Heterogeneous Gold Catalysts and Catalysis

RSC Catalysis Series

Titles in the Series:

1: Carbons and Carbon Supported Catalysts in Hydroprocessing
2: Chiral Sulfur Ligands: Asymmetric Catalysis
3: Recent Developments in Asymmetric Organocatalysis
4: Catalysis in the Refining of Fischer–Tropsch Syncrude
5: Organocatalytic Enantioselective Conjugate Addition Reactions: A Powerful Tool for the Stereocontrolled Synthesis of Complex Molecules
6: *N*-Heterocyclic Carbenes: From Laboratory Curiosities to Efficient Synthetic Tools
7: *P*-Stereogenic Ligands in Enantioselective Catalysis
8: Chemistry of the Morita–Baylis–Hillman Reaction
9: Proton-Coupled Electron Transfer: A Carrefour of Chemical Reactivity Traditions
10: Asymmetric Domino Reactions
11: C–H and C–X Bond Functionalization: Transition Metal Mediation
12: Metal Organic Frameworks as Heterogeneous Catalysts
13: Environmental Catalysis Over Gold-Based Materials
14: Computational Catalysis
15: Catalysis in Ionic Liquids: From Catalyst Synthesis to Application
16: Economic Synthesis of Heterocycles: Zinc, Iron, Copper, Cobalt, Manganese and Nickel Catalysts
17: Metal Nanoparticles for Catalysis: Advances and Applications
18: Heterogeneous Gold Catalysts and Catalysis

How to obtain future titles on publication:
A standing order plan is available for this series. A standing order will bring delivery of each new volume immediately on publication.

For further information please contact:
Book Sales Department, Royal Society of Chemistry, Thomas Graham House, Science Park, Milton Road, Cambridge, CB4 0WF, UK
Telephone: +44 (0)1223 420066, Fax: +44 (0)1223 420247
Email: booksales@rsc.org

Visit our website at www.rsc.org/books

Heterogeneous Gold Catalysts and Catalysis

Edited by

Zhen Ma
Fudan University, Shanghai, PR China
Email: zhenma@fudan.edu.cn

Sheng Dai
Oak Ridge National Laboratory, TN, USA
Email: dais@ornl.gov

ROYAL SOCIETY OF CHEMISTRY

THE QUEEN'S AWARDS FOR ENTERPRISE: INTERNATIONAL TRADE 2013

RSC Catalysis Series No. 18

Print ISBN: 978-1-84973-917-7
PDF eISBN: 978-1-78262-164-5
ISSN: 1757-6725

A catalogue record for this book is available from the British Library

Published by The Royal Society of Chemistry,
Thomas Graham House, Science Park, Milton Road,
Cambridge CB4 0WF, UK

Registered Charity Number 207890

For further information see our web site at www.rsc.org

Printed in the United Kingdom by CPI Group (UK) Ltd, Croydon, CR0 4YY, UK

Preface

Heterogeneous catalysts are ubiquitous, typically including metal oxides, zeolites, and supported metal catalysts. They have found many applications in the synthesis of chemicals and materials, the conversion of fossil, hydrogen and biomass fuels, and the protection of environments. Supported metal catalysts are at the heart of heterogeneous catalysis. These catalysts include platinum, palladium, rhodium, and silver supported on oxide supports and have played key roles in many important chemical processes. For instance, platinum catalysts are useful for hydrotreating and complete oxidation, whereas palladium and rhodium are components in catalysts used for ablating car emissions, silver catalysts are useful in selective oxidation.

Gold was initially regarded as an inactive metal in catalysis. Gold catalysts prepared by traditional impregnation of $HAuCl_4$ followed by calcination usually have big gold particles, thus being inactive in reactions such as CO oxidation. However, in the 1980s, Haruta and co-workers in Japan have found that gold catalysts prepared by coprecipitation or deposition-precipitation have small gold nanoparticles and are highly active for CO oxidation. In the 1990s, Goodman and co-workers have discovered that the CO oxidation activity correlates nicely with the size of gold nanoparticles, being the highest when the size of gold nanoparticles is about 3 nm.

Since then, there have been a number of papers dealing with the synthesis of gold catalysts. Attention has been paid to the synthesis methods and synthesis details, including pH values during synthesis, gold loadings, aging temperature and duration, calcination conditions, and pretreatment conditions. These catalysts were often characterized by a number of techniques, such as X-ray diffraction (XRD), transmission electron microscopy (TEM), X-ray photoelectron spectroscopy (XPS), etc. CO oxidation has been chosen as a probe to compare the activity of different gold catalysts prepared. Often

RSC Catalysis Series No. 18
Heterogeneous Gold Catalysts and Catalysis
Edited by Zhen Ma and Sheng Dai

Published by the Royal Society of Chemistry, www.rsc.org

a "good" gold catalyst can exhibit high activity in CO oxidation well below room temperature.

As the research proceeds, there have been more papers reporting the application of gold catalysts in other reactions, such as selective oxidation or hydrogenation of organic molecules, elimination of other environmental pollutants in air and water, and purification of hydrogen fuel by selective oxidation of CO or the water-gas shift reaction. In addition, attention has been paid to the nature of active sites and reaction mechanisms. They can be studied by spectroscopic methods using supported catalysts under operando conditions, or using model catalysts such as single-crystal gold surfaces or gold nanoparticles supported on planer films. Finally, DFT calculations can help understand the nature of active sites and reaction mechanisms. These are really tough questions and the conclusions in different papers are often controversial due to different preparation details and properties of different catalysts.

Bond, Louis, and Thompson published an outstanding book "Catalysis by Gold" in 2006. That book summarized the various aspects of gold catalysis, and is particularly suitable for a new researcher of the field. Several years have passed by, and the field has undergone significant development. More and more new gold catalysts have been developed, and gold catalysts have found more applications in reactions other than CO oxidation. It is time-consuming to search through such a body of work, especially when new papers are published all the time.

The current book presents a collection of overviews, written by experts in the field, on the recent development of heterogeneous gold catalysts and catalysis. The first three chapters are devoted to the synthesis and design of new gold catalysts. Chapter 1 by Ma and Dai summarizes the several strategies for the design of new gold catalysts with enhanced thermal stability. Chapter 2 by Li and Jin overviews the synthesis, structure, thermal stability and catalytic applications of well defined gold nanoclusters and their supported forms. Chapter 3 by Murray and co-workers describes the preparation, characterization, and properties of nanocrystal superlattices as a new class of model materials for obtaining fundamental insights in catalysis.

The next four chapters are related to the application of gold catalysts in ablating environmental pollutants. In Chapter 4, Chen and Sasirekha summarize various metal oxide-based gold catalysts for preferential oxidation of CO in H_2 stream. In Chapter 5, Ivanova and co-workers review the development of various metal oxide-based gold catalysts for the water-gas shift reaction. In Chapter 6, Ma and Hao briefly summarize the application of gold catalysts in complete oxidation of CO, ethylene, and formaldehyde, as well as in the decomposition of ozone, N_2O, and NO. In Chapter 7, Zanella and co-workers review the modification of various semiconductors by gold nanoparticles for photocatalytic removal of water pollutants and for photocatalytic hydrogen production.

Chapters 8–11 are about the application of gold catalysts in the synthesis of chemicals via selective oxidation or hydrogenation. Han and co-workers summarize the application of heterogeneous gold catalysts in selective oxidation of alkanes, alkenes, alcohols, and aldehydes, and briefly introduce Fenton-like reaction and photocatalytic degradation. Zhong and co-workers furnish a long review on the application of heterogeneous gold catalysts for selective oxidation reactions. Simakova and Murzin concisely report the selective oxidation of biomass-derived secondary alcohols using gold catalysts. Finally, Keane and co-workers describe the selective hydrogenation of various organic substrates using supported gold catalysts.

Chapters 12–15 focus more on the fundamental aspects of heterogeneous gold catalysts and catalysis. Wu and Liotta summarize the interaction between gold and various reducible oxide supports, which is important for achieving high catalytic activity. Centeno and co-workers describe the oxygen vacancies presented in gold catalysts based on reducible and non-reducible supports, together with their role in oxidation catalysis. Wu and Overbury summarize their infrared spectroscopic study toward understanding nature of active sites and reaction mechanisms on gold catalysts based on non-reducible supports such as SiO_2, $FePO_4$, and $LaPO_4$. Finally, Huang summarizes their fundamental studies based on single crystalline and supported gold catalysts.

Overall, this book furnishes a diversified collection of overviews on various aspects of heterogeneous gold catalysts and catalysis. We wish to thank all the authors and publishing staff for their hard work.

Zhen Ma
Fudan University, P.R. China

Sheng Dai
Oak Ridge National Laboratory, TN, USA

Contents

RSC Catalysis Series No. 18
Heterogeneous Gold Catalysts and Catalysis
Edited by Zhen Ma and Sheng Dai

Published by the Royal Society of Chemistry, www.rsc.org

CHAPTER 1

Stabilizing Gold Nanoparticles by Solid Supports

ZHEN MA*[a] AND SHENG DAI[b]

[a] Department of Environmental Science and Engineering, Fudan University, Shanghai, 200433, PR China; [b] Chemical Sciences Division, Oak Ridge National Laboratory, Oak Ridge, TN 37831, USA
*Email: zhenma@fudan.edu.cn

1.1 Introduction

Catalysis by nanostructured materials has attracted tremendous interest recently.[1–4] Nanostructured catalysts may have interesting catalytic properties associated with their small sizes and geometric/electronic structures. In particular, Haruta and co-workers found that gold nanoparticles finely dispersed on some metal oxide supports have excellent activities in low-temperature CO oxidation.[5–11] This finding has been followed by thousands of studies on supported gold catalysts and their catalytic applications in environmental catalysis and chemical synthesis.[12–26]

Gold nanoparticles may be synthesized *via* a traditional colloidal chemistry approach,[27–30] in which $AuCl_4^-$ ions are reduced by sodium citrate, tetrakis(hydroxymethyl)phosphonium chloride (THPC) or sodium borohydride ($NaBH_4$), and the formed gold nanoparticles can be stabilized by polyvinylalcohol (PVA), polyvinylpyrrolidone (PVP), polydiallyldimethylammonium chloride (PDDA), or cetyltrimethylammonium bromide (CTAB).[13,31] These gold nanoparticles (colloids) can be used either directly as catalysts in liquid phase or deposited onto solid supports.

RSC Catalysis Series No. 18
Heterogeneous Gold Catalysts and Catalysis
Edited by Zhen Ma and Sheng Dai

Published by the Royal Society of Chemistry, www.rsc.org

Alternatively, gold nanoparticles can be formed on solid supports by loading a gold precursor (usually a gold salt or complex) onto solid supports followed by reduction or calcination.[13] During the reduction or calcination process, the gold cations are reduced into gold atoms that aggregate into gold nanoparticles. The extent of agglomeration during this stage depends on many factors such as the ambient, temperature, and duration of the process as well as the nature of solid supports.

Supported metal catalysts are usually composed of metal nanoparticles and solid supports. Solid supports may provide a platform for dispersing and stabilizing gold nanoparticles so as to expose more surface gold atoms to the reactants, thus increasing catalytic activity. They may tune the oxidation state of gold by charge transferring or by mediating the reducing degree of gold precursors upon calcination or reduction. Some supports may undergo phase transformation or structural collapse under high temperatures, thus aggravating the sintering of gold nanoparticles on these supports or leading to the encapsulation of gold nanoparticles by these supports. Figure 1.1(c) shows a schematic diagram illustrating the phase transformation of a support at high temperatures.[32] Besides the facets related to metal–support interactions mentioned above, solid supports may participate in catalysis by adsorbing and activating reactants as well as supplying active oxygen. They may also, of course, strongly adsorb some reaction intermediates or products, leading to catalyst deactivation.

Although solid supports can disperse gold nanoparticles, the sintering of gold nanoparticles at elevated temperatures is often inevitable because of their low melting points and high surface free energies.[13,33] Figure 1.1(a) and

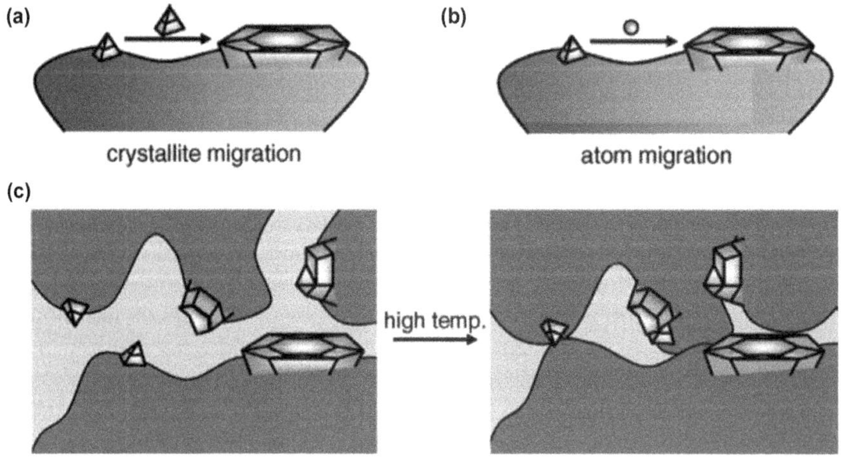

Figure 1.1 Several mechanisms for the sintering of supported metal catalysts.[32] (a) Crystalline migration. (b) Atom migration. (c) Phase transformation of the support at high temperatures.
Reproduced with permission of Wiley-VCH from G. Rothenberg, *Catalysis: Concepts and Green Applications*, Wiley-VCH, Weinheim, 2008.

Figure 1.1(b) show two models (crystalline migration and atom migration) proposed for the sintering of metal nanoparticles on supports.[32] In the first model, entire metal crystallites migrate, collide, and coalesce on the support surface. In the second model, metal atoms migrate from one crystallite to another *via* the surface or the gas phase, making big crystallite bigger and small crystallites smaller. The phase transformation or structural collapse of supports under elevated temperatures, as shown in Figure 1.1(c), may exacerbate the sintering or encapsulation of gold nanoparticles.[32]

Because the catalytic activities of supported gold catalysts often decrease sharply as the gold nanoparticle agglomerate under elevated temperatures but a high temperatures is often encountered during the calcination, operation, and regeneration of catalysts,[34] it is necessary to enhance the thermal stability of supported gold nanoparticles. This can be achieved by improving synthesis details, *e.g.*, thoroughly washing away residual chloride that may facilitate sintering.[35,36] However, a more versatile way is to tune the structural environment surrounding gold nanoparticles, *e.g.*, by strengthening the metal–support interaction and by designing sturdy inorganic shells that encapsulate gold nanoparticles.[19–21,33] These strategies rely on synthesis and modification of catalytic materials.

Most of the publications relevant to gold catalysis deal with the conventional synthesis, characterization, and applications of supported gold nanoparticles, as well as the elucidation of the nature of active sites and reaction mechanisms.[13,16,26] Only a small portion of publications have addressed the thermal stability and stabilization of gold nanoparticles on solid supports. The Dai group at the US Oak Ridge National Laboratory has been interested in designing new-structured gold nanocatalysts with enhanced properties, including catalytic activity, stability on stream, and thermal stability.[37–70] It is known from these studies that catalytic performance and thermal stability of supported gold catalysts depend critically on their composition and catalyst structure. Below we first summarize some recent advances in the stabilization of gold nanoparticles by solid supports, and then furnish our perspectives on future development.

1.2 Catalysts with Complex Interfacial Structures

1.2.1 General Considerations

Supported gold catalysts are usually prepared by loading gold onto supports (*e.g.*, TiO_2, ZrO_2, SiO_2, Al_2O_3, Fe_2O_3, and CeO_2) *via* impregnation, deposition–precipitation, co-precipitation, and colloidal deposition.[13] When supported gold catalysts are prepared by impregnation, deposition–precipitation, or co-precipitation methods, the gold precursors (gold cations) reduce to metallic gold atoms that migrate and aggregate into gold nanoparticles upon reduction or calcination. When gold catalysts are prepared by loading preformed gold nanoparticles (colloids) onto supports, it is not necessary to reduce the metallic gold nanoparticles again, but a calcination or pre-treatment

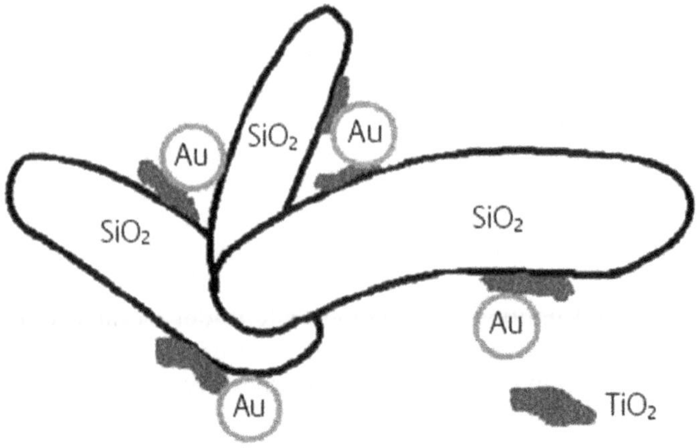

Figure 1.2 Scheme showing the arrangement of particles and various interfaces ($Au-TiO_2$, $Au-SiO_2$, TiO_2-SiO_2) in $Au/TiO_2/SiO_2$ catalyst.[71]
Reproduced with permission of Springer from L. Guczi, A. Beck and K. Frey, *Gold Bull.*, 2009, **42**, 5.

step may still be needed to remove organic capping agents that may influence catalytic activity and to enhance the metal–support interaction.

Regardless of preparation methods, the formed supported gold catalysts usually have simple metal–support interfaces (*e.g.*, $Au-TiO_2$). The sintering of gold nanoparticles on these neat metal oxide supports is a common problem under elevated temperatures. Attempts have been made to build up more complex interfaces for enhancing the thermal stability. For instance, Figure 1.2 illustrates the structural feature of $Au/TiO_2/SiO_2$ (*i.e.*, gold nanoparticles supported on TiO_2-modified SiO_2 support), highlighting the additive–support (TiO_2-SiO_2) and metal–additive ($Au-TiO_2$) interfaces in addition to the metal–support ($Au-SiO_2$) interface.[71] The presence of additional interfaces or complex structures may mitigate the sintering of gold nanoparticles due to the enhanced metal–support interaction.

1.2.2 Pre-modification of Supports before Loading Gold

Au/SiO_2 catalysts are usually not very active for CO oxidation and gold nanoparticles on SiO_2 can sinter easily, unless the preparation method is carefully chosen.[48,49,72] To enhance the thermal stability, one idea is to modify the SiO_2 support by another metal oxide before loading gold. For instance, Tai and co-workers developed $Au/TiO_2/SiO_2$ catalysts.[73] SiO_2 wet-gel was prepared by the hydrolysis of $Si(OCH_3)_4$ in the presence of NH_4OH, and was subsequently soaked in a toluene solution of Ti(*iso*-$OC_3H_7)_4$. Dodecanethiol-capped gold nanoparticles (2.1 nm) were then deposited onto TiO_2/SiO_2. For comparison, a TiO_2 support was prepared using Ti(*iso*-$OC_3H_7)_4$ as the precursor, and was used to load gold nanoparticles. As shown in Figure 1.3,[73] gold nanoparticles in $Au/TiO_2/SiO_2$ were still small (average

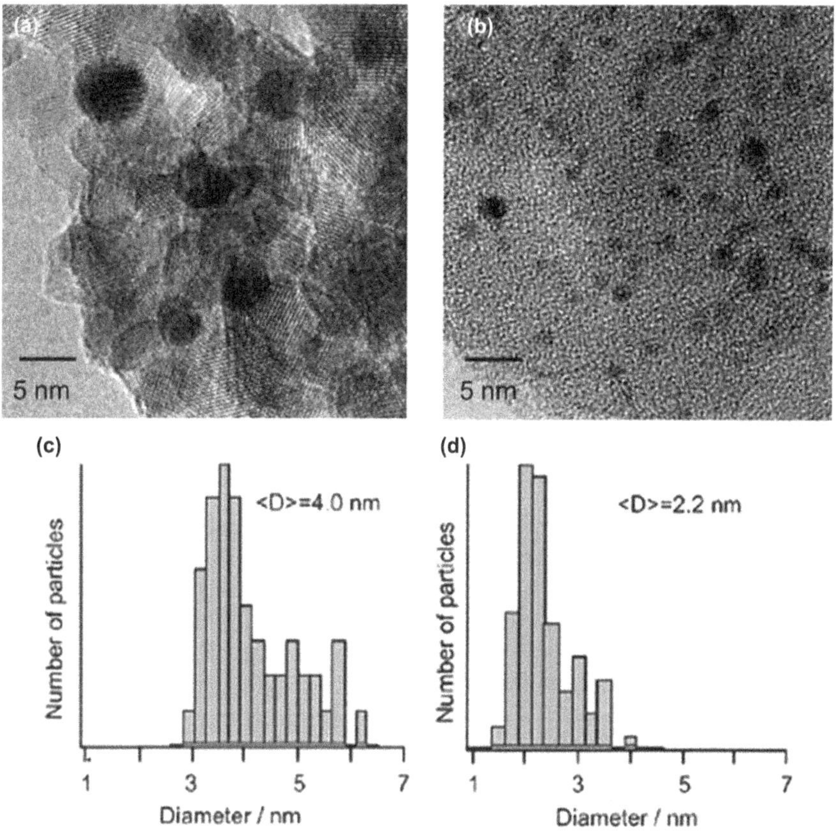

Figure 1.3 TEM images and size distributions of gold nanoparticles on TiO_2 (a, c) and TiO_2-coated SiO_2 (b, d) composite aerogels. These catalysts were calcined at 400 °C.[73]
Reproduced with permission of Elsevier from Y. Tai, J. Murakami, K. Tajiri, F. Ohashi, M. Daté and S. Tsubota, *Appl. Catal. A*, 2004, **268**, 183.

diameter 2.2 nm) after calcination in air at 400 °C, whereas those in Au/TiO_2 grew obviously (average diameter 4.0 nm). Although the authors did not show the sintering behavior of gold nanoparticles on a neat SiO_2 support, this study nicely showed that gold nanoparticles exhibit high thermal stability on TiO_2-modified SiO_2 gel.

In another work, Yan *et al.* developed Au/TiO_2/mesoporous SiO_2.[39] The mesoporous SiO_2 (SBA-15) surface was functionalized by amorphous TiO_2 *via* a surface-sol-gel method, using $Ti(OC_4H_9)_4$ as the precursor. Gold was then loaded onto the support *via* a deposition–precipitation method. For comparison, Au/P25 TiO_2 was also prepared. Here the 'P25 TiO_2' refers to a commercial TiO_2 furnished by Degussa. Although the as-synthesized Au/P25 TiO_2 showed high activity in CO oxidation when the reaction temperature was below −20 °C, the Au/P25 TiO_2 calcined at 300 °C was much less active due to the aggregation of gold nanoparticles. For comparison, the activities of the as-synthesized and 300 °C calcined Au/TiO_2/mesoporous SiO_2 were similar, due to the preservation of small gold nanoparticles at 300 °C.[39] The authors additionally showed that it was difficult to load gold onto mesoporous SiO_2 *via* deposition–precipitation due to the low isoelectric point of SiO_2, and the obtained gold nanoparticles were usually big.

Above, we have highlighted two examples for the pre-modification of SiO_2 supports by TiO_2 before loading gold. The presence of TiO_2 species not only stabilizes gold nanoparticles on the modified supports, but also increases the isoelectric point of supports (note that the isoelectric point of TiO_2 is higher than that of SiO_2), thus increasing the gold loading when the deposition–precipitation method is used to load gold. In addition, the additional Au-TiO_2 interface leads to high activity in CO oxidation.[74–81] SiO_2 supports can also be modified by other metal oxides (*e.g.*, CoO_x,[82] ZnO,[83] CeO_2,[83,84] CuO[69]) to increase the dispersion of gold nanoparticles on supports. The role of these metal oxide additives is similar to that of the TiO_2 additive mentioned above.

Au/TiO_2 is the most studied gold catalyst. If prepared properly, it should be active for low-temperature CO oxidation, but the sintering of gold nanoparticles on TiO_2 is still a problem. Yan *et al.* developed a new catalyst, $Au/Al_2O_3/TiO_2$, for CO oxidation.[44] First, the P25 TiO_2 support was modified by amorphous Al_2O_3 *via* surface-sol-gel processing of Al(sec-$OC_4H_9)_3$ followed by controlled hydrolysis. Gold was then loaded onto Al_2O_3/TiO_2 *via* deposition–precipitation using $HAuCl_4$ as the precursor. Interestingly, the gold nanoparticles on the Al_2O_3/TiO_2 support showed excellent thermal stability upon aging at 500 °C, whereas gold nanoparticles on neat TiO_2 sintered significantly.[44]

Ma *et al.* subsequently prepared $Au/M_xO_y/TiO_2$ catalysts.[51] In the preparation, the surface–sol-gel method was not used. Instead, a traditional impregnation method was used to load metal nitrates onto TiO_2. The metal nitrates/TiO_2 were calcined to form M_xO_y/TiO_2 supports, and gold was loaded *via* deposition–precipitation.[51] The use of impregnation instead of the surface–sol-gel method to functionalize TiO_2 support was based on several considerations. First, it was thought that a catalyst (*e.g.*, $Au/Al_2O_3/TiO_2$) prepared involving decomposing a metal nitrate (*e.g.*, $Al(NO_3)_3$) on TiO_2 support followed by loading gold should exhibit a similar performance compared with its counterpart prepared using a surface–sol-gel method. Second, the nitrate decomposition method is less demanding and therefore suitable for large-scale preparation. Third, the synthesis *via* the surface–sol-gel method is constrained by the availability, expensiveness, and storage stability of metal alkoxide precursors.

It was found that $Au/M_xO_y/TiO_2$ (M = Ca, Ni, Zn, Ga, Y, Zr, La, Pr, Nd, Sm, Eu, Gd, Dy, Ho, Er, Yb) retained significant activity in CO oxidation even after thermal treatment at 500 °C.[51] This was explained by the enhanced thermal stability of gold nanoparticles caused by the surface modification of TiO_2 support by certain metal oxides, as demonstrated by X-ray diffraction (XRD) and transmission electron microscopy (TEM) data.[51,63] In addition, it was speculated that the presence of metal oxide additives with redox properties adjacent to gold nanoparticles may change the oxidation state of gold and the redox property of the support. More experiments should be performed to better understand the promotional effects of the metal oxide additives on TiO_2 support.[85–88]

The empirical observations mentioned above have led to further fundamental research. Liu and co-workers shed light on the stabilizing effect of the amorphous Al_2O_3 by means of density functional theory (DFT) calculations.[89] Figure 1.4 shows the models proposed for the local structures of a gold atom or a two-layer gold strip on an Al_2O_3/TiO_2 surface.[89] These models were used for their theoretical calculations. The authors found that the binding of gold on Al_2O_3/TiO_2 was much stronger than that on TiO_2, and the stronger binding was valid for some other metals (*i.e.*, Ag, Cu, Pt, Pd, Ir) on Al_2O_3/TiO_2 compared with these metals on TiO_2. This finding explained the enhanced thermal stability of gold nanoparticles on Al_2O_3/TiO_2 *versus* TiO_2. A further idea would be to extend this DFT approach to the $Au/M_xO_y/TiO_2$ system to obtain deeper insights.

The pre-modification strategy is not limited to the modification of oxide supports by another metal oxide. For instance, TiO_2 was treated by an aqueous H_3PO_4 solution before loading gold.[52] That treatment was found to

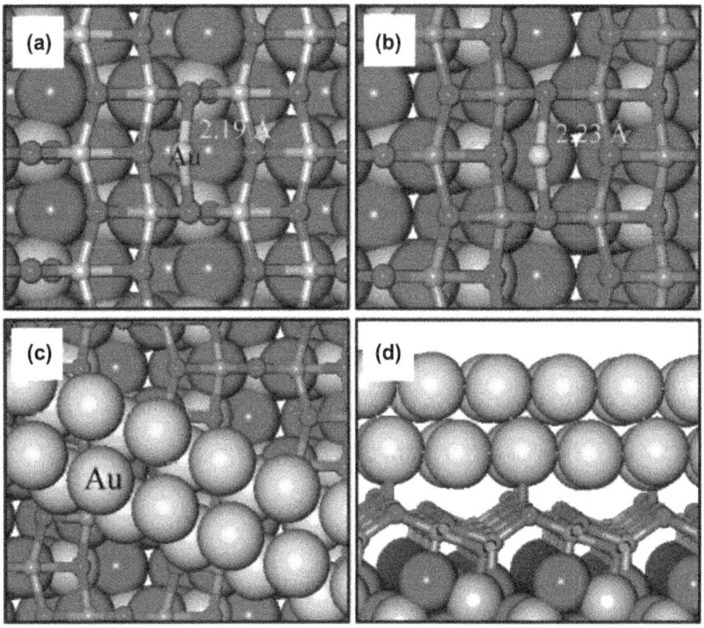

Figure 1.4 Optimized structures of Au atom on (a) $TiO_2(101)$ and (b) Al_2O_3/TiO_2, and a two-layer Au strip on Al_2O_3/TiO_2 [(c) top view and (d) side view].[89] The distances of Au–O bonds are also labeled.
Reproduced with permission of the American Chemical Society from C.-M. Wang, K.-N. Fan and Z.-P. Liu, *J. Phys. Chem. C*, 2007, **111**, 13539.

enhance the thermal stability of gold nanoparticles. A similar stabilization effect was found with $Au/H_3PO_4\text{-}Al_2O_3$[90] and $Ag/H_3PO_4\text{-}TiO_2$.[91] However, the presence of phosphorus species also suppressed the catalytic activities. In addition, it is not clear why the H_3PO_4 treatment can help stabilize gold or silver nanoparticles.

1.2.3 Post-modification of Supported Gold Catalysts

Supported gold catalysts are usually prepared by loading gold onto neat metal oxide supports such as TiO_2. However, few attempts have been made to additionally modify supported gold catalysts when gold was already loaded onto solid supports.[92–95] Ma *et al.* developed $PO_4^{3-}/Au/TiO_2$ catalysts.[52] In the synthesis, Au/TiO_2 was prepared by deposition–precipitation, reduced at 150 °C in H_2–Ar, and soaked in a diluted H_3PO_4 solution, followed by washing and drying. This treatment can stabilize gold nanoparticles on TiO_2, as evidenced by the fact that $PO_4^{3-}/Au/TiO_2$ calcined at 500 °C still had small gold nanoparticles and appreciable activity in CO oxidation at room temperature. However, overloaded phosphate ions may lead to low catalytic activity due to the blockage of active sites. In addition, it is not clear why the H_3PO_4 treatment can stabilize gold nanoparticles.

Zhu *et al.* developed $SiO_2/Au/TiO_2$ catalysts.[50] In the synthesis, pre-formed Au/TiO_2 was treated in $Si(OCH_3)_4$, $H_2N(CH_2)_3Si(OC_2H_5)_3$, or $((CH_3)_3CO)_3\text{-}SiOH$ solution, followed by calcination (Figure 1.5).[50] The silicon precursors

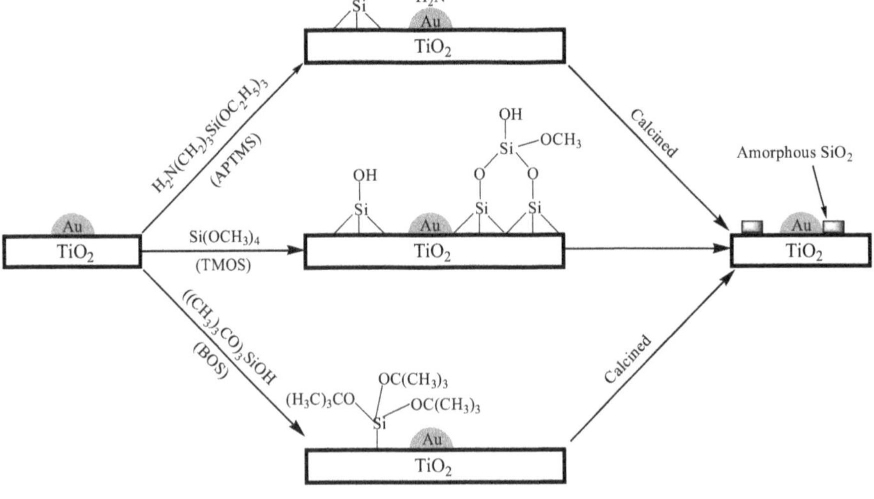

Figure 1.5 Scheme for the synthesis of $SiO_2/Au/TiO_2$.[50] Pre-formed Au/TiO_2 was treated in $Si(OCH_3)_4$, $(C_2H_5O)_3Si(CH_2)_3NH_2$, or $(CH_3)_3CO)_3SiOH$ solution, followed by calcination to form amorphous SiO_2.
Reproduced with permission of Springer from H. G. Zhu, Z. Ma, S. H. Overbury and S. Dai, *Catal. Lett.*, 2007, **116**, 128.

decomposed to amorphous SiO_2 near gold nanoparticles upon calcination, therefore mitigating the sintering of gold nanoparticles and maintaining high catalytic activity. Although XRD and TEM data demonstrated the maintenance of small gold nanoparticles upon thermal aging,[19,50] there is no direct information on the location and morphology of the amorphous SiO_2 additive.

Ma *et al.* developed $SiO_2/Au/TiO_2$ catalysts using an alternative method.[56] In the synthesis, SiO_2 was loaded onto Au/TiO_2 *via* atomic layer deposition (ALD). $Si(OCH_3)_4$ vapor was used as the precursor to the amorphous SiO_2 overlayer, and the SiO_2 content could be adjusted by varying the number of ALD cycles or $Si(OCH_3)_4$ exposures. High-resolution TEM images showed the presence of amorphous SiO_2 adjacent to the gold nanoparticles and TiO_2 support. The amorphous SiO_2 may mitigate the sintering of gold nanoparticles.[56] Although the ALD method is expected to provide more uniform coating of gold nanoparticles and TiO_2 support, it is more demanding in terms of the apparatus and control of the experimental conditions. In addition, the uniform and dense coating of gold nanoparticles makes the reactants less accessible to the active sites, thus leading to lower catalytic activities in CO oxidation.

The stabilization effect of SiO_2 coatings that physically isolate and encapsulate gold nanoparticles is understandable. But sometimes the loading of SiO_2 was too low to allow for complete coating, whereas the stabilization effect of the amorphous SiO_2 was still observed.[50] Rashkeev and co-workers conducted DFT calculations, and concluded that the deposition of SiO_2 on TiO_2 support may cause lattice-mismatch instabilities and lead to the formation of strong anchoring sites for gold nanoparticles, even if the coverage of SiO_2 is below monolayer.[96] Moreno and co-workers provided direct imaging evidence for the stabilization effect caused by the SiO_2 decoration attached to gold nanoparticles (Figure 1.6).[97] Note that the SiO_2 decoration was formed by the transformation of $HS(CH_2)_3Si(CH_3)_3$ modified gold nanoparticles supported on SiO_2 substrate but not by post-modification.

The structures of $SiO_2/Au/TiO_2$ highlighted above are relatively straightforward. Further attempts have been made to design catalysts with more complex architectures.[98,99] For instance, Yin and co-workers prepared core-satellite nanocomposite catalysts protected by a porous SiO_2 shell (Figure 1.7).[98] In the synthesis, a superparamagnetic Fe_3O_4 core was coated by a SiO_2 shell, and gold nanoparticles were loaded onto SiO_2/Fe_3O_4. The resulting material was coated with another layer of SiO_2 to fix the position of the gold nanoparticles, and a 'surface-protected etching' technique was used to make the outer SiO_2 shell porous. The $SiO_2/Au/SiO_2/Fe_3O_4$ catalyst was still active after running the catalytic reduction of 4-nitrophenol for several times, whereas $Au/SiO_2/Fe_3O_4$ without the porous SiO_2 shell suffered from agglomeration of gold nanoparticles and loss of activity during the reaction.

To put this work in perspective, it should be mentioned that the post-modification strategy was not limited to the modification of supported gold catalysts. Other modified catalysts reported include $SiO_2/Pt/zeolite$,[100]

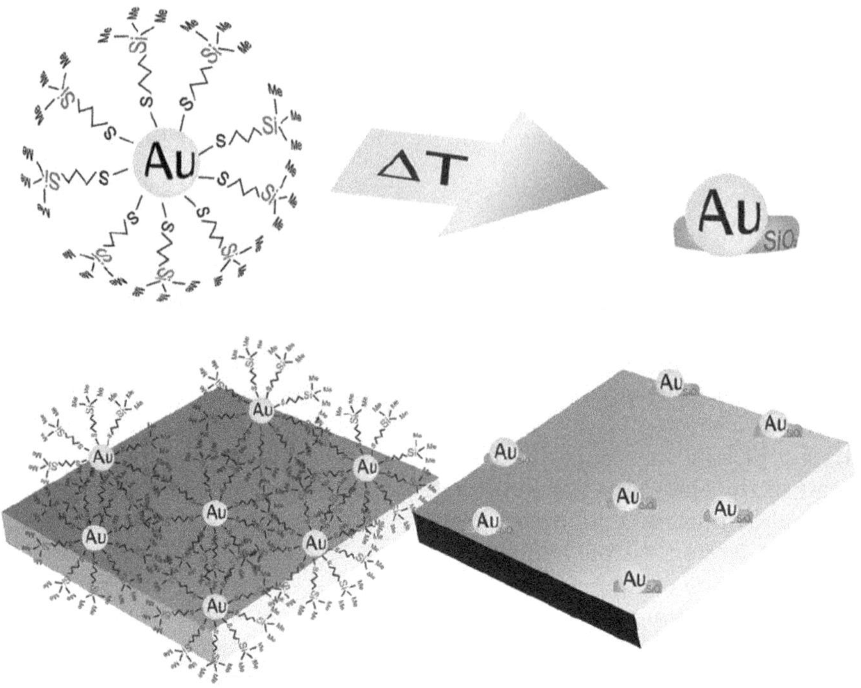

Figure 1.6 Proposed scheme for the stabilization of gold nanoparticles on top of SiO_2 and $SrTiO_3$ substrates *via* ligand-assisted pinning.[97]
Reproduced with permission of the Royal Society of Chemistry from C. Moreno, N. J. Divins, J. Gázquez, M. Varela, I. Angurell and J. Llorca, *Nanoscale*, 2012, **4**, 2278.

SiO_2/Pt/C,[101–104] SiO_2/Pt/Fe_2O_3,[105] SiO_2/Pt/SiO_2,[106,107] TiO_2/Pt/SiO_2,[108] Al_2O_3/Pd/SiO_2,[109] Al_2O_3/Pd/Al_2O_3/SiO_2,[110] Al_2O_3/Pd/Al_2O_3,[111] and HfO_x/Pd/SiO_2.[112] The thermal stability of metal nanoparticles in these catalysts has been significantly enhanced. The modifier of choice was SiO_2 due to the relative easiness of preparation and high thermal stability of SiO_2, but post-modification by metal oxides has not been demonstrated widely. It is expected that post-modification of supported metal catalysts by metal oxides (especially transition metal oxides) will bring more functionalities to the catalyst systems. However, the synthesis of metal oxide coatings is more difficult and the thermal stability of metal oxide coatings may not be that good.

1.2.4 Dispersion of Au-Fe_3O_4 Dumbbell Structures on Supports

Delicate materials synthesis has lead to new materials with interesting structures or morphologies, thus providing new opportunities for designing better catalysts. The dumbbell-like Au-Fe_3O_4 composite is such a material.[113–115] The dumbbell structure is composed of a small gold

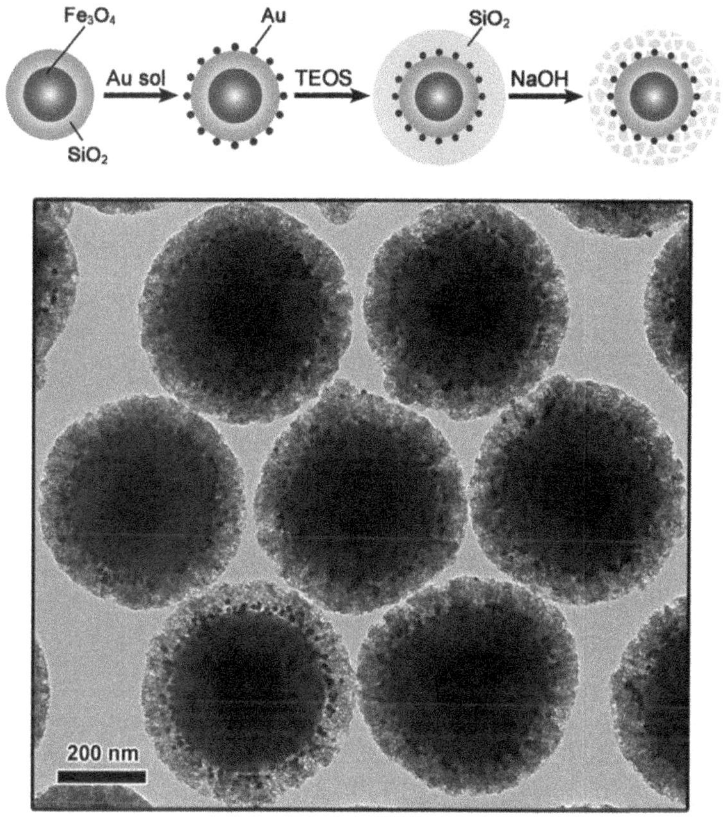

Figure 1.7 Synthesis procedure and a typical TEM image of porous SiO_2 protected $Au/SiO_2/Fe_3O_4$ composite catalyst.[98]
Reproduced with permission of Wiley-VCH from J. P. Ge, Q. Zhang, T. R. Zhang and Y. D. Yin, *Angew. Chem., Int. Ed.*, 2008, **47**, 8924.

nanoparticle at one end and a bigger Fe_3O_4 nanoparticle at the other end. These two components interact with each other through an Au-Fe_3O_4 interface. Yin *et al.* prepared Au-Fe_3O_4 dumbbell nanoparticles *via in situ* decomposition of $HAuCl_4$ and $Fe(CO)_5$, followed by oxidation of the Fe component, and then dispersed the Au-Fe_3O_4 dumbbell nanoparticles on several supports *via* colloidal deposition (Figure 1.8).[59] Figure 1.9 shows the TEM images of unsupported and SiO_2-supported Au-Fe_3O_4 dumbells.[59] The presence of Fe_3O_4 attached to gold nanoparticles was proposed to be responsible for the high activity in CO oxidation and thermal stability of the resulting Au-Fe_3O_4/SiO_2 catalyst.

Zheng and co-workers studied supported Au-Fe_3O_4 dumbbell catalysts in more detail.[116] Gold nanoparticles (6.7 nm) were epitaxially grown on Fe_3O_4 nanoparticles (15.2 nm) to form dumbbell particles, and the dumbbell particles were then dispersed onto TiO_2. For comparison, Au/TiO_2 was prepared by dispersing gold nanoparticles (6.7 nm) onto TiO_2. The thermal

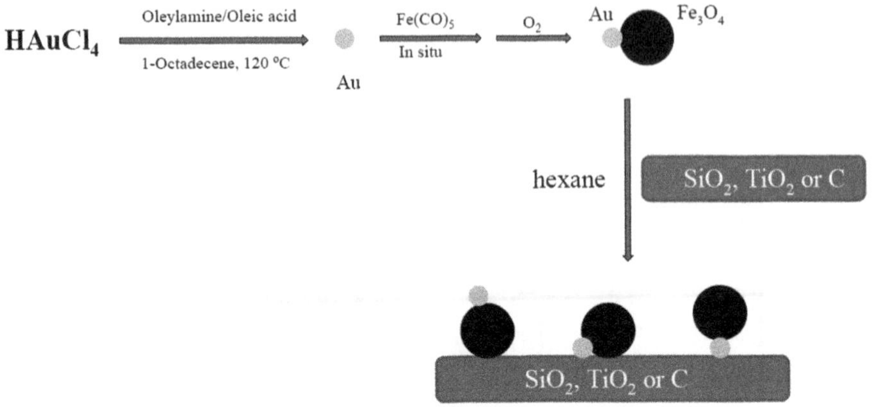

Figure 1.8 Schematic diagram showing the synthesis of supported Au-Fe_3O_4 dumbbell structures.[59]
Courtesy of Dr Hongfeng Yin.

stability was studied by calcining these catalysts at 350, 450, and 550 °C and recording TEM images afterwards. The sintering of gold nanoparticles in Au-Fe_3O_4/TiO_2 was less obvious than that in Au/TiO_2.

To summarize, dumbbell-structured Au-Fe_3O_4 particles have been used to make supported gold catalysts. The advantage of the catalyst design is that active interfaces can be installed on sturdy supports, thus mitigating sintering and enhancing catalytic activity. In addition, the sizes of gold particles and Fe_3O_4 can be tuned, and the solid supports can be chosen according to desired needs. However, the types of catalysts are limited by the availability of dumbbell structures; many other dumbbell structures have not been developed. In addition, the local contact structures of the catalysts were not made clear in previous studies. High angle annular dark field (HAADF) imaging may provide more detailed information.

1.2.5 *In situ* Transformation of Supported Bimetallic Catalysts into Metal Oxide Modified Supported Gold Catalysts

Gold-containing bimetallic particles have been used to prepare supported catalysts.[117–119] In most cases, the supported particles are assumed to be metallic before and after thermal treatments or catalytic reactions, especially considering that it is difficult to oxidize some noble metal components (*e.g.*, Pt) with oxygen. However, the transformation of supported bimetallic particles under certain conditions is often inevitable.[67,70,120–126] In particular, some metals (*e.g.*, Cu, Ni, Fe) can be oxidized at elevated temperatures. This situation has led to the development of catalysts with enhanced performance.

Zhou *et al.* developed Au-NiO/SiO_2 catalysts.[60,61] First, NiAu alloy nanoparticles were prepared *via* the reduction of mixed Ni and Au salts by

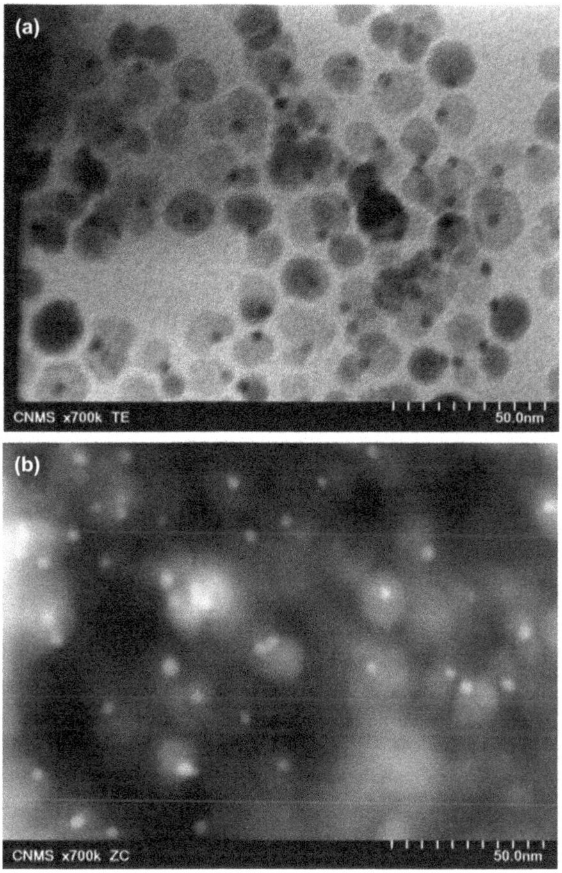

Figure 1.9 (a) Bright-field TEM image of Au-Fe_3O_4 dispersed in hexane. (b) Dark-field TEM image of Au-Fe_3O_4 dumbbells supported on SiO_2.[59] The scale bars represent 50 nm.
Reproduced with permission of the Royal Society of Chemistry from H. F. Yin, C. Wang, H. G. Zhu, S. H. Overbury, S. H. Sun and S. Dai, *Chem. Commun.*, **2008**, 4357.

butyl-lithium. The NiAu nanoparticles were then deposited onto SiO_2 support. Upon treating NiAu/SiO_2 in H_2 at 600 or 720 °C and then in O_2 at 300 °C, the catalyst transformed into Au-NiO/SiO_2, the working catalyst. The amorphous NiO was demonstrated to locate near gold nanoparticles, thus mitigating the sintering of gold nanoparticles. As shown in Figure 1.10, gold nanoparticles in Au/SiO_2 (a reference catalyst prepared by colloidal deposition) sintered obviously after thermal treatment, whereas those in Au-NiO/SiO_2 were quite small.[60] This *in situ* transformation strategy may prompt the development of new advanced catalysts with better chemical and structural properties.

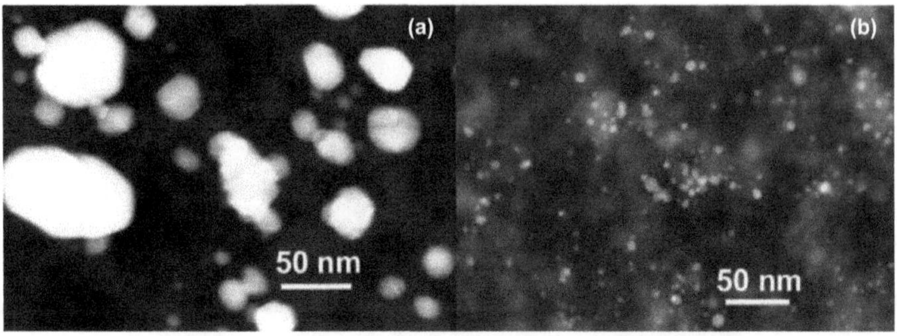

Figure 1.10 Dark-field TEM images of Au/SiO_2 (a) and $NiAu/SiO_2$ (b) catalysts collected after pre-treatment in H_2 at 720 °C, in O_2 at 300 °C, and after testing in CO oxidation.[60] The as-prepared $NiAu/SiO_2$ transformed to $Au-NiO/SiO_2$ (the working catalyst) during the pre-treatment. Reproduced with permission of Wiley-VCH from S. H. Zhou, H. F. Yin, V. Schwartz, Z. L. Wu, B. Eichhorn, S. H. Overbury and S. Dai, *ChemPhysChem*, 2008, **9**, 2475.

1.3 Catalysts with Single Metal–Support Interfaces

1.3.1 General Considerations

So far, we have highlighted some gold catalysts with complex metal–support, metal–modifier, and modifier–support interactions. The thermal stability of gold nanoparticles can be enhanced in this way. The question now arises as whether gold catalysts with single metal–support interfaces can still have good thermal stability.

1.3.2 Enhancing the Metal–Support Contact

Li and co-workers designed mesoporous multicomponent nanocomposite colloidal spheres (MMNCSs) in four steps: (1) synthesis of gold nanoparticles and CeO_2 nanoparticles; (2) mixing of these nanoparticles in a solvent at a certain ratio; (3) assembly of the mixed colloidal nanoparticles according to an emulsion-based bottom-up self-assembly strategy; and (4) calcination at 200 °C to form mesoporous structures.[127] Figure 1.11 shows the N_2 adsorption–desorption isotherm, pore size distribution, and the corresponding TEM image of $Au-CeO_2$ MMNCs.[127] The resulting $Au-CeO_2$ composite catalyst showed higher activity and better thermal stability than a conventional Au/CeO_2 catalyst. Similar observations were obtained for $Au-TiO_2$ MMNCSs.[127]

Murray and co-workers prepared $Au-Fe_3O_4$ binary nanoparticle superlattices (BNSLs) through self-assembly of gold and Fe_3O_4 nanocrystals at a liquid–air interface, and then transferred the BNSLs onto Si wafers.[128] $Au-Fe_3O_4$ BNSLs with different compositions and geometric arrangements can be fabricated in this way. In addition, the size of gold nanoparticles as

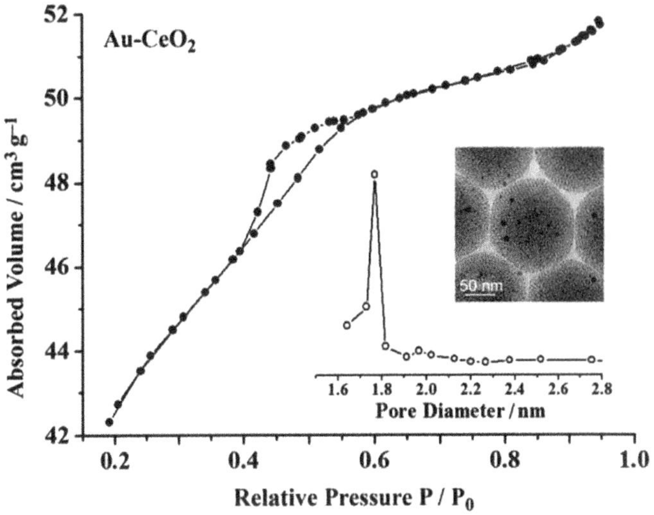

Figure 1.11 N_2 adsorption–desorption isotherm, pore size distribution curve, and the corresponding TEM image (inset) of Au-CeO_2 MMNCs.[127] Reproduced with permission of Wiley-VCH from C. Chen, C. Y. Nan, D. S. Wang, Q. Su, H. H. Duan, X. W. Liu, L. S. Zhang, D. R. Chu, W. G. Song, Q. Peng and Y. D. Li, *Angew. Chem., Int. Ed.*, 2011, **50**, 3725.

well as the thickness of Au-Fe_3O_4 BNSLs on Si wafers can be selected. This type of catalysts provides an excellent model for studying the contribution of interfacial sites for catalysis as well as thermal stability of gold catalysts. The authors demonstrated a strategy of keeping gold nanocrystals separated to enhance the thermal stability of gold catalysts.

1.3.3 Au@oxide Core-shell or Yolk-shell Structures

The thermal stability of gold nanoparticles may be improved by encapsulating gold nanoparticles in oxide shells. The inorganic shells can physically confine the gold nanoparticles, thus mitigating their sintering. In addition, they may provide interfacial sites active for catalysis. Normally there is one gold nanoparticle in one shell, not multiple gold nanoparticles in one shell. Depending on whether there is any space between gold nanoparticle and the shell, core-shell and yolk-shell structures can be distinguished.[129–133] These catalysts have metal–support interfaces without other additives, but the catalyst structures are different from conventional supported metal catalysts. To make the catalyst structure more complex, supported core-shell structures may be prepared,[66] and another component (*e.g.*, a metal oxide) can be added into the oxide shell.[134]

Yin and co-workers developed the Au@porous SiO_2 catalyst.[135] Au@SiO_2 core-shell particles were synthesized first *via* a modified sol-gel procedure using gold sol as the core and tetraethylorthosilicate as the silicon source for the formation of SiO_2 shells. The surface of SiO_2 shells was then functionalized by PVP, and etched by NaOH solution. The interior of the Au@SiO_2 was selectively etched to create porous structures, whereas gold nanoparticles were still encapsulated into the SiO_2 shells. The Au@porous SiO_2 catalyst demonstrated excellent stability in catalytic reduction of 4-nitrophenol. In contrast, uncoated gold nanoparticles aggregated immediately in the reaction mixture.

Schüth and co-workers developed a yolk-shell 'Au,@ZrO_2' catalyst.[136] Gold nanoparticles (15–17 nm) were prepared by sodium citrate reduction, and were coated with SiO_2 shells. The Au@SiO_2 spheres were then coated by ZrO_2 shells, and the SiO_2 in between the gold nanoparticles and the ZrO_2 shells was removed by NaOH solution. The ZrO_2 shells not only protected the gold cores, but also created Au-ZrO_2 interfaces active for CO oxidation. In addition, the thin (20 nm) porous ZrO_2 shells make the encapsulated gold nanoparticles accessible to the reactants.

Schüth and co-workers also developed a yolk-shell 'Au,@ZrO_2' catalyst with smaller gold particle sizes (Figure 1.12).[137] Gold nanoparticles (15 nm) were coated by SiO_2 shells. The size of the encapsulated gold cores was decreased to 10 nm by treating Au@SiO_2 core-shell particles with NaCN solution, during which some metallic gold atoms reacted with NaCN to form soluble $[Au(CN)_2]^-$. The material was then coated by ZrO_2 shells (thickness 7–10 nm), and the SiO_2 shells in between the gold cores and ZrO_2 shells were

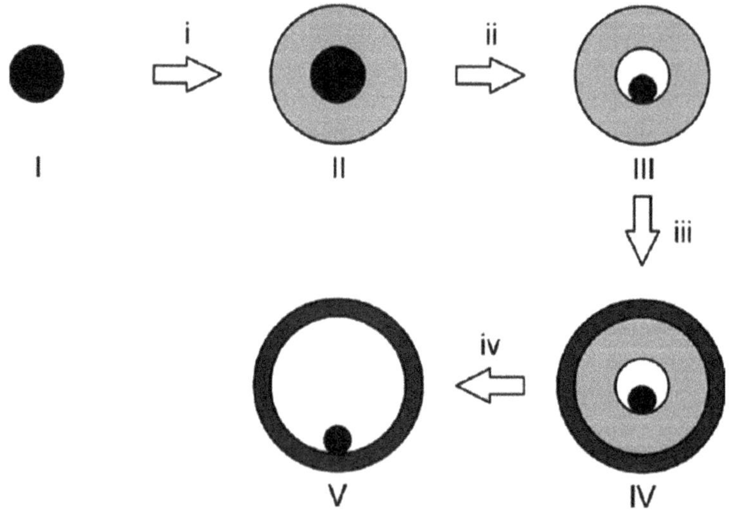

Figure 1.12 Scheme showing the preparation steps for Au,@ZrO_2 yolk-shell catalysts with reduced gold core size: (I) gold colloid; (II) SiO_2-covered gold nanoparticles; (III) SiO_2-covered gold cores with reduced size; (IV) ZrO_2-covered sphere; and (V) Au,@ZrO_2 yolk-shell structure.[137] Reproduced with permission of the Royal Society of Chemistry from R. Güttel, M. Paul and F. Schüth, *Chem. Commun.*, 2010, **46**, 895.

leached by aqueous NaOH. The resulting catalyst could withstand calcination at 800 °C, with no loss of catalytic activity in CO oxidation. Nevertheless, it was difficult to reduce the size of gold particles by NaCN further, and an alternative aqua regia treatment allowed fine-tuning of gold particle sizes down to 5 nm but the catalyst was deactivated by residual chlorine.

Zheng and co-workers prepared Au@hollow mesoporous-ZrO_2 (Au@*hm*-ZrO_2).[138] Hydrophobic dodecanethiol-capped gold nanoparticles (6.3 nm) underwent ligand exchange with mercaptoundecanoic acid, and were treated with ammonia to be water soluble. The water-soluble gold nanoparticles were then treated by $Si(OC_2H_5)_4$ to form SiO_2 shells (120 nm). The Au@SiO_2 spheres were coated with ZrO_2 *via* the hydrolysis of $Zr(OC_4H_9)_4$ in the presence of Brij 30 surfactant and calcined at 850 °C. The SiO_2 shells were removed by leaching with NaOH solution. The catalyst was active for the reduction of 4-nitrophenol, and showed excellent thermal stability.

Tang and co-workers prepared Au@CeO_2 *via* a facile 'self-templating' method.[139] $HAuCl_4$ and $CeCl_3$ were mixed with aqueous glucose and urea, and the solution was subjected to hydrothermal treatment to generate a dark brown solution. The suspension was then calcined at 600 °C to obtain Au@CeO_2. The role of glucose is to reduce cationic gold and to form amorphous carbon submicrospheres encapsulating the formed gold nanoparticles and adsorbing Ce^{3+} ions. The final calcination step removes the carbon submicrospheres and forms CeO_2 shells. Au@CeO_2 showed much higher activity than Au/CeO_2 due to the better stabilization of gold nanoparticles and more contact areas between gold nanoparticles and CeO_2.

To summarize, gold containing core-shell structures have been synthesized and enhanced thermal stability has been demonstrated. The studies mentioned above aimed at demonstrating protocols, and the structures of the catalysts developed had many versions, but there is still no universal method to prepare a variety of core-shell or yolk-shell catalysts and it was not discussed in the literature which catalyst structure is ideal. In addition, the synthesis often involves leaching, which is not good for environmental and safety reasons. Convenient ways for large-scale production of catalysts are still needed.

1.3.4 Encapsulation of Gold Nanoparticles in a Solid Matrix

Laursen *et al.* prepared zeolite-encapsulated gold nanoparticles.[140] This invention involves the immobilization of gold nanoparticles in an amorphous silica matrix, followed by hydrothermal treatment to crystallize the amorphous silica to a zeolite (silicalite-1) phase (Figure 1.13).[140] Most of the gold nanoparticles were trapped in the zeolite crystal. They exhibited high thermal stability and are resistant to acid leaching. On the other hand, some gold nanoparticles on the external surface of the zeolite crystal can still agglomerate upon heating, and they can be leached by aqua regia easily. In a subsequent work,[141] the authors developed a two-step method in which gold nanoparticles were immobilized in amorphous SiO_2 followed by

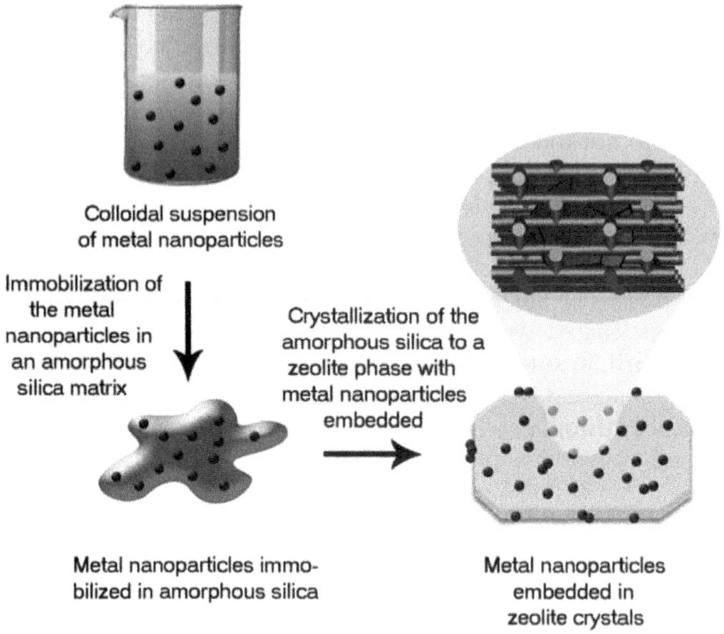

Figure 1.13 Schematic illustration of the encapsulation of gold nanoparticles in zeolite silicalite-1 crystals.[140]
Reproduced with permission of Wiley-VCH from A. B. Laursen, K. T. Højholt, L. F. Lundegaard, S. B. Simonsen, S. Helveg, F. Schüth, M. Paul, J.-D. Grunwaldt, S. Kegnæs, C. H. Christensen, and K. Egeblad, *Angew. Chem., Int. Ed.*, 2010, **49**, 3504.

hydrothermal treatment to create gold-embedded silicalite-1 seeds. The resulting silicalite-1 seeds were then suspended in a ZSM-5 growth medium containing aluminum to form gold-embedded ZSM-5.

1.4 Conclusions

Here we have summarized some strategies and examples for the stabilization of gold nanoparticles on solid supports. These strategies include: (1) pre-modification of supports before loading gold; (2) post-modification of supported gold catalysts; (3) dispersion of Au-Fe_3O_4 dumbbell structures on supports; (4) *in situ* transformation of supported alloy catalysts into metal oxide modified supported gold catalysts; and (5) development of Au@oxide core-shell or yolk-shell structures. In these cases, the interfacial structures have been modified. The presence of additional additives may enhance the interaction between metal and support, thus mitigating the sintering of gold nanoparticles. These oxide additives (especially transition metal oxides) may also tune the oxidation state of gold and tune the redox properties of the support, thus enhancing catalytic activity. Alternatively, the encapsulation of gold nanoparticles by oxide shells may avoid the sintering of gold nanoparticles. Considering that the catalytic activities of supported gold catalysts depend critically on the size of the gold nanoparticles, the stabilization of gold nanoparticles is meaningful for maintaining their performance under high temperature conditions.

There are still some deficiencies in previous research, which may provide new opportunities for further investigation.

(1) XRD and TEM data can provide direct evidence for the stabilization of gold nanoparticles (as judged by the sizes of gold nanoparticles), but they are not enough to pin down the location and morphologies of the additives that stabilize gold nanoparticles. High-resolution TEM and HAADF images may help address this issue and provide direct evidence.[64,66,124,142] For instance, Figure 1.14 shows an appealing high-resolution TEM image of the microscopic structure of Au/IrO_2/TiO_2 obtained from oxidative transformation of Au-Ir/TiO_2 at 400 °C.[142]
(2) First principle DFT calculations may be able to tell us why the presence of certain additives may enhance the thermal stability of gold nanoparticles,[89,96,143] and to predict the compositions and structures of better catalysts. It is also worthwhile to use DFT methods to understand the nature of active sites and reaction mechanisms on new catalysts,[144] although care should be taken when interpreting the results because of different conditions in different systems.
(3) Surface science studies, based on model catalyst systems, could be carried out to better understand the sintering behavior and support effect.[145–147]

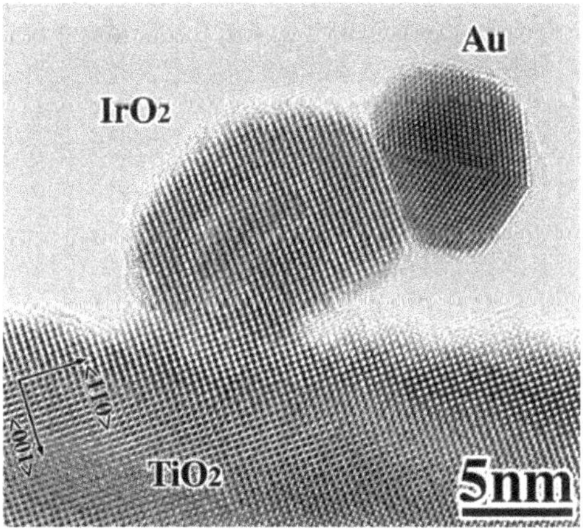

Figure 1.14 TEM image of gold nanoparticle-IrO_2 pillar structure on TiO_2 support.[142] This structure was formed by calcining Au-Ir/TiO_2 at 400 °C. Reproduced with permission of Oxford University Press from T. Akita, M. Okumura, K. Tanaka, S. Tsubota and M. Haruta, *J. Electron Microsc.*, 2003, **52**, 119.

(4) The synthesis methodologies highlighted above have not been demonstrated systematically, and most of the studies rely on simple test reactions such as CO oxidation and reduction of 4-nitrophenol. It is worthwhile to synthesize a wide variety of catalysts following these synthesis methodologies, and to explore the applications of these catalysts in a wide variety of catalytic reactions.

Acknowledgements

Z. Ma thanks the National Natural Science Foundation of China (Grant Nos. 21007011 and 21177028), the PhD programs foundation of the Ministry of Education in China (Grant No. 20100071120012), and the overseas returnees start-up research fund of the Ministry of Education in China for the financial support. S. Dai thanks the Office of Basic Energy Sciences, US Department of Energy for financial support. Oak Ridge National Laboratory is managed by UT-Battelle, LLC for the US DOE under Contract DE-AC05-00OR22725.

References

1. K. B. Zhou and Y. D. Li, *Angew. Chem., Int. Ed.*, 2012, **51**, 602.
2. H. Tüysüz and F. Schüth, *Adv. Catal.*, 2012, **55**, 127.
3. K. An and G. A. Somorjai, *ChemCatChem*, 2012, **4**, 1512.
4. F. Zaera, *Chem. Soc. Rev.*, 2013, **42**, 2746.
5. M. Haruta, T. Kobayashi, H. Sano and N. Yamada, *Chem. Lett.*, 1987, **16**, 405.
6. M. Haruta, N. Yamada, T. Kobayashi and S. Iijima, *J. Catal.*, 1989, **115**, 301.
7. M. Haruta, S. Tsubota, T. Kobayashi, H. Kageyama, M. J. Genet and B. Delmon, *J. Catal.*, 1993, **144**, 175.
8. M. Haruta, *Catal. Today*, 1997, **36**, 153.
9. M. Haruta and M. Daté, *Appl. Catal. A*, 2001, **222**, 427.
10. M. Haruta, *CATTECH*, 2002, **6**, 102.
11. M. Haruta, *Chem. Rec.*, 2003, **3**, 75.
12. G. C. Bond and D. T. Thompson, *Catal. Rev. Sci. Eng.*, 1999, **41**, 319.
13. G. C. Bond, C. Louis and D. T. Thompson, *'Catalysis by Gold'*, Imperial College Press, London, 2006.
14. A. S. K. Hashmi and G. J. Hutchings, *Angew. Chem., Int. Ed.*, 2006, **45**, 7896.
15. M. C. Kung, R. J. Davis and H. H. Kung, *J. Phys. Chem. C*, 2007, **111**, 11767.
16. S. A. C. Carabineiro and D. T. Thompson, in *Nanocatalysis*, ed. U. Heiz and U. Landman, Springer, Berlin, 2007, p. 377.
17. C. Della Pina, E. Falletta, L. Prati and M. Rossi, *Chem. Soc. Rev.*, 2008, **37**, 2077.
18. A. Corma and H. Garcia, *Chem. Soc. Rev.*, 2008, **37**, 2096.

19. Z. Ma, S. H. Overbury, and S. Dai, in *Nanomaterials: Inorganic and Bioinorganic Perspectives*, ed. C. M. Lukehart and R. A. Scott, John Wiley & Sons, Chichester, 2009, p. 247.
20. Z. Ma and S. Dai, *Nano Res.*, 2011, **4**, 3.
21. Z. Ma and S. Dai, *ACS Catal.*, 2011, **1**, 805.
22. L.-F. Gutiérrez, S. Hamoudi and K. Belkacemi, *Catal.*, 2011, **1**, 97.
23. M. Haruta, *Faraday Discuss.*, 2011, **152**, 11.
24. L. Parti and A. Villa, *Catal.*, 2012, **2**, 24.
25. M. Stratakis and H. Garcia, *Chem. Rev.*, 2012, **112**, 4469.
26. T. Takei, K. Akita, I. Nakamura, T. Fujitani, M. Okumura, K. Okazaki, J. H. Huang, T. Ishida and M. Haruta, *Adv. Catal.*, 2012, **55**, 1.
27. M.-C. Daniel and D. Astruc, *Chem. Rev.*, 2004, **104**, 293.
28. D. Astruc, F. Lu and J. R. Aranzaes, *Angew. Chem., Int. Ed.*, 2005, **44**, 7852.
29. H. Bönnemann and K. S. Nagabhushana, in *Surface and Nanomolecular Catalysis*, ed. R. Richards, CRC Press, Boca Raton, 2006, p. 63.
30. C.-J. Jia and F. Schüth, *Phys. Chem. Chem. Phys.*, 2011, **13**, 2457.
31. P. X. Zhao, *Coord. Chem. Rev.*, 2012, **257**, 638.
32. G. Rothenberg, *Catalysis: Concepts and Green Applications*, Wiley-VCH, Weinheim, 2008.
33. A. M. Cao, R. W. Lu and G. Veser, *Phys. Chem. Chem. Phys.*, 2010, **12**, 13499.
34. G. Pattrick, E. van der Lingen, C. W. Corti, R. J. Holliday and D. T. Thompson, *Top. Catal.*, 2004, **30–31**, 273.
35. Q. Xu, K. C. C. Kharas and A. K. Datye, *Catal. Lett.*, 2003, **85**, 229.
36. S. Ivanova, C. Petit and V. Pitchon, *Appl. Catal. A*, 2004, **267**, 191.
37. H. G. Zhu, B. Lee, S. Dai and S. H. Overbury, *Langmuir*, 2003, **19**, 3974.
38. W. F. Yan, B. Chen, S. M. Mahurin, S. Dai and S. H. Overbury, *Chem. Commun.*, 2004, 1918.
39. W. F. Yan, B. Chen, S. M. Mahurin, E. W. Hagaman, S. Dai and S. H. Overbury, *J. Phys. Chem. B*, 2004, **108**, 2793.
40. H. G. Zhu, Z. W. Pan, B. Chen, B. Lee, S. M. Mahurin, S. H. Overbury and S. Dai, *J. Phys. Chem. B*, 2004, **108**, 20038.
41. B. Lee, H. G. Zhu, Z. T. Zhang, S. H. Overbury and S. Dai, *Micropor. Mesopor. Mater.*, 2004, **70**, 71.
42. W. F. Yan, V. Petkov, S. M. Mahurin, S. H. Overbury and S. Dai, *Catal. Commun.*, 2005, **6**, 404.
43. W. F. Yan, B. Chen, S. M. Mahurin, V. Schwartz, D. R. Mullins, A. R. Lupini, S. J. Pennycook, S. Dai and S. H. Overbury, *J. Phys. Chem. B*, 2005, **109**, 10676.
44. W. F. Yan, S. M. Mahurin, Z. W. Pan, S. H. Overbury and S. Dai, *J. Am. Chem. Soc.*, 2005, **127**, 10480.
45. W. F. Yan, S. M. Mahurin, B. Chen, S. H. Overbury and S. Dai, *J. Phys. Chem. B*, 2005, **109**, 15489.
46. W. F. Yan, S. Brown, Z. W. Pan, S. M. Mahurin, S. H. Overbury and S. Dai, *Angew. Chem., Int. Ed.*, 2006, **45**, 3614.

47. W. F. Yan, S. M. Mahurin, S. H. Overbury and S. Dai, *Top. Catal*, 2006, **39**, 199.
48. H. G. Zhu, C. D. Liang, W. F. Yan, S. H. Overbury and S. Dai, *J. Phys. Chem. B*, 2006, **110**, 10842.
49. H. G. Zhu, Z. Ma, J. C. Clark, Z. W. Pan, S. H. Overbury and S. Dai, *Appl. Catal. A*, 2007, **326**, 89.
50. H. G. Zhu, Z. Ma, S. H. Overbury and S. Dai, *Catal. Lett.*, 2007, **116**, 128.
51. Z. Ma, S. H. Overbury and S. Dai, *J. Mol. Catal. A*, 2007, **273**, 186.
52. Z. Ma, S. Brown, S. H. Overbury and S. Dai, *Appl. Catal. A*, 2007, **327**, 226.
53. Z. Ma, C. D. Liang, S. H. Overbury and S. Dai, *J. Catal.*, 2007, **252**, 119.
54. W. F. Yan, B. Chen, S. M. Mahurin, S. H. Overbury and S. Dai, *Stud. Surf. Sci. Catal.*, 2007, **170**, 1065.
55. W. F. Yan, Z. Ma, S. M. Mahurin, J. Jiao, E. W. Hagaman, S. H. Overbury and S. Dai, *Catal. Lett.*, 2008, **121**, 209.
56. Z. Ma, S. Brown, J. Y. Howe, S. H. Overbury and S. Dai, *J. Phys. Chem. C*, 2008, **112**, 9448.
57. H. F. Yin, Z. Ma, S. H. Overbury and S. Dai, *J. Phys. Chem. C*, 2008, **112**, 8349.
58. Z. Ma, H. F. Yin, S. H. Overbury and S. Dai, *Catal. Lett.*, 2008, **126**, 20.
59. H. F. Yin, C. Wang, H. G. Zhu, S. H. Overbury, S. H. Sun and S. Dai, *Chem. Commun.*, 2008, 4357.
60. S. H. Zhou, H. F. Yin, V. Schwartz, Z. L. Wu, D. R. Mullins, B. Eichhorn, S. H. Overbury and S. Dai, *ChemPhysChem*, 2008, **9**, 2475.
61. S. H. Zhou, Z. Ma, H. F. Yin, Z. L. Wu, B. Eichhorn, S. H. Overbury and S. Dai, *J. Phys. Chem. C*, 2009, **113**, 5758.
62. B. Lee, Z. Ma, Z. Zhang, C. Park and S. Dai, *Micropor. Mesopor. Mater.*, 2009, **122**, 160.
63. Z. Ma, H. F. Yin and S. Dai, *Catal. Lett.*, 2010, **136**, 83.
64. H. F. Yin, Z. Ma, M. F. Chi and S. Dai, *Catal. Lett.*, 2010, **136**, 209.
65. Z. Ma, H. F. Yin and S. Dai, *Catal. Lett.*, 2010, **138**, 40.
66. H. F. Yin, Z. Ma, M. F. Chi and S. Dai, *Catal. Today*, 2011, **160**, 87.
67. J. C. Bauer, D. Mullins, M. J. Li, Z. L. Wu, E. A. Payzant, S. H. Overbury and S. Dai, *Phys. Chem. Chem. Phys.*, 2011, **13**, 2571.
68. C. C. Tian, S. H. Chai, X. Zhu, Z. L. Wu, A. Binder, J. C. Bauer, S. Brown, M. F. Chi, G. M. Veith, Y. L. Guo and S. Dai, *J. Mater. Chem.*, 2012, **22**, 25227.
69. G. C. Ma, A. Binder, M. F. Chi, C. Liu, R. C. Jin, D. E. Jiang, J. Fan and S. Dai, *Chem. Commun.*, 2012, **48**, 11413.
70. J. C. Bauer, G. M. Veith, L. F. Allard, Y. Oyola, S. H. Overbury and S. Dai, *ACS Catal.*, 2012, **2**, 2537.
71. L. Guczi, A. Beck and K. Frey, *Gold Bull.*, 2009, **42**, 5.
72. G. M. Veith, A. R. Lupini, S. Rashkeev, S. J. Pennycook, D. R. Mullins, V. Schwartz, C. A. Bridges and N. J. Dudney, *J. Catal.*, 2009, **262**, 92.

73. Y. Tai, J. Murakami, K. Tajiri, F. Ohashi, M. Daté and S. Tsubota, *Appl. Catal. A*, 2004, **268**, 183.
74. A. M. Venezia, F. L. Liotta, G. Pantaleo, A. Beck, A. Horvath, O. Geszti, A. Kocsonya and L. Guczi, *Appl. Catal. A*, 2006, **310**, 114.
75. M. Bandyopadhyay, O. Korsak, M. W. E. van den Berg, W. Grunert, A. Birkner, W. Li, F. Schüth and H. Gies, *Micropor. Mesopor. Mater.*, 2006, **89**, 158.
76. A. Beck, A. Horváth, G. Stefler, R. Katona, O. Geszti, G. Tolnai, L. F. Liotta and L. Guczi, *Catal. Today*, 2008, **139**, 180.
77. Y. Tai and K. Tajiri, *Appl. Catal. A*, 2008, **342**, 113.
78. Y. Tai, W. Yamaguchi, K. Tajiri and H. Kageyama, *Appl. Catal. A*, 2009, **364**, 143.
79. A. Beck, A. Horváth, G. Stefler, M. S. Scurrel and L. Guczi, *Top. Catal.*, 2009, **52**, 912.
80. V. S. Narkhede, A. De Toni, V. V. Narkhede, M. Guraya, J. W. Niemantsverdriet, M. W. E. van den Berg, W. Grünert and H. Gies, *Micropor. Mesopor. Mater.*, 2009, **118**, 52.
81. C. L. Peza-Ledesma, L. Escamilla-Perea, R. Nava, B. Pawelec and J. L. G. Fierro, *Appl. Catal. A*, 2010, **375**, 37.
82. K. Qian, W. X. Huang, Z. Q. Jiang and H. X. Sun, *J. Catal.*, 2007, **248**, 137.
83. K. Qian, W. X. Huang, J. Fang, S. S. Lv, B. He, Z. Q. Jiang and S. Q. Wei, *J. Catal.*, 2008, **255**, 269.
84. L. F. Liotta, G. Pantaleo, F. Puleo and A. M. Venezia, *Catal. Today*, 2012, **187**, 10.
85. N. Sahu, K. M. Parida, A. K. Tripathi and V. S. Kamble, *Appl. Catal. A*, 2011, **399**, 110.
86. J. J. Plata, A. M. Márquez, J. F. Sanz, R. S. Avellaneda, F. Romero-Sarria, M. I. Domínguez, M. A. Centrno and J. A. Odriozola, *Top. Catal.*, 2011, **54**, 219.
87. R. Zanella, V. Rodríguez-González, Y. Arzola and A. Moreno-Rodriguez, *ACS Catal.*, 2012, **2**, 1.
88. Y.-W. Chen, D.-S. Lee and H.-J. Chen, *Int. J. Hydrogen Energy*, 2012, **37**, 15140.
89. C.-M. Wang, K.-N. Fan and Z.-P. Liu, *J. Phys. Chem. C*, 2007, **111**, 13539.
90. C. X. Qi, H. J. Su, R. G. Guan and X. F. Xu, *J. Phys. Chem. C*, 2012, **116**, 17492.
91. N. Comsup, J. Panpranot and P. Praserthdam, *Catal. Commun.*, 2010, 11.
92. B. Solsona, M. Conte, Y. Cong, A. Carley and G. Hutchings, *Chem. Commun.*, 2005, 2351.
93. A. Horváth, A. Beck, A. Sárkány, G. Stefler, Z. Varga, O. Geszti, L. Tóth and L. Guczi, *J. Phys. Chem. B*, 2006, **110**, 15417.
94. F. Wang and G. X. Lu, *Catal. Lett.*, 2007, **115**, 46.
95. A. Horváth, A. Beck, G. Stefler, T. Benkó, G. Sáfrán, Z. Varga, J. Gubicza and L. Guczi, *J. Phys. Chem. C*, 2011, **115**, 20388.

96. S. N. Rashkeev, S. Dai and S. H. Overbury, *J. Phys. Chem. C*, 2010, **114**, 2996.
97. C. Moreno, N. J. Divins, J. Gázquez, M. Varela, I. Angurell and J. Llorca, *Nanoscale*, 2012, **4**, 2278.
98. J. P. Ge, Q. Zhang, T. R. Zhang and Y. D. Yin, *Angew. Chem., Int. Ed.*, 2008, **47**, 8924.
99. Y. H. Deng, Y. Cai, Z. K. Sun, J. Liu, C. Liu, J. Wei, W. Li, C. Liu, Y. Wang and D. Y. Zhao, *J. Am. Chem. Soc.*, 2010, **132**, 8466.
100. T. Kanazawa, *Catal. Lett.*, 2006, **108**, 45.
101. S. Takenaka, H. Matsumori, K. Nakagawa, H. Matsune, E. Tanabe and M. Kishida, *J. Phys. Chem. C*, 2007, **111**, 15133.
102. S. Takenaka, T. Arike, H. Matsune, E. Tanabe and M. Kishida, *J. Catal.*, 2008, **257**, 345.
103. K. Nakagawa, Y. Tanimoto, T. Okayama, K.-I. Sotowa, S. Sugiyama, S. Takenaka and M. Kishida, *Catal. Lett.*, 2010, **136**, 71.
104. K. Nakagawa, T. Okayama, Y. Tanimoto, K.-I. Sotowa, S. Sugiyama, T. Moriga, S. Takenaka and M. Kishida, *Appl. Catal. A*, 2012, **419**, 13.
105. P. Zhang, M. F. Chi, S. Sharma and E. W. McFarland, *J. Mater. Chem.*, 2010, **20**, 2013.
106. Q. Zhang, I. Lee, J. P. Ge, F. Zaera and Y. D. Yin, *Adv. Funct. Mater.*, 2010, **20**, 2201.
107. I. Lee, Q. Zhang, J. P. Ge, Y. D. Yin and F. Zaera, *Nano Res.*, 2011, **4**, 115.
108. A. S. Reddy, S. Kim, H. Y. Jeong, S. Jin, K. Qadir, K. Jung, C. H. Jung, J. Y. Yun, J. Y. Cheon, J.-M. Yang, S. H. Joo, O. Terasaki and J. Y. Park, *Chem. Commun.*, 2011, **47**, 8412.
109. X. H. Liang, J. H. Li, M. Yu, C. N. McMurray, J. L. Falconer and A. W. Weimer, *ACS Catal.*, 2011, **1**, 1162.
110. H. Feng, J. L. Lu, P. C. Stair and J. W. Elam, *Catal. Lett.*, 2011, **141**, 512.
111. J. F. Lu, B. S. Fu, M. C. Kung, G. M. Xiao, J. W. Elam, H. H. Kung and P. C. Stair, *Science*, 2012, **335**, 1205.
112. A. Binder and M. Seipenbusch, *Appl. Catal. A*, 2011, **396**, 1.
113. H. Yu, M. Chen, P. M. Rice, S. X. Wang, R. L. White and S. H. Sun, *Nano Lett.*, 2005, **5**, 379.
114. C. Wang, C. J. Xu, H. Zeng and S. H. Sun, *Adv. Mater.*, 2009, **21**, 3045.
115. F.-H. Lin, W. Chen, Y.-H. Liao, R.-A. Doong and Y. D. Li, *Nano Res.*, 2011, **4**, 1223.
116. B. H. Wu, H. Zhang, C. Chen, S. C. Lin and N. F. Zheng, *Nano Res.*, 2009, **2**, 975.
117. A.-Q. Wang, C.-M. Chang and C.-Y. Mou, *J. Phys. Chem. B*, 2005, **109**, 18860.
118. D. I. Enache, J. K. Edwards, P. Landon, B. Solsona-Espriu, A. F. Carley, A. A. Herzing, M. Watanabe, C. J. Kiely, D. W. Knight and G. J. Hutchings, *Science*, 2006, **311**, 362.
119. X. Y. Liu, A. Q. Wang, X. D. Wang, C.-Y. Mou and T. Zhang, *Chem. Commun.*, 2008, 3187.

120. S. Albonetti, R. Bonelli, J. E. Mengou, C. Femoni, C. Tiozzo, S. Zacchini and F. Trifirò, *Catal. Today*, 2008, **137**, 483.
121. S. Albonetti, R. Bonelli, R. Delaigle, C. Femoni, E. M. Gaigneaux, V. Morandi, L. Ortolani, C. Tiozzo, S. Zacchini and F. Trifirò, *Appl. Catal. A*, 2010, **372**, 138.
122. R. Bonelli, S. Albonetti, V. Morandi, L. Ortolani, P. M. Riccobene, S. Scirè and S. Zacchini, *Appl. Catal. A*, 2011, **395**, 10.
123. X. Y. Liu, A. Q. Wang, L. Li, T. Zhang, C.-Y. Mou and J. F. Lee, *J. Catal.*, 2011, **278**, 288.
124. R. Bonelli, S. Zacchini and S. Albonetti, *Catal.*, 2012, **2**, 1.
125. X. Li, S. S. S. Fang, J. Teo, Y. L. Foo, A. Borgna, M. Lin and Z. Y. Zhong, *ACS Catal.*, 2012, **2**, 360.
126. X. D. Wang, H. B. Yu, D. Y. Hua and S. H. Zhou, *J. Phys. Chem. C*, 2013, **117**, 7294.
127. C. Chen, C. Y. Nan, D. S. Wang, Q. Su, H. H. Duan, X. W. Liu, L. S. Zhang, D. R. Chu, W. G. Song, Q. Peng and Y. D. Li, *Angew. Chem., Int. Ed.*, 2011, **50**, 3725.
128. Y. J. Kang, X. C. Ye, J. Chen, L. Qi, R. E. Diaz, V. Doan-Nguyen, G. Z. Xing, C. R. Kagan, J. Li, R. J. Gorte, E. A. Stach and C. B. Murray, *J. Am. Chem. Soc.*, 2013, **135**, 1499.
129. L. M. Liz-Marzan and P. Mulvaney, *J. Phys. Chem. B*, 2003, **107**, 7312.
130. S. H. Liu and M. Y. Han, *Chem. Asian J.*, 2010, **5**, 36.
131. J. C. Park and H. Song, *Nano Res.*, 2011, **4**, 33.
132. Z. W. Seh, S. H. Liu and M.-Y. Han, *Chem. Asian J.*, 2012, **7**, 2174.
133. S. H. Liu, S.-Q. Bai, Y. G. Zheng, K. W. Shah and M. Y. Han, *ChemCatChem*, 2012, **4**, 1462.
134. R. Güttel, M. Paul and F. Schüth, *Catal. Sci. Eng.*, 2011, **1**, 65.
135. Q. Zhang, T. R. Zhang, J. P. Ge and Y. D. Yin, *Nano Lett.*, 2008, **8**, 2867.
136. P. M. Arnal, M. Comotti and F. Schüth, *Angew. Chem., Int. Ed.*, 2006, **45**, 8224.
137. R. Güttel, M. Paul and F. Schüth, *Chem. Commun.*, 2010, **46**, 895.
138. X. Q. Huang, C. Y. Guo, J. Q. Zuo, N. F. Zheng and G. D. Stucky, *Small*, 2009, **5**, 361.
139. J. Qi, J. Chen, G. D. Li, S. X. Li, Y. Gao and Z. Y. Tang, *Energy Environ. Sci.*, 2012, **5**, 8937.
140. A. Laursen, K. T. Højholt, L. F. Lundegaard, S. B. Simonsen, S. Helveg, F. Schüth, M. Paul, J.-D. Grunwaldt, S. Kegnæs, C. H. Christensen and K. Egeblad, *Angew. Chem., Int. Ed.*, 2010, **49**, 3504.
141. K. T. Hojhølt, A. B. Laursen, S. Kagnæs and C. H. Christensen, *Top. Catal.*, 2011, **54**, 1026.
142. T. Akita, M. Okumura, K. Tanaka, S. Tsubota and M. Haruta, *J. Electron Microsc.*, 2003, **52**, 119.
143. Z.-P. Liu, S. J. Jenkins and D. A. King, *Phys. Rev. Lett.*, 2004, **93**, 156102.

144. J. Graciani, J. J. Plata, J. F. Sanz, P. Liu and J. A. Rodriguez, *J. Chem. Phys.*, 2010, **132**, 104703.
145. B. K. Min, W. T. Wallace and D. W. Goodman, *J. Phys. Chem. B*, 2004, **108**, 14609.
146. W. T. Wallace, B. K. Min and D. W. Goodman, *J. Mol. Catal. A*, 2005, **228**, 3.
147. R. Si, J. Tao, J. Evans, J. B. Park, L. Barrio, J. C. Hanson, Y. M. Zhu, J. Hrbek and J. A. Rodriguez, *J. Phys. Chem. C*, 2012, **116**, 23547.

CHAPTER 2

Well-defined Gold Nanoclusters for Catalytic Applications

GAO LI AND RONGCHAO JIN*

Department of Chemistry, Carnegie Mellon University, 4400 Fifth Ave, Pittsburgh, PA, 15213, USA
*Email: rongchao@andrew.cmu.edu

2.1 Introduction

Metal nanoparticles occupy an important niche in catalysis, such as selective oxidation and hydrogenation, and carbon–carbon coupling reactions.[1–4] In fundamental studies of the catalytic reaction mechanism, the polydispersity of conventional nanocatalysts poses some major issues: the size-dependent catalytic activities of nanoparticles are averaged out in polydispersed catalysts, and it is difficult to relate the catalytic performance with the structure and intrinsic properties of nanocatalysts. To overcome these obstacles, well-defined nanoparticle catalysts are highly desirable. The ultimate level is to prepare atomically precise nanoparticles, which resemble molecular compounds as they possess well-defined formulas and their structures can be solved by growing single crystals followed by X-ray crystallography. These truly monodispersed, molecularly pure nanoparticles (often called nanoclusters) are of paramount importance for fundamental studies, not only for catalysis, but also for many other topics in nanoscience research.

In recent years, remarkable advances have been achieved in solution-phase synthesis of atomically precise nanoclusters by a 'size-focusing' methodology,

RSC Catalysis Series No. 18
Heterogeneous Gold Catalysts and Catalysis
Edited by Zhen Ma and Sheng Dai

Published by the Royal Society of Chemistry, www.rsc.org

notably thiolate-protected gold nanoclusters.[5–7] These thiolate-protected Au nanoclusters are well-defined to the ultimate atomic level (*i.e.*, molecular purity), rather than the usual nanometre precision in conventional nanoparticle synthesis. A number of specific-size, thiolate-protected $Au_n(SR)_m$ nanoclusters have become available; each of them is composed of a precise number of metal atoms (n) and of ligands (m). $Au_n(SR)_m$ with n from a dozen to about 200 atoms (equivalently 1–2 nm core diameter)[7] are found to be non-metallic, manifested in their sizable bandgaps. Thus, such nanoclusters are particularly interesting in terms of probing how the evolution of the electronic properties of nanoclusters influences their catalytic performance.

Compared with conventional metallic nanoparticle catalysts, non-metallic $Au_n(SR)_m$ nanoclusters possess several distinct features that are of particular interest to catalysis. The small sized gold nanoclusters (core size 1–2 nm) exhibit distinct electron-energy quantization,[7,8] which is in constrast to the continuous conduction band in metallic gold nanoparticles or bulk gold. Thus, gold nanoclusters become semiconducting and possess a sizable bandgap (*e.g.* ~1.3 eV for $Au_{25}(SR)_{18}$[9]). In addition, $Au_n(SR)_m$ nanoclusters can be doped with a single foreign atom[10–12] (*e.g.*, Pd and Pt), which is of particular interest to nanocatalysis, as the catalytic properties of gold nanoclusters can be tuned on an atom-by-atom basis.

In this chapter we introduce $Au_n(SR)_m$ nanoclusters as a new promising class of model catalyst. These nanoclusters can be either used as a homogeneous catalyst for solution phase reactions, or loaded onto a solid support to become a heterogeneous catalyst. Correlation between catalytic performance and the structure of gold nanoclusters is also discussed.

2.2 Synthesis of Atomically Precise Gold Nanoclusters $Au_n(SR)_m$

The atomically precise $Au_n(SR)_m$ nanocluster catalysts are synthesized *via* a size focusing process. In this process, a proper distribution of size-mixed nanoclusters is first made by kinetically controlling the reduction of gold precursor, typically Au(I)-SR polymer, with $NaBH_4$. The as-prepared size-mixed gold nanoclusters are then subjected to size focusing under appropriate conditions (*e.g.* at 80 °C overnight and in the presence of excess thiol). During the size focusing process, the unstable gold nanoclusters decompose or convert to the thermodynamically stable gold nanoclusters, and by controlling the initial size range of the starting size-mixed nanoclusters, one can obtain single-sized, thermodynamically stable gold nanoclusters of a specific size. By tuning the initial size range of the gold nanoclusters in the first step, the subsequent size-focusing step can give rise to a series of size discrete $Au_n(SR)_m$ nanoclusters, such as $Au_{25}(SR)_{18}$,[9,13] $Au_{38}(SR)_{24}$,[14] $Au_{144}(SR)_{60}$,[15] $Au_{333}(SR)_{79}$,[16] *etc.* Herein, the $Au_{25}(SR)_{18}$ nanocluster is chosen as a typical example for detailed discussion of the size-focusing processes (*vide infra*).

2.2.1 The Case of the $Au_{25}(SR)_{18}$ Nanocluster

The $Au_{25}(SR)_{18}$ nanocluster was the first case found in our early work using the 'size-focusing' methodology.[9,13] The synthesis procedure of $Au_{25}(SR)_{18}$ nanoclusters consists of two steps. In the first step, Au(III) salt (typically $HAuCl_4$) was reduced with excess thiol to form an Au(I)-SR polymeric intermediate. Then, the Au(I)-SR intermediate was further reduced by a strong reducing agent (typically $NaBH_4$). After prolonged aging of the reaction product, three prominent absorption peaks (670, 450, and 400 nm) emerged in the ultraviolet-visible (UV-vis) spectrum of the crude product [Figure 2.1], indicating a spontaneous size focusing process under the reaction conditions.[9,13] The observed optical absorption peaks are characteristic of $Au_{25}(SR)_{18}$ nanoclusters and can be conveniently used to identify or trace the size-focusing process. The intriguing focusing process of $Au_{25}(SR)_{18}$ nanoclusters was later found to be quite universal, *i.e.* irrespective of the type of thiol (*e.g.*, HSC_2H_4Ph, HSC_6H_{13}, $HSC_{12}H_{25}$, glutathione, $HSC_{10}H_{22}COOH$, *etc.*). The UV-vis spectra of $Au_{25}(SR)_{18}$ nanoclusters with different types of thiolates were also found to be the same, implying a common structure adopted in all $Au_{25}(SR)_{18}$ protected by different thiolates.[13,17]

The case of $Au_{25}(SR)_{18}$ synthesis illustrates the basic principle of the 'size-focusing' process. This process is based upon the stability of different sized

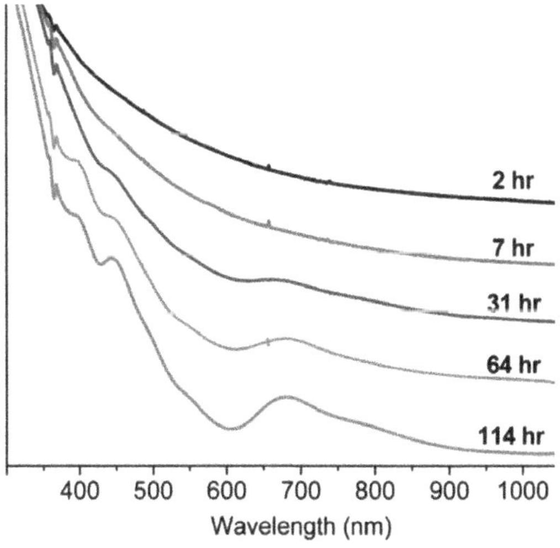

Figure 2.1 UV-vis spectral monitoring of the 'size-focusing' process for the growth of $Au_{25}(SR)_{18}$ nanoclusters.
Adapted with permission from ref. 13.

gold nanoclusters. Controlling the size distribution of the starting $Au_n(SR)_m$ mixture plays a critical role in finally obtaining the atomic monodispersity of nanoclusters. In addition, O_2 is probably involved in the size-focusing process, though details remain to be unraveled in future work.

2.3 Crystallographic Structure of the $Au_{25}(SR)_{18}$ Nanocluster

Among the reported $Au_n(SR)_m$ nanoclusters, the $Au_{25}(SR)_{18}$ nanocluster has been extensively investigated for nanocatalysis (*vide infra*). Thus, we briefly discuss the structure and electronic properties of the $Au_{25}(SR)_{18}$ nanocluster (where, R = CH_2CH_2Ph), which serve as the basis of the catalytic studies.[8,18] The $Au_{25}(SR)_{18}$ nanocluster comprises a 13 atom Au_{13} icosahedral core and a $Au_{12}(SR)_{18}$ shell [Figure 2.2(a)]. In the space-filling model [Figure 2.2(b)], an external, triangular Au_3 motif and even some of the core Au atoms are indeed accessible to reactant molecules in catalytic reactions; thus, the triangular Au_3 motif may be associated with the catalytically active site. The electronic property of $Au_{25}(SR)_{18}$ exhibits discrete energy levels, which is caused by the quantum-size effect. The molecular orbitals[18] can be roughly divided into the Au core orbitals (*i.e.*, primarily contributed by the Au_{13} core) and the shell orbitals (*i.e.*, primarily contributed by the $Au_{12}(SR)_{18}$ shell), reflecting the geometric core/shell structure. In addition, the $Au_{25}(SR)_{18}$ nanoclusters [*e.g.*, R = C_6H_{13}, $C_{12}H_{25}$, and glutathione (G)] exhibit the same geometric structure regardless of the thiolate ligand.[19] In the case of the $Au_{25}(SR)_{18}$ nanoclusters with foreign atom(s), the electronic and catalytic properties (*vide infra*) were found to be largely influenced by the foreign atom(s), albeit the structure of the bimetal nanocluster is the same.[10–12]

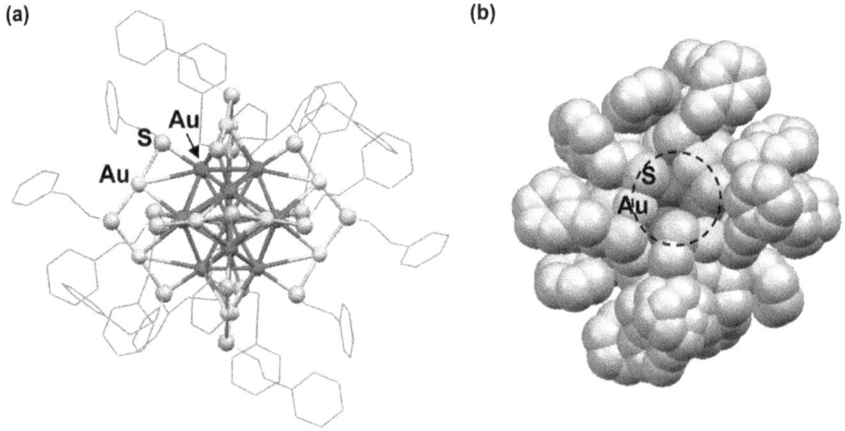

Figure 2.2 Crystallographic structure of $Au_{25}(SCH_2CH_2Ph)_{18}$ nanoclusters [magenta: gold atoms in the core, cyan: gold atoms in the shell (or surface)]. (a) Ball-stick model; (b) space-filling model.
Adapted with permission from ref. 18.

2.4 Thermal Stability of $Au_n(SR)_m$ Nanoclusters

The thermal stability of $Au_n(SR)_m$ nanoclusters is important in the catalytic application since most catalytic reactions are carried out under thermal conditions (*e.g.*, <150 °C). The thiolate-protected gold nanoclusters are particularly robust as they are synthesized *via* the size-focusing process (*e.g.* at 80 °C overnight in the presence of excess thiol). The gold nanoclusters can be stable for months or even years under ambient conditions (*e.g.*, stored in a refrigerator). Thermogravimetric analysis (TGA) shows that $Au_n(SR)_m$ (*e.g.*, $Au_{25}(SR)_{18}$, $Au_{38}(SR)_{24}$, $Au_{144}(SR)_{60}$, and $Au_{25}(SG)_{18}$) nanoclusters start to lose thiolate ligands at ~200 °C in a N_2 atmosphere [Figure 2.3(a)].[20] It is worth noting that the thermal decomposition of the Au nanoclusters is not affected by the atmosphere (*e.g.*, N_2, air, O_2, or H_2). Furthermore, the isothermal stability test of $Au_{25}(SR)_{18}$ nanoclusters (*i.e.*, maintained at 150 °C in air atmosphere for 60 min) indicates that the gold nanoclusters remain intact after the annealing process [Figure 2.3(b)]. The thermal test confirms that the $Au_{25}(SR)_{18}$ nanoclusters are robust and thermally stable (<150 °C).[21]

2.5 Reactivity and Catalytic Properties of $Au_n(SR)_m$ Nanoclusters

2.5.1 The Reversible Conversion between $[Au_{25}(SR)_{18}]^0$ and $[Au_{25}(SR)_{18}]^-$

It is known that the electron transfer between nanocatalyst and reactant(s) is key in catalytic reactions. The $Au_{25}(SR)_{18}$ nanocluster catalyst indeed exhibits good capability in the electron transfer. The native $Au_{25}(SR)_{18}$ nanocluster is

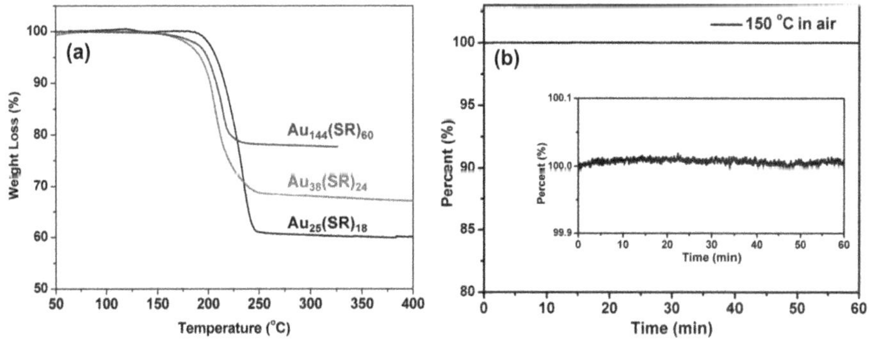

Figure 2.3 (a) Thermogravimetric analysis of $Au_n(SR)_m$ nanoclusters (under N_2 atmosphere). (b) Isothermal stability analysis of pure $Au_{25}(SR)_{18}$ nanoclusters (maintained at 150 °C in air atmosphere for 60 min). No discernable loss of ligands (starting: 100%) was observed. The inset shows a zoom-in of the TGA curve at a constant 150 °C.
Adapted with permission from ref. 20.

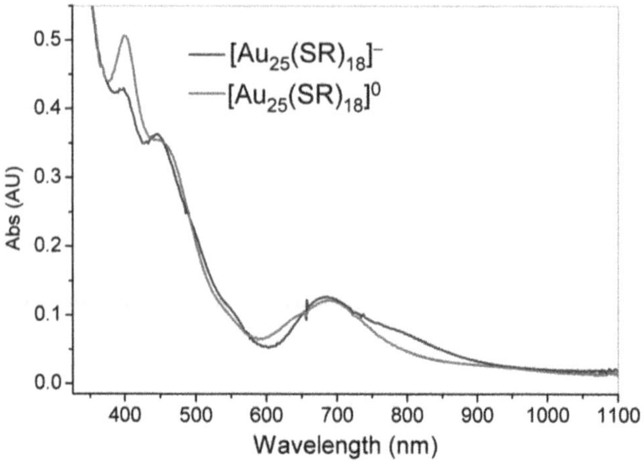

Figure 2.4 Reversible conversion between $[Au_{25}(SR)_{18}]^0$ and $[Au_{25}(SR)_{18}]^-$. Adapted with permission from ref. 23.

anionic (*i.e.*, $[Au_{25}(SCH_2CH_2Ph)_{18}]^-[N(C_8H_{17})_4]^+$). UV-vis spectral changes were observed when a solution of $[Au_{25}(SCH_2CH_2Ph)_{18}]^-$ was exposed to air or O_2 (Figure 2.4). The product was identified to be charge-neutral $[Au_{25}(SCH_2CH_2Ph)_{18}]^0$ nanoclusters.[22] It is interesting that the gold core, instead of the thiolate ligands on the cluster surface, loses one electron. Single crystal X-ray crystallography revealed that the $[Au_{25}(SCH_2CH_2Ph)_{18}]^0$ structure[23] shares the same framework as that of $[Au_{25}(SCH_2CH_2Ph)_{18}]^-$. The redox process between $[Au_{25}(SCH_2CH_2Ph)_{18}]^0$ and $[Au_{25}(SCH_2CH_2Ph)_{18}]^-$ is completely reversible, which is also found in the aqueous-soluble $Au_{25}(SG)_{18}$ nanocluster (SG: glutathione). The charge-state conversion of $[Au_{25}(SR)_{18}]^q$ (where $R=CH_2CH_2Ph$, C_6H_{13}, and G) indicates that the gold nanoclusters can be applied in catalytic oxidation or reduction processes, as well as other reactions.

2.5.2 Catalytic Oxidation

The O_2 activation is one of the key steps in selective oxidation catalysis.[1,24] As discussed above, the $Au_{25}(SR)_{18}$ nanocluster can interact with O_2 even at room temperature.[23] The O_2 activation should be much more efficient at elevated temperatures.

2.5.2.1 Catalytic Oxidation of CO to CO_2

We first discuss CO oxidation, as this reaction has been extensively investigated in nanogold catalysis.[1,25] The TiO_2-supported nanogold catalysts are generally found to be the most effective catalysts in the oxidation of CO to CO_2. However, in the case of thiolate ligand-protected gold nanocluster catalysts, Nie *et al.*[26] surprisingly found that the $Au_{25}(SR)_{18}/TiO_2$ catalyst had

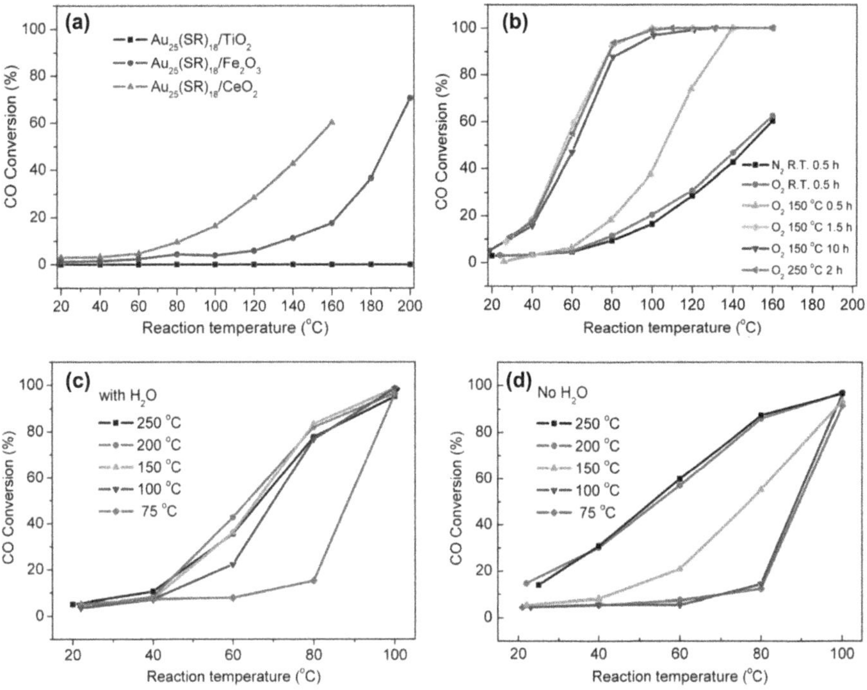

Figure 2.5 (a) Catalytic activity of different oxide supported $Au_{25}(SR)_{18}/MO_x$ catalysts for CO oxidation. Catalyst pretreatment condition: N_2 at room temperature (R.T.) for 0.5 h. Reaction conditions: gas hourly space velocity ~7500 mL g^{-1} h^{-1}; catalyst 0.1 g. (b) Catalytic activity of $Au_{25}(SR)_{18}/CeO_2$ after different pre-treatments. (c, d) Effect of water vapor on CO conversion over various pre-treated $Au_{25}(SR)_{18}/CeO_2$ catalysts under different pre-treatment temperatures: (c) feed gases with water vapor; and (d) without vapor.
Adapted with permission from ref. 26.

no catalytic activity even up to 200 °C [Figure 2.5(a)], whereas $Au_{25}(SR)_{18}/CeO_2$ exhibited moderate activity (onset temperature 60 °C, 62% conversion at 160 °C). The catalytic results imply striking differences between conventional (bare) Au/TiO_2 catalysts[25] and ligand-on $Au_{25}(SR)_{18}/TiO_2$ catalyst.[26]

Significantly, pre-treatment of the $Au_{25}(SR)_{18}/CeO_2$ catalyst in O_2 for 1.5–2 h at 150 °C (T_{pre}) led to a drastic increase in catalytic activity (onset temperature shifted to room temperature), and 100% CO conversion was reached at 100 °C [Figure 2.5(b)]. Prolonged O_2 pretreatment at 150 °C (*e.g.*, >2 h) did not lead to further enhancement in catalytic activity [Figure 2.5(b)], nor in the case of the T_{pre} increasing to 250 °C (above the thiolate desorption). It is worth noting that the drastic effect of thermal O_2 pre-treatment was not observed in the $Au_{25}(SR)_{18}/TiO_2$ system and the $Au_{25}(SR)_{18}/$oxide catalysts pre-treated in N_2 (as opposed to O_2). Characterization of the $Au_{25}(SR)_{18}/CeO_2$ catalyst showed that the ligands remained on the cluster after the 150 °C pre-treatment in O_2. We further found that the presence of

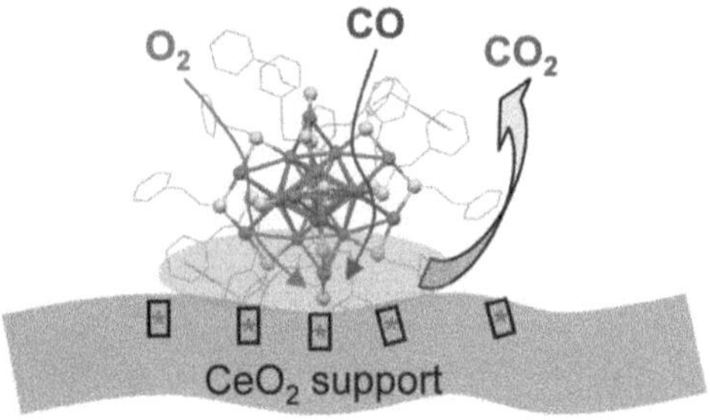

Scheme 2.1 Proposed CO oxidation at the perimeter sites of $Au_{25}(SR)_{18}/CeO_2$. Adapted with permission from ref. 26.

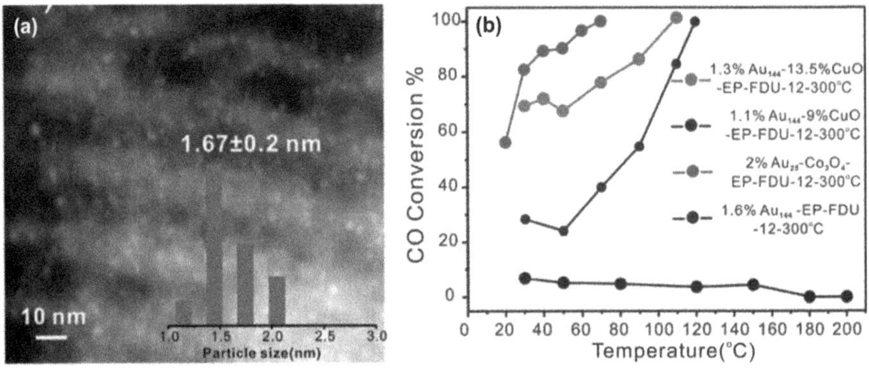

Figure 2.6 (a) Z-contrast STEM image of Au_{144} nanoclusters supported on CuO-EP-FDU-12 after calcination at 300 °C. (b) Oxidation of CO as a function of temperature over supported gold cluster catalysts.
Adapted with permission from ref. 28.

water vapor in the feed gases exhibited a promotional effect on catalyst performance [Figure 2.5(c) and 2.5(d)]. For example, in the case of feed gas containing vapor, O_2 pre-treatment at an even lower temperature (*e.g.*, $T_{pre} = 100$ °C) could lead to the same drastic enhancement in activity.[26]

The above results imply that the interface between $Au_{25}(SR)_{18}$ and CeO_2 is critical and should constitute the catalytic active site.[27] We speculate that O_2 first converts to O_2^- by withdrawing an electron from the $Au_{25}(SR)_{18}$ nanocluster. This oxygen species then migrates to the nanocluster–CeO_2 interface and converts to hydroperoxide species on CeO_2; the activated CO should be oxidized at the perimeter sites of the catalyst (Scheme 2.1). The detailed mechanism of the catalytic CO oxidation by $Au_{25}(SR)_{18}/CeO_2$ catalyst requires more detailed investigation.

Ma *et al.*[28] also reported the CO oxidation reaction using 'ligand-off' Au_{25} and Au_{144} clusters as catalysts. The protecting ligands on $Au_{25}(SCH_2CH_2Ph)_{18}$ and $Au_{144}(SCH_2CH_2Ph)_{60}$ nanoclusters can be removed *via* thermal treatment at 300 °C in air (Figure 2.6; note: the ligand desorption temperature ~200 °C,

Figure 2.3) without sintering the Au_{25} and Au_{144} nanoclusters because the 'ligand-off' Au_{25} and Au_{144} nanoclusters can be effectively stabilized on ordered heterostructured mesoporous material supports (*e.g.*, CuO-EP-FDU-12 and Co_3O_4-EP-FDU-12). The sizes of the ligand-off Au_{25} and Au_{144} clusters are characterized by Z-contrast scanning transmission electron microscopy (STEM, where Z refers to the atomic number of the element). The average size of 'ligand-off' Au_{144} clusters is determined to be 1.67 ± 0.2 nm which is consistent with the expected size (metal core) of the $Au_{144}(SCH_2CH_2Ph)_{60}$ nanoclusters [Figure 2.6(a)]. The ligand-off Au_{25} and Au_{144} nanoclusters supported on both CuO-EP-FDU-12 and Co_3O_4-EP-FDU-12 are more active than ligand-on catalysts in CO oxidation at relatively low temperatures [*e.g.*, 56% CO conversion at 20 °C and 100% conversion at 70 °C for the Au_{25} cluster supported on Co_3O_4-EP-FDU-12, Figure 2.6(b)]. Thus, this thermal treatment method can be applied to the nanogold catalysts when ligand-off gold cluster catalysts are preferred.

2.5.2.2 *Selective Oxidation of Styrene*

Zhu *et al.*[29] investigated the solution-phase styrene oxidation with O_2 catalyzed by $Au_n(SR)_m$ nanocluster catalysts (free gold nanoclusters and SiO_2-supported gold nanoclusters, respectively), Scheme 2.2. The catalytic reaction was carried out at 80–100 °C for 12–24 h and gave rise to benzaldehyde (up to ~70% selectivity), styrene epoxide (selectivity ~25%), and acetophenone (<5% selectivity), Scheme 2.2. A strong size dependence was observed in the catalytic reaction using free $Au_n(SR)_m$ nanocluster catalysts: the smaller $Au_n(SR)_m$ nanoclusters gave rise to a much higher catalytic activity (on a mass basis). A 35% conversion of styrene was obtained even with a parts per million (ppm) level of catalyst in solution when using the $Au_{25}(SR)_{18}$ nanocluster catalyst. For SiO_2-supported $Au_{25}(SR)_{18}$ (without calcination pre-treatment prior to catalytic reaction), the catalyst gave rise to a performance comparable with that of the unsupported catalyst. The advantage of the supported $Au_{25}(SR)_{18}/SiO_2$ catalyst is its recyclability as the collected catalyst is re-usable in catalytic reaction;[29] no apparent deterioration in the activity and selectivity of the re-used supported catalyst was observed after multiple cycles.

Liu *et al.*[30] investigated a ligand-off Au_{25}/HAP catalyst where HAP is hydroxyapatite [$Ca_{10}(PO_4)_6(OH)_2$)] for selective oxidation of styrene with

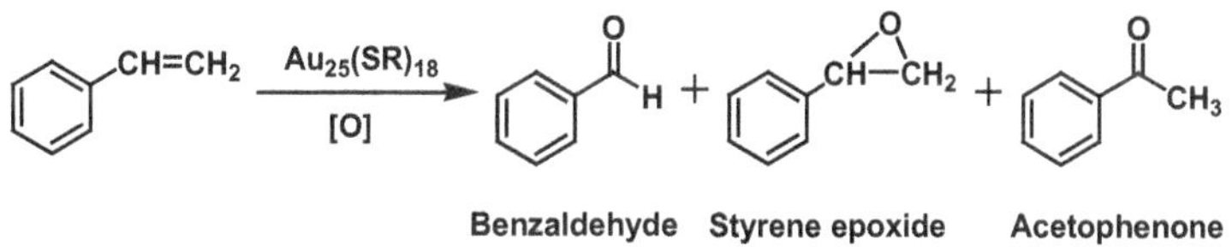

Scheme 2.2 Selective oxidation reaction of styrene catalyzed by $Au_{25}(SR)_{18}$ nanocluster catalysts.

anhydrous *tert*-butyl hydroperoxide (TBHP) as the oxidant. The thiolate ligand on the $Au_{25}(SR)_{18}$ nanocluster was removed by heat treatment at 300 °C. Styrene epoxide was obtained as the major product in the Au_{25}/HAP catalyzed oxidation run at 80 °C in toluene (Scheme 2.2). A 92% yield of styrene oxide was obtained, which is much higher than conventional nanocatalysts (*ca.* 50–60% yield) with the same Au loading (0.5 wt%). They deemed that anhydrous toluene and TBHP were critical to obtain a high yield of styrene epoxide.

Zhu *et al.*[20] further investigated and compared three oxidant systems: (1) TBHP as the oxidant; (2) TBHP as the initiator and O_2 as the main oxidant; and (3) O_2 as the oxidant (without initiator). Since TBHP is more reactive than O_2, the reaction with only TBHP [*i.e.* system (1)] catalyzed by $Au_{25}(SR)_{18}/SiO_2$ (without calcination pre-treatment) gave high conversion of styrene (*e.g.* 86%) with 100% selectivity for benzaldehyde. System (2) gave rise to a much lower activity (25% conversion with 100% selectivity for benzaldehyde), while the conversion and selectivity in system (3) (solely O_2) were even lower (18% conversion with 80% selectivity for benzaldehyde). A similar phenomenon was also found in the three oxidant systems using $Au_{38}(SR)_{24}/SiO_2$ and $Au_{144}(SR)_{60}/SiO_2$ catalysts.[20] These results indicate that the ingredients of the oxidant play a critical role in the oxidation reaction, and that O_2 activation is a key step for achieving high conversion of styrene. The results also imply that the active oxygen species in the catalytic cycles may be a peroxide-like or hydroperoxide-like species.[20] In addition, a distinct size dependence of $Au_n(SR)_m$ nanocluster catalysts is observed in the selective oxidation of styrene, with the order of catalytic activity: $Au_{25}(SR)_{18} > Au_{38}(SR)_{24} > Au_{144}(SR)_{60}$.

To compare the catalytic activity of doped nanoclusters with undoped $Au_{25}(SR)_{18}$, Qian *et al.*[31] further synthesized the monoplatinum-doped $Pt_1Au_{24}(SR)_{18}$ nanocluster and investigated the performance of the cluster in the selective oxidation of styrene using $PhI(OAc)_2$ as oxidant. The catalytic reaction was carried out at 70 °C for 10 h in acetonitrile under N_2 atmosphere. The $Pt_1Au_{24}(SR)_{18}/TiO_2$ catalyst performed with a higher catalytic activity (including conversion of styrene and selectivity for benzaldehyde) than the $Au_{25}(SR)_{18}/TiO_2$ catalyst. The different catalytic performance was mainly attributed to the electronic properties of the clusters.

2.5.2.3 *Aerobic Alcohol Oxidation*

Xie *et al.*[32] investigated the catalytic activity of the ligand-off 25-atom Pd_1Au_{24} and Au_{25} supported on multiwalled carbon nanotubes (CNTs) in the selective aerobic oxidation of benzyl alcohol, Scheme 2.3. Benzaldehyde, benzoic acid, and benzyl benzoate as the major products appeared in the aerobic oxidation reaction. Au_{25}/CNT and Pd_1Au_{24}/CNT catalysts were synthesized *via* two main steps: the thiolate-protected $Au_{25}(SC_{12}H_{25})_{18}$ and $Pd_1Au_{24}(SC_{12}H_{25})_{18}$ nanoclusters were first obtained following the procedures described in Section 2.2, and then the thiolate-protected clusters were loaded on CNTs. Finally, the thiolate ligand was removed by heat treatment at 300 or 450 °C for

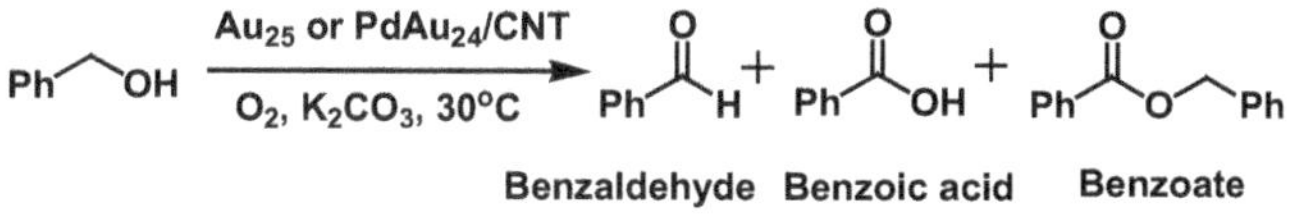

Scheme 2.3 Selective oxidation reaction of benzyl alcohol catalyzed by ligand-off Pd_1Au_{24}/CNT and Au_{25}/CNT catalysts.

2 h in vacuum. The transmission electron microscopy (TEM) images indicate that no aggregation occurred when the cluster loading was <0.2 wt%, whereas the cluster aggregation was obvious at a higher loading. It is very interesting that the catalytic performance of the Pd_1Au_{24}/CNT catalyst (74% conversion of benzyl alcohol) was much better than that of the Au_{25}/CNT catalyst (22% conversion). Meanwhile, the selectivity of the aerobic oxidation was altered from benzaldehyde and benzoate (when catalyzed by Au_{25}/CNT catalyst) to benzoic acid (when catalyzed by Pd_1Au_{24}/CNT catalyst). The improvement of the catalytic performance of the single-atom doping cluster catalyst was deemed to be caused by electron transfer from the central Pd atom to surface gold atoms. The ligand-off Au_{25}/CNT and Pd_1Au_{24}/CNT catalysts were found to be robust and durable.

2.5.2.4 *Aerobic Oxidation of Cyclohexane*

Tsukuda and co-workers[33] made HAP-supported $Au_n(SG)_m$ ($n = 10$, 18, 25, and 39) nanocluster catalysts with 0.2 wt% loading of nanoclusters. Diffuse reflectance UV-vis spectra and TEM measurements confirmed that the $Au_n(SG)_m$ nanoclusters adsorbed on HAP were intact. Subsequent thermal calcination at 300 °C for 2 h *in vacuo* completely removed the –SG ligands on the surface of the gold nanoclusters. TEM analysis showed no aggregation of the ligand-off Au nanoclusters to larger nanoparticles; this is attributed to the protection of HAP (*i.e.*, the strong interaction between the Au clusters and PO_4^{3-} moiety).[30] It should be noted that TEM cannot measure the precise number of Au atoms in the resultant ligand-off Au_n/HAP catalysts (originally $n = 10$, 18, 25, and 39) and that ligand-off gold nanoparticles might have a different structure from the original structure of ligand-protected particles. The catalytic performance of these ligand-off Au_n/HAP catalysts was evaluated for the selective oxidation of cyclohexane using O_2 as oxidant (under 1 MPa and solvent-free conditions at 150 °C for 4 h). Cyclohexanol and cyclohexanone were obtained as the primary products with nearly equal selectivity in the oxidation reaction (Figure 2.7). It is interesting that the turnover frequency (TOF) values exhibited an optimal cluster size around $n \sim 40$ (Figure 2.7). The TOF value of the Au_{39}/HAP catalyst reached 18 500 h^{-1} Au $atom^{-1}$.[33]

2.5.2.5 *Selective Oxidation of Sulfide*

It is known that sulfides can bind to the surface of gold nanoclusters *via* the interaction between the sulfur atom (–S–) and the gold surface, while

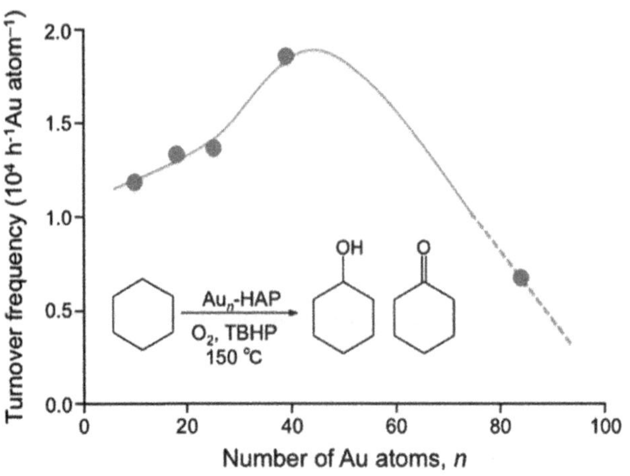

Figure 2.7 Selective oxidation of cyclohexane to cyclohexanol and cyclohexanone with O_2 as the oxidant catalyzed by ligand-off Au_n/HAP. Adapted with permission from ref. 33.

sulfoxides [–S(O)–] only weakly bind to the gold surface; hence the conversion from sulfide to sulfoxide on the clusters surface should be feasible. Li *et al.*[34] investigated the oxide-supported ∼1 wt% $Au_{25}(SR)_{18}$ nanocluster catalyst for selective oxidation of sulfides to sulfoxides using PhIO as oxidant. The TiO_2-supported $Au_{25}(SR)_{18}$ nanocluster catalyst gave rise to high catalytic reactivity (∼97% conversion of sulfide with ∼92% selectivity for sulfoxide).

Meanwhile, different types of substrates have been investigated to test the effects of electronic and steric factors in selective oxidation (Table 2.1). Sulfides with electron-rich groups were oxidized at a higher conversion than those with electron-deficient groups, and sulfides with greater steric hindrance showed lower conversion than those with less steric hindrance (Table 2.1). The support effect was also investigated, and it was found that the order of the catalyst activity of various $Au_{25}(SR)_{18}$/oxide catalyst was as follows: $TiO_2 > Fe_2O_3 > CeO_2 > MgO$. The oxide-supported $Au_{25}(SR)_{18}$ catalysts showed excellent recyclability in the sulfoxidation process.[34]

2.5.2.6 *Proposed Mechanism for Catalytic Oxidation Processes*

The above results (sections 2.5.2.1 to 2.5.2.5) of the catalytic selective oxidation processes (*e.g.* CO, alcohol, styrene, cyclohexane, sulfide) over $Au_n(SR)_m$ catalysts allow some valuable insight to be gained at the atomic level. Herein we use the Au_{25} as a catalyst model for a brief discussion. Under the assumption that thiolate localizes one electron of Au(6s) in the Au-SR bond, the Au_{13} core of $[Au_{25}(SR)_{18}]^q$ formally possesses eight (when $q = -1$) or seven (when $q = 0$) delocalized valence electrons—which originate from Au(6s) and are primarily distributed within the Au_{13} core,[22] while the Au_{12} shell bears positive charges due to charge transfer from gold to sulfur of the

Table 2.1 Oxidation of sulfides using $Au_{25}(SR)_{18}/TiO_2$ catalyst. Adapted with permission from ref. 34.

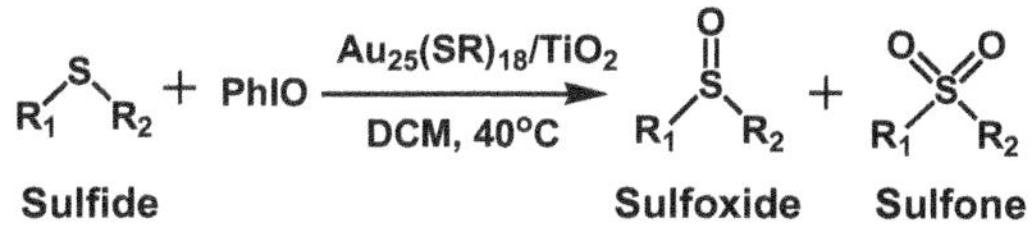

Entry	Eubstrates	Conversion (%)	Yields (%) Sulfoxide	Sulfone
1	C_6H_5–S–Me	96.8	89.6	7.2
2	o-O_2N-C_6H_4–S–Me	68.3	67.7	0.6
3	o-Br-C_6H_4–S–Me	90.9	90.5	0.4
4	o-Me-C_6H_4–S–Me	91.1	90.9	0.2
5	o-OHC-C_6H_4–S–Me	83.5	83.4	0.1
6	C_6H_5–S–$CH_2CH{=}CH_2$	98.9	98.2	0.7
7	o-Ac-C_6H_4–S–C_6H_5	72.7	71.5	1.2
8	n-C_8H_{17}–S–n-C_8H_{17}	73.4	73.0	0.4
9	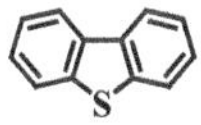	56.6	40.6	16.0

thiolate ligands. We propose that the electron-rich Au_{13} core should facilitate [O] (*e.g.*, O_2, TBHP, PhIO, *etc.*) activation by electron transfer to [O], accompanied by $[Au_{25}(SR)_{18}]^-$ conversion to neutral $[Au_{25}(SR)_{18}]^0$. Of note, this oxidation readily occurs at elevated temperatures (*e.g.*, under catalytic reaction conditions). A similar [O] activation mechanism *via* electron transfer was also identified in PVP-protected gold nanoclusters for aerobic alcohol oxidation.[17] For selective oxidation on $Au_{25}(SR)_{18}$, activation of reactants (*e.g.*, styrene, sulfide, CO, and alcohol, *etc.*) should be facilitated by the partial positive charge ($Au^{\delta+}$, $\delta = \sim 0.3$)[35] on the triangular gold atom configuration (*i.e.* Au_3). Theoretical work is still needed to obtain a deeper understanding of the adsorption geometry and the structure of intermediates.

2.5.3 Catalytic Selective Hydrogenation

2.5.3.1 Selective Hydrogenation of Aldehydes and Ketones

Beside the oxidation reactions, the $Au_n(SR)_m$ nanocluster catalysts were also demonstrated to be capable of catalyzing hydrogenation reactions, such as the chemoselective hydrogenation of α,β-unsaturated ketones to α,β-unsaturated alcohols under mild conditions (60 °C, in mixed solvents (1 : 1 toluene/acetonitrile)), Scheme 2.4.[36] The $Au_{25}(SR)_{18}$ nanoclusters were found

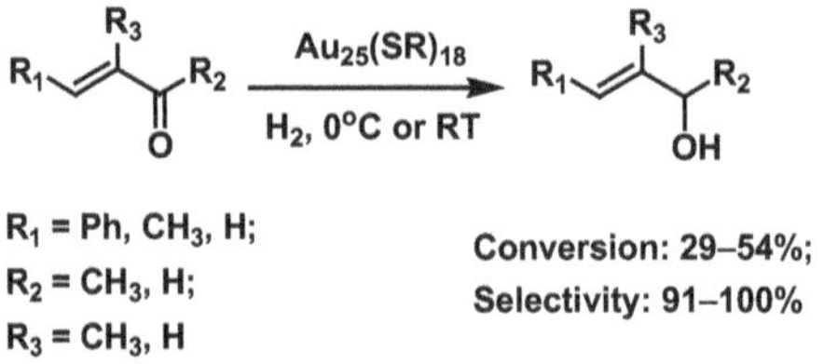

Scheme 2.4 Chemoselective hydrogenation reaction of α,β-unsaturated ketones to α,β-unsaturated alcohols catalyzed by $Au_{25}(SR)_{18}$ nanocluster catalysts.

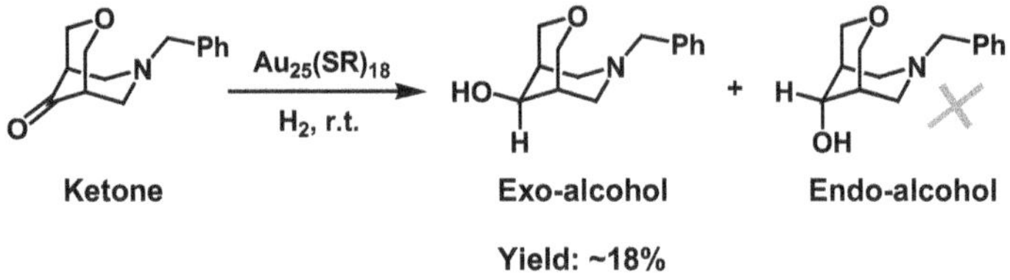

Scheme 2.5 Stereoselective hydrogenation reaction of ketone to *exo*-alcohol catalyzed by $Au_{25}(SR)_{18}$ nanocluster catalysts.

to hydrogenate preferentially the C=O bond of benzalacetone against the C=C bond, and the C=O hydrogenated product (*i.e.*, unsaturated alcohol) was obtained with 76% selectivity compared with 14% selectivity for unsaturated ketone and 10% for saturated alcohol.[31] UV-vis spectra and mass spectrometry (MS) analysis of the free $Au_{25}(SR)_{18}$ nanocluster catalyst confirm that the $Au_{25}(SR)_{18}$ nanoclusters remained intact during the hydrogenation reaction. Further optimization of the reaction led to a nearly complete selectivity for α,β-unsaturated alcohol in a mixed toluene/ethanol (1 : 1) solvent at room temperature.[36] The oxide-supported catalysts, including $Au_{25}(SR)_{18}/Fe_2O_3$, $Au_{25}(SR)_{18}/TiO_2$, and $Au_{25}(SR)_{18}/SiO_2$, were also investigated, and it was found that the catalytic activity was remarkably improved with the exception of the $Au_{25}(SR)_{18}/SiO_2$ catalyst. In addition, the chemoselective hydrogenation was established including a range of substituted α,β-unsaturated ketones and aldehydes with moderate yields and high selectivities (Scheme 2.4).

Further, Zhu *et al.*[37] expanded the intact $Au_{25}(SR)_{18}$ nanocluster-catalyzed reaction to stereoselective hydrogenation of a bicyclic ketone, *i.e.*, 7-(phenylmethyl)-3-oxa-7-azabicyclo [3.3.1] nonan-9-one, to *exo*-alcohol (Scheme 2.5). The stereoselective hydrogenation was run at room temperature in an ethanol–toluene mixture (5 : 1, v/v) solvent and using H_2 as hydrogen source; ∼18% conversion of the bicyclic ketone with 100% selectivity for the *exo*-alcohol was obtained; note that the selectivity was determined by nuclear magnetic resonance (NMR) spectroscopy.[37] The high stereoselectivity (∼100%) result was correlated with the atomic spatial construction and the unique core-shell nature of the $Au_{25}(SR)_{18}$ nanoclusters.

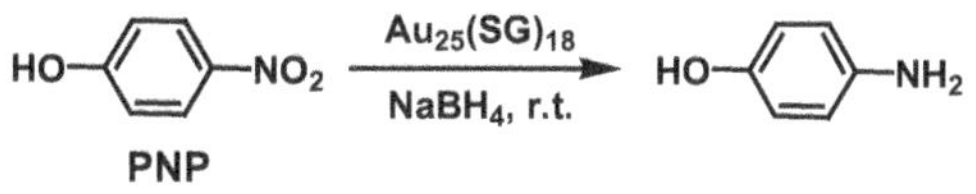

Scheme 2.6 Hydrogenation of 4-nitrophenol to 4-aminophenol catalyzed by $Au_{25}(SR)_{18}$ nanocluster catalysts.

2.5.3.2 Hydrogenation of Nitrophenol

Kawasaki and co-workers[38] investigated and compared the catalytic properties of $Au_{25}(SG)_{18}$ nanoclusters and *N,N*-dimethylformamide (DMF) stabilized gold nanoclusters [not of atomic precision compared with $Au_{25}(SG)_{18}$ nanoclusters] in the reduction of 4-nitrophenol (PNP) to 4-aminophenol by $NaBH_4$ in water at room temperature, Scheme 2.6. The reduction process was monitored by UV-vis spectra. The $Au_{25}(SG)_{18}$ nanoclusters exhibited higher catalytic activity than DMF-capped clusters in the PNP reduction [pseudo-first-order rate constant 8×10^{-3} s^{-1} for $Au_{25}(SG)_{18}$ *vs.* 3×10^{-3} s^{-1} for DMF-capped Au clusters]. High catalytic activity of the $Au_{25}(SG)_{18}$ nanocluster catalyst can be achieved even at a low catalyst concentration (*e.g.*, 1.0 μM). No induction time was observed in the case of $Au_{25}(SG)_{18}$ despite the fact that–SG is a strongly binding ligand due to less steric hindrance, while the DMF-stabilized Au clusters exhibited an induction time ($\sim$4000 s), which was attributed to impedance by the surface DMF layer on the clusters when reactants accessed the cluster surface.[38] For $Au_{25}(SG)_{18}$ nanoclusters, the unique core-shell structure seems to pose less steric hindrance and thus renders the catalytic active sites more accessible to reactants. Scott and co-workers[39] recently also demonstrated intact $Au_{25}(SR)_{18}$ (where, $R=C_6H_{13}$, C_8H_{17}, and $C_{12}H_{25}$) nanoclusters for the reduction of PNP by $NaBH_4$ in a mixture of tetrahydrofuran–water and under N_2 atmosphere. The $Au_{25}(SR)_{18}$ nanoclusters were more stable and kept their structural integrity towards $NaBH_4$ treatment compared with the larger $Au_{\sim180}(SC_6H_{13})_{\sim100}$ nanoclusters.

2.5.4 Catalytic Carbon–Carbon Coupling Reaction

Gold nanoclusters also hold great promise for homogeneous and heterogeneous catalysis in the carbon–carbon coupling reaction, such as Ullmann-type homo-coupling and Sonogashira cross-coupling reactions. The catalytic application of gold nanoclusters in C–C coupling reaction was due to the good capability of gold nanocluster catalysts in activating the C–I bond of iodobenzene and the terminal C–H bond of phenylacetylene.

2.5.4.1 Ullmann Homo-coupling Reaction

The gold nanoclusters are good catalysts for activating the C–I bond of the iodobenzene in the carbon–carbon homo-coupling process. Li *et al.*[40] investigated the catalytic activity of the oxide-supported $Au_{25}(SR)_{18}$ (where

Table 2.2 Homo-coupling of aryl iodides using Au_{25}/CeO_2 catalyst. (A) Crystalline structure of $Au_{25}(SR)_{18}$ (ball-stick model); (B) space-filling model and the triangular catalytic active sites (Au_3). Adapted with permission from ref. 40.

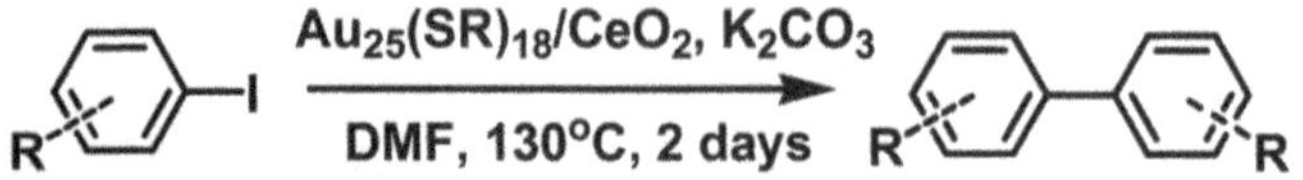

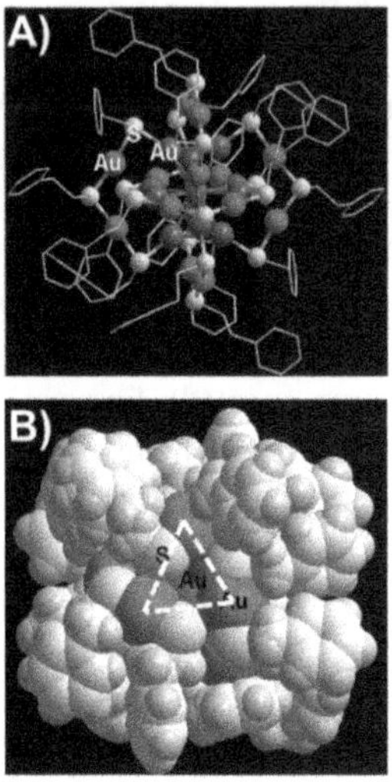

Entry	Substrates	Conversion[%]
1	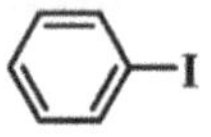	99.8
2	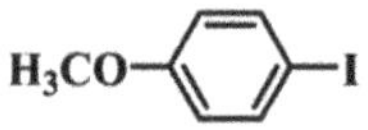	99.5
3	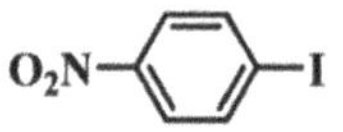	67.5
4	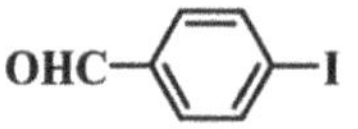	78.2
5	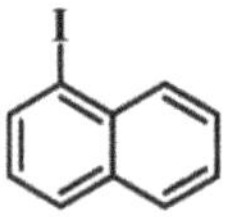	99.7

$R = CH_2CH_2Ph$) catalysts for the Ullmann-type homo-coupling reactions of aryl iodides. The oxide-supported catalyst was prepared by impregnation of oxide powders in a CH_2Cl_2 solution of $Au_{25}(SR)_{18}$ (~1 wt% loading). The $Au_{25}(SR)_{18}/CeO_2$ catalyst gave rise to 99.8% conversion of iodobenzene under N_2 atmosphere at 130 °C using DMF as solvent and K_2CO_3 as base. The application of the $Au_{25}(SR)_{18}$ catalyzed homo-coupling reaction was tested with a variety of substituents. The electron-rich substrates performed better than electron-deficient substrate in this reaction (Table 2.2). The support effect was also investigated; no distinct effect of the oxide supports was observed (*e.g.*, CeO_2, SiO_2, TiO_2, and Al_2O_3).

2.5.4.2 Sonogashira Cross-coupling Reaction

Li *et al.*[41] expanded the catalytic application of the oxide-supported $Au_{25}(SR)_{18}$ nanocluster catalysts to the Sonogashira cross-coupling of iodobenzene and phenylacetylene. The supported catalyst was made by impregnation of oxide powders (including CeO_2, TiO_2, MgO, and SiO_2) in a CH_2Cl_2 solution of $Au_{25}(SR)_{18}$ (~1 wt% loading), and then annealed at 150 °C (1 h) in a vacuum oven. STEM and thermogravimetric analysis confirmed that the protecting thiolate ligand remained on the surface of gold nanoclusters after the annealing. The $Au_{25}(SR)_{18}$/oxide catalysts after thermal treatment were applied in the Sonogashira cross-coupling between iodobenzene and phenylacetylene; the optimized reaction was carried out under N_2 atmosphere at 160 °C using DMF as solvent and K_2CO_3 as base. A high conversion of *p*-iodoanisole (up to 96.1%), with excellent selectivity for the cross-coupling product (up to 88.1%), was obtained when the reaction was catalyzed by $Au_{25}(SR)_{18}/CeO_2$ (Table 2.3, entry 1). The $Au_{25}(SR)_{18}$ nanocluster catalyst performed much better than both the larger sized 2–3 nm Au nanoclusters (Table 2.3, entry 1 *vs.* 5) and the CeO_2-supported Au nanoparticles (bare, ~20 nm).[42] The support effects were also investigated in the coupling reaction, and no distinct effect of the oxide supports was observed (*i.e.*, CeO_2, SiO_2, TiO_2, and MgO). In the recyclability test of the catalyst, the conversion of *p*-iodoanisole exhibited no significant loss, while the selectivity decreased from 88.1% to 64.5% after five cycles. The drop in selectivity could be caused by the gradual degradation of nanoclusters in the multiple recycling tests, as larger Au particles give a much lower selectivity. Moreover, density functional theory (DFT) calculations of the reactant adsorption showed that both reactants (*i.e.*, iodobenzene and phenylacetylene) prefer to adsorb on the open facet with the phenyl ring facing a surface gold atom. A total adsorption energy reaches −0.90 eV when the two reactants co-adsorb on the $Au_{25}(SR)_{18}$ catalyst. The DFT results suggested that the

Table 2.3 Catalytic performance of $Au_{25}(SR)_{18}$ ($R=CH_2CH_2Ph$) nanoclusters supported on various oxides as catalysts for Sonogashira cross-coupling of *p*-iodoanisole and phenylacetylene. (a) Top view of one of the two open facets of the $Au_{25}(SR)_{18}$ nanoclusters where three external gold atoms are exposed; the other facet is on the back side. (b) Top view of the co-adsorption of phenylacetylene and iodobenzene on the surface of the $Au_{25}(SR)_{18}$ nanoclusters. $R'=OCH_3$; MPEP = 1-methoxy-4-(2-phenylethynyl) benzene; DMBP = 4,4′-dimethoxy-1,1′-biphenyl; Au, yellow; S, blue; C, gray; H, white; I, green. Adapted with permission from ref. 41.

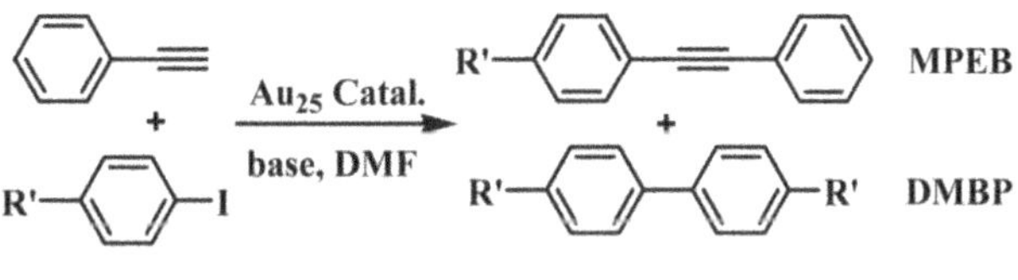

entry	catalyst	conv. (%)	selectivity (%) MPEB	selectivity (%) DMBP
1	$Au_{25}(SR)_{18}/CeO_2$	96.1	88.1	11.9
2	$Au_{25}(SR)_{18}/TiO_2$	92.8	82.9	17.1
3	$Au_{25}(SR)_{18}/SiO_2$	90.8	79.3	20.7
4	$Au_{25}(SR)_{18}/MgO$	93.3	80.6	19.4
5	AuNC 2-3 nm $(SC_6H_{13})/CeO_2$	65.5	57.2	42.8

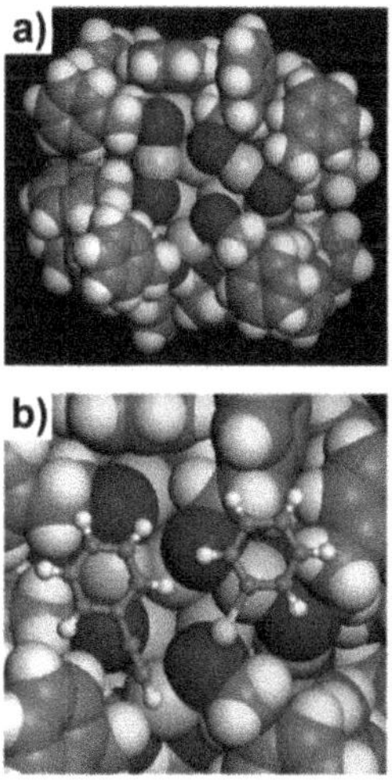

catalytic active sites is associated with the $Au_{25}(SR)_{18}$, which is consistent with the experimental results.

2.6 Conclusions

The atomically precise $Au_n(SR)_m$ nanoclusters have been successfully applied in a number of reactions, including: (1) oxidation processes such as CO to CO_2, styrene to benzaldehyde and styrene epoxide, cyclohexane to cyclohexanol and cyclohexanone, benzyl alcohol to benzaldehyde and benzoic acid and benzoate, sulfide to sulfoxide, *etc.*; (2) selective hydrogenation processes such as aldehydes or ketones to alcohols, nitrophenol to aminophenol, *etc.*; and (3) carbon–carbon coupling reactions such as the Ullmann homo-coupling of iodobenzene, Sonogashira cross-coupling of iodobenzene and phenylacetylene, *etc.*

This new class of $Au_n(SR)_m$ nanocluster catalysts is particularly promising as model catalysts. They are expected to bridge the gap between the bulk single crystal surface model catalysts and the realistic nanocatalysts. Results from the above catalytic reactions strongly indicate that the catalytic performance (including the catalytic activity and selectivity) are closely associated with the unique structure and electronic properties of $Au_n(SR)_m$ nanoclusters. Using the structurally characterized $Au_{25}(SR)_{18}$ nanocluster as a paradigm, we discussed some correlations between structure and catalytic reactivity. Although much work remains to be carried out in the future, these well-defined gold nanoclusters will provide new opportunities for achieving fundamental understanding of nanocatalysis at the atomic level, such as the insight into the size dependence and deep understanding of the molecular activation, catalytically active sites, and catalytic mechanism by correlation with the atomic structures of gold nanoclusters. Future research on atomically precise nanocluster catalysts will contribute largely to fundamental catalysis studies and the design of new, highly efficient catalysts for other specific chemical processes.

Acknowledgements

The work was supported by the US Department of Energy-Office of Basic Energy Sciences (Grant DE-FG02-12ER16354).

References

1. G. C. Bond, C. Louis and D. T. Thompson, *Catalysis by Gold*, Imperial College Press, London, 2006.
2. U. Heiz and U. Landman, *Nanocatalysis*, Springs, New York, 2007.
3. G. Li and R. Jin, *Acc. Chem. Res.*, 2013, **46**, 1749.
4. R. Jin, *Nanotechnol. Rev.*, 2012, **1**, 31.
5. R. Jin, H. Qian, Z. Wu, Y. Zhu, M. Zhu, A. Mohanty and N. Garg, *J. Phys. Chem. Lett.*, 2010, **1**, 2903.

6. R. Jin, *Nanoscale*, 2010, **2**, 343.
7. H. Qian, M. Zhu, Z. Wu and R. Jin, *Acc. Chem. Res.*, 2012, **45**, 1470.
8. D.-E. Jiang, *Nanoscale*, 2013, **5**, 7149.
9. M. Zhu, E. Lanni, N. Garg, M. E. Bier and R. Jin, *J. Am. Chem. Soc.*, 2008, **130**, 1138.
10. E. Gottlieb, H. Qian and R. Jin, *Chem.Eur. J.*, 2013, **19**, 4238.
11. D.-E. Jiang and S. Dai, *Inorg. Chem.*, 2009, **48**, 2720.
12. Y. Negishi, K. Munakata, W. Ohgake and K. Nobusada, *J. Phys. Chem. Lett.*, 2012, **3**, 2209.
13. Z. Wu, J. Suhan and R. Jin, *J. Mater. Chem.*, 2009, **19**, 622.
14. H. Qian, M. Zhu, U. N. Andersen and R. Jin, *J. Phys. Chem. A*, 2009, **113**, 4281.
15. H. Qian and R. Jin, *Nano Lett.*, 2009, **9**, 4083.
16. H. Qian, Y. Zhu and R. Jin, *Proc. Natl. Acad. Sci. U. S. A.*, 2012, **109**, 696.
17. Y. Shichibu, Y. Negishi, T. Tsukuda and T. Teranishi, *J. Am. Chem. Soc.*, 2005, **127**, 13464.
18. M. Zhu, C. M. Aikens, F. J. Hollander, G. C. Schatz and R. Jin, *J. Am. Chem. Soc.*, 2008, **130**, 5883.
19. Z. Wu, C. Gayathri, R. Gil and R. Jin, *J. Am. Chem. Soc.*, 2009, **131**, 6535.
20. Y. Zhu, H. Qian and R. Jin, *Chem. Eur. J.*, 2010, **16**, 11455.
21. Z. Wu and R. Jin, *ACS Nano*, 2009, **3**, 2036.
22. M. Zhu, C. M. Aikens, M. P. Hendrich, R. Gupta, H. Qian, G. C. Schatz and R. Jin, *J. Am. Chem. Soc.*, 2009, **131**, 2490.
23. M. Zhu, W. T. Eckenhoff, T. Pintauer and R. Jin, *J. Phys. Chem. C*, 2008, **112**, 14221.
24. B. K. Min and C. M. Friend, *Chem. Rev.*, 2007, **107**, 2709.
25. M. Haruta, S. Tsubota, T. Kobayashi, H. Kageyama, M. J. Genet and B. Delmon, *J. Catal.*, 1993, **144**, 175.
26. X. Nie, H. Qian, Q. Ge, H. Xu and R. Jin, *ACS Nano*, 2012, **6**, 6014.
27. M. Haruta, *Faraday Discuss.*, 2011, **152**, 11.
28. G. Ma, A. Binder, M. Chi, C. Liu, R. Jin, D.-E. Jiang, J. Fan and S. Dai, *Chem. Commun.*, 2012, **48**, 11413.
29. Y. Zhu, H. Qian, M. Zhu and R. Jin, *Adv. Mater.*, 2010, **22**, 1915.
30. Y. Liu, H. Tsunoyama, T. Akita and T. Tsukuda, *Chem. Commun.*, 2010, **46**, 550.
31. H. Qian, D.-E. Jiang, G. Li, C. Gayathri, A. Das, R. R. Gil and R. Jin, *J. Am. Chem. Soc.*, 2012, **134**, 16159.
32. S. Xie, H. Tsunoyama, W. Kurashige, Y. Negishi and T. Tsukuda, *ACS Catal.*, 2012, **2**, 1519.
33. Y. Liu, H. Tsunoyama, T. Akita, S. Xie and T. Tsukuda, *ACS Catal.*, 2011, **1**, 2.
34. G. Li, H. Qian and R. Jin, *Nanoscale*, 2012, **4**, 6714.
35. D.-E. Jiang, M. L. Tiago, W. Luo and S. Dai, *J. Am. Chem. Soc.*, 2008, **130**, 2777.
36. Y. Zhu, H. Qian, B. A. Drake and R. Jin, *Angew. Chem., Int. Ed.*, 2010, **49**, 1295.

37. Y. Zhu, Z. Wu, G. C. Gayathri, H. Qian, R. R. Gil and R. Jin, *J. Catal.*, 2010, **271**, 155.
38. H. Yamamoto, H. Yano, H. Kouchi, Y. Obora, R. Arakawa and H. Kawasaki, *Nanoscale*, 2012, **4**, 4148.
39. A. Shivhare, S. J. Ambrose, H. Zhang, R. W. Purves and R. W. J. Scott, *Chem. Commun.*, 2013, **49**, 276.
40. G. Li, C. Liu, Y. Lei and R. Jin, *Chem. Commun.*, 2012, **48**, 12005.
41. G. Li, D.-E. Jiang, C. Liu, C. Yu and R. Jin, *J. Catal.*, 2013, **306**, 177.
42. V. K. Kanuru, G. Kyriakou, S. K. Beaumont, A. C. Papageorgiou, D. J. Watson and R. M. Lambert, *J. Am. Chem. Soc.*, 2010, **132**, 8081.

CHAPTER 3

Nanocrystal Superlattices: Novel Model Materials for Study of Catalysis

YIJIN KANG,[a,b] VOJISLAV R. STAMENKOVIC[b] AND CHRISTOPHER B. MURRAY*[a,c]

[a] Department of Chemistry, University of Pennsylvania, 231 South 34th Street, Philadelphia, PA 19104, USA; [b] Materials Science Division, Argonne National Laboratory, 9700 South Cass Avenue, Argonne, IL 60439, USA; [c] Department of Materials Science and Engineering, University of Pennsylvania, 231 South 34th Street, Philadelphia, PA 19104, USA
*Email: cbmurray@sas.upenn.edu

3.1 Introduction and Background

Well-defined surfaces of oriented single (or poly-) crystals (*i.e.* extended surfaces) have been used for decades to understand fundamental physical and chemical processes. An example is shown in Figure 3.1, in which the correlation between surface structure and CO adsorption is studied on different surfaces.[1] The insights gained from similar systems have been extrapolated to elucidate the properties of the real heterogeneous catalysts, which are usually nanoparticles dispersed on high surface area supports.[2,3] The extended surface experiments are usually performed in an ultra-high vacuum (UHV) environment (*i.e.* under highly controlled conditions), as this would allow more precise identification of specific mechanism that may operate in a nanoparticulate heterogeneous catalytic system. However, the

RSC Catalysis Series No. 18
Heterogeneous Gold Catalysts and Catalysis
Edited by Zhen Ma and Sheng Dai

Published by the Royal Society of Chemistry, www.rsc.org

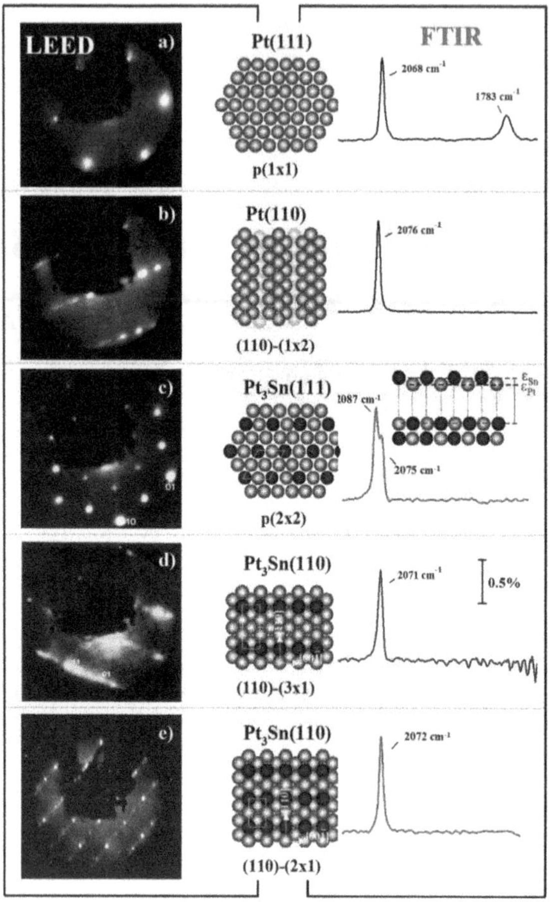

Figure 3.1 Low-energy electron diffraction (LEED) patterns, corresponding top-view atomic models and 'signature' spectra of adsorbed CO_L and CO_B on different surfaces recorded at 0.05 V and calculated as $(-\Delta R/R)$: (a) Pt(111); (b) Pt(110); (c) $Pt_3Sn(111)$; (d) $Pt_3Sn(110)$-(3×1); and (e) $Pt_3Sn(110)$-(2×1). FTIR = Fourier transform infrared. Reproduced by permission of Elsevier.[1]

knowledge gained from extended surface models is not always directly applicable to real catalysts because distributions in particle size, morphology, surface states and variations in the interactions between catalyst particles or between catalyst particles and the support are all influential.[4]

Although challenging, the preparation of well-controlled nano-materials which may bridge conventional models and real catalysts, is of great interest in research.[5–17] For instance, a layered synthetic microstructure was fabricated as a model catalyst to isolate the support effects or the non-equilibrium surface structures from the size effects in catalysis studies.[18,19] Similarly, electron beam lithography was employed to fabricate Pt nanoparticle arrays as model catalysts, allowing control of the metal catalyst surface structure, the oxide–metal interface area, and the interparticle distances.[20,21] In another example, a multi-component nanocrystal (NC) architecture may operate as a tandem catalyst, displaying catalytic properties that have not been observed in a simple mixture of the same components.[22]

Recent developments in self-assembly allowed large-scale fabrication of multi-component superlattices.[23] This in turn thus enabled the chemical and physical properties of different materials to be combined in a single periodic architecture (colloidal crystal). This offered a unique opportunity to exploit synergistic interactions and collective properties that arise as NCs couple in these more complex systems.[23–29] Specifically, the self-assembly of two types of NCs into a binary superlattice (BNSL) has attracted great attention for the design of functional metamaterials. With their highly ordered structures and precisely controlled stoichiometry, BNSLs could be valuable models for investigation of catalytic processes, providing a unique approach to connect information gained from the extended surfaces to real catalysts.

3.2 Supercrystals, Superlattices and Binary Superlattices: Preparation, Characterization and Properties

While atoms and molecules can form crystals, colloidal particles can form crystal-like ordered structures (superlattices) through properly balanced attractive and repulsive interparticle forces. Naturally occurring opal is an example of a colloidal crystal (supercrystal) resulting from ordered array of tiny silica particles.[30,31] Since half a century ago, the colloidal superlattices and supercrystals have attracted ever increasing attention.[23–26,28,29,32–45] Supercrystals and superlattices bear some resemblance to atomic crystals, such as the existence of terraces, steps, kinks, and vacancies [Figure 3.2(a)], as well as line-defect caused grain boundaries [Figure 3.2(b)]. Superlattices with a long-range order can be prepared with two distinct morphologies: faceted colloidal supercrystals [Figure 3.2(c)] and ordered thin films [Figure 3.2(d)].

The ligand-capped NCs readily form a stable dispersion in non-polar solvents such as alkanes, but reversibly aggregate in polar solvents such as alcohols.[34] Destabilizing NC dispersion ('solution') by adding polar solvents causes recrystallization of colloidal NCs into supercrystals. During the recrystallization process, the NCs self-organize into ordered structure through a 'self-cleaning' effect.[39]

The conventional strategy to assemble superlattices involves precisely controlled recrystallization and deposition of superlattice thin films on a substrate.[24,34–36,46–49] To achieve the controlled assembly, one can carefully manipulate the evaporation process by tuning pressure and temperature. However, the limited control in such conventional methods limits the size and uniformity of superlattices. In contrast, a recently developed method based on the liquid–air interfacial assembly of NCs helps to overcome these

Figure 3.2 (a) Scanning electron microscopy (SEM) image of Pt–Pd BNSLs having terraces, steps, kinks, and vacancies; (b) transmission electron microscopy (TEM) image of Au-FeO_x BNSLs showing line-defect caused grain boundaries; (c) SEM image of Pt supercrystal having faceted surfaces; and (d) TEM image of ordered thin film of Au-FeO_x BNSLs. Scale bars: (a–c) 200 nm, (d) 50 nm.
Reproduced by permission of American Chemical Society.[41]

limitations, allowing integration of superlattices onto any substrate [Figure 3.3(a)].[23] This liquid–air interfacial assembly also enables fabrication of large area superlattice membranes (*e.g.* ∼ cm).

By employing the liquid–air interface self-assembly method, a group of high-quality superlattices and BNSLs were prepared. For instance, AlB_2-type BNSL membranes were self-assembled from 15 nm Fe_3O_4 and 6 nm FePt NCs, and were transferred to a 7 mm × 7 mm SiO_2/Si wafer [Figure 3.3(b–d)]. Atomic force microscopy was used to characterize the thickness, which was found to be ∼40 nm, corresponding to the thickness of approximately one unit cell (Figure 3.4).

Moreover, a great diversity of compositions and structures of BNSLs have been prepared, including Pt-Pd,[40] Pt-Au,[41] Pt-Ni,[41] Pt-MnO,[41] Pt-FeO_x,[41] FePt-FeO_x,[23,38] $CoPt_3$-FeO_x,[23] FePt-MnO,[26] Au-Cu,[41] Ni-Cu,[41] Au-Ni,[41] Au-FeO_x,[38,41] Cu-FeO_x,[41] Pd-FeO_x,[41] and MnO-FeO_x,[41] all in a variety of crystal structures. A demonstration of the range of BNSLs is presented in Figure 3.5. In addition to the variety of composition, the superlattices or even BNSLs have been built with building blocks beyond hard spheres.[13,39,40,50–54] Figure 3.6 exhibits the rod-sphere BNSLs self-assembled from $NaYF_4$ nanorods and

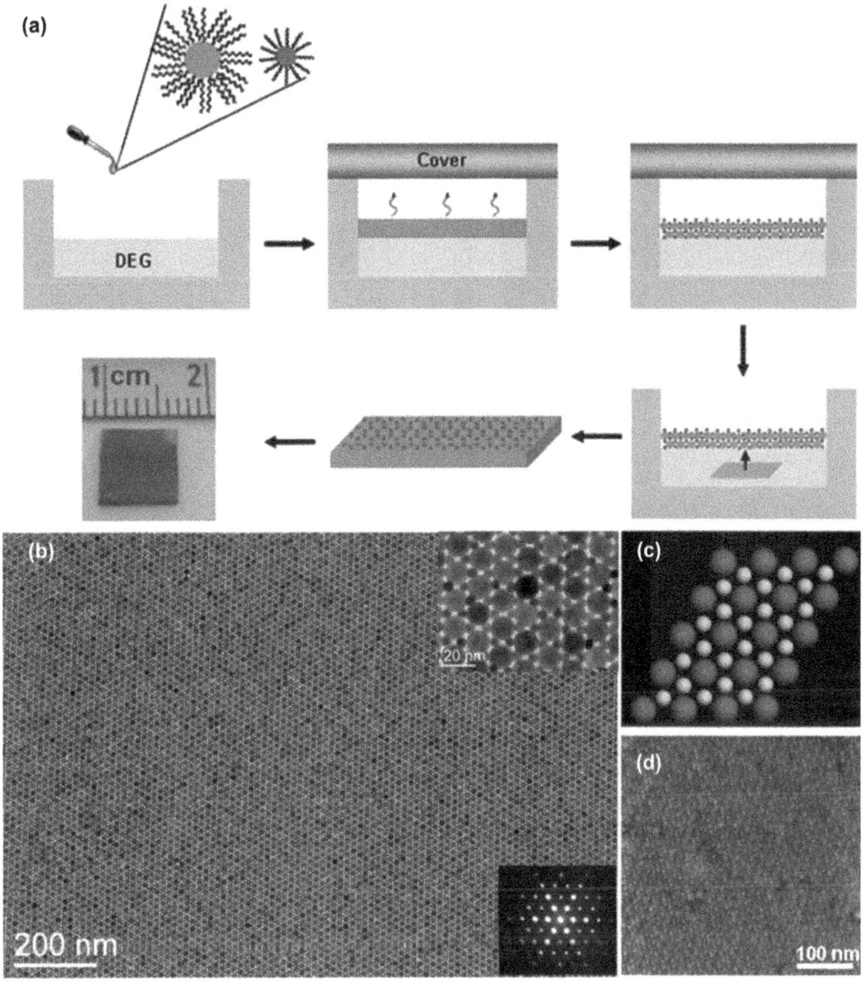

Figure 3.3 (a) Schematics of the superlattice membrane growth and transfer processes by the liquid–air interface self-assembly method. The photograph shows a typical BNSL membrane transferred to a SiO_2–Si wafer. Mechanical damage from tweezers in the membrane's upper right corner helps visualize the scale. (b–d) AlB_2-type BNSL membranes self-assembled from 15 nm Fe_3O_4 and 6 nm FePt NCs: (b) TEM image of the (001) lattice projection (upper inset, magnified view; lower inset, small-angle electron diffraction pattern); (c) crystallographic model; and (d) high-resolution scanning electron microscopy (HRSEM). Reproduced by permission of Nature Publishing Group.[23]

11.0 nm Fe_3O_4 nanospheres, presenting a 'shape alloy' of binary NCs.[53] Figure 3.7 shows two-dimensional (2D) superlattices self-assembled from lanthanide fluoride nanoplates (DyF_3, TbF_3, and EuF_3) with various aspect ratios,[52] and Figure 3.8 demonstrates BNSLs having Pt octahedra and Pd spheres as the building blocks.[40]

Figure 3.4 (a) Atomic force microscopy (AFM) height image (scan size, 1 μm × 1 μm) of an AlB_2-type BNSL membrane consisting of 15 nm Fe_3O_4 and 6 nm FePt NCs. Inset: height analysis of the profile indicated in the AFM image. (b) AFM phase image of the same membrane, showing a (100) projection in-plane view. (c) Side view of a crystallographic model of the membrane, showing that the membrane is one unit cell thick. Reproduced by permission of Nature Publishing Group.[23]

From all the cases shown above, one can see that the microscopy is the primary approach to characterize the supercrystals and superlattices. However, TEM and STEM images are two-dimensional (2D) projection of three-dimensional (3D) structures, and SEM images are limited at the surface of selected orientation. Therefore, these conventional electron microscopy images are not sufficient to ascertain the structure of BNSLs.[37,40] To unambiguously assign the structure, a crystallographic analysis procedure using a dual-axis tomography TEM holder was developed.[37] For example, a tilt series of BNSLs composed of 5.6 and 10.5 nm Fe_3O_4 NCs were obtained on a TEM equipped with a dual-axis tomography holder allowing full-angle rotation [Figure 3.9(a)]. With the tilt series, an ico-AB_{13} structure was unambiguously identified. In contrast, if the same BNSLs were examined solely by regular TEM 2D projection images, at certain angle the ico-AB_{13} structure [Figure 3.9(b)] might be incorrectly taken as the AlB_2 structure [Figure 3.9(c)]. By the crystallographic analysis employing the tomography holder, the Pt-Pd BNSLs shown in Figure 3.8 were confirmed to have the ico-AB_{13} type

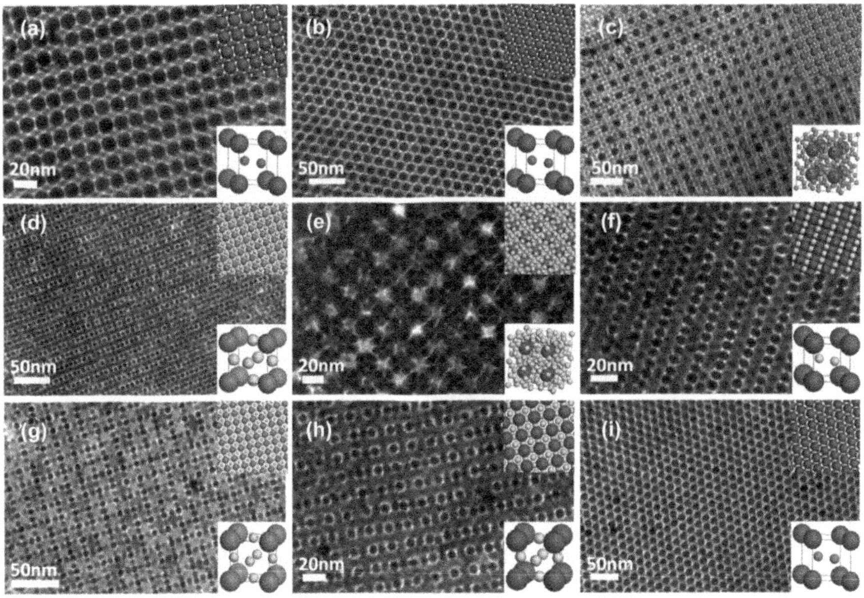

Figure 3.5 A selection of available BNSLs: (a, b) AB_2 Cu-FeO_x; (c) AB_{13} Cu-FeO_x; (d) AB_3 Au-Ni; (e) AB_{13} Cu-Au; (f) AB_2 Cu-Au; (g) AB_3 Au-FeO_x; (h) AB_3 Cu-Au; and (i) AB_2 Cu-MnO. The insets are cartoon structures and unit cells. Scale bars: (a, e, f, h) 20 nm; (b–d, g, i) 50 nm.
Reproduced by permission of American Chemical Society.[41]

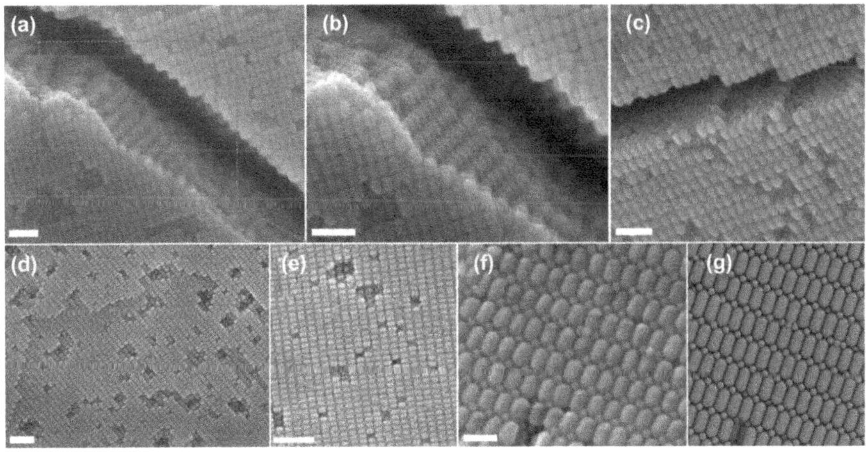

Figure 3.6 SEM images of the rod-sphere BNSLs self-assembled from $NaYF_4$ nanorods and 11.0 nm Fe_3O_4 nanospheres close to a cracked region. The crack separates the superlattice (a), magnified in (b) almost perpendicular, and (c) almost parallel to the hexagonal layers revealing terraces. (d–f) Voids are often observed on outside surfaces as seen in the TEM and (g) compared with a theoretical reconstruction. Scale bars: (a–c) 100 nm; (d,e) 200 nm; (f) 50 nm.
Reproduced by permission of American Chemical Society.[53]

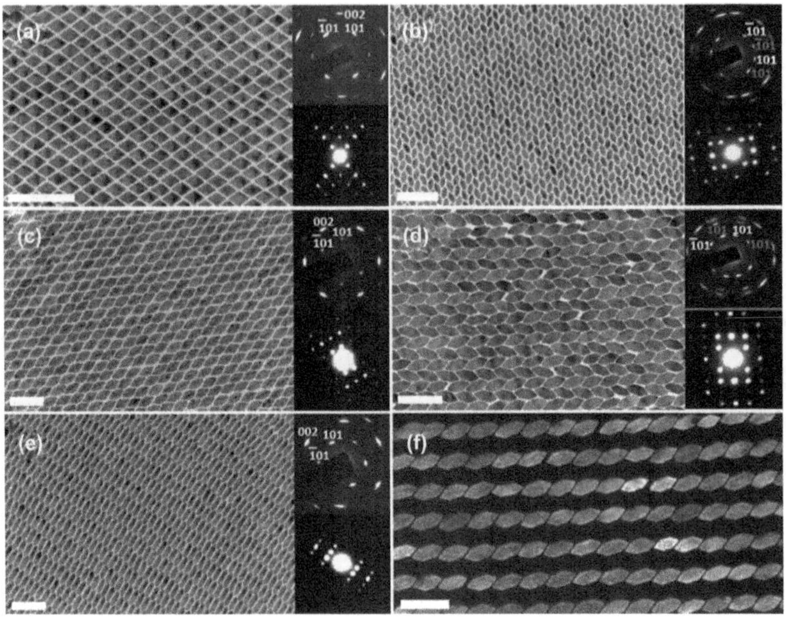

Figure 3.7 2D superlattices self-assembled from lanthanide fluoride nanoplates. (a) Parallel arrangements of DyF_3 rhombohedral nanoplates. (b) Small aspect ratio TbF_3 hexagonal nanoplates and (c) EuF_3 large aspect ratio hexagonal nanoplates. Alternating arrangements of intermediate aspect ratio hexagonal nanoplates of composition (d) DyF_3 and (e) TbF_3; for each, TEM images are shown on the left, and wide-angle and small-angle (upper right and lower right, respectively) electron diffraction patterns. (f) Dark-field TEM image of the same area as shown in (e). Scale bars: 100 nm.
Reproduced by permission of Nature Publishing Group.[52]

structure analogous to $NaZn_{13}$ (Figure 3.10).[40] These examples emphasize that systematic 3D structural characterization is essential for further exploration of structure–property relationships in multi-component nanoscale assemblies.

NC superlattices not only provide an opportunity to study fundamental interparticle interactions, but also offer a great opportunity to capture and harness emergent properties. For instance, large area (1.5 mm $\times$ 2.5 mm) BNSL membranes were used to construct magneto-resistive devices and their magneto-transport measurements clearly showed that device magneto-resistance was dependent on the structure (stoichiometry) of the BNSLs.[23] In a magnetics study, BNSL membranes exhibited a single phase like magnetization alignment process because of the collective interparticle dipolar interactions, which was not observed in the phase-separated NC mixtures that have the same stoichiometry.[25] In a study of the plasmonic properties, it was found that the strength of near-field couplings between adjacent plasmonic NCs can be systematically engineered by varying the NC size,

Figure 3.8 (a) TEM image, (b) high angle annular dark field scanning transmission electron microscopy (HAADF-STEM) image, and (c) SEM image of Pt-Pd AB_{13}-type BNSL. (f–h) STEM energy dispersive X-ray (EDX) mapping (green: Pt, red: Pd; (f) shows overlaps of STEM image and EDX-mappings) of Pt-Pd BNSLs. Inset of (c) proposed structural model: a unit cell of ico-AB_{13} BNSL (A, Pt octahedron in green; B, Pd sphere in red). Scale bars: (a, b, f, g, h) 50 nm; (c) 100 nm.
Reproduced by permission of American Chemical Society.[40]

composition, and the lattice symmetry of BNSLs, leading to broadband spectral tunability of the collective plasmonic response of BNSLs across the entire visible spectrum.[55] In addition to the examples above, superlattices have been explored in many other areas, such as electronics and catalysis.[36,40,41,47,49,51]

3.3 BNSLs: Novel Model Materials for Study of Catalysis

The interfacial interaction could be phenomenal in catalysis as well. For instance, catalyst–support interactions (*e.g.* metal–oxide interaction) have attracted great attention, because such interaction may result in significant enhancement in activity and/or selectivity.[10,11,56–61] Scientists use various model materials such as layered microstructure,[18,19] lithography fabricated arrays,[20,21] and multi-component NC architecture,[22,40,41] to try to isolate the support effects from other effects (*e.g.* size, structure, *etc.*). In addition to these model materials, BNSLs offer a class of novel materials for study of catalysis, such as searching for the active sites and investigating the thermal stability.[26,40,41]

3.3.1 Searching for the Active Sites

BNSLs offer great flexibility for searching for the active sites in catalytic processes, because their diversified structures and compositions can be

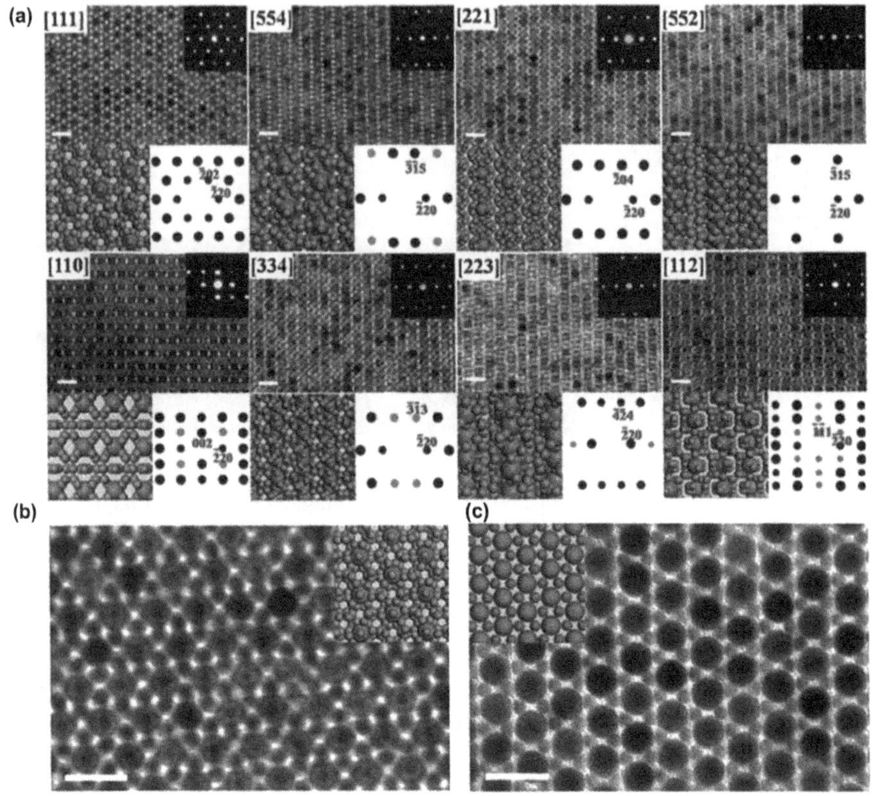

Figure 3.9 Tilt series of ico-AB_{13} BNSLs composed of 5.6 and 10.5 nm Fe_3O_4 NCs and structural comparison between ico-AB_{13} and AlB_2 type BNSLs. (a) Tilt series starts from [111] zone axis and tilts around $[\bar{1}10]$ direction. Tilting sequence: [111]→[554]→[221]→[552]→[110] and [111]→[334]→[223]→[112] zone axes. Below each TEM image (inset of each is the corresponding small-angle electron diffraction (ED pattern) is the corresponding structural model (left) and simulated ED pattern (right). The blue spots of the simulated ED patterns are kinematically forbidden reflections but could appear experimentally due to secondary diffraction. (b) TEM image of [111] projection of ico-AB_{13} Fe_3O_4 BNSLs. The inset is the structural model of the [111] projection of ico-AB_{13}. (c) TEM image of the [001] projection of AlB_2 BNSLs assembled from 5.6 and 10.5 nm Fe_3O_4 NCs. The inset is the structural model of the [001] projection of AlB_2. Scale bars: 20 nm.
Reproduced by permission of American Chemical Society.[37]

carefully controlled. Here, the classic reaction of CO oxidation on Au is used as model system to demonstrate the benefits of the BNSLs in catalytic studies.

The Langmuir–Hinshelwood mechanism is the most widely accepted mechanism for CO oxidation, in which both CO and O_2 need to adsorb on catalyst surface to facilitate CO oxidation. The rate limiting step is O_2

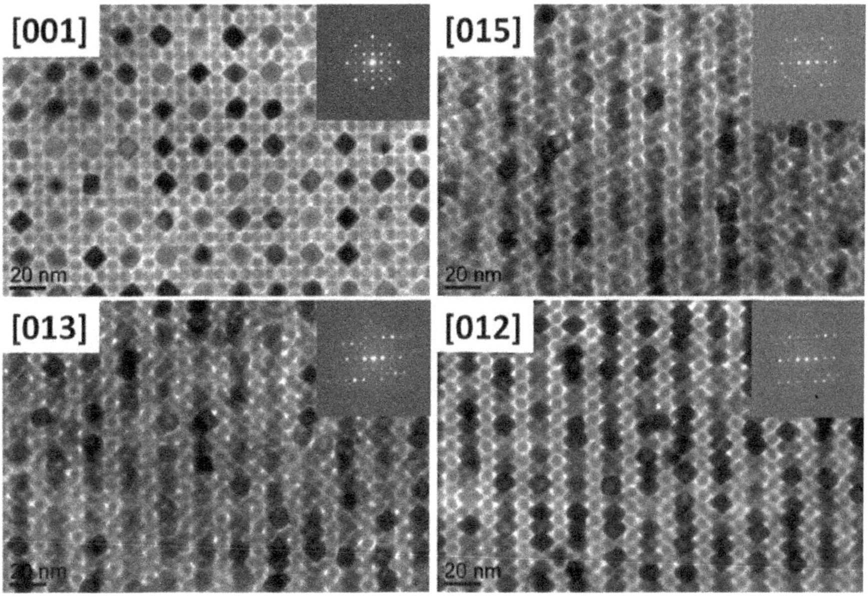

Figure 3.10 Crystallographic analysis of Pt-Pd BNSLs confirms the ico-AB_{13} type structure analogous to $NaZn_{13}$. Tilting sequence: [001] → [015] → [013] → [012] zone axes.
Reproduced by permission of American Chemical Society.[40]

adsorption and dissociation.[62] However, the adsorption strength of O_2 molecules on a Au surface is negligible at typical conditions used in practical CO oxidation.[62–65] Thus, bulk Au had long been considered to be catalytically inactive toward CO oxidation. However, some oxide-supported Au catalysts, especially Au/iron oxide catalysts, have been found to be highly active for CO oxidation, even at ambient temperature.[66–69] The origin of such high activity has been under debate for decades.[63,64,67,69–73] One common argument was that only small Au particles below a size called 'critical size' are active.[67,69,72] Such critical size was reported to be 5 nm, whereas Au particles larger than 5 nm were generally regarded as inactive for low-temperature CO oxidation. However, Au-Fe_3O_4 dumbbell nanoparticles (heterodimers) composed of 5 nm Au attached to 12 nm Fe_3O_4 were identified as a highly active catalyst for CO oxidation, even below 0 °C.[74] Catalytic CO oxidation was also observed on a random mixture of 6 nm Au NCs and 16 nm Fe_3O_4 NCs (Figure 3.11), while either 16 nm Fe_3O_4 NCs or 6 nm Au NCs alone showed negligibly low activity toward CO oxidation under the same reaction condition.[41] These two results imply that Au might not necessarily need to be smaller than 5 nm to be active in Au-FeO_x catalysts. Instead, the interaction between Au and FeO_x may be responsible for the extraordinary activity.

Due to the advantage of the structural control that is available in BNSLs, BNSLs were employed to experimentally support the proposal that the active sites correspond to the interparticle contacts. BNSLs have two major

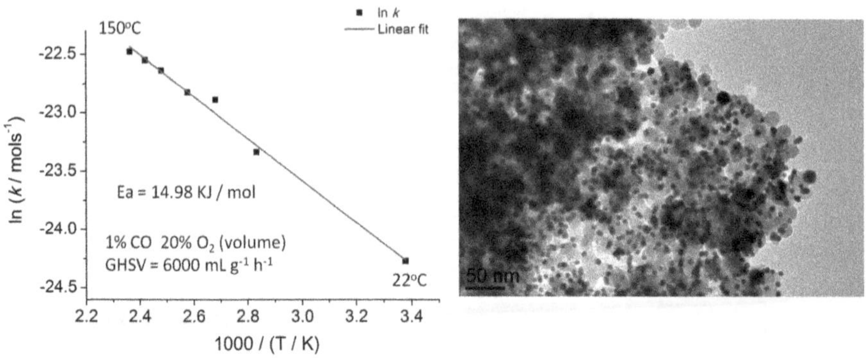

Figure 3.11 CO oxidation on a random mixture of 6 nm Au NCs and 16 nm Fe_3O_4 NCs. GHSV = gas hourly space velocity.
Reproduced by permission of American Chemical Society.[41]

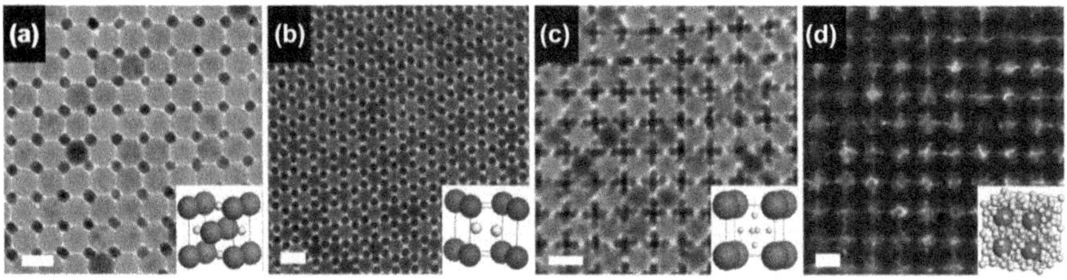

Figure 3.12 TEM images of Au-FeO_x BNSLs with structures analogous to (a) CuAu, (b) AlB_2, (c) CaB_6 and (d) $NaZn_{13}$. Scale bars: 20 nm.
Reproduced by permission of American Chemical Society.[41]

advantages. First, the size of building blocks can be precisely controlled, eliminating any uncertainties arising from polydispersity of the catalyst or support. In this study, the Au NCs were controlled at a size of 6 nm, ensuring that the individual Au NCs would not contribute significantly to the overall activity unless they were in contact with FeO_x, simply because Au NCs in a size that is greater than 5 nm is supposed to be inactive. Second, the tunable BNSL structures allow the stoichiometry of Au : Fe to be adjusted, and the number of contacts between Au and FeO_x NCs can be rigorously controlled.

NC superlattices can be induced to form a variety of crystal structures by adjusting the relative size and concentration of the constituent NCs. In the Au-FeO_x system (here, FeO_x is A and Au is B), AB [Figure 3.12(a)], AB_2 [Figure 3.12(b)], AB_6 [Figure 3.12(c)], and AB_{13} [Figure 3.12(d)] structures, which are analogous to the atomic crystals of CuAu, AlB_2, CaB_6, and $NaZn_{13}$, respectively, are all accessible. Moreover, it is possible to prepare an A_2B_3 structure which does not have an atomic counterpart.[38,75] Among these available structures, AB, AB_2, and AB_{13} were chosen for the catalysis study because they were able to be produced in a high phase purity with large-size grains on the order of μm^2 and the formed continuous films were able to

Figure 3.13 HAADF-STEM images of pre-treated (a) AB, (b) AB_2, and (c) AB_{13} Au-FeO_x BNSLs. The cartoon below each HAADF-STEM image represents the corresponding structure. Scale bars: 20 nm.
Reproduced by permission of American Chemical Society.[41]

cover areas on the order of cm^2. As shown in the HAADF-STEM images, each Au NC in the AB structure [Figure 3.13(a)] was directly in contact with four neighboring FeO_x NCs, and *vice versa*. For the AB_2 monolayer BNSL [Figure 3.13(b)], the number of contacts is six per FeO_x NC and three per Au NC, while the AB_{13} structure [Figure 3.13(c)] has 24 contacts per FeO_x and two per Au NC. The effective contact number per Au for AB_{13} is 1.85 (*i.e.* $2 \times 12/13$), because the Au NCs are arranged in an icosahedron, in which 12 Au NCs are on the exterior and the 13th Au NC sits in the interior with no contact to FeO_x.

The BNSLs with the structures shown in Figure 3.13 were fabricated through a liquid–air interfacial self-assembly method[23,41] transferred onto Si wafers and then loaded into a batch reactor for catalytic studies. The overall reaction rates were proportional to the loadings, indicating that the diffusion of gas molecules was not the rate limiting step. The reaction temperature was chosen at 150 °C. At this relatively high temperature relatively high reaction rates were able to compensate the low catalyst loading, ensuring that the concentration of product (*i.e.* CO_2) was sufficiently high to be detected by gas chromatography (GC); this temperature was sufficiently low so that the activity of Au or FeO_x alone was negligible. Additionally and importantly, the BNSL morphology did not change at 150 °C. The HAADF-STEM images in Figure 3.13 demonstrate the structure of BNSL catalysts, and the atomic resolution HAADF-STEM image in Figure 3.14(b) confirms

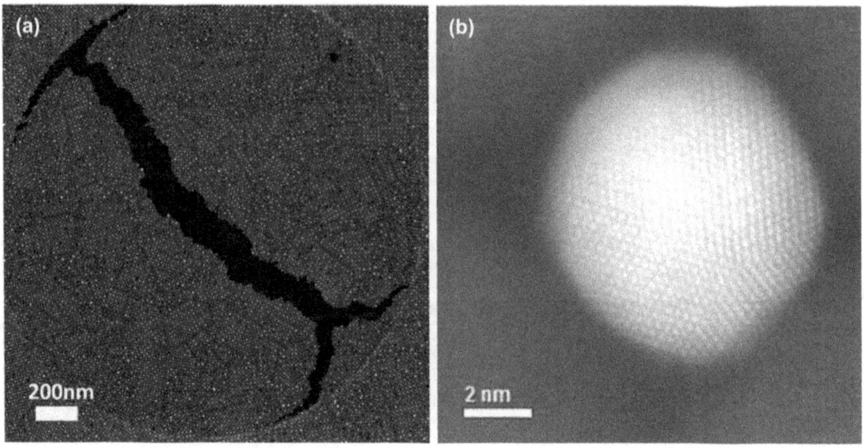

Figure 3.14 HAADF-STEM images of AB Au-FeO_x BNSL showing that (a) the BNSLs can be free-standing and (b) one Au NC is in contact with four adjacent FeO_x NCs.
Reproduced by permission of American Chemical Society.[41]

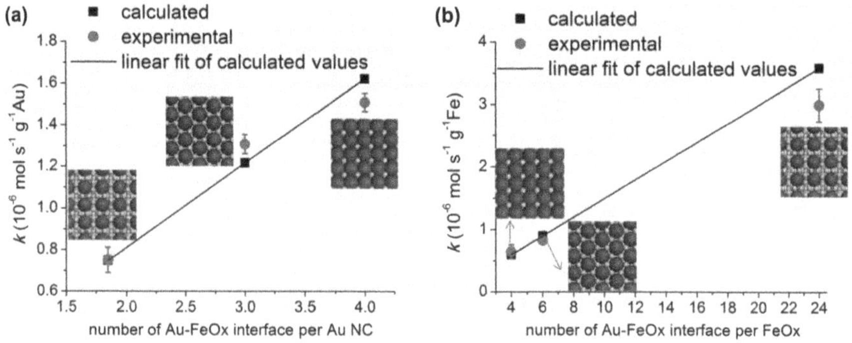

Figure 3.15 The activity (reaction rate of CO oxidation expressed by the rate of CO_2 production) normalized to the mass of (a) Au and (b) Fe, is a function of the number of the Au-FeO_x contacts (a) per Au NC and (b) per FeO_x NC.
Reproduced by permission of American Chemical Society.[41]

that the Au NC are in direct contact with neighboring FeO_x NCs after organic stabilizers were removed by a thermal treatment.[41]

When the size of Au and FeO_x NCs were fixed, the sizes (*i.e.* area or perimeter) of each contact between Au and FeO_x were regarded constant. Therefore, according to the hypothesis that the active site is the contact, it was predicted that the overall activity should be proportional to the number of catalytic contacts. The measurements on Au-FeO_x BNSLs provided direct evidence for this hypothesis (Figure 3.15). The overall specific rates were normalized to the mass of Au or Fe, respectively, and then plotted *versus* the number of Au-FeO_x contacts in each structure. Because the AB_2 BNSL had the largest domain sizes, the rates obtained on AB_2 Au-FeO_x were averaged to provide a relative rate per Au-FeO_x contact, and then this reference was used to derive the expected activity of AB and AB_{13} based on their respective number of Au-FeO_x contacts. The calculation generated a linear correlation: the overall reaction rate was proportional to the number of catalytic contacts. The experimental rates of CO oxidation measured on AB and AB_{13} were consistent with this linear correlation. As the Au and FeO_x NCs alone showed negligible activity under the same reaction conditions, this linear correlation between the overall catalytic activities and the numbers of Au-FeO_x contacts was strong evidence that the contacts of Au-FeO_x were the active sites for CO oxidation. Moreover, if the size of FeO_x NCs in the BNSLs changed, the activity per FeO_x NC subsequently changed, but the activity per Au NC did not significantly change (Table 3.1). This observation thus provided further confirmation of the hypothesis. Additionally, disordered mixtures of Au and FeO_x NCs showed significantly lower activities (except mixture 5) compared with the ordered BNSLs (Table 3.2 and Figure 3.16). This was able to be explained by the reduced density of Au-FeO_x contacts due to local phase segregation. Although the intermediate and long range orders were absent in mixture 5, the composition and size ratio of the constituents in mixture 5 was close to that in AB_5 BNSL (equivalent to the atomic $CaCu_5$ crystal structure). The absence of phase separation and the high density of Au-FeO_x contacts in mixture 5 in turn resulted in a high reaction rate per FeO_x particle. When the same Au and FeO_x NCs were loaded onto opposite sides of a

Table 3.1 CO oxidation on Au-FeO_x BNSLs. Reproduced by permission of American Chemical Society.[41]

	Loading/μg		Size/nm		k/ml $min^{-1} \times 10^5$	k/mol $s^{-1} \times 10^{11}$	k_{Au}/mol s^{-1} g^{-1}Au $\times 10^6$	k_{Fe}/mol s^{-1} g^{-1}Fe $\times 10^6$
	Au	Fe	Au	Fe				
AB_2-type	41.16	65.38	6	16	7.330	5.454	1.325	0.834
AB_2-type	52.89	44.12	6	12	9.467	7.044	1.332	1.597

Table 3.2 CO oxidation on Au-FeO_x BNSLs and random mixtures. Reproduced by permission of American Chemical Society.[41]

	Loading (Au)/μg	Loading (Fe)/μg	k/ml $min^{-1} \times 10^5$	k/mol $s^{-1} \times 10^{11}$	k_{Au}/mol s^{-1} g^{-1}Au $\times 10^6$	k_{Fe}/mol s^{-1} g^{-1}Fe $\times 10^6$
AB-type	22.4	59.34	4.663	3.467	1.548	0.584
AB_{13}-type	179.67	49.17	19.660	14.583	0.812	2.966
AB_2-type	41.16	65.38	7.330	5.454	1.325	0.834
mixture1	52.89	44.12	1.564	1.164	0.220	0.264
mixture2	53.36	98.26	2.178	1.620	0.304	0.165
mixture3	50.8	102.21	2.602	1.936	0.381	0.189
mixture4	0.37	95.93	0.000	0.000	0.000	0.000
mixture5	187.08	83.65	23.210	17.269	0.923	2.064

Note: 1. Only one (of three) measurement for each BNSL is listed. 2. The activity for mixture 4 is too low to be detected by gas chromatography–thermal conductivity detector (GC-TCD).

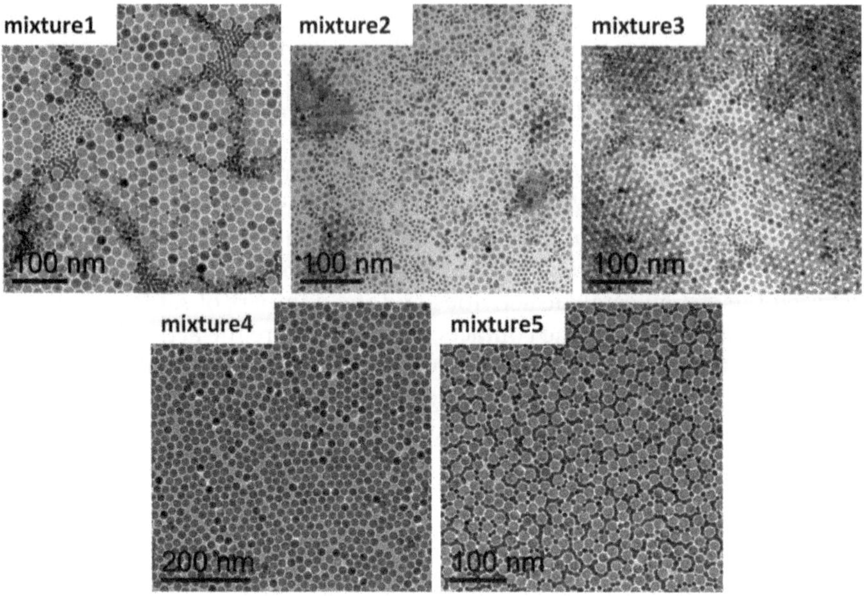

Figure 3.16 TEM images of Au-FeO_x random mixtures.
Reproduced by permission of American Chemical Society.[41]

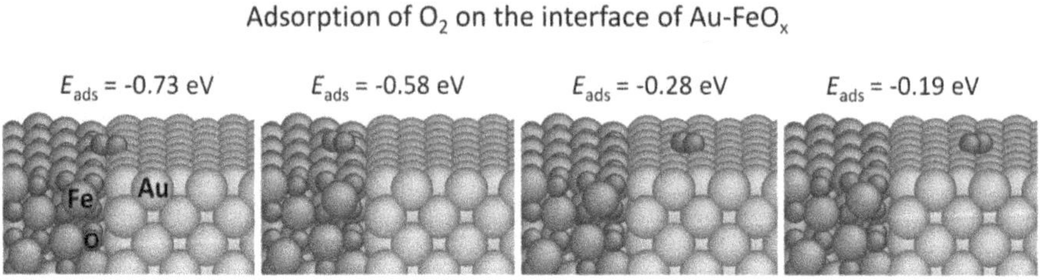

Figure 3.17 DFT calculated adsorption of O_2 on the interface of Au-FeO_x.
Reproduced by permission of American Chemical Society.[41]

Si Wafer (*i.e.* when no Au-FeO_x contact existed), the samples were tested in equivalent conditions but showed negligible activity toward CO oxidation. This again strongly suggested that the Au-FeO_x interaction at the contacts was responsible for the high catalytic activity.

As discussed earlier, the poor catalytic activity on pure Au is attributed to the weak adsorption of O_2 on Au surfaces. A density functional theory (DFT) study of the adsorption at the contact between Au and FeO_x NCs indicated that the O_2 adsorption was strongest at the contact ($E_{ads} = -0.73$ eV), and the adsorption strength decreased dramatically as the adsorption site moved away from the Au-FeO_x contact (Figure 3.17).[41] While CO is adsorbed on the Au surface,[57] the enhanced O_2 adsorption is essential to facilitate CO oxidation. Moreover, as O_2 adsorbs and dissociates readily on FeO_x, at the temperature of reaction, the O adsorbed on FeO_x may migrate to Au or the Au-FeO_x interface to promote CO oxidation. Many mechanisms for CO oxidation at Au–oxide interfaces have been discussed.[57,59,61,76–79] Nevertheless, in addition to many works which study the effect of the Au–oxide interface that eliminates Au particle size effects,[10,19,59,61,78,80] the observations shown in this chapter provide direct evidence that the Au-FeO_x contact is responsible for the high activity, instead of the small Au particles.

The overall activities measured on BNSLs were about two orders of magnitude lower than the reported best catalysts. However, the turnover frequency (TOF) normalized to total number of contacts ($\sim$1 s^{-1}) was in the same order of magnitude as the reported high TOFs.[69] This was ascribed to the actual number of active sites. Assuming the NCs are near spherical, the atoms involved in the contacts are highlighted in Figure 3.18. For 6 nm Au NCs, only 0.48% of the total Au atoms are at the Au-FeO_x contact. However, for the 0.5 nm Au cluster, the percentage of interfacial Au atoms over total Au atoms is 50% which means that, for the same Au loading, the 0.5 nm Au NCs have $\sim$100 times more 'active' Au atoms than 6 nm Au NCs. The number of active sites at the Au-FeO_x contact can be increased by decreasing the Au NC size. The results thus indicate that it is not uniquely the small sizes for Au particle, but rather the maximized number of active sites (*i.e.* the Au-FeO_x contact), that is desirable to obtain high activity.

3.3.2 Investigating the Thermal Stability

In addition to enabling the engineering of catalytic activity, BNSLs also hold the potential to improve the stability of catalysts. Before discussing the investigation of the thermal stability, the background is introduced.

The NC has a significantly reduced melting temperature compared with the bulk material. The melting point T(d) is inversely proportional to the particle size *d*, as described by the Gibbs–Thompson equation:

$$T(d) = T_b\left(1 - \frac{4\sigma_{sl}}{H_f \rho d}\right) \tag{3.1}$$

where T_b is the bulk melting temperature, σ_{sl} is the solid–liquid interface energy, H_f is the bulk enthalpy of fusion, and ρ is the solid density.

Thermodynamically, the reduction in surface energy drives sintering; kinetically, the temperature is the key to sintering.

In the thin film of single component NCs (or the random binary mixture), metal NCs contact with each other as shown in Figure 3.19. Mass transfer between NCs can be described by equation (3.2):

$$\Gamma = \nu e^{-E/k_B T} \tag{3.2}$$

where Γ is the jump or hopping rate, ν is the attempt frequency, E is the potential energy barrier for mass transfer, k_B is the Boltzmann constant, and T is the temperature.

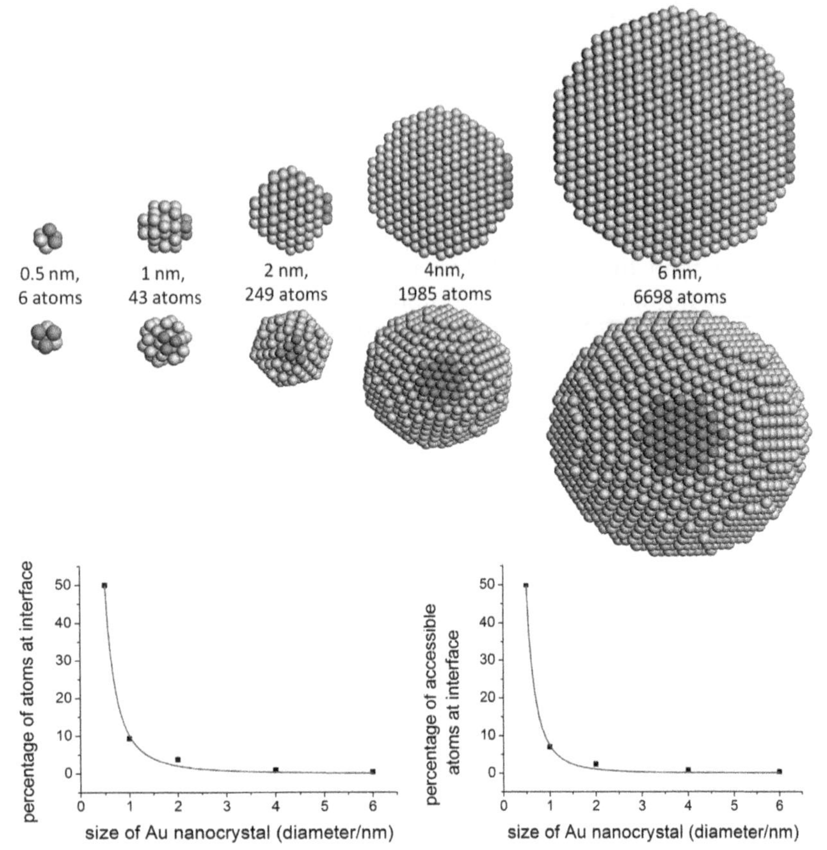

Size of Au nanocrystal (diameter)	Atoms in contact with FeOx via Au(111)	Atoms in contact with FeOx via Au(110)	Atoms in contact with FeOx via Au(100)	Total number of atoms	Percentage of atoms at interface	Percentage of accessible atoms at interface
6	43	17	37	6698	0.48	0.36
4	21	12	24	1985	0.96	0.76
2	6	9	13	249	3.74	2.41
1	3	4	5	43	9.30	6.98
0.5	3	3	3	6	50	50

Figure 3.18 Correlation between size of Au NCs and the effective Au atoms in the active Au-FeO_x catalysts. The cartoons show the side view and top view of Au NCs at the size of 0.5–6 nm; the atoms involved in the contact with FeO_x are highlighted by orange color. The term 'accessible atoms' is used when it assumes that only atoms at the periphery may access to CO molecules.
Reproduced by permission of American Chemical Society.[41]

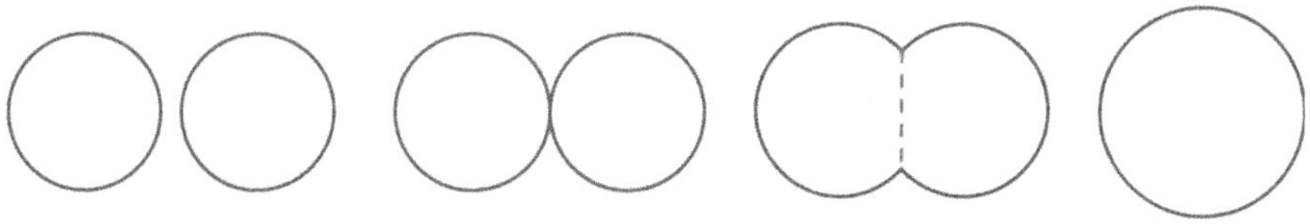

Figure 3.19 Illustration of the sintering process occurred between spherical nanocrystals.
Reproduced by permission of American Chemical Society.[41]

When NCs are in contact with each other, the capillary stress acts as the driving force for the mass transfer in the sintering process. This stress, σ, at a curved surface can be described by the Laplace equation:

$$\sigma = \gamma\left(\frac{1}{R_1} + \frac{1}{R_2}\right) \tag{3.3}$$

where γ is the energy associated with the curved surface, and R_1 and R_2 are the radii of curvature. When a neck formed, σ in the neck region is different from σ in the region next to the neck. This gradient in curvature generates a thermodynamic gradient that drives mass transfer. Metal atoms on the NCs flow to reach equilibrium. Metal atoms have higher mobility at higher temperature and on smaller sized NCs.

Metal usually exhibits substantially reduced melting points when its size drops into the nanometre regime – a phenomenon known as the melting point depression.[81] For instance, Au NCs having diameter of 2.5 nm were found to have their melting point reduced to around 300 °C.[81] Furthermore, the metal NCs start to sinter at the temperature well below their reduced melting point, because of elevated mass transfer on the surface. In a metal NC system, it is very common that one NC is in direct contact with the other one, forming a neck (Figure 3.19). The great change in curvature over distance at the neck generates a capillary stress, which acts as a driving force to move the metal atoms for sintering. Additionally, a liquid phase is also possibly involved in the sintering process, providing another major source of sintering stress.[82–84]

As an example of sintering at low temperature, 6 nm Au and 12 nm Cu NCs start to sinter at 200 °C (in N_2).[41] This behavior of metal sintering at a relatively low temperature causes the problem of losing catalytic surface area of metal catalysts, thus leading to a reduction in their overall activities. However, in the catalysis studies involving BNSLs, it was found that metal–oxide BNSLs with ordered structures exhibited substantially enhanced stability at elevated temperature.

The observations are summarized in Figure 3.20. Au-FeO_x AB_2 BNSLs had their long-range ordered structures retained after a thermal treatment at 200 °C (in N_2). In contrast, the single component thin film of Au NCs and the

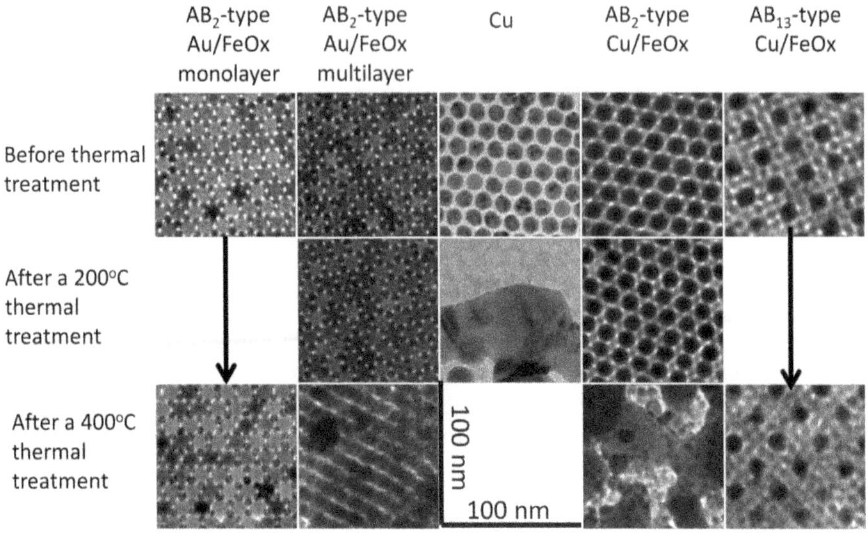

Figure 3.20 Structure-dependent thermal stability of NCs and NC superlattice. All TEM images are 100 nm × 100 nm.
Reproduced by permission of American Chemical Society.[41]

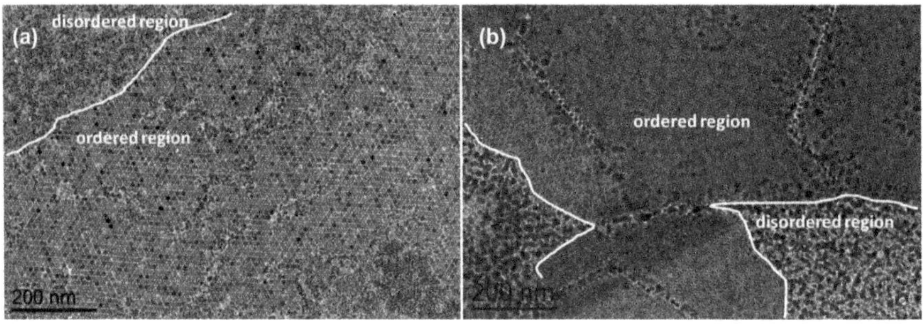

Figure 3.21 TEM images of AB_2-type Au-FeO_x (a) before and (b) after annealing at 200 °C for 30 min (in N_2).
Reproduced by permission of American Chemical Society.[41]

disordered Au-FeO_x mixtures were severely sintered. As a demonstration, a thin film, in which the ordered and disordered domains co-existed, was treated at 200 °C for 30 min [Figure 3.21(a)]. After the thermal treatment, the disordered domains were sintered while the ordered domains were preserved [Figure 3.21(b)]. When the temperature of thermal treatment was further increased to 400 °C (in N_2) in an AB_2 Au-FeO_x BNSL that had multilayers, Au NC sintered in the interstices formed by the lattice of larger FeO_x NCs, but the FeO_x skeleton was preserved (Figure 3.20). In contrast, a monolayer of AB or AB_2 structure of Au-FeO_x changed little in morphology. This enhanced stability was not limited to Au-containing BNSLs. As seen in the case of Cu-FeO_x, the AB_2 (here, Cu is A, FeO_x is B) structure retained at

200 °C, but sintered at 400 °C. However, the AB_{13} Cu-FeO_x was stable up to 400 °C, which is a sufficiently high operating temperature for most industrially important catalytic reactions.

These results indicated that the stability of a catalyst can be substantially improved for certain BNSL structures if the metal NCs are well isolated from each other by more thermally stable oxide NCs. At high temperature, adjacent Au NCs tended to coalesce (Figure 3.22). In multi-layers of AB_2 Au-FeO_x, an Au NC shared interstices (from packing of FeO_x) with

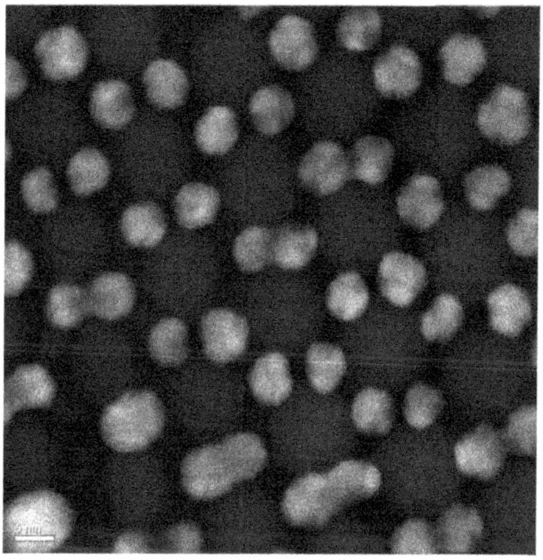

Figure 3.22 TEM images of AB_2-type Au-FeO_x BNSL after annealing at 200 °C for 30 min. In the lower left region, the defected FeO_x NC with smaller size cannot confine the Au NCs in spatial, causing the coalescence of Au NCs; in the upper right region, the Au NCs are well-confined in spatial by FeO_x NCs. Scale bar: 5 nm.
Reproduced by permission of American Chemical Society.[41]

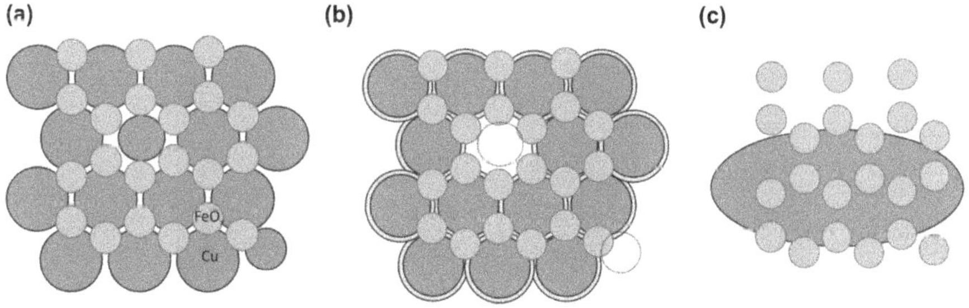

Figure 3.23 Illustration of (a) as prepared AB_2 Cu-FeO_x, (b) AB_2 Cu-FeO_x heated at certain temperature which is below the melting point of most Cu nanocrystals (*e.g.* 200 °C), and (c) Cu NCs fully sintered and coalesced at high temperature (*e.g.* 400 °C). Due to the size distribution of Cu NCs (although it is small), some small sized Cu NCs exist. Such small Cu NCs melted at relatively low temperature (*e.g.* 200 °C), wetting the rest Cu NCs.
Reproduced by permission of American Chemical Society.[41]

neighboring Au NCs in both lateral and vertical directions; in the monolayer, Au NCs were isolated by FeO_x NCs in one plane. Therefore, the probability of coalescence in a multi-layer was higher than that in a monolayer. Moreover, the wetting liquid from melted Au might also increase the probability of coalescence to form a neck, and thus facilitates the sintering process. However, in the AB_2 structures, especially in monolayer AB_2, the large FeO_x NCs blocked mass transfer between the Au NCs and the wetting of liquid Au, alleviating the sintering. In AB_2 Cu-FeO_x, Cu was the larger one of the two building blocks and thus the distances between two Cu NCs were insufficient to prevent coalescence. Specifically, the liquid Cu spread over the surface (wetting) and thereby provided an easy mass-transport route between

Figure 3.24 Structural model of the (111)-oriented NaCl-type BNSL viewed from (a) the top and (b) the side. TEM images of (c) the as-made NaCl-type FePt-MnO BNSL membrane with the (111) lattice projection oriented parallel to the TEM grid, and (d) the FePt-MnO BNSL after annealing at the temperature up to 650 °C.
Reproduced by permission of American Chemical Society.[26]

Cu NCs (Figure 3.23). In contrast, in an AB_{13} Cu-FeO_x BNSL the metal NCs were separated spatially and thus they did not coalesce.

By using BNSLs as models, a general strategy of keeping metal NCs precisely separated to enhance the stability of metal catalysts was demonstrated, consistent with recent reports.[26,85] One of recent results showed that, in FePt-MnO, the well-confined FePt NCs were stable up to 650 °C (Figure 3.24).[26]

3.4 Conclusions

NC superlattices, especially BNSLs, offer great opportunities to study catalysis because of the highly programmable metal–metal, metal–oxide, or oxide–oxide interactions. The Au-FeO_x BNSLs with well-defined structures have been demonstrated to be particularly useful for searching for the active sites in Au-FeO_x catalysts. BNSL architectures have also been exploited to enhance the thermal stability of catalysts by suppressing sintering and coalescence. As the variety of available building blocks increases, it is expected that NC superlattices will provide a powerful new platform for research in the field of catalysis that bridges traditional studies of single crystals and supported particulate catalysts.

References

1. V. R. Stamenkovic, M. Arenz, B. B. Blizanac, K. J. J. Mayrhofer, P. N. Ross and N. M. Markovic, *Surf. Sci.*, 2005, **576**, 145.
2. N. Mizuno and M. Misono, *Chem. Rev.*, 1998, **98**, 199.
3. G. A. Somorjai, *Chem. Rev.*, 1996, **96**, 1223.
4. P. L. J. Gunter, J. W. Niemantsverdriet, F. H. Ribeiro and G. A. Somorjai, *Catal. Rev. Sci. Eng.*, 1997, **39**, 77.
5. Y. Li and G. A. Somorjai, *Nano Lett.*, 2010, **10**, 2289.
6. I. Lee, F. Delbecq, R. Morales, M. A. Albiter and F. Zaera, *Nat. Mater.*, 2009, **8**, 132.
7. R. Narayanan and M. A. El-Sayed, *Nano Lett.*, 2004, **4**, 1343.
8. K. M. Bratlie, H. Lee, K. Komvopoulos, P. D. Yang and G. A. Somorjai, *Nano Lett.*, 2007, **7**, 3097.
9. S. E. Habas, H. Lee, V. Radmilovic, G. A. Somorjai and P. Yang, *Nat. Mater.*, 2007, **6**, 692.
10. Q. Fu, H. Saltsburg and M. Flytzani-Stephanopoulos, *Science*, 2003, **301**, 935.
11. R. Si and M. Flytzani-Stephanopoulos, *Angew. Chem., Int. Ed.*, 2008, **47**, 2884.
12. M. R. Buck, J. F. Bondi and R. E. Schaak, *Nat. Chem.*, 2012, **4**, 37.
13. Y. J. Kang and C. B. Murray, *J. Am. Chem. Soc.*, 2010, **132**, 7568.
14. Y. J. Kang, *et al.*, *ACS Nano*, 2012, **6**, 2818.
15. Y. J. Kang, J. B. Pyo, X. Ye, T. R. Gordon and C. B. Murray, *ACS Nano*, 2012, **6**, 5642.

16. C. Wang, M. Chi, D. Li, D. Strmcnik, D. van der Vliet, G. Wang, V. Komanicky, K.-C. Chang, A. P. Paulikas, D. Tripkovic, J. Pearson, K. L. More, N. M. Markovic and V. R. Stamenkovic, *J. Am. Chem. Soc.*, 2011, **133**, 14396.
17. C. Wang, D. van der Vliet, K. L. More, N. J. Zaluzec, S. Peng, S. H. Sun, H. Daimon, G. F. Wang, J. Greeley, J. Pearson, A. P. Paulikas, G. Karapetrov, D. Strmcnik, N. M. Markovic and V. R. Stamenkovic, *Nano Lett.*, 2011, **11**, 919.
18. I. Zuburtikudis and H. Saltsburg, *Science*, 1992, **258**, 1337.
19. Z. Zhou, S. Kooi, M. Flytzani-Stephanopoulos and H. Saltsburg, *Adv. Funct. Mater.*, 2008, **18**, 2801.
20. P. W. Jacobs, F. H. Ribeiro, G. A. Somorjai and S. J. Wind, *Catal. Lett.*, 1996, **37**, 131.
21. J. Grunes, J. Zhu, E. A. Anderson and G. A. Somorjai, *J. Phys. Chem. B*, 2002, **106**, 11463.
22. Y. Yamada, C. K. Tsung, W. Huang, Z. Y. Huo, S. E. Habas, T. Soejima, C. E. Aliaga, G. A. Somorjai and P. D. Yang, *Nat. Chem.*, 2011, **3**, 372.
23. A. G. Dong, J. Chen, P. M. Vora, J. M. Kikkawa and C. B. Murray, *Nature*, 2010, **466**, 474.
24. E. V. Shevchenko, D. V. Talapin, N. A. Kotov, S. O'Brien and C. B. Murray, *Nature*, 2006, **439**, 55.
25. J. Chen, A. G. Dong, J. Cai, X. C. Ye, Y. J. Kang, J. M. Kikkawa and C. B. Murray, *Nano Lett.*, 2010, **10**, 5103.
26. A. G. Dong, J. Chen, X. C. Ye, J. M. Kikkawa and C. B. Murray, *J. Am. Chem. Soc.*, 2011, **133**, 13296.
27. B. A. Korgel, *Nat. Mater.*, 2007, **6**, 551.
28. D. K. Smith, B. Goodfellow, D. M. Smilgies and B. A. Korgel, *J. Am. Chem. Soc.*, 2009, **131**, 3281.
29. H. Friedrich, C. J. Gommes, K. Overgaag, J. D. Meeldijk, W. H. Evers, B. de Nijs, M. P. Boneschanscher, P. E. de Jongh, A. J. Verkleij, K. P. de Jong, A. van Blaaderen and D. Vanmaekelbergh, *Nano Lett.*, 2009, **9**, 2719.
30. M. J. Murray and J. V. Sanders, *Philos. Mag. A*, 1980, **42**, 721.
31. J. V. Sanders, *Philos. Mag. A*, 1980, **42**, 705.
32. S. Hachisu and S. Yoshimura, *Nature*, 1980, **283**, 188.
33. P. Bartlett, R. H. Ottewill and P. N. Pusey, *Phys. Rev. Lett.*, 1992, **68**, 3801.
34. C. B. Murray, C. R. Kagan and M. G. Bawendi, *Science*, 1995, **270**, 1335.
35. C. B. Murray, C. R. Kagan and M. G. Bawendi, *Annu. Rev. Mater. Sci.*, 2000, **30**, 545.
36. F. X. Redl, K. S. Cho, C. B. Murray and S. O'Brien, *Nature*, 2003, **423**, 968.
37. J. Chen, X. C. Ye and C. B. Murray, *ACS Nano*, 2010, **4**, 2374.
38. A. G. Dong, X. C. Ye, J. Chen and C. B. Murray, *Nano Lett.*, 2011, **11**, 1804.
39. Y. J. Kang, M. Li, Y. Cai, M. Cargnello, R. E. Diaz, T. R. Gordon, N. L. Wieder, R. R. Adzic, R. J. Gorte, E. A. Stach and C. B. Murray, *J. Am. Chem. Soc.*, 2013, **135**, 2741.

40. Y. J. Kang, X. Ye, J. Chen, Y. Cai, R. E. Diaz, R. R. Adzic, E. A. Stach and C. B. Murray, *J. Am. Chem. Soc.*, 2013, **135**, 42.
41. Y. J. Kang, X. Ye, J. Chen, L. Qi, R. E. Diaz, V. Doan-Nguyen, G. Xing, C. R. Kagan, J. Li, R. J. Gorte, E. A. Stach and C. B. Murray, *J. Am. Chem. Soc.*, 2013, **135**, 1499.
42. F. Dumestre, B. Chaudret, C. Amiens, P. Renaud and P. Fejes, *Science*, 2004, **303**, 821.
43. C. Desvaux, C. Amiens, P. Fejes, P. Renaud, M. Respaud, P. Lecante, E. Snoeck and B. Chaudret, *Nat. Mater.*, 2005, **4**, 750.
44. M. I. Bodnarchuk, M. V. Kovalenko, W. Heiss and D. V. Talapin, *J. Am. Chem. Soc.*, 2010, **132**, 11967.
45. J. Henzie, M. Gruenwald, A. Widmer-Cooper, P. L. Geissler and P. Yang, *Nat. Mater.*, 2012, **11**, 131.
46. S. H. Sun and C. B. Murray, *J. Appl. Phys.*, 1999, **85**, 4325.
47. S. H. Sun, C. B. Murray, D. Weller, L. Folks and A. Moser, *Science*, 2000, **287**, 1989.
48. C. B. Murray, S. H. Sun, H. Doyle and T. Betley, *MRS Bull.*, 2001, **26**, 985.
49. D. V. Talapin and C. B. Murray, *Science*, 2005, **310**, 86.
50. X. Ye, J. E. Collins, Y. J. Kang, J. Chen, D. T. N. Chen, A. G. Yodh and C. B. Murray, *Proc. Natl. Acad. Sci. U. S. A.*, 2010, **107**, 22430.
51. D. C. Reifsnyder, X. Ye, T. R. Gordon, C. Song and C. B. Murray, *Acs Nano*, 2013, **7**, 4307.
52. X. Ye, J. Chen, M. Engel, J. A. Millan, W. Li, L. Qi, G. Xing, J. E. Collins, C. R. Kagan, J. Li, S. C. Glotzer and C. B. Murray, *Nat. Chem.*, 2013, **5**, 466.
53. X. Ye, J. A. Millan, M. Engel, J. Chen, B. T. Diroll, S. C. Glotzer and C. B. Murray, *Nano Lett.*, 2013, **13**, 4980.
54. X. Ye, C. Zheng, J. Chen, Y. Gao and C. B. Murray, *Nano Lett.*, 2013, **13**, 765.
55. X. Ye, J. Chen, B. T. Diroll and C. B. Murray, *Nano Lett.*, 2013, **13**, 1291.
56. M. Valden, X. Lai and D. W. Goodman, *Science*, 1998, **281**, 1647.
57. M. S. Chen and D. W. Goodman, *Science*, 2004, **306**, 252.
58. M. Cargnello, V. V. T. Doan-Nguyen, T. R. Gordon, R. E. Diaz, E. A. Stach, R. J. Gorte, P. Fornasiero and C. B. Murray, *Science*, 2013, **341**, 771.
59. J. A. Rodriguez, S. Ma, P. Liu, J. Hrbek, J. Evans and M. Perez, *Science*, 2007, **318**, 1757.
60. J. A. Rodriguez, X. Wang, P. Liu, W. Wen, J. C. Hanson, J. Hrbek, M. Perez and J. Evans, *Top. Catal.*, 2007, **44**, 73.
61. J. B. Park, J. Graciani, J. Evans, D. Stacchiola, S. Ma, P. Liu, A. Nambu, J. Fernandez Sanz, J. Hrbek and J. A. Rodriguez, *Proc. Natl. Acad. Sci. U. S. A.*, 2009, **106**, 4975.
62. X. Y. Deng, B. K. Min, A. Guloy and C. M. Friend, *J. Am. Chem. Soc.*, 2005, **127**, 9267.
63. B. K. Min, A. R. Alemozafar, D. Pinnaduwage, X. Deng and C. M. Friend, *J. Phys. Chem. B*, 2006, **110**, 19833.
64. R. Meyer, C. Lemire, S. K. Shaikhutdinov and H. Freund, *Gold Bull.*, 2004, **37**, 72.

65. N. Lopez and J. K. Norskov, *J. Am. Chem. Soc.*, 2002, **124**, 11262.
66. M. Haruta and M. Date, *Appl. Catal. A*, 2001, **222**, 427.
67. M. Haruta, *Catal. Today*, 1997, **36**, 153.
68. M. Haruta, S. Tsubota, T. Kobayashi, H. Kageyama, M. J. Genet and B. Delmon, *J. Catal.*, 1993, **144**, 175.
69. A. A. Herzing, C. J. Kiely, A. F. Carley, P. Landon and G. J. Hutchings, *Science*, 2008, **321**, 1331.
70. N. Lopez, T. V. W. Janssens, B. S. Clausen, Y. Xu, M. Mavrikakis, T. Bligaard and J. K. Norskov, *J. Catal.*, 2004, **223**, 232.
71. A. S. K. Hashmi and G. J. Hutchings, *Angew. Chem., Int. Ed.*, 2006, **45**, 7896.
72. Y. Liu, C. J. Jia, J. Yamasaki, O. Terasaki and F. Schuth, *Angew. Chem., Int. Ed.*, 2010, **49**, 5771.
73. B. K. Min and C. M. Friend, *Chem. Rev.*, 2007, **107**, 2709.
74. C. Wang, H. F. Yin, S. Dai and S. H. Sun, *Chem. Mater.*, 2010, **22**, 3277.
75. X. C. Ye, J. Chen and C. B. Murray, *J. Am. Chem. Soc.*, 2011, **133**, 2613.
76. I. N. Remediakis, N. Lopez and J. K. Norskov, *Angew. Chem., Int. Ed.*, 2005, **44**, 1824.
77. M. M. Schubert, S. Hackenberg, A. C. van Veen, M. Muhler, V. Plzak and R. J. Behm, *J. Catal.*, 2001, **197**, 113.
78. T. Yan, D. W. Redman, W.-Y. Yu, D. W. Flaherty, J. A. Rodriguez and C. B. Mullins, *J. Catal.*, 2012, **294**, 216.
79. I. X. Green, W. Tang, M. Neurock and J. T. Yates, Jr., *Science*, 2011, **333**, 736.
80. Z. Zhou, M. Flytzani-Stephanopoulos and H. Saltsburg, *J. Catal.*, 2011, **280**, 255.
81. P. Buffat and J. P. Borel, *Phys. Rev. A*, 1976, **13**, 2287.
82. P. R. Couchman and W. A. Jesser, *Nature*, 1977, **269**, 481.
83. J. W. M. Frenken and J. F. Vanderveen, *Phys. Rev. Lett.*, 1985, **54**, 134.
84. H. Sakai, *Surf. Sci.*, 1996, **351**, 285.
85. S. H. Joo, J. Y. Park, C.-K. Tsung, Y. Yamada, P. Yang and G. A. Somorjai, *Nat. Mater.*, 2009, **8**, 126.

CHAPTER 4

Preferential Oxidation of Carbon Monoxide in Hydrogen Stream over Gold Catalysts

YU-WEN CHEN*[a] AND NATARAJAN SASIREKHA[b]

[a] Department of Chemical and Materials Engineering, National Central University, Chung-Li, 32054, Taiwan; [b] Department of Chemistry, Anna University, Bharathidasan Institute of Technology, Tiruchirappalli, 620 024, Tamil Nadu, India
*Email: ywchen@ncu.edu.tw

4.1 Introduction

Gold was historically considered as an inert transition metal for catalytic reactions, because of its deep-lying valence *d* band and very diffuse *s, p* orbitals.[1,2] In earlier days, Bond[3,4] and Parravano[5] and co-workers attempted to investigate the catalytic properties of gold, and reported some interesting results about the efficiency of very small gold particles supported on silica for hydrogenation reactions.[6] Yet, the earlier results failed to prove its practical application in heterogeneous catalysis.[7–9]

The turning point was the ground-breaking work by Haruta and co-workers using gold nanoparticles supported on metal oxides for CO oxidation at sub-ambient temperature in the 1980s,[10,11] which initiated the interest to exploit the chemical properties of nanosized gold particles for catalytic reactions. The most important breakthrough in gold catalysis revealed that clusters of gold atoms in the size range of 2–5 nm can act as a selective catalyst, especially for reactions involving oxygen.[12] Apart from the

RSC Catalysis Series No. 18
Heterogeneous Gold Catalysts and Catalysis
Edited by Zhen Ma and Sheng Dai

Published by the Royal Society of Chemistry, www.rsc.org

oxidation of carbon monoxide, Hutchings and co-workers[13,14] advanced the catalytic potential of gold nanoparticles for various reactions such as direct formation of hydrogen peroxide from H_2/O_2,[15] selective oxidation of glycerol to glyceric acid,[16] selective hydrocarbon oxidation under mild conditions,[17,18] hydrochlorination of acetylene,[19,20] oxidation of glycerol,[21] hydrogenation of crotonaldehyde,[22] *etc.*

In the past two decades, significant progress has been made in the field of gold catalysis for numerous applications such as de-NOx, catalytic combustion of volatile organic compounds, water–gas shift reaction, photocatalysis and chemical synthesis.[23–28] In particular, preferential oxidation of carbon monoxide in a hydrogen-rich stream (PROX) for the CO removal from the feed gas stream for polymer–electrolyte membrane fuel cells (PEMFCs) has attracted the attention of many scientists in the recent past, as hydrogen is expected to play a major role in the future.

Fuel cells are the potential source of pollution-free energy that can replace existing conventional fuels. The presence of carbon monoxide at 1000–10 000 parts per million (ppm) in reformate hydrogen gas poisons the platinum electrodes, and hence it is necessary to effectively oxidize CO without oxidizing H_2 at the operating temperature of PEMFCs (80–100 °C) to less than 10 ppm. In the presence of conventional pure Pt electrocatalysts, the CO concentration should be less than 5 ppm, whereas for more CO tolerant bimetallic alloy catalysts (Pt–Ru), the CO concentration should be less than 100 ppm. Metal oxide supported nanosized gold catalysts are, in contrast to Pt metal group catalysts, intrinsically more active for CO oxidation[29–31] than for H_2 oxidation and found to be almost insensitive to CO_2, which makes them a suitable candidate for PEMFCs. Although numerous works have been reported on PROX using supported gold catalysts,[32–38] further improvement is still needed in catalytic activity, selectivity and durability of catalysts.

This chapter presents a critical review of some of the recent research on gold nanoparticles on metal oxide supports for the preferential oxidation of carbon monoxide in a hydrogen-rich stream. Also, we have integrated some of our own results and present our perspective of PROX over gold-supported catalysts. The discussion focuses on various preparation methods to achieve gold nanoparticles, the effect of the support and the effect of promoters on the support.

4.2 Gold Nanoparticles Supported on Metal Oxides

Low-temperature CO oxidation has been studied intensively using supported gold catalysts. The selective oxidation of CO is one of the simplest and most cost-effective methods for the purification of hydrogen-rich reformate gas that has been used in PEMFCs. The suitable catalyst should possess a high CO oxidation capability at PEMFC temperatures. Also, it should prevent oxidation of hydrogen selectively along with sufficient resistance towards deactivation in the presence of CO_2 and H_2O. Conventional Pt-based catalysts cannot protect hydrogen from oxidation; however, Au-based catalysts display extraordinary activity towards CO oxidation selectively, even compared with non-precious metal based catalysts such as CuO-CeO_2 catalysts.

The support plays a decisive role in modifying the catalytic activity of the gold catalysts. The metal oxide support can be classified into either an active or reducible support such as TiO_2,[39,40] Fe_2O_3,[41,42] MnO_2,[43] and CeO_2,[44–47] or an inert support such as SiO_2[48] and Al_2O_3.[49] The reducible supports not only stabilize the small gold nanoparticles, but also provide oxygen atoms during the PROX reaction and improve the oxidation of CO. However, non-reducible supports facilitate high dispersion of gold nanoparticles and the oxygen adsorption occurs on metal sites.[50]

Despite numerous works on various supports,[51–56] the attainment of the best performing supported gold catalyst remains a challenge. The oxygen donating property of the reducible oxide support has been correlated with the catalytic activity of metal oxide supported gold particles, but the multiplicity of other factors critically affects their performance in CO oxidation. Each support has certain unique properties, which facilitates high CO conversion and selectivity.

One of the industrially used catalysts for CO oxidation is Hopcalite, which is a mixture of manganese dioxide and copper oxide in different ratios. An interesting reducible oxide that has not been explored completely for CO oxidation is MnO_x, even though it is environmentally benign and has an advantage of low cost and natural abundance. The oxidation state of manganese is most influential when using manganese oxide as a support. MnO_x compounds have a typical benthollide structure containing labile lattice oxygen. The labile oxidation state facilitates Mn to act either as a reducing agent or an oxidizing agent. The oxygen storage capacity in the crystalline lattice and the capability of manganese to form oxides with variable oxidation states determines its catalytic properties. Gardner *et al.*[57] indicated that Au and MnO_x interact synergistically, exhibiting long-term CO oxidation near ambient temperature with negligible activity decay. Hoflund and co-workers[58,59] demonstrated Au on MnO_x as one of the best candidates for selective CO oxidation in a H_2-rich stream. They investigated a number of reducible oxide supported metals and found that gold supported on MnO_x was highly efficient. Nieuwenhuys and co-workers[60] investigated Au/MgO/MnO_x/Al_2O_3 catalyst with an Au particle size of 5 nm for CO oxidation at low temperature using MnO_x as co-catalyst. Chang and McCarthy[61] reported that MnO_x has high oxygen storage along with faster oxygen adsorption and oxide reduction rates, which can be used to reduce the amount of CO in the exhaust gas of a natural gas vehicle. MnO_x has good O_2 uptake (oxidation) and O_2 release (reduction) rates at temperatures below 600 °C. Au/MnO_2 has high selectivity of CO oxidation in a hydrogen stream, but low activity in CO conversion for the PROX reaction.[33] The crystalline nature of MnO_2 was compared with amorphous MnO_2 for CO oxidation and it was found that the crystalline MnO_2 had higher CO conversion than amorphous MnO_2,[62] which could be due to different oxygen activated species formed on MnO_2. In Au/MnO_2 catalysts, only the metallic gold species is the active species for CO oxidation. Our results for the PROX reaction revealed high activity of Au/MnO_2-CeO_2 catalysts due to the synergistic effects of CeO_2-MnO_2 composite oxides with minor portion of Ce^{3+} species and co-existence of metallic and non-metallic gold species within the nanogold particle.[63]

The oxygen mobility from MnO_x structure is greatly affected by the presence of other oxides such as CeO_2. At low temperatures, cerium provides oxygen to manganese and increases the activity of MnO_x; whereas at elevated temperatures, cerium withdraws oxygen and decreases its activity. Mn^{3+} can provide effective adsorption sites for CO and supply oxygen from its oxide for CO oxidation, which results in the formation of Mn^{2+}.

CeO_2 has been of wide interest for decades because of its important applications including three-way catalysts for automotive emission control. CeO_2 is one of the most thermally stable compounds; under various redox conditions, the oxidation state of the cation may vary between +3 and +4. Its distinct defect chemistry and the ability to exchange lattice oxygen with the gas phase result in an oxide with unique catalytic properties,[64] including the promotion of metal dispersion, enhancement of the catalytic activity at the metal–support interface sites, and promotion of CO removal through oxidation using lattice oxygen. Zhou *et al.*[65] reported the synthesis of nano single crystalline CeO_2 particles by the precipitation method and Carrettin *et al.*[66] reported that 2.8% Au loaded on nanocrystalline CeO_2 exhibited extra high activity by two orders of magnitude over 1.5% Au/TiO_2 and 5% Au/Fe_2O_3 catalysts. Nanosized gold supported on CeO_2 has been reported to be a good catalyst for PROX, owing to its high activity and selectivity to CO oxidation, and resistance to H_2O and CO_2 in the temperature range of 303–403 K. The presence of ionic Au^{3+}/Au^+ species strongly interacting with ceria would increase CO conversion. A high gold dispersion can lead to high reducibility of the ceria support, and surface oxygen reducibility may be enhanced through a lattice substitution effect. The vacant Ce^{4+} sites may be occupied by Au^+ or Au^{3+}, which will lead to the formation of oxygen vacancies, thereby increasing oxygen mobility and reducibility.[67]

Transition metals supported on some oxides are good substitute catalysts because of their low price and widespread use. Among them, copper or copper oxide can exhibit higher activities per unit surface area than those of noble metal catalysts such as platinum in CO oxidation. Copper or copper oxides supported on TiO_2 are active in catalyzing selective CO oxidation in the presence of O_2. Friedman *et al.*[68] pointed out that, beyond a critical quantity, any additional copper loading was in the form of bulk crystalline oxide. Liu and Flytzani-Stephanopoulos[69] suggested that the activity of catalyst derived primarily from the combination of finely dispersed copper support systems and bulk copper oxide had negligible contributions. Similar results reported by Yu *et al.*[70] indicated that the highest activity was achieved when the amount of CuO_x was near the monolayer dispersion threshold, whereas a higher amount of CuO_x loading would have reverse effect.

Iron oxide has been reported to be a good support for gold catalysts towards CO oxidation at low temperature. Li *et al.*[71] reported higher activity of Fe_2O_3 nanoparticles over Fe_2O_3 powders for CO oxidation at temperatures above 210 °C. This was attributed to the small particle size and the presence of the hydroxylated phase of iron oxide. Our result[72] demonstrated that nanosized iron oxide had high activity for CO oxidation even at ambient conditions. The higher activity was attributed to the small particle size, high concentration of hydroxyl groups, and more densely populated surface co-ordination unsaturated sites. Haruta[30] obtained the best results when 5% Au was supported on Fe_xO_y. Besides a reducible support, Fe_xO_y strengthens the catalyst to tolerate CO_2 in the PROX mixture. The disordered structure facilitates the formation of anion vacancies near gold particles and enhances the catalytic activity. We developed Au/Fe_xO_y catalyst for low temperature CO oxidation by the deposition–precipitation method.[73] A Fe_xO_y sample prepared by the precipitation method provided amorphous iron oxide with abundant hydroxyl groups on the surface. Fe_xO_y had a large surface area of about 406 $m^2\ g^{-1}$ with a small and uniform particle size. The pH value and calcination temperature play key roles in creating high activity for Au/Fe_xO_y catalysts. Moreover, the catalyst was highly resistant towards moisture and high concentration of CO_2.[74]

Co/TiO_2 catalyst was considered to have a strong metal–support interaction (SMSI) and showed high activity in CO hydrogenation.[75,76] This interaction is an important factor for determining the properties of Co/TiO_2 catalyst such as cobalt dispersion and reduction behavior.[77] Indeed, the synthesis of highly dispersed cobalt on TiO_2 support requires a strong interaction between cobalt and the support. However, too strong an interaction can produce the cobalt–support compound as a sub-oxide at an interface that is highly resistant to reduction.[78,79] Chen *et al.*[80] investigated nanoparticles of Au/CoO_x towards CO oxidation and reported that the deposition of gold crystallites on the surface of CoO_x not only promoted the catalytic activity but also altered the kinetic order of CO from $m=1$ to $m=0$. Co_3O_4 powders are n-type semiconductors and tend to lose lattice oxygen on the surface on interaction with the adsorbed CO. The advantage of n-type oxides was attributed to an enhanced ability in oxygen adsorption on their defect sites. Cobalt oxide is particularly active and possesses a high activity for CO oxidation.

ZnO does not belong to the reducible oxides, but Au/ZnO catalyst has been demonstrated to be active in low-temperature CO oxidation.[81–84] The study by Boccuzzi *et al.*[83] on gold supported on ZnO and TiO_2 showed the existence of two kinds of metallic gold sites, which are able to adsorb both oxygen and carbon monoxide at the same time. Nevertheless, little information is available concerning the synthesis of Au/ZnO. Castillejos *et al.*[85] reported that 3 nm gold nanoparticles supported on ZnO tetrapods showed high activity for PROX; the catalytic activity was at least twice the values observed for any conventional support with Au for PROX reaction.

Compared with other supports, MgO is considered as inactive,[50] since it is an unreducible oxide. Unreducible oxides generally have low ability to adsorb or store oxygen at low temperatures. MgO is relatively thermally stable, which makes it a suitable candidature as a catalyst support for noble metals. It tends to stabilize the small gold nanoparticles, and hence the activity seems to depend very critically on the gold particle size. Moreover, the surface of MgO is positively charged due to its strong basicity, and its structural simplicity makes it a suitable support to study the oxidation of CO at MgO supported gold nanoparticles by density functional theory (DFT) calculations.[86] Au/MgO had a low CO conversion and high selectivity to CO oxidation due to a low hydrogen oxidation rate.[87] The CO oxidation also occurs through direct interaction of CO with hydroxyl radicals present at the interface between gold and $Mg(OH)_2$.[88]

4.3 Gold Nanoparticles Supported on Titania

Among the supports discussed so far, the most extensively studied catalyst for CO oxidation is Au/TiO_2 because it is one of the most active catalysts for this reaction at low temperature. It should be noted that neither gold nor titania is active for CO oxidation, but their combination generates surprisingly high catalytic activity. On Au/TiO_2 catalysts, the reaction was observed even at temperatures as low as 90 K.[61]

The improved catalytic activities for CO oxidation can be attributed to the interaction between gold and TiO_2, because the perimeters of Au-TiO_2 interfaces are assumed to act as the important active site,[89] which in turn depends on the preparation methods. It is known that catalysts prepared by the incipient wetness method are generally less active than those prepared by the deposition–precipitation (DP) method for low-temperature CO oxidation. The DP method yields hemispherical gold particles, which are strongly attached to TiO_2 by epitaxial contact with their flat planes, Au(111) to anatase TiO_2(112) and rutile TiO_2(110), yielding the longest Au/TiO_2 perimeter interface apart from fine dispersion with small gold particles. However, the impregnation method yields spherical gold particles loaded on TiO_2 support with comparatively larger gold particles than the DP method.[90] Consequently, the activation of supported gold can be explained from the limited size of the active gold particles to the effects of metal–support interactions.[91–98] Even though several attempts have been made to study the role of support on Au nanoparticles, it is still under debate.

4.3.1 Effect of Preparation Methods

Gold supported on titania is one of the most efficient catalysts for CO oxidation and the optimum gold particle size was found to be 2–3 nm. Studies on gold catalysts so far showed that gold in nanodispersed form leads to high catalytic activity due to strong metal–support interaction;[31,99] the nature of support is also important. The particle size and catalytic activity of the gold catalysts are greatly influenced by the preparation methods and variable parameters. Several techniques such as coprecipitation,[12] deposition–precipitation,[39] chemical vapour deposition,[100] impregnation,[101] photodeposition,[102] plasma deposition,[103] and laser vapourisation[104] methods have been used for the synthesis of gold nanoparticles on metal oxides.

4.3.1.1 Deposition–Precipitation

The studies made so far highlighted that the reactivity of gold is highly structure sensitive and that preparation techniques play a major role in the synthesis of highly active gold catalysts.[91] The DP method has been regarded as the most simple and cost-effective method to prepare nanogold catalysts supported on several metal oxides. The preparation parameters, such as the control of pH value, loading period, precursor choice and calcination temperature, have great influences on the particle size and the catalytic activity of the gold catalysts. In contrast, the DP method is very tedious and only a part of gold could be deposited on the support.[90] The gold metal has exhibited catalytic activity only with a particle size of <5 nm and the bilayer structure is more active than the monolayer one as evidenced by previous reports.[50,105,106] It is vital to select suitable preparation conditions, and to restrict the particle size of gold to within 5 nm, while preparing active gold catalysts by the available preparation methods.

Mostly the supported gold catalysts are prepared by chloride-containing chloroauric precursors. Moreau *et al.*[39] reported that the gold complexes present in solution vary with pH as shown in Equations (4.1) to (4.6). An increase in the pH of the solution results in displacement of chloride ions from $AuCl_4^-$ by water, which will lead to loss of a proton from a neutral hydrated ion and hydrolysis due to the replacement of Cl^- by OH^-. Furthermore, the increase in temperature of the $HAuCl_4$ solution also increases the hydrolysis of gold species. At the same time, the charge on the surface of supports also depends on pH value, being negative above the isoelectric point and positive below it. Therefore, $AuCl_x(OH)_{4-x}^-$ adsorbed easily on the positive surface of supports as shown in Equation (4.7). Moreover, the deposited gold particle size decreases with an increase in pH value.

It can concluded that gold particles can be dispersed on the metal oxide supports by adjusting the pH value of the gold precursor solution to the isoelectric point of the metal oxide supports at a temperature of 47–87 °C. This will facilitate the deposition of $Au(OH)_3$ on the support without precipitating in the liquid. The pH of precursor solution is selected primarily based on the isoelectric point of the metal oxide supports as given by Haruta.[90] In contrast, it was also suggested that the optimum pH for high activity of CO oxidation was proved to be 9 for Au/TiO_2 catalyst, even though the isoelectric point of TiO_2 is at pH 6.[39]

$$HAuCl_4^- + H_2O \Leftrightarrow AuCl_3(H_2O) + Cl^- \tag{4.1}$$

$$AuCl_3(H_2O) \Leftrightarrow AuCl_3(OH)^- + H^+ \tag{4.2}$$

$$AuCl_3(OH)^- + H_2O \Leftrightarrow AuCl_2(H_2O)(OH) + H^+ + Cl^- \quad (4.3)$$

$$AuCl_2(H_2O)(OH) \Leftrightarrow AuCl_2(OH)_2^- + H^+ \quad (4.4)$$

$$AuCl_2(OH)_2^- + H_2O \Leftrightarrow AuCl(OH)_3^- + H^+ + Cl^- \quad (4.5)$$

$$AuCl(OH)_3^- + H_2 \Leftrightarrow Au(OH)_4^- + H^+ + Cl^- \quad (4.6)$$

$$Ti(OH) + AuCl(OH)_3^- \rightarrow Ti - O - Au(OH)_3^- + H^+ + Cl^- \quad (4.7)$$

At extremely low pH value (<2) in the precursor solution, gold complexes retain more chlorine; the deposited Au particles will consequently be large and that reduces the catalytic activity. However, under basic conditions, the TiO_2 surface will be negatively charged and repulse the $Au(OH)_4^-$ ion from the surface.[39] The low concentration of Au anions on the TiO_2 surface leads to the deposition of small gold particles with a narrow particle size distribution; however, there will be a decrease in the exact loading percentage of gold during the minimal duration of the preparation process. In general, pH <6 results in gold particles larger than 10 nm. The amount of gold deposited also depends on the pH of the solution. Nanoparticle gold on TiO_2 can be achieved by suitable preparation methods and careful control of the preparation conditions.

To study the effect of base agents in controlling the gold particle size and catalytic activity of Au/TiO_2, we prepared Au/TiO_2 at pH 6 using different base reagents, such as urea, NaOH and NH_4OH to control the pH value.[35] The order of gold loading for Au/TiO_2 catalysts prepared at pH 6 was $NaOH \geq NH_4OH \gg$ urea. It is apparent that not only the pH value, but also the base agent influences the amount of Au deposit on the support. The real gold loading decreased with a decrease in strength of OH^- by decomposition of base. Zanella *et al.*[107] identified that the gold content of Au/TiO_2 catalysts prepared by the DP method with urea was independent of loading time, but a longer aging time (4, 16, 90 h) resulted in re-dispersion of small gold particles. They also mentioned that the urea dissociated faster at 90 °C than at 80 °C. With a relatively short loading period, 2 h in preparation, and lower aging temperature (65 °C), the urea is not enough to dissociate the hydroxide group in solution for the formation of gold surface complexes.

Figure 4.1 shows the CO conversion and selectivity of Au/TiO_2 catalysts prepared by different base agents at pH 6 for the PROX reaction. At the same preparation condition of pH 6, NaOH helps to deposit the maximum amount of gold on the TiO_2 surface (0.81%), which is responsible for its higher catalytic activity. Thus as the Au loading increases on the support, CO conversion also increases. The CO conversion is higher for Au/TiO_2 catalysts prepared by NaOH, whereas the CO selectivity of Au/TiO_2 catalysts prepared by urea was higher when compared with other base agents. This is mainly attributed to the lower amount of deposition of gold on the TiO_2 surface

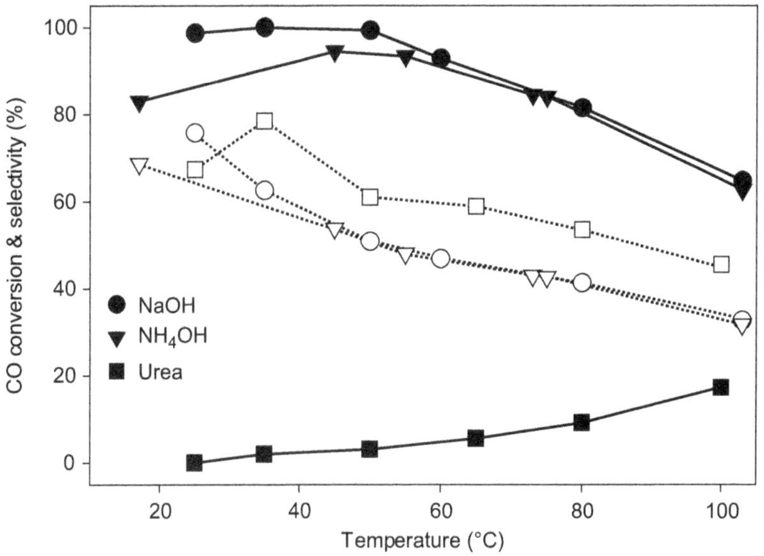

Figure 4.1 PROX reaction on gold catalysts prepared by different bases at pH 6: CO selectivity (dotted lines) and CO conversion (solid line). Reactant: 1.33% CO, 1.33% O_2, and 65.33% H_2, helium balance, with a total space velocity (SV) = 30 000 ml $h^{-1} \cdot g^{-1}$ catalyst.
From ref. 35.

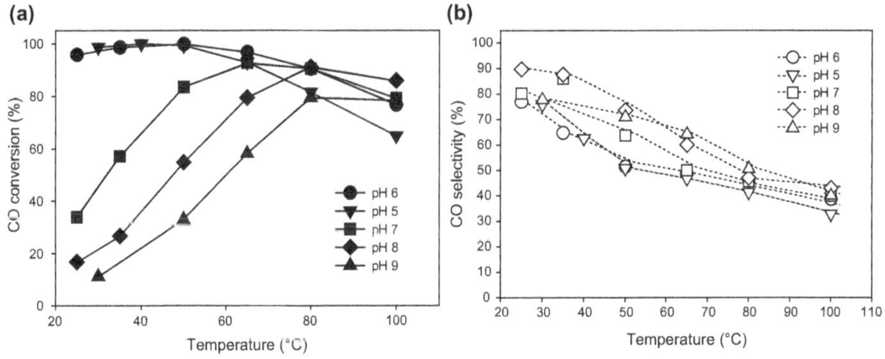

Figure 4.2 PROX reaction on gold catalysts prepared by NaOH with different pH value: (a) CO conversion; and (b) CO selectivity. Reactant: 1.33% CO, 1.33% O_2, and 65.33% H_2, helium balance, with a total SV = 30 000 ml $h^{-1} \cdot g^{-1}$.
From ref. 35.

(0.11%), and thus CO conversion is also low. Moreover, residual chlorine also reduces the catalytic activity.

The exact gold loading decreased from 0.91 to 0.42 wt% as the pH value increased for gold catalysts prepared using NaOH. The CO conversion (Figure 4.2) was found to decrease at room temperature when the pH value decreased, mainly due to the low content of gold on the support. When the reaction temperature increased, the CO and H_2 reacted simultaneously with O_2. All the catalysts showed an increase in CO conversion with an increase in reaction temperature until a situation arose when O_2 was totally consumed by CO or H_2; at this point, the CO conversion started to decrease with increasing reaction temperature. Especially at higher temperature (above 80 °C), H_2 competed with CO to react with O_2 dramatically. In this situation, only 0.59% gold (pH 8) was enough to approach CO conversion to 90% without sacrificing redundant CO selectivity. Moreau and Bond[108] also reported that the gold content passes through a maximum when the catalyst was prepared at about the isoelectric point (pH ~6), but maximum specific rates for CO oxidation occur at pH 8–9, because the Au particle size becomes smaller as the pH is further increased.[107–109] The reason for the lower selectivity for gold catalysts may be due to agglomeration of gold particles during the reaction, which leads to a reduction in the catalytic activity. Au/TiO_2 catalyst prepared by the DP method with base agent NaOH at pH 6 has an advantage of higher gold deposition on the support compared with other base agents, and the gold loading of the catalyst can be easily controlled. It is suitable to enhance the catalytic activity for the preferential oxidation of CO in H_2 stream.

4.3.1.2 *Photodeposition*

Gold nanoparticles can be dispersed on metal oxide semiconductors by the photodeposition (PD) method. Both DP and PD could successfully produce gold nanoparticles with a narrow particle size distribution. However, the amount of gold deposited is higher and the gold particle size is smaller with the PD method than with the DP method. The heat treatment in the case of DP becomes unnecessary with PD, because gold is reduced by ultraviolet (UV) irradiation itself. This method is employed mainly to modify the photocatalytic properties of semiconductors by depositing noble metals.

The size distribution of metal particles and the absence of agglomerates were influenced by the concentration of precursor, pH value, irradiation time, and wavelength of light source. Chan and Barteau[110] reported that the size distribution of gold particles decreased significantly with decreasing precursor concentration from 10^{-3} to 10^{-5} M, and the highest Au loading obtained was ~10.75 wt% with an average Au particle diameter of ~1 nm. In our investigation,[111] we found that the exact gold loading on photodeposited catalysts prepared by a 400 W lamp increased when the irradiation time increased from 3 min to 10 min. However, an increase in the average gold particle size is detrimental as bigger gold particles reduce CO conversion. Under UV light irradiation with a wavelength less than 387 nm, the illuminated TiO_2 (anatase type; band gap energy, $E_g = 3.2$ eV) could generate photo-excited electrons and positive holes in aqueous medium. $AuCl_4^-$ ions adsorbed on the surface of TiO_2 particles can react with the photogenerated e^- to form Au^{3+}, Au^+, and Au^0.[112]

$AuCl_3$ is occasionally used as a precursor in the PD method. When $HAuCl_4$ and $AuCl_3$ were selected as precursors, it was found that there is no significant difference in exact gold loading; both $HAuCl_4$ and $AuCl_3$ exhibited almost similar CO conversion and selectivity.[34] The results demonstrated that the gold precursor did not have a significant influence on the physicochemical properties of the supported gold catalyst prepared by the PD method with preparation variables.

Moreau *et al.*[39] studied in great detail the effect of the pH value on the preparation of Au/TiO_2 catalysts. At low pH values, the surface of the TiO_2 is positively charged and the species in solution anionic; hence an electrostatic mechanism can explain the deposition of Au. However, the uptake is rapid, large particles form and the catalysts have low catalytic activity. As the pH value increased, the amount of gold deposited on the support decreased but the activity increased. Moreau and Bond[113] noted that, above pH 8, the adsorbed gold complex is substantially chlorine-free which leads to a catalyst containing little chlorine (which is an effective poison for CO oxidation) and consisting of small particles; hence the catalysts display high activity.

Table 4.1 shows the effect of pH value on performance for PROX over gold catalysts. When the reaction temperature increased, both CO and H_2 reacted simultaneously with O_2. All the catalysts showed an increase in CO conversion with an increase in reaction temperature until the temperature reached up to 80 °C, when the CO conversion started to decrease with increasing reaction temperature. At high reaction temperature (>80 °C), H_2 competed with CO to react with O_2. The CO conversion of Au/TiO_2 (pH 10) had a higher CO conversion than other catalysts from room temperature to 120 °C. Au/TiO_2 prepared at (pH 10) had the highest CO conversion of 95% at 80 °C. This indicates that H_2 competed with CO to react with oxygen.

Table 4.1 CO conversion and selectivity of Au/TiO_2 catalysts using 8.53×10^{-4} M Au precursor solution at a reaction temperature of 80 °C. From ref. 36.

Light Power	pH Value in Preparation	Irradiation Time/min	CO Conversion/%	CO Selectivity/%
16 W	6	10	20	28
	9	10	71	43
	10	10	95	47
	10	30	56	51
	10	60	31	56
	11	10	56	32
10 W	6	30	49	39
	6	60	62	37
	6	120	76	47
	6	180	55	51
	9	120	90	45
	10	120	88	44
400 W	6	3	7	25
	6	6	18	42
	6	10	14	34

Thermodynamically, the exothermic enthalpy of H_2 reacting with oxygen is lower than that of CO. At high temperature, it is expected that H_2 will keep CO from adsorbing on the Au active site.

In the PD method, the rate of reduction of ionic gold to the metallic state is influenced by the power of light source. It was observed that, when irradiated by a 400 W and a 10 W lamp, a similar gold loading, 0.211% (400 W) and 0.292% (10 W), was uploaded on TiO_2. However, the time of irradiation required for 400 W and 10 W was different, *i.e.*, 6 min and 120 min, respectively. At the same time, the small gold particles continued to aggregate to form larger gold particles during irradiation by the 400 W lamp. Even though Au/TiO_2 was irradiated for 3 min by the 400 W lamp, the mean size (2.7 nm) of the gold particles was still larger (Figure 4.3) than those prepared by the 10 W lamp (1.7 nm). By comparing two UV lamps with different power sources, it can be concluded that the lower power source lamp can deposit smaller gold particles on the support than the high power lamp, and that this is responsible for its high catalytic activity. The catalyst prepared by the 10 W lamp had the highest CO selectivity of 98% at 25 °C; this decreased rapidly from 25 to 65 °C. Therefore, increasing the irradiation time (from 3 min to 10 min) leads to agglomeration of gold particles (>5 nm), and the gold particles are widely distributed from 1 nm to 5 nm. However, large particles of gold are not formed on the surface of TiO_2 during the irradiation

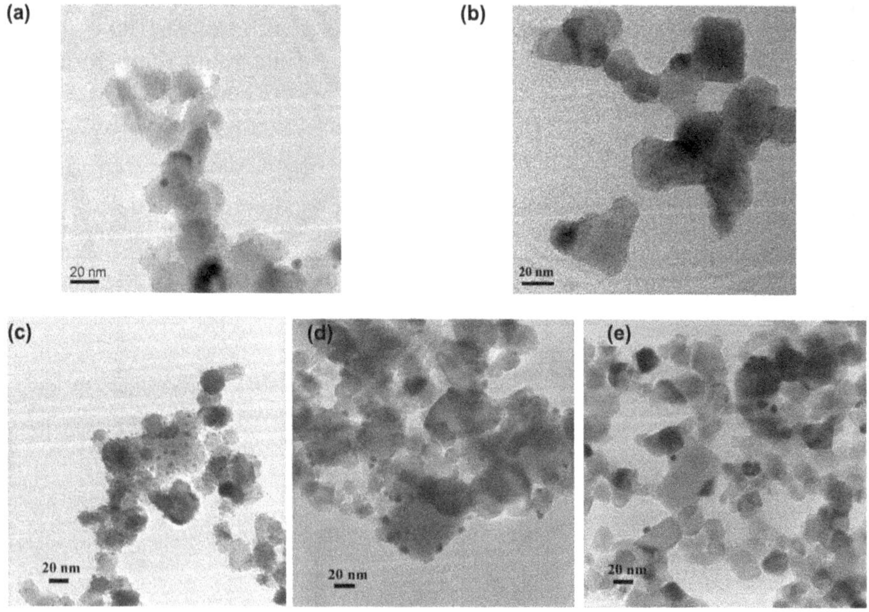

Figure 4.3 TEM images of Au/TiO_2 by 10 W lamps at (a) pH 6 and (b) pH 9 for 120 min, and by 400 W lamp at pH 6 for (c) 3 min; (d) 6 min and (e) 10 min.
From refs. 34 and 111.

by the 10 W lamp. The catalytic activity of Au/TiO_2 prepared using a 400 W lamp decreased with an increase in gold particle size.

The highest CO conversion was found on the Au/TiO_2 catalyst prepared by the PD method at pH 9. The differences in CO selectivities are within 10% among all catalysts. The main reason behind the distinct CO conversions was the presence of small gold particles; the gold particles prepared at pH 6 were larger than those prepared at pH 9 or 10 which resulted in very low CO conversion. The exact gold loadings directly affect the catalytic performance. The catalyst prepared at pH 10 had low loading of gold content, which exhibited lower CO conversion than those prepared at pH 9.

It is well-known that the exact gold loading depends strongly on the pH value in the deposition–precipitation method, especially for a TiO_2 support. Similar results were observed for those prepared by the PD method. The results also demonstrated that the gold loading at pH 9 was only half that of the catalyst prepared at pH 6. But in this method, the pH value of the gold solution did not dramatically affect the content of gold deposition within the same range of pH values. The gold loading and particle size played a vital role, as expected, were related to the catalytic activity for the PROX reaction. The catalyst prepared by the low power lamp (10 W) demonstrated a high CO conversion and high selectivity of O_2 reacting with CO in the PROX reaction.

The effect of the DP and PD (10 W, 1.5×10^{-3} M, pH 6) methods can be compared by investigating the catalytic activities of gold catalysts prepared with similar gold contents in the temperature range of 17–100 °C. The CO conversion (Figure 4.4) increased from room temperature to 50 °C and decreased when the temperature was higher than 50 °C. CO selectivity

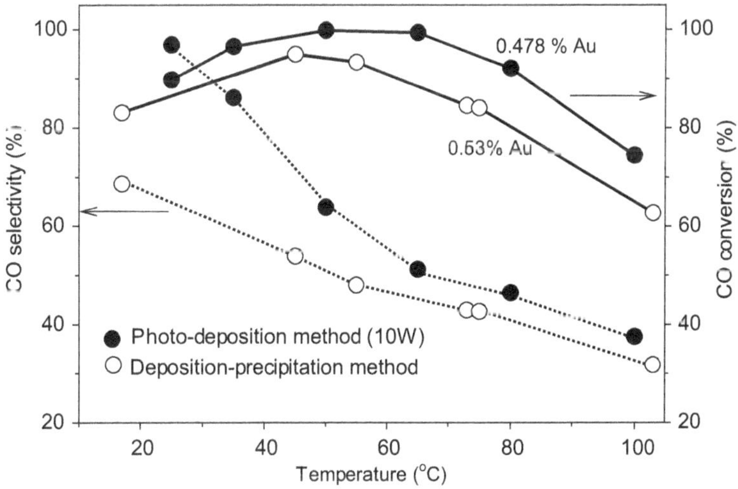

Figure 4.4 PROX reaction on gold catalysts: CO selectivity (dashed line) and CO conversion (solid line). Reactant: 1.33% CO, 1.33% O_2, and 65.33% H_2, helium balance, with a total GHSV = 30 000 ml $h^{-1}\cdot g^{-1}$. From ref. 35.

decreased with increasing reaction temperature, because hydrogen competed to react with oxygen at high temperature. Hydrogen oxidation has a higher activation energy than CO oxidation. Since only limited oxygen was used in the reaction, CO conversion and selectivity decreased at high temperatures. The catalyst prepared by the PD method shows higher CO conversion and selectivity than that prepared by the DP method. The photodeposited catalyst (prepared by the 10 W lamp) had the highest CO selectivity of 98% at room temperature, and the selectivity decreased rapidly between 25 and 65 °C. Since the photodeposited catalyst was not calcined, the color of the catalysts changed from light blue to purple when the reaction temperature increased. This indicated that the 1.5 nm Au particles sintered to ~3 nm at high temperature (>50 °C). Besides, the hydrogen competed with CO to adsorb on the gold surface and oxidize with oxygen at high temperatures. However, comparing the catalysts prepared by the DP and PD methods, the catalysts obtained by the PD method showed relatively higher CO selectivity below 65 °C than those prepared by the DP method, due to the difference in chemical state of gold particles and the particle size of gold. In a long time test, the catalysts were stable at 80 °C for more than 60 h. However, the PD method can be exclusively used to prepare gold nanoparticles on semiconductors only.

4.4 Gold Nanoparticles Supported on Mixed Metal Oxides

There are numerous reports on the catalytic activities of Au/TiO_2 towards low-temperature CO oxidation and preferential oxidation of CO. The improved catalytic activities for the CO oxidation can be ascribed to the interaction between gold and TiO_2. This is because the perimeters of the Au-TiO_2 interfaces are assumed to act as the important active site,[89] which in turn depends on the preparation method. It is known that catalysts prepared by the incipient wetness method are generally less active than those prepared by the DP method for low-temperature CO oxidation. The DP method yields hemispherical gold particles, which strongly attached to TiO_2 by epitaxial contact with their flat planes, Au(111) to anatase TiO_2(112) and rutile TiO_2(110), yielding the longest Au/TiO_2 perimeter interface apart from fine dispersion and small gold particles. However, the impregnation method yields spherical gold particles loaded on a TiO_2 support with comparatively larger gold particles than the DP method.[90] Consequently, the activation of supported gold can be explained from the limited size of the active gold particles to the effects of metal–support interactions.

Au/TiO_2 has been reported to have high activity in CO oxidation, but the selectivity of O_2 reacting with CO was not high due to the rapid reaction of O_2 with H_2 at higher reaction temperature. It can be overcome by the addition of another metal oxide to suppress H_2 oxidation and maintain the CO conversion as Au/TiO_2. Less attention has been paid to mixed or composite oxide supports for the PROX reaction using gold catalysts. To unravel the synergistic effect of a second metal oxide on Au/TiO_2 for the PROX reaction, we have used some of the active and inactive metal oxides to prepare mixed and composite oxide supports.

4.4.1 Au/MnO_2-TiO_2

Au/MnO_2 has high selectivity towards CO oxidation in a hydrogen stream, but low activity in CO conversion. It was expected that Au/MnO_2-TiO_2 with a certain composition and preparation technique might have high CO conversion and high selectivity of CO oxidation in a hydrogen-rich stream.

The preparation of gold catalysts was accomplished by the DP method using chloroauric acid as precursor and NH_4OH as base agent. In our study[33] on Au/MnO_2-TiO_2, we also investigated the influence of pH (5, 6, 9) on the preparation of gold catalyst by the DP method. Our results revealed that the average gold particle size was minimal at pH 6 and that there was insignificant influence of pH over the gold composition in the range used in the study. However, with an increase in the Mn/Ti ratio, gold loading on a MnO_2-TiO_2 support increased from 0.58% to 0.69%. The MnO_2 support prepared by the incipient wetness impregnation method formed large particles with a very low surface area. Consequently, the surface area of the MnO_2-TiO_2 composite oxide decreased with increasing manganese content. Addition of MnO_2 to TiO_2 from Mn : Ti = 2 : 98 to Mn : Ti = 4 : 96 did not change the surface area. However, a further increase in the Mn to Ti ratio exhibited a lower Braunauer–Emmett–Teller (BET) surface area of 46 $m^2\ g^{-1}$. Qi and Yang[114] reported that, with an increase in manganese content, the surface area of Mn/TiO_2 decreased from 42.74 to 35.28 $m^2\ g^{-1}$. The addition of small amounts of MnO_2 can result in the formation of homogeneous solid solution as the closeness of the ionic radii and co-ordination number of $Mn^{3+/4+}$ and Ti^{4+} would indeed allow for the dissolution of $Mn^{3+/4+}$ within the TiO_2 lattice.

Transmission electron microscopy (TEM) photographs and the size distribution of gold particles uphold the presence of highly dispersed and fine gold particles (mean particle size of 2.37 nm) on the support. Based on the TEM results, one can conclude that the gold particles were highly dispersed on the MnO_2-TiO_2 support with particles of smaller size and narrow size distribution. Nevertheless, addition of MnO_2 to the TiO_2 support slightly broadened the distribution of gold particle size. Especially for $Au/Mn_{13}Ti_{87}$, the mean gold particle size has been found to be 2.63 nm but the gold particle size varied up to 6 nm. The particle size of TiO_2 P-25 support from Degussa was in the range of 20–40 nm, which can constrain the size of gold nanoparticles deposited on it. Furthermore, it can be observed that the percentage gold loading increased with MnO_2 loading. This is beneficial for preparing the gold catalyst by the DP method. With an increase in pH from 6 to 9 in DP preparation process, the percentage composition of metallic gold decreased with a corresponding increase in non-metallic gold ($Au^{\delta+}$) and Au_2O_3. The metallic gold content showed a maximum at pH 6, even higher than that of Au/TiO_2. Increase in MnO_2 loading leads to an increase in the percentage of Au_2O_3 in the catalyst. However, Au/Mn_4Ti_{96} has a lower metallic Au content than $Au/Mn_{13}Ti_{87}$ and Au/Mn_2Ti_{98}, combined with an increase in non-metallic gold ($Au^{\delta+}$).

The CO conversion of Au/TiO_2 catalysts (Figure 4.5) increased with an increase in reaction temperature from 25 °C to 50 °C, and then decreased after 50 °C along with temperature [Figure 4.5(a)]. The selectivity of oxidation of CO [Figure 4.5(b)] also followed the same trend. In addition, Au/MnO_2-TiO_2 exhibited a much higher selectivity than Au/TiO_2. Moreover, the selectivity of CO oxidation on Au/TiO_2 decreased to 38% while the reaction

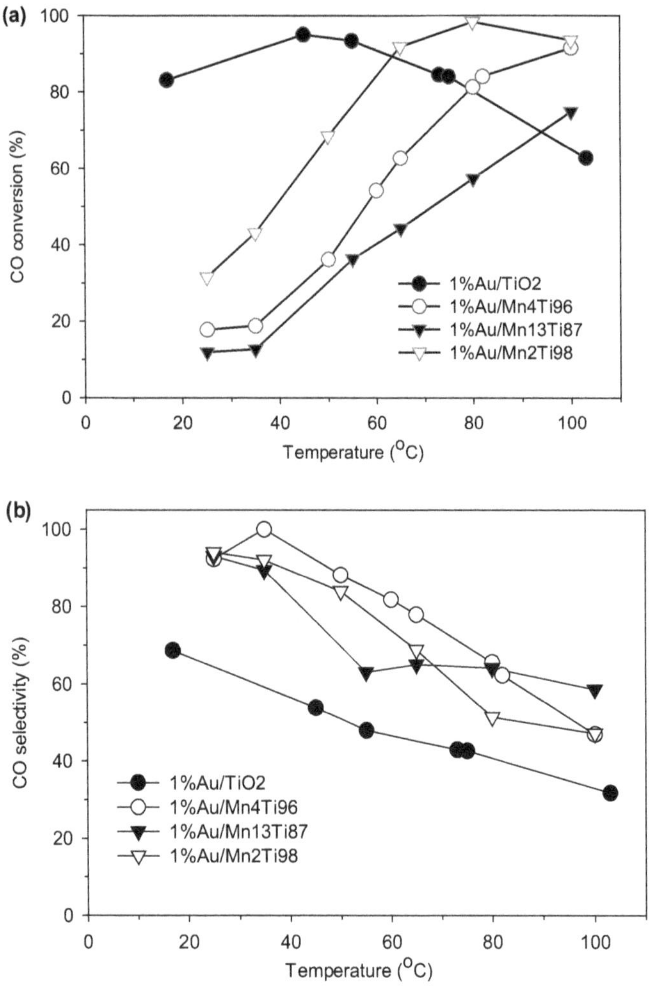

Figure 4.5 PROX reaction on gold catalysts: (a) CO conversion; and (b) CO selectivity. Reactant: 1.33% CO, 1.33% O_2, and 65.33% H_2, helium balance, with a total SV = 30 000 ml $h^{-1} \cdot g$.
From ref. 33.

temperature increased to 80 °C. On the other hand, a high CO conversion on Au/MnO_2–TiO_2 catalysts was observed even at the expected fuel cell temperature (80 °C), with an exception of $Au/Mn_{13}Ti_{87}$ catalyst. In spite of that, a decreasing trend has been observed in CO selectivity.

The astonishing fact is the high selectivity of all Au/MnO_2-TiO_2 catalysts compared with Au/TiO_2. Au/MnO_2-TiO_2 catalyst with a Mn to Ti ratio of 2 : 98 demonstrated excellent performance on both CO conversion as well as selectivity of CO oxidation at 80 °C. However, a decrease in CO conversion at higher temperatures was observed. This suggests that hydrogen is easier to react with oxygen at higher temperature in the presence of metallic gold. The electron-deficient gold species seems to play an important role in the oxidation of CO. However, the active state of gold is still a subject of controversy. Haruta and co-workers[100,115] observed that metallic gold clusters are responsible for high activity of supported gold catalysts. Several authors have reported the presence of cationic gold clusters.[116–119] Hutchings *et al.*[120] reported that, among the existence of both metallic and cationic form of gold species in Au/Fe_2O_3, the presence of a significant fraction of cationic gold is essential for CO oxidation. Kung and co-workers[121] proposed that CO oxidation requires the presence of ensembles of Au^0 and Au^I–OH sites located at the metal–surface interface of Au/γ-Al_2O_3 catalysts. Some authors have postulated that anionic gold species may be a possible cause of high activity.[122–124] On the other hand, Hoflund and co-workers[59,125] concluded that non-metallic gold present in α-Fe_2O_3 and Co_3O_4 influenced low-temperature CO oxidation. After detailed X-ray photon spectroscopy (XPS) investigations, Park and Lee[126] concluded that an oxidic gold species acts as the active species for Au/TiO_2, Au/Al_2O_3 and Au/Fe_2O_3. Most of the literature reports suggest that metallic gold is an active species for Au/TiO_2 or $Au/Ti(OH)_3$.[127,128]

In our study, the correlation of catalytic activity with oxidation state of gold for Au/MnO_2-TiO_2 catalysts supports the influence of metallic gold on CO oxidation. It can be observed that the CO conversion increases with an increase in the content of metallic gold. This is in accordance with the results of Bamwenda *et al.*[129] that metallic gold is the active species in Au/TiO_2 catalysts, as a clear correlation between catalytic activity and the proportion of metallic gold. Comparing different preparation pH values and the same support (Mn_2Ti_{98}), the sample prepared by pH 6 had the highest metallic gold content and the lowest non-metallic gold content; this demonstrated the highest catalytic activity and selectivity.

In all the catalysts investigated in our study, Au/Mn_2Ti_{98} having the maximum metallic gold content was found to be the most efficient catalyst for preferential oxidation of CO in a hydrogen-rich stream. Although non-metallic gold also has been regarded as the active site in the reaction, there is no confirmed correlation with the catalytic performance for Au_2O_3 species. Nevertheless, we hypothesized that metallic gold is active for both CO and H_2 oxidation. In order to suppress the oxidation of H_2 in the PROX reaction, the presence of electron-deficient gold is required. In other words, an optimum combination of metallic and electron-deficient gold species is the way to obtain high CO conversion, high selectivity of oxidation of CO and low selectivity of oxidation of H_2. The high catalytic activity of Au/MnO_2-TiO_2 over Au/TiO_2 has been attributed to the synergistic effect of gold particle size, optimum combination of metallic, and electron-deficient gold species, as well as to Au-support interactions.

4.4.2 Au/CeO_2-TiO_2

Combination of two metal oxides can produce novel structural and electronic properties. TiO_2 and CeO_2 are the most widely used metal oxide supports in the field of catalysis.

Ceria is in the most stable face centered cubic fluorite phase. CeO_2 has been of wide interest for decades because of its important applications, including three-way catalysts for automotive emission control. CeO_2 is one of the most thermally stable compounds; under various redox conditions, the oxidation state of the cation may vary between +3 and +4. Its distinct defect chemistry and the ability to exchange lattice oxygen with the gas phase result in an oxide with unique catalytic properties,[130] including the promotion of metal dispersion, enhancement of the catalytic activity at the metal–support interface sites, and promotion of CO removal through oxidation using lattice oxygen. Zhou *et al.*[65] reported the synthesis of nano single-crystalline CeO_2 particles by the precipitation method and Carrettin *et al.*[131] reported that 2.8% Au loaded on nanocrystalline CeO_2 exhibited higher activity by two orders of magnitude than 1.5% Au/TiO_2 and 5% Au/Fe_2O_3 catalysts.

Our results[132] using Au/CeO_2-TiO_2 for PROX showed that CeO_2 particle size grew and aggregated at high CeO_2 loading. The average gold particle sizes were 2.3–2.5 nm; they were nearly the same at various Ce to Ti ratios. The nanosize support was beneficial for gold species to diffuse into the pore of the support during preparation. It should be noted that when the gold particle size reached about 2 nm, the metallic gold species would become electron deficient to some extent. It would be difficult to differentiate the oxidation state of gold particles. This is why different researchers arrived at different conclusions on the active state of gold. The intensity ratio of I_{Au}/I_{Ti} observed from XPS was greater than 0.04, indicating that most gold particles were dispersed on the surface of the support. The metallic gold species was about 93% on the CeO_2-TiO_2 catalyst surface and only 7% of Au^+ dispersed on the support. Although the accurate mechanism of CO oxidation by gold was unclear, that the presence of metallic gold improved CO oxidation was a fact beyond any doubt.

The influence of Ce to Ti ratio on the conversion and selectivity of the PROX reaction on Au/CeO_2-TiO_2 is shown in Figure 4.6. Au/CeO_2-TiO_2 (1 : 9) showed the maximum CO conversion at room temperature and increased with temperature to attain 100% conversion. However, there was a slight decrease in conversion above 50 °C. With an increase in the Ce to Ti ratio from 1 : 9 to 3 : 7, there was a decrease in the CO conversion at room

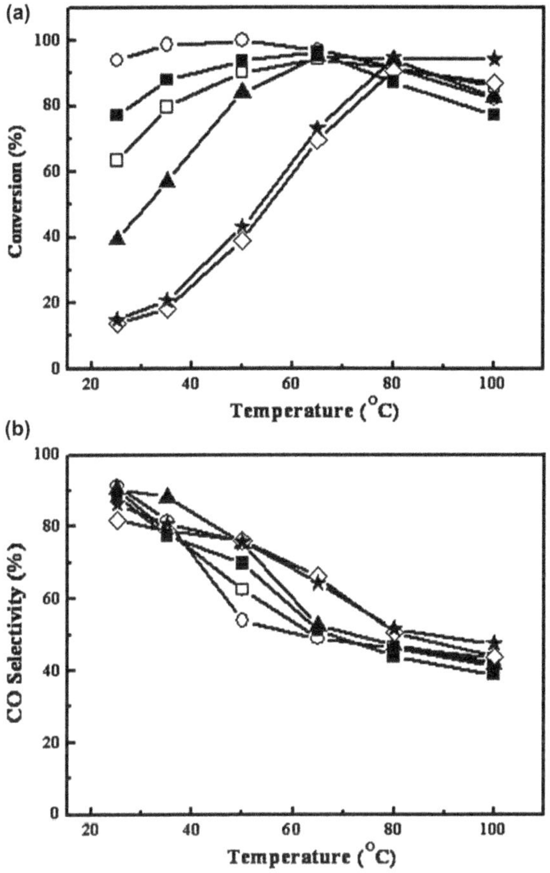

Figure 4.6 Influence of Ce to Ti ratio on Au/CeO_2-TiO_2 catalyst: (a) CO conversion; and (b) selectivity for PROX reaction. □ = CeO_2-TiO_2 (1 : 1); ○ = CeO_2-TiO_2 (10 : 90); ▲ = CeO_2-TiO_2 (20 : 80); ◇ = CeO_2-TiO_2 (30 : 70); ■ = TiO_2 (Degussa); ★ = CeO_2 (Degussa).
From ref. 132.

temperature. Since the gold particle size was similar, it could be suggested that the intensity of oxygen supplied from the support weakened. The Au/TiO_2 catalyzed CO oxidation more easily than Au/CeO_2 when the temperature was below 70 °C. Increasing the Ce to Ti ratio accurately affected the oxidizing ability of the catalyst; a Ce to Ti ratio of 1 : 9 could reach optimum activity for PROX reaction. However, the CO conversion of Au/CeO_2-TiO_2 at a Ce to Ti ratio of 1 : 1 was observed to be higher than those with 2 : 8 and 3 : 7, which may be due to the formation of a mixed oxide phase. Furthermore, CO conversion increased with rising temperature until 65 °C. All catalysts showed nearly the same CO conversion above 80 °C.

The limited rate of CO oxidation was not due to the oxygen supplied by the support at high temperature because there was enough energy to induce oxygen on the support to dissociate and then react with CO, which was adsorbed on the surface of gold particle. Au is not known to be an effective metal for dissociative adsorption in general. The determining step for CO oxidation was the rate of CO adsorbed on gold surface. This is why the conversion fell slightly at high temperature. For all catalysts, the selectivity decreased with increased temperature. Au/CeO_2-TiO_2 (1 : 9) showed the highest CO conversion of 94% and selectivity of 91% at 25 °C. It had better catalytic performance than other Ce/Ti catalysts, which may be due to its fine gold particle size and the nature of Ce^{4+} at low levels of loading. CeO_2 plays the role of promoter and increases the catalytic activity of gold nanoparticles supported on titania.

4.4.3 Au/CuO_x-TiO_2

CuO is also active for CO oxidation. Wu and co-workers[133] reported that CuO_x-TiO_2 could oxidize CO completely above 100 °C. They suggested that CuO_x-TiO_2 catalyst is an active and stable catalyst for CO oxidation. In that case, TiO_2 particles acted as the support and the CuO_x acted as an active component above 100 °C. In our study,[134] a series of Au/CuO_x-TiO_2 catalysts with various Cu to Ti atomic ratios were prepared by the DP method. The preferential oxidation of CO in a hydrogen-rich stream on these catalysts was carried out in a fixed bed microreactor with a gas mixture ($CO : O_2 : H_2 : He = 1.33 : 1.33 : 65.33 : 32.01$ by volume ratios) fed with a space velocity of 30 000 mL g^{-1} h^{-1}. The literature results demonstrated that the presence of CuO could influence the surface properties of TiO_2, thereby resulting in higher gold deposition on the CuO-TiO_2 support. The mean particle diameter of gold for Au/CuO_x-TiO_2 was about 2.39–2.57 nm. When the loading of CuO_x increased, a slight increase in gold particle size was observed. It means that gold was highly dispersed and formed small particles. Moreover, a small amount of 4–5 nm gold particles appeared after the reaction, but most particles were still less than 3 nm with a hemispherical shape on the support. The concentration of oxidic gold species on the surface of the support decreased with the increasing Cu to Ti ratio. It is proposed that both Au^0 and Au^{3+} were required for higher activity in PROX.[135] Au^0 is considered to adsorb CO, while Au^{3+} cemented the metallic particle on the support and activated surface hydroxyl groups, which could react with the adsorbed CO to form adsorbed carboxylate. The highly dispersed gold particles (Au^0 and Au^{3+}) are responsible for the superior catalytic performance. The catalytic activity decreased with decreasing Au^{3+} content.

Figure 4.7 shows the effect of Cu to Ti molar ratios for the preferential oxidation of CO over Au/CuO_x-TiO_2 and Au/TiO_2 catalysts. Addition of a suitable amount of CuO_x improved the CO conversion. In the operating temperature for fuel cell (~80 °C), Au/CuO_x-TiO_2 catalysts showed superior activity compared with Au/TiO_2. The activity of Au/TiO_2 increased with increasing temperature until it reached the maximum CO conversion of 97% at 65 °C. Then the activity decreased to 78% at 100 °C. It is well-known that, although TiO_2 is an effective support for Au to selectively oxidize CO, the

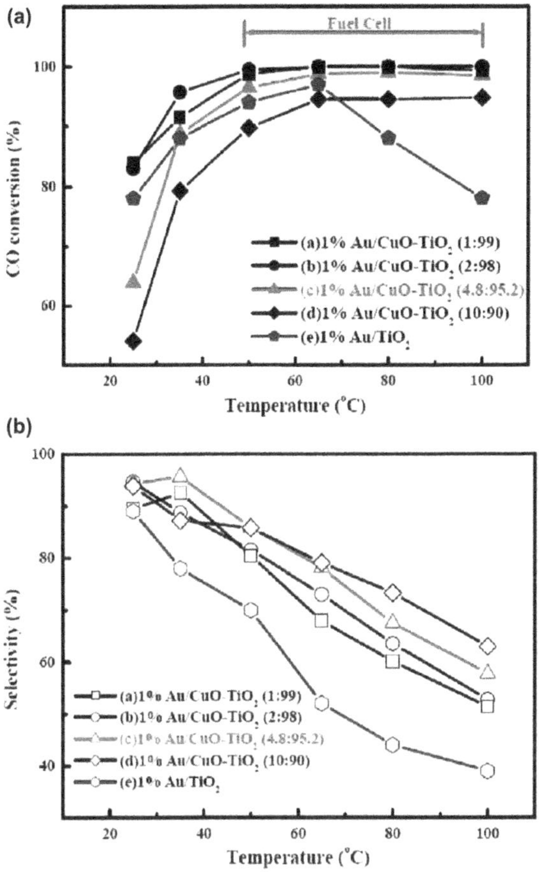

Figure 4.7 Effect of Cu to Ti mole ratios over Au/CuO_x_TiO_2 catalysts for: (a) CO conversion; and (b) for CO selectivity.
From ref. 134.

conversion will decrease appreciably. It was observed that there was no significant difference in the Cu to Ti ratios of 1 : 99, 2 : 98, and 4.8 : 95.2 in Au/CuO_x-TiO_2 catalysts when the temperature was above 35 °C. The CO conversions of these catalysts achieved ~99%, and the activity could be maintained at high temperature. Generally, the CO conversion fell slightly with increasing CuO_x content. Although Au/CuO_x-TiO_2 (10 : 90) had a lower activity, the CO conversion could reach 95% at 80 °C. It indicates that CuO_x promoted Au at high temperature. CuO_x played the role of a promoter but not as an active site. If CuO_x offers the same active site as Au, the higher CuO_x content should induce higher CO conversion. Therefore, CuO_x-TiO_2 was proposed to be a supplier or storage of oxygen. Suitable CuO_x content formed new support properties to increase catalytic activity of CO oxidation. CuO_x can inhibit the oxidation of H_2 to H_2O. Hence CO selectivity increased on addition of CuO_x. The selectivity of Au/TiO_2 decreased sharply, due to the nature of TiO_2. When the temperature was higher than 65 °C, the selectivity was less than 50%. This indicates that H_2 competed with CO to capture oxygen. In thermodynamics, the exothermic enthalpy of H_2 reacting with oxygen is lower than that with CO. At high temperature, it is estimated that H_2 would stop CO from adsorbing on the Au active site.

Our results showed that Au/CuO_x-TiO_2 is a highly effective and thermally stable catalyst for the PROX reaction.

4.4.4 Au/Fe_2O_3-TiO_2

Haruta[30] reported that the best results were obtained with Fe_2O_3 as a support with a gold loading of 5 wt%. Gold on α-Fe_2O_3 has been investigated by Schubert *et al.*[136] to remove CO while not oxidizing H_2. The synthesis of nanoscale Au/Fe_2O_3 catalysts *via* the deposition–precipitation technique has been investigated for CO oxidation.[137] The selective oxidation of CO in a hydrogen-rich stream was studied on iron oxide–supported gold catalysts prepared using DP or coprecipitation.[138] We attempted to study catalytic oxidation of CO over gold/iron hydroxide catalyst under ambient conditions.[41] In our study on PROX,[139] we used FeO_x-TiO_2 as a support to deposit gold nanoparticles. FeO_x is an easily reducible oxide support; thus, the amorphous oxide and the promoting effect of FeO_x on TiO_2 turns out to be an effective support to deposit gold for the preferential oxidation of CO in a hydrogen-rich stream.

Low-temperature CO oxidation over a supported catalyst has a size effect of gold nanoparticles. Generally, smaller gold nanoparticles show higher catalytic activity toward CO oxidation. A strong interaction between gold and the support, such as TiO_2 and Fe_2O_3, leads to a higher catalytic activity than that of unsupported gold catalysts. Active catalysts are prepared by precipitation, and many of the active catalysts are typically found to comprise small gold crystallites, 2–4 nm in diameter, supported on an oxide. There has been much debate concerning the nature of the active site of these catalysts. Bond and Thompson[1] proposed a model where Au atoms at the interface between the gold particle and the oxide are the active oxidation centers. Okazaki *et al.*[140] concluded that the oxygen-rich surface of TiO_2 is important in the consideration of real Au/TiO_2 catalysts prepared using the DP method and calcined in air as the final preparative step. They concluded that electron transfer and orbital hybridization at the interface may be the source of the catalytic activity. In our study, the correlation of catalytic activity with electronic state of gold for Au/FeO_x-TiO_2 catalysts supports the influence of metallic gold on CO oxidation.

The morphology of promoting oxide comprises: (1) crystalline or amorphous character; (2) the type of crystalline structure that may affect the oxygen binding energy and the abundance of the oxygen vacancies; and (3) the size and shape of the oxide phase affecting the length of the perimeter of active Au/promoter oxide interfaces.[141] The small size of gold nanoparticles, amorphous active oxide FeO_x, TiO_2, and the synergistic effect of FeO_x-TiO_2 support favor higher activity in the preferential oxidation of CO in a hydrogen stream. In our present study, amorphous active oxide FeO_x plays the role of promoter; thus the promoter along with TiO_2 not only stabilizes the gold particle size, but the gold supported on the promoter interface may form a perimeter and increase activity.

Figure 4.8 shows the effect of the Fe to Ti molar ratios on the catalytic behaviors of PROX reaction over gold catalysts. The addition of a suitable amount of FeO_x could improve the CO conversion. The catalytic activity of Au/TiO_2 catalyst increased with an increase in reaction temperature until it reached the maximum CO conversion of 85%, from 65 °C to 80 °C; then the CO conversion decreased to 70% at 100 °C. The Au/FeO_x-TiO_2 (10 : 90) catalyst had a higher CO conversion than the other catalysts when the temperature was >55 °C. Au/FeO_x (Fe : Ti = 10 : 90) had a CO conversion of 100% at 80 °C. With regard to these catalysts, the CO conversions of >85% were achieved at 80 °C. When the temperature was increased, the CO

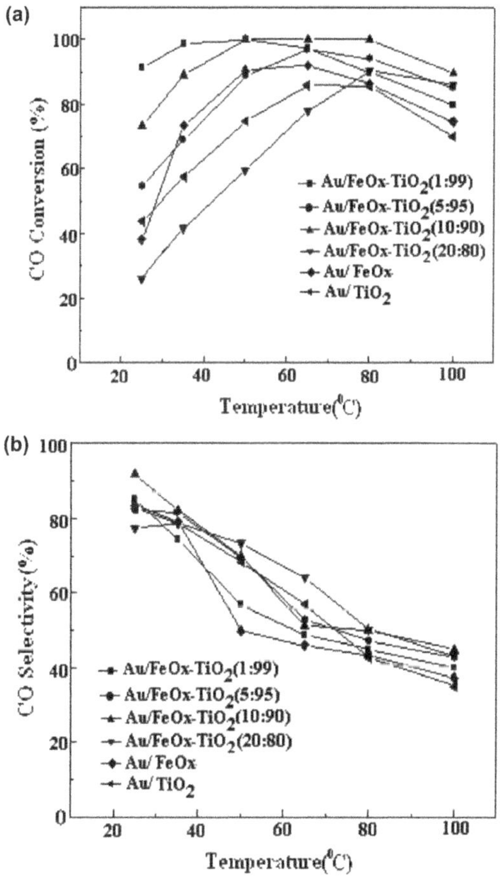

Figure 4.8 CO conversion (a) and CO selectivity (b) of Au/FeO_x-TiO_2 catalysts with various Fe to Ti molar ratios.
From ref. 139.

selectivity decreased. This indicates that H_2 competed with CO to react with oxygen at high reaction temperatures. It can be observed that FeO_x modified the support and prevented comparative H_2 oxidation. Au/FeO_x-TiO_2 with a molar ratio of 10 : 90 was determined to be the best catalyst, when compared with other catalysts for PROX, and it showed high CO conversion (100%) and high CO selectivity (48%) at 80 °C for fuel cell applications. The catalytic activity of the gold/oxide perimeter is dependent not only on the size of the gold particles, but also on the size and morphology of the oxide component (likely to be an amorphous structure), regardless of whether it is supporting gold nanoparticles or decorating them. The activity in CO oxidation over gold catalysts is strongly affected by the length of the Au/'active' oxide perimeter, which is regarded as the 'active interface'.[141] Thus, gold supported on FeO_x-TiO_2 with a molar ratio of 10 : 90 showed higher activity for preferential oxidation of CO, mainly because of the size of the gold particles, strong metal–support interactions, the amorphous nature of FeO_x, and the promoting effect and synergistic effect of FeO_x and TiO_2.

4.4.5 Au/Co_3O_4-TiO_2

Au/Co_3O_4 was recognized as one of the most active catalysts for methane oxidation, among a series of co-precipitated Au on several transition metal oxides.[142] The structural promotion by CeO_2 of a Co_3O_4 catalyst was observed in methane total oxidation.[143,144] The positive effect was attributed to improved thermal stability, and increased dispersion of Au metal. Cobalt oxide is particularly active and possesses a high activity for CO and hydrocarbon oxidation.[145] There are several literature reports on Au/TiO_2 and Au/Co_3O_4 for the PROX reaction, but the results on Au/Co_3O_4-TiO_2 are scarce in the literature. We tried to study the effects of Co_3O_4 additive on Au/TiO_2 for the PROX reaction.[146]

The results infer that adding Co_3O_4 to TiO_2 did not change the isoelectric point of TiO_2. The low loading of cobalt on TiO_2 did not show any XRD peaks corresponding to cobalt oxide. The weaker acidic surface of TiO_2 favors better dispersion of cobalt species, since Co^{2+} cations can be considered as a Lewis acid. Therefore, the Co_3O_4 phase is almost not observable on Au/CoO_x-TiO_2 (1 : 99) and Au/CoO_x-TiO_2 (5 : 95). However, when the Co to Ti atomic ratio increased from 5 : 95 to 1 : 9, the appearance of (311) plane of spinel Co_3O_4 can be observed. Au/CoO_x-TiO_2 (2 : 8) and Au/CoO_x-TiO_2 (3 : 7) showed well-defined Co_3O_4 phase, which may be attributed to an increase in crystallite size of Co_3O_4 due to increase in cobalt loading. The strong metal–support interaction of TiO_2 with cobalt oxide makes the cobalt oxide species difficult to reduce due to a strong Co–O interaction with the support or the formation of stable compound like cobalt titanate. The well-defined pattern of anatase TiO_2, even after loading cobalt oxide, suggests that cobalt oxide did not enter into the anatase phase of TiO_2, but instead formed a crystalline cobalt oxide species on the surface of TiO_2. A uniform dispersion of gold nanoparticles with diameter less than 4 nm and narrow size distribution was observed. The isoelectric point of CoO_x is higher than that of TiO_2, which subsequently enhances the effective loading of gold on TiO_2 rather than on CoO_x. This is in agreement with the TEM images.

Although the cobalt oxide species is not detectable by XRD technique for Au/CoO_x-TiO_2 (5 : 95), some dark spots shown in high-resolution TEM (Figure 4.9) should be the cobalt oxide species according to the Z-contrast, because cobalt is a heavier metal than titanium. Co_3O_4 formed very small particles without definite shape on TiO_2 surface. At high magnification, two different types of lattices can be distinguished and Co_3O_4 particles were observed by the presence of ordered fringes with a space distance between crystal face of 0.243 nm typical of crystalline Co_3O_4. The good interaction between TiO_2 and CoO_x can be confirmed by the absence of distinct boundaries between the phases. We found that, on Au/TiO_2 catalysts, gold nucleates probably at oxygen defect sites of titania, and remains highly dispersed and in good contact with the support. The gold atoms directly in contact with titania are positively charged and highly stable. Doping with

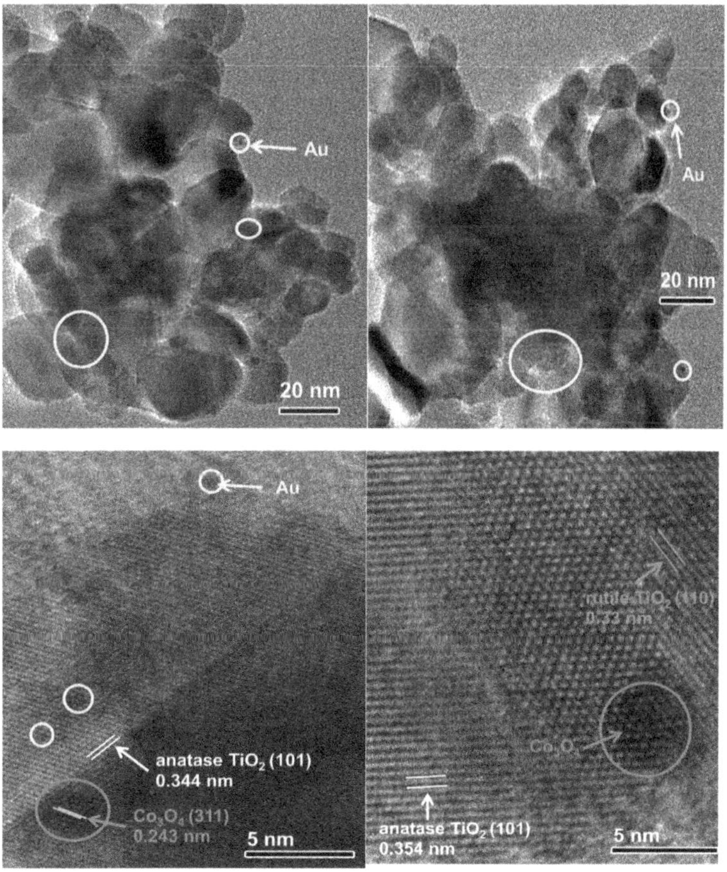

Figure 4.9 High-resolution TEM images at different magnification of Au/Co_3O_4-TiO_2 (5 : 95) prepared at pH 7. From ref. 146.

Co_3O_4 increases the number of oxygen defects in titania, thus leading to a higher concentration of strongly bound gold atoms.

The catalytic activity of Au/TiO_2 catalyst increased with an increase in temperature until it reached the maximum CO conversion at 80 °C; the CO conversion then decreased at 100 °C. Au/Co_3O_4-TiO_2 (5 : 95) had a higher CO conversion than the other catalysts when the temperature was >65 °C. The catalyst was very stable if the reaction temperature was below 100 °C. Both CO conversion and selectivity did not change after 400 h on stream. It is well-known that CO could adsorb on Co_3O_4 and Au. However, CO is more strongly adsorbed on Au particles. Most researchers[145] have concluded that Au is in the semispherical shape on the support if the particle size is around 3 nm and CO is strongly adsorbed on the perimeter of the Au and support.

4.4.6 Au/ZnO-TiO_2

We also tried to investigate the influence of inactive supports such as ZnO on Au/TiO_2.[147] Adding ZnO to TiO_2 was expected to change the properties of TiO_2, such as isoelectric point and hydroxyl groups, and therefore influence the catalytic properties of Au. The isoelectric points of TiO_2 and ZnO are 6 and 9, respectively.[128,148] The gold uptake depends on the isoelectric point of the supports and slightly increases at high pH is due to a change in the charge of the support, becoming increasingly more negative above the isoelectric point.[149–155] This results in an electrostatic repulsion of gold-containing anions. It has been reported that the amount of gold loading on TiO_2 decreased with increase in the pH value of preparation solution when the preparation pH value was higher than 6.[108,156] The catalyst prepared at pH 9 deposited the highest amount of gold on ZnO-TiO_2 (5 : 95) support due to the presence of ZnO. The isoelectric point of mixed oxides corresponds to pure ZnO when the concentration of ZnO increases. The amount of gold loading decreased with an increase in ZnO content. The actual Zn content in each catalyst was slightly less than the nominal Zn loading, possibly because some of the ZnO and/or $Zn(OH)_2$ leached out during the DP process. We observed that the originally white ZnO-TiO_2 became violet or pink upon Au modification. The change in color indicates that Au nanoparticles become deposited on the metal oxide surfaces. The colors of the Au/ZnO-TiO_2 catalysts also changed with gold content, which was affected by the Zn to Ti ratio in the support and the preparing pH value; the more the gold particles deposited on supports, the more the purple of the color. Therefore, one could easily judge the amount of Au loading by the color of the catalyst.

The average particle size of these gold catalysts was in the range of 1–2 nm. Au/ZnO-TiO_2 (5 : 95) had a narrower particle size distribution as shown in Figure 4.10. The TEM results confirmed the existence of fine Au particles deposited on ZnO-TiO_2 support. They showed a narrow range of gold particle size (1.2–1.6 nm). The addition of ZnO decreased the average gold particle size from 2.5 to 1.3 nm. The ZnO formed very thin islands without any definite shape on TiO_2 surface as shown in Figure 4.11. Most of the gold

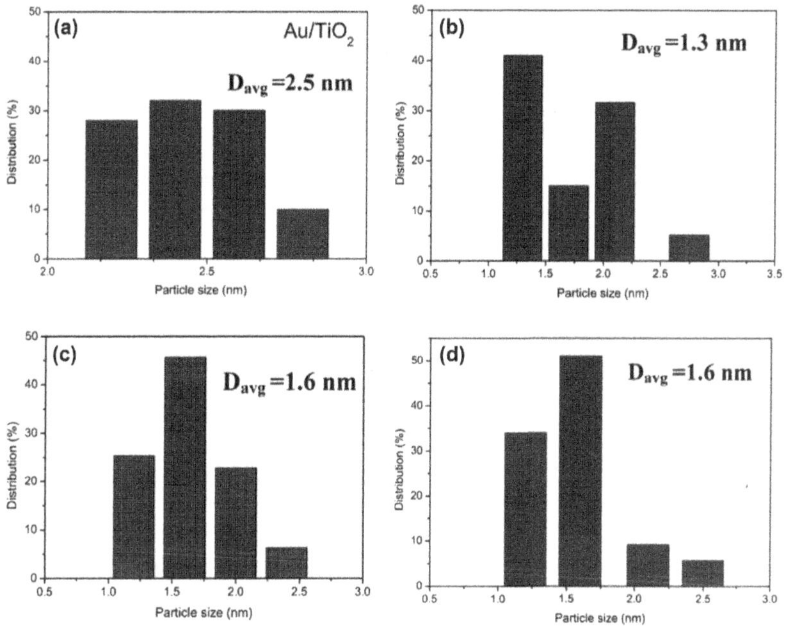

Figure 4.10 Particle size distributions of Au/ZnO-TiO_2 prepared at pH 7: (a) Au/TiO_2; (b) Au/ZnO–TiO_2 (1 : 99); (c) Au/ZnO-TiO_2 (1 : 9); and (d) Au/ZnO. From ref. 147.

particles were deposited on the surface of anatase TiO_2 in semispherical shape with particle size around 2.5 nm. P25 TiO_2 contains both anatase and rutile phases. Anatase TiO_2 has more hydroxyl groups than rutile TiO_2 and Zn species easily attached on it. High-resolution TEM analyses of a heterostructured Au/ZnO-TiO_2 nanocomposite showed that a single ZnO nanoparticle was connected with an Au nanoparticle in semispherical shape. The gold lattice was clearly shown in the high-resolution TEM image, and the d-space of Au (112) was 0.23 nm, consistent with the XRD result. Since the surface area of TiO_2 was not large and the concentration of ZnO was much higher than that of Au, most of the gold particles had intimate contact with ZnO. However, the exact situation depends on the Zn to Ti ratio and the preparation pH value during the DP process.

The preparation parameters play a major role in determining the properties of the catalyst and the dispersion of gold nanoparticles. In this study, the DP method was used to prepare gold-supported catalysts. There is a wide difference in the isoelectric point of ZnO and TiO_2, and hence pH value is a vital parameter in the dispersion and catalytic activity of gold particles. The influence of Zn to Ti ratio (1 : 99, 5 : 95, and 1 : 9) on the catalytic activities of Au/ZnO-TiO_2 catalysts prepared at pH 7 was investigated at temperatures between 25 and 100 °C. Selectivity on Au/ZnO at 25 °C was found to achieve 100%, whereas for Au/TiO_2 it was only 83%. Au/ZnO showed higher selectivity of CO oxidation than Au/TiO_2 at all temperatures. Therefore, under the

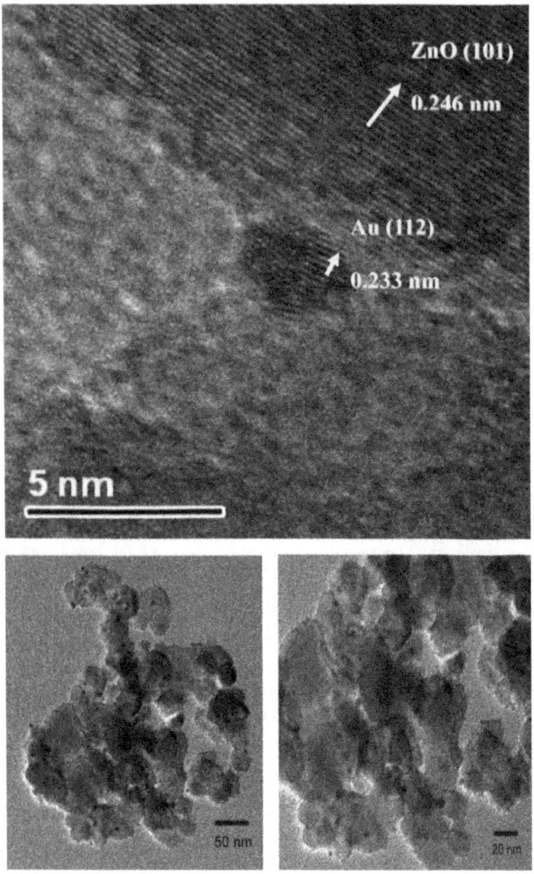

Figure 4.11 High resolution TEM images at different magnifications of Au/ZnO-TiO_2 (5 : 95) prepared at pH 6.
From ref. 147.

PROX reaction with limited oxygen concentration, Au/TiO_2 has high CO conversion and low CO selectivity. In contrast, Au/ZnO has low CO conversion and high CO selectivity. When ZnO was added on Au/TiO_2 (Zn : Ti = 5 : 95), there was an increase in the CO conversion at 25 °C from 44% to 64%. However, a further increase in the Zn to Ti ratio (1 : 9) led to a decrease in CO conversion. Increase in temperature resulted in improved CO conversion. At 50 °C, all the catalysts reached 100% CO conversion. Further increase in temperature reduced CO conversion. The decrease in CO conversion at high temperatures is due to the increased competition between CO and H_2 oxidation, because O_2 concentration was limited in the reaction.

Au/ZnO-TiO_2 (5:95) showed a better performance than Au/ZnO-TiO_2 (1 : 9) at high temperatures. At 25 °C, Au/ZnO-TiO_2 (5 : 95) had better selectivity to CO oxidation than other catalysts. Au/ZnO-TiO_2 (5 : 95) had optimum CO conversion and selectivity to CO oxidation in the PROX reaction. Au/ZnO-TiO_2

has a better performance than both Au/ZnO and Au/TiO_2 in the PROX reaction. The addition of ZnO stabilizes the gold particles, and improves the CO conversion and selectivity to CO oxidation.

In our opinion, the charge of the support cations is not the most relevant parameter. The most important one is the presence of oxygen vacancies at the interface between Au and the support, which enables an electron transfer. The mechanism over Au/ZnO-TiO_2 is not similar to CO oxidation over Au/TiO_2. For Au/TiO_2, the oxygen adsorbed on the interface between gold and TiO_2 reacts with CO on the same active site. Due to the presence of $Zn(OH)_2$ and nano Au particles, Au has a strong interaction with $Zn(OH)_2$ and is in Au^+ state. The presence of $Zn(OH)_2$ could stabilize this Au^+ species. CO and O_2 could adsorb on this site and react with each other. Hydrogen is less strongly adsorbed on it. Therefore, the presence of $Zn(OH)_2$ could suppress hydrogen oxidation.

With an increase in the preparation pH value, the CO conversion decreased at room temperature. The high catalytic activity of Au/ZnO-TiO_2 over Au/TiO_2 and Au/ZnO has been attributed to the synergistic effect of gold particle size, optimum combination of metallic and electron-deficient gold species, as well as to Au–support interactions. Au had a strong interaction with $Zn(OH)_2$ and was in Au^+ state. The presence of $Zn(OH)_2$ could stabilize this Au^+ species. CO and O_2 could adsorb on this site and react with each other. Hydrogen was less strongly adsorbed on it. Therefore, the presence of $Zn(OH)_2$ could suppress hydrogen oxidation. Hence, the oxidation state of the gold species and pH played an important role in determining CO conversion and selectivity of CO oxidation in a hydrogen-rich stream. Moreover, the catalysts were stable at 80 °C for more than 80 h.

4.4.7 Au/MgO-TiO_2

Another important non-reducible metal oxide is MgO. Because MgO is a good additive, it can stabilize transition metals and prevents sintering and evaporation.[86] Margitfalvi and co-workers[157] used MgO to modify Au/Al_2O_3 catalysts. Au/MgO-Al_2O_3 showed high activity in the sub-ambient and ambient temperature range, and the authors suggested that most of the gold particles were in contact with MgO. Au/MgO-TiO_2 was also tried for the PROX reaction.[32] The isoelectric points of MgO and TiO_2 are 12 and 6, respectively. Gold uptake depends on the isoelectric point of the supports and decreases at higher pH due to a change in the charge of the support, becoming increasingly more negative above the isoelectric point. This results in an electrostatic repulsion of gold-containing anions. It has been reported that the amount of gold loading on TiO_2 decreased with increase in the pH value of preparation solution when this was higher than 6.[108] Hence, the actual loading of gold deposited on MgO_x-TiO_2 mixed oxides was less than that on pure TiO_2 support.

The isoelectric point of mixed oxides corresponds to pure MgO when the concentration of MgO increases. The amount of gold loading increased with an increase in MgO_x content. In addition, the catalyst prepared at pH 9 deposited the highest amount of gold on MgO_x-TiO_2 (2 : 8) support. The colors of the Au/MgO_x-TiO_2 catalysts also changed with different gold content, which was affected by the Mg to Ti ratio in the support. The more the gold particles deposited on supports, the purpler the color is. When MgO_x-TiO_2 was prepared by the impregnation method, $Mg(OH)_2$ was deposited on the TiO_2 surface, and hence the particles must be smaller than the particle size of TiO_2. The addition of MgO to TiO_2 resulted in a good dispersion of gold particles and a narrow particle size distribution. The average particle size of these gold catalysts was found to be in the range of 2–3 nm. The TEM results confirmed the existence of fine Au particles deposited on the MgO_x-TiO_2 support. The addition of magnesium oxide reduced the size of gold particles from 3 to 2 nm. Grisel and Nieuwenhuys[158] also reported that the addition of MgO on Au/Al_2O_3 resulted in a highly dispersed Au phase with a narrow particle size distribution. The MgO_x forms very thin islands without any definite shape on TiO_2 surface. Most of the gold particles were deposited on the surface of TiO_2 in semispherical shape with particle size around 2.5 nm.

To study the influence of preparation pH value in the DP process to deposit gold on the catalytic activities of gold catalysts in PROX reaction, Au/MgO_x-TiO_2 (2 : 8) prepared at various pH values *viz.*, 8, 9, and 10.4, were tested. It was observed that, at higher pH, there was a decrease in the conversion and selectivity. The highest CO conversion was obtained when the prepared pH value was 9. The optimum pH for higher CO conversion was found to be 9, which is due to smaller Au particle size and higher gold content. Hence, a pH value of 9 was used to prepare Au/MgO_x-TiO_2 in the study.

The influence of Mg to Ti ratio (1 : 9, 2 : 8, and 3 : 7) on the catalytic activities of Au/MgO_x-TiO_2 catalysts prepared at pH 9 was investigated. Figure 4.12 shows the CO conversion and selectivity of oxygen reacting with CO, respectively, for Au/TiO_2, Au/MgO_x, and Au/MgO_x-TiO_2 catalysts at temperatures between 25 and 100 °C. Both Au/MgO_x and Au/TiO_2 showed nearly the same CO conversion at 25 °C. However, the selectivity on Au/MgO_x at 25 °C was found to be 100%, whereas for Au/TiO_2 it was only 72%. The conversion of CO using Au/MgO_x and Au/TiO_2 increased gradually with increasing temperature. Above 50 °C, the CO conversion on Au/MgO_x increased rapidly and reached 100% at 100 °C. Au/TiO_2 reached the maximum CO conversion (72%) at 80 °C and the conversion remained the same at 100 °C. Au/MgO_x showed better selectivity than Au/TiO_2 at all temperatures.

The results demonstrate that Au/TiO_2 is very active for both CO oxidation and H_2 oxidation. Therefore, under the PROX reaction with limited oxygen concentration, Au/TiO_2 has high CO conversion and low CO selectivity. In contrast, Au/MgO_x has low H_2 conversion and high CO selectivity. When MgO_x (Mg : Ti = 1 : 9) was deposited on TiO_2, there was an increase in the CO conversion at 25 °C from 17% to 40%. With an increase in the Mg to Ti ratio from 1 : 9 to 2 : 8, there was a further increase in CO conversion to 80% at 25 °C. However, a further increase in the Mg to Ti ratio (3 : 7) led to a

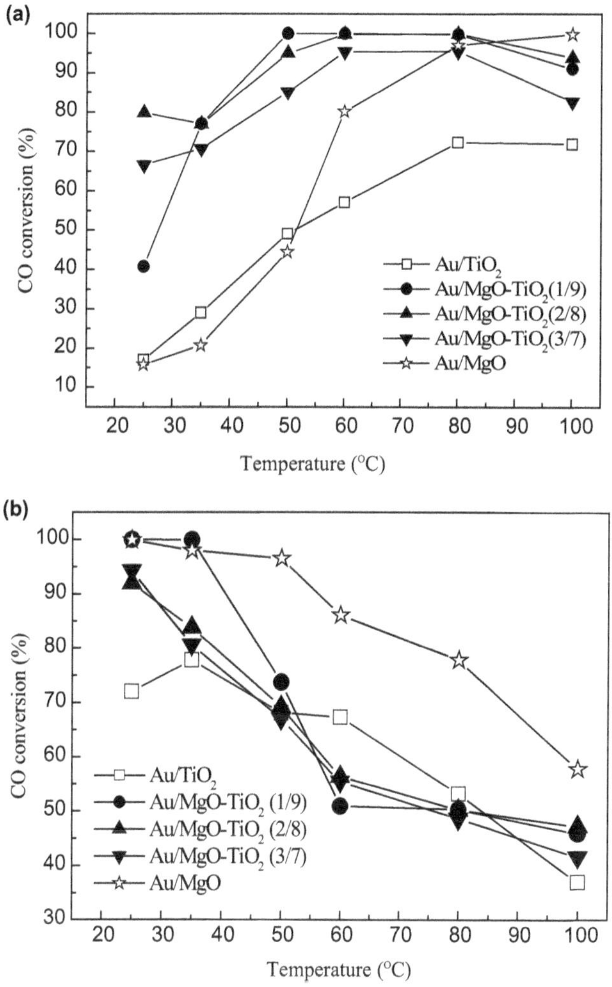

Figure 4.12 CO conversion (a) and CO selectivity (b) of Au/MgO_x-TiO_2 prepared at pH 9. Reactant gas: 1.33% CO, 1.33% O_2, 65.33% H_2, 32.01% He for balance, 50 mL min^{-1}, GHSV = 30 000 h^{-1}.
From ref. 32.

decrease in CO conversion, which was due to larger gold particles (2.64 nm) and lower gold loading compared with Au/MgO_x-TiO_2 (2 : 8).

Increase in temperature results in improved CO conversion. Above 50 °C, all the catalysts reached more than 95% CO conversion. At 80 °C, Au/MgO_x-TiO_2 (1 : 9) and Au/MgO_x-TiO_2 (2 : 8) reached 100% CO conversion. A further increase in temperature drastically reduced the CO conversion. However, Au/MgO_x-TiO_2 (2 : 8) showed a better performance than Au/MgO_x-TiO_2 (1 : 9) at high temperatures. The decrease in CO conversion at high temperatures is due to the increased competition between CO and H_2 oxidation because O_2 concentration was limited in the reaction. At 25 °C, Au/MgO_x-TiO_2 (1 : 9) had better selectivity to CO oxidation than other catalysts. However, the selectivity to CO oxidation of all the catalysts decreased with reaction temperature as H_2 oxidation became intense as the temperature increased, thereby retarding the oxidation of CO. Au/MgO_x-TiO_2 (2 : 8) had the optimum CO conversion and selectivity to CO oxidation in the PROX reaction.

The addition of MgO_x stabilizes the gold particles and improves the CO conversion and selectivity to CO oxidation. The results demonstrated that adding suitable amount of MgO_x on Au/TiO_2 can increase CO oxidation and suppress H_2 oxidation in the PROX reaction. Au/MgO_x-TiO_2 has a better performance than both Au/MgO_x and Au/TiO_2 in the PROX reaction.

4.5 Conclusions

Fuel cells are potential battery replacements; however, platinum electrodes are often poisoned by the presence of CO. Clean hydrogen fuel is necessary to improve the efficiency of the fuel cell. In order to remove CO, highly dispersed gold nanoparticles on metal oxides have been extensively studied for preferential oxidation of CO in a hydrogen-rich stream. Little attention has been paid to mixed or composite oxide supports, in which the second metal oxide can improve the selectivity of oxidation of CO on Au/TiO_2. We have studied the influence of active supports such as MnO_x, CeO_2, CuO_x, Fe_2O_3 and CoO_x, and inactive supports such as ZnO and MgO on the catalytic activity of Au/TiO_2 for the PROX reaction. Irrespective of reducible or non-reducible supports, all the second metal oxides enhanced the selectivity of CO oxidation without sacrificing CO conversion. Depending on the nature of the supports, the metal–support interaction varies and plays a major role in determining the conversion and selectivity. The preparation methods and pH modify the nature of mixed metal oxide supports. The results confirmed that gold nanoparticles supported on mixed or composite metal oxides are better than Au/TiO_2 for the PROX reaction.

References

1. G. C. Bond and D. T. Thompson, *Catal. Rev. Sci. Eng.*, 1999, **41**, 319.
2. B. Hammer and J. K. Norskov, *Adv. Catal.*, 2000, **45**, 71.
3. G. C. Bond, *Gold Bull.*, 1972, **5**, 11.
4. P. A. Sermon and G. C. Bond, *J. Chem. Soc. Faraday I*, 1978, **74**, 385.
5. S. Galvagno and G. Parravano, *J. Catal.*, 1978, **55**, 178.
6. P. A. Sermon, G. C. Bond and G. Webb, *J. Chem. Soc., Faraday Trans. I*, 1979, **75**, 395.
7. I. E. Wachs, *Gold Bull.*, 1983, **16**, 98.
8. J. Schwank, *Gold Bull.*, 1983, **16**, 103.
9. J. Schwank, *Gold Bull.*, 1985, **18**, 2.
10. M. Haruta and H. Sano, *Abstr. Pap. Am. Chem. Soc.*, 1985, **189**, 171.
11. M. Haruta, T. Kobayashi, H. Sano and N. Yamda, *Chem. Lett.*, 1987, **2**, 405.

12. M. Haruta, N. Yamada, T. Kobayashi and S. Iijima, *J. Catal.*, 1989, **115**, 301.
13. G. J. Hutchings, *Catal. Today*, 2005, **100**, 55.
14. A. Hashmi and G. J. Hutchings, *Angew. Chem., Int. Ed.*, 2006, **45**, 7896.
15. P. Landon, P. J. Collier, A. J. Papworth, C. J. Kiely and G. J. Hutchings, *Chem. Commun.*, 2002, 2058.
16. S. Carrettin, P. McMorn, P. Johnston, K. Griffin and G. J. Hutchings, *Chem. Commun.*, 2002, 696.
17. M. D. Hughes, Y.-J. Xu, P. Jenkins, P. McMorn, P. Landon, D. I. Enache, A. F. Carley, G. A. Attard, G. J. Hutchings, F. King, E. H. Stitt, P. Johnston, K. Griffin and C. J. Kiely, *Nature*, 2005, **437**, 1132.
18. Y.-J. Xu, P. Landon, D. Enache, A. F. Carley, M. W. Roberts and G. J. Hutchings, *Catal. Lett.*, 2005, **101**, 175.
19. B. Nikosi, N. J. Coville and G. J. Hutchings, *J. Chem. Soc., Chem. Commun.*, 1988, 71.
20. M. Conte, A. F. Carley, C. Heirene, D. J. Willock, P. Johnston, A. A. Herzing, C. J. Kiely and G. J. Hutchings, *J. Catal.*, 2007, **250**, 231.
21. S. Carrettin, P. McMorn, P. Johnston, K. Griffin, C. J. Kiely, G. A. Attard and G. J. Hutchings, *Top. Catal.*, 2004, **27**, 131.
22. J. E. Bailie and G. J. Hutchings, *Chem. Commun.*, 1999, 2151.
23. T. V. Choudhary and D. W. Goodman, *Top. Catal.*, 2002, **21**, 25.
24. C. D. Pina, E. Falletta, L. Prati and M. Rossi, *Chem. Soc. Rev.*, 2008, **37**, 2077.
25. A. Corma and H. Garcia, *Chem. Soc. Rev.*, 2008, **37**, 2096.
26. W. Yan, B. Chen, S. M. Mahurin, S. Dai and S. H. Overbury, *Chem. Commun.*, 2004, 1918.
27. G. Ma, A. Binder, M. Chi, C. Liu, R. Jin, D. Jiang, J. Fan and S. Dai, *Chem. Commun.*, 2012, **48**, 11413.
28. Z. Ma and S. Dai, *Nano Res.*, 2011, **4**, 3.
29. M. Haruta, G. Bamwenda, S. Tsubota and T. Nakamura, *Catal. Lett.*, 1997, **44**, 83–87.
30. M. Haruta, *Catal. Today*, 1997, **36**, 153.
31. M. Haruta, S. Tsubota, T. Kobayashi, M. Kageyama, M. J. Genet and B. Delmon, *J. Catal.*, 1993, **144**, 175.
32. L.-H. Chang, Y.-W. Chen and N. Sasirekha, *Ind. Eng. Chem. Res.*, 2008, **47**, 4098.
33. L.-H. Chang, N. Sasirekha and Y.-W. Chen, *Catal. Commun.*, 2007, **8**, 1702.
34. P. Sangeetha, L.-H. Chang and Y.-W. Chen, *Ind. Eng. Chem. Res.*, 2009, **48**, 5666.
35. P. Sangeetha, L.-H. Chang and Y.-W. Chen, *Mater. Chem. Phys.*, 2009, **118**, 181.
36. Y.-F. Wang, P. Sangeetha and Y.-W. Chen, *Int. J. Hydrogen Energy*, 2009, **34**, 8912.
37. Y. Kuwauchi, H. Yoshida, T. Akita, M. Haruta and S. Takeda, *Angew. Chem., Int. Ed.*, 2012, **51**, 7729.

38. A. Longo, L. F. Liotta, G. Pantaleo, F. Giannici, A. M. Venezia and A. Martorana, *J. Phys. Chem. C*, 2012, **116**, 2960.
39. F. Moreau, G. C. Bond and A. O. Taylor, *J. Catal.*, 2005, **231**, 105.
40. F. Boccuzzi, A. Chiorino and M. Manzoli, *Surf. Sci*, 2000, **454–456**, 942.
41. K.-C. Wu, Y.-L. Tung, Y.-L. Chen and Y.-W. Chen, *Appl. Catal. B: Environ.*, 2004, **53**, 111.
42. H.-Y. Lin and Y.-W. Chen, *Ind. Eng. Chem. Res.*, 2005, **44**, 4569.
43. L.-H. Chang, N. Sasirekha, Y.-W. Chen and W.-J. Wang, *Ind. Eng. Chem. Res.*, 2006, **45**, 4927.
44. Z. Chen and Q. Gao, *Appl. Catal. B: Environ.*, 2008, **84**, 790.
45. X. S. Huang, H. Sun, L. C. Wang, Y. M. Liu, K. N. Fan and Y. Cao, *Appl. Catal. B: Environ.*, 2009, **90**, 224.
46. O. H. Laguna, F. Sarria, M. A. Centeno and J. A. Odriozola, *J. Catal.*, 2010, **276**, 360.
47. D. Widmann, R. Leppelt and R. J. Behm, *J. Catal.*, 2007, **251**, 437.
48. S. D. Lin, M. Bollinger and M. A. Vannice, *Catal. Lett.*, 1993, **17**, 245.
49. E. Quinet, F. Morfin, F. Diehl, P. Avenier, V. Caps and J. L. Rousset, *Appl. Catal. B: Environ.*, 2008, **80**, 195.
50. M. M. Schubert, S. Hackenberg, A. C. van Veen, M. Muhler, V. Plzak and R. J. Behm, *J. Catal.*, 2001, **197**, 113.
51. S. S. Pansare, A. Sirijaruphan and J. G. Goodwin Jr., *J. Catal.*, 2005, **234**, 151.
52. O. H. Laguna, W. Y. Hernández, G. Arzamendi, L. M. Gandía, M. A. Centeno and J. A. Odriozola, *Fuel*, 2014, **118**, 176.
53. J. Fonseca, S. Royer, N. Bion, L. Pirault-Roy, M. C. Rangel, D. Duprez and F. Epron, *Appl. Catal. B: Environ.*, 2012, **128**, 10.
54. O. H. Laguna, F. R. Sarria, M. A. Centeno and J. A. Odriozola, *J. Catal.*, 2010, **276**, 360.
55. N. F. P. Ribeiro, F. M. T. Mendes, C. A. C. Perez, M. M. V. M. Souza and M. Schmal, *Appl. Catal. A: Gen.*, 2008, **347**, 62.
56. M. A. Kipnis, E. A. Volnina, A. A. Ezhov and V. K. Ivanov, *Kinet. Catal.*, 2013, **54**, 358.
57. S. D. Gardner, G. B. Hoflund, B. T. Upchurch, D. R. Schryer and E. J. Kielin, *J. Catal.*, 1991, **129**, 114.
58. G. B. Hoflund, S. D. Gardner, D. R. Schryer, B. T. Upchurch and E. J. Kielin, *Appl. Catal. B: Environ.*, 1995, **6**, 117.
59. S. D. Gardner, G. B. Hoflund, M. R. Davidson, H. A. Laitinen, D. R. Schryer and B. T. Upchurch, *Langmuir*, 1991, **7**, 2140.
60. R. Griesel, K. J. Weststrate, A. Gluhoi and B. E. Nieuwenhuys, *Gold Bull.*, 2002, **35**, 39.
61. Y.-F. Chang and J. G. McCarthy, *Catal. Today*, 1996, **30**, 163.
62. L.-H. Chang, N. Sasirekha, B. Rajesh and Y-W. Chen, *Sep. Purif. Technol.*, 2007, **58**, 211.
63. B. M. Reddy, In *Metal Oxides: Chemistry and Applications*, ed. J. L. G. Fierro, CRC Press, Florida, 2006, ch. 8, pp. 215–246.

64. A. Trovarelli, *Catal. Rev. Sci. Eng.*, 1996, **38**, 439.
65. X. D. Zhou, W. Huebner and H. U. Anderson, *Appl. Phys. Lett.*, 2002, **80**, 3814.
66. S. Carrettin, P. Concepcion, A. Corma, J. M. L. Nieto and V. F. Puntes, *Angew. Chem., Int. Ed.*, 2004, **43**, 2538.
67. A. M. Venezia, G. Pantaleo, A. Longo, G. Di Carlo, M. P. Casaletto, F. L. Liotta and G. Deganello, *J. Phys. Chem. B*, 2005, **109**, 2821.
68. R. M. Friedman, J. J. Freeman and F. W. Lytle, *J. Catal.*, 1978, **55**, 10–28.
69. W. Liu and M. Flytzani-Stephanopoulos, *Chem. Eng. J.*, 1996, **64**, 283.
70. X. F. Yu, N. Z. Wu, Y. C. Xie and Y. Q. Tang, *J. Mater. Chem.*, 2000, **10**, 1629.
71. P. Li, D. E. Miser, S. Rabiei, T. T. Yadav and M. R. Hajaligol, *Appl. Catal. B: Environ.*, 2003, **43**, 151.
72. H.-Y. Lin, Y.-W. Chen and W.-J. Wang, *J. Nanopart. Res.*, 2005, **7**, 249.
73. H.-Y. Lin and Y.-W. Chen, *Ind. Eng. Chem. Res.*, 2005, **44**, 4569.
74. K.-C. Wu, Y.-L. Tung, Y.-L. Chen and Y.-W. Chen, *Appl. Catal. B: Environ.*, 2004, **53**, 111.
75. J. Li and N. J. Coville, *Appl. Catal. A: Gen.*, 1999, **181**, 201.
76. D. J. Duvenhage and N. J. Coville, *Appl. Catal. A: Gen.*, 2002, **233**, 63.
77. N. J. Coville and J. Li, *Catal. Today*, 2002, **71**, 403.
78. S. Sun, N. Tsubaki and K. Fujimoto, *Appl. Catal. A: Gen.*, 2000, **202**, 121.
79. Y. Zhang, D. Wei, S. Hammache and J. G. Goodwin Jr., *J. Catal.*, 1999, **188**, 281.
80. Y. J. Chen, D.-E. Wu and C.-T. Yeh, *Res. Adv. Mater. Sci.*, 2003, **5**, 41.
81. J. D. Grunwald, C. Kiner, C. Wogerbauer and A. Baiker, *J. Catal.*, 1999, **181**, 223.
82. G. Y. Wang, W. X. Zhang, H. L. Lian, D. Z. Jiang and T. H. Wu, *Appl. Catal. A: Gen.*, 2003, **239**, 1.
83. F. Boccuzzi, A. Chiorino, S. Tsubota and M. Haruta, *Catal. Lett.*, 1994, **29**, 225.
84. N. S. Phala, G. Klatt, E. van Steen, S. A. French, A. A. Sokol and C. R. A. Catlow, *Phys. Chem. Chem. Phys.*, 2005, **7**, 2440.
85. E. Castillejos, R. Bacsa, A. Guerrero- Ruiz, I. Rodriguez-Ramos, L. Datas and P. Serp, *Nanoscale*, 2011, **3**, 929.
86. L. M. Molina and B. Hammer, *Phys. Rev. Lett.*, 2003, **90**, 206102.
87. M. M. Schubert, V. Plzak, J. Garvhe and R. J. Behm, *Catal. Lett.*, 2001, **76**, 143.
88. D. A. H. Cunningham, W. Vogel and M. Haruta, *Catal. Lett.*, 1999, **63**, 43.
89. J. D. Grunwaldt and A. Baiker, *J. Phys. Chem. B*, 1999, **103**, 1002.
90. M. Haruta, *J. New Mater. Electrochem. Systems*, 2004, **7**, 163.
91. J. M. C. Soares, P. Morrall, A. Crossley, P. Harris and M. Bowker, *J. Catal.*, 2003, **219**, 17.
92. B. Schumacher, Y. Denkwitz, V. Plzak, M. Kinne and R. J. Behm, *J. Catal.*, 2004, **224**, 449.

93. M. Date, Y. Ichihashi, T. Yamashita, A. Chiorino, F. Boccuzzi and M. Haruta, *Catal. Today*, 2002, **72**, 89.
94. A. Luengnaruemitchaia, S. Osuwana and E. Gularib, *Int. J. Hydrogen Energy*, 2004, **29**, 429.
95. M. M. Schubert, A. Venugopal, M. J. Kahlich, V. Plzak and R. J. Behm, *J. Catal.*, 2004, **222**, 32.
96. A. Luengnaruemitchaia, D. T. K. Thos, S. Osuwana and E. Gularib, *Int. J. Hydrogen Energy*, 2005, **30**, 981.
97. L. M. Liu, B. McAllister, H. Q. Ye and P. Hu, *J. Am. Chem. Soc.*, 2006, **128**, 4017.
98. K. Y. Ho and K. L. Yeung, *J. Catal.*, 2006, **242**, 131.
99. M. Valden, X. Lai and D. W. Goodman, *Science*, 1998, **281**, 1647.
100. M. Okumura, S. Nakamura, S. Tsubota, T. Nakamura, M. Azuma and M. Haruta, *Catal. Lett.*, 1998, **51**, 53.
101. W. C. Li, M. Comotti and F. Schuth, *J. Catal.*, 2006, **237**, 190.
102. G. R. Bamweda, S. Tsubota, T. Nakamura and M. Haruta, *Catal. Lett.*, 1997, **44**, 83.
103. S. H. Kim, C.-H. Jung, N. Sahu, D. Park, J. Y. Yun, H. Ha and J. Y. Park, *Appl. Catal. A: Gen*, 2013, **454**, 53.
104. C. Rossignol, S. Arrii, F. Morfin, L. Piccolo, V. Caps and J. L. Rousset, *J. Catal.*, 2005, **230**, 476.
105. M. S. Chen and D. W. Goodman, *Catal. Today*, 2006, **111**, 22.
106. N. Lopez, T. V. W. Janssens, B. S. Clausen, Y. Xu, M. Mavrikakis, T. Bligaard and J. K. Nørskov, *J. Catal.*, 2004, **223**, 232.
107. R. Zanella, S. Giorgio, C. R. Henry and C. Louis, *J. Phys. Chem. B*, 2002, **106**, 7634.
108. F. Moreau and G. C. Bond, *Catal. Today*, 2007, **122**, 215.
109. R. Zanella, S. Giorgio, C.-H. Shin, C. R. Henry and C. Louis, *J. Catal.*, 2004, **222**, 357.
110. S. C. Chan and M. A. Barteau, *Langmuir*, 2005, **21**, 5588.
111. L.-H. Chang, Y.-L. Yeh and Y. W. Chen, *Int. J. Hydrogen Energy*, 2008, **33**, 1965.
112. C. Y. Wang, C. Y. Liu, X. Zheng, J. Chen and T. Shen, *Colloids Surf. A.*, 1998, **131**, 271.
113. F. Moreau and G. C. Bond, *Appl. Catal. A: Gen.*, 2006, **302**, 110.
114. G. Qi and R. T. Yang, *Appl. Catal. B: Environ.*, 2003, **44**, 217.
115. M. Okumura, S. Tsubota and M. Haruta, *J. Mol. Catal. A*, 2003, **199**, 73.
116. J. Guzman and B. C. Gates, *J. Am. Chem. Soc.*, 2004, **126**, 2672.
117. J. Guzman and B. C. Gates, *J. Phys. Chem. B*, 2002, **106**, 7659.
118. J. C. Fierro-Gonzalez, V. A. Bhirud and B. C. Gates, *Chem. Commun.*, 2005, **42**, 5275.
119. D. C. Koningsberger, M. K. Oudenhuijzen, J. de Graaf, J. A. van Bokhoven and D. E. Ramaker, *J. Catal.*, 2003, **216**, 178.
120. G. J. Hutchings, M. S. Hall, A. F. Carley, P. Landon, B. E. Solsona, C. J. Kiely, A. Herzing, M. Makkee, J. A. Moulijn, A. Overweg, J. C. Fierro-Gonzalez, J. Guzman and B. C. Gates, *J. Catal.*, 2006, **242**, 71.

121. C. K. Costello, J. H. Yang, H. Y. Law, Y. Wang, J.-N. Lin, L. D. Marks, M. C. Kung and H. H. Kung, *Appl. Catal. A: Gen.*, 2003, **243**, 15.
122. Z. Yan, S. Chinta, A. A. Mohamed, J. P. Fackler, Jr. and D. W. Goodman, *J. Am. Chem. Soc.*, 2005, **127**, 1604.
123. L. D. Socaciu, J. Hagen, T. M. Bernhardt, L. Woste, U. Heiz, H. Hakkinen and U. Landman, *J. Am. Chem. Soc.*, 2003, **125**, 10437.
124. B. Yoon, H. Hakkinen, U. Landman, A. S. Norz, J.-M. Antonietti, S. Abbet, K. Judai and U. Heiz, *Science*, 2005, **307**, 403.
125. W. E. Epling, G. B. Hoflund, J. Weaver, S. Tsubota and M. Haruta, *J. Phys. Chem.*, 1996, **100**, 9929.
126. E. D. Park and J. S. Lee, *J. Catal.*, 1999, **186**, 1.
127. M. A. P. Dekkers, M. J. Lippits and B. E. Nieuwenhuys, *Catal. Lett.*, 1998, **56**, 195.
128. J. D. Grunwaldt, M. Maciejewski, O. S. Becker, P. Fabrizioli and A. Baiker, *J. Catal.*, 1999, **186**, 458.
129. G. R. Bamwenda, S. Tsubota, T. Nakamura and M. Haruta, *Catal. Lett.*, 1997, **44**, 83.
130. A. Trovarelli, *Catal. Rev. Sci. Eng.*, 1996, **38**, 439.
131. S. Carrettin, P. Concepcion, A. Corma, J. M. L. Nieto and V. F. Puntes, *Angew. Chem., Int. Ed.*, 2004, **43**, 2538.
132. P. Sangeetha and Y.-W. Chen, *Int. J. Hydrogen Energy*, 2009, **34**, 7342.
133. J. S. Huang, Y. Wang, X. Zhao, S. Wang, S. Wu and W. Huang, *Catal. Commun.*, 2006, 7, 1029.
134. P. Sangeetha, B. Zhao and Y.-W. Chen, *Ind. Eng. Chem. Res.*, 2010, **49**, 2096.
135. F. Boccuzzi, A. Chiorino, M. Manzoli, P. Lu, T. Akita, S. Ichikawa and M. Haruta, *J. Catal.*, 2001, **202**, 256.
136. M. M. Schubert, M. J. Kahlich, H. A. Gasteiger and R. J. Behm, *J. Power Sources*, 1999, **84**, 175.
137. M. Khoudiakov, M. C. Gupta and S. Deevi, *Nanotechnology*, 2004, **15**, 987.
138. S. Scire, C. Crisafulli, S. Minico, G. G. Condorelli and A. D. Mauro, *J. Mol. Catal. A: Chem.*, 2008, **284**, 24.
139. Y.-F. Yang, P. Sangeetha and Y.-W. Chen, *Ind. Eng. Chem. Res.*, 2009, **48**, 10402.
140. K. Okazaki, S. Ichikawa, Y. Maeda, M. Haruta and M. Kohyama, *Appl. Catal. A: Gen.*, 2005, **291**, 45.
141. L. Guczi, A. Beck and K. Frey, *Gold Bull.*, 2009, **42**, 5.
142. R. D. Waters, J. J. Weimer and J. E. Smith, *Catal. Lett.*, 1995, **30**, 181.
143. L. F. Liotta, G. Di Carlo, G. Pantaleo, A. M. Venezia and G. Deganello, *Appl. Catal. B: Environ.*, 2006, **66**, 217.
144. L. F. Liotta, G. Di Carlo, G. Pantaleo and G. Deganello, *Catal. Comm.*, 2005, **6**, 329.
145. P. Thormahlen, M. Skoglundh, E. Fridell and B. Andersson, *J. Catal.*, 1999, **188**, 300.

146. Y.-W. Chen, H.-J. Chen and D.-S. Lee, *J. Mol. Catal. A: Chem.*, 2012, **363**, 470.
147. Y.-W. Chen, D.-S. Lee and H.-J. Chen, *Int. J. Hydrogen Energy*, 2012, **37**, 15140.
148. P. Benezeth, D. A. Palmer and D. J. Wesolowski, *Geochim. Cosmochim. Acta*, 1999, **63**, 1571.
149. C. R. Lemire, K. Meyer, K. Shaikhutdinov and H. J. Freund, *Surf. Sci.*, 2004, **552**, 27.
150. H. Yoshida, T. Shimizu, C. Murata and T. Hattori, *J. Catal.*, 2003, **220**, 226.
151. G. Shan, M. Zhong, S. Wang, Y. Li and Y. Liu, *J. Colloid Interface Sci.*, 2008, **326**, 392.
152. M. Manzoli, A. Chiorino and F. Boccuzzi, *Appl. Catal. B: Environ.*, 2004, **52**, 259.
153. M. Manzoli, A. Chiorino and F. Boccuzzi, *Surf. Sci.*, 2003, **532–535**, 377.
154. M. P. Casaletto, A. Longo, A. Martorana, A. Prestianni and A. M. Venezia, *Surf. Interface Anal.*, 2006, **38**, 215.
155. M. P. Casaletto, A. Longo, A. M. Venezia, A. Martorana and A. Prestianni, *Appl. Catal. A: Gen.*, 2006, **302**, 309.
156. T. Tabakova, V. Idakiev, D. Andreeva and I. Mitov, *Appl. Catal. A: Gen.*, 2000, **202**, 91.
157. E. G. Szabo, M. Hegedus, A. Szegedi, I. Sajo and J. L. Margitfalvi, *React. Kinet. Catal. Lett.*, 2005, **86**, 339.
158. R. J. H. Grisel and B. E. Nieuwenhuys, *J. Catal.*, 2001, **199**, 48.

CHAPTER 5

Twenty Years of Golden Future in the Water Gas Shift Reaction

T. RAMÍREZ REINA, M. GONZÁLEZ CASTAÑO, S. PALMA, S. IVANOVA* AND J. A. ODRIOZOLA

Department of Inorganic Chemistry, University of de Seville and Institute of Materials Science of Seville, Mixed Center US-CSIC Avda. Américo Vespucio 49, 41092, Seville, Spain
*Email: svetlana@icmse.csic.es

5.1 Short Overview

The water gas shift (WGS) reaction $CO + H_2O \leftrightarrow H_2 + CO_2$ was first described by the Italian scientist Felice Fontana in 1780. However, it was not until the beginning of the 20th century with the development of the Haber–Bosch process for ammonia production that it started to be of real practical use. In this process, the WGS reaction is normally combined with the gasification of coal or natural gas reforming to produce the hydrogen needed for the synthesis of ammonia. From this moment, the WGS reaction gained importance and became an inseparable part of various industrial processes, such as ammonia and methanol production, hydrogen supply for the petrochemical industry, the Fisher–Tropsch reaction and numerous reforming reactions.

The most important aspect of this hydrogen production process is that the WGS reaction is an exothermic reaction and the equilibrium conversion decreases with increasing temperature.[1] Therefore, a significantly better

RSC Catalysis Series No. 18
Heterogeneous Gold Catalysts and Catalysis
Edited by Zhen Ma and Sheng Dai

Published by the Royal Society of Chemistry, www.rsc.org

performance can be achieved by operating at a relatively high temperature, thereby exploiting the reaction kinetics when the gas composition is far from equilibrium, and then lowering the temperature as thermodynamics begin to limit the CO conversion.[2] Hence, the lowest temperature at which maximum CO conversion can be achieved is determined by the catalyst's kinetics and the dew point of the gas.[3] The first known catalysts were iron–chromium oxides as a high temperature shift catalyst which converts CO to CO_2 producing hydrogen in the temperature range 400–500 °C.[1] Later, the Cu-based low-temperature catalysts (250–400 °C) were discovered and applied in a series of adiabatic high and low temperature shift reactors for hydrogen supply under industrial conditions.

Although the WGS reactors and catalysts applied industrially have remained unchanged since the application of this technology, the emerging fuel cell technology for portable electronic devices rewrites the requisites for the WGS catalyst application. The exigency of a smaller volume and weight, non-pyrophoric and low-temperature operating catalyst triggered studies for possible application of noble metal based catalysts for generation of pure hydrogen on-board a vehicle. Nevertheless despite the diversity of the catalytic system reported as active in the WGS reaction, Pt- and Au-based catalysts appeared to be the most promising systems to substitute the traditional catalysts for small-scale applications.[4] In particular, the gold-based catalysts received particular attention due to their high activity and absence of conditioning before the reaction. The gold-based catalysts are now entering their third decade of special interest aimed at on-board hydrogen generation technology. An irrefutable proof is the ever increasing number of papers published since the first report in the late 1990s (Figure 5.1).[5]

In an attempt to gather all the existing information, this chapter is organized over five-year periods in which the most relevant aspects of the gold-based materials and their application in the WGS reaction are described. Within the groups of years, the achievements are organized around the support type, pointing to the extreme importance of its nature and the metal–support interaction for the production of gold-based catalysts for the low-temperature WGS reaction with good performance.

5.2 1996–1999

Andreev's group in the late 1990s presented the very first report on gold-based catalysts for the WGS reaction.[5] Based on the high catalytic activity of gold-supported catalysts for low temperature CO and H_2 oxidation and combustion of hydrocarbons and the enhanced activity in the presence of moisture previously demonstrated by Haruta and co-workers,[6,7] the group used Au/α-Fe_2O_3 catalysts for the WGS reaction. The support was carefully selected based on the well-known activity of the iron oxide for the high-temperature WGS reaction.

Hence the idea was simple and very clever: the use of gold supported on Fe_2O_3 should produce a promising system for the low-temperature WGS

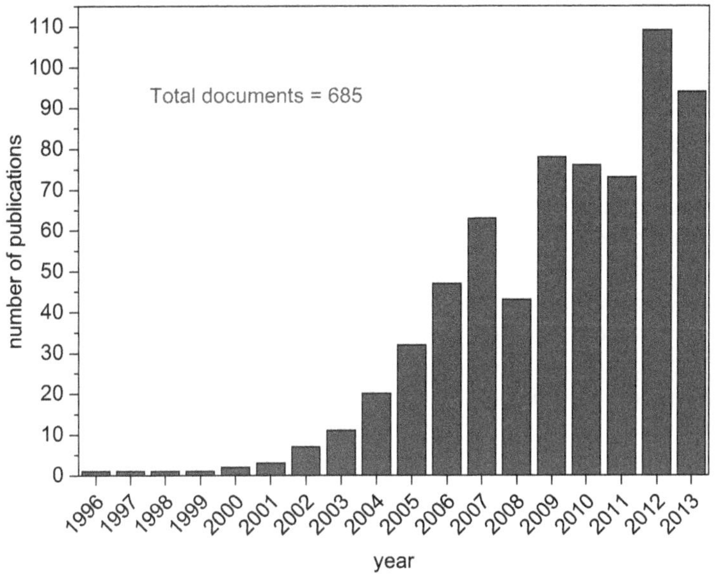

Figure 5.1 Number of published reports in the period 1998–2013 in Science Direct using 'gold catalyst' and 'WGS' as search strings.

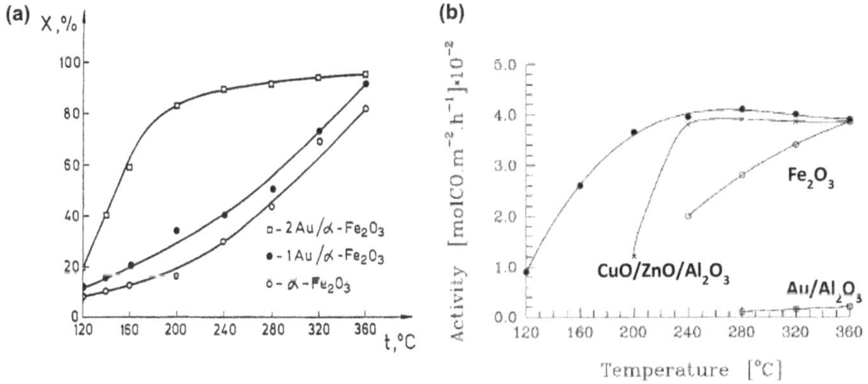

Figure 5.2 (a) Activity of the first Au catalyst, adapted from ref. 5. (b) Comparison between various catalysts from ref. 8.

reaction. Indeed, an excellent activity was achieved as shown in Figure 5.2(a). The sample displayed different Fe_2O_3 reducibility and the presence of gold clearly reflected the catalytic activity. Au/α-Fe_2O_3 presented a very high activity in the low temperature range where the iron oxide is hardly active. The successful performance of this material was ascribed as a consequence of a specific gold–support interaction, which affects the mechanism of the reaction. The reaction was reported to proceed *via* an associative mechanism where the intermediates could be either formates or carbonates.

The formation of surface intermediates through the interaction between the CO molecule and one OH group is the key step in the process, underlining the importance of the high concentration of OH groups in the catalysts, in order to achieve a high activity in the WGS reaction. The essential aspects of this mechanism are the dissociative adsorption of water on ultrafine gold nanoparticles followed by the spillover of active hydroxyl groups onto adjacent sites of iron oxide. The formation and decomposition of the intermediate involves the redox transfer ($Fe^{3+} \rightarrow Fe^{2+}$) and the reverse step, re-oxidation ($Fe^{2+} \rightarrow Fe^{3+}$) occurring during the water splitting step.

Some years later, the same group reported that the activity of Au/α-Fe_2O_3 material was even superior to that of the industrial CuO/ZnO/Al_2O_3 catalyst under the same conditions [Figure 5.2(b)].[8] In the same work, it was also demonstrated that Au/Al_2O_3 catalyst with a similar gold particles size is almost inactive, thus reinforcing the idea that the high activity of the gold–iron oxide catalyst is related to the metal–support interaction. To elucidate the possible mechanism, Ilieva *et al.*[9] carried out some temperature-programmed desorption–temperature-programmed reduction (TPD-TPR) experiments with the Au/α-Fe_2O_3 catalyst. These experiments revealed that the addition of gold facilitates the hematite–magnetite transition (magnetite is the active phase of iron oxide for the high-temperature WGS reaction) and involves the dissociative adsorption of water onto finely divided gold nanoparticles.

An improvement of the method of preparation, known to be crucial for the gold catalysis, was also proposed by this group in order to improve the performance of the Au/α-Fe_2O_3 catalysts for the WGS reaction.[10] A complete Fourier transform infrared (FTIR) study of the low-temperature WGS reaction on Au/Fe_2O_3 and Au/TiO_2 was reported by Boccuzzi *et al.*[11] Titania by itself was almost inactive and iron oxide only showed some activity in the high-temperature range. However, both gold catalysts presented similar high activities in the WGS reaction. The comparable performance was justified considering that the active sites for hydrogen dissociation and for water–CO interactions are located at the surface of metallic gold nanoparticles.

5.3 2000–2005

Compared to the previous period where only a few papers are found, a remarkable increase of the number of publications on this topic took place. The tendency was to obtain a deeper understanding of the promising Au/Fe_2O_3 catalyst developed by Andreev's group and to try some new types of support, especially TiO_2 and CeO_2, which started to gain importance.

5.3.1 Au/Fe_2O_3

Using the results of Andreev's group described above for Au/Fe_2O_3 as a starting point, some authors studied important aspects of the water gas shift with this type of catalysts. For example, Hua *et al.*[12] studied the influence of the calcination temperature on the structure and catalytic performance of Au/iron oxide catalysts. They found that the catalytic activity of the samples depended heavily on the calcination temperature, due to the change in the gold dispersion, phase state and reductive properties of the support. The better performance at the lower calcination temperatures was assigned to the large amount of highly dispersed oxidized gold species, synergized with the hydrated support, and the ease of its reducibility.

Venugopal and Scurrell[13] employed different reductive pre-treatments on the Au/Fe_2O_3 catalyst prior to the WGS measurements. The addition of 2,6-dimethylpyridine to the water–gas mixture decreased the WGS activity, suggesting the involvement of surface hydroxyl groups in the reaction; addition of ethanol seemed to be even more efficient than 2,6-dimethylpyridine. From TPR, temperature-programmed oxidation (TPO) and TPR analysis, gold re-dispersion for the 5% Au/Fe_2O_3 catalyst was observed when treated with H_2 prior to the O_2.

In an attempt to improve the catalytic performance of what became the classical Au/Fe_2O_3 catalyst, Venugopal *et al.*[14] introduced ruthenium as a second metal in order to compare the performance of the bimetallic Au-Ru/iron oxide catalysts with the monometallic ones (Au/iron oxide and Ru/iron oxide). The bimetallic system exhibited the highest activity in the low-temperature WGS reaction due to the synergistic effect between gold and ruthenium, and the textural stabilization of the system.

5.3.2 Au/TiO_2

Some valuable works using titania as a support for gold nanoparticles also appeared in this period. Idakiev *et al.*[15] presented for the first time a series of gold catalysts supported on mesoporous titania for the low-temperature WGS reaction. The mesoporous titania with high surface area and uniform pore size distribution was obtained using a surfactant template method through a neutral $[C_{13}(EO)_6\text{-}Ti(OC_3H_7)_4]$ assembly pathway.

The influence of gold content (1–5 wt%) and particle size on the catalytic behavior was addressed. They commented that the catalytic activity, and especially the catalyst stability, depended strongly on both the structure and nature of the support, and on the specific gold–support interaction. It was observed by transmission electron microscopy (TEM) that gold tended to epitaxially contact the anatase instead of rutile phase, implying that the locations of the Ti cations around the gold nanoparticles are more regular on the anatase surface. There was no measured activity for the pure mesoporous anatase, meaning that the reaction evolves in a different way on Au/TiO_2 than on pure TiO_2. A relevant observation was that the best performance was achieved with the sample loaded with 2.5 wt% of gold (and not higher), probably because the gold was homogeneously dispersed in the form of smaller hemispherical particles. A higher gold content results in bigger particles covering more surface and therefore a reduction in the active sites on the surface compared with the 2.5% Au/TiO_2 was reported.

The influence of the space velocity and H_2O to CO ratio was also evaluated. The increase of the space velocity from 2000 h^{-1} up to 8000 h^{-1} remarkably reduced the activity of the 5% Au/TiO_2 sample but not that of the 2.5% Au/TiO_2 sample. The latter manifested high and very stable (two weeks under operation) CO conversion at low H_2O to CO ratios, making the mesoporous titania a very promising support for gold nanoparticles in a WGS system.

The same authors also proposed the use of TiO_2 nanotubes (TNTs) for supporting gold nanoparticles as a good catalyst for the low-temperature WGS reaction.[16] The inner diameter of the nanotubes was similar to the size of the deposited gold nanoparticles and a fraction of Au was inserted inside, limiting the transport of the reactants to the active sites and leading to a reduction in the contact surface. In fact, these samples presented lower activity than the gold supported on mesoporous titania.[15] However, the catalytic activity of Au/TNTs might be enhanced if anatase-based nanotubes are employed instead of titanate-based ones.

In this period, Boccuzzi *et al.*[17] carried out a comparative study of the activity in the WGS reaction of Au, Cu and Ag supported on TiO_2. The following activity trend was observed: no activity (Ag), intermediate (Cu) and high activity (Au). An FTIR study confirmed that the gold catalyst activates CO molecules on the surface step sites and that the lack of activity in the silver-based system is due to the incapability of silver to adsorb and activate CO molecules.

5.3.3 Au/CeO_2

The first reports on using cerium oxide as a support for gold in the WGS reaction appeared in these five years. Ceria then rapidly became the most popular and employed support for gold nanoparticles in the low-temperature WGS reaction.

In general, the suitability of cerium oxide for oxidation catalysis is related to its redox behavior (Ce^{4+}/Ce^{3+}) and oxygen storage/release capacity resulting from the oxygen mobility in its lattice.[18] Under moderate temperatures and reductive conditions such as the WGS atmosphere, oxygen vacancies can be created in CeO_2, leading to the formation of nonstoichiometric oxides CeO_{2-x} (where $0<x<0.5$).[19] The concentration of these structural defects has been often targeted for excellent catalytic activity in CO elimination reactions.

The most cited paper so far regarding WGS and Au/CeO_2 catalysts in the 2000–2005 period comes from the group of Flytzani-Stephanopulos.[20] In this paper, the importance of the preparation method is underlined. Ceria was prepared by two common routes, co-precipitation and deposition–precipitation, with gold deposited by the urea gelation precipitation technique described in ref. 21. Moreover lanthanum (around 10 atom%) was added to the ceria as a promotor. In these studies an interesting conclusion was reached: the gold particle size is not the most important factor in the WGS reaction; meanwhile, there is a supplier of oxygen like ceria for the reaction. However, gold particle size is relevant for oxygen transfer. The presence of gold is necessary because CO adsorption occurs on the gold nanoparticles, since CO adsorption on hydroxylated ceria is completely hindered. From this observation, a cooperative redox mechanism for the WGS reaction on Au-CeO_2 was suggested. Another critical point appears to be control of the ceria crystal size and its influence on the amount of available oxygen, the oxide reducibility, and the catalytic activity of the Au/CeO_2.[22] Despite the fact that the amount of available surface oxygen mainly depends on the ceria crystal size, the presence of gold greatly weakens the Ce–O bonds, thus easing oxygen mobility and its interaction with CO at low temperatures.

A high and stable activity for a series of Au/CeO_2 catalysts varying the gold content from 1 wt% to 5wt% was found by Andreeva *et al.*[23] An increase in the CO conversion proportional to gold loading up to a certain limit (3 wt% of gold) was observed. 3Au/CeO_2 and 5Au/CeO_2 catalysts exhibited comparable activity and stability for the WGS reaction; 3Au/CeO_2 was the most stable sample maintaining high conversion over 3 weeks under operation. Two interesting conclusions should be pointed out from this paper: (1) Au promotes ceria reducibility and these redox properties are associated with the catalytic activity; and (2) gold particle size decreases under reaction conditions.

One of the first studies *via* FTIR of the WGS reaction mechanism on the Au/CeO_2 system was carried out by Tabakova *et al.*[24] Some chemisorption experiments of the molecules involved in the WGS reaction were performed to obtain clues on the active sites and reaction pathways. Apparently, nanosized metallic gold clusters in intimate contact with the oxygen vacancies of ceria play an essential role for the genesis of the high activity. In this type of material, ceria is not a simple support. Its ability to store oxygen related to the fast $Ce^{4+} \rightarrow Ce^{3+}$ transfer, together with the capacity to create anionic vacancies is of great importance for the high activity showed by this catalyst. The main achievements of this work can be summarized in three points: (1) the reduction of Au/CeO_2 in hydrogen at 373 K produces oxygen vacancies in the ceria surface; (2) an electron transfer from the support to the small gold cluster occurs and negatively charged gold sites appeared on the surface; and (3) the metal–support interface plays a decisive role in the surface structure of the catalysts modulating the oxidation state of the both ceria and gold. As for the mechanism of the WGS reaction, it seems to proceed at the boundary between the small metallic gold cluster and ceria, with CO adsorption taking place on the gold and water dissociation on the ceria defect sites. The presence of some bands associated with formate species confirms their role as intermediates in the water–gas shift.

In contrast to the idea that metallic gold is the active species in the WGS process, some works from the laboratory of Flytzani-Stephanopoulos note that gold can be present in the ionic state during the reaction.[25] The activity and stability of low content (0.2 and 0.9 atom%) gold-cerium oxide catalysts were studied. As a completely new approach, the authors proposed to remove the weakly bound gold species by leaching with NaCN. The starting materials were with gold loading of 2 and 8 atom%, which resulted in the above mentioned 0.2 and 0.9 atom% samples, respectively. As a principal observation, it was reported that the leached fraction mainly contained metallic gold and a small fraction of cationic gold. Based on this, the authors claimed that, during the WGS reaction, their catalysts were free of metallic gold and gold was present in an oxidized form. The leached materials displayed the same WGS activity as the high content gold ones and remained free of metallic gold species during the WGS reaction after working under a simulate reforming outlet pf 11% CO, 26% H_2O, 26% H_2 and 7% CO_2 balanced in He. Presumably this gas mixture controlled the oxidation state of gold and modulated the reduction–oxidation behavior of ceria. Good results of stability were also reported but the shutdown operation involves the apparition of hydroxycarbonate species, limiting the activity.

In the same period, several interesting studies by Jacobs and co-workers on active sites, intermediates and mechanistic aspects, and a comparison between Au/CeO_2 and Pt/CeO_2 systems were published and must be discussed at this point.[26–29] Using steady-state diffuse reflectance infrared Fourier transform spectroscopy (DRIFTS), formates were observed on the reduced ceria surface, formed by the reaction of CO with geminal OH groups, present only after reduction of the ceria surface.[27] Under high H_2O to CO ratios, CO exhibits a first-order rate dependency, and therefore, the active sites are expected to move to lower coverage of CO, suggesting that the WGS mechanism is likely proceeded *via* formates.

Despite the fact that some authors proposed a ceria-mediated redox mechanism,[20] the results of Jacobs and co-workers disagreed. X-ray absorption near-edge structure (XANES) data showed no evidence of ceria re-oxidation in the presence of water under a hydrogen atmosphere.[27] This process of ceria re-oxidation is a fundamental step in the redox mechanism (Figure 5.3); according to this, the formate pathway must be considered as

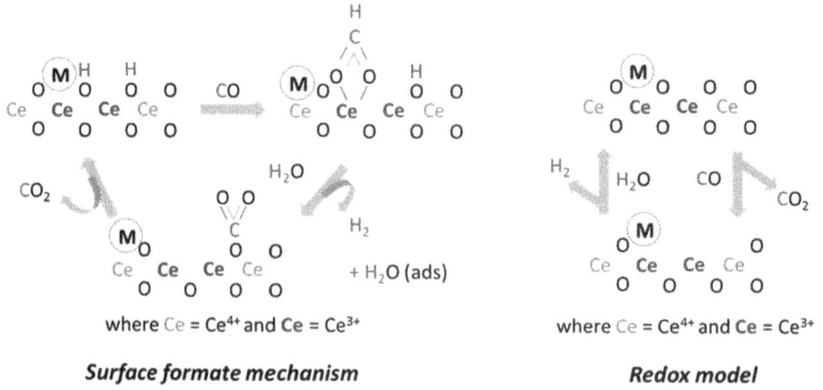

Figure 5.3 Proposed mechanisms for WGS reaction on gold/ceria catalysts. Adapted from ref. 28.

the dominant mechanism in this kind of system. Comparing the Au-ceria catalyst to a Pt-based homologue, Jacobs and co-workers found that the activity of the latter is about 20 times higher and that the metallic function participates in the reaction mechanism.[26]

Supporting the idea that the WGS reaction is indifferent to the metal (Au or Pt) and proceeds under the formate mechanism, the same group proposed a study devoted to the decomposition of the formate species, generated through the CO reaction with type II OH bridging groups, to yield the WGS products using different metal loadings (0.5 and 2.5 wt%) for both metals.[29] A faster and higher degree of formate decomposition over Pt-based samples was observed, with the increase being proportional to the metal loading. Based on these data, rupture of the formate C–H bond as a rate-limiting step in the WGS reaction was suggested, with the ability to overcome this step depending on the metal promoter, with platinum being better than gold.

These findings seemed controversial compared with those presented by Flytzani-Stephanopoulos and co-workers,[25] where cationic gold species were suggested to participate in the WGS reaction.

Similar results are reported in the work of Luengnaruemitchai *et al.*[30] where a comparison of the activity of Pt/CeO_2, Au/CeO_2 and Au/Fe_2O_3 catalysts for the low-temperature WGS was presented. Their results confirmed that the activity of the catalysts depends on the type of metal, with platinum being more active than gold, and on the nature of the support, with CeO_2 being a better choice for the WGS reaction than Fe_2O_3.

However, the superiority of Pt/CeO_2 over Au/CeO_2 catalysts was challenged by work presented by Haruta's group,[31] where an Au (4.0 wt%)/CeO_2 sample showed much higher activity than Au(2.4 wt%)/TiO_2, Pt(2.8 wt%)/CeO_2 and the commercial low-temperature shift catalyst, $Cu/ZnO/Al_2O_3$. The diversity of the results concerning these metals is often caused by the extreme importance of the method of preparation for the gold catalysts, where very precise control of the parameters is needed to produce active catalysts. An important finding was that the Au/CeO_2 catalyst did not produce methane and maintained 100% selectivity to CO_2 up to 623 K, while Pt/CeO_2 produced methane and resulted in a decrease in H_2 yield above 523 K.

The deactivation of the Au/CeO_2 catalyst is a point to consider in view of a possible practical application. This issue was addressed by Kim and Thompson.[32] Contrary to previous reports where the deactivation of gold supported on reducible metal oxides was attributed to over-reduction of the support and nanoparticles sintering, in this paper the formation of carbonate and formate species on the catalyst surface was reported as a main reason for deactivation. The authors proposed that oxygen-deficient sites on the catalyst surface facilitate the deposition of carbonates and/or formates. This implies that deactivation could be mitigated by conditioning the CeO_2 surface or by the addition of constituents to minimize oxygen deficiency. The initial activity for the Au/CeO_2 catalyst was fully recovered by calcination of the spent catalyst in flowing air at elevated temperatures.

5.4 2006–2010

WGS catalysts in general, and gold-based systems in particular, were studied intensively during this period due to the increased interest in hydrogen and fuel cell technologies. Hydrogen and fuel cells together have strong potential to play a significant role in the energy system that must be defined in the future. The interest was driven by the desire to generate hydrogen on-board a vehicle using a fuel processor. Fuel processors convert liquid hydrocarbons, where the volumetric energy content of these liquids is much larger than that of bottled hydrogen, into an almost CO-free hydrogen-rich product through the coupling of several catalytic reactions such as a reforming (usually steam reforming), water–gas shift and preferential CO oxidation in the presence of hydrogen (PrOx) and/or methanation (Figure 5.4).[4,33–35] The WGS reaction ($CO + H_2O \rightarrow CO_2 + H_2$) is a critical step in fuel processors for preliminary CO cleanup and additional hydrogen generation prior to the CO preferential oxidation or methanation step.[33–36] WGS units are placed downstream of the reformer to lower the CO content and improve the hydrogen yield.

The original work of Andreev's group demonstrated for the first time the promising character of oxide-supported gold catalysts, in particular Au/Fe_2O_3, as a very efficient catalyst for the WGS reaction.[5,8] However, as discussed above, in the 2000–2005 period many papers claimed that Au/CeO_2 catalysts were clearly more active and more interesting to analyze. On the basis of the acquired knowledge, the tendency during 2006–2010 was to investigate in depth some aspects that had generated much debate in the literature such as the nature of the active sites, the stability of gold-based systems, the kinetics of the reaction, and the mechanism. In addition the comparison between the relative activities of gold and platinum catalysts, which had emerged in the previous years, is worth considering. The most

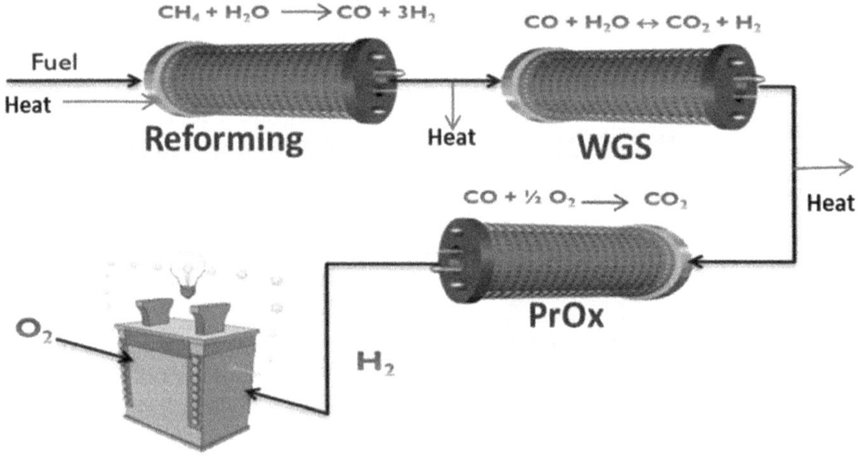

Figure 5.4 Fuel processor reactions.

studied systems in this period were again Au/CeO_2 materials, but also some studies using Au/TiO_2, Au/ZrO_2, *etc.* Not for the last time, the doping strategies appearing in the literature converted the CeO_2 doped systems to the most important systems to consider.

5.4.1 Au/CeO_2

Kim *et al.*[37] evaluated the importance of nanocrystalline gold on the Au/CeO_2 WGS catalyst. The study deals with some Au/CeO_2 catalysts, having gold loadings up to 8.3% and crystallites size ranging from 4 to 7 nm. Gold was leached from one of these materials. The rates for the unleached, nanocrystalline gold-containing catalysts were significantly higher than those for the leached catalysts, suggesting that the active sites associated with nanocrystalline gold domains are much more active than those associated with the atomically dispersed gold. These results challenge the observations obtained by the group of Flytzani-Stephanopoulos that reported no significant changes in the WGS activity of Au/CeO_2 catalysts after leaching. Based on the latter, they proposed that the WGS activity is associated with atomically dispersed cationic gold species. As explained by Burch,[38] the conclusion that only non-metallic gold is active for the WGS reaction should be treated with caution. It is likely that the leached catalysts contained most of the gold in the active form, probably as a small metallic cluster, and the parent catalysts before leaching (with 8% of gold) contained a large proportion of the inactive gold species.

The role of cationic Au^{3+} and metallic Au^0 on Au/CeO_2 in the WGS reaction was also examined by Karpenko *et al.*[39] who compared the reaction behavior of cyanide leached and non-leached catalysts. They observed that, despite the leaching process being selective for the removal of non-ionic Au species, the resulting catalyst did not contain only Au^{3+}. The abundance of the gold species present in the system depends strongly on the post-treatment of the material; high temperatures and reductive atmospheres favor the presence of metallic Au^0 nanoparticles, and leaching led to a considerable loss in activity. Despite the diversity in the authors' opinions regarding the nature of the active gold species, there is a general agreement that bulk metallic gold by itself is not efficient for catalytic reaction. However, gold nanoparticles supported in appropriate manner on an oxide can constitute highly active and relatively stable catalysts.

At this point some key questions appear. Is metallic gold the active form? Is gold activated due to the interaction with the support? Is the gold–oxide interface the place where the reaction takes place?

In this regard, Rodriguez's group has undertaken very intensive work.[40–44] In a smart paper, they proved that the high performance of Au-CeO_2 and Au-TiO_2 catalysts in the WGS process relies mainly on the direct participation of the oxide in the catalytic process.[40] Practically the same CO conversion was obtained for the inverse CeO_{2-x}/Au(111) and TiO_{2-x}/Au(111) systems, thus underlining the role of the interface. In CeO_{2-x}/Au(111) and TiO_{2-x}/Au(111),

CO is absorbed on gold-located sites, water splitting occurs on the oxygen vacancies of the oxide, and the subsequent reaction takes place on the metal–oxide interface. Density functional theory (DFT) calculations show that the TiO_2/Au (111) system can readily perform the WGS reaction with minimum energy barriers, thus confirming that the metal–support interface seems to be crucial for the process.

Deeper insights into the nature of the gold species presented during the low-temperature WGS reaction were published by Deng *et al.*[45] by combining XANES and extended X-ray absorption fine structure (EXAFS) studies over a low content (<1% Au) gold-ceria material. The authors observed that, under WGS conditions, reduction of the oxidized gold species and progressive gold agglomeration occurs. The Au–Au coordination number is zero for the fresh sample, but increases with the reaction temperature to 6.5 ± 2.4 (after use at 100 °C) and to 8.7 ± 1.5 (after 200 °C) in a feed stream composed by 5% CO and 3% H_2O in helium. An influence of the reaction gas composition on the degree of Au-O reduction was also detected. The lower the reduction potential of the mixture, the more oxidized gold is present in the used catalyst and the higher its activity. Nevertheless, photoemission results for the oxidation and reduction of gold nanoparticles supported on rough ceria (111) single crystal indicate that cationic gold species cannot be the key sites responsible for the WGS activity at high temperatures.[44] This study suggests that the active sites in gold-ceria catalysts involve pure gold nanoparticles in contact with the O vacancies of ceria.

Using DFT calculations, Jenssens *et al.*[46] also observed that the activity of gold catalysts increased when the gold particle size decreased due to the increasing number of low-coordinated Au atoms. The effect of the gold coordination number on the adsorption strength of CO and oxygen was found to be larger than other electronic effects and a critical parameter for the catalytic activity. Among the low-coordinated atoms, those placed on the corner sites of Au crystallites were the most active. The authors concluded that the synthesis procedure and the properties of the support surface influenced the size, dispersion and morphology of gold nanoparticles, and hence the concentration of very active low-coordinated gold atoms.

Further relevant information regarding the nature of gold and ceria during the WGS reaction has been obtained from a broad group of techniques such as TEM, TPR, FTIR and X-ray photon spectroscopy (XPS). A good summary of the most valuable information extracted from these experiments can be found in ref. 38. TEM evidences that the gold nanoparticles sizes depend heavily on the synthesis method and the pre-treatment conditions. For example, vapor grafting methodology leads to big Au clusters (close to 30 nm), while some optimized impregnation methods could generate nanogold of 1 nm. Fu *et al.*[47] carefully studied the CO and H_2-TPR of Au-CeO_2 catalysts and observed that gold reduction happens at temperature *ca.* 50 °C, and CeO_2 reduction normally begins at approximately 100 °C. Considering that CO and H_2 are typical components of the WGS mixture, a high level of reduction of gold and ceria is expected at the reaction temperatures (conventionally between 160 and 350 °C). The degree of reduction was confirmed by the infrared CO adsorption experiments reported by Tabakova *et al.*[48] Some bands at about 2060 cm^{-1} related to $Au^{\delta-}$ and a band at 2100 cm^{-1} associated with Au^0 likely to be adjacent to Ce^{4+} were found. The presence of oxygen vacancies on the ceria lattice was correlated to the presence of Ce^{3+}, as reflected in a well-defined band at 2127 cm^{-1}, corresponding to the electronic transition $2F_{5/2}$ to $2F_{7/2}$ of the Ce^{3+}. Furthermore, an XPS experiment detected a high concentration of Ce^{3+}, reinforcing the idea that the active phase in the WGS reaction is constituted of a gold cluster in contact with the oxygen vacancies of ceria. From here it can be indicated that, although gold-based catalysts are known to be effective even without activation, maybe it would be more adequate to state that the reductive atmosphere of the WGS activates the catalysts by reduction.

DFT calculations complete the information from experimental techniques. Some of the results have shown that the presence of oxygen defects surrounding the metal particles is of crucial relevance in the reaction mechanism.[41] Considering experimental and theoretical data together, it can be concluded that metallic gold is the active form of gold in the WGS reaction. Nevertheless, the intimate Au-oxide contact and the presence of oxygen vacancies in the oxide structure is required for a high catalytic performance.

Another important factor controlling WGS activity is the average size of the ceria particles.[45] Ceria particle size can be modulated using an adequate preparation method. Andreeva *et al.*[49] investigated the influence of the preparation method on nanosized gold supported on ceria and ceria–alumina systems. A comparison between samples, prepared by mechanochemical activation and conventional precipitation was reported. The former catalyst presented higher activity, attributed to the smaller size of Au and ceria crystallites in those samples. The addition of alumina to ceria results in smaller ceria crystallites, and therefore an increase in the concentration of the oxygen vacancies and oxygen storage capacity of the ceria was obtained. The main role of alumina was to act as a textural promoter, stabilizing Au and ceria, and avoiding agglomeration of the metal and oxide particles during the WGS process.

Chen and Goodman[50] considered the multi-role support effect in activating nanosized Au particles as a source of nucleating sites and/or charge transfer, and directly related to the reactants and their behavior. This concept reinforces the active role of ceria as a support and it was employed as an initiating point to investigate further different ways to promote ceria features in relevant reactions such as the WGS. In this way some works where ceria is doped or mixed with several promoters for the WGS came out during this five-year period.

Tabakova and co-workers[51] made some contribution in this sense. For example, they presented a quite efficient WGS catalyst based on a series of Au/CeO_2-TiO_2 catalysts prepared by the deposition–precipitation method. The existence of small cluster and ultra-small gold nanoparticles was observed by infrared, and their relative amount and nature were reported to depend strongly on the support composition and the pre-treatment process. Samarium oxide was also employed as a ceria promoter and compared with alumina as a dopant.[52] It was reported that the nature of the created oxygen vacancies depends on the dopant: Al_2O_3 creates vacancies in the bulk ceria, while the SmO_x sample presented vacancies located around SmO_x, and the ceria structure seems to be better ordered than the Al_2O_3 doped one. In addition, the concentration of Ce^{3+} was higher when SmO_x was used as a promoter and the amount of formate species was also superior in a parallel way with the activity in the WGS reaction. They proposed a higher concentration of $Au^{\delta+}V_oCe^{3+}$ sites on the surface of the SmO_x doped catalyst, which facilitates the dissociation of water and enhances the activity.

An Au-mixed lanthanum/cerium oxide catalyst was proposed by Wang *et al.*[53] as promising systems for the WGS reaction. The authors' attention was focused on the impact of La addition on the structure and reducibility of CeO_2 and on the reactivity of Au-based $La_xCe_{1-x}O_{2-0.5x}$ mixed oxides in the WGS reaction. Both structure and reducibility could be tailored by adjusting the LaO_x content. The reducibility of the Au-free sample was improved by the addition of LaO_x, with 25 atom% being the optimum loading to achieve the highest reducibility and therefore the best performance in the WGS reaction.

Zirconia has been also employed as a ceria promoter with even better results than the above mentioned dopants. The addition of Zr to the ceria lattice exercises a remarkable effect on both the oxygen storage capacity and the ability to maintain it. Zirconia stimulates the storage and release oxygen not only in the surface layers but also in the bulk lattice structure.[54] Ameiro-Fonseca *et al.*[55] published a very interesting paper where they presented $Au/CeZrO_4$ systems as excellent catalysts for the low-temperature WGS reaction. It should be noted that some activity was detected at temperatures as low as 100 °C. The effect of gold loadings from 0.19 wt% to 3 wt% was also studied; the highest specific reaction rate was reached for the lowest gold content and dropped steadily as the loading increased. Another interesting point is that 1.7 wt% $Au/CeZrO_4$ turned out to be much more active than a 4 wt% $Pt/CeZrO_4$ material, attributed to higher reducibility of the gold-based material.

Among the studied formulations, ceria attracted the greatest attention during 2006–2010, and it is clearly the most published support for gold catalyst in the WGS reaction. Nevertheless, some other types of metal oxides supports were also considered by several authors and it is worth briefly summarizing them.

5.4.2 Au/ZrO_2

Gold supported on zirconia can also be an adequate catalyst for the low-temperature WGS reaction, as demonstrated by several authors.[56–58] For example, Idakiev *et al.*[58] prepared nanosized gold catalysts by using mesoporous zirconia with a high surface area as a support. This sample reached high CO conversion under a model WGS mixture (4.49% CO, 30% of H_2O

under He). It was concluded that the structure of mesoporous zirconia facilitates the formation of well-dispersed and stable gold particles upon calcination and reduction, and thus greatly improves WGS activity. Additionally, they observed that as long as the gold content decreased, a bigger gold cluster appeared leading to a reduction of the CO oxidation rate. Some deactivation due to the formation of carbonates was observed and attributed to the ability of the materials to adsorb and accumulate CO. However, this catalyst deactivation was found to be reversible since the activity could be recovered after a simple thermal treatment.

The crystal phase of zirconia employed as a gold support is another matter to consider. The performance of Au/ZrO_2 samples for the WGS reaction using two types of zirconia structures, monoclinic (m) and tetragonal (t), was investigated by Li *et al.*[59] It was found that the Au/m-ZrO_2 catalyst showed much higher activity than the Au/t-ZrO_2 catalyst, attributed basically to the higher CO adsorption capacity of the Au/m-ZrO_2 catalyst. Moreover, using FTIR measurements, it was identified that the interaction between CO adsorbed on gold nanoparticles and OH groups on the surface of m-ZrO_2 produced formate species, which acted as the reaction intermediates and finally decomposed into CO_2 and H_2 in the presence of water.

Similarly to the case of ceria, the zirconia particle size plays a role and induces some changes in the catalytic behavior of the Au/ZrO_2 systems. This issue was addressed by Shen's group which examined the effect of both gold and zirconia particle sizes in the low-temperature WGS reaction.[59] As expected, the Au-ZrO_2 contact boundary appears to play an essential role in determining the catalytic activity. Maintaining the Au particle size at about 3 nm, an increase in the ZrO_2 particle size from 7 to 55 nm caused a significant detrimental effect on the activity. In a similar way, by fixing the ZrO_2 particle size at about 20 nm, the CO conversion decreased greatly when the size of the gold nanoparticles increased from 2.9 to 6.2 nm. In good agreement with the interpretation of Rodriguez's group these authors proposed that, in the WGS reaction, the CO is adsorbed on the Au species and H_2O is activated in the support (this time in the surface of ZrO_2).

The use of sulfates during gold deposition can enhance the activity of Au/ZrO_2 catalysts in the WGS reaction.[57] The addition of sulfates increases the surface acidity, retards the crystallization, and leads to materials with a higher surface area. Charging zirconium hydroxide with 2–8% of sulfates allows an improvement in the dispersion of gold nanoparticles, which is immediately reflected in a positive effect on CO conversion.

5.4.3 Au/TiO_2

As discussed earlier, the deposition of gold nanoparticles on TiO_2 produces very good catalysts for the WGS reaction. This is remarkable because neither bulk metallic gold nor titania are known as WGS catalysts. Rodriguez's group has contributed significantly to our understanding of the behavior of Au-TiO_2 systems for the WGS reaction.[40,42]

The high performance of Au-TiO_2 materials relies heavily on the direct participation of the metal–oxide interface in the catalytic process. The oxide assists the dissociation of water, a process that the extended surfaces and nanoparticles of gold cannot perform.[43] Although the performance of this system is good, one of the main drawbacks reported is its deactivation due to the sintering of gold nanoparticles at high temperatures.[60] Promoting TiO_2 with different oxides can modulate the gold–support interaction and prevent the gold particles sintering. Following this idea, a complete catalytic screening of the WGS reaction behavior of a series of promoted titania systems was reported by Ma *et al.*[61] Within the prepared Au/M_xO_y/TiO_2 catalysts with M = Al, Ca, Fe, Zn, Ga, Y, Zr, La, Ce, Pr, Nd, Sm, Eu, Gd, Dy, Ho, Er or Yb, the best specific rate at 200 °C was reached with a catalyst promoted with La_2O_3. At 300 °C the use of holmium oxide leads to better performance. Regarding the stability, small drops of CO conversion after 120 h on stream were observed for the Au/Eu_2O_3/TiO_2 system.

Mesoporous titania, and particularly ceria-modified mesoporous titania, has been successfully used in the low-temperature WGS reaction by Idakiev *et al.*[62] The mesoporous titania was prepared using a surfactant templating method through a neutral [$C_{13}(EO)_6$-Ti($OC_3H_7)_4$] assembly pathway, and ceria was added as promoter by the deposition–precipitation method (DP). Several factors are responsible for the promising performance observed for these catalysts: (1) ceria as an additive increases the crystallinity of titania and induces a decrease in the titania particles size; (2) the beneficial effect on the reducibility of the support due to the interaction of ceria with mesoporous TiO_2; and (3) the high degree of synergistic interaction between ceria and mesoporous titania. The good activity and excellent stability of the resulting materials was attributed to larger number of active sites located at the Au/CeO_2-TiO_2 interface.

5.4.4 Au/Miscellaneous Supports

Ceria, titania and zirconia, and multiple combinations of them, were the most broadly employed supports for gold nanoparticles in the WGS reaction between 2006 and 2010. However, some other works during this period using different supports are worth mentioning. For example, a very promising Au/ThO_2 catalyst was presented by Tabakova *et al.*[63] The catalyst showed extremely high activity at the lowest temperature range: the observed CO conversion was 84% at 120 °C. This system exhibited superior activity compared with commercial CuO-ZnO-Al_2O_3 and the Au/TiO_2 reference catalyst (World Gold Council). Apparently, the excellent activity was associated with the presence of highly dispersed gold clusters in a strong interaction with thoria.

Zhang *et al.*[64] also presented a very interesting work regarding Au/Fe_2O_3 catalysts promoted with ZrO_2 and Nb_2O_5. This was the first report in the literature where Au/Fe_2O_3 material had been simultaneously doped with two promoters. The doping process led to a progressive increase in the surface area and gold dispersion. In addition, the magnetite crystal size decreased, and its enrichment with ZrO_2 and Nb_2O_5 on the surface induced a strong interaction between the oxides that avoided the particles sintering, leading to improved stability and high activity for the low-temperature WGS reaction under hydrogen-rich conditions.

5.4.5 Au *vs.* Other Metals

Although not a specified subject of the chapter, a brief comparison between gold and other noble metals should be briefly discussed to clarify the importance of the future of gold metal in the low-temperature WGS reaction. When thinking about a future application, a truthful description of its benefits and disadvantages is necessary. For example, the activity of Au is often compared with the activity of Pt-, Rh- and Pd-based catalysts in the low-temperature WGS range (*i.e.* below 250 °C). The catalytic system is always designed to take advantage of the favorable thermodynamics at low temperatures. In this sense, gold catalysts can offer some advantages in the range of 160–250 °C where the Pt group metals are not active enough.[1] One of the best comparisons was presented in the review by Burch[38] (Figure 5.5). In this work the performance of gold catalysts is compared to that of Pt-based materials; the former are shown as being more effective WGS catalysts than the latter.

However, as commented above, some authors like Jacobs *et al.*[28] and Tibilleti *et al.*[65] found better catalytic performances for the Pt-based materials compared with the gold ones. In the paper by Jacobs and co-workers,

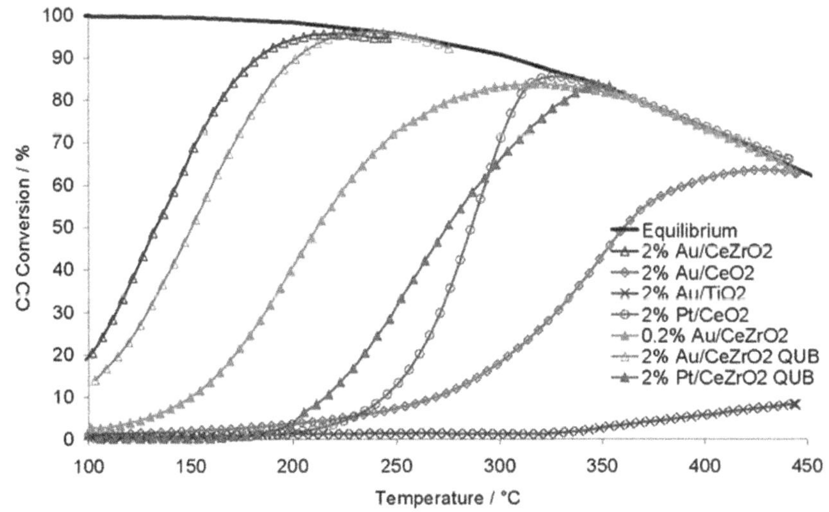

Figure 5.5 Comparison between Pt- and Au-based catalysts in the WGS reaction. Reprinted from ref. 38 with permission from the Royal Society of Chemistry.

a 5% Pt/CeO_2 catalyst exhibited almost 20 times higher activity than the equivalent 5% Au/CeO_2 material. The authors claimed that the reports, in which Au presented a higher activity than Pt, are ones where the Pt systems were tested under conditions of incomplete activation. However, the importance of the preparation method should always be kept in mind when comparing this kind of material. In this work, the catalysts were prepared using a vapor-deposition procedure, which is a technique currently not considered the most efficient way to achieve a strong contact between gold and the support, and leading to less active gold catalysts.

In contrast with the over-enhanced activity of the Pt-based system found by Jacobs *et al.*,[28] the Pt based system in the work of Tibilleti *et al.*[65] and in the paper of Luengnaruemitchai *et al.*[30] was only two times more efficient than gold-based catalysts. These results reinforce the importance of the preparation method and the reaction conditions (reactant mixture and pre-treatment). Overall the idea that can be extracted from all these works is that Au/CeO_2 and Pt/CeO_2 catalysts present comparable activity in the WGS reaction (with the exception of Jacobs' systems) with a factor of two that can oscillate considering the sensitivity of gold to the preparation method. In any case, it is difficult to establish direct comparison between Au- and Pt-based systems since most of the authors use broad options of preparation procedures, pre-treatments, input streams, *etc.* Nevertheless, these experimental conditions are of crucial relevance and must not be ignored to properly compare this type of system since both metals respond differently to changes in the concentration of the reactants.

Extensive research has been conducted looking for better alternatives to the well-known commercial low-temperature shift catalysts $Cu/ZnO/Al_2O_3$. Gold-based systems, a priori, present some advantages over this commercial system. For instance, they are not pyrophoric if exposed to air, they do not require special pre-treatment before use, and they show better thermal stability than the CuO/ZnO formulations. A meaningful comparison of nano-sized Au/CeO_2 home-made and Au/TiO_2 (World Gold Council) with CuO/Al_2O_3 (BASF) and $CuO/ZnO/Al_2O_3$ (REB Research & Consulting) was presented by Mendes *et al.*[66] It was found that the presence of the reaction products in the feed stream had the most adverse negative effect on the activity of the Cu-based catalysts at 150 °C. Under realistic reformate conditions (4.74% CO; 35.39% H_2O; 28.46% H_2; 10.06% CO_2 in N_2), the Au/CeO_2 sample was the most promising catalyst, particularly in the temperature range 150–200 °C. On the other hand, the commercial $CuO/ZnO/Al_2O_3$ catalyst exhibited the best balance of activity and stability. Moreover, a big effect on the CO concentration was observed. Depending on the reaction temperature, changes in the CO concentration may affect the catalytic activity positively or negatively. A positive effect of water for all the catalysts, apart from the Au/TiO_2 material, was also reported. The authors also noted that the operating temperature is a key parameter to take into account for correct catalyst selection. At temperatures ≥250 °C, Au/CeO_2 is clearly the better option since it does not become deactivated by the reaction intermediates and presents a higher CO conversion than $CuO/ZnO/Al_2O_3$. Nevertheless, at lower temperatures, stability is a negative factor for its selection and $CuO/ZnO/Al_2O_3$ seems to be the most suitable option.

An interesting comparative study of Cu, Ag and Au (IB group metals) supported on ceria was carried out by Tabakova *et al.*[48] The influence of the method of preparation was extensively studied and the influences on the activity of the IB metal/ceria catalysts in the WGS reaction can be summarized as follows: (1) regarding gold, different methods of synthesis lead to different gold particles size modulating the activity; (2) concerning copper, the particle size is almost unaffected by the preparation route and the key factor is the availability of active copper species in contact with ceria defects; and (3) for silver, Ag/ceria catalysts were almost inactive in the WGS reaction due to the very low adsorption energy of CO on metallic silver.

5.5 2011 to the Present

Despite the vast work carried out up to 2010, in very recent years and nowadays, gold catalysts are still extensively studied and their potential for the WGS reaction keeps attracting the attention of many research groups. The use of cerium oxide as a support for gold nanoparticles maintains the edge over the other supports as the most promising and most reported material, although the ceria promoted systems are much more frequent. To conserve the scheme of the chapter, the main results obtained in this period are organized considering the choice of the support.

5.5.1 Au/CeO_2 and CeO_2-promoted Systems

As stated above, the success of CeO_2 in several applications is mainly due to its unique combination of an elevated oxygen storage capacity (OSC) coupled with the ability to shift easily between reduced and oxidized state (Ce^{3+}/Ce^{4+}). A widely held opinion is that potentiating the oxygen vacancies formation should result in a higher possibility for binding the catalytically active species to ceria. When ceria is used as a support for nanosized gold particles, the resulting catalyst shows a high activity in the WGS reaction over a wide temperature range, but the stability could become very problematic.[67] That is why the role of ceria in the WGS reaction has also been the subject of many theoretical investigations. In particular, DFT methods have been employed to investigate the water dissociation step into OH and H fragments. It has been concluded that the ceria support is directly involved in the WGS reaction, because it lowers the barrier of water dissociation. However, as a perfect (111) CeO_2 surface is not able to dissociate water, it is inferred that this ability is displayed only when O vacancies are present.[68]

An interesting approach to study the ceria's electronic properties is the use of Raman spectroscopy. It is well-known that the shape and position of the main Raman bands of ceria are strongly influenced by particle morphology and the oxygen dosage. Moreover, it is demonstrated that not only oxygen, but also water can be employed to investigate the structure and the reactivity of the surface defects by means of *in situ* Raman spectroscopy. For example, Vindigni *et al.*[69] studied the influence of the pre-treatment conditions on the type of oxygen vacancies produced. It was found that the presence of defects could be related to the simultaneous and synergic contribution of the smaller ceria particle size and the presence of highly dispersed gold strongly interacting with the ceria surface. Using O_2 as a surface probe molecule in Raman experiments, a characterization of the surface defects was carried out.[70] The adsorption of oxygen on surface defect sites could result in the formation of surface superoxide (O^{2-}) and peroxide ($O_2{}^{2-}$) species; all of them are Raman active and hence can be easily revealed by means of spectroscopic experiments under a controlled atmosphere. The existence of three types of defects in ceria was affirmed, but only those located in the surface are able to react with H_2O.

The demonstrated positive role of the ceria defects in the catalytic activity motivated all future developments to look for a way of creating these defects in order to enhance the WGS performance. The inclusion of heterocations on the ceria lattice causes structural distortions inside ceria, thus provoking strain on the oxide lattice and favoring the formation of oxygen vacancies.[71] In this sense, the use of different metals and metal oxides has been considered.[72–75] The use of iron oxide as a ceria promoter is particularly interesting. In an *in situ* XANES study during the WGS reaction of an Au/CeO_2-FeO_x/Al_2O_3 system, Ramirez-Reina *et al.*[73] revealed the relevance of the presence of Ce^{3+} for obtaining high activities. CO conversion can be then directly correlated to the Ce valence, as presented in Figure 5.6.

It was found that the CO conversion increased when Ce valance decreased until an optimum value of 3.3. Moreover, the Ce–Fe interaction modifies the covalent character of the Ce–O bond, altering the stability of the hydroxide ions on the support and increasing the CO oxidation activity. The formation of a Ce–Fe solid solution is a key point to consider when iron is employed as a ceria promoter.[74] If iron oxide is present as a segregated phase, Fe^{3+} reduction during the WGS reaction results in divalent iron ions that stabilize

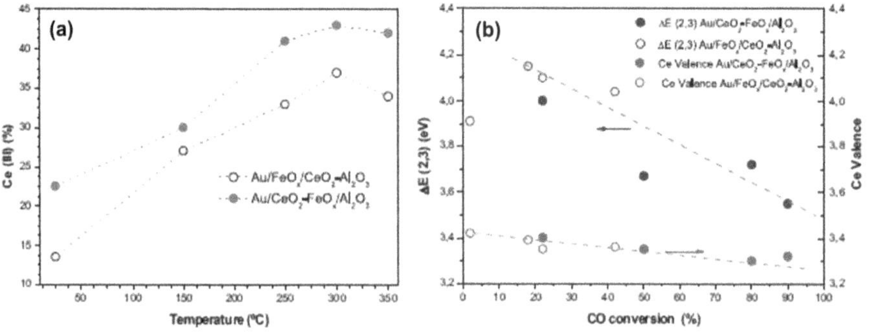

Figure 5.6 WGS activity correlation to the Ce oxidation state. Reprinted from ref. 73.

by crystallizing as Fe_3O_4 spinel structures and the Ce–Fe synergy is restricted to the perimeter of the spinel structure. In this way, the reducibility of the support is smaller than in the case of the solid solution. The redox properties of the Fe-doped systems are superior than the unmodified gold-ceria based catalysts. Recently, an enhancement of the oxygen storage complete capacity (OSCC) and oxygen storage capacity (OSC) has been proposed as one of the main reason explaining the activity promotion when iron is added to ceria.[75] The OSC and the number of atomic oxygen layers (NL) denote that the redox properties increased when iron was used as a promoter, leading to more active materials. Iron can be definitely considered as an efficient ceria promoter for the design of a highly effective WGS catalyst.

The promising nature of the ceria-based gold catalyst in this period encounters in this period a problem related to the deactivation with time-on-stream and/or shutdown restart operations. This deactivation process was commonly attributed to the blockage of the active sites by carbonate and/or formate formed during the WGS reaction.[32] Since the deposition of carbonate and/or formate is facilitated by oxygen-deficient sites on the catalyst surface, this process can be modulated by the addition of constituents, such as ZrO_2, to control oxygen deficiency and decrease the formation of carbonate species, and the ceria-zirconia based systems were carefully studied.[76] A higher tendency for the formation of carbonaceous species on pure ceria with respect to ceria-zirconia mixtures was found.[77,78] Vindigni *et al.*[77] reported that $Ce_{50}Zr_{50}$ and $Ce_{80}Zr_{20}$ mixtures present similar acidities, with both being two times higher than that of pure ceria. Therefore, zirconia addition modifies the acid–base properties of the support, but it is $Ce_{50}Zr_{50}$ where the less stable carbonate-like species appear. This feature, along with the presence of a high gold dispersion, makes $Au/Ce_{50}Zr_{50}$ the best catalyst in terms of catalytic activity and stability within the studied systems.

Other studies on gold-based catalyst supported on ceria-zirconia mixtures have been published by Idakiev *et al.*[79] In order to shed some light on the reason of the enhanced activity of the ceria-zirconia mixtures, they focused their efforts on examining the textural properties. From adsorption–desorption isotherms, it was observed that the samples contained an appreciable amount of secondary mesoporosity or macroporosity. This complementary mesoporosity and macroporosity makes the transport of reagents to framework reaction centers more efficient, and it was thus reported to favor the catalytic activity.

The impact of the preparation method on the activity and stability of gold supported on ceria–zirconia low-temperature WGS catalysts was also investigated by Pilasombat *et al.*[80] The influence of the gold deposition method, nature of the gold precursor, nature of the washing solution and drying method was considered.

Lanthanum has been also evaluated and successfully employed as a ceria promoter.[81] The preparation of $La_xCe_{1-x}O_y$ nanorods ($x = 0$–0.5) with uniform aspect ratio, and their evaluation as supports for Au in the WGS reaction was reported. By independently controlling aspect ratio and composition, it was demonstrated that the reducibility of the nanorods is determined by the preferential exposure of highly active {110} planes and the La to Ce ratio. The La doping was found to act as an inhibiter for the growth of $La_xCe_{1-x}O_y$ nanorods along the [110] direction and correlated with the preferential exposure of {110} crystal planes for CeO_2 nanorods, resulting in a lower reduction temperature and an increased number of reducible sites in the temperature range of the WGS reaction. All of this led to strongly increased WGS activity compared with nanoparticle-based Au/ceria catalysts. However, unlike nanoparticle-based catalysts, the reducibility of CeO_2 nanorods cannot be enhanced significantly by La doping, suggesting that the already low reduction energy of Ce^{4+} along the {110} planes cannot be further decreased by La. As a result, the WGS activity of pure Au/CeO_2 and Au/$La_xCe_{1-x}O_y$ with $x = 25\%$ is essentially unchanged, with conversions significantly higher than those for Au/ceria nanoparticle catalysts and commercial Cu-based catalyst.

Based on the best known support, some authors studied bimetallic Au-M systems supported on ceria as another catalyst design approach. Çağlayan and Aksoylu[82] reported a comparative study among Au-Re/ceria, Au/ceria and Re/ceria materials. Au-Re/ceria catalysts showed better performances, especially at higher H_2O to O ratios, while the monometallic catalysts suffered from steam inhibition. The influence of the preparation method, especially the application sequence of the impregnation and deposition–precipitation methods which were used in Re and Au addition, respectively, was found to be crucial; gold addition by the DP technique on impregnated Re/ceria catalysts led to a higher dispersion and stronger interaction between Au and Re particles and, consequently, higher CO conversion values. On the other hand, the impregnation of Re on Au/ceria catalysts resulted in blockage of the active sites, agglomeration of gold particles, and a reduction in the Ce^{3+} content.

Another point worth mentioning regarding a bimetallic system supported on ceria was reported by Gamboa-Rosales *et al.*[83] The Co-Au/CeO_2 bimetallic system was prepared by incipient wetness impregnation (IWI) and the DP method to check the proficiency of cobalt oxide as an alternative for copper oxide. From CO TPD, it was observed that bimetallic samples (gold- and cobalt-containing samples), desorbed a larger amount of CO_2 in comparison with the monometallic samples (cobalt–ceria samples), suggesting that CO is adsorbed on both the cobalt and gold surface, although some CO could be adsorbed onto ceria surface in the form of carbonates desorbing at higher temperatures. The ability of AuCoCe catalysts to supply lattice oxygen was related to gold dispersion, cobalt oxide dispersion, and the gold–cobalt oxide interface. Also, the gold addition enhanced the synergistic interaction between both metals, improving their activity. In this study, the application of oxygen-assisted water gas shift (OWGS) was proposed; it was shown that O_2 addition clearly improved the activity of the catalyst CoCe and AuCoCe, correlated with the highest OSC capacity of AuCoCe catalysts and its easier reducibility.

5.5.2 Au/TiO_2

Research on gold supported on titania also continued in this period. For instance, Sakurai *et al.*[84] confirmed that gold on titania, prepared by the DP method, exhibited very high activity for the low-temperature WGS reaction and comparable with that of the conventional $Cu/ZnO/Al_2O_3$ catalyst. Recently, Williams *et al.*[68] found a strong size effect for the Au-TiO_2 powder catalyst; that is, the WGS reaction rate per mol% Au significantly increased with the reduction in average gold particle size, more importantly for sizes less than 1 nm.

The deposition of ceria nanoparticles on TiO_2 produces an oxide–oxide interface that has unique structural and catalytic properties.[70,85–88] The CeO_x nanoparticles have an unusual coordination mode, which stabilizes Ce^{3+} ions. The increase in the stability of the Ce(III) oxidation state leads to an enhancement in the chemistry and catalytic activity of the ceria nanostructures. In a fundamental study of the WGS reaction on model $Au/CeO_x/TiO_2(110)$ and powder $Au/CeO_x/TiO_2$ catalysts, Si *et al.*[67] paid particular attention to the effect of ceria on the activity of the gold–titania systems. The main effect of the ceria addition was to enhance the concentration of oxygen vacancies in the catalysts, and thus help the dissociation of water during the reaction. CeO_x nanoparticles deposited on $TiO_2(110)$ acted as anchoring sites for gold, improving the dispersion of the admetal on the oxide support. The chemical properties showed by the cerium–titania support and the improvement of gold dispersion lead to a substantial reduction in the apparent activation energy for the WGS reaction, which decreases from 18 kcal mol^{-1} on Cu(111) to 7 kcal mol^{-1} on $Au/CeO_x/TiO_2(110)$.

Regarding the oxidation state of gold nanoparticles, results of *in situ* XAS (X-ray absorption spectroscopy) experiments indicated that the gold oxidation state in the $Au/CeO_x/TiO_2$ powder catalysts changed from ionic ($Au^{\delta+}$) to metallic (Au^0) when the WGS began. From *in situ* characterization results, it was concluded that the active phase for these powder catalysts contains gold particles with average sizes of $\sim$2 nm, which are preferentially stabilized by the oxygen vacancies in the CeO_x/TiO_2 surfaces. Based on these results, Rodriguez's group proposed that a complex and promising configuration for WGS catalysts is a mixed metal oxide array in which nanoparticles of a metal and an oxide can interact with the reactants. Extremely active WGS catalysts have been found after co-adsorbing nanoparticles of gold and ceria on a $TiO_2(110)$ substrate.[89]

Yang *et al.*[90] carried out further works regarding Au/TiO_2, leading to the publication of the first demonstration of the application of ultraviolet (UV) treatment to activate Au/TiO_2 for the WGS reaction with or without subsequent cyanide leaching. The catalytic activity of the UV-treated samples was significantly enhanced, and this was attributed to a large fraction of atomically dispersed and stable gold produced in the UV-irradiated sample. The slightly better activity of the UV-irradiated sample after leaching is attributed to the removal of gold particles, which uncovered additional anchored gold atoms from the basal planes of the particles. From CO TPR studies, the gold–titania interface was associated with a facile H_2O dissociation initiating the WGS reaction at close to ambient temperatures. High angle annular dark field–scanning electron transmission spectroscopy (HAADF-STEM) results evidenced that atomically dispersed gold species with surrounding extra surface –OH groups are the active sites for the WGS reaction on titania, as well as on other oxide supports. In the authors' opinion, the active gold sites are the isolated atoms stably anchored on the titania sites.

5.5.3 Au/ZrO_2

A few studies on gold supported on zirconia also appeared, thus closing the circle of using supports around ceria, zirconia and titania. The effect of the synthesis temperature of the hydrothermal zirconia on the structural properties and low-temperature WGS catalytic activity of Au/ZrO_2 catalysts was investigated.[91] The catalytic activity of the gold-zirconia catalysts for the low-temperature WGS reaction decreased as follows: Au/ZrO_2 – 150 > Au/ZrO_2 – 170 > Au/ZrO_2 – 190 > Au/ZrO_2 – 130. The catalytic behavior correlates well with the crystallite size of the ZrO_2 particles for Au/ZrO_2 – 150, Au/ZrO_2 – 170 and Au/ZrO_2 – 190, *i.e.* the smaller the zirconia nanoparticles the higher the catalytic activity, consistent with the results reported by Li *et al.*[56] and Zhang *et al.*[92] In both studies, the improvement of the activity on reducing the ZrO_2 crystallite size was attributed to the increased Au-ZrO_2 contact boundaries in the catalysts, which are regarded as active sites in the WGS reaction. It was also observed that a larger pore volume and larger average pore diameter lead to higher catalytic activity of the Au/ZrO_2 catalyst. This benefit in catalytic activity was explained by Liao *et al.*[93] They suggested that a benefit of the large pore diameter was the promotion of gold dispersion on the surface of the supports. In addition, Tabakova *et al.*[94] showed that the deactivation of the Au/ZrO_2 catalysts for the low-temperature WGS reaction was mainly attributed to a reduction in pore volume, suggesting the importance of a large pore volume for high catalytic activity.

The effect of the calcination temperature was studied by Zhang *et al.*[95] for gold-based zirconia catalysts. The support and catalyst were synthetized *via* the hydrothermal and DP methods, respectively. The increase in the calcination temperature from 350 °C to 450 °C led to an enlargement in crystal size of ZrO_2, while changes to the gold particles were not observed.

5.6 Final Remarks

All the 20 years of gold catalyst development for the WGS reaction have not been enough to conclude on the utility and applicability of those systems in practical applications. The benefits are clear, no needs of pre-treatment, no pyrophoricity and a relatively high activity in the low-temperature range. Nevertheless, the negative aspects such as extreme sensibility to the preparation procedure (existence or not of gold nanoparticles), the importance of the support nature (especially the development of an active gold–support interface), and the relatively low stability of the catalysts, mainly attributed to the support deactivation, also need to be considered. From the evolution of the gold systems over the years, it is clear that the ceria-supported catalysts present the highest activity among all candidates; however, high activity usually corresponds to a high deactivation rate. A common strategy in the past few years is to dope ceria with some elements to look for the activity improvement. However, one important thing that should be taken into account is that the catalyst design should include not only promotion of the activity but also improvement of its stability. The near future lies in a finely balanced catalytic system for which the activity to stability ratio is the highest possible.

It should also not be forgotten that a potential application in fuel cell technology calls for research on the design of structured catalysts regarding the volume and weight decrease of the WGS catalytic unit and an increase in efficiency in low-residence time working systems.

There is still much work to be done, challenges to be overcome, the gold catalysts to design.

References

1. C. Ratnasamy and J. P. Wagner, *Catal. Rev.*, 2010, **51**, 325.
2. A. R. Dubrovskiy, E. V. Rebrov, S. A. Kuznetsov and J. C. Schouten, *Catal. Today*, 2009, **147S**, S198.
3. T. Giroux, S. Hwang, Y. Liu, W. Ruettinger and L. Shore, *Appl. Catal., B*, 2005, **56**, 95.
4. S. Ivanova, O. H. Laguna, M. A. Centeno, A. Eleta, M. Montes and J. A. Odriozola, in *Renewable Hydrogen Technologies*, ed. L. M. Gandia, G. Arzamendi and P. M. Dieguez, Elsevier, Amsterdam, 2013, p. 225–243.
5. D. Andreeva, V. Idakiev, T. Tabakova, A. Andreev and R. Giovanoli, *J. Mater. Sci.*, 1996, **31**, 1101.
6. M. Haruta, N. Yamada, T. Kobayashi and S. Iijima, *J. Catal.*, 1989, **115**, 301.
7. S. Tsubota, A. Ueda, H. Sakurai, T. Kobayashi and M. Haruta, *Environ. Catal., Am. Chem. Soc. Symp. Ser.*, 1994, **552**, 420.
8. D. Andreeva, V. Idakiev, T. Tabakova and A. Andreev, *J. Catal.*, 1996, **158**, 354.
9. I. Ilieva, D. Andreeva and A. Andreev, *Thermochim. Acta*, 1997, **292**, 169.
10. D. Andreeva, T. Tabakova, V. Idakiev, P. Christov and R. Giovanoli, *Appl. Catal., A*, 1998, **169**, 9.
11. F. Boccuzzi, A. Chiorino, M. Manzoli, D. Andreeva and T. Tabakova, *J. Catal.*, 1999, **188**, 176.
12. J. Hua, K. Wei, Q. Zheng and X. Lin, *Appl. Catal., A*, 2004, **259**, 121.
13. A. Venugopalv and M. S. Scurrell, *Appl. Catal., A*, 2004, **258**, 241.

14. A. Venugopal, J. Aluha, D. Moganov and M. S. Scurrell, *Appl. Catal., A*, 2003, **245**, 149.
15. V. Idakiev, T. Tabakova, Z. Y. Yuan and B. L. Su, *Appl. Catal., A*, 2004, **270**, 135.
16. V. Idakiev, Z. Y. Yuan, T. Tabakova and B. L. Su, *Appl. Catal., A*, 2005, **281**, 149.
17. F. Boccuzzi, A. Chiorino, M. Manzoli, D. Andreeva, T. Tabakova, L. Ilieva and V. Idakiev, *Catal. Today*, 2002, **75**, 169.
18. A. Trovarelli, *Catal. Rev. Sci. Tech.*, 1996, **38**, 439.
19. A. Penkova, O. H. Laguna, M. A. Centeno and J. A. Odriozola, *J. Phys. Chem. C*, 2012, **116**, 5747.
20. Q. Fu, A. Weber and M. Flytzani-Stephanopuolos, *Catal. Lett.*, 2001, **77**, 87.
21. Y. Li, Q. Fu and M. Flytzani-Stephanopoulos, *Appl. Catal., B*, 2000, **27**, 179.
22. Q. Fu, S. Kudriavtseva, H. Saltsburg and M. Flytzani-Stephanopoulos, *Chem. Eng. J.*, 2003, **93**, 41.
23. D. Andreeva, V. Idakiev, T. Tabakova, L. Ilieva, P. Falaras, A. Bourlinos and A. Travlos, *Catal. Today*, 2002, **72**, 51.
24. T. Tabakova, F. Boccuzzi, M. Manzoli and D. Andreeva, *Appl. Catal., A*, 2003, **252**, 385.
25. Q. Fu, W. Deng, H. Saltsburg and M. Flytzani-Stephanopoulos, *Appl. Catal., B*, 2005, **56**, 57.
26. G. Jacobs, E. Chenu, P. M. Patterson, L. Williams, D. Sparks, G. Thomas and B. H. Davis, *Appl. Catal., A*, 2004, **258**, 203.
27. G. Jacobs, E. Chenu, P. M. Patterson, L. Williams, D. Sparks, G. Thomas and B. H. Davis, *Appl. Catal., A*, 2004, **262**, 177.
28. G. Jacobs, S. Ricote, P. M. Patterson, U. M. Graham, A. Dozier, S. Khalid, E. Rhodus and B. H. Davis, *Appl. Catal., A*, 2005, **292**, 229.
29. G. Jacobs, S. Ricote, U. M. Graham, P. M. Patterson and B. H. Davis, *Catal. Today*, 2005, **106**, 259.
30. A. Luengnaruemitchai, S. Osuwan and E. Gulari, *Catal. Commun.*, 2003, **4**, 215.
31. H. Sakurai, T. Akita, S. Tsubota, M. Kiuchi and M. Haruta, *Appl. Catal., A*, 2005, **291**, 179.
32. C. H. Kim and L. T. Thompson, *J. Catal.*, 2005, **230**, 66.
33. C. Song, *Catal. Today*, 2002, **77**, 17.
34. G. Kolb, *Fuel Processing for Fuel Cells*, Wiley-VCH, Weinheim, 2008.
35. O. Sanz, F. J. Echave, F. Romero-Sarria, J. A. Odriozola, M. Montes, L. M. Gandia, G. Arzamendi and P. M. Dieguez, in *Renewable Hydrogen Technologies*, Elsevier, Amsterdam, 2013, p. 201–224.
36. A. F. Ghenciu, *Curr. Opin. Solid State Mater. Sci.*, 2002, **6**, 389.
37. C. H. Kim and L. T. Thompson, *J. Catal.*, 2006, **244**, 248.
38. R. Burch, *Phys. Chem. Chem. Phys.*, 2006, **8**, 5483.
39. A. Karpenko, R. Leppelt, V. Plzak and R. J. Bhem, *J. Catal.*, 2007, **252**, 231.

40. J. A. Rodriguez, S. Ma, P. Liu, J. Hrbek, J. Evans and M. Pérez, *Science*, 2007, **318**, 1757.
41. J. A. Rodriguez, M. Perez, J. Evans, G. Liu and J. Hrbek, *J. Chem. Phys.*, 2005, **122**, 241101.
42. J. A. Rodriguez, J. Evans, J. Graciani, J. B. Park, P. Liu, J. Hrbek and J. F. Sanz, *J. Phys. Chem. C*, 2009, **113**, 7364.
43. P. Liu and J. A. Rodriguez, *J. Chem. Phys.*, 2007, **126**, 164705.
44. J. A. Rodriguez, X. Wang, P. Liu, W. Wen, J. C. Hanson, J. Hrbek, M. Pérez and J. Evans, *Top. Catal.*, 2007, **44**, 73.
45. W. Deng, A. I. Frenkel, R. Si and M. Flytzani-Stephanopoulos, *J. Phys. Chem. C*, 2008, **112**, 12834.
46. T. V. W. Janssens, B. S. Clausen, B. Hvolbæk, H. Falsig, C. H. Christensen, T. Bligaard and J. K. Nørskov, *Top. Catal.*, 2007, **44**, 15.
47. Q. Fu, H. Saltsburg and M. Flytzani-Stephanopoulos, *Science*, 2003, **301**, 935.
48. T. Tabakova, F. Boccuzzi, M. Manzoli, J. W. Sobczak, V. Idakiev and D. Andreeva, *Appl. Catal., A*, 2006, **298**, 127.
49. D. Andreeva, I. Ivanov, L. Ilieva, J. W. Sobczak, G. Avdeev and T. Tabakova, *Appl. Catal., A*, 2007, **333**, 153.
50. M. S. Chen and D. W. Goodman, *Catal. Today*, 2006, **111**, 22.
51. M. Manzoli, F. Vindigni, A. Chiorino, T. Tabakova, V. Idakiev and F. Boccuzzi, *Reac. Kinect. Catal. Lett.*, 2007, **2**, 213.
52. D. Andreeva, M. Kantcheva, I. Ivanov, L. Ilieva, J. W. Sobczak and W. Lisowski, *Catal. Today*, 2010, **158**, 66.
53. Y. Wang, S. Liang, A. Cao, R. L. Thompson and G. Veser, *Appl. Catal., B*, 2010, **99**, 89.
54. D. Duprez and C. Descorme, in *Catalysis by Ceria and Related Materials*, ed. A. Trovarelli, Imperial College Press, London, 2002, Vol. 2, p. 242–279.
55. A. Ameiro-Fonseca, J. M. Fisher, D. Ozkaya, M. D. Shannon and D. Thompsett, *Top. Catal.*, 2007, **44**, 223.
56. J. Li, N. Ta, W. Song, E. Zhan and W. Shen, *Gold Bull.*, 2009, **42**, 48.
57. M. Manzoli, F. Boccuzzi, V. Trevisan, F. Menegazzo, M. Signoretto and F. Pinna, *Appl. Catal., B*, 2010, **96**, 28.
58. V. Idakiev, T. Tabakova, A. Naydenov, Z. Y. Yuan and B. L. Su, *Appl. Catal., B*, 2006, **63**, 178.
59. J. Li, J. Chen, W. Song, J. Liu and W. Shen, *Appl. Catal., A*, 2008, **334**, 321.
60. Z. Ma, S. H. Overbury and S. Dai, *J. Mol. Catal. A*, 2007, **273**, 97.
61. Z. Ma, H. Yin and S. Dai, *Catal. Lett.*, 2010, **136**, 83.
62. V. Idakiev, T. Tabakova, K. Tenchev, Z. Y. Yuan, T. Z. Ren and B. L. Su, *Catal. Today*, 2006, **128**, 223.
63. T. Tabakova, V. Idakiev, K. Tenchev, F. Boccuzzi, M. Manzoli and A. Chiorino, *Appl. Catal., B*, 2006, **63**, 94.
64. F. Zhang, Q. Zheng, K. Wei, X. Lin, H. Zhang, J. Li and Y. Cao, *Catal. Lett.*, 2006, **118**, 131.

65. D. Tibilleti, A. Ameiro-Fonseca, R. Burch, Y. Chen, J. M. Fisher, A. Goguet, C. Hardacre, P. Hu and D. Thompsett, *J. Phys. Chem. B*, 2005, **109**, 22553.
66. D. Mendes, H. Garcia, V. B. Silva, A. Mendes and L. M. Madeira, *Ind. Eng. Chem Res.*, 2008, **48**, 430.
67. R. Si, J. Tao, J. Evans, J. B. Park, L. Barrio, J. C. Hanson, Y. Zhu, J. Hrbek and J. A. Rodriguez, *J. Phys. Chem. C*, 2012, **116**, 23547.
68. W. D. Williams, M. Shekhar, W. S. Lee, V. Kispersky, W. N. Delgass, F. H. Ribeiro, S. M. Kim, E. A. Stach, J. T. Miller and L. F. Allard, *J. Am. Chem. Soc.*, 2010, **132**, 14018.
69. F. Vindigni, M. Manzoli, A. Damin, T. Tabakova and A. Zecchina, *Chem. Eur. J.*, 2011, **17**, 4356.
70. J. B. Park, J. Graciani, J. Evans, D. Stacchiola, S. D. Senanayake, L. Barrio, P. Liu, J. Fernández-Sanz, J. Hrbek and J. A Rodriguez, *J. Am. Chem. Soc.*, 2009, **132**, 356.
71. G. Avgouropoulos, M. Manzoli, F. Boccuzzi, T. Tabakova, J. Papavasiliou, T. Ioannides and V. Idakiev, *J. Catal.*, 2008, **256**, 237.
72. T. Tabakova, L. Ilieva, I. Ivanov, R. Zanella, J. W. Sobczak, W. Lisowski, Z. Kaszkur and D. Andreeva, *Appl. Catal., B*, 2013, **136**, 70.
73. T. Ramirez Reina, W. Xu, S. Ivanova, M. A. Centeno, J. Hanson, J. A. Rodriguez and J. A. Odriozola, *Catal. Today*, 2013, **205**, 41.
74. T. Ramirez Reina, S. Ivanova, V. Idakiev, J. J. Delgado, I. Ivanov, T. Tabakova, M. A. Centeno and J. A. Odriozola, *Catal. Sci. Technol.*, 2013, **3**, 779.
75. T. Ramirez Reina, S. Ivanova, M. A. Centeno and J. A. Odriozola, *Front. Chem.*, 2013, **1**, 12.
76. C. H. Kimand, L. T. Thompson, Role of zirconia in stabilizing ceria supported gold water gas shift catalysts, presented at *GOLD 2006: New Industrial Application for Gold*, Limerick, Ireland, 2006.
77. F. Vindigni, M. Manzoli, T. Tabakova, V. Idakiev, F. Boccuzzi and A. Chiorino, *Appl. Catal., B*, 2012, **125**, 507.
78. S. Zeng, X. Zhang, X. Fu, L. Zhang, H. Su and H. Pan, *Appl. Catal., B*, 2013, **136–137**, 308.
79. V. Idakiev, T. Tabakova, K. Tenchev, Z. Y. Yuan, T. Z. Ren and B. L. Su, *J. Porous Mater.*, 2012, **19**, 15.
80. R. Pilasombat, H. Daly, A. Goguet, J. P. Breen, R. Burch, C. Hardacre and D. Thompsett, *Catal. Today*, 2012, **180**, 131.
81. S. Liang and G. Veser, *Catal. Lett.*, 2012, **142**, 936.
82. B. S. Çağlayan and A. E. Aksoylu, *Catal. Commun.*, 2011, **12**, 1206.
83. N. K. Gamboa-Rosales, J. L. Ayastuy, A. Iglesias-González, M. P. González-Marcos and M. A. Gutiérrez-Ortiz, *Chem. Eng. J.*, 2012, **207–208**, 49.
84. H. Sakurai, A. Ueda, T. Kobayashi and M. Haruta, *Chem. Commun.*, 1997, **271–272**.
85. A. Bruix, J. A. Rodriguez, P. J. Ramírez, S. D. Senanayake, J. Evans, J. B. Park, D. Stacchiola, P. Liu, J. Hrbek and F. Illas, *J. Am. Chem. Soc.*, 2012, **134**, 8968.

86. S. Kundu, J. Ciston, S. D. Senanyake, D. A. Arena, E. Fujita, D. Tacchiola, L. Barrio, R. M. Navarro, J. L. G. Fierro and J. A. Rodriguez, *J. Phys. Chem. C*, 2012, **116**, 14062.
87. I. D. González, R. M. Navarro, W. Wen, N. Marinkovic, J. A. Rodriguéz, F. Rosa and J. L. G. Fierro, *Catal. Today*, 2010, **149**, 372.
88. L. Barrio, G. Zhou, I. D. Gonzalez, M. Estrella, J. Hanson, J. A. Rodriguez, R. M. Navarro and J. L. G. Fierro, *Phys. Chem. Chem. Phys.*, 2012, **14**, 2192.
89. J. A. Rodriguez, *Catal. Today*, 2011, **160**, 3.
90. M. Yang, L. F. Allard and M. Flytzani-Stephanopoulos, *J. Am. Chem. Soc.*, 2013, **135**, 3768.
91. Y. Zhang, Y. Zhan, Y. Cao, C. Chen, X. Lin and Q. Zheng, *Chin. J. Catal.*, 2012, **33**, 230.
92. X. Zhang, H. Wang and B. Q. Xu, *J. Phys. Chem. B*, 2005, **109**, 9678.
93. W. P. Liao, Y. Y. Dong, M. Jin, T. He and H. Suo, *Chin. J. Catal*, 2008, **29**, 134.
94. T. Tabakova, V. Idakiev, D. Andreeva and I. Mitov, *Appl. Catal., A*, 2000, **202**, 91.
95. Y. Zhang, Y. Zhan, C. Chen, Y. Cao, X. Lin and Q. Zheng, *Int. J. Hydrogen Energy*, 2012, **37**, 12292.

CHAPTER 6

Gold Catalysis in the Complete Oxidation or Decomposition of Small Molecule Pollutants

CHUNYAN MA* AND ZHENGPING HAO*

Department of Environmental Nanomaterials, Research Center for Eco-Environmental Sciences, Chinese Academy of Sciences, Beijing, 100085, PR China
*Email: cyma@rcees.ac.cn; zpinghao@rcees.ac.cn

6.1 Introduction

Gold has been the source of jewelry, decoration and medical material. Gold itself has value and is readily traded. Indeed the gold price and gold reserve are still indicators of the financial status of the world and national economies. Since nanoscale gold was found to be exceptionally active as a catalyst, much research has focused on the synthesis, surface properties, theoretical calculation and catalytic performance of gold catalysts. The World Gold Council publishes *Gold Bulletin* and *CatGold News* on a regular basis, and has held five conferences including Gold 2001, Gold 2003, Gold 2006, Gold 2009 and Gold 2012. There were 14 oral presentations totally on gold catalysts at the 14th International Congress on Catalysis in 2008, and G. Hutchings gave an award lecture (Heina Heinemann Award) on catalysis

RSC Catalysis Series No. 18
Heterogeneous Gold Catalysts and Catalysis
Edited by Zhen Ma and Sheng Dai

Published by the Royal Society of Chemistry, www.rsc.org

using gold nanoparticles at the 15th International Congress on Catalysis in 2012.

A number of topics related to gold catalysis are supported financially by National Natural Science Foundation of China every year. In Japan, government-funded research organizations also support studies on gold catalysis. These investigations are concerned with the synthesis, structure analysis, theoretical calculations, surface science and catalytic applications of gold materials. It is obvious that the fascination with gold catalysis and gold catalysts has prompted an avalanche of research activity.

Over the past 30 years, there have been a number of discoveries that nanogold is a good catalyst for a lot of reactions including: selective oxidation, selective hydrogenation and complete oxidation in the gas phase; and oxidation, hydrogenation, deoxygenation, coupling and addition reaction in the liquid phase. The complete oxidation of small molecule gases at low temperatures over gold catalysts is an important discovery, which has many advantages such as lower energy consumption and mild reaction conditions.

Among them, the elimination of CO at ambient temperature is very important for air purification systems and breathing apparatus. The catalyst used at the present in breathing apparatus is Hopcalite (a mixed oxide of copper and manganese). Since lots of research results indicated that elimination of CO at ambient temperature can be achieved by oxidation over supported gold catalysts, some researchers tried to use gold catalysts in breathing apparatus. However, they realized that the poor stability of gold catalysts prevented their application of CO elimination in breathing apparatus. Air purification systems and breathing apparatus are applied in submarine, military, firefighting and mining operations, *etc.* in which the effectiveness of the catalyst can mean the difference between life and death. Thus, before Hopcalite could be replaced by supported gold catalysts, repeated testing of gold catalysts is required for security. Hopcalite is still so far the catalyst used in air purification systems and breathing apparatus.

Gold catalysts are still far from being usable for practical application of CO oxidation. However, this does not discourage researchers from searching for methods to promote the activity and stability of gold catalysts for CO oxidation, and from studying the reaction mechanism of CO oxidation over gold catalysts. The oxidation of other small molecules such as ethylene and formaldehyde over gold catalysts has also been investigated at room temperature or lower.

The decomposition of small molecule gases is another pathway to eliminate waste gas pollutants. The decomposition of gaseous pollutants such as ozone, N_2O and NO has been achieved over gold catalysts at a relative low temperature. These studies help us to understand the dissociating function of the small molecule on gold catalysts, and to develop effective gold catalysts for the elimination of waste gases by decomposition. In this chapter, we summarize reported results of the catalytic oxidation or decomposition of small molecule pollutants.

6.2 Complete Oxidation of Small Molecule Pollutants

6.2.1 Complete Oxidation of Carbon Monoxide

In 1987, Haruta *et al.*[1] reported that supported nanogold catalysts are highly effective for CO oxidation at low temperature. Since then, the surprising high activity of nanogold catalysts has spurred much research interest in gold catalyst applications, including selective oxidation of hydrocarbons,[2,3] the water gas shift reaction,[4] and liquid phase selective oxidation.[5,6] CO oxidation over gold catalysts, in particular, has been extensively studied.

6.2.1.1 The Effect of Gold Particle Size on Activity

The effect of gold particle size on the activity of gold catalysts is one aspect that has been studied. For bulk catalysts, peak activity is typically found for particle sizes in the range of 2–5 nm, which are composed of hundreds of atoms.[7,8] Boyen *et al.*[9] concluded that gold particles 1.4 nm in diameter, containing 55 atoms, are extraordinarily stable and these 1.4 nm gold particles could be active sites for CO oxidation. Goodman and co-workers[10] investigated the unusual size dependence of the low-temperature catalytic oxidation of CO by preparing gold clusters ranging in diameter from 1 to 6 nm supported on TiO_2. They found that Au clusters near 3.5 nm in size exhibited the maximum reactivity, and that a further decrease in cluster diameter below 3 nm leads to a decrease in the activity of the Au.

Furthermore, CO oxidation activities were typically found to peak for gold clusters in the sub-nanometer size range—composed of several gold atoms or even a single gold atom. Anderson and co-workers[11] reported room temperature CO oxidation on planar model Au_n/TiO_2 catalysts prepared by deposition of size-selected Au_n^+, $n=1$, 2, 3, 4, 7. Activity is strongly dependent on deposited cluster size, with substantial activity for Au_n as small as three atoms. Heiz and co-workers[12] used deposition of size-selected cluster cations to study CO oxidation on Au_n/MgO, observing significant reactivity for clusters as small as Au_8.

6.2.1.2 The Effect of Support and Preparation Method on Activity

Oxides,[13–15] carbon[16,17] and zeolite[18] supported gold materials have been investigated as effective CO oxidation catalysts under room temperature or even lower temperatures. It is observed that the catalytic activities of CO oxidation depend on the support.[19] Many researchers have summed up the CO oxidation activity over different supported gold catalysts. For example, Haruta[20] summarized the influence of the support acidity/basicity on CO oxidation activity over supported gold catalysts. Previous results also demonstrated that Au nanoparticles dispersed on reducible oxide supports are better than those supported on nonreducible supports. The high activity of

CO oxidation over Au/TiO_2 is due to Ti^{3+}, which is more capable of forming an active oxygen-rich Au-TiO_2 interface.[21]

Here, we prefer to emphasize that, even with the same support, different activities can be obtained using different preparation methods. The loading and dispersion of Au nanoparticles on different supports depends on the preparation method. For example, it is difficult to support gold nanoparticles on carbon by impregnation method due to the hydrophobicity of carbon. Thus, carbon was usually treated by nitric acid in the preparation of Au/C catalysts. Oxygen-containing functional groups generated on the carbon after treatment by nitric acid due to the oxidizability of nitric acid. Carbon that contains more oxygen-containing functional groups will be more hydrophilic. Even though, the sol-gel method is suitable for the preparation of Au/C catalysts due to the better control of gold dispersion. The impregnation method is one of the most used methods. The size of gold nanoparticles depends on the pH value of the gold precursor, which is adjusted by adding hydrochloric acid, nitric acid, sodium chloride or sodium hydroxide to the $HAuCl_4$ solution as required. When the pH value is in the range of 7–8, the hydrolysis of $AuCl_4^-$ proceeds, generating $[AuCl(OH)_3]^-$.[22] The supported gold catalysts may have relatively smaller gold nanoparticles (<5 nm), prepared by the impregnation method from the $[AuCl(OH)_3]^-$ precursor.[22] The incipient wetness method, an improved impregnation method, requires no extra precursor solution after support saturation adsorption. Uniformly dispersed gold particles are easily obtained for gold catalysts prepared by the incipient wetness method. There are several other preparation methods reported for supported gold catalysts such as gas-phase grafting,[23] deposition–precipitation[24] and deposition–reduction methods.[25] The purpose of all the preparation methods is to obtain supported gold catalysts with uniformly dispersed gold particles smaller than 5 nm.

It is reported that unsupported nanoporous gold is active for CO oxidation at temperatures as low as −20 °C[26] or −30 °C.[27] The nanoporous gold is made by selective dissolution (de-alloying) of silver from silver/gold alloy.[26] Furthermore, the nanoporous gold catalyst has been found to be active and highly selective for the aerobic oxidation of D-glucose to D-gluconic acid.[28] Is unsupported gold catalytically active? Haruta[29] wrote a special highlight article to discuss this question; he postulated that the unsupported gold is not really unsupported but an inversely supported gold catalyst. The unsupported gold was prepared from an Au-Ag alloy by leaching Ag, and the presence of residual Ag measured by X-ray photoelectron spectroscopy (XPS) may play an important role.

6.2.1.3 The Reaction Mechanism of CO Oxidation over Supported Gold Catalysts

The reaction mechanism and active sites of supported gold catalysts for CO oxidation have been studied for many years. Arguments continue concerning the nature of the active gold species and the catalytic sites. The hypotheses for the catalytic activity of gold were considered to stem from more than one of the following four factors: (1) size of gold clusters; (2) the support effect; (3) charge transfer between the support and gold; and (4) the junction interface between the support and gold.

When starting to review the CO oxidation over supported gold in this chapter, the size of the gold particles and the support effect were discussed. In fact, the interaction between gold and support leads to the charge transfer and modifies the valence state of gold clusters. Metallic gold,[30] cationic gold[31,32] or electron-rich gold particles[33] have been suggested to play an important role in the high activity for CO oxidation. For cationic gold, Au^+,[34] Au^{3+},[35] and $Au^{\delta+}$ [36] were considered to be responsible for the catalytic activity for CO oxidation. The debate about the electronic structure of active gold particles suggests that there is no absolute conclusion on the active state of gold nanoparticles.

The junction interface between gold nanoparticles and support has been regarded as providing reaction sites for CO oxidation. This was proposed first by Haruta, who postulated that the CO and O_2 molecules were adsorbed on the interface of Au and Fe_2O_3, and that the dissociation of O_2 occurred on the interface.[37] Then Hu and co-workers[38] studied CO oxidation on Au/TiO_2 using density functional theory (DFT) calculation; they identified that CO oxidation occurs at the interface between Au and the oxide with a very small barrier, and that O_2 adsorption at the interface is the key step in the CO oxidation reaction. More recently, Green *et al.*[39] reported that the CO oxidation over Au/TiO_2 catalyst takes place at the perimeter zone between Au and TiO_2.

Haruta[40] proposed the reaction mechanism of CO oxidation over Au/TiO_2 catalysts (Figure 6.1) in which CO is thought to be probably adsorbed on the gold particle edge or step site, and the oxygen is envisioned to adsorb on the support or the interface between gold particle and support.

Discussion of the reaction mechanism of CO oxidation focuses mainly on the oxygen species. The earlier reaction mechanism of CO oxidation proposed by Bond and Thompson[41] incorporated the essential features of cationic gold and the hydroxyl group (Figure 6.2). They mentioned the hydroxyl group as the active oxygen species, which were adsorbed on the cationic gold sites.[41]

Next, oxygen vacancies on the support were thought to play an important role in CO oxidation, but the oxygen vacancies may result in the formation of a negatively charged gold cluster. The electron-rich gold is expected to result in enhanced binding of molecular oxygen to the gold surface and the formation of further active oxygen species, such as oxygen molecular anion O_2^- or oxygen atomic anion O^-.[42] Another viewpoint is that oxygen attacks neighboring anion defects (F centers containing trapped electrons existed in reducible support), forming O_2^- species.[43]

Wang and Hammer[44] proposed reaction mechanisms of CO oxidation over Au/oxide catalysts. The first one is CO attacks O at the Au-oxide; CO_2 is

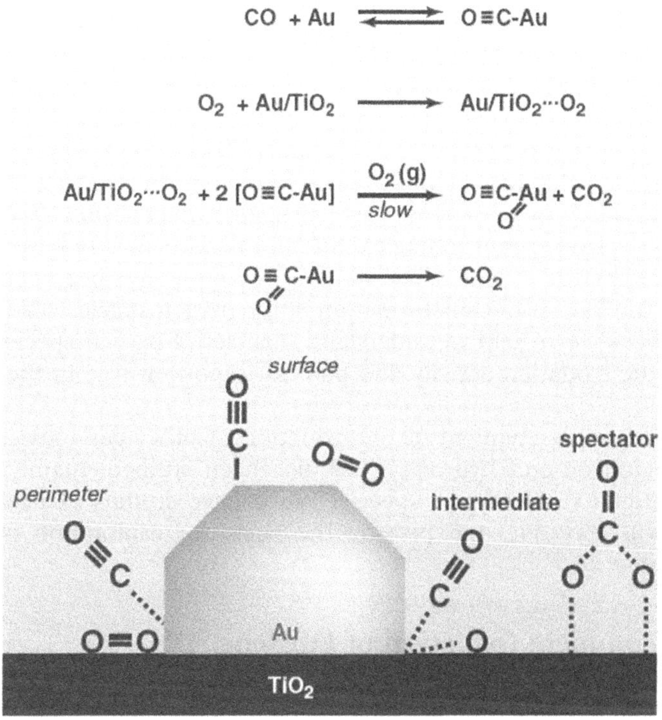

Figure 6.1 Schematic representation for CO oxidation pathways over Au/TiO_2.[40]

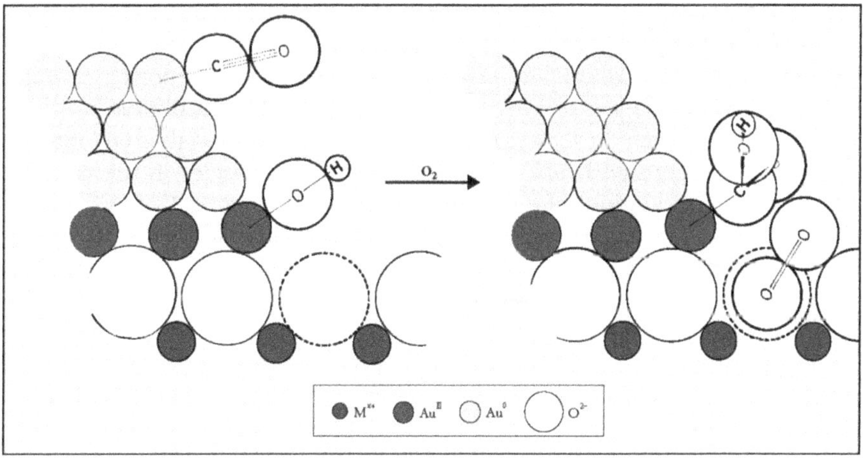

Figure 6.2 Schematic representation of the mechanism of CO oxidation.[41]

formed together with a vacancy in the Au-oxide. The second one is that O_2 adsorbs molecularly in the vacancy, CO attacks the O_2, and then CO_2 is formed.

Min and Friend[45] concluded that the rate of CO oxidation with atomic oxygen is limited by dissociation of O_2, but that the oxidation step is rapid; for CO oxidation with molecular oxygen, the oxidation step is rate limiting, but the oxygen adsorption is quick. Actually, in CO oxidation, both molecular and atomic oxygen exist in most cases.

Since the active oxygen species is thought to be an important species in CO oxidation, considerable attention has focused on the characterization and theoretical study of oxygen species. XPS, ultraviolet–visible (UV-vis) spectroscopy, electron spin resonance (ESR) and Raman spectra have been used to characterize the oxygen species and a DFT study has been performed to calculate the barrier of transformation of the oxygen species. The studies of oxygen species can help to understand how active oxygen species improve the catalytic oxidation activity and play an important role in the oxidation reactions.

The reaction mechanism of the oxidation of other gas molecules (such as formaldehyde and hydrogen) has also been studied, mainly based on the investigation of oxygen species. We believe studies of active oxygen species will continue and provide the basis for exploration of reaction mechanisms.

6.2.2 Complete Oxidation of Ethylene

Ethylene is one of the harmful volatile organic compounds (VOCs) that possess both C–C σ and C–C π bonds, enhancing photochemical pollution. In fruit stores, ethylene is released from fruits and then accelerates their premature spoilage. To maintain freshness, the efficient elimination of trace ethylene in the fruit storage environment is required. The complete oxidation of ethylene is a thermodynamically favored process, but this oxidation reaction is difficult to carry out unless using catalysts. In 2001, Park *et al.*[46] reported complete photocatalytic oxidation of ethylene over Pt/TiO_2, but the photocatalytic reactivity was low even under extra UV light. They concluded that OH radicals, O_2^- and O_3^- play significant roles as the key active species in the complete photocatalytic oxidation of ethylene. In 2006, Ahn *et al.*[47] reported that Au/Al_2O_3 catalyzed ethylene oxidation to CO_2 and H_2O above 100 °C. The activity is not sufficient to meet the demand of fruit freshness, because the fruits are usually stored in refrigerated warehouses at temperature near 0 °C.

Hao and co-workers[48] investigated different supports, including Fe_2O_3, Co_3O_4, TiO_2 and ZnO loaded Au catalysts, to catalyze the decomposition of ethylene to CO_2 and H_2O. They found that Au/Co_3O_4 prepared by the precipitation method was the most active, showing 7.4% ethylene conversion at 20 °C. The influence of ethylene concentration in the initial gas on the stability was investigated. It was found that the Au/Co_3O_4 catalyst exhibited a rapid deactivation trend with increasing time under condition of high ethylene concentration, and the deactivation rate of the catalyst slowed down when the concentration of ethylene fell to 5 parts per million (ppm) (Figure 6.3).

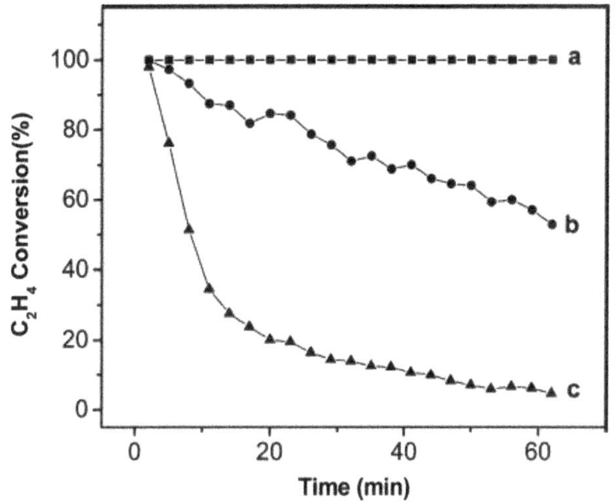

Figure 6.3 Reaction tests for ethylene (C_2H_4) conversion with time-on-stream over the Au/Co_3O_4 catalyst: (a) 5 ppm ethylene, 0 °C; (b) 50 ppm ethylene, 20 °C; and (c) 1050 ppm ethylene, 50 °C.[48]

Nanoporous Co_3O_4 and Au/Co_3O_4 materials with a three-dimensional (3D) structure were successfully synthesized by the hard templating method. Nanoporous Co_3O_4 and Au/Co_3O_4 catalysts with highly reactive {110} facets showed 30% and 76% conversion of ethylene (50 ppm in air) at 0 °C, respectively. The stable, highly dispersed and exposed gold nanoparticles in the more active catalyst abide in the pore wall of the nanoporous Co_3O_4 support, making the generation of surface active oxygen species easy for the ethylene oxidation. Both the C–C π and σ bonds of pollutant molecule of ethylene are broken on these catalysts with exceptional activities.[49]

Further research results revealed that Co_3O_4 with a different morphology supported gold catalyst showed different activities of ethylene oxidation. For supported Au/Co_3O_4 catalysts, the Co_3O_4 nanorods selectively exposing {110} facets were more active than Co_3O_4 nanopolyhedra and Co_3O_4 nanocubes exposing predominantly {011} and {001} facets, respectively.[50] Recently, 5% Au/MCM-41 was prepared by the impregnation method and the ethylene conversion over the catalyst was 50% at 25 °C, which is lower than that achieved on 5% Pt/MCM-41 prepared by the same method. In fact, supported gold catalysts prepared by impregnation method using $HAuCl_4$ aqueous solution as precursor without improvement usually contain large gold particles. The gold particle size of Au/MCM-41 was too large (35 nm calculated from X-ray diffraction data using the Scherrer equation) to achieve good activity. Thus, compared with Pt/MCM-41, the lower activity of ethylene oxidation over Au/MCM-41 may be influenced not only by active metal element, but also by the preparation method.[51]

The complete oxidation of ethylene at low temperature was achieved using the supported gold catalyst. However, the catalytic stabilities of ethylene oxidation over supported gold catalysts still need to be improved for the sake of their practical application. The reaction mechanism of ethylene oxidation in terms of the detail of the intermediate species has not been explored.

6.2.3 Complete Oxidation of Formaldehyde

Formaldehyde is a typical indoor air pollutant emitted from widely used decorative materials in airtight buildings. Nasopharyngeal or blood cancer could be triggered by prolonged exposure to formaldehyde with concentrations exceeding safe limitations (0.08 mg m^{-3} according to the Chinese standard).

Methods for the elimination of formaldehyde include adsorption, photocatalysis, and thermocatalysis. However, adsorption saturation is an existing problem which troubles the users. Compared with photocatalysis, thermocatalysis does not need extra light; an effective catalyst should be used to eliminate formaldehyde at room temperature.

Recently, some results were reported on the catalytic oxidation of formaldehyde over oxide-supported noble metal catalysts. Au/α-Fe_2O_3 exhibited catalytic activity comparable with that of Pd/γ-Al_2O_3 and Pt/γ-Al_2O_3, giving a formaldehyde conversion of 50% at 80 °C.[52] Zhu and co-workers[53] reported that 7.1% Au/Fe-O can catalyze the conversion of 20% HCHO to CO_2 at 20 °C, although the gold loading of this catalyst is higher. The Au/CeO_2 catalysts prepared by co-precipitation did not show any activity at room temperature.

Zhang *et al.*[54] continued to improve the CeO_2 support and synthesized three-dimensionally ordered macroporous Au/CeO_2 catalysts, which exhibited better catalytic activity with 10% formaldehyde conversion at room temperature. However, better activity for catalytic formaldehyde oxidation is still a desirable technical goal.

Hao and co-workers[55] reported that two-dimensional mesoporous Au/Co_3O_4-CeO_2 prepared by the hard template method oxidized 50% formaldehyde to CO_2 at 25 °C. This is an enhanced activity of formaldehyde oxidation over supported gold catalysts at room temperature. They also studied the reaction mechanism of formaldehyde oxidation over mesoporous Au/Co_3O_4-CeO_2. They proposed that only C–H bond activation is required, and that the gold nanoparticles could make the surface active oxygen desorb more easily and thus raise the activity of formaldehyde oxidation. A microreaction process for formaldehyde oxidation over mesoporous Au/Co_3O_4-CeO_2 is proposed in Figure 6.4.

6.3 Decomposition of Small Molecule Pollutants

6.3.1 Decomposition of Ozone

Ozone is an important component of the upper atmosphere that protects both humans and ecosystems from harmful UV radiation. However, ground-level ozone that exceeds the safety level (0.1 ppm as proposed by the

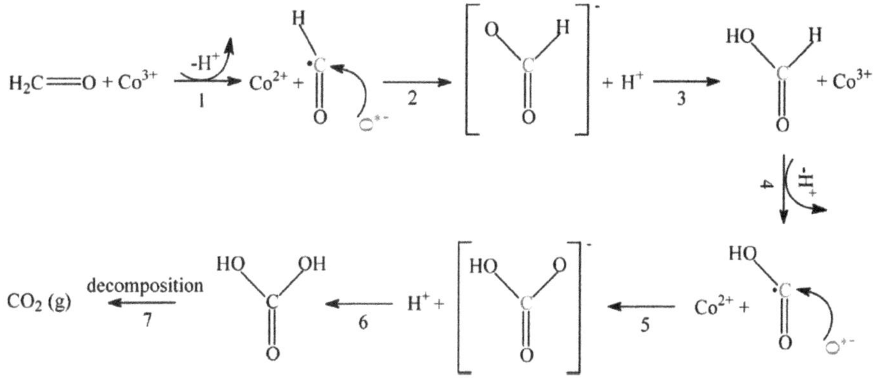

Figure 6.4 Microreaction process for formaldehyde oxidation over mesoporous Au/Co_3O_4-CeO_2.[55]

International Ozone Association) is harmful to health and enhances photochemical pollution.

Ozone is useful as a powerful oxidizing and disinfection agent, but its release into the environment should be avoided. Although the half-life of ozone is 16 hours, it is necessary to accelerate its decomposition. Hao *et al.*[56] reported that supported gold catalysts are superior to Ag and Ni catalysts in terms of both catalytic activity and stability for ozone decomposition; the ozone ($\sim$3000 ppm O_3/O_2 mixture gas) decomposition conversion over Au/Fe_2O_3 is up to 98% at 0 °C. They proposed that ozone adsorption can produce some active oxygen species (O_2^-) on the supported gold catalysts in ozone decomposition.

There are several patents focused on ozone decomposition. Axel *et al.*[57] invented a material whose surface has regions made of CeO_2 and regions made of MnO_2. This material can be used to reduce the ozone content. Takeshi *et al.*[58] provided a simple and compact-sized equipment for decomposing and discharging an ozone mixture.

Although ground-level ozone in air exceeds the standard level in some cities and regions, effective measures have been taken to reduce the ozone content only in some small and special environments.

6.3.2 Decomposition of Nitrous Oxide and Nitric Oxide

Nitrous oxide (N_2O) has been identified as a strong greenhouse effect gas with a global warming potential per molecule 300 times that of carbon dioxide. The direct catalytic decomposition of N_2O into N_2 and O_2 is an attractive solution to reduce N_2O.

A considerable amount of work on N_2O decomposition catalysts has concentrated on Fe/zeolite or transition metal oxides, but some research has focused on supported noble metal catalysts. Suo and co-workers[59] reported that 90% N_2O decomposition over Au/Co_3O_4 catalyst was achieved at a temperature of 265 °C, and the temperature was raised to 270 °C in the presence of excess O_2. Zhang and co-workers[60] reported that the temperature at which N_2O is completely decomposed over Au/SBA-15 is 370 °C, and that introduction of Pd to the Au/SBA-15 catalyst reduced the activity of N_2O decomposition. Limtrakul and co-workers[61] reported a density functional study of the activity of Au/ZSM-5 for N_2O decomposition. They found that the charge state of Au played an important role in N_2O decomposition and that ZSM-5 facilitates charge transfer between gold and N_2O, a key process for the reaction.

The reported activities of N_2O decomposition over supported gold catalysts are not lower than some reported activities over Fe/zeolite catalysts. However, there are no explicit data on the stabilities of N_2O decomposition over supported gold catalysts. The deactivation of supported gold catalysts at a relatively higher reaction temperature (about 350 °C) is easier. The surface atoms become significantly mobile above 350 °C, which is the Hütting temperature of gold (about 0.3 times the melting temperature).[62] The surface atoms shift and gather together, leading to the aggregation and enlargement of gold nanoparticles, which is deduced to cause the deactivation. Thus, the stabilities of gold catalysts should be investigated when supported gold catalysts are used in a high-temperature reaction.

Ichikawa and co-workers[63] reported the decomposition of nitric oxide (NO) over Au/NaY catalysts; the NO conversion to N_2 and to N_2O is 80.48% and 0.56% at 400 °C, respectively. Huang and co-workers[64] studied the adsorption and decomposition of NO on Au(997) and Au(110)-(1×2) surfaces. They revealed that the $(NO)_2$ dimer species, which bonded with the lowest-coordinated Au atoms *via* the N atom, is the active surface species for NO decomposition into O adatom and N_2O. Nieuwenhuys and co-workers[65] reported that NO decomposed to N_2O over Au(310) at temperatures as low as 80 K. They proposed that two mechanisms may lead to NO decomposition to N_2O:

(1) dimer mechanism:

$$NO_g \rightarrow NO_{ads}$$

$$2NO_{ads} \rightarrow N_2O_{ads} + O_{ads}$$

(2) *via* N–O bond breaking

$$NO_{ads} \rightarrow N_{ads} + O_{ads}$$

$$2N_{ads} \rightarrow N_{2g}$$

$$NO_{ads} + N_{ads} \rightarrow N_2O_{ads}$$

The above two reports (references 64 and 65) studied the model gold catalyst and helped to improve understanding of the low-temperature NO decomposition over nanogold catalysts.

6.4 Conclusions

Catalysis by gold is now a mature field and synthesis of supported nano-size gold catalysts is easy for experienced researchers. The complete oxidation or decomposition of small molecule pollutants over supported gold catalysts has been studied and the obtained low-temperature activities are significant for practical applications.

As a good model for theoretical studies and calculations, the reaction mechanism of CO oxidation over gold catalyst has also been investigated.

(1) Although different size of gold nanoparticles were investigated for catalysis, researchers agree that gold particles smaller than 5 nm are more active than those bigger than 5 nm.
(2) Selection of the support affects the catalytic performance of the supported gold catalyst for the same reaction. However, the preparation method influences the gold dispersion on the selected support, and then leads to different catalytic activity. Thus, when a substance is selected as support, the correct method should be chosen to prepare supported gold catalyst.
(3) The nature of the active sites for supported nanogold is a highly controversial topic and a number of key debates focused on the electronic nature of the active gold sites. Which valence of nanogold is the most active for the supported gold catalysts—metallic gold, cationic gold or anionic gold?
(4) The interface between gold particles and support was thought to provide the site for reaction. Synthesis of model gold catalysts with a specific structure and interface, and tests of their catalytic efficacy, will lead the way in a new era of theoretical studies of gold catalysts.

Except for CO oxidation, studies of the oxidation mechanism of ethylene and formaldehyde are possible and worthy. This will help us to understand the activation mechanism of simple pollutants over supported gold catalysts, because the changes of chemical bonds of these simple pollutant molecules are easy to analyze by modern characterization techniques. It is worth pointing out that we need to obtain more stabilized gold catalysts which can be used in the actual environment, such as for the elimination of CO, formaldehyde and ozone in air purification systems, and the elimination of ethylene in fruit storage warehouses. Catalytic technologies based on nanogold materials have the potential to significantly impact on environmental remediation. Nanogold catalysts could be used to control some emissions from different sources and to develop environmentally friendly technologies.

References

1. M. Haruta, T. Kobayashi, H. Sano and N. Yamada, *Chem. Lett.*, 1987, **16**, 405.

2. G. C. Bond and D. T. Thompson, *Catal. Rev. Sci. Eng.*, 1999, **41**, 319.
3. G. J. Hutchings, *Gold Bull.*, 2004, **37**, 3.
4. Q. Fu, W. Deng, H. Saltsburg and F. Flytzani-Stephanopoulos, *Appl. Catal. B: Environ.*, 2005, **56**, 3.
5. D. I. Enache, J. K. Edwards, P. Landon, B. Solsona-Espriu, A. F. Carley, A. A. Herzing, M. Watanabe, C. J. Kiely, D. W. Knight and G. J. Hutchings, *Science*, 2006, **311**, 362.
6. C. Y. Ma, W. J. Xue, J. J. Li, W. Xing and Z. P. Hao, *Green Chem.*, 2013, **15**, 1035.
7. H. Okatsu, N. Kinoshita, T. Akita, T. Ishida and M. Haruta, *Appl. Catal. A: Gen.*, 2009, **369**, 8.
8. T. Ishida, H. Watanabe, T. Bebeko, T. Akita and M. Haruta, *Appl. Catal. A: Gen.*, 2010, **377**, 42.
9. H. G. Boyen, G. Kästle, F. Weigl, B. Koslowski, G. Dietrich, P. Ziemann, J. P. Spatz, S. Rietmüller, T. Hartmann, M. Nöller, G. Smid, M. Garnier and P. Oelhafen, *Science*, 2002, **297**, 1533.
10. M. Valden, X. Lai and D. W. Goddman, *Science*, 1998, **281**, 1647.
11. S. Lee, C. Fan, T. Wu and S. L. Anderson, *J. Am. Chem. Soc.*, 2004, **126**, 5682.
12. A. Sanchez, S. Abbet, U. Heiz, W. D. Schneider, H. Häkkinen, R. N. Barnett and U. Landman, *J. Phys. Chem. A*, 1999, **103**, 9573.
13. S. A. C. Carabineiro, N. Bogdanchikova, M. Avalos-Borja, A. Pestryakov, P. B. Tavares and J. L. Figueiredo, *Nano Res.*, 2011, **4**, 180.
14. J. Huang, C. Xue, B. Wang, X. Guo and S. Wang, *React. Kinet. Mech. Cat.*, 2013, **108**, 403.
15. A. A. Herzing, C. J. Kiely, A. F. Carley, P. Landon and G. J. Hutchings, *Science*, 2008, **321**, 1331.
16. C. Galeano, R. Güttel, M. Paul, P. Arnal, A. H. Lu and F. Schüth, *Chem. Eur. J.*, 2011, **17**, 8434.
17. E. Castillejos, R. Chico, R. Bacsa, S. Coco, P. Espinet, M. Pérez-Cadenas, A. Guerrero-Ruiz, I. Rodríguez-Ramos and P. Serp, *Eur. J. Inorg. Chem.*, 2010, **32**, 5096.
18. J. C. Fierro-Gonzalea, B. G. Anderson, K. Ramesh, C. P. Vinod, J. W. Niemantaverdriet and B. C. Gates, *Catal. Lett.*, 2005, **101**, 265.
19. T. V. Choudhary and D. W. Goodman, *Top. Catal.*, 2002, **21**, 25.
20. M. Haruta, *Chem. Rec.*, 2003, **3**, 75.
21. N. Weiher, A. M. Beesley, N. Tsapatsaris, L. Delannoy, C. Louis, J. A. van Bokhoven and S. L. M. Schroeder, *J. Am. Chem. Soc.*, 2007, **129**, 2240.
22. C. Baata, N. Decker and U. Prüβe, *J. Catal.*, 2008, **258**, 165.
23. S. Hermes, M. K. Schröter, R. Schmid, L. Khodeir, M. Muhler, A. Tissler, R. W. Fischer and R. A. Fischer, *Angew. Chem., Int. Ed.*, 2005, **44**, 6237.
24. F. Moreau, G. C. Bond and A. O. Taylor, *J. Catal.*, 2005, **231**, 105.
25. Y. Sunagawa, K. Yamamoto, H. Takahashi and A. Muramatsu, *Catal. Today*, 2008, **132**, 81.
26. V. Zielasek, B. Jurgens, C. Schulz, J. Biener, M. M. Biener, A. V. Hamza and M. Bäumer, *Angew. Chem., Int. Ed.*, 2006, **45**, 8241.

27. C. Xu, J. Su, X. Xu, P. Liu, H. Zhao, F. Tian and Y. Ding, *J. Am. Chem. Soc.*, 2007, **129**, 42.
28. H. Yin, C. Zhou, C. Xu, P. Liu, X. Xu and Y. Ding, *J. Phys. Chem. C*, 2008, **112**, 9673.
29. M. Haruta, *ChemPhysChem*, 2007, **8**, 1911–1913.
30. F. Porta, L. Prati, M. Rossi and G. Scari, *J. Catal.*, 2002, **211**, 464.
31. G. J. Hutching, M. S. Hall, A. F. Carley, P. Landon, B. E. Solsona, C. J. Kiely, A. Herzing, M. Makkee, J. A. Moulijn, A. Overweg, J. C. Fierro-Gonzalez, J. Guzman and B. C. Gates, *J. Catal.*, 2006, **242**, 71.
32. V. Aguilar-Guerrero and B. C. Gates, *Chem. Commun.*, 2007, 3210.
33. M. S. Chen and D. W. Goodman, *Science*, 2004, **306**, 252.
34. J. Guzman and B. C. Gates, *Angew. Chem., Int. Ed.*, 2003, **42**, 690.
35. F. Wagner, S. Galvagno, C. Milone, A. Visco, L. Stievano and S. Calogero, *J. Chem. Soc. Faraday Trans.*, 1997, **93**, 3403.
36. M. Valden, X. Lai and D. Goodman, *Science*, 1998, **281**, 1647.
37. M. Haruta, *Catal. Today*, 1997, **36**, 153.
38. Z. P. Liu, X. Q. Gong, J. Kohanoff, C. Sanchez and P. Hu, *Phys. Rev. Lett.*, 2003, **91**, 266102.
39. X. Green, W. Tang, M. Neurock and J. T. Yates, *Science*, 2011, **333**, 736.
40. M. Haruta, *CATTECH*, 2002, **6**, 102.
41. G. C. Bond and D. T. Thompson, *Gold Bull.*, 2000, **33**, 41.
42. J. D. Stiehl, T. S. Kim, S. M. McClure and C. B. Mullins, *J. Am. Chem. Soc.*, 2004, **126**, 1606.
43. B. Yoon, H. Häkkinen, U. Landman, A. S. Wörz, J. M. Antonietti, S. Abbet, K. Judai and U. Heiz, *Science*, 2005, **307**, 403.
44. J. G. Wang and B. Hammer, *Top. Catal.*, 2007, **44**, 49.
45. B. K. Min and C. M. Friend, *Chem. Rev.*, 2007, **107**, 2709.
46. D. R. Park, B. J. Ahn, H. S. Park, H. Yamashita and M. Anpo, *Korean J. Chem. Eng.*, 2001, **18**, 930.
47. H. G. Ahn, B. M. Choi and D. J. Lee, *J. Nanosci. Nanotechnol.*, 2006, **6**, 3599.
48. J. J. Li, C. Y. Ma, X. Y. Xu, J. J. Yu, Z. P. Hao and S. Z. Qiao, *Environ. Sci. Technol.*, 2008, **42**, 8947.
49. C. Y. Ma, Z. Mu, J. J. Li, Y. G. Jin, J. Cheng, G. Q. Lu, Z. P. Hao and S. Z. Qiao, *J. Am. Chem. Soc.*, 2012, **132**, 2608.
50. W. J. Xue, Y. F. Wang, P. Li, Z. W. Liu, Z. P. Hao and C. Y. Ma, *Catal. Commun.*, 2011, **12**, 1265.
51. C. Jiang, K. Hara and A. Fukuoka, *Angew. Chem., Int. Ed.*, 2013, **52**, 6265.
52. C. Li, Y. Shen, M. Jia, S. Sheng, M. O. Adebajo and H. Zhu, *Catal. Commun.*, 2008, **9**, 355.
53. Y. Shen, X. Yang, Y. Wang, Y. Zhang, H. Zhu, L. Gao and M. Jia, *Appl. Catal. B-Environ.*, 2008, **79**, 142.
54. J. Zhang, Y. Jin, C. Li, Y. Shen, L. Han, Z. Hu, X. Di and Z. Liu, *Appl. Catal. B-Environ.*, 2009, **91**, 11.
55. C. Y. Ma, D. H. Wang, W. J. Xue, B. J. Dou, H. L. Wang and Z. P. Hao, *Environ. Sci. Technol.*, 2011, **45**, 3628.

56. Z. P. Hao, D. Y. Cheng, Y. Guo and Y. H. Liang, *Appl. Catal. B-Environ.*, 2001, **33**, 217.
57. A. Axel, D. Christian, K. Ursus and P. Uwe, Component having a catalytic surface, method for producing same and use of said component. *US Pat.*, PCT/EP2011/057949, 2013.
58. M. Takeshi and A. Tomonobu, Method for decomposing and discharging ozone mixture, and device for decomposing and discharging ozone mixture. JP 2010201462, 2012.
59. L. Yan, X. Zhang, T. Ren, H. Zhang, X. Wang and J. Suo, *Chem. Commun.*, 2002, 860.
60. W. Xing, X. F. Yang, A. Q. Wang, L. Li, X. Y. Liu, T. Zhang, C. Y. Mou and J. Li, *J. Phys. Chem. C*, 2012, **116**, 6222.
61. T. Maihom, S. Wannakao, B. Boekfa and J. Limtrakul, *Chem. Phys. Lett.*, 2013, **556**, 217.
62. C. N. Satterfield, *Heterogeneous Catalysis in Industrial Practice*, McGraw Hill, New York, 2nd edn, 1991.
63. T. M. Salama, T. Shido, R. Ohnishi and M. Ichikawa, *J. Chem. Soc., Chem. Commun.*, 1994, 2749.
64. Z. Wu, L. Xu, W. Zhang, Y. Ma, Q. Yuan, Y. Jin, J. Yang and W. Huang, *J. Catal.*, 2013, **304**, 112.
65. C. P. Vinod, J. W. Niemantsverdriet and B. E. Nieuwenhuys, *Appl. Catal. A: Gen.*, 2005, **291**, 93.

CHAPTER 7

Superficial Modification of Semiconductors with Gold Nanoparticles for Photocatalytic Applications

JUAN CARLOS DURÁN-ÁLVAREZ, RODOLFO ZANELLA* AND SOCORRO OROS-RUÍZ†

Centro de Ciencias Aplicadas y Desarrollo Tecnológico, Universidad Nacional Autónoma de México, Circuito Exterior S/N, Ciudad Universitaria, A. P. 70-186, Coyoacán, México D.F., 04510, Mexico
*Email: rodolfo.zanella@ccadet.unam.mx

7.1 Introduction

This chapter reviews recent work related to the surface modification of semiconductors with gold nanoparticles to cause plasmon-enhanced photocatalytic degradation of organic molecules, water splitting, removal of air pollutants, and enhancement of photocatalytic production of fine chemicals.

As is well known, photocatalysis can convert the energy of light into chemical energy.[1] Direct solar-to-chemical energy conversion has several advantages over solar-to-electric energy conversion. As an intermittent light source, there is no way of storing large amounts of electricity to be used

†Present Address: Departamento de Química, ECOCATAL, Universidad Autónoma Metropolitana-Iztapalapa, Avenida San Rafael Atlixco 186, 09340, México, D.F., Mexico.

RSC Catalysis Series No. 18
Heterogeneous Gold Catalysts and Catalysis
Edited by Zhen Ma and Sheng Dai

Published by the Royal Society of Chemistry, www.rsc.org

during the night.[2] The photocatalytic production of fuels provides an alternative method for storing the energy from the Sun in chemical bonds which can be released later without producing harmful byproducts.[2] Moreover, in photocatalytic processes, the energy of photons can be used to drive many other useful chemical reactions, including water purification and the production of valuable fine chemical products.

The efficiency of solar conversion in photocatalytic devices depends on four processes: light absorption; charge separation; charge migration; and charge recombination.[3–5] Semiconductors used in photocatalytic devices have a low solar energy conversion efficiency because of limitations in one or more of these four processes. For example, many commonly used semiconductors such as TiO_2, ZnO, SnO_2, and perovskites such as $La_2Ti_2O_7$ absorb only ultraviolet (UV) light[6–8] which accounts for about 4% of the total solar radiation reaching the Earth.[5] In contrast to the UV range, the visible region of the solar spectrum accounts for nearly 43% of the solar energy incident on the surface of the Earth.

Recently it has been shown that the superficial modification of semiconductors with plasmonic metal nanostructures, such as gold, enhances the photocatalytic activity toward the reactions mentioned above.[1,2,9–19] Surface plasmon resonance (SPR) improves the solar energy conversion efficiency by: (1) extending light absorption to longer wavelengths; (2) increasing light scattering; and (3) exciting electron–hole pairs in the semiconductor by transferring the plasmonic energy from the metal to the semiconductor.[8] The metal that modifies the semiconductor does not form part of the solid framework, but it is in a separate phase in interfacial contact with the semiconductor.[1] In order to obtain stable and durable materials, the metal used in this approach should be chemically inert, especially towards (photo) oxidation. For this reason, noble metals are most suitable for this purpose. Gold is one of the important noble metals and does not undergo corrosion in photocatalytic processes. Gold nanoparticles provide not only a convenient way to direct the flow of photo-induced charge carriers because of irradiation by UV light, which helps to hinder electron–hole pair recombination, but also to display a characteristic plasmon band in the visible region due to the collective excitation of electrons in gold nanoparticles.[1,2,10,20–23]

7.2 General Aspects

7.2.1 Photocatalysis

Many semiconductors, such as TiO_2, ZnO, CdS, ZnS, and Cu_2O can act as photocatalysts to carry out oxidation or reduction reactions.[2,18,24,25] When a photocatalyst absorbs a photon of energy equal to or greater than its bandgap, an electron may be promoted from the valence band to the conduction band leaving behind an electron vacancy or 'hole' in the valence band. If the charge separation is maintained, the electron and hole may migrate to the catalyst's surface where they participate in photocatalytic redox reactions with adsorbed species.[26–32] When positive holes oxidize water molecules or OH^- ions, highly reactive hydroxyl radicals are formed. Hydroxyl radicals are extremely aggressive species and can act as strong electrophiles towards unsaturated double bonds; for aromatics, they can abstract hydrogen from hydrocarbons to form water and can oxidize electron-rich compounds by removing a single electron from a substrate. On the other hand, the reaction of electrons with adsorbed oxygen molecules produces superoxide anions which, in the following steps, can be converted to hydrogen peroxide and also to hydroxyl radicals, both able to oxidize stable organic molecules to CO_2 and H_2O or fragments of lower molecular weight.[26,33,34] However, fast recombination of the photogenerated electrons and holes decreases the photocatalytic efficiency, limiting the applications of semiconductors in photocatalysis.[35,36]

7.2.2 Semiconductors Commonly used in Photocatalysis

One of the prerequisites of a solid to be a photocatalyst is stability and durability under irradiation.[1] Among photocatalysts, TiO_2 has gained much attention because of its superior photoreactivity, nontoxicity, long-term stability, relative abundance, low cost, chemical inertness, and environmentally friendly features.[37,38] Nonetheless, the photocatalytic efficiency of pure TiO_2 under solar light is still low, mainly because of the fast recombination of photogenerated electron–hole pairs and the inability to utilize visible light owing to its large bandgap (3.2 eV). Only UV light can be utilized to activate the photocatalytic activity of pure TiO_2. Since UV light only accounts for 4% of solar radiation energy while visible light contributes 43–45%, the inability to utilize visible light limits the efficiency of solar photocatalytic activities of pure TiO_2. These fundamental problems prevent TiO_2 from practical application.[39,40] Therefore, the development of effective and highly sensitive visible light-responsive TiO_2 photocatalytic systems is desirable.

Titania (TiO_2) has three most commonly encountered crystalline polymorphs: rutile (tetragonal, space group P42/mnm); anatase (tetragonal, space group I41/amd); and brookite (orthorhombic, space group Pbca). Among the three phases, rutile has the highest density and refractive index, and is thought to be the most thermally stable. Anatase is generally catalytically and photocatalytically more active than rutile. The brookite phase is the least studied because of the difficulties encountered in obtaining its pure form.[41,42]

Even if TiO_2 is the most widely used photocatalyst,[43–45] its limited activity under visible light irradiation has motivated the quest for modified titania materials absorbing visible light[1] (see Section 7.5). Hematite (α-Fe_2O_3) is another well-known photocatalyst,[46–48] with a smaller bandgap (2–2.3 eV); Fe_2O_3 can capture approximately 40% of the incident sunlight. Although Fe_2O_3 is highly stable over a wide range of pH environments, it typically requires the use of other reagents, such as H_2O_2, to initiate the photocatalytic evolution reaction.[2,46] ZnO is a wide-bandgap oxide semiconductor with a direct energy gap of about 3.3 eV and, as a consequence, it absorbs UV radiation because of the band-to-band transitions.[49] Cu_2O is a p-type semiconductor with a bandgap of 2.0–2.2 eV and has been used for solar energy conversion because of its absorption in the visible region.[50] It has also been used as a photocatalyst for the electrochemical splitting of water using visible light irradiation and the degradation of organic materials.[51,52] Tin dioxide (SnO_2) has been utilized as an important n-type semiconductor with a bandgap of 3.6 eV. Many metal oxide semiconductors such as ZrO_2,[53] SiO_2,[53] CeO_2,[54] Al_2O_3,[55] $KNbO_3$,[56] $BiVO_4$,[57,58] Bi_2O_3,[59,60] $AgNBO_3$,[61] $ZnWO_4$,[62] and WO_3[63,64] have been developed as photocatalysts with visible light activities. Among them, $BiVO_4$, with a bandgap of $E_g \approx 2.40$ eV, has been recognized as a potentially suitable visible light photocatalyst for pollutant decomposition and water oxidation.[65] Other compounds that have been used as photocatalysts include sulfides such as CdS and ZnS, and other compounds such as zeolite Y,[18,53] silver halides (*e.g.* AgCl,[66,67] AgBr,[68] AgI[69]), and graphene oxide.[70,71]

7.2.3 Optical Properties of Gold Nanoparticles

Metallic gold nanoparticles absorb light in the visible range because of SPR, which consists on the collective oscillation of electrons on the surface of gold nanoparticles, resulting in the that resonance with the electromagnetic field of the incident light. Usually, gold nanoparticles exhibit an absorption band in the visible region around 560 nm which is responsible for the different colors visually observed in gold colloids.[72] In fact, one of the old uses of gold nanoparticles was as pigment.[1]

The position and shape of the surface plasmon (SP) band depend on many factors including the dielectric constant of the medium, the particle size, the shape of the particles, and the Coulombic charge of the nanoparticle, among others.[1,73] The SP band is the most important characteristic property of gold nanoparticles in order to identify their shape and size. The shape or size of the gold nanoparticle changes the surface geometry that leads to a shift in the electric field density on the surface.[1,2] This causes a change in the oscillation frequency of the electrons, generating different cross-sections for the optical properties including absorption and scattering. Because the SPR is greatly influenced by the size of the nanoparticles, the efficiency of electron transfer from gold to TiO_2 may be tuned and optimized by altering the size of the gold nanoparticles.[74]

As previously mentioned, gold is a noble metal that has the advantage of not undergoing corrosion under photocatalytic conditions;[1] moreover, it can be strongly anchored to semiconductor surfaces, especially to TiO_2 semiconductor surfaces.[11,75–81] It has been evidenced that the electron transfer from gold to TiO_2 is viable under excitation of the plasmon band because of the strong interaction between the environment and oscillating electrons in gold nanoparticles.[82,83] As supported gold nanoparticles can absorb visible light *via* SPR, they can enhance photocatalytic activity; the SPR absorption may also cause rapid heating of the nanoparticles,[84] which can induce oxidation of organic compounds in air.[85] In addition, supported gold nanoparticles also show considerable UV light absorption *via* electron interband transitions from 5d to 6sp.[53,86]

7.2.4 Superficial Modification of Semiconductors with Gold Nanoparticles

As mentioned previously, one of the main problems of most of the semiconductors used in photocatalysis is the rapid recombination rate of the photogenerated electrons and holes before they migrate to the surface of the photocatalyst. In this regard, an effective way to facilitate charge separation involves the deposition of noble metal nanoparticles with a large work function (*e.g.* Au and Ag) onto metal oxide supports. These metal–semiconductor combinations have proven to be efficient in inhibiting electron–hole pair recombination with the metal particles acting as irreversible electron sinks, thus helping efficient charge separation and resulting in a higher photocatalytic reaction rate. Moreover, in several semiconductors such as TiO_2, it is necessary to enhance the visible light absorption. Among these metals, gold is a popular choice because of its high chemical stability and its ability to generate surface plasmons at the desired wavelength.[87–89]

The presence of gold nanoparticles on the surface of TiO_2 helps to avoid electron–hole pair recombination because this combination results in the formation of an efficient electron trap (Schottky junction) and improves the photocatalyst's response to illumination.[90–92] This is because, upon UV irradiation, the photogenerated electrons migrate to the Fermi energy level of the gold nanoparticles. At the same time, the formed Schottky barrier hinders the inverse migration and, hence, charge recombination is avoided. When a metal and a semiconductor are joined to form a metal–semiconductor interface, a significant redistribution of charge is expected to take place because of the overlap of wave functions from two sides.[93] Depositing gold nanoparticles on semiconductor photocatalysts was proposed to enhance the photocatalytic activity because of their different Fermi levels characterized by the work function of the metals and the band structure of the semiconductors.[13,58,94–97]

As shown in Figure 7.1, the Fermi level and electron-accepting states of noble metals are located at an energy level just below the conduction band (CB) of TiO_2.[98] Therefore, upon UV light irradiation, the e^- in the TiO_2 conduction band can be effectively transferred to the metal nanoparticles deposited on the surface, while the h^+ remains in the valence band (VB) of TiO_2, eventually forming surface reactive radicals and thus avoiding recombination.[99] It is worth mentioning that gold nanoparticles alone are also potential photocatalysts since they also can absorb UV light *via* interband

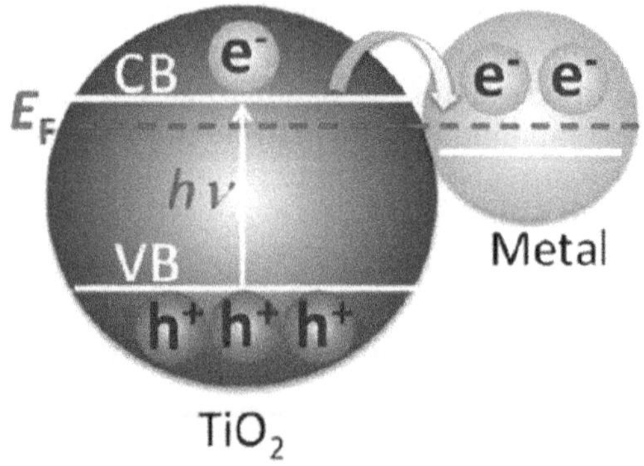

Figure 7.1 Schematic Illustration of Electron Transfer from TiO_2 to Noble Metal Nanoparticles Upon Irradiation of UV Light.
Reproduced with Permission.[99] Copyright 2012, American Chemical Society.

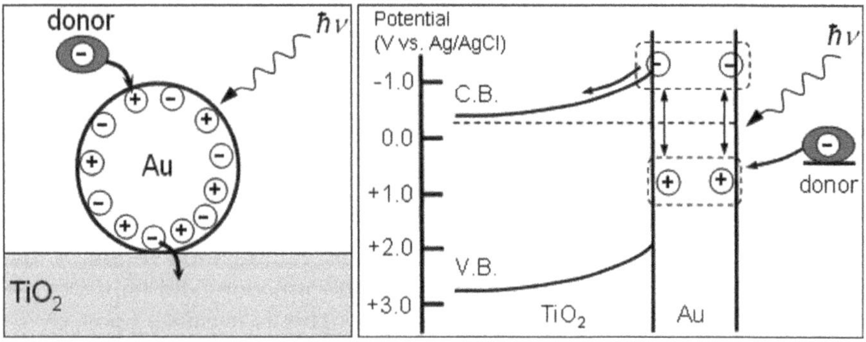

Figure 7.2 Schematic Illustration of the Proposed Charge Transfer Mechanism.
Reproduced with Permission.[106] Copyright 2005, American Chemical Society.

transitions.[96] Then, noble metal–semiconductor hybrids not only possess the catalytic properties of the noble metal, but also inherit the intrinsic photocatalytic capacity of TiO_2.[100–102]

In addition to the formation of electron traps, the SPR of gold nanoparticles is expected to enhance the absorption of incident photons, which will assist semiconductors to work as visible light active catalysts and enhance their photocatalytic efficiency.[25,74,103]

Two main mechanisms have been discussed in the literature regarding plasmonic enhancement of photocatalysis under visible illumination: charge transfer and local electric field enhancement. In the case of the charge transfer mechanism, it is proposed that the plasmon resonance excites electrons in Au, which are transferred to the conduction band of the adjacent TiO_2,[2] as shown schematically in Figure 7.2. Surface plasmons consist of a charge density wave on the surface of the metal. All the plasmonic charge resides at the Fermi energy of the metal and, therefore, it is not able to drive the reduction and oxidation half reactions, which are only driven for the semiconductors.[2] In addition, irradiating metal nanoparticles near their plasmon resonance frequency can generate intense local electric fields near the surface of the nanoparticles. It has been shown that the electric field intensity of local plasmonic 'hot spots' can reach as much as a thousand times that of the incident electric field.[2,104,105] In these 'hot spot' regions, the electron–hole pair generation rate is a thousand times higher than that of the incident electromagnetic field.[2] Furthermore, because this field is confined within a few nanometers of the TiO_2 surface, most of the plasmon-induced electron–hole pairs diffuse to the photocatalytic surface and contribute to the catalytic process.[2]

Additionally, it has been proposed that as the Fermi energy of the metal particles increases with decreasing size (due to a quantum size effect), gold particles with an appropriate size can possess an energy level between that of the conduction band of titania and that of the adsorbed oxygen.[107] Thus, the photoelectrons can be captured only by the gold nanoparticles and subsequently transferred to the adsorbed oxygen, thus leading to an effective separation of charges, as well as an increase of the photocatalytic activity of titania. When the size of the gold particles becomes too large, their Fermi energy will be lower than that of the adsorbed oxygen and, as a consequence, the photoelectrons cannot be transferred to the adsorbed oxygen. Also, when the size of gold particles is too small, the photoelectrons cannot be transferred from the bottom of the titania conduction band to the gold particles because the Fermi energy of gold particles is higher than that observed in the conduction band of titania. Therefore, it has been concluded that only gold nanoparticles with an appropriate size are effective for the enhancement of the photocatalytic activity of titania.[107]

7.3 Preparation of Semiconductors Modified with Gold Nanoparticles

It has been shown that parameters such as particle size, particle shape, dispersion (percentage of support surface covered by gold), and loading (percentage of gold weight on the material) also play an important role in the photocatalytic activity of this type of gold/semiconductor photocatalysts. The method used for the deposition of gold nanoparticles on the semiconductor surface can have an important impact on these parameters, thus affecting the photocatalytic activity. Several methods have been developed to synthesize gold nanoparticles on metal oxides. Here we review only the most important ones, as it is not the aim of this chapter to furnish a deep review about preparation methods. For more details readers can refer to excellent works on this topic.[108–110]

Gold–metal oxide systems have been prepared by both physical and chemical routes, including the radio frequency (RF) sputtering approach,[107,111–113] physical vapor deposition,[43,114,115] sol-gel dip coating,[116,117] deposition–precipitation,[75–77,118,119] assembling of pre-synthesized colloidal gold nanoparticles formed by chemical reduction onto the surface of the metal oxide,[44,55,86,120] photoreduction method,[121] ion exchange method,[66,75,76,122] impregnation method,[67,75,83,123,124] hydrothermal method,[102,125–130] sonochemical synthesis,[131,132] and encapsulation method.[9,46–48,133–135]

The most commonly used gold precursor is $AuCl_4^-$. If the metal oxide is porous, the gold precursor is first transferred into the pores in advance through adsorption, ion exchange, or ultrasonication. The gold precursors are then reduced *in situ* with UV irradiation,[136] chemical methods,[137–139] electrochemical deposition,[140] or heating the material in the presence of a reducing gas flow.[11,75–77,141] Then the reduced gold nanoparticles are incorporated into pore channels or on the external surface of the metal oxide. Usually it is possible to control the morphology, size, and distribution of deposited gold nanoparticles, but sometimes this becomes a difficult task because of the influence of oxide supports.[142]

The most used methods for preparation of plasmonic photocatalysts are described below.

7.3.1 Adsorption of Preformed Gold Nanoparticles onto Semiconductors

Gold nanoparticles are synthesized *via* the reduction of Au(I) or Au(III) precursors in aqueous or organic media in the presence of surface stabilizers.[108,109] Appropriate selection of synthetic method can provide control of various factors including particle size, shape, and size distribution. Reduction of gold salts is a simple process that only requires mixing of reactants under well-controlled external conditions, which can affect the final morphology of the particles.[75–77,109,118,119]

Apart from the strength of the reductant, the action of a stabilizer is critical in liquid phase synthesis. The reaction temperature is one of the main factors in determining particle size because the oxidation potential and related kinetics of the reductant are normally dependent on the temperature.

The representative reduction agents are citrate ions, $NaBH_4$, tetrakis(hydroxymethyl) phosphonium chloride (THPC), cetyltrimethylammonium bromide (CTAB),[143] citrate ions, poly(vinylpyrrolidone) (PVP), poly(vinylalcohol) PVA, among others.[109] Usually, deposition of gold colloids on the semiconductor is performed by mere immersion of the semiconductor into the colloidal solution, followed by washing with water and drying. This method has been used to prepare gold photocatalysts supported on TiO_2,[74,144–152] WO_3,[153] CeO_2,[154] Cu_2O,[25] SnO_2,[155] ZnO,[20,156] $BiVO_4$,[17] $SiO_2@TiO_2$,[157] and $SiO_2@Cu_2O$.[5]

7.3.2 Photodeposition

Photodeposition is based on the principle that metal cations with appropriate redox potentials can be reduced by the photoelectrons created by bandgap illumination of semiconductors such as TiO_2, ZnO, and WO_3. UV irradiation of de-aerated solutions containing $HAuCl_4$ and TiO_2 or ZnO led to both the deposition of gold and its reduction.[109] The average gold particle sizes obtained by this method are typically in the range of 5–15 nm.[108,109,121]

Usually it is much easier to control the size and monodispersity of gold nanoparticles in colloidal solutions than *in situ* reduction on TiO_2 film or powders.[158] This method has been used to prepare gold photocatalysts supported on TiO_2[103,159–161] and CeO_2.[19]

7.3.3 Deposition–Precipitation

The deposition–precipitation method involves the gradual and homogeneous basification of a suspension of metal oxide in a solution containing the metal precursor (usually $HAuCl_4$).[75–81] The aim of that procedure is to perform the interaction and precipitation of gold species on the surface of the metal oxide and not in the solution. This method is certainly the most used for the preparation of gold catalysts and photocatalysts since it readily leads to the formation of small gold particles (2–3 nm).

The method was first proposed by Haruta and co-workers[118,119,162] using NaOH or Na_2CO_3 or NH_3 to adjust the pH of the solution containing both $HAuCl_4$ and the oxide support to 7–8; after agitation for some time, the catalyst is washed with water, dried, and calcined in air or reduced in hydrogen. The gradual rise in pH required in the deposition–precipitation procedure can be obtained by the addition of urea [$CO(NH_2)_2$], which at temperatures higher than 60 °C in the presence of water, decomposes and gradually increases the pH.[75,76]

$$CO(NH_2)_2 + 3H_2O \rightarrow CO_2 + 2NH_4^+ + 2OH^-$$

An extensive study performed by Zanella and co-workers[75–81] showed that the small particles ($\sim$2 nm) could be obtained on various metal oxides. Figure 7.3 shows typical transmission electron microscopy (TEM) images of Au/TiO_2 prepared by deposition–precipitation with urea.

In contrast to deposition–precipitation at fixed pH (Haruta method), in the deposition–precipitation method using urea, all the gold of the solution can be deposited onto the metal oxide within the first hour of preparation. This method has been used to prepare gold photocatalysts supported on TiO_2,[10,11,14,16,107,141,163–171] ZnO,[172] and $KNbO_3$.[173]

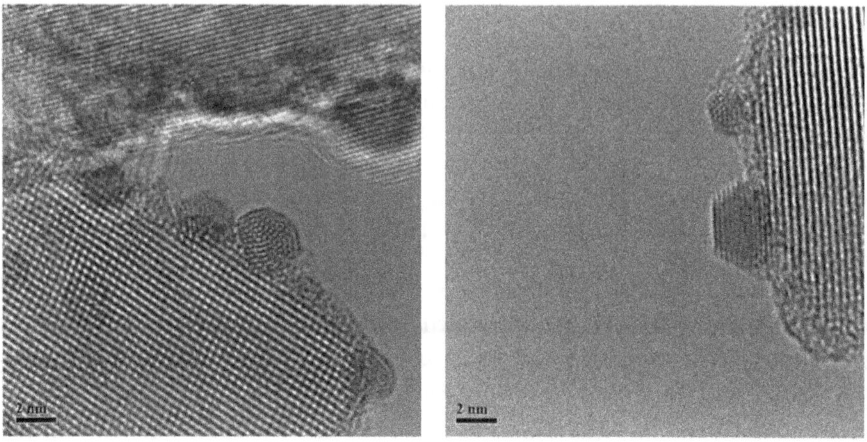

Figure 7.3 Typical High Resolution TEM Images of Au/TiO_2 Prepared by Deposition–Precipitation with Urea.

7.3.4 Impregnation and Ion Adsorption

Impregnation is the simplest method of preparation of metal/metal oxide catalysts. It consists of wetting the support with aqueous solution containing the gold precursor, usually $HAuCl_4$ or $AuCl_3$. Afterwards, the sample is dried. Then, a thermal treatment is performed to reduce the precursor to metallic particles. The drawback of this method is that the chlorides of the precursor are also present on the metal oxide and induce gold particle sintering during thermal treatment as well as the formation of large particles (>10–20 nm).[75] This method has been used to prepare gold photocatalysts supported on TiO_2[12] and WO_3.[28]

In the so-called anion adsorption method, the deposition of gold is performed by specific interactions between the gold precursor and the oxide support, so it is necessary to wash the sample after deposition and remove the chloride without leaching gold.[75,76,103,108,109] The preparation is based on the principle that the hydroxyl groups present on the oxide surface may be protonated or deprotonated on the solution pH in which the oxide is immersed. Adsorption of gold anions (as $AuCl_4^-$) on oxide supports is possible only in the case the pH of the solution is lower than the point of zero charge (PZC) of the metal oxide. If the pH of the solution is higher than the PZC, gold cations such as $Au(en)_2^{3+}$ can be used to synthesize Au/metal oxide photocatalysts.[75,76,174] After ion adsorption, the sample is usually washed with water, dried under vacuum at temperatures between 25 and 100 °C, and then thermally treated to reduce the precursor into metallic particles.[174]

There is a myriad of other techniques to deposit gold nanoparticles on metal oxides; for more information about their characteristics, procedures, principles, and results, the reader can refer to relevant references.[1,18,108,109]

7.4 Use of Semiconductors Modified with Gold Nanoparticles for Photocatalytic Removal of Water Pollutants

Heterogeneous photocatalysis using gold-modified semiconductors is an emerging technique for the removal of organic and inorganic pollutants in water. A large number of studies may be found in the literature aimed at testing the degradation of pollutants in water using this approach. Most studies testing the dissipation of pollutants in water are carried out to test the photocatalytic activity of a given gold-modified semiconductor. Such degradation tests are regularly performed in pure water, and therefore further investigations should focus on the degradation of contaminants in natural waters as well as in reclaimed wastewaters (*i.e.* municipal or industrial treated wastewater).

One of the most promising potential advantages of gold-modified semiconductors for heterogeneous photocatalysis is in obtaining not only a high level of degradation, but also the complete mineralization of pollutants (conversion to CO_2 and H_2O). Such degradation and mineralization rates may be higher than those obtained using non-modified semiconductors, as reported by several authors.[18,54,121,141,160,175–184] A high level of mineralization is the most desirable when pollutants are recalcitrant or have the potential to degrade into more dangerous substances. Examples of such pollutants are polyphenols (*e.g.* triclosan), which may produce dioxins upon photolysis,[185] or antibiotic agents that may retain their antibacterial properties after photocatalysis using non-modified semiconductors (*e.g.* ciprofloxacin).[186] In addition to organic substances, photocatalysis using gold-modified semiconductors is used to reduce ions in natural waters from harmful to harmless forms; for instance, Cr^{6+} to Cr^{3+} and NO_3 to N_2.[187–192] Photocatalytic reduction of inorganic pollutants occurs more efficiently when oxidizable species are used as scavenger agents for the photo-produced holes, resulting in improved charge separation.[191,192] Small carboxylic acids (*e.g.* oxalic, formic, and malonic acids) provide a trap for photo-produced holes upon UV or visible light irradiation, facilitating the persistence of electrons in the conduction band of the semiconductor.

When implemented, gold-modified semiconductor-based photocatalysis is capable of remediating water polluted with both heavy metals and dissolved organic matter. In such cases, the evaluation of the catalysts is based on two aspects: (1) the change in concentration of the pollutants in the aqueous phase as well as CO_2 production due to mineralization; and (2) the potential of the catalyst to be recycled through several stages of photocatalysis. On the basis of the aforementioned characteristics, the best materials for photocatalysis are those that on the one hand can transform pollutants in the shortest time possible, generating the fewest degradation byproducts, and on the other hand, can be re-used many times without losing catalytic activity.

Currently, a large number of gold-modified semiconductors are used for the photocatalytic removal of organic and inorganic contaminants in water; however the latter is used to a considerably lesser extent. Examples include WO_3, Fe_2O_3, CeO_2, and Bi_2O_3, although TiO_2 is the semiconductor used in most cases due to its ease of handling, low cost and low toxicity as well as its resistance to corrosion. Anatase is the TiO_2 crystallographic phase most commonly used to deposit gold nanoparticles due to its high photoactivity. Other phases such as rutile, brookite and even the amorphous form are also used for gold deposition and further degradation of pollutants and inactivation of bacteria in water.[103,149,184,193–195] When used for bacterial inactivation, silver nanoparticles have proven to be more active than gold nanoparticles, which is attributable to the inherent disinfecting properties of silver.[196]

The aim of modifying the semiconductor surface by deposition of gold nanoparticles is to exploit visible light for photodegradation and mineralization of dissolved pollutants. A large number of semiconductors display photoactivity to a greater extent in the UV region compared with visible light, either because of their wide bandgap or due to the increased potential for recombination of electron–hole pairs; *e.g.* TiO_2, ZnO, SiO_2, $BiVO_4$, and Fe_2O_3. The use of UV irradiation in photocatalysis causes water treatment costs to increase significantly, making it unaffordable for sites in developing countries. Therefore it is necessary to move photocatalytic degradation towards the visible region of the spectrum and ultimately to sunlight.

Reduction and oxidation reactions by superoxide and hydroxyl radicals formed by the splitting of water molecules may be an important pathway in the degradation of dissolved organic pollutants.[1,197] SPR is expressed in the visible light spectrum between 500 and 550 nm,[1,5,73,147,198,199] which implies that the full solar spectrum can be used for driving reactions with gold-modified semiconductors. As mentioned previously, in addition to the SPR effect, gold nanoparticles act as sinks for excited electrons in the conduction band, thereby preventing the recombination of the photo-produced electron-hole pair.[73] Metals displaying a higher work function than that of the semiconductor, such as Au, Ag and Cu (5.1, 4.6 and 4.9 eV compared with 4.2 eV for TiO_2) are effective as electron sinks.[200] Excited electrons can occur in gold nanoparticles due to the SPR effect; which may result in the degradation of pollutants in the surface of such nanoparticles. This occurs *via* the following mechanism: (1) the incident photons are absorbed by gold nanoparticles through the SPR effect; (2) the excited electrons may be injected from gold nanoparticles into the conduction band of the semiconductor, producing holes within the gold nanoparticles; and (3) electron deficient gold nanoparticles are able to oxidize organics in order to achieve the original metallic state.[201] The overall process occurring in the photocatalysis of water pollutants using gold-modified semiconductors is shown in Figure 7.4.

One of the most important factors affecting the photocatalytic degradation of dissolved pollutants is the dispersion and size of the gold nanoparticles deposited on the surface of the semiconductor.[149,160,199] Large loads and size

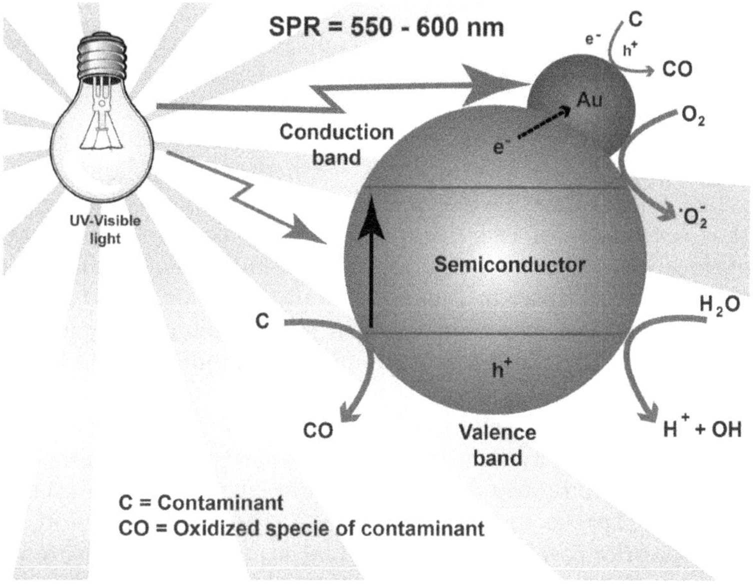

Figure 7.4 General Mechanism Proposed for the Degradation of Dissolved Pollutants in Water by Photocatalysis Using Gold-Modified Semiconductors.

of gold nanoparticles can result in a drop in photodegradation efficiency, while a uniform distribution of nanoparticles results in increased degradation of the compound.[160,161,163,202,203] Large loads of gold nanoparticles on the semiconductor surface lead to an increase in the size of the metal nanoparticles, and thus to a smaller separation of the electron–hole pair, as well as to the formation of recombination centers.[1]

The specific surface area of the catalyst has a significant effect on the photocatalytic conversion of pollutants in water, since this property is related to the adsorption of the compound onto both the semiconductor and gold nanoparticles, which in turn results in greater exchange of electrons from the catalyst to the compound. Deposition of gold nanoparticles on the surface of the semiconductor increases its specific surface area and thus the efficiency of photocatalysis.[1]

The ionization state of dissolved contaminants should also be taken into account, since it may determine the extent of adsorption onto the gold-modified semiconductor; for instance, ionized carboxylic acids (carboxylates) have been shown to be adsorbed onto Au/TiO_2 to a greater extent than their protonated forms.[204] Even though gold deposition on the semiconductor surface has been proven to increase the adsorption of organic molecules, when gold loads are significantly high the opposite effect is observed.[1,205–209] The photocatalytic efficiency of the gold-modified semiconductor can be tuned by varying the shape and size of both semiconductor and gold nanoparticles; thus a variety of shapes of both elements such as nanocups,[210] nanotubes,[161,211–213] aerogels,[149] urchin-like forms,[202] spheres,[129,135,214,215] nanorods,[144,150,216–218] nanodumbells,[219,220] flower-like structures[221] as well as nanoparticles coupled to carbon nanotubes,[100] in addition to many others, have been tested in the photocatalytic degradation of pollutants in water.

One important factor to be considered when photocatalysis is used for removing contaminants from water is the nature of the liquid matrix. In wastewater, for instance, a wide variety of dissolved compounds, which may be either scavengers of the electron–hole pairs or sacrificial agents preventing recombination, can be found.[222–224] Moreover, dissolved organic matter in water may react by sequestering the photo-produced electrons in both semiconductors and gold nanoparticles, producing excited species capable of degrading the target contaminants, although mineralization is not achieved.[223,225]

Dissolved organic matter in water can be a serious impediment for photocatalysis of target pollutants since it may: (1) act as a screening agent for irradiation; (2) adsorb onto the semiconductor surface; and/or (3) scavenge the photoactive species in the water.[222,225,226] The implementation of water pretreatment prior to photocatalysis is crucial, not only to ensure optimal performance of photocatalysis but also to obtain acceptable catalyst regeneration and thereby promote recycling. So far, few studies have been carried out to investigate the performance of gold-modified semiconductors used to degrade pollutants in natural waters;[187,227] moreover, no work aimed at treating wastewater using these materials was found in the present review. Table 7.1 shows a summary of the research studies to date using gold-modified semiconductors for the removal of pollutants in water.

In most studies testing the photocatalytic efficiency of gold-modified semiconductors, the catalyst is added as a powder and is commonly applied at concentrations ranging from 0.5 to 2 g L^{-1}.[135,202,205,213,228,229] The use of powdered catalyst for purifying water introduces the problem of its removal and management as a residue after water treatment, for which different filtering techniques are used. Due to the small size of the catalyst particles, separation techniques such as nanofiltration are commonly used for removing the sludge produced after water treatment. Recently, nanomaterial residues have been classified as water pollutants since they may be able to generate harmful effects in exposed organisms, such as oxidative stress and pathological changes in aquatic species due to the facilitated entry of metals into cells.[230]

In order to replace powdered catalyst in heterogeneous catalytic processes, and thus avoid the need for filtering and the risk of future contamination, some studies have aimed at removing contaminants from water using gold-modified semiconductors fixed to a surface, *e.g.* thin films or the semiconductor attached to the reactor wall. Some studies report the use of gold-modified semiconductor thin films for removing dissolved pollutants using UV or UV–visible light.[12,116,211,231–237] Arabatzis *et al.*[236] studied the photocatalytic performance of non-crystalline rough thin films of titania surface-modified with gold nanoparticles. Deposition of gold nanoparticles

Table 7.1 Some Studies Using Gold Modified TiO_2 Catalyst for Degradation of Pollutants in Water.

Pollutant	Catalyst	Preparation Method	Characteristics of Photocatalytic Experiments	Performance of Photocatalyst	Photocatalytic Activity Comparison Experiments	Enhancement of Photocatalytic Activity Using the Gold-Modified Material	Ref.
Phenol	Au/TiO_2	Photodeposition	Visible light (300 nm cutoff) irradiation of 200 μM phenol solution. 0.02 g of 1 wt.% Au/TiO_2 used	90% removal after 250 min of irradiation 25% mineralization was determined.	Unmodified TiO_2	1.2 upon deposition of gold at 0.5 wt.%	245
Phenol	Gold on mesoporous TiO_2	Impregnation	UV light irradiation (365 nm) applied for 3 h to O_2 saturated pollutants in solution 5–15 nm gold nanoparticles were loaded onto TiO_2 surface at 0.1–5.0 wt.%	Removals from 15 to 80% when gold load decreased from 5.0 to 0.5 wt.%	Unmodified TiO_2	3 and 4 for gold loads of 0.1 and 0.5 wt.%, respectively 2 and 1.5 for gold loads of 1 and 2 wt.%, respectively 1.5 for gold load of 5 wt.%	205
Chromium				Chromium reduction from 60 to 95% when gold load decreased from 5.0 to 0.5 wt.%		1.4 and 1.5 for gold loads of 0.1 and 0.5 wt.%, respectively 1.3 and 1.2 for gold loads of 1 and 2 wt.%, respectively 1.1 for gold load of 5 wt.%	
Methyl orange	Au/TiO_2	Impregnation	UV light (254 nm) irradiated to the 10 mg L^{-1} methyl orange air saturated solution	100% after 1 h of irradiation	Unmodified (as synthesized and P25) TiO_2	9 and 5 for as synthesized and P25 TiO_2, respectively	246
Rhodamine B	Gold supported in TiO_2 nanotubes	Photodeposition	Simulated sunlight irradiation applied to rhodamine B solution using a 36 W Xe lamp Gold loads of 0.1, 0.4, 0.8, and 1.6 wt.% were tested	Removals of 75% after 4 h of irradiation.	Unmodified TiO_2	1.1	213

Table 7.1 *(Continued)*

Pollutant	Catalyst	Preparation Method	Characteristics of Photocatalytic Experiments	Performance of Photocatalyst	Photocatalytic Activity Comparison Experiments	Enhancement of Photocatalytic Activity Using the Gold-Modified Material	Ref.
Methyl orange	TiO_2 nanotubes co-loaded with Au and reduced graphene nanoparticles (Au/RGO/TiO_2)	Electrochemical deposition	Simulated solar irradiation (280–2000 nm) supplied to 20 mg L^{-1} methylene orange solution using a 500 W Xe lamp	100% after 140 min irradiation	Unmodified TiO_2 nanotubes	2.5	212
					RGO/TiO_2	1.6	
					Au/TiO_2	1.25	
Methylene blue	Gold supported in TiO_2 thin films	RF reactive magnetron co-sputtering method	Irradiation of a 2 mg L^{-1} methylene blue solution was conducted with UV and visible light using 6 W and 500 W halogen lamp, respectively	80% of degradation efficiency after 200 min of UV irradiation and 65% of degradation after 200 min of visible light irradiation	Unmodified TiO_2	6 and 4.3 for UV and visible light irradiation, respectively	200
					Ag/TiO_2	1.6 for both UV and visible light irradiation	
					Cu/TiO_2	− 1.5 for UV light irradiation (degradation using Cu/TiO_2 was higher than that using Au/TiO_2) 1.2 for visible light irradiation	
Methyl blue	Urchin-like TiO_2 decorated with gold nanoparticles	Wet chemical process	UV–visible light irradiation (centered at 365, 405, 436, 547, and 578 nm) was applied to 40 mg L^{-1} methyl blue solution using a 20 W low pressure Hg lamp. Au/TiO_2 catalyst was used at 0.33 g L^{-1}	70% of disappearance after 180 min of irradiation	Unmodified as synthesized TiO_2	1.75	202
					P25 TiO_2	2.3	
					Ag/TiO_2	− 1.14	

Methyl orange	TiO_2/ZnO nanofibers decorated with gold nanoparticles	Wet chemical process	UV irradiation (313 nm) supplied using a 50 W high pressure Hg lamp to the solution containing 10 mg L^{-1} of both compounds Catalyst was used at 1 g L^{-1}	Near 100% of disappearance after 30 h of irradiation	Unmodified TiO_2 and ZnO nanofibers	2	247
4-Chlorophenol				Nearly 100% removal after 40 h of irradiation	TiO_2/ZnO nanofibers	1.2	
Rhodamine B	Gold deposited in TiO_2 nanosheets with a high percentage of exposed facets	Sulfide mediated polyol process	Visible light (cutoff of 420 nm) supplied to 20 μM rhodamine B solution using a 300 W Xe lamp. Catalyst concentration was 1 g L^{-1}	100% degradation after 80 min of irradiation using a gold load of 1.0 wt.% Lower degradation was observed using gold loads of 0.5 and 2.0 wt.%	Unmodified TiO_2 N-doped TiO_2	2.2 2.5	209
Salicylic acid	Au/TiO_2 gold nanoparticles of various shapes namely small, medium and large sized nanospheres (SSN, MSN, LSN), aggregated nanospheres (ANS), photodeposited particles (PAP) and nanorods (NR)	Photodeposition	UV light (254 nm) applied to 1 mM salicylic acid solution using a 125 W Hg arc lamp. Catalyst added at 10 g L^{-1}	Degradation after 60 min of irradiation of 85% using SNS, 75% for ANS, LNS and PAP, and 70% for NR	Unmodified TiO_2	From 1.4 to 1.7	151

Table 7.1 (*Continued*)

Pollutant	Catalyst	Preparation Method	Characteristics of Photocatalytic Experiments	Performance of Photocatalyst	Photocatalytic Activity Comparison Experiments	Enhancement of Photocatalytic Activity Using the Gold-Modified Material	Ref.
Methylene blue	Gold clusters coated with phenylethyl mercaptan on TiO_2 anatase thin films	Electron beam sputtering of preformed gold nanoclusters Coverage of gold clusters on TiO_2 was 3.8, 15.6, 41.9, and 100%	UV light irradiation was carried out upon a 0.1 mM methylene blue solution	$k = 0.076\ h^{-1}$ was obtained for phenylethyl mercaptan modified Au/TiO_2 at 41.9 wt.% Lower values were obtained for other treatments	Unmodified TiO_2 thin films.	From 1.05 to 1.8 for different treatments tested	244
Methyl orange	Gold nanoparticles supported on butterfly wing reticular hierarchical structured TiO_2	Biomorphed Au/TiO_2 was synthesized using butterflies as a template Gold nanoparticles deposited by the deposition–precipitation method	Visible light irradiation (cutoff at 400 nm) supplied to 10 μM methyl orange solution using a 1000 W Xe lamp Catalyst used at 1 g L^{-1} and gold load was 8 wt.%	100% of disappearance after 80 min of visible light irradiation	Unmodified control TiO_2 Unmodified biomorphic TiO_2	2.5 1.8	229
Acid orange 7	Gold nanotubes embedded into pores of anodic TiO_2 nanotubes	Electrodeposition	Simultaneous photocatalytic removal of Cr^{6+} (10 mg L^{-1}) and AO7 (20 mg L^{-1}) carried out by visible light irradiation using a 500 W Xe lamp	100% after 2 h of irradiation at pH value of 2.5 Highest removals observed when both compounds were present	Unmodified TiO_2	3.3	190
Chromium				90% under the same conditions		9	

Methyl orange	Au/TiO_2 nanocomposites	Photodeposition	20 mg L^{-1} methyl orange solution irradiated with visible light Catalyst used at 0.04 g L^{-1}	87% after 24 h of visible light (590 nm) irradiation	Unmodified TiO_2	14.5	248
1,4-dioxane	Au/TiO_2	Purchased	Simulated sunlight was used to irradiate 500 mg L^{-1} 1,4-dioxane solution using a 2000 W Xe lamp Three different concentrations of dissolved oxygen (DO) tested: 0.1, 8, and 35 mg L^{-1}, at 5 °C Gold contents were 0.37 and 1.2 wt.%	50% degradation after 8 h of simulated sunlight irradiation, DO concentration of 35 mg L^{-1} and gold load of 1.2 wt.% The lowest removal was observed when DO and gold concentrations were 0.4 mg L^{-1} and 0.37 wt.%, respectively	Unmodified TiO_2	– 2 (higher removal when TiO_2 tested) when DO and gold concentrations were 0.4 mg L^{-1} and 0.37 wt.%, respectively 1.4 when DO concentration was 0.4 mg L^{-1}	208
4-chlorophenol	Au/TiO_2	Deposition–precipitation	UV irradiation (200–280 nm) supplied to 4-chorophenol solution (0.5 mM) using an 11 W low pressure Hg lamp Catalyst concentration 0.5 g L^{-1}	100% degradation and 20% mineralization after 1 h of irradiation using a gold load of 0.1 wt.% Byproducts were determined in this experiment	Unmodified TiO_2	1 and 1.38 for 0.1 wt.% Au/TiO_2 and 0.1 wt.% Au/0.1 wt.% Ag/TiO_2, respectively. Based on mineralization of 4-chorophenol after 6 h of irradiation	206
	$Au–Ag/TiO_2$			100% degradation and 25% mineralization after 1.5 h of irradiation, using 0.1 wt.% Au–0.1 wt.% Ag/TiO_2			

Table 7.1 (*Continued*)

Pollutant	Catalyst	Preparation Method	Characteristics of Photocatalytic Experiments	Performance of Photocatalyst	Photocatalytic Activity Comparison Experiments	Enhancement of Photocatalytic Activity Using the Gold-Modified Material	Ref.
Sulforhodamine B	Au/TiO_2–bentonite composites	Deposition–precipitation	UV (365 nm) and visible (cutoff over 450 nm) light irradiation were carried out using a 100 W low pressure Hg lamp and a 500 W Xe lamp, respectively	100% degradation after 2 h of UV irradiation and 70% of degradation after 20 h of irradiation using visible light Measurements carried out using the chemical oxygen demand parameter	Unmodified TiO_2	− 1.1 for UV light and −1.3 for visible light (degradation was higher using unmodified TiO_2)	249
					TiO_2–bentonite	1.3 for UV light and −1.15 for visible light irradiation	
Methylene blue	Gold nanoparticles supported on mesoporous TiO_2	Deposition–precipitation	Methylene blue solution (20 mg L^{-1}) irradiated with UV light ($\lambda > 365$ nm) Catalyst concentration 1 g L^{-1}	85% degradation after 3 h of irradiation Gold load used was 0.25 wt.% Higher gold loads resulted in lower degradation	Photolysis	3.4	101
					Unmodified TiO_2	1.4	
Malathion	Co-supported Au–Pd on TiO_2 nanotubes film	Photodeposition	UV light ($\lambda = 365$ nm) irradiation supplied to the 10 mg L^{-1} malathiol solution using a 450 W high pressure Hg lamp	100% herbicide degradation after 4 h of irradiation	Photolysis	No degradation found after 4 h of irradiation	250
					Unmodified TiO_2	2.5	

Rhodamine B	$Au@TiO_2$ spheres@SiO_2 spheres	Citrate reduction of Au preformed nanoparticles	Natural sunlight irradiation	Degradation >93% for all the compounds within 40 min. Highest degradation obtained using a gold load of 0.1 wt.%	Photolysis	No degradation detected	251
Rhodamine 6G					Anatase TiO_2	3.3	
Methylene blue					TiO_2 P25	2.5	
2,4-dichlorophenol					SiO_2/TiO_2	6.3	
Methylene blue	Au/TiO_2 Janus-like nanoparticles	Spin coating followed by photodeposition	Piece of the Janus-like Au/TiO_2 put in the 10 μM methylene blue solution and irradiated using UV light at 254 nm Average size of TiO_2 and Au nanoparticles was 17 and 7.5 nm, respectively	48% degradation after 400 min of UV irradiation	Unmodified TiO_2	1.25	252
					Au/TiO_2 composites	1.15	
Methyl orange	$Au/N–TiO_2$	Deposition–precipitation	UV and visible light supplied using a 300W low pressure Hg lamp and a 1000 W Xe lamp, respectively Concentrations of catalyst and compound were 0.8 g L^{-1} and 20 mg L^{-1}, respectively	85% degradation after 5 h of visible light irradiation and 95% disappearance using UV light irradiation for 75 min The highest degradation observed when gold loads of 4.0 wt.% used	Unmodified TiO_2 P25	1	253
					$N–TiO_2$	2.4 for visible light irradiation and 2.26 when using UV light irradiation	

Table 7.1 (*Continued*)

Pollutant	Catalyst	Preparation Method	Characteristics of Photocatalytic Experiments	Performance of Photocatalyst	Photocatalytic Activity Comparison Experiments	Enhancement of Photocatalytic Activity Using the Gold-Modified Material	Ref.
2,4-diclorophenol (DCP)	Gold nanoparticles inserted into TiO_2 photonic crystals	Infiltration of preformed gold nanoparticles into TiO_2 photo crystals by hydrothermal reduction	Visible light ($\lambda > 420$ nm) supplied to 10 mg L^{-1} 2,4-DCP solution using a high pressure Xe lamp	Photodegradation rate constant of 0.57 min^{-1}	Photolysis Unmodified TiO_2 nanocrystals	7.1 4.75	130
Acid orange 7	Mesoporous Au/TiO_2 composite microspheres	Microemulsion–based bottom–up self-assembly method	Visible light irradiation supplied to 200 ppm AO7 azo dye solution Concentration of catalyst 0.5 g L^{-1}	Total degradation after 4 h of irradiation The highest removals observed when gold loads of 1 and 2.5 wt.% used	TiO_2 P25 TiO_2 nanocomposites TiO_2 nanospheres	Close to 10 5.8 1.5	215
Rhodamine B	Mesoporous Au/TiO_2 composites	Photodeposition	Visible light irradiation ($\lambda > 420$ nm) applied to a 10 μM rhodamine B solution using a 15 W daylight lamp	Total disappearance of color after 180 min of visible light irradiation The highest degradation observed using gold load of 0.015 wt.%	Unmodified TiO_2	3.4 by comparing photodegradation rate constants	254
Methylene blue	SiO_2@TiO_2–Au core@shell nanostructures	Citrate reduction of preformed gold nanoparticles	Visible light ($\lambda < 420$ nm) supplied to the 10 ppm organic compound solution using a 450 W Xe lamp	85% of degradation after 4 h of visible light irradiation The highest disappearance observed when using 15 nm gold nanoparticles	SiO_2@TiO_2	4 (comparison only for methylene blue)	74

Methyl orange				100% disappearance after 4 h of irradiation			
p-nitrophenol				100% degradation after 3 h of irradiation			
Methyl orange	Gold nanocrystal/ titanate nanobelt heterojunctions	Hydrothermal method to produce TiO_2 nanobelts followed by brust method to load gold	UV and visible light irradiation applied to 0.1 mM methyl orange solution 0.05 g of catalyst added	100% degradation after 4 h of UV irradiation and >90% of degradation using visible light irradiation for 5 h The highest degradation using visible light observed when gold loads of 3–5 wt.% used	Unmodified titanate nanobelts	1.9 for UV irradiation No degradation observed when visible light applied	255
p-chlorophenol	Au@TiO_2 spheres	Hydrothermal treatment	Visible light irradiation ($\lambda > 360$ nm) applied to 0.1 mM *p*–cholorophenol solution Amount of catalyst added was 0.05 g	100% degradation after 4 h of irradiation	Core@shell TiO_2 Crushed Au/TiO_2	1.17 1.4	135
Methyl orange	Au/TiO_2 films	Evaporation	UV light ($\lambda > 365$ nm) was supplied using a Hg lamp	13% removal after 1 h of irradiation	Unmodified TiO_2	9	43
Phenol	Au–Cu/TiO_2	Chemical reduction	UV–visible light irradiation applied to 0.2 mM phenol solution Concentration of catalyst 1 g L^{-1}	80% phenol degradation after 3 min of irradiation The highest degradation rate obtained using the Au–Cu (1 : 3)/ TiO_2 catalyst	Unmodified TiO_2 P25 Au/TiO_2 Cu/TiO_2	1.47 1.4 1.15	256

Table 7.1 (*Continued*)

Pollutant	Catalyst	Preparation Method	Characteristics of Photocatalytic Experiments	Performance of Photocatalyst	Photocatalytic Activity Comparison Experiments	Enhancement of Photocatalytic Activity Using the Gold-Modified Material	Ref.
Methyl orange	Au/Fe–TiO_2	Deposition–precipitation	UV and visible light illumination provided to 20 mg L^{-1} methyl orange solution using a high pressure Hg and tungsten halogen lamp, respectively	75% degradation after 85 min of visible light irradiation The highest degradation rate observed using gold loads of 2.0 and 3.0% wt for visible and UV light, respectively	Unmodified TiO_2 P25 0.5Fe–TiO_2	4.4 for visible light and 2.5 for UV light 1.7 for visible light and 0.5 for UV light	207
4-chlorophenol	Au/TiO_2	Deposition–precipitation	Visible light ($\lambda > 420$ nm) used to irradiate 200 mg L^{-1} 4-chlorophenol solution	90% degradation after 4.5 h of irradiation The highest degradation rate obtained using gold load of 0.42% wt	Unmodified TiO_2	2	257
Brilliant red X-3B	Au/TiO_2 nanoflakes	Photodeposition	UV light ($\lambda > 365$ nm) applied to 0.11 mM X-3B dye solution using a Hg lamp Concentration of catalyst 1 g L^{-1}	Total degradation after 20 min of UV irradiation The highest degradation rates obtained when Au/TiO_2 nanoflakes were calcined at 500 °C	Photolysis Unmodified TiO_2 nanoflakes Unmodified TiO_2 P25 Au/TiO_2 P25	No degradation observed 4.8 2.25 1.25	37

Nitrate (reduction to N_2)	Au/TiO_2	Deposition–precipitation	UV light applied to 100 ppm NO_3 and 8 mM oxalic acid solutions using a 400 W lamp Anaerobic conditions maintained Gold load set at 1 wt.%	Conversions of 60% achieved after 150 min of irradiation at 30 °C and when oxalic acid (8 mM) used as a hole scavenger			188
Oxalic acid				100% removal after 150 min of irradiation			
3,4-Dichloro phenylurea	Au/TiO_2	Deposition–precipitation	Simulated solar irradiation applied to 20 ppm 3,4-dichloro phenylurea solution Concentration of catalyst was 100 ppm and gold load set as 0.83% wt Aerobic conditions maintained in the solution	84% degradation after 100 min of simulated solar light irradiation Photocatalytic byproducts were determined using liquid chromatography tandem mass spectrometry (LC/MS/MS) techniques	Photolysis Unmodified TiO_2	2.4 1.3	258
Methylene blue	Gold nanorods supported in TiO_2 P25 Silver-coated gold nanorods supported in TiO_2 P25	Chemical deposition of preformed metal nanoparticles	Visible light ($\lambda = 400–750$ nm) irradiated to 10 μM methylene blue solution Catalyst added as a thin film on the surface of a glass slide and immersed into the compound solution; then irradiation carried out	50% of degradation after 3 h of irradiation $>$85% of degradation after 3 h of irradiation The highest degradation observed when wavelength of 680 nm used for both catalysts	Unmodified TiO_2 P25	1.25 and 1.7 for Au and Au/Ag/TiO_2 catalysts, respectively	150

Table 7.1 *(Continued)*

Pollutant	Catalyst	Preparation Method	Characteristics of Photocatalytic Experiments	Performance of Photocatalyst	Photocatalytic Activity Comparison Experiments	Enhancement of Photocatalytic Activity Using the Gold-Modified Material	Ref.
Methyl orange	Au/TiO_2	Photodeposition	UV light (350–400 nm) irradiated to 20 ppm dye solution	Total degradation after 30 min of irradiation			228
Phenol	Gold nanoparticles supported on sulfate treated TiO_2	Photodeposition Chemical reduction	UV light ($\lambda = 365$ nm) used to irradiate the 50 ppm phenol solution Concentration of catalyst set as 1 g L^{-1} Deposited gold content 1% wt	Mineralization rates of 7.0 and 2.8 g organic C L^{-1} s^{-1} for sulfate treated and non-treated TiO_2, respectively	Unmodified TiO_2	1.6 and 1.8 for sulfate-treated and non-treated TiO_2, respectively	259
Methyl *tert*-butyl ether (MTBE)	Au/TiO_2	Deposition–precipitation	Artificial sunlight supplied to 200 mg L^{-1} MTBE solution	90% degradation after 5 h of irradiation The highest degradation rate reached when solution was O_2 saturated	Unmodified TiO_2	2.8	171

(Au^0 and Au^{3+}) resulted in significant enhancement of methyl orange photocatalysis; the activity was twice than that obtained using the pristine material. The optimal concentration of gold nanoparticles on a titania surface in terms of degradation yield was 0.8 mg cm^{-2}. The authors reported no deactivation of the catalyst after five consecutive photocatalytic experiments. With regard to the photocatalytic degradation of colorless pollutants, Orlov *et al.*[238] reported the total photocatalytic degradation of 4-chlorophenol and methyl *tert*-butyl ether (MTBE) after 7.5 h and 20 h of UV light irradiation, respectively, using gold-modified thin TiO_2 films at loads of 0.55% wt. Removal efficiencies obtained with the gold-modified catalyst were twofold higher than those obtained using the bare catalyst.

The photocatalytic performance of gold-modified semiconductor thin films may depend on factors additional to those mentioned for catalyst powders, for instance the thickness of semiconductor films,[231] dispersion of gold nanoparticles,[233] and the porosity of semiconductor films[235,236] as well as the method used for gold deposition. The use of gold-modified semiconductor thin films may result in pollutant degradation mechanisms that are different from those observed when powdered catalyst is used. For instance, Yogi *et al.*[239] observed decreased demethylation of the azo dye, methyl orange, upon UV irradiation (365 nm) in testing Au/TiO_2 thin films compared with observations when Au/TiO_2 powder was used. According to Subramanian *et al.*,[240] it is possible that gold nanoparticles in Au/semiconductor thin films act as recombination centers for photo-produced electron–hole pairs, which would result in lower rates of degradation compared with the powdered catalyst. The studies of Arabatzis *et al.*[235] support such a hypothesis as they observed reduced degradation of 3,5-dichlorophenol using Au/TiO_2 thin films compared with those obtained using powdered catalyst.

Gold–semiconductor ternary composites have been synthesized as thin films showing a remarkable increase in the potential of photocatalysis of dissolved organic pollutants. Very thin films (300 nm) of TiO_2 modified with $Au@SiO_2$ core@shell nanostructures showed efficient degradation of methylene blue compared with unmodified Au/TiO_2 thin films.[241] Such a behavior was attributable to the SPR increasing the intensity of the surrounding electric field; which was nearly nine times higher compared with Au/TiO_2 alone. The SiO_2 coating significantly increased the SPR effect of gold nanoparticles, leading to the enhanced photocatalytic activity.

ZnO/TiO_2 thin films surface-modified with gold and silver nanoparticles were tested for the degradation of dilute solutions of the antibiotic agent tetracycline.[242] The highest degradation rate was found when the gold-loaded composite was tested, which was attributed not only to the greater SPR effect of gold nanoparticles, but also to the higher adsorption capacity of the gold-modified composite compared with the silver-modified one. The authors investigated byproduct formation throughout the reaction; they reported the notably higher byproduct formation on the surface of the thin sheets rather than in the solution. Similarly, mineralization of the compound was found to be higher on the surface of the gold-modified thin layer composite.

Some methods for increasing the photocatalytic potential of gold-modified semiconductors, either in powdered form or as thin films, have been reported. Adding nanoparticles of a second noble metal to a gold-modified semiconductor can significantly increase its photocatalytic performance in degrading dissolved pollutants. Bimetallic Au–Pd composites supported on a TiO_2 surface have demonstrated increased efficiency in removing azo dyes using visible light compared with monometallic composites.[146] The addition of oxidizing agents such as peroxomonosulfate (PMS), peroxodisulfate (PDS), and hydrogen peroxide (H_2O_2) to the solution during photocatalytic tests increases photocatalysis yield since such oxidizing agents act as hole scavengers, preventing recombination of photo-produced species. Increases of up to 100% in the degradation of azo dyes have been reported upon addition of oxidizing agents using Au/TiO_2 photocatalysts.[182]

Sensitization of gold nanoparticles using organic molecules, notably azo dyes, increases the catalytic efficiency of the composite. This is clearly shown in the study by Bae and Choi,[243] which investigated the degradation of trichloroacetate and carbon tetrachloride using sensitized Au/TiO_2 irradiated with visible light. Accordingly, Kao *et al.*[244] assessed the photocatalytic efficiency of TiO_2 anatase films modified with phenylethylmercaptan (PEM) chemisorbed gold clusters. The authors found that degradation of azo dyes using PEM (3.8%) gold clusters/TiO_2 was 52% higher than that observed for unmodified Au/TiO_2. PEM chemisorbed to gold nanoparticles causes photo-produced electrons in the conduction band of the semiconductor to flow onto gold nanoparticles and then to the delocalized electronic structure of the benzene ring in the organic molecule, thereby promoting prolonged charge separation and thus an increase in photocatalytic activity.

With regard to contaminant molecules used to evaluate the photocatalytic capability of the gold-modified semiconductors, the change in concentration of colored substances, such as methyl orange, methylene blue and rhodamine B, is quite simple to follow by quantifying the disappearance of the color using spectrophotometric methods.[103,115,161,163,182,200,213,231,236,239,249,252,260–267] Some of the byproducts generated by the photocatalyst may still have color, but most are colorless.[239] Colorless molecules are used to evaluate photocatalytic performance to a lesser extent since sample preparation and measurement methods tend to be more complex.

Phenol and simple fatty acids, such as oxalic acid, succinic acid or malonic acid, are commonly used to elucidate the reaction mechanism leading to the degradation and potential mineralization of more complex organic compounds, and are therefore referred to as model molecules. It is known that in photocatalytic systems using gold-modified semiconductors, succinic acid is converted to malonic acid, which is then converted to oxalic acid and finally mineralized to CO_2. Photodecomposition of malonic acid *via* oxalate is only one possible pathway, accounting for about 50% of the malonic acid conversion.[160] Degradation of malonic acid using UV light begins with the production of the hole in the semiconductor, which results in the elimination of CO_2 from the dicarboxylic acid and thus the generation of oxalic acid. The photocatalytic mineralization of oxalate occurs mainly through the hole oxidation mechanism. The lack of evidence pointing to malonic acid degradation into oxalic acid under visible light irradiation of Au/TiO_2 indicates that holes formed in the gold nanoparticles are unable to oxidize malonic acid efficiently, and therefore this reaction occurs on the surface of the semiconductor rather than on gold nanoparticles.

The reactions resulting in the degradation and mineralization of simple fatty acids using gold-modified TiO_2 can be summarized as follows:[121]

Formation of the excited species

$$TiO_2 \underset{h\nu}{\longrightarrow} e^-{}_{cb} + h^+{}_{vb} \longleftrightarrow \text{recombination}$$

$$h^+{}_{vb} + H_2O_{adsorbed} \rightarrow HO^\bullet{}_{adsorbed} + H^+$$

$$h^+{}_{vb} + HO^-{}_{adsorbed} \rightarrow HO^\bullet{}_{adsorbed}$$

$$h^+{}_{vb} + C_{adsorbed} \rightarrow C^{\bullet+}{}_{adsorbed}$$

$$HO^\bullet{}_{adsorbed} + C_{adsorbed} \rightarrow (P\text{–}OH^\bullet)_{adsorbed} \leftrightarrow P^{\bullet+}{}_{adsorbed} + OH^-{}_{adsorbed}$$

$$e^-{}_{cb} + O_{2adsorbed} \rightarrow O_2{}^{\bullet-}{}_{adsorbed}$$

where h^+ and e^- are the photo-produced hole–electron pair, respectively; vb and cb refer to the valence and conduction bands, respectively; and, C refers to the pollutant molecule.

In the absence of electron acceptors, such as dissolved oxygen (O_2), electron–hole recombination is possible. Dissolved oxygen impedes recombination by trapping electrons and forming superoxide radicals. Oxidation by holes is favored by molecules that adsorb easily onto the photocatalyst. On the other hand, $HO^\bullet$ is able to react either while it is adsorbed or dispersed in the solution, thereby reacting with the dissolved molecules. The degradation of simple carboxylic acids may occur according to the following reactions:

$$R\text{–}COOH + h^+{}_{vb} + OH^\bullet{}_{adsorbed} \rightarrow RCOO^\bullet + H^+ \rightarrow CO_2 + R^\bullet$$

In an acidic medium, catalysis can be carried out by the reaction with the superoxide radical. Since oxalic acid's mineralization occurs through a series of chain reactions, it is possible that the formation of organic peroxides takes place:

$$R^\bullet + O_2{}^{\bullet-} \rightarrow R\text{–}O_2{}^- + H^+ \leftrightarrow R\text{–}O_2H$$

$$R\text{–}O_2H + h^+{}_{vb} + OH^\bullet{}_{adsorbed} \rightarrow CO_2$$

Lastly, the final reaction of oxalic acid mineralization can be described as:

$$2H_2C_2O_4 + O_2 + 2h\nu \rightarrow 4CO_2 + 2H_2O$$

Morphology of gold nanoparticles supported on a semiconductor has been shown to impact upon the degradation routes of organic contaminants in water. In this respect, Hu and Bürgi[204] compared the degradation pathways of dicarboxylic acids using TiO_2 modified with gold nanorods and nanospheres. The authors noted a higher number of intermediates when using nanospheres than those observed in photocatalysis experiments using gold nanorods supported on TiO_2.

7.4.1 Use of Semiconductors other than TiO_2 in Removal of Pollutants from Water

Even though gold-modified TiO_2 is the most common catalyst used to degrade pollutants in water streams, other gold-modified semiconductors have also been developed for such purposes. Such semiconductors may have a smaller bandgap than titania or lower cost, as well as better morphological properties.

7.4.1.1 Zinc Oxide

ZnO is a low cost alternative to TiO_2; this semiconductor displays a similar bandgap to that of TiO_2 (3.37 eV), so that similar photocatalytic efficiencies for water pollutant degradation can be expected when using this oxide. ZnO, as an alternative nanomaterial, has a similar photocatalytic activity, greater in some cases, than TiO_2. This property has been observed for discoloration of reactive blue 19, a textile anthraquinone dye, in aqueous suspensions.[267]

As with TiO_2, photocatalytic performance of the semiconductor under visible light can increase significantly on surface modification with gold nanoparticles. Li *et al.*[268] synthesized Au–ZnO nanocomposites composed of hexagonal pyramid shaped ZnO (25 nm of slide edge length) and a spherical gold nanoparticle of 8 nm in diameter at the base of the pyramid. Composites showed maximum absorption at 519 nm, due to SPR, and thus high photocatalytic activity resulting in complete degradation of rhodamine B after 10 min of UV–visible irradiation, half of the time needed for unmodified ZnO.

A wide variety of Au/ZnO microstructures have been tested to degrade water pollutants. For example, three-dimensional (3D) flower-like microstructures decorated with gold were capable of degrading azo dyes at higher rates than bare ZnO.[221] Furthermore, gold nanoparticle on ZnO nanorods have been shown to produce complete degradation of rhodamine B within 15 min using UV light, substantially surpassing the performance of other TiO_2 and ZnO based photocatalysts.[269] Other studies reported that the synthesis of glutathione bifunctionalized Au/ZnO resulted in the optimal monodispersion of gold nanoparticles on the surface of ZnO,[270] producing complete degradation of rhodamine 6G after just 30 min of irradiation with UV light (320 nm) at an optimal gold loading of 2% wt. When noble metals, such as gold or platinum, are simultaneously deposited onto the surface of ZnO particles, a high electron flow oscillates from the conduction band of the semiconductor to noble metal nanoparticles. This is because the work function of ZnO (Φ_m of 4.3 eV) tends to be much lower than those of gold or platinum (Φ_s Au: 5.1 eV and Pt: 5.65 eV).[271] In this respect, Pawinrat *et al.*[262] compared the photocatalytic efficiency of Au–Pt modified ZnO composites synthesized by flame spray pyrolysis for methylene blue degradation. The highest dissipation of the azo dye after 1 h irradiation with UV light (365 nm) was obtained with the Au–Pt/ZnO composite, showing conversion rates of 1.14 h^{-1} with Au loads of 3% wt., compared with Pt/ZnO and bare ZnO.

On the subject of the use of thin layer gold-modified ZnO, Chen *et al.*[232] tested the photocatalytic performance of immobilized bimetallic Au–Ag/ZnO thin films on indium tin oxide (ITO). The Au–Ag/ZnO film showed nearly twice the photocatalytic efficiency of pure ZnO, and much higher efficiency than that observed for the Ag/ZnO film; this can be explained by the rapid oxidation of silver in the aqueous medium. A feasible approach to protect silver in bimetallic composites could be based on the protection of silver nanoparticles by coating the semiconductor with gold nanoparticles. Pyne *et al.*[272] reported an increase in the catalytic activity for photodegradation of methylene blue using silver nanoparticles coated with ZnO nanowires, supporting such a hypothesis. Au/ZnO composites have proven to be efficient in the catalytic degradation of organic compounds in water using both visible light and UV light, as demonstrated by the disappearance of azo dyes.[273] Moreover, the addition of oxidizing agents such PMS, PDS and H_2O_2 significantly facilitated not only the degradation, but also the mineralization of organic compounds using visible light.[273]

7.4.1.2 Silicon Oxide

SiO_2 is one of the most commonly used semiconductors for the degradation of pollutants in water, either as a powder, thin layers or core@shell structures. However, there is a general lack of information regarding the photocatalytic activity of the gold-modified SiO_2.

Studies by Garron *et al.*[187] testing the photocatalytic reduction of nitrate and nitrite in groundwater are of interest since they are among the few works dealing with environmental matrices. The authors used surface-modified SiO_2 with three noble metals: Pd, Sn, and Au. The highest activity and selectivity were obtained when using the Pd(5 wt.%)–Sn(1.5)–Au(3 wt.%)/SiO_2 catalyst, which presented the highest selectivity in N_2 production, *i.e.* 97.5%, compared with the bimetallic alloys. More recent works have tested LaB_6@SiO_2/Au nanocomposite as a novel catalyst with near-infrared (NIR) photothermally enhanced activity.[274] Under NIR irradiation, hot spots on the catalyst surface were shown to cause an increase in the reduction rate of 4-nitrophenol. The resulting LaB_6@SiO_2/Au nanocomposite not only retained the NIR absorption (around 1000 nm) of the LaB_6 cores, but also exhibited an additional characteristic absorption at 520 nm, owing to the SPR of gold nanoparticles. Photoconversion of 4-nitrophenol using this approach can be explained not only by photocatalytic activity but also by the increase in the temperature of the solution due to the NIR irradiation, given that the highest degradation rates were found at 45 °C (0.251 min^{-1}). The catalytic ability of LaB_6@SiO_2/Au composite nanoparticles was mainly due to the SRP exerted by the deposited gold nanoparticles.

7.4.1.3 *Ceria*

Cerium oxide has been shown to possess significant capacity as a photocatalyst in the visible light spectrum. In fact, gold-modified CeO_2 displays higher photoactivity in the visible light region than other gold-modified semiconductors, such as Au/TiO_2, Au/ZrO_2, and Au/SiO_2.[54] Total mineralization of formic acid has been observed after 28 h of visible light irradiation using 1 wt.% Au/CeO_2, as reported by Kominami *et al.*[54] The absorption of incident photons by gold nanoparticles through their localized SPR, followed by the injection of excited electrons from gold nanoparticles into the conduction band of the semiconductor, seems to be the key process controlling the rate of mineralization of formic acid. Given the potential of CeO_2 to absorb visible light, Kominami *et al.*[179] tested the photocatalytic performance of Au/CeO_2 through the mineralization of simple fatty acids, namely formic, oxalic and acetic acids, using visible and green light. Mineralization of formic acid using visible light was dramatically higher when gold loads of 1% wt were tested (total removal was achieved after 28 h), although it was negligible when the reaction was carried out under anaerobic conditions, indicating that oxygen plays an important role in the mineralization of the organic compound. The apparent quantum efficiency (determined as the ratio of CO_2 generated and the absorbed photons per molecule) was greater when green light (wavelength 550 nm) was used in photocatalytic tests, although the transformation rate was half of that observed with visible light. Even when the photocatalysis rate is lower using green light, the use of this kind of light for the mineralization of organic compounds instead of UV light can significantly reduce the operating costs of photocatalysis treatment systems.

7.4.1.4 *Tungsten Oxide*

WO_3 is a photocatalyst with a narrow bandgap (2.8 eV) which is activated by visible light. However, recombination of photo-induced electron–hole pairs tends to be greater for this oxide compared with other semiconductors. Furthermore, due to its low specific surface area, its photocatalytic activity is expected to be low when the bare semiconductor is used. Similarly to CeO_2, WO_3 absorbs light up to *ca.* 480 nm (yellow color), remaining stable and harmless in acidic and oxidative conditions. The phototransformation of organic pollutants in water, namely methyl orange and 2,4-dichlorophenoxyacetic acid (2,4-D), using Au/WO_3 powder has demonstrated high efficiency compared with unmodified oxide under sunlight irradiation.[28] This, in addition to its resilience to corrosion, leads to consideration of WO_3 as a promising visible light-driven photocatalyst.

Even though the photocatalytic performance of WO_3 is improved when gold nanoparticles are supported on the semiconductor's surface, the size of WO_3 particles, which is at the level of nanometres, makes the deposition of large gold loads unfeasible. This is because it will cause the total coating of the semiconductor nanoparticles, which in turn results in a drop in photocatalysis performance. In order to achieve higher loadings of gold on the semiconductor's surface, and thus increase the photocatalytic degradation of organic pollutants in water, an alternative is the synthesis of ternary composites containing WO_3 and another semiconductor. This method results in an increase in the specific surface area of the composite and a decrease in the recombination of the photo-induced electron–hole pairs.

Recently, the ternary composite system formed by TiO_2, WO_3, and gold nanoparticles has gained attention in terms of the enhancement of photocatalytic response to UV (absorbed by TiO_2 up to 380 nm, in the case of anatase) and visible light (absorbed by WO_3 and gold). TiO_2(75 wt.%)/WO_3(24 wt.%)/Au(1 wt.%) ternary composites synthesized by impregnation of WO_3 soluble precursors on TiO_2 surface, followed by photoreduction of gold nanoparticles have been used to degrade phenol upon irradiation with fluorescent light ($\lambda = 365$ nm).[275] The presence of gold on the catalyst surface was shown to facilitate the degradation of phenol by a factor of two compared with the composite without gold. The increased photocatalytic efficiency of the ternary composite is attributed to the transfer of photogenerated electrons from TiO_2 into the lower energy conduction band of WO_3, whereas the generated holes in the valence band of WO_3 are moved into the higher energy valence band of TiO_2.[275] In this way an efficient charge separation is assured, and thus visible light harvesting is possible.

When ternary composites are used, the site where gold nanoparticles are deposited (*i.e.* in one or in both catalysts) has a determining effect on the photodegradation of organic compounds. In this regard, the results reported by Karácsonyi *et al.*[276] evidenced the decrease in the bandgap of the Au/WO_3/TiO_2 composite as the WO_3 loading increased, and gold nanoparticles remained on TiO_2, from 2.88 eV, when the load of WO_3 was 1 wt.%, to 2.66 eV, for WO_3 loads of 33 wt.%. On the other hand, an increase in the gold loading on WO_3 leads to a negligible increase in the mineralization rate, as demonstrated when degradation of oxalic acid was tested.[276] The addition of oxidizing agents during the photocatalytic process has been shown to significantly enhance the photocatalytic degradation of organic contaminants in solution when using Au/WO_3/TiO_2 ternary composites. For example, the use of ozone increased the degradation efficiency of 2,4,6-trinitrotoluene by 6.2 times compared with bare TiO_2 using UV–visible light.[183]

The use of Au/WO_3 thin films for degradation of pollutants in water is a field that has just begun to be explored. The absorption edge of the WO_3 in thin film is reported to be red-shifted in the presence of the gold-buffered layer, indicating a reduction of the optical bandgap.[237] The photocatalytic performance of the gold-buffered WO_3 films to degrade methylene blue dye has been shown to be 50% greater than that observed for bare WO_3 thin films, although the photocatalytic activity of WO_3 films was found to be negligible after repeated use, and therefore recycling potential of Au/WO_3 films can be considered to be low.[237] When Au/WO_3/TiO_2 thin layers are tested, high rates of photo-induced degradation of contaminants have been observed using simulated sunlight.[178] This can be attributed to the decrease in the bandgap in the ternary composite (from 2.75 eV for TiO_2/WO_3 to 2.64 eV for Au/TiO_2/WO_3). The highest photocatalytic activity using ternary composite thin films has been found for the composite Au(0.4 wt.%)/WO_3(4 wt.%)/WO_3(95.6 wt.%).[178]

Studies focused on monitoring the byproducts resulting from the photo-catalyzed degradation of organic pollutants in water are very scarce. One example of this kind of studies is that reported by Kovács *et al.*,[275] in which four phenol intermediates, namely hydrochinon, pyrocatechol, resorcinol, and 1,2,4-trihydroxy benzene, were followed throughout photocatalytic degradation using Au/TiO_2/WO_3 ternary composites. Results showed that the increase in the WO_3 load within the ternary composite favored the formation of pyrocatechol, resorcinol, and 1,2,4-trihydroxy benzene, while the distribution of the intermediates was altered by changes in the gold load on the composites.

7.4.1.5 Bismuth Oxides

Similarly to WO_3 and CeO_2, bismuth oxides (notably $BiVO_4$ and Bi_2O_3) exhibit photocatalytic activity in the visible light spectrum. The lower values of bismuth oxide bandgap compared to that of TiO_2 (E_g 2.4 eV for $BiVO_4$ and 2.8 eV for Bi_2O_3) have encouraged the study of the photocatalytic potential for different morphologies of this semiconductor. Nanosheets, nanotubes, and hierarchical nanostructures have been synthesized to carry out photo-catalytic degradation of water pollutants.

Scheelite monoclinic is the crystallographic phase of $BiVO_4$ displaying the largest photoactivity, due to its smaller bandgap in comparison with other crystallographic phases, *e.g.* scheelite and tetragonal ones (E_g 3.1 eV). Despite the aforementioned advantages, the photoreactivity of $BiVO_4$ is restricted by its low specific surface area, which in some cases has been determined as 10 $m^2\ g^{-1}$, as well as by its high rate of electron–hole pair recombination.[277] Therefore, modification of the surface of the bismuth oxides with deposition of gold nanoparticles is seen as a plausible option to both increase the specific surface area of semiconductor and reduce the recombination of the photo-induced pairs. $BiVO_4$ by itself exhibits no light absorption beyond 550 nm; in contrast, the gold-modified catalyst exhibits light absorption at wavelengths of 600–615 nm due to the SRP effect.[278]

Chen *et al.*[279] reported a high efficiency of degradation of crystal violet and methylene blue due to low recombination of the electron–hole pair using a crystalline peanut-like form of $BiVO_4$. Regarding the use of gold-modified nanosheets and nanotube $BiVO_4$ structures, Cao *et al.*[17] reported total removal of methyl orange after 50 min of irradiation with visible light using a gold load at 3.5–3.7% wt. On the other hand, unmodified $BiVO_4$ showed no photocatalytic activity, even when the energy supplied was slightly higher than the bandgap of such $BiVO_4$ structures (2.4 eV). Other studies report a similar behavior for phenol[278] and for mineralization of simple fatty acids.[280]

In comparing the pollutants removal for nanosheets and nanotubes, increased activity has been found for the former. Using electrochemical determinations, it has been demonstrated that a complex transfer of electrons occurs when Au/$BiVO_4$ catalyst is irradiated with visible light—either the trapping of the photo-produced electron in the conduction band into the gold nanoparticle or the migration of the excited electrons in the gold nanoparticles (*d* and *f* orbitals) to the conduction band of semiconductor.[278] A red shift in the absorption peak is observed in $BiVO_4$ when high loadings of gold are deposited on the semiconductor, probably because of the agglomeration of gold particles on the semiconductor surface. Additionally, the specific surface area of scheelite-monoclinic $BiVO_4$ has been shown to increase after gold nanoparticles deposition at 1 wt.% from 23 to 31 $m^2 g^{-1}$. Regarding gold-modified Bi_2O_3, the study by Pugazhenthiran *et al.*[181] reported a decrease in the bandgap of gold nanorod-modified α-scheelite bismuth trioxide, depending on the method of preparation of the composite, from 2.81 eV for sonochemically processed material to 2.73 eV for hydrothermally prepared Au/Bi_2O_3. Photocatalytic degradation of acid orange 10 using Au/Bi_2O_3 resulted in a mineralization yield of 47% over 30 min using ozone to assist photocatalysis. The use of ozone resulted in a 3.5-fold enhancement in the degradation rate compared with gold-modified Bi_2O_3, which is explained by the increase in the local electric field in the vicinity of gold nanoparticles by ozone molecules.

7.4.1.6 Alumina

Al_2O_3 can improve the textural properties of TiO_2 by increasing the specific surface area when it is used in the form of a ternary composite with gold. An increase in the surface area of TiO_2 P25 from 48–65 $m^2 g^{-1}$ to 220–240 $m^2 g^{-1}$ has been observed for the Au/Al_2O_3/TiO_2 ternary composite, resulting in increased potential for adsorption of organic pollutants onto the catalyst surface.[165]

Regarding the photocatalytic activity of the ternary composite, total removal of MTBE has been reported by Rodríguez-González *et al.*,[165] with gold nanoparticles smaller than 7 nm being shown to be the most active for MTBE removal. Ternary composites including gold and ceria (Au/Al_2O_3/CeO_2) have been tested for the photocatalytic decomposition of diuron herbicide using UV light (247 nm).[176] The highest degradation rate (91% and $k = 0.0162\ s^{-1}$) was obtained using a photocatalyst with 50 wt.% CeO_2 and 2% wt of gold. However, 85% of diuron decomposition was obtained using Au/Al_2O_3 while Au/Al_2O_3/CeO_2 composite was capable of mineralizing up to 77% of the herbicide, showing the highest photocatalytic activity.

Composites constituted by bimetallic nanoparticles uniformly dispersed onto mesoporous alumina (Au–Ag–AgI/Al_2O_3) were used for the photocatalytic degradation of the anti-inflammatory drug naproxen with visible light irradiation;[281] total degradation was achieved within 20 min at pH 5.6. It is interesting to note that, as pH values increased, photocatalytic rates decreased to less than 50% at a pH value of 9.1. The interaction between gold and silver nanoparticles resulted in more electrons participating in the collective oscillation of SPR. This, in turn, enhanced the efficiency of visible light in catalysis. Moreover, the authors reported that the photocatalytic activity of Au–Ag–AgI/Al_2O_3 did not decrease markedly after six successive cycles of photocatalysis.

In bimetallic composites where silver is used, the release of silver ions from the composite can pose a risk to water quality. However, in the case of composite Au–Ag–AgI/Al_2O_3 this problem is negated when the reaction is carried out in water with a certain salt content (*i.e.* natural waters). This is because, while the plasmon-induced holes on gold nanoparticles are able of trap electrons from pollutants, NO_3, HCO_3 and SO_4 ions, these electrons are donated to the photoexcited holes forming $HCO_3^{\bullet}$ or $NO_3^{\bullet}$ radicals. This increases the transfer of electrons to silver, suppressing its release from the composite as Ag^+ ions. However, this phenomenon may reduce photocatalytic activity since anion radicals created in metal nanoparticles and released to solution are less reactive than holes for degrading pollutants.[281]

7.4.1.7 Hematite

Au/Fe_2O_3 is one of the earliest gold catalysts studied intensively; this material has been extensively used in the water gas shift reaction as well as in selective hydrogenation of unsaturated aldehydes or ketones. However, there is a general lack of information regarding the use of hematite modified with gold nanoparticles for degradation of pollutants in water.

As occurs with alumina, hematite has been proven to be efficient in the adsorption of organic compounds in aqueous matrices. Additionally, this material can be magnetically recovered after photocatalytic processes and reused.[282] Zhang *et al.*[283] synthesized nanocomposites composed of gold nanoparticles, polyaniline, and hematite with a core@shell structure. The photocatalytic activity of this material, expressed as the reduction of rhodamine B in the presence of $NaBH_4$, was shown to be high as complete reduction of the compound achieved after 7.5 min using UV light. The highest activity was found for the material displaying the smallest deposition of gold nanoparticles. Cao *et al.*[284] used α-Fe_2O_3 nanocrystals for degradation of rhodamine B upon visible light irradiation. The highest photocatalytic activity was observed for the 11 nm Au/α-Fe_2O_3 hetero-nanoparticles, resulting in nearly 82% of rhodamine B degradation within 8 h compared with only 35% of rhodamine B degradation when bare hematite was tested.

The improved photocatalytic efficiency of Au/α-Fe_2O_3 hetero-nanoparticles is attributed to both the transfer of electrons from the conduction band of hematite to gold nanoparticles and the SPR expressed by the gold nanoparticles. More importantly, SPR excitation of gold nanoparticles can generate enhanced localized electric fields, which may concentrate the light energy incident on the surface of composite and generate more photoexcited electrons nearby, increasing in turn the photocatalytic reaction on the surface of the gold-modified semiconductor.

7.4.1.8 Other Oxides

The surfaces of other oxides such as ZrO_2, Ag_2O, Cu_2O, NbO_3, and SnO_2 have been modified by deposition of gold nanoparticles and used for the dissipation of pollutants in water. However, given that many of these investigations are still in progress, few reports in this respect are found in the literature.

Zhu *et al.*[285] deposited preformed gold nanoparticles by impregnation onto semiconductors that display broad bandgaps (higher than 5 eV), namely zeolite Y, ZrO_2, and SiO_2. The bandgap of the three semiconductors decreased upon the deposition of gold nanoparticles. The gold-modified materials exhibited better catalytic performance in degrading the dye sulforhodamine B under UV irradiation than under blue light (450 nm) irradiation. Photocatalytic activity observed was attributed to the SPR displayed by gold nanoparticles.

Regarding copper oxides, Kong *et al.*[286] synthesized octahedral Au@Cu_2O nanocrystals which displayed remarkable photocatalytic performance for methyl orange degradation due to improved charge separation from the core@shell interaction. The catalytic activity was higher for composites with the smallest sizes of nanoparticles (127 and 92 nm). Similarly to other catalysts, the enhanced photocatalytic efficiency was ascribed to enhanced light absorption in the visible region as well as to the electron–sink effect of the gold nanoparticles. In addition, photocatalytic efficiency of the gold-modified Cu_2O depends on the coverage density of the gold nanoparticles. Higher efficiency is achieved by increasing the coverage of gold nanoparticles, but this then decreases as the surface of Cu_2O becomes too densely covered, as reported by Pan *et al.*[25] for Au/Cu_2O nanowires. The use of Au/Cu_2O nanoframes has been reported as a novel method for the removal of dissolved contaminants in water.[50] Using this array, the lifetime of the photo-induced electron–hole pairs tends to increase as the Cu_2O cage thickness increases, which leads to an augment in the hydroxyl radical concentration and thus an enhanced rate of photocatalytic degradation. This same kind of arrangement has also been used for Ag_2O. Yen *et al.*[287] synthesized porous hollow gold nanocages using gold nanoparticles as a template. In this arrangement, gold replaced silver within the hollow nanoparticles, while silver remained as a shell inside the nanocages, with the remaining silver oxidized as Ag_2O. The use of Au/Ag_2O nanocages has resulted in the rapid degradation of methyl orange after 50 min of UV light irradiation.[287]

Recently, gold-modified $KNbO_3$ has been used for the removal of contaminants in water. $KNbO_3$ is a perovskite displaying low toxicity and high stability under light illumination, which makes it attractive for use as a photocatalyst in water treatment. Lan *et al.*[173] anchored gold nanoparticles onto one-dimensional orthorhombic potassium niobate nanowires. Photocatalytic activity expressed as rhodamine B degradation using Au/$KNbO_3$ appeared to be much greater than that of $KNbO_3$ nanowires, nanorods, and commercial $KNbO_3$ from Alfa Aesar. The use of UV excitation (365 nm) in photocatalytic experiments resulted in greater degradation of azo dye than that observed under visible light (420 nm) irradiation. Additionally, increasing the size of the gold nanoparticles from 5 to 10 nm showed significantly improved reactivity. The use of gold nanoparticles supported on $KNbO_3$ nanowires increased the degradation of azo dye by a factor of 15 compared with unmodified $KNbO_3$. Given the relatively large bandgap (3.3–3.8 eV) of $KNbO_3$, it is clear that visible light catalysis is exclusively occurring on gold nanoparticles *via* SPR.

Lastly, other works have reported the deposition of gold nanoparticles in ultrafine SnO_2 nanoclusters assembled to form highly ordered hollow structures. The gold-modified SnO_2 structures presented enhanced photocatalytic performance in the photodegradation of rhodamine 6G, owing largely to the high specific surface area of the composites as well as the effective electron–hole separation at the Au/SnO_2 interfaces.[155]

7.4.1.9 Cadmium Sulfide

CdS is considered as a suitable substitute for titania due to its small bandgap, 2.4 eV, which can result in the absorption of light in the visible region at wavelengths of 400–517 nm, *i.e.* green and violet light. Au/CdS composites are promising plasmonic materials that can be useful in photocatalytic applications. In Au/CdS structures, gold acts as an electron sink since the Fermi energy level of gold nanoparticles [+0.5 V *vs.* normal hydrogen electrode (NHE)] is lower than the conduction band of the CdS (−1.0 V *vs.* NHE). A study by Singh and Pal[288] demonstrated that the degradation of dissolved pollutants using Au/CdS structures can be tuned by varying the shape of both semiconductor and gold nanoparticles. The authors synthesized nanosphere and nanorod CdS modified on their surfaces with 9.1 nm spherical and rod-shaped gold nanoparticles. The bandgap of CdS spheres and nanorods (2.81 and 2.64 eV, respectively) was found to be significantly red-shifted to 2.44 and 2.56 eV, respectively, after impregnation with gold nanospheres, and to 2.38 and 2.56 eV, respectively, after nanorod deposition. Photooxidation of salicylic acid by bare CdS nanorods was found to be 62% after 4 h of UV irradiation. Subsequently, when gold nanospheres were loaded onto CdS nanorods, the degradation rate of salicylic acid was increased by 12%. The highest degradation of salicylic acid (*ca.* 86%) was found when gold nanorods were supported on CdS rods. The deposition of gold nanoparticles on the surface of CdS resulted in a red-shift of the plasmon band towards wavelengths of 620–655 nm, depending on the content of gold on the semiconductor surface.[289]

Ternary composites using CdS and TiO_2 have been used to degrade organic pollutants in water.[290] Lv *et al.*[291] tested the photocatalytic performance of Au/CdS/TiO_2 by measuring the degradation of methyl orange under visible light irradiation. Degradation efficiency using the ternary composite (98%) was shown to be higher than with unmodified TiO_2 (43%) and CdS/TiO_2 (80%) composite. The highest mineralization degree was observed when the gold content in the composite was 0.1% wt.

7.4.1.10 Zinc Sulfide

ZnS is an important II-VI semiconductor material displaying a wide bandgap energy (3.7 eV). Large specific surface area of ZnS (up to 173.14 $m^2\ g^{-1}$) reflects the high adsorption capacity of the composite and the potential to carry out photocatalytic reactions. As a consequence, this semiconductor is a candidate for noble metal deposition in order to catalyze the photodegradation of dissolved pollutants.

Degradation of 4-nitrophenol using UV light irradiation was tested using ZnS particles modified by gold and platinum deposition.[214] Photodegradation efficiency of 4-nitrophenol was higher when the catalyst modified with platinum was used (78% of the initially added concentration) compared with that observed for gold (72% of the initially added concentration). This is explained by the fact that better Fermi level equilibration between the conduction band of ZnS and platinum (4.2 eV) was achieved compared with gold (5.5 eV).

7.4.2 Use of Gold-modified Semiconductors to Remove Pollutants of Emerging Concern in Water

As shown above, semiconductors with surfaces modified by the deposition of gold nanoparticles have been demonstrated not only to efficiently degrade but to mineralize a broad spectrum of water pollutants, from azo dyes and simple fatty acids to pollutants commonly found in wastewater streams. This is of particular interest to those concerned with the quality of water sources, especially water bodies constantly receiving persistent organic substances.

Examples of this type of pollutants are the so-called organic pollutants of emerging concern (OPECs), which are chemical substances used in everyday consumer products, such as pharmaceutically active compounds, personal care products, plasticizers, flame retardants and detergents, among many others.[292] The common way to dispose of these pollutants is *via* discharge to wastewater and thus any water body receiving this wastewater will present a concentration of emerging contaminants. Endocrine disruption in mammals, birds, fish, and reptiles as well as the development of antibiotic resistance by pathogenic microorganisms are the two effects causing most concern.[293] A broad range of emerging pollutants have been shown to be resistant to conventional water treatment systems,[294] and because of this heterogeneous photocatalysis is considered to be a good alternative to remove these persistent pollutants from water sources or effluent.

Research on the removal of OPECs using gold-modified semiconductors has just begun. Results obtained so far indicate acceptable and consistent removal of pharmaceutical compounds and plasticizers. However, mineralization is not fully achieved and byproducts have not been fully characterized either in terms of their structure or toxicology. An exception of this is the antibiotic compound ciprofloxacin, whose photocatalysis has been documented and byproducts characterized.[186] Interestingly, many of these byproducts retain their antibiotic properties after irradiation with UVA–visible light for two hours. A study by Ye *et al.*[157] showed that the non-steroidal anti-inflammatory naproxen was degraded using SiO_2@Au@TiO_2 core@shell nanostructures with visible light ($\lambda > 420$ nm) provided by a 250 W Xe lamp. Gold decoration on the SiO_2@TiO_2 nanoparticles increased the potential for degradation of naproxen with visible light. The highest removal rates were obtained using a gold loading of 0.1 wt.%, while higher loads of gold resulted in a decrease in photocatalytic activity.

Monitoring the photocatalytic degradation of antibiotics has become important as an environmental application of gold-modified semiconductors. In this regard, photocatalytic degradation of commonly used substances, such as tetracycline, by these means has been reported in the literature.[242] Oros-Ruiz *et al.*[141] studied the degradation of the antibiotic trimethoprim using Au, Ag and Ni modified TiO_2 P25 (2 nm average particle size and metal loading of 0.5% wt). The authors reported mineralization rates greater than 80% using UV irradiation ($\lambda = 254$ nm). Gold supported on titania was the most active catalyst compared with the other noble metals deposited (Ag, Cu, Ni) on TiO_2. In another study, the photocatalytic removal of the third generation antibiotic, ceftiofur sodium, under UV–visible light was reported using Au/TiO_2. Removal efficiency of 95% were achieved when Au/TiO_2 was tested; lower efficiencies (by a factor of 10) were observed when TiO_2 was tested.[180] The compound removal rate was increased when oxidizing agents such as PMS, PDS and H_2O_2 were added, and the Au/TiO_2–PMS combination resulted in 95% of ceftiofur sodium degradation within 30 min, as well as the complete mineralization after 10 h of irradiation. The increase in the photocatalytic degradation rate with the addition of oxidizing agents can be attributed to the immediate trapping of the photogenerated electrons by the oxidants, leading to a decrease in the charge recombination rate and eventually the release of more free radicals.

The study by Alvaro *et al.*[96] is noteworthy as they are pioneers in the degradation of the neurotoxin soman. The authors used Au/TiO_2 synthesized in a one-step procedure in which gold nanoparticles were formed in the sol-gel reaction with titania and pluronic surfactant was used as a structuring agent. The catalyst showed a high efficiency resulting in the complete degradation of the organic compound upon 30 min of visible light irradiation.

Degradation byproducts of detergents in wastewater such as nonylphenol have been photocatalytically removed using Au/TiO_2. According to results reported by Naya *et al.*,[295] rutile showed a higher removal efficiency than the anatase phase and unmodified $BiVO_4$. Total removal and mineralization of nonylphenol was observed after 2 h of visible light irradiation. The greater efficiency of adsorption of nonylphenol on rutile than on anatase might explain the high degradation efficiency of the compound when rutile is used. The decomposition mechanism of nonylphenol begins with the decomposition of the aromatic ring by electron transfer from the gold nanoparticles. Subsequently, intermediary groups migrate from gold nanoparticles to TiO_2 and are then chemisorbed to undergo a series of oxidations that finally result in CO_2.

With regard to plasticizers, Li *et al.*[296] studied the decomposition of bisphenol A by photoelectrocatalysis using Au(2 wt.%)/TiO_2 anchored to a thin sheet of ITO. Thin layers were irradiated with UV light and an anodic current was applied. Results showed that greater photoactivity was achieved using the gold-modified catalyst than with bare TiO_2. The photoelectrocatalysis reaction was more efficient for oxidation of bisphenol A (30% mineralization) than photocatalysis by itself (25% mineralization). Greater degradation of bisphenol A was observed using indium tin oxide (ITO) thin films than for powdered catalyst in the batch tests; this was attributed to the greater penetration of light into Au/TiO_2–ITO thin layers. No deactivation was observed over 10 cyclical tests.

7.5 Comparison of Photocatalytic Activity of Doped and Gold-modified Semiconductors for Degradation of Water Pollutants

In heterogeneous photocatalysis aimed at treating water, one of the main objectives is to use sunlight in photocatalytic reactions. In order to achieve this goal, either the deposition of noble metal nanoparticles on the surface of semiconductors or the inclusion of metal ions (Pt, Au, Ag, Cu, Fe, Ce), metal oxides (SnO_2, WO_3, MnO_2, V_2O_5), and non-metals (C, N, F, S, P) within the semiconductor confers the possibility of using solar energy for degradation of organic compounds in water through the use of visible light.

Doping with transition metals has resulted in both positive and negative effects. Metal ion doping decreases the bandgap of semiconductors, notably TiO_2; however, the metal ions may also serve as recombination centers for electrons and holes, thus reducing the overall photocatalytic activity of the semiconductor.[297] Introducing non-metals within the semiconductors can overcome this obstacle. Substitution of oxygen atoms in a metal oxide semiconductor lattice, for instance TiO_2, with non-metals, such as nitrogen, carbon or boron, results in the subtraction of one, two or three electrons in the valence band of the semiconductor, respectively. Nitrogen is one of the most effective dopants as the mixing of nitrogen 2p states with oxygen 2p states causes a significant decrease in the bandgap. Doping semiconductors with boron, carbon, and nitrogen introduces defects into semiconductors which may induce absorption in the visible region.[298] Introducing small amounts of nitrogen impurities into the TiO_2 lattice resulted in a significant improvement in its photocatalytic activity with regard to the decomposition of organic pollutants. However, excessive quantities of N impurities within the structure of semiconductors may act as recombination centers for photo-induced electrons and holes, as occurs with metal oxide doping. In the case of N-doped TiO_2, the reduction in the bandgap energy has been determined from 3.2 eV in unmodified TiO_2 to 2.4 eV for doped titania, which leads to the enhanced utilization of visible light for the N-doped photocatalysts.[177]

To date, the reactivity and quantum efficiency of N-doped TiO_2 materials remain too low for them to be used in water treatment.[148] To overcome these limitations, the use of co-doped materials as well as the combination of gold-doping modification and non-metal doping may increase the use of sunlight for the photocatalytic degradation of pollutants in water. N and C co-doping of titania, for instance, may lead to an increase in the Braunauer–Emmett–Teller (BET) surface area of the catalyst and improved photoactivity in the visible and UV regions as both carbon and nitrogen may act as photosensitizers. Wang and Lim[299] described the degradation of bisphenol A by visible light-emitting diode (LED) light in the presence of N and C co-doped TiO_2 prepared by solvothermal synthesis. Although non-metal doping can extend the photocatalytic activity of semiconductor into the visible light region, it often reduces the quantum yield of TiO_2, and recombination of electron–hole pairs has been observed.

On the other hand, the use of gold nanoparticles on the surface of semiconductors can improve the separation of photo-induced electron-hole pairs, but it has almost no effect on extending the photocatalytic response of TiO_2 to the visible light region. Therefore, a promising modification of semiconductors to harvest sunlight avoiding the recombination of the photo-induced pair is the concomitant use of doping and gold deposition on the surface of the semiconductor. Tian *et al.*[300] reported the use of combined noble metal loaded and gold-modified semiconductors for the decomposition of methylene blue and 2,4-dichlorophenol.

Doping might be useful to improve the TiO_2 photocatalytic response in the visible light region. Although gold nanoparticles are able to improve the visible light activity of TiO_2 they are not able to improve the light harvesting achieved by the N-doped semiconductor. A 2 wt.%Au/N–TiO_2 catalyst showed the highest removal of methylene blue using visible light compared with N–TiO_2, Au/TiO_2 and pure TiO_2 catalyst. In this study, it was observed that the improvement in the photocatalytic activity of the semiconductor was due to the deposition of gold, leading to the increased lifetime of photo-induced electron-hole pairs. This was evidenced by the greater number of electrons generated using Au/N–TiO_2 compared with N–TiO_2 and bare TiO_2. In a study carried out by Iliev *et al.*,[177] it was observed that the photooxidation rate of oxalic acid in water using UV light decreases in the following order: Au/N–TiO_2 > Au/TiO_2 > N–TiO_2 > TiO_2, while the reactivity using visible light was: Au/N–TiO_2 > N–TiO_2 >>> Au/TiO_2 > TiO_2. These results confirm the aforementioned concept regarding the combination of gold modification and N-doping semiconductor as an enhancing factor of the photocatalytic performance. The photocatalytic activity of gold modified N-doped TiO_2 is increased approximately twofold when using UV or visible light irradiation compared with both gold-modified and N-doped TiO_2.

The non-metal boron has been found to significantly improve the visible light absorption of TiO_2. Moreover, boron doping can effectively retard the crystal growth of anatase (450 °C) and phase transition from anatase to rutile (650 °C), and modify the bandgap to allow for visible light absorption. However, boron doping does not always result in significant visible light photocatalytic activity. This is caused by the promotion of structural defects with doping (*e.g.* formation of Ti^{3+}), which leads to charge recombination. In these cases, gold deposition on the surface of the doped semiconductor can overcome this problem due to its ability to hinder the charge recombination. Wang *et al.*[170] reported that the deposition of gold nanoparticles on titania increased the charge separation within B/TiO_2. A higher yield in photodecomposition of the azo dye methylene blue using Au/B–TiO_2 composite was achieved (2 wt.% wt gold and 0.25 wt% boron, calcined at 650 °C) compared with B–TiO_2 and pure TiO_2.

For metal ion doping, Xiao *et al.*[207] tested the photocatalytic performance of Au/Fe–TiO_2 expressed as the removal of methyl orange on UV and visible light irradiation. The use of gold nanoparticles on Fe-doped TiO_2 increased the photoactivity of the material by a factor of 4.4 compared with unmodified TiO_2 and by a factor of 1.7 compared with the Fe-doped catalyst. In spite of the combination of Fe-doping and gold deposition, the highest photocatalytic performance was achieved using UV light. Santiago-Morales *et al.*[227] used Ce-doped TiO_2 for the degradation of the β-blocker propranolol with visible light. The authors found mineralization rates of the contaminant of up to 17% after 6 h of irradiation. It is noteworthy that these authors studied the removal of pollutants in treated wastewater, where they found a decline of photodegradation due to the matrix effect.

7.6 Modification of Semiconductors by Gold Nanoparticles for Photocatalytic Hydrogen Production

At the present time, the search and development of alternatives to fossil fuels has become an urgent matter because of several factors: the effects of global warming; the growth of world population, and consequently a higher and increasing worldwide consumption of energy; and the closer perspective of a time-limited durability of oil reserves. All these have drawn the attention to the use and eventual substitution of fossil fuels oil by fuels obtained from clean and renewable sources such as tidal, geothermal, hydraulic, and solar energy.[301,302]

Hydrogen represents an important and feasible fuel option since its energy capacity and clean combustion are very attractive characteristics for its use. Hydrogen can be obtained from water, one of the most abundant natural resources on the Earth, by photocatalytic means using light, another clean and constant source of energy supplied by the Sun. Photocatalysis has drawn attention as it is a viable way to obtain hydrogen by water splitting and interest in it has risen over the past few decades.[303–307]

The photocatalytic water splitting process involves the absorption of light to split water and generate hydrogen fuel by irradiating oxide semiconductors. The first studies of this reaction were performed by Honda and Fujishima at the beginning of the 1970s,[308,309] they reported the decomposition of water and estimated the redox potential for this reaction (1.2 eV). In their experiment, a photoelectrochemical cell, consisting of one electrode of TiO_2 and one of Pt, was irradiated by UV light producing the decomposition of water; the oxidation reaction was performed on the TiO_2 electrode and the reduction of H^+ to H_2 was performed on the Pt electrode.

This reaction can be performed on semiconductors having a bandgap larger than 1.23 eV corresponding to the water splitting potential. The conduction and valence band levels should satisfy this energy requirement to conduct the reduction and oxidation potentials for H_2O, respectively (Figure 7.5).[310,311] The bottom level of the conduction band has to be more negative than the redox potential of H^+/H_2 (0 V *vs.* NHE), while the top level of the valence band be more positive than the redox potential of O_2/H_2O (1.23 V).[312]

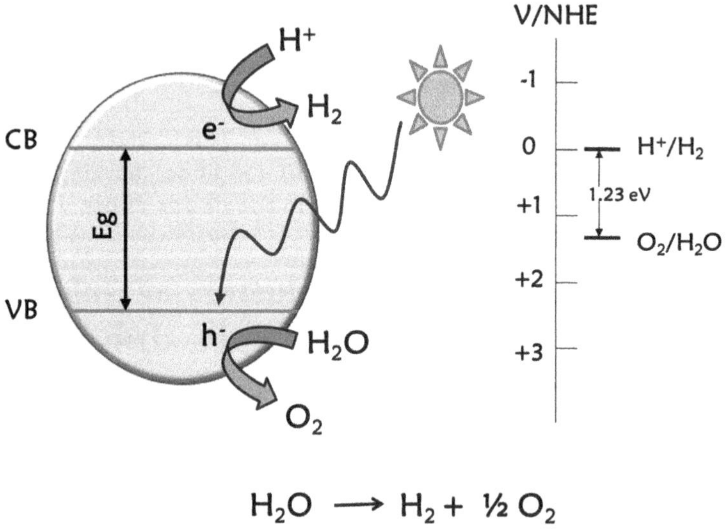

Figure 7.5 Schematic Representation of the Photocatalytic Water Splitting Mechanism.

A number of studies on this topic have been performed through the years,[305,310,312,313] and many semiconductors have been reported for the production of hydrogen and oxygen under UV light. They can be classified into four groups:[305,312] (1) d^0 metal (Ti^{4+}, Zr^{4+}, Nb^{5+}, Ta^{5+}, W^{6+}, Mo^{6+}) oxide photocatalysts; (2) d^{10} metal (In^{3+}, Ga^{3+}, Ge^{4+}, Sn^{4+}, Sb^{5+}) oxide photocatalysts; (3) f^0 metal (Ce^{4+}) oxide photocatalysts; and (4) non-oxide photocatalysts. These materials have been studied and reported to produce hydrogen under UV light. However, as has been mentioned previously, only 4% of the solar spectrum corresponds to UV light, while 46% of the solar light corresponds to visible light irradiation (400–700 nm).[312]

For the last 20 years, the target of producing hydrogen by using visible light has been studied,[201,310,314–316] and semiconductors capable of harvesting solar energy with low quantum efficiencies have been reported. To achieve this goal, it is necessary to build new materials to perform the photocatalytic water splitting reaction in the presence of visible light. Several techniques have been developed to improve the visible light absorption of semiconductors such as: (1) metal or nonmetal loading;[304,307,317] (2) deposition of transition metal oxides;[304] (3) metal or non-metal doping,[312] for example non-metals such as C, N, and S that shift the photoresponse to the visible region;[307] (4) semiconductor coupling;[307,312] and (5) dye surface sensitization.[305]

The photocatalytic systems for water splitting can be classified into two main kinds: sacrificial reagent-containing water-splitting systems and overall water-splitting systems.[312] These sacrificial electron donors reagents can irreversibly consume photogenerated holes, reducing undesirable charge recombination.[317] Organic molecules such as alcohols, organic acids, aldehydes, glucose, and dyes, among others, can be used as electron donors for photocatalytic hydrogen generation; inorganic sacrificial reagents such as S^{2-}/SO_3^{2-}, Fe^{3+}, Ce^{4+}, and I_3^- have been used.[312]

The photocatalytic applications of gold nanoparticles supported on different materials[315,316] are numerous and outstanding. For hydrogen production, gold nanoparticles used as co-catalysts have been proved to increase activity by several orders of magnitude on a wide variety of semiconductors.[318] Gold nanoparticles possess the ability to absorb visible light, because of the SPR, to increase the separation of the electron–hole by acting as a sink of electrons, reducing the possibility of recombination;[307] the electrons accumulate on the gold nanoparticles and the Fermi levels shift closer to the conduction band of the support,[319] resulting in more negative energy levels. Then the accumulated electrons on the metal particle can be transferred to protons adsorbed on the surface resulting in hydrogen, molecules (Figure 7.6).

Gold nanoparticles supported on different semiconductors have been applied for different catalytic reactions, among them, photocatalytic water splitting.[1,106,316,317] The most studied and reported support for gold nanoparticles in photocatalytic water splitting is TiO_2;[10,11,318–326] nevertheless, some other semiconductors have also been reported as supports: $NaTaO_3$,[327]

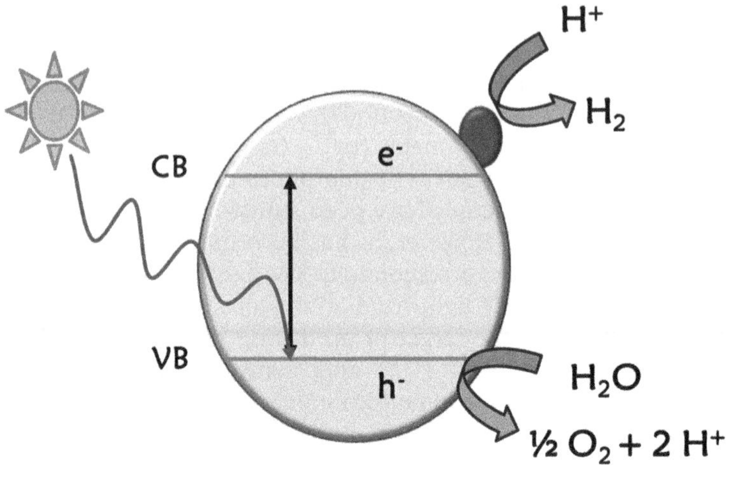

Figure 7.6 Effect of Gold Nanoparticles on the Surface of the Semiconductor.

$K_4Nb_6O_{17}$, $Sr_2Nb_2O_7$, $KTaO_3$, $NaTaO_3$, Na–TaO_3 : La,[328,329] ZnS,[330] Fe_2O_3,[48] TiO_2/WO_3,[276] and $LaInO_3$.[331]

The first photocatalytic studies of gold nanoparticles (from 2–5 nm) highly dispersed on TiO_2 for hydrogen production were reported by Bamwenda *et al.*[332] using an aqueous solution of ethylene glycol, they reported an optimal gold loading of 2 wt.%, a higher activity at neutral pH and an optimal thermal treatment of 673 K. Bamwenda *et al.*[322] also studied the production of hydrogen using platinum and from gold nanoparticles in ethanol–water solution, evaluating the deposition method and metal loading, and finding a production of hydrogen 30% higher for platinum nanoparticles than that observed when using gold nanoparticles; the most active photocatalysts for both metallic nanoparticles were generated by the photodeposition method obtaining methane, carbon dioxide, and traces of acetic acid as byproducts of the reaction.

Chiarello and co-workers[325] evaluated photocatalytic hydrogen production using gold nanoparticles on a series of TiO_2 samples synthesized by spray pyrolysis. The effect of the different crystalline phases anatase and rutile was compared with that of commercial TiO_2 P25, as well as the presence of methanol as a hole scavenger for the water splitting reaction under UV light using a mercury lamp of 250 W (330–450 nm). Their results showed that anatase is the most active crystalline phase for water splitting, and that the presence of methanol increased the rate of hydrogen production by a factor of three, whereas the addition of gold nanoparticles on the TiO_2 surface increased the rate of hydrogen generation by a factor of 30. In a later work, Chiarello *et al.*[326] compared the reaction rate for photocatalytic methanol/methanol reforming using flame spray pyrolysis Au/TiO_2 materials in a bubbling reservoir containing a 20 vol.% solution of methanol–water, where the photocatalyst was placed as a bed. The main advantages found in the gas-phase photocatalytic system were the elimination of the mass transfer limitations since the vapor gas cell ensures a more homogeneous irradiation of the photocatalyts and a high reproducibility of the rate of hydrogen production for the same reused photocatalyst. The vapor phase photoreactor increased the reaction rate for hydrogen evolution by over 30% compared to the photoreactor for liquid reaction.

The effect of gold particle size on hydrogen production has been evaluated by several authors. Wu *et al.*[167] evaluated gold nanoparticles supported on TiO_2 P25 by the deposition–precipitation method using NaOH, varying the metallic particle size, methanol concentration, and pH for optimal hydrogen production, and using a 300 W Xe lamp as a light source and a filter to exclude the infrared (IR) radiation. Their results showed that the rate of hydrogen production increased with decreasing the gold particle size from 10 to 3 nm. The authors concluded that the smaller gold particles deposited on the titania surface induced a better charge separation, generating more holes, depressing the CO production and increasing the oxidation of the formic acid intermediate to CO_2; the optimal pH for the photocatalytic reforming of methanol solutions was obtained at a neutral pH of 7, where a more favorable interaction between CH_3OH and the photocatalyst surface is achieved. Oros-Ruiz *et al.*[11] reported the effect of gold particle size as well as an evaluation for optimal gold-loading thermal treatment for hydrogen production from a 1 : 1 methanol-water solution under UV light using a low energy mercury lamp of 2 W, with a primary emission at 254 nm. In that work, the deposition of gold nanoparticles was performed on TiO_2 P25 by the deposition method using urea. It was observed that the optimal gold loading was 0.5 wt.% and that the optimal thermal treatment was between 150 and 300 °C, generating average particle sizes of 2.8 ± 0.2 nm.

Naldoni *et al.*[333] evaluated the presence of Pt and Au/TiO_2 for photocatalytic reforming of methanol to hydrogen; the materials were prepared by an impregnation–reduction method for metal loadings of 1 wt% and evaluated *in situ* by electron paramagnetic resonance (EPR). The system consisted of a bubbling system filled with a 20 vol.% aqueous methanol solution. A carrier gas N_2 was saturated by methanol–water vapors and fed to the catalyst bed. The irradiation source was a xenon arc lamp emitting UV–Vis light. The photoactivity was greatly improved in the presence of the metal nanoparticles by increasing the hydrogen production rate by more than one order of magnitude; the platinum catalyst was more active than the gold co-catalyst. It was corroborated by EPR that, when gold or platinum particles were added, several cationic surface sites were available for the stabilization of the superoxide radical and the photogenerated electrons were easily transferred to the metal particles, reducing the recombination rate. Platinum nanoparticles on the TiO_2 surface acted as electron sink more efficiently than gold nanoparticles, probably because of a larger difference between their work function compared with that of TiO_2.

Nadeem *et al.*[334] studied hydrogen evolution from ethanol by Au/TiO_2 materials varying the TiO_2 anatase particle sizes from nano (<10 nm) to micro (0.15 μm) size, using similar gold nanoparticles size under UV light. The metallic nanoparticles were deposited by deposition precipitation using urea, for nominal gold loadings of 2 wt.%. The photocatalysts were pre-treated under H_2 flow up to 400 °C for 15 h prior to the photocatalytic tests. The irradiation sources were six lamps of 15 W black UV light of 350 nm wavelength. The main products obtained were hydrogen, acetaldehyde, methane, carbon dioxide, and carbon monoxide. The nanosized Au/TiO_2 particles showed to be one order of magnitude more active for hydrogen production compared with the micro-sized Au/TiO_2 photocatalysts.

Murdoch *et al.*[320] studied gold nanoparticles supported on sol-gel TiO_2 (anatase and rutile); the resulting photoactivity was studied for hydrogen production under UV light irradiation supplied by six 150 W UV lamps of wavelength 350 nm, using an ethanol solution and varying the crystalline phase of TiO_2 between anatase and rutile, and the gold loadings between 1 and 8 wt.%. Their results showed that the most active crystalline phase was anatase, presenting a photocatalytic production of hydrogen two orders of magnitude higher than that obtained from Au/TiO_2 rutile.

In another study, the effect of gold nanoparticles on the crystalline composition of TiO_2 P25 for the photocatalytic hydrogen production under UV light was reported by Jovic *et al.*[335] The gold nanoparticles were supported on TiO_2 P25 by deposition–precipitation, varying the nominal gold loadings between 0 and 10 wt.%, obtaining similar gold particle sizes of about 5 nm for each loading. The reactions were carried out in ethanol–water solutions (80% ethanol–20% water) using a 200 W lamp with an emission of 365 nm as irradiation source. The higher rates for hydrogen production of Au/TiO_2 P25 were obtained for gold loadings between 0.5 wt.% and 2 wt.% showing yields in the order of 32–34 mmol g^{-1} h^{-1}. By normalizing the rate of hydrogen production with the total amount of gold, it was observed that the 0.5 wt.% Au/TiO_2 sample was the most active photocatalyst for hydrogen production.

To determine the effect of gold/rutile and gold/anatase on the electron transfer for hydrogen production, anatase and rutile phases were obtained from TiO_2 P25 by selective chemical dissolution. Gold nanoparticles on each crystalline phase were deposited using a 3 wt.% loading and photoactivity was compared for the two loaded phases using the same gold loading. It was observed that the lowest activity was obtained by pure TiO_2 P25 (0.4 mmol g^{-1} h^{-1}), whereas almost the same activity was observed for bare rutile and anatase phases (1.4 and 1.3 mmol g^{-1} h^{-1}, respectively). It was proposed that the rapid electron–hole pair recombination limits hydrogen production in the absence of co-catalyst. The addition of gold as co-catalyst increased the hydrogen production for gold/anatase and gold/rutile to 21.7 and 10.3 mmol g^{-1} h^{-1}, respectively. When a physical blend of gold/anatase and gold/rutile displaying a similar anatase to rutile ratio than that of TiO_2 P25 (85 : 15) was evaluated, hydrogen production was half that obtained for Au/TiO_2 P25, indicating that gold nanoparticles at the close interface of rutile to anatase enhance the hydrogen yield.

Gärtner *et al.*[318] studied the effect of different gold precursors such as $NaAuCl_4 \cdot 2H_2O$, $AuCl_3$, AuCl, $[Au(PPh_3)]Cl$, and $NaAuCl_4 \cdot 2H_2O$ in basic and acidic media for hydrogen production using the *in situ* synthesis by photodeposition of 1 wt.% Au/TiO_2 obtaining particle sizes of gold in the range of 3–15 nm. The most active precursors were $NaAuCl_4.2H_2O$ and $AuCl_3$. Methanol was the most efficient hole scavenger compared with other organic molecules such as glucose, isopropyl alcohol, and glycerin. Experiments were performed under UV light irradiation using a 1.5 W Xe lamp. These materials were tested under visible light irradiation but no activity was observed with light of $\lambda > 400$ nm. The long-term stability of these materials was investigated under UV light (320 nm $< \lambda < 400$ nm) and the reaction over two days evolved with no decreasse in activity.

Bowker *et al.*[336] evaluated hydrogen production by methanol reforming using Au/TiO_2 and Pd/TiO_2. It was observed that the use of glycerol produced three times more hydrogen than methanol.[337] More recently Bowker *et al.*[338] proposed hydrogen production from alcohols, evaluating the activities of gold, palladium, and Au–Pd as co-catalysts on TiO_2. These materials were prepared by incipient wetness impregnation (IW). Additionally, another two colloidal methods (CM) for the deposition of Au/TiO_2 were evaluated; using citrate (Turkevich method) and $NaBH_4$. The photocatalytic reactions were performed under UV light using a 400 W Xe lamp as irradiation source. It was observed that the IW method produced the most active Au/TiO_2 photocatalyst calcinated at 400 °C. Under the same conditions, the resulting activity of Pd/TiO_2 was 30% higher than that of the Au/TiO_2 photocatalyst. By analyzing the metal loading, it was found that the optimal value for Pt/TiO_2 was 0.1 wt.% for hydrogen photoproduction, being 1.5 times higher than that of Au/TiO_2. When the sample Au–Pt/TiO_2 by CM was evaluated at 0.5 wt.% of each metal, a synergetic effect was observed duplicating the photoactivity obtained by the Pd/TiO_2 sample, suggesting that an Au–Pt alloy is formed which traps the photoexcited electrons more efficiently reducing the recombination process. Also, it is proposed that Au–Pd alloys destabilize CO, avoiding the deactivation of the photocatalyst. Au/TiO_2 was proved to be efficient for hydrogen production from different alcohols, in the following order: methanol > propanol > ethanol > butanol > isopropyl alcohol > tert-butanol.

Hydrogen production by *in situ* photodeposition of different metals on TiO_2 P25 using glucose solutions was studied under UV light by Gomathisankar *et al.*[339] The irradiation source was a LED lamp of 365 nm. Metal ions and their redox potentials *versus* standard hydrogen electrode (SHE) were $Ag^+ = 0.799$ V, $Au^{3+} = 1.002$V, $Cu^{2+} = 0.3419$ V, $Ni^{2+} = -0.257$ V, $Pd^{2+} = 0.951$V, $Rh^{3+} = 0.758$ V, and $Zn^{2+} = -0.7618$ V. To reduce a metal ion by photocatalysis, the conduction band of the semiconductor must be more negative than the reduction potential of the metal ion. At pH = 7, the conduction and valence bands of the TiO_2 are -0.46 and $+2.7$ V, respectively. Therefore, the photodeposition of Zn^{2+} did not occur. The photocatalytic activities obtained for the different metal ion deposition might be related to the work function of the deposited metal. The reported work functions of metals Pd, Au, Rh, Cu, and Ag are 5.12, 5.1, 4.98, 4.65, and 4.26 eV, respectively; all of them are larger than that of TiO_2 (4.2 eV). The only metals capable of performing hydrogen production in decreasing order were Pd, Au, and Rh, which is consistent with the order of the work functions of these metals. It was observed that hydrogen production increased by increasing the concentration of glucose, following a Langmuir–Hinshelwood mechanism where the concentration of glucose found the optimal value of 1 wt.% in aqueous solution; the metal ion concentration was 5 ppm. The photoproduction of hydrogen using Au/TiO_2 was 203 times higher than that observed for TiO_2, whereas for Pd production was 362 fold more efficient than that of bare catalyst.

7.6.1 Photocatalytic Water Splitting Reactions under Visible Light

The use of gold nanoparticles as co-catalysts increases the light absorption of semiconductors in the visible region, since the former present light absorption because of the SPR. The mechanism proposed for photocatalytic hydrogen production under visible light is presented in Figure 7.7;[1,10,340] when gold nanoparticles are photoexcited, electrons from them are injected into the TiO_2 conduction band, leading to the generation of holes on the gold nanoparticles and electrons into the semiconductor's conduction band. The latter are known to provoke hydrogen generation, and the holes will be quenched by the sacrificial electron donor.

The effect of the excitation wavelength of the SPR under UV and visible light was studied by Gomes Silva *et al.*[10] using the resonance of gold nanoparticles as light harvesters to perform photocatalytic water splitting for

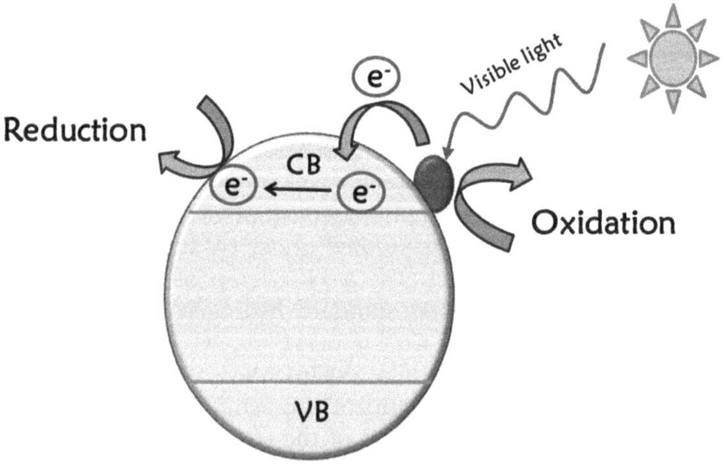

Figure 7.7 Photoexcitation of a Semiconductor Under Visible Light.

hydrogen and oxygen production. The synthesis of 1.5 wt.% Au/TiO_2 P25 catalysts was conducted by deposition–precipitation using NaOH from a 3 : 1% v/v water–methanol solution and 0.02 M of ethylene diamine tetra-acetic acid (EDTA) as hole scavenger. Photoactivity of TiO_2 under UV light was performed using a 200 W Hg–Xe mercury lamp of $\lambda < 380$ nm, the photocatalytic water splitting proceeded by direct activation of titania, whereas the generated holes in the valence band were quenched by EDTA, and electrons on the conduction band migrated to the gold nanoparticles acting as catalytic sites for hydrogen generation. The studies under visible light were performed by polychromatic light using an Hg–Xe lamp ($\lambda > 400$ nm) and also by using a monochromatic light from a laser with wavelength of $\lambda = 532$ nm. TiO_2 did not present activity under these visible light irradiation conditions. On the other hand, the materials containing 1.5 wt.% Au/TiO_2 presented hydrogen production. The use of a monochromatic wavelength of 532 nm from the laser pulse showed a three-fold increase of the volume of hydrogen evolved compared with that obtained using the polychromatic light. In both cases, using visible light, the mechanism was explained by a direct photo-excitation of electrons in gold nanoparticles, which were then injected into the TiO_2 conduction band, leading in turn to the generation of holes on the gold nanoparticles.

Rosseler *et al.*[341] evaluated photocatalytic hydrogen production under solar light simulation using platinum and gold nanoparticles supported on TiO_2 by varying the anatase–rutile phases, methanol–water ratios (0–25% v/v), and the metal loading. The optimal platinum loading was 0.3 wt.% on a 50 : 50 anatase–rutile phase mixture, whereas for Au/TiO_2 it was 2 wt.% on a 80 : 20 anatase–rutile TiO_2 P25. Under the same reaction conditions the optimal Pt/TiO_2 photoactivity was enhanced by several orders of magnitude compared to that observed for Au/TiO_2 photocatalyst, and for this last catalyst, the presence of 1% v/v of methanol showed good efficiency for hydrogen production. Under optimized conditions, hydrogen production could be maintained over 10 days without deactivation of the photocatalysts and without the addition of extra methanol.

Ilie *et al.*[342] proposed the re-use of wastewaters to simultaneously help water decontamination and the photoproduction of hydrogen. The photocatalytic hydrogen production obtained from sugar solutions as sacrifice molecules was studied for gold deposited on TiO_2 and TiO_2/SiO_2, TiO_2, and TiO_2/SiO_2 modified by doping metals such as Cr, Mn, and V, and also by non-metals such as N and S to expand the light absorption of titania to the visible region. For all the TiO_2 dopants, the nominal loading was fixed at 1 wt.%, whereas for gold nanoparticles it was 0.5 wt.%. The reactions were carried out by the irradiation of commercial sugar solutions (10 wt.%) with visible light using a 200 W Ne lamp. The results showed that anion doping and metal doping of TiO_2 presented a very similar activity; however, the best performance was achieved by S–TiO_2 followed closely by V/TiO_2 and 0.5 wt.%Au/1 wt.%TiO_2/SiO_2.

The effect of sub-nanometre gold particles on CdS was studied by Shen *et al.*[343] using sub-nanometre particles of gold synthesized from $Au(PPh_3)Cl$

and the reducing agent, borane–*tert*-butylamine complex. Varying the Au/CdS ratios, they found an optimal gold loading of 1 wt.% with average gold particle sizes of 0.95 ± 0.1 nm on CdS. These photocatalysts produced about 404 μmolH_2 h^{-1}, increasing up to 35 times the hydrogen production obtained by bare CdS nanoparticles. The reactions were carried out using Na_2S and Na_2SO_3 as sacrificial reagents under visible light using a 300 W Xe lamp with a cutoff filter allowing only $\lambda > 420$ nm and a water filter for $\lambda < 800$ nm. The materials proved to be stable for 18 h of reaction. When gold nanoparticles of about 3 nm were synthesized at the same metal loading (1 wt.%) and evaluated under the same conditions, the photoactivity for hydrogen production decreased, generating only 30 μmolH_2 h^{-1}. The effect of gold as co-catalyst was compared with other noble metals such as Pt, Ru, Pd and Rh, synthesized by photodeposition to generate similar particle sizes of 3 nm. It was observed that the order of the photoactivities for hydrogen production was Pt/CdS > Ru/CdS > Au/CdS > Pd/CdS > CdS ≈ Rh/CdS.

Zhang *et al.*[330] studied the photocatalytic activity for hydrogen production under UV and visible light irradiation of gold nanoparticles supported on ZnS flower-shaped nanostructures. The photocatalytic reactions for hydrogen production were carried out in aqueous solutions containing 0.35 M Na_2S and 0.25 M Na_2SO_3 as sacrificial reagents and evaluating different gold loadings (0, 2, 4, and 6 wt.%/ZnS) to optimize hydrogen production. These structures were also compared with the activity obtained by nanometrical ZnS particles of about 3 nm. The studies under UV light showed a high efficiency for hydrogen generation by Au/ZnS nanostructures using a 300 W Xe arc lamp as irradiation source, where bare ZnS flower-shaped nanostructures reported an activity of 1602 $\mu mol \cdot h^{-1} \cdot g^{-1}$ being about six times more active than 3 nm ZnS particles. When the gold loading increased on the ZnS flower-shaped nanostructures, an optimal value of 4 wt.% was reached, duplicating the activity up to 3306 $\mu mol \cdot h^{-1} \cdot g^{-1}$, showing that these metallic particles act as efficient attraction centers for electrons, avoiding their recombination. A further increase of gold loading to 6 wt.% resulted in a detrimental effect on activity. The stability of the Au/ZnS flower-shaped structures was also studied for the optimal gold loading. After four cycles, the photocatalyst did not exhibit any loss of activity, proving to be stable under the photocatalytic reaction. When these structures were tested under visible light irradiation ($\lambda > 420$ nm), however, no activity was observed despite the fact that these materials presented absorption at 553 nm SPR. These results were explained as a highly negative reduction potential of the ZnS conduction band that prevented the excited energetic electrons from the SPR state to be injected from the conduction band to the ZnS.

Recent works have reported a major improvement for photocatalytic hydrogen production using Au/TiO_2 and visible light. These studies present a promissory scenario for performing water splitting by using only visible light, which is known as plasmonic photocatalysis. Seh *et al.*[344] synthesized and reported efficient non-centrosymmetric Au–TiO_2 photocatalysts under visible light irradiation by using the localized SPR. The Janus Au_{50nm}–TiO_2 particles' photocatalytic performance was compared with that obtained by core@shell Au_{50nm}–TiO_2. The reactions were performed in a 1 : 2 v/v isopropanol–water solution under visible light irradiation using a 500 W tungsten halogen lamp, with a cutoff filter for $\lambda > 400$ nm. The rate of visible light hydrogen generation was significantly higher for the Janus Au_{50nm}–TiO_2 photocatalyst because of a stronger localization of plasmonic near-fields close to the Au–TiO_2 interface by a synergetic interaction generating 1.7 times the hydrogen photoproduction by using a core@shell Au_{50nm}–TiO_2 photocatalyst. The gold particle size was optimized and the most active photocatalyst under visible light irradiation for the generation of hydrogen was Janus Au_{70nm}–TiO_2, showing an efficient visible light plasmonic photocatalysis generating up to 170 mL of H_2 in 3 h. These results are very promising for solar light harvesting since they overcome the low hydrogen production previously reported even using Au/TiO_2 under UV light irradiation.

7.6.2 Alternative Supports to TiO_2 Reported for Hydrogen Production

The activity of gold nanoparticles have been reported using semiconductors other than TiO_2. Lin and Chang[327] studied the effect of gold nanoparticles on potassium titanoniobiate ($KTiNbO_5$), comparing the resultant photoactivity with TiO_2 P25 and $NaTaO_3$: La semiconductors. The photoactivity for hydrogen production of Au/$KTiNbO_5$ was also evaluated using different deposition methods: deposition–precipitation, impregnation, and photodeposition using nominal gold loadings of 1.0 wt.%. The experiments were performed under UV light by means of a 400 W medium pressure halide lamp emitting at $\lambda_{max} = 360$ nm in a methanol–water 1 : 5 solution. The photoactivities for water splitting were obtained as follows: DPAu/$KTiNbO_5$ > IMPAu/$KTiNbO_5$ > FDAu/$KTiNbO_5$ > $NaTaO_3$: La > TiO_2 P25 > $KTiNbO_5$. The DPAu/$KTiNbO_5$ sample containing 0.63 wt.% gold showed an activity 47 times higher than that obtained for pure $KTiNbO_5$.

Puangpetch *et al.*[345] reported hydrogen production using gold nanoparticles on $SrTiO_3$ and the effect of different hole scavengers such as methanol, formic acid, acetic acid, and propanoic acid under UV light irradiation. The nominal gold loading for these experiments was 1 wt.%. Hydrogen production using only water splitting without the presence of any hole scavenger was observed. A positive effect of hole scavengers was observed by using formic acid at 2.5 vol.%, increasing the hydrogen production rate compared with the pure system. A negative effect of hole scavengers occurred when using propanoic acid. The catalytic activity decreased significantly with increasing concentration. Methanol and acetic acid were reported as non-effective hole scavengers for this Au/$SrTiO_3$ system since the hydrogen production rate tended to remain unchanged. The most effective pH for hydrogen generation occurred at natural pH. From these results, the authors concluded that both the small molecular size as well as the complete dissociation of formic acid were the characteristics responsible of the higher hydrogen production with the Au/$SrTiO_3$ photocatalyst.

Iwase *et al.*[328] evaluated the photocatalytic water splitting activity of $K_4Nb_6O_{17}$, $Sr_2Nb_2O_7$, $KTaO_3$, $NaTaO_3$, and $NaTaO_3$ doped with La using either pure water or a 10% v/v methanol–water solution, comparing their activity with the modified supports with 1 wt.% of gold nanoparticles synthesized by photodeposition, obtaining gold particle sizes between 22 and 48 nm. The reactions were carried out under UV light using a 300 W Xe lamp. The results showed that bare $Sr_2Nb_2O_7$ and $KTaO_3$ were not active for hydrogen generation. $K_4Nb_6O_{17}$, $NaTaO_3$, and $NaTaO_3$: La presented activity for hydrogen production. These activities were improved at least by four times when gold nanoparticles were deposited on their surface, resulting in active sites for hydrogen production. Iwase *et al.*[329] studied the effect of deposition method of gold nanoparticles on La:$NaTaO_3$ in the photocatalytic water splitting. The gold (1 wt.%) co-catalyst on La:$NaTaO_3$ prepared by impregnation improved the activity of bare La:$NaTaO_3$ by 17 times, whereas the same catalyst (1 wt.% Au) prepared by photodeposition increased the hydrogen production reported by the bare support by 7 times. The impregnated materials also showed higher and more stable activity. The effect was attributed to the periphery of the interface between gold and the La:$NaTaO_3$ support. The photodeposited gold nanoparticles presented a larger periphery than the impregnated ones; this contact structure might affect the ability of the catalyst for oxygen activation.

Dhanasekaran and Gupta[331] studied the effect of gold nanoparticles on doped $LaInO_3$ with Ga and N for the photoreduction of water and hydrogen generation. The materials were prepared *via* co-precipitation and solid state reaction. The metal/oxide nanocomposites were fixed at a metal loading of 1 wt.% of Pt, Au or NiO by incipient impregnation. The photoactivities were measured under UV light (250 nm $< \lambda <$ 420 nm) without the presence of any sacrificial reagent; they were also evaluated under visible light irradiation using a 300 W Xe lamp with a filter for $\lambda >$ 395 nm using a methanol–water solution (2:1 v/v). The undoped $LaInO_3$ prepared by co-precipitation showed poor activity under UV light irradiation, increasing as the content of Ga increased (to 0.1 and 0.2); doping with N increased the activity of $LaInO_3$ significantly. Hydrogen evolution increases consistently with loading co-catalysts, showing a better performance for Au (130 $\mu mol \cdot h^{-1} \cdot g^{-1}$) > Pt > NiO. For the experiments under visible light irradiation, however, no water splitting was observed for Ga : $LaInO_3$, and only small amounts of hydrogen were generated using the N : $LaInO_3$ doped sample (about 1 $\mu mol \cdot h^{-1} \cdot g^{-1}$); this activity increased three-fold by using gold as co-catalyst, showing a higher hydrogen production under visible light than that observed for the other co-catalysts evaluated (Au > Pt > NiO). The results obtained for this system showed that doping with N led to a significant increase in the catalytic activity of $LaInO_3$ while gold was the best co-catalyst compared with Pt and NiO at similar loadings, revealing that the outstanding presence of gold nanoparticles as co-catalysts may be associated to their high dispersion, metal–support interfaces, and optical properties.

7.7 Use of Semiconductors Modified with Gold Nanoparticles for Photocatalytic Removal of Air Pollutants

Although most of the work on photocatalysis has been focused on the degradation of organic pollutants in aqueous media,[29,346–348] photocatalysis is also of interest for the abatement of pollution in the gas phase (mainly atmospheric pollution) because of carbon monoxide, nitrogen oxides (NO_X), volatile organic compounds (VOCs) and other air pollutants,[1,16,107,159,166,171,172] and for the conversion of carbon dioxide for fuel production.[349]

Gold nanoparticles supported on transition metal oxides have an extraordinary catalytic performance for CO oxidation at room temperature (in the dark). Only a very few works have studied the photocatalytic activity of gold/semiconductors in CO oxidation. Zhang *et al.*[172] studied the photocatalytic activity in CO oxidation under UV light illumination ($\lambda = 365$ nm) at room temperature of pure and gold-deposited ZnO nanoparticles prepared by deposition–precipitation with differing gold content. These authors showed that gold deposition can enhance the photocatalytic activity of CO oxidation; this was attributed to the increased amounts of $^{\bullet}O_2^{-}$ and the formation of O (or O −) on the Au/ZnO surface which can react directly with CO on the Au/ZnO surface under UV light illumination. Moreover, they showed that there are certain relationships between photoluminescence (PL) spectra and photocatalytic activity, namely, the higher the gold-deposited content, the weaker the PL intensity; thus the photocatalytic activity of CO oxidation is higher, which according to the authors, confirmed the effective transfer of photogenerated electrons from ZnO to gold nanoparticles under UV light illumination.

NO_X are harmful air pollutants because of their role not only in the generation of photochemical smog and acid rain, but also in the promotion of ozone formation.[350] The use of TiO_2 and Au/TiO_2 as a photocatalyst for NO elimination has been scarcely studied despite the dual benefit that can be obtained with this system, namely, the photocatalytic activity of TiO_2 and NO adsorption on the gold nanoparticles.[16,351–353] The photocatalytic activity of Au/TiO_2 has been evaluated by the photocatalytic decomposition of NO using an ultraviolet light source of 365 nm.[16,351] TiO_2 was prepared by gelling Ti(IV) isopropoxide, and gold nanoparticles were added by the deposition–precipitation method with urea.[16] The gold particle size was distributed between 4.4 and 6.7 nm, and the bandgap energy of Au/TiO_2 was lower than the one observed for bare TiO_2. A NO + O_2 mixture (150 ppm) was used to evaluate the photocatalytic activity *in situ*, at room temperature, under atmospheric pressure. The Au/TiO_2 materials showed an enhanced photocatalytic activity when compared with the reference TiO_2. The highest activity was obtained for Au/TiO_2 with 0.3 wt.% of gold; a further increase in gold loading reduced the photoactivity. The analysis of the stoichiometric balance indicated that the main reaction deals with the NO reduction to form N_2 and O_2; some formation of N_2O was detected, whereas formation of NO_2 was not observed. The role of the gold nanoparticles was associated with the decrease of the e^-/h^+ pair recombination rate, a higher interaction of NO with the catalyst surface, and the displacement of the bandgap to lower energy values. The photocatalytic oxidation of NO has been also studied with gold supported on mesoporous TiO_2 films deposited on a tin-doped ITO layer. Incorporation of gold resulted in a significant improvement in the photonic efficiency and a significant increase in the photocatalytic oxidation of NO.[12] Moreover, it was reported that ultra-small sub-nanometer 1 nm gold nanoparticles supported on TiO_2 synthesized by a grafting method using $Au(PPh_3)Cl$ dissolved in $CHCl_3$, using $P(PPh)_2(CH_2)(PPh)_2$ as stabilizer, exhibited a remarkable photocatalytic activity for the NO_2 conversion. It was shown that the activity of Au/TiO_2 was six times higher than that of unmodified catalysts.[354]

Other type of air pollutants are VOCs, which originate largely from industrial processes, internal sources such as outgassing of building materials and furnishings, cleaning products, and daily activities. VOCs lead to lower air quality, global warming, and climate change with a consequent negative impact on environmental health. This impact occurs both in indoor and outdoor environments. Indoor environments typically contain higher levels of VOCs than outdoor environments with exposure to these pollutants potentially causing headaches, nausea, fatigue, and other symptoms. VOCs include formaldehyde (HCHO), acetaldehyde (CH_3CHO), benzene (C_6H_6), acetone (CH_3COCH_3), and aromatic compounds, among others.[18]

Removal of VOCs by gold photocatalysis has been reported. For example, Kowalska *et al.*[201] reported the visible light induced oxidation of 2-propanol on gold-modified titania through the excitation of the gold surface plasmon by using action spectrum analyses. They speculated that an electron may be injected from the gold particles into the conduction band of TiO_2 and then molecular oxygen adsorbed on the surface of TiO_2 could be reduced. The resultant electron-deficient gold could oxidize organic compounds such as 2-propanol, and be recovered to the original metallic state. The photocatalytic activities of core@shell $Au@TiO_2$ nanoparticles were investigated by oxidation degradation of gaseous acetaldehyde.[355] This structures presented high photocatalytic activity when exposed to UV or visible light irradiation.[128] Lee *et al.*[159] studied the photocatalytic degradation of gas phase toluene (C_7H_8) by deposition of Ag, Au, Pt, Pd, and Rh metals on Degussa P25 TiO_2 surface, at a nominal loading of 0.5 wt.%. An enhancement of the photocatalytic performance and inhibition of deactivation by a factor of two were observed for Pd, Rh, and Au.

The photocatalytic oxidation of acetone was studied using Au/TiO_2 catalysts prepared by direct current (DC) magnetron sputtering and by deposition–precipitation.[107] The results showed that the photocatalytic activity of these Au/TiO_2 samples depended strongly on the preparation method. The catalysts prepared using DC reactive sputtering with particle sizes between 6.2 and 15 nm showed activity under both UV and visible light irradiation conditions, whereas those prepared using the deposition–precipitation technique with an average particle size of about 3.4 nm showed activity only under UV activation.[107] Moreover, it has been shown that photo-induced degradation of acetone takes place more rapidly than the decomposition of intermediate species adsorbed on the surface.[356]

A new possibility that has been recently studied is the photocatalytic reduction of CO_2 for fuel production. Photocatalytic conversion of CO_2 into hydrocarbons is of great interest for its potential to convert an abundant greenhouse gas to useful hydrocarbon fuels. It has been shown that deep UV activation of CO_2 can lead to the selective formation of CO or even to the reduction to methane by hydrogen or water. Photochemical CO_2 reduction requires the presence of hydrogen, water, or other renewable reducing agents as a source of hydrogen.

As has been mentioned in Section 7.6, photocatalytic hydrogen generation from water is particularly efficient when metal nanoparticles are deposited on the surface of a semiconductor to act as co-catalysts.[349] Methane formation by the reduction of CO_2 with H_2O ($CO_2 + 2H_2O \rightarrow CH_4 + 2O_2$) is a more complicated process than the water splitting reaction because the reduction of CO involves eight photons to reduce CO_2 to CH_4. In addition to methane, methanol and formaldehyde are also possible products. The conduction band edge and photovoltage of TiO_2 are barely high enough to carry out direct CO_2 reduction and water oxidation. Because of this, an additional CO_2 reduction catalyst is typically needed in order to accomplish reasonable CO_2 reduction directly on a semiconductor surface.[349] The reasons for the lower efficiency and higher difficulty of the photocatalytic CO_2 reduction are both thermodynamic and kinetic.[349]

It has been proposed that TiO_2 is one of the most promising photocatalysts for carbon dioxide reduction.[43] The use of co-catalysts such as gold seems very promising for a plasmonic photocatalytic enhancement. Recently, Hou *et al.*[43] found that the quantum efficiency was improved by a factor of 24 under visible illumination by integrating gold nanoparticles onto TiO_2 (2.1×10^{-5}% for Au/TiO_2 and 8.8×10^{-7}% for bare TiO_2).[43] This enhancement was attributed to the strong electric fields created by the SPR of the gold nanoparticles, which excite electron–hole pairs locally in the TiO_2 at a rate several orders of magnitude higher than the normal incident light. They proposed that the plasmon-excited electrons in the gold nanoparticles cannot transfer directly from the gold to the TiO_2. Only when the photon energy is high enough to excite the d-band electrons of gold to a conduction band that lies above the conduction band of TiO_2 direct charge transfer occur between these two materials. The authors proposed that, when the incident photon energy was high enough (254 nm UV), an additional mechanism involving the interband electric transitions in gold produced a number of additional photocatalytic reaction products. In this wavelength range, both the excited electrons in Au and TiO_2 contribute to the reduction of CO_2 with H_2O vapor.[43]

7.8 Use of Gold-modified Semiconductors in Production of Fine Chemicals

In recent years gold-modified semiconductors have emerged as efficient catalysts for fine chemical synthesis either in gas or liquid phase. Oxidation reactions based on molecular oxygen as oxidant have been extensively reported to produce: (1) carboxylic acids, esters, imines, *N*-formylamine, benzymidazole, tricarbonyl compounds, ketones and oxime by oxidation of alcohols; (2) silanol through the oxidation of silane; (3) aryl compounds by oxidation of alkylaromatic compounds; and (4) azo compounds from amines.[357] To give some examples, small gold nanoparticles supported on CeO_2, Al_2O_3, TiO_2, and SiO_2 have been used for aldehyde production from primary alcohols,[358] imines synthesis *via* oxidation of secondary amines,[359] oxidation of amines into amides and lactams[360,361] and oxidative cyclization of aryl-substituted anilines with aldehydes, respectively.[362] However, the production of fine chemicals using photoreactions catalyzed by gold-modified semiconductors is a poorly explored area to date.

Since in photocatalysis using a semiconductor both oxidation and reduction occur simultaneously at the adjacent sites on the catalyst surface, these processes can be used for organic synthesis. Furthermore, the production of free radicals from the oxidation of adsorbed water and oxygen molecules on catalyst may increase the performance, although not necessarily the selectivity, of the fine chemical reactions. As mentioned above, visible light irradiation of gold/semiconductor composites results in the transfer of electrons from gold nanoparticles to the semiconductor's conduction band due to the SPR effect; this may result in the occurrence of mild oxidizing gold nanoparticles, which may be promising materials for fine chemicals production *via* oxidation.

Kowalska *et al.*[363] have reported the selective oxidation of isopropanol into acetone in well-aerated Au/TiO_2 suspensions. The excited electrons in gold nanoparticles by the SPR are transferred to the semiconductor and thus oxidation of organic compounds takes place as gold recovers to the original metallic state. More recent studies report the selective oxidation of cinnamyl alcohol to cinnamaldehyde using gold nanoparticles supported on several semiconductors (*e.g.* ZnO, TiO_2, $BiVO_4$, WO_3, In_2O_3).[364] The authors pointed out the strong dependence of oxidation on the semiconductor nature and size of the gold particles; they found the $Au/SrTiO_3$ to be the best catalyst in terms of aldehyde production. Naya *et al.*[365] studied the Au SPR-induced visible light photocatalytic oxidation of thiol to disulfide using Au/TiO_2

catalyst. Visible light ($\lambda > 420$ nm) led to selective oxidation of thiol to disulfide, whereas the reaction could be reversed by UV light irradiation ($\lambda > 300$ nm). As SPR is the key factor in this oxidation reaction, it is expected that the size and dispersion of gold nanoparticles impact both in oxidation and reduction.

The photoactivity of gold nanoparticles through the SPR under UV or visible light may be manipulated for the synthesis of fine chemicals *via* reduction reactions. Examples of this can be found in the reduction of nitroaromatics into azo compounds, the hydrogenation of azobenzenes to hydroazobenzenes, the reduction of ketones to alcohols, and the reduction of epoxides to alkenes.[366] Au/ZrO_2 has been reported as the most effective catalyst for the photocatalytic reduction of nitrobenzenes to azobenzenes,[367] while Zhao *et al.*[368] set the optimal conditions for the photo-induced hydroamination of alkynes using visible light at room temperature. According to the authors, this reaction is carried out by either the nucleophilic attack of C≡C by amines or the direct addition of N–H bonds to the C≡C bonds.

7.9 Concluding Remarks

Research on gold-modified semiconductors in photocatalysis is growing over the past two decades. Deposition of gold nanoparticles on the surface of some semiconductors makes possible the use of sunlight in heterogeneous catalysis. Surface modification of semiconductors with gold nanoparticles can prove advantageous: in semiconductors with high bandgaps, *e.g.* TiO_2, ZnO_2, and Al_2O_3, the SPR of gold nanoparticles results in light harvesting at higher wavelengths, from 515 to 580 nm, depending on the material; in addition, the recombination of photo-produced electron–hole pairs decreases in semiconductors because of the electron sink effect caused by gold nanoparticles. The latter is of considerable relevance for the use of low-bandgap semiconductors that have a high recombination rate, such as WO_3, CeO_2, and $BiVO_4$.

The role of gold nanoparticles in extending the lifetime of the photo-produced electron–hole pairs, harvesting light in the visible region, and increasing the specific surface area of the catalyst lead us to consider these materials as promising photocatalysts for: (1) the removal of pollutants in water and air; and (2) the production of hydrogen and fine chemicals.

Several methods for the synthesis and deposition of gold nanoparticles on semiconductors aimed at producing small and well-dispersed nanoparticles (between 2 and 20 nm) on the surface of semiconductors have been developed. The most used methods are deposition–precipitation, adsorption of preformed gold nanoparticles (colloids), and photoreduction. Other kinds of modification developed to reduce the bandgap of semiconductors are based on doping with different chemical species (metal and non-metal). However, doping does not necessarily result in a decrease in the rate of recombination of the photo-produced electron–hole pairs; in this sense, future research should be aimed at combining doping and gold-modification to take advantage of the features achieved using both modifications and thus shift the photocatalytic activity of modified semiconductors to visible or sunlight.

Regarding the degradation of pollutants, an extensive body of research has been conducted to determine the photocatalytic potential of a large number of gold-modified semiconductors to remove water pollutants. It is in this kind of study in which important improvements in the photocatalytic activity caused by the gold deposition have been observed. So far, most of the studies aimed at removing water pollutants have dealt with conventional contaminants (*i.e.* azo dyes, phenol, MTBE, Cr^{6+}). Nevertheless, more recent studies have investigated the removal of pollutants of emerging concern in water using gold-modified semiconductors. Such pollutants raise concern because they are not completely removed in conventional treatment systems and may cause subtle but harmful effects to the exposed organisms. Upcoming studies should use the state-of-the-art so far reached to develop heterogeneous photocatalysis systems based on gold-modified materials and sunlight for the removal of conventional and emerging contaminants in natural waters and effluents from wastewater treatment plants.

Contrary to that observed for water pollutants, studies on the removal of air pollutants by photocatalysis using gold-modified semiconductors are scarce. This may be because, as occurs for oxidation of CO with Au/TiO_2, heterogeneous catalysis with no supply of light has shown promising results. Similarly, there is a significant lack of information on the removal of NO_X in air streams by heterogeneous photocatalysis, while the state-of-the-art is more developed for the removal of VOCs using this method. Future prospects should aim at using heterogeneous photocatalysis with gold-modified semiconductors for artificial photosynthesis. Initially, research on the production of hydrogen has been developed, while future studies should aim at producing methanol by CO_2 reduction.

Gold-modified semiconductors have been intensively used for photocatalytic water splitting in recent years; in their use as co-catalysts on different supports, gold nanoparticles have proved to be effective in reducing the energy requirements to split water on systems such as high-power lamps or low-energy sources. Moreover, new materials based on gold nanoparticles have proved to be efficient at working under visible light by enhancing the SPR and reducing the recombination at the semiconductor's surface. Since it has been observed from the reported studies on the topic, the optimal gold loading, morphology, and the design of the metal–semiconductor interface as well as the proper selection of work function of the materials involved in the photo reaction may induce synergetic effects and enhance the photoactivity of these materials.

Acknowledgements

The authors acknowledge CONACyT-Mexico for providing funding in the framework of the 130407 and 163098 projects, as well as DGAPA-UNAM for the funding supplied through the 103513 project.

References

1. A. Primo, A. Corma and H. Garcia, *Phys. Chem. Chem. Phys.*, 2011, **13**, 886.
2. W. Hou and S. B. Cronin, *Adv. Funct. Mater.*, 2013, **23**, 1612.
3. M. A. Green, *Prog. Photovoltaics*, 2009, **17**, 183.
4. N. Q. Wu, J. Wang, D. Tafen, H. Wang, J. G. Zheng, J. P. Lewis, X. Liu, S. S. Leonard and A. Manivannan, *J. Am. Chem. Soc.*, 2010, **132**, 6679.
5. S. K. Cushing, J. Li, F. Meng, T. R. Sentry, S. Suri, M. Zhi, M. Li, A. D. Bristow and N. Wu, *J. Am. Chem. Soc.*, 2012, **134**, 15033.
6. D. Tafen, J. Wang, N. Q. Wu and J. P. Lewis, *Appl. Phys. Lett.*, 2009, **94**, 93101.
7. F. Meng, Z. Hong, J. Arndt, M. Li, M. Zhi, F. Yang and N. Q. Wu, *Nano Res.*, 2012, **5**, 213.
8. M. Li, S. Zhou, Y. Zhang, G. Chen and Z. Hong, *Appl. Surf. Sci.*, 2008, **254**, 3762.
9. K. Awazu, M. Fujimaki, C. Rockstuhl, J. Tominaga, H. Murakami, Y. Ohki, N. Yoshida and T. Watanabe, *J. Am. Chem. Soc.*, 2008, **130**, 1676.
10. C. Gomes Silva, R. Juárez, T. Marino, R. Molinari and H. García, *J. Am. Chem. Soc.*, 2010, **133**, 595.
11. S. Oros-Ruiz, R. Zanella, R. López, A. Hernández-Gordillo and R. Gómez, *J. Hazard. Mater.*, 2013, **263**, 2.
12. I. Bannat, K. Wessels, T. Oekermann, J. Rathousky, D. Bahnemann and M. Wark, *Chem. Mater.*, 2009, **21**, 1645.
13. A. A. Ismail, D. W. Bahnemann, I. Bannat and M. Wark, *J. Phys. Chem. C*, 2009, **113**, 7429.
14. D. Tsukamoto, Y. Shiraishi, Y. Sugano, S. Ichikawa, S. Tanaka and T. Hirai, *J. Am. Chem. Soc.*, 2012, **134**, 6309.
15. F. J. López-Tenllado, A. Marinas, F. J. Urbano, J. C. Colmenares, M. C. Hidalgo, J. M. Marinas and J. M. Moreno, *Appl. Catal. B*, 2012, **128**, 150.
16. J. Hernández-Fernández, R. Zanella, A. Aguilar-Elguezabal, R. D. Arizabalo, S. Castillo and M. Moran-Pineda, *Mater. Sci. Eng. B*, 2010, **174**, 13.
17. S.-W. Cao, Z. Yin, J. Barber, F. Y. C. Boey, S. C. J. Loo and C. Xue, *ACS Appl. Mater. Interfaces*, 2011, **4**, 418.
18. X. Zhou, G. Liu, J. Yu and W. Fan, *J. Mater. Chem.*, 2012, **22**, 21337.
19. A. Tanaka, K. Hashimoto and H. Kominami, *J. Am., Chem. Soc.*, 2012, **134**, 14526.
20. H. Yu, H. Ming, J. Gong, H. Li, H. Huang, K. Pan, Y. Liu, Z. Kang, J. Wei and D. Wang, *Bull. Mater. Sci.*, 2013, **36**, 367.
21. H. Yu, H. Ming, H. Zhang, H. Li, K. Pan, Y. Liu, F. Wang, J. Gong and Z. Kang, *Mater. Chem. Phys.*, 2012, **137**, 113.
22. J. L. Zhang, C. Yang and C. He, *J. Am., Chem. Soc.*, 2006, **128**, 1798.
23. I. Tanahashi, H. Iwagishi and G. Chang, *Mater. Lett.*, 2008, **62**, 2714.

24. V. Augugliaro, V. Loddo, M. Pagliaro, G. Palmisano and L. Palmisano, *Clean by Light Irradiation: Practical Applications of Supported TiO_2*, RSC Publishing, Cambridge, UK, 2010.
25. Y. Pan, S. Deng, L. Polavarapu, N. Gao, P. Yuan, C. H. Sow and Q.-H. Xu, *Langmuir*, 2012, **28**, 12304.
26. C. S. Turchi and D. F. Ollis, *J. Catal.*, 1990, **122**, 178.
27. R. W. Mathews and S. R. MacEvoy, *J. Photochem. Photobiol. A*, 1992, **64**, 231.
28. M. Qamar, Z. H. Yamani, M. A. Gondal and K. Alhooshani, *Solid State Sci.*, 2011, **13**, 1748.
29. M. R. Hoffman, S. T. S. T. Martin, W. Choi and D. W. Bahnemann, *Chem. Rev.*, 1995, **95**, 69.
30. M. Muneer, D. Bahnemann, M. Qamar, M. A. Tariq and M. Faisal, *Appl. Catal. A*, 2005, **289**, 224.
31. J. Tang, J. R. Durrant and D. R. Klug, *J. Am., Chem. Soc.*, 2008, **130**, 13885.
32. P. V. Kamat, *J. Phys. Chem. C*, 2007, **111**, 2834.
33. U. Cernigoj, U. L. Stangar, P. Trebse and P. R. Ribic, *Acta Chim. Slov.*, 2006, **53**, 19.
34. D. Daniel and I. G. R. Gutz, *Electrochem. Commun.*, 2007, **9**, 522.
35. A. Dawson and P. V. Kamat, *J. Phys. Chem. B*, 2001, **105**, 960.
36. S. Hotchandani and P. V. Kamat, *J. Phys. Chem.*, 1992, **96**, 6834.
37. Y. Liu, L. Chen, J. Hu, J. Li and R. Richards, *J. Phys. Chem. C*, 2010, **114**, 1641.
38. A. Kafizas, S. Kellici, J. A. Darr and I. P. Parkin, *J. Photochem. Photobiol.*, 2009, **204**, 183.
39. Y. Wu, J. Zhang, L. Xiao and F. Chen, *Appl. Catal., B*, 2009, **88**, 525.
40. J. Zhu, J. Xie, M. Chen and D. Jiang, *J. Nanosci. Nanotechnology*, 2010, **10**, 1.
41. M. Addamo, V. Augugliaro, M. B. A. Di-Paola, V. Loddo, G. Palmisano, L. Palmisano and S. Yurdakal, *Catal. Lett.*, 2008, **126**, 58.
42. J. Xie, L. Xiaomeng, L. Jun and S. Huoming, *Pure Appl. Chem.*, 2009, **81**, 2407.
43. W. Hou, W. H. Hung, P. Pavaskar, A. Goeppert, M. Aykol and S. B. Cronin, *ACS Catal.*, 2011, **1**, 929.
44. D. B. Ingram, P. Christopher, J. L. Bauer and S. Linic, *ACS Catal.*, 2011, **1**, 1441.
45. Q. Zhang, D. Q. Lima, I. Lee, F. Zaera, M. F. Chi and Y. D. Yin, *Angew. Chem., Int. Ed.*, 2011, **50**, 7088.
46. I. Thomann, B. A. Pinaud, Z. B. Chen, B. M. Clemens, T. F. Jaramillo and M. L. Brongersma, *Nano Lett.*, 2011, **11**, 3440.
47. H. W. Gao, C. Liu, H. E. Jeong and P. D. Yang, *ACS Nano*, 2012, **6**, 234.
48. E. Thimsen, F. Le Formal, M. Grätzel and S. C. Warren, *Nano Lett.*, 2011, **11**, 35.
49. A. B. Djurisic, Y. Chan and E. Herbert-Li, *Mater. Sci. Eng. R*, 2002, **38**, 237.

50. M. A. Mahmoud, W. Qian and M. A. El-Sayed, *Nano Lett.*, 2011, **11**, 3285.
51. C. A. N. Fernando, P. H. C. de-Silva, S. K. Wethasinha, I. M. Dharmadasa, T. Delsol and M. C. Simmonds, *Renew. Energy*, 2002, **26**, 521.
52. M. Hara, T. Kondo, M. Komoda, S. Ikeda, K. Shinohara, J. N. Kondo and K. Domen, *Chem. Commun.*, 1998, **20**, 2185.
53. H. Y. Zhu, X. Chen, Z. F. Zheng, X. B. Ke, E. Jaatinen, J. C. Zhao, C. Gou, T. T. Xie and D. J. Wang, *Chem. Commun.*, 2009, 7524.
54. H. Kominami, A. Tanaka and K. Hashimoto, *Chem. Commun.*, 2010, **46**, 1287.
55. R. G. Nair, A. M. Tripathi and S. K. Samdarshi, *Sol. Energy Mater. Sol. C.*, 2011, **94**, 2379.
56. J. Y. Lan, X. M. Zhou, G. Liu, J. G. Yu, J. C. Zhang, L. J. Zhi and G. J. Nie, *J. Mater. Chem.*, 2011, **22**, 21337.
57. A. Iwase and A. Kudo, *J. Phys. Chem.*, 2010, **17**, 2607.
58. D. Wang, H. F. Jiang, X. Zong, Q. Xu, Y. Ma, G. L. Li and C. Li, *Chem. Eur. J*, 2011, **17**, 1275.
59. T. Saison, N. Chemin, C. Chanéac, O. Durupthy, V. Ruaux, L. Mariey, F. Maugué, P. Beaunier and J. P. Jolivet, *J. Phys. Chem. C*, 2011, **115**, 5657.
60. L. S. Zhang, W. Z. Wang, J. Yang, Z. G. Chen, W. Q. Zhang, L. Zhou and S. W. Liu, *Appl. Catal. A*, 2006, **308**, 105.
61. G. Li, S. Yan, Z. Wang, X. Wang, Z. Li, J. Ye and Z. Zou, *Dalton Trans.*, 2009, **2009**, 8519.
62. G. Huang and Y. Zhu, *Mater. Sci. Eng. B*, 2007, **139**, 201.
63. D. Chen and J. H. Ye, *Adv. Funct. Mater.*, 2008, **18**, 1922.
64. T. Arai, M. Yanagida, Y. Konishi, A. Ikura, Y. Iwasaki, H. Sugihara and K. Sayama, *Appl. Catal. B*, 2008, **84**, 42.
65. G. C. Xi and J. H. Ye, *Chem. Commun.*, 2010, **46**, 1893.
66. P. Wang, B. Huang, X. Qin, X. Zhang, Y. Dai, J. Wei and M. H. Whangbo, *Angew. Chem., Int. Ed.*, 2008, **47**, 7931.
67. J. G. Yu, G. P. Dai and B. B. Huang, *J. Phys. Chem. C*, 2009, **113**, 16394.
68. Y. Zhang, Z. R. Tang, X. Z. Fu and Y. J. Xu, *Appl. Catal. B*, 2011, **106**, 445.
69. C. Hu, T. W. Peng, X. X. Hu, Y. L. Nie, X. F. Zhou, J. H. Qu and J. H. He, *J. Am. Chem. Soc.*, 2011, **132**, 857.
70. H. Zhang, X. F. Fan, X. Quan, S. Chen and H. T. Yu, *Environ. Sci. Technol.*, 2011, **45**, 5731.
71. M. S. Zhu, P. L. Chen and M. H. Liu, *ACS Nano*, 2011, **5**, 4529.
72. C. J. Orendorff, T. K. Sau and C. J. Murphy, *Small*, 2006, **2**, 636.
73. L. Du, A. Furube, K. Yamamoto, K. Hara, R. Katoh and M. Tachiya, *J. Phys. Chem. C*, 2009, **113**, 6454.
74. S. T. Kochuveedu, D.-P. Kim and D. H. Kim, *J. Phys. Chem. C*, 2011, **116**, 2500.
75. R. Zanella, S. Giorgio, C. R. Henry and C. Louis, *J. Phys. Chem. B*, 2002, **106**, 7634.
76. R. Zanella, L. Delannoy and C. Louis, *Appl. Catal. A*, 2005, **291**, 62.

77. R. Zanella and C. Louis, *Catal. Today*, 2005, **107–108**, 768.
78. X. Bokhimi and R. Zanella, *J. Phys. Chem. C*, 2007, **111**, 2525.
79. X. Bokhimi, R. Zanella and A. Morales, *J. Phys. Chem. C*, 2007, **111**, 15210.
80. X. Bokhimi, R. Zanella and A. Morales, *J. Phys. Chem. C*, 2008, **112**, 12463.
81. X. Bokhimi, R. Zanella, A. Morales, V. Maturano and C. Ángeles-Chávez, *J. Phys. Chem. C*, 2011, **115**, 5856.
82. A. Furube, L. Du, K. Hara, R. Katoch and M. Tachiya, *J. Am. Chem. Soc.*, 2007, **129**, 14852.
83. Z. Zheng, B. Huang, X. Qin, X. Zhang, Y. Dai and M. H. Whangbo, *J. Mater. Chem.*, 2011, **21**, 9079.
84. A. Takami, H. Kurita and S. Koda, *J. Phys. Chem. B*, 1999, **103**, 1226.
85. X. Chen, H. Y. Zhu, J. C. Zhao and Z. F. Gao, *Angew. Chem. Int. Ed.*, 2008, **47**, 5353.
86. H. Zhu, X. Ke, X. Yang, S. Sarina and H. Liu, *Angew. Chem. Int. Ed.*, 2010, **49**, 9657.
87. Q. C. Xu, Y. H. Ng, Y. Zhang, J. S. C. Loo, R. Amal and T. T. Y. Tan, *Chem. Commun.*, 2011, **47**, 8641.
88. P. Zhang, C. Shao, Z. Zhang, J. Mu, Z. Gou and Y. Liu, *Chem. Commun.*, 2011, **21**, 17746.
89. P. Zhang, C. Shao, Z. Zhang, M. Zhang, J. Mu, Z. Guo and Y. Liu, *Nanoscale*, 2011, **3**, 3357.
90. V. Subramanian, E. E. Wolf and P. V. Kamat, *J. Am. Chem. Soc.*, 2004, **126**, 4943.
91. A. L. Linsebigler, G. Q. Lu and J. T. Yates, *Chem. Rev.*, 1995, **95**, 735.
92. Y. Tian and T. Tatsuma, *J. Am. Chem. Soc.*, 2005, **127**, 7632.
93. R. T. Tung, *Mater. Sci. Eng. R*, 2001, **35**, 1.
94. X. Wang and R. A. Caruso, *J. Mater. Chem.*, 2011, **21**, 20.
95. A. A. Ismail, D. W. Bahnemann and S. A. Al-Sayari, *Appl. Catal. A*, 2012, **431–432**, 62.
96. M. Alvaro, B. Cojocaru, A. A. Ismail, N. Petrea, B. Ferrer, F. A. Harraz, V. I. Parvulescu and H. García, *Appl. Catal. B*, 2010, **99**, 191.
97. N. Lakshminarasimhan, E. Bae and W. Choi, *J. Phys. Chem. C*, 1996, **111**, 15244.
98. M. Jakob, H. Levanon and P. V. Kamat, *Nano Lett.*, 2003, **3**, 353.
99. Z. Bian, T. Tachikawa, W. Kim, W. Choi and T. Majima, *J. Phys. Chem.*, 2012, **116**, 25444.
100. H. Wang, S. Dong, Y. Chang and J. L. Faria, *J. Hazard. Mater.*, 2012, **235–236**, 230.
101. H. Wang, J. L. Faria, S. Dong and Y. Chang, *Mater. Sci. Eng. B*, 2012, **177**, 913.
102. H. Wang, T. You, W. Shi, J. Li and L. Guo, *J. Phys. Chem. C*, 2012, **116**, 6490.
103. P. Osei Bonsu, X. Lü, J. Xie, D. Jiang, M. Chen and X. Wei, *React. Kinet. Mech. Catal.*, 2012, **107**, 487.

104. F. Le, D. W. Brandl, Y. A. Urzhumov, H. Wang, J. Kundu, N. J. Halas, J. Aizpurua and P. Nordlander, *ACS Nano*, 2008, **2**, 707.
105. S. I. Zou and G. C. Schatz, *Chem. Phys. Lett.*, 2005, **403**, 62.
106. Y. Tian and T. Tatsuma, *J. Am. Chem. Soc.*, 2005, **127**, 7632.
107. B. Cojocaru, Ş. Neaţu, E. Sacaliuc-Pårvulescu, F. Lévy, V. I. Pårvulescu and H. García, *Appl. Catal. B*, 2011, **107**, 140.
108. G. C. Bond, C. Louis and D. T. Thompson, *Catalysis by Gold*, Imperial College Press, London, 2006.
109. C. Louis and O. Pluchery, *Gold Nanoparticles for Physics, Chemistry and Biology*, Imperial College Press, London, 2012.
110. M.-C. Daniel and D. Astruc, *Chem. Rev.*, 2003, **104**, 293.
111. J. W. Yoon, T. Sasaki and N. Koshizaki, *Electrochem. Solid State Lett.*, 2002, **5**, A256.
112. L. Armelao, D. Barreca, G. Bottaro, A. Gasparotto and E. Tondello, *Chem. Mater*, 2004, **16**, 3331.
113. L. Armelao, D. Barreca, G. Bottaro, A. Gasparotto, C. Maccato, C. Maragno, E. Tondello, U. L. Stangar, M. Bergant and D. Mahne, *Nanotechnology*, 2007, **18**, 375709.
114. Z. Liu, W. Hou, P. Pavaskar, M. Aykol and S. B. Cronin, *Nano Lett.*, 2011, **11**, 1111.
115. W. Hou, Z. Liu, P. Pavaskar, W. H. Hung and S. B. Cronin, *J. Catal.*, 2011, **277**, 149.
116. P. Fu and P. Zhang, *Appl. Catal. B*, 2010, **96**, 176.
117. R. S. Sonawane and M. K. Dongare, *J. Mol. Catal. A*, 2006, **243**, 68.
118. M. Haruta, *Catal. Today*, 1997, **36**, 153.
119. M. Haruta, S. Tsubota, T. Kobayashi, H. Kageyama, M. J. Genet and B. Delmon, *J. Catal.*, 1993, **144**, 175.
120. D. B. Ingram and S. Linic, *J. Am. Chem. Soc.*, 2011, **133**, 5202.
121. V. Iliev, D. Tomova, R. Todorovska, D. Oliver, L. Petrov, D. Todorovsky and M. Uzunova-Bujnova, *Appl. Catal. A*, 2006, **313**, 115.
122. P. Wang, B. Huang, Q. Zhang, X. Zhang, X. Qin, Y. Dai, J. Zhan, J. Yu, H. Lin and Z. Lou, *Chem. Eur. J.*, 2010, **16**, 10042.
123. X. Chen, H. Y. Zhu, J. C. Zhao, Z. F. Zheng and X. P. Gao, *Angew. Chem. Int. Ed.*, 2008, **47**, 5353.
124. H. G. Yu, R. Liu, X. F. Wang, P. Wang and J. G. Yu, *Appl. Catal. B*, 2012, **111–112**, 326.
125. Q. J. Xiang, J. G. Yu, B. Chen and H. C. Ong, *Chem. Asian J.*, 2010, **5**, 1466.
126. S. M. Sun, W. Z. Wang, L Zhang, M. Shang and L. Wang, *Catal. Commun.*, 2009, **11**, 290.
127. Y.-C. Pu, Y.-C. Chen and Y.-J. Hsu, *Appl. Catal. B*, 2010, **97**, 389.
128. X.-F. Wu, H.-Y. Song, J.-M. Yoon, Y.-T. Yu and Y.-F. Chen, *Langmuir*, 2009, **25**, 6438.
129. J. Yu, L. Yue, S. Liu, B. Huang and X. Zhang, *J. Colloid Interface Sci.*, 2009, **334**, 58.
130. Y. Lu, H. Yu, S. Chen, X. Quan and H. Zhao, *Environ. Sci. Technol.*, 2012, **46**, 1724.

131. S. Anandan and M. Ashokkumar, *Ultrason. Sonochem.*, 2009, **16**, 316.
132. D. Yang, S.-E. Park, J.-K. Lee and S.-W. Lee, *J. Cryst. Growth*, 2009, **311**, 508.
133. T. Torimoto, H. Horibe, T. Kameyame, K. Okazaki, S. Ikeda, M. Matsumura, A. Ishikawa and I. Ishihara, *J. Phys. Chem. Lett.*, 2011, **2**, 2057.
134. K. Mori, M. Kawashima, M. Che, Q. Zhang, D. Q. Lima, I. Lee, F. Zaera, M. F. Chi and Y. D. Yin, *Angew. Chem. Int. Ed.*, 2011, **49**, 8598.
135. Z. Bian, J. Zhu, F. Cao, Y. Lu and H. Li, *Chem. Commun.*, 2009, **25**, 3789.
136. J. C. Yu, X. C. Wang, L. Wu, W. K. Ho, L. Z. Zhang and G. T. Zhou, *Adv. Funct. Mater.*, 2004, **14**, 1178.
137. J. H. He and T. Kunitake, *Chem. Mater.*, 2004, **16**, 2656.
138. J. H. He, I. Ichinose, T. Kunitake and A. Nakao, *Langmuir*, 2002, **18**, 10005.
139. N. Chandrasekharan and P. V. Kamat, *Nano Lett.*, 2001, **1**, 67.
140. M. D. Pérez, E. Otal, S. A. Bilmes, G. J. A. A. Soler-Illia, E. L. Crepaldi, D. Grosso and C. Sanchez, *Langmuir*, 2004, **20**, 6879.
141. S. Oros-Ruiz, R. Zanella and B. Prado, *J. Hazard. Mater.*, 2013, **263**, 28.
142. M. N. Patel, R. D. Williams, R. D. May, H. Uchida, J. Stevenson and K. P. Johnston, *Chem. Mater.*, 2008, **20**, 6029.
143. M. C. Hidalgo, M. Maicu, J. A. Navío and G. Colón, *Appl. Catal. A*, 2011, **397**, 112.
144. P. D. Cozzoli, M. L. Curri, C. Giannini and A. Agostiano, *Small*, 2006, **2**, 413.
145. A. Snyder, Z. Bo, R. Moon, J.-C. Rochet and L. Stanciu, *J. Colloid Interface Sci.*, 2013, **399**, 92.
146. R. Su, R. Tiruvalam, Q. He, N. Dimitratos, L. Kesavan, C. Hammond, J. A. Lopez-Sanchez, R. Bechstein, C. J. Kiely, G. J. Hutchings and F. Besenbacher, *ACS Nano*, 2012, **6**, 6284.
147. Y. H. Tseng, I. G. Chang, Y. Tai and K. W. Wu, *J. Nanosci. Nanotechnology*, 2012, **12**, 416.
148. Z. Zheng, J. Zhao, Y. Yuan, H. Liu, D. Yang, S. Sarina, H. Zhang, E. R. Waclawika and H. Zhu, *Chem. Eur. J*, 2013, **19**, 5731.
149. Z. Pap, A. Radu, I. J. Hidi, G. Melinte, L. Diamandescu, T. Popescu, L. Baia, V. Danciu and M. Baia, *Chin. J. Catal.*, 2013, **34**, 734.
150. N. Zhou, L. Polavarapu, N. Gao, Y. Pan, P. Yuan, Q. Wang and Q.-H. Xu, *Nanoscale*, 2013, **5**, 4236.
151. R. Kaur and B. Pal, *J. Mol. Catal. A*, 2012, **355**, 39.
152. Y. Ide, N. Nakamura, H. Hattori, R. Ogino, M. Ogawa, M. Sadakane and T. Sano, *Chem. Commun.*, 2011, **47**, 11531.
153. Q. Xiang, G. F. Meng, H. B. Zhao, Y. Zhang, H. Li, W. J. Ma and J. Q. Xu, *J. Phys. Chem. C*, 2010, **114**, 2049.
154. A. Nakajima, T. Kobayashi, T. Isobe and S. Matsushita, *Mater. Lett.*, 2011, **65**, 3051.
155. H. You, R. Liu, C. Liang, S. Yang, F. Wang, X. Lu and B. Ding, *J. Mater. Chem. A*, 2013, **1**, 4097.
156. S. Sarkar, A. Makhal, T. Bora, S. Baruah, J. Dutta and S. K. Pal, *Phys. Chem. Chem. Phys.*, 2011, **13**, 12488.

157. M. Ye, H. Zhou, T. Zhang, Y. Zhang and Y. Shao, *Chem. Eng. J.*, 2013, **226**, 209.
158. Z. Y. Zhong, J. Y. Lin, S. P. Teh, J. Teo and F. M. Dautzenberg, *Adv. Funct. Mater.*, 2007, **17**, 1402.
159. S. L. Lee, J. Scott, K. Chiang and R. Amal, *J. Nanopart. Res.*, 2009, **11**, 209.
160. V. Iliev, D. Tomova, L. Bilyarska and G. Tyuliev, *J. Mol. Catal. A*, 2007, **263**, 32.
161. A. Pearson, H. Zheng, K. Kalantar-zadeh, S. K. Bhargava and V. Bansal, *Langmuir*, 2012, **28**, 14470.
162. M. Haruta and M. Daté, *Appl. Catal. A*, 2001, **222**, 427.
163. S. Oros-Ruiz, R. Gómez, R. López, A. Hernández-Gordillo, J. A. Pedraza-Avella, E. Moctezuma and E. Pérez, *Catal. Commun.*, 2012, **21**, 72.
164. S. Oros-Ruiz, J. A. Pedraza-Avella, C. Guzmán, M. Quintana, E. Moctezuma, G. del Angel, R. Gómez and E. Pérez, *Top. Catal.*, 2011, **54**, 519.
165. V. Rodríguez-González, R. Zanella, G. del-Angel and R. Gómez, *J. Mol. Catal. A*, 2008, **281**, 93.
166. T. Marino, R. Molinari and H. García, *Catal. Today*, 2013, **206**, 40.
167. G. Wu, T. Chen, W. Su, G. Zhou, X. Zong, Z. Lei and C. Li, *Int. J. Hydrogen Energy*, 2008, **33**, 1243.
168. N. Wang, T. Tachikawa and T. Majima, *Chem. Sci.*, 2011, **2**, 891.
169. J. T. Carneiro, T. J. Savenije, J. A. Moulijn and G. Mul, *J. Photochem. Photobiol. A*, 2011, **217**, 326.
170. X. Wang, M. Blackford, K. Prince and R. A. Caruso, *ACS Appl. Mater. Interfaces*, 2012, **4**, 476.
171. A. Orlov, D. A. Jefferson, M. Tikhov and R. M. Lambert, *Catal. Commun.*, 2007, **8**, 821.
172. Z. Zhang, P. He, J. Yang, S. Dai and M. I. N. Zhang, *Surf. Rev. Lett.*, 2005, **12**, 749.
173. J. Lan, X. Zhou, G. Liu, J. Yu, J. Zhang, L. Zhi and G. Nie, *Nanoscale*, 2011, **3**, 5161.
174. R. Zanella, A. Sandoval, P. Santiago, V. A. Basiuk and J. M. Saniger, *J. Phys. Chem. B*, 2006, **110**, 8559.
175. S. V. Awate, R. K. Sahu, M. D. Kadgaonkar, R. Kumar and N. M. Gupta, *Catal. Today*, 2009, **141**, 144.
176. F. Galindo-Hernández, J. A. Wang, R. Gómez, X. Bokhimi, L. Lartundo and A. Mantilla, *J. Photochem. Photobiol. A*, 2012, **243**, 23.
177. V. Iliev, D. Tomova and S. Rakovsky, *Desalination*, 2010, **260**, 101.
178. V. Iliev, D. Tomova, S. Rakovsky, A. Eliyas and G. LiPuma, *J. Mol. Catal. A*, 2010, **327**, 51.
179. H. Kominami, A. Tanaka and K. Hashimoto, *Appl. Catal. A*, 2011, **397**, 121.
180. N. Pugazhenthiran, S. Murugesan, P. Sathishkumar and S. Anandan, *Chem. Eng. J.*, 2014, **241**, 401.
181. N. Pugazhenthiran, P. Sathishkumar, S. Murugesan and S. Anandan, *Chem. Eng. J.*, 2011, **168**, 1227.

182. P. S. Sathish Kumar, R. Sivakumar, S. Anandan, J. Madhavan, P. Maruthamuthu and M. Ashokkumar, *Water Res.*, 2008, **42**, 4878.
183. D. Tomova, V. Iliev, S. Rakovsky, M. Anachkov, A. Eliyas and G. LiPuma, *J. Photochem. Photobiol. A*, 2012, **231**, 1.
184. G. Veréb, Z. Ambrus, Z. Pap, Á. Kmetykó, A. Dombi, V. Danciu, A. Cheesman and K. Mogyorósi, *Appl. Catal. A*, 2012, **417–418**, 26.
185. D. E. Latch, J. L. Packer, B. L. Stender, J. VanOverbeke, W. A. Arnold and K. McNeill, *Environ. Toxicol. Chem.*, 2005, **24**, 517.
186. T. Paul, M. C. Dodd and T. J. Strathmann, *Water Res.*, 2010, **44**, 3121.
187. A. Garron, K. Lázar and F. Epron, *Appl. Catal. B*, 2006, **65**, 240.
188. J. A. Anderson, *Catal. Today*, 2012, **181**, 171.
189. J. A. Anderson, *Catal. Today*, 2011, **175**, 316.
190. S. Luo, Y. Xiao, L. Yang, C. Liu, F. Su, Y. Li, Q. Cai and G. Zeng, *Sep. Purif. Technol.*, 2011, **79**, 85.
191. M. V. Dozzi, A. Saccomanni and E. Selli, *J. Hazard. Mater.*, 2012, **211–212**, 188.
192. A. Pandikumar and R. Ramaraj, *J. Hazard. Mater.*, 2012, **203–204**, 244.
193. G. Veréb, L. Manczinger, G. Bozsó, A. Sienkiewicz, L. Forró, K. Mogyorósi, K. Hernádi and A. Dombi, *Appl. Catal. B*, 2013, **129**, 566.
194. S. F. Chen, J. P. Li, K. Qian, W. P. Xu, Y. Lu, W. X. Huang and S. H. Yu, *Nano Res.*, 2010, **3**, 244.
195. L. Armelao, D. Barreca, G. Bottaro, A. Gasparotto, C. Maccato, C. Maragno, E. Tondello, U. L. Stangar, M. Bergant and D. Mahne, *Nanotechnology*, 2007, **18**, 375709.
196. M. Rai, A. Yadav and A. Gade, *Biotechnol. Adv.*, 2009, **27**, 76.
197. A. Sobczyński and A. Dobosz, *Pol. J. Environ. Stud.*, 2001, **10**, 195.
198. L. Du, A. Furube, K. Hara, R. Katoh and M. Tachiya, *Thin Solid Films*, 2009, **518**, 861.
199. B. K. Min, J. E. Heo, N. K. Youn, O. S. Joo, H. Lee, J. H. Kim and H. S. Kim, *Catal. Commun.*, 2009, **10**, 712.
200. P. Sangpour, F. Hashemi and A. Z. Moshfegh, *J. Phys. Chem. C*, 2010, **114**, 13955.
201. E. Kowalska, O. O. Prieto Mahaney, R. Abe and B. Ohtani, *Phys. Chem. Chem. Phys.*, 2010, **12**, 2344.
202. L. Xiang, X. Zhao, C. Shang and J. Yin, *J. Colloid Interface Sci.*, 2013, **403**, 22.
203. E. Grabowska, H. Remita and A. Zaleska, *Physicochem. Probl. Mi.*, 2010, **45**, 29.
204. X. Hu and T. Bürgi, *Appl. Catal. A*, 2012, **449**, 139.
205. H. Li, Z. Bian, J. Zhu, Y. Huo, H. Li and Y. Lu, *J. Am. Chem. Soc.*, 2007, **129**, 4538.
206. P. Wongwisate, S. Chavadej, E. Gulari, T. Sreethawong and P. Rangsunvigit, *Desalination*, 2011, **272**, 154.
207. L. Xiao, J. Zhang, C. Ye, B. Tian, F. Chen and M. Anpo, *Catal. Lett.*, 2006, **111**, 207.
208. N. K. Youn, J. E. Heo, O. S. Joo, H. Lee, J. Kim and B. K. Min, *J. Hazard. Mater.*, 2010, **177**, 216.

209. S. Zhu, S. Liang, Q. Gu, L. Xie, J. Wang, Z. Ding and P. Liu, *Appl. Catal. B*, 2012, **119–120**, 146.
210. J. Lu, P. Zhang, A. Li, F. Su, T. Wang, Y. Liu and J. Gong, *Chem. Commun.*, 2013, **49**, 5817.
211. I. Paramasivam, J. M. Macak and P. Schmuki, *Electrochem. Commun.*, 2008, **10**, 71.
212. Y. Chen, Y. Tang, S. Luo, C. Liu and Y. Li, *J. Alloys Compd.*, 2013, **578**, 242.
213. S. Sun, C. Chen and J. Sun, *Procedia Environ. Sci.*, 2013, **18**, 620.
214. S. Singla and B. Pal, *Mater. Res. Bull.*, 2013, **48**, 4867.
215. G. Wang, X. Wang, J. Liu and X. Sun, *Chem. Eur. J.*, 2012, **18**, 5361.
216. J. W. Chiou, S. C. Ray, H. M. Tsai, C. W. Pao, F. Z. Chien, W. F. Pong, C. H. Tseng, J. J. Wu, M.-H. Tsai, C.-H. Chen, H. J. Lin, J. F. Lee and J.-H. Guo, *J. Phys. Chem. C*, 2011, **115**, 2650.
217. G. Menagen, J. E. Macdonald, Y. Shemesh, I. Popov and U. Banin, *J. Am. Chem. Soc.*, 2009, **131**, 17406.
218. J.-J. Wu and C.-H. Tseng, *Appl. Catal. B*, 2006, **66**, 51.
219. R. Costi, G. Cohen, A. Salant, E. Rabani and U. Banin, *Nano Lett.*, 2009, **9**, 2031.
220. R. Costi, A. E. Saunders, E. Elmalem, A. Salant and U. Banin, *Nano Lett.*, 2008, **8**, 637.
221. C. Chen, Y. Lu, H. He, K. Wu and Z. Ye, *Appl. Phys. A*, 2013, **110**, 47.
222. A. Haarstrick, O. M. Kut and E. Heinzle, *Environ. Sci. Technol.*, 1996, **30**, 817.
223. C. Minero, E. Pelizzetti, M. Sega, S. E. Friberg and J. Sjóblom, *J. Disper. Sci. Technol.*, 1999, **20**, 643.
224. C. Qin, S. Yang, C. Sun, M. Zhan, R. Wang, H. Cai and J. Zhou, *Front. Environ. Sci. Eng.*, 2010, **4**, 321.
225. T. E. Doll and F. H. Frimmel, *Water Res.*, 2005, **39**, 403.
226. I. K. Konstantinou and T. A. Albanis, *Appl. Catal. B*, 2004, **49**, 1.
227. J. Santiago-Morales, A. Agüera, M. M. Gómez, A. R. Fernández-Alba, J. Giménez, S. Esplugas and R. Rosal, *Appl. Catal. B*, 2012, **129**, 13.
228. D. Zhang, *Russ. J. Phys. Chem. A*, 2012, **86**, 498.
229. J. Chen, H. Su, F. Song, W. J. Moon, Y. S. Kim and D. Zhang, *J. Colloid Interface Sci.*, 2012, **370**, 117.
230. G. Federici, B. J. Shaw and R. D. Handy, *Aquat. Toxicol.*, 2007, **84**, 415.
231. F. Barka-Bouaifel, K. Makaoui, P. Y. Jouan, X. Castel, N. Bezzi, R. Boukherroub and S. Szunerits, *RSC Adv.*, 2012, **2**, 12482.
232. L. Chen, T. Tran, C. A. Huang, J. Li, L. Yuan and Q. Cai, *Appl. Surf. Sci.*, 2013, **273**, 82.
233. Z.-Y. Chen, Y. Hu, T.-C. Liu, C.-L. Huang and T.-S. Jeng, *Thin Solid Films*, 2009, **517**, 4998.
234. R. S. Sonawane and M. K. Dongare, *J. Mol. Catal. A*, 2006, **243**, 68.
235. I. M. Arabatzis, S. Antonaraki, T. Stergiopoulos, A. Hiskia, E. Papaconstantinou, M. C. Bernard and P. Falaras, *J. Photochem. Photobiol. A*, 2002, **149**, 237.

236. I. M. Arabatzis, T. Stergiopoulos, D. Andreeva, S. Kitova, S. G. Neophytides and P. Falaras, *J. Catal.*, 2003, **220**, 127.
237. H. W. Choi, E. J. Kim and S. H. Hahn, *Chem. Eng. J.*, 2010, **161**, 285.
238. A. Orlov, M. S. Chan, D. A. Jefferson, D. Zhou, R. J. Lynch and R. M. Lambert, *Environ. Technol.*, 2006, **27**, 747.
239. C. Yogi, K. Kojima, N. Wada, H. Tokumoto, T. Takai, T. Mizoguchi and H. Tamiaki, *Thin Solid Films*, 2008, **516**, 5881.
240. V. Subramanian, E. Wolf and P. V. Kamat, *J. Phys. Chem. B*, 2001, **105**, 11439.
241. J.-J. Chen, J. C. S. Wu, P. C. Wu and D. P. Tsai, *J. Phys. Chem. C*, 2012, **116**, 26535.
242. A. Eremenko, N. Smirnova, I. Gnatiuk, O. Linnik, N. Vityuk, I. Mukha and A. Korduban, in *Nanocomposites and Polymers with Analytical Methods*, ed. J. Cuppoletti, *InTech*, Riijeka, Croatia, 2011, ch. 3 . p. 51.
243. E. Bae and W. Choi, *Environ. Sci. Technol.*, 2002, **37**, 147.
244. C.-Y. Kao, J.-D. Liao, I. W. Huang and C.-W. Chang, *Surf. Coat. Tech.*, 2012, **206**, 4887.
245. A. Dobosz and A. Sobczyński, *Monatsh. Chem.*, 2001, **132**, 1037.
246. J. Li and H. C. Zeng, *Chem. Mater.*, 2006, **18**, 4270.
247. P. Zhang, C. Shao, X. Li, M. Zhang, X. Zhang, Y. Sun and Y. Liu, *J. Hazard. Mater.*, 2012, **237–238**, 331.
248. M. M. Khan, S. A. Ansari, J. Lee and M. H. Cho, *J. Ind. Eng. Chem.*, 2013, **19**, 1845.
249. J. Li, Suyoulema, W. Wang and Sarina, *Solid State Sci.*, 2009, **11**, 2037.
250. H. Yu, X. Wang, H. Sun and M. Huo, *J. Hazard. Mater.*, 2010, **184**, 753.
251. Q. Zhang, D. Q. Lima, I. Lee, F. Zaera, M. Chi and Y. Yin, *Angew. Chem., Int. Ed.*, 2011, **123**, 7226.
252. X. Fu, J. Liu, H. Yang, J. Sun, X. Li, X. Zhang and Y. Jia, *Mater. Chem. Phys.*, 2011, **130**, 334.
253. Y. Wu, H. Liu, J. Zhang and F. Chen, *J. Phys. Chem. C*, 2009, **113**, 14689.
254. M. Zhou, J. Zhang, B. Cheng and H. Yu, *Int. J. Photoenergy*, 2012, **2012**, 1.
255. Y. Liu, W. Shu, Z. Peng, K. Chen and W. Chen, *Catal. Today*, 2013, **208**, 28.
256. Z. Hai, N. El Kolli, D. B. Uribe, P. Beaunier, M. J. Yacaman, J. Vigneron, A. Etcheberry, S. Sorgues, C. Colbeau-Justin, J. Chen and H. Remita, *J. Mater. Chem. A*, 2013, **1**, 10829.
257. A. Orlov, D. A. Jefferson, N. Macleod and R. M. Lambert, *Catal. Lett.*, 2004, **92**, 41.
258. S. Chusaksri, J. Lomda, T. Saleepochn and P. Sutthivaiyakit, *J. Hazard. Mater.*, 2011, **190**, 930.
259. M. C. Hidalgo, M. Maicu, J. A. Navío and G. Colón, *J. Phys. Chem. C*, 2009, **113**, 12840.
260. A. Pearson, H. Jani, K. Kalantar-zadeh, S. K. Bhargava and V. Bansal, *Langmuir*, 2011, **27**, 6661.

261. B. Tian, J. Zhang, T. Tong and F. Chen, *Appl. Catal. B*, 2008, **79**, 394.
262. P. Pawinrat, O. Mekasuwandumrong and J. Panpranot, *Catal. Commun.*, 2009, **10**, 1380.
263. G. K. Naik, P. M. Mishra and K. Parida, *Chem. Eng. J.*, 2013, **229**, 492.
264. Y. Badr, M. G. Abd El-Wahed and M. A. Mahmoud, *J. Hazard. Mater.*, 2008, **154**, 245.
265. Y. Badr and M. A. Mahmoud, *J. Phys. Chem. Solids*, 2007, **68**, 413.
266. A. Barras, M. R. Das, R. R. Devarapalli, M. V. Shelke, S. Cordier, S. Szunerits and R. Boukherroub, *Appl. Catal. B*, 2013, **130–131**, 270.
267. C. Lizama, J. Freer, J. Baeza and H. D. Mansilla, *Catal. Today*, 2002, **76**, 235.
268. P. Li, Z. Wei, T. Wu, Q. Peng and Y. Li, *J. Am. Chem. Soc.*, 2011, **133**, 5660.
269. Q. Wang, B. Geng and S. Wang, *Environ. Sci. Technol.*, 2009, **43**, 8968.
270. N. Udawatte, M. Lee, J. Kim and D. Lee, *ACS Appl. Mater. Interfaces*, 2011, **3**, 4531.
271. M.-K. Lee and H.-F. Tu, *J. Electrochem. Soc.*, 2008, **155**, 758.
272. S. Pyne, G. P. Sahoo, D. K. Bhui, H. Bar, P. Sarkar, S. Samanta, A. Maity and A. Misra, *Spectrochim. Acta Part A*, 2012, **93**, 100.
273. P. S. Sathish Kumar, M. Ruby Raj, S. Anandan, D. Zhou and M. Ashokkumar, *Water Sci. Technol.*, 2009, **60**, 1586.
274. B.-H. Lai, Y.-R. Lin and D.-H. Chen, *Chem. Eng. J.*, 2013, **223**, 418.
275. G. Kovács, L. Baia, A. Vulpoi, T. Radu, A. Dombi, K. Hernádi, V. Danciu, S. Simon and Z. Pap, *Appl. Catal. B*, 2013, **147**, 508.
276. É. Karácsonyi, L. Baia, A. Dombi, V. Danciu, K. Mogyorósi, L. C. Pop, G. Kovács, V. Cosoveanu, A. Vulpoi, S. Simon and Z. Pap, *Catal. Today*, 2013, **208**, 19.
277. Y.-H. Xu, C.-J. Liu, M.-J. Chen and Y.-Q. Liu, *Int. J. Nanopart.*, 2011, **4**, 268.
278. M. Long, J. Jiang, Y. Li, R. Cao, L. Zhang and W. Cai, *Nano-Micro Lett.*, 2011, **3**, 171.
279. L. Chen, S.-F. Yin, R. Huang, Q. Zhang, S.-L. Luo and C.-T. Au, *CrystEngComm*, 2012, **14**, 4217.
280. K. Pingmuang, N. Wetchakun, W. Kangwansupamonkon, K. Ounnunkad, B. Inceesungvorn and S. Phanichphant, *Int. J. Photoenergy*, 2013, **2013**, 1.
281. T. Peng, C. Hu, X. Hu, X. Zhou and J. Qu, *Catal. Lett.*, 2012, **142**, 646.
282. X. Deng, K. Lin, X. Chen, Q. Guo and P. Yao, *Chem. Eng. J.*, 2013, **225**, 656.
283. X. Zhang, X. Zhang, R.-P. Feng, L.-H. Liu and H. Meng, *Mater. Chem. Phys.*, 2012, **136**, 555.
284. S.-W. Cao, J. Fang, M. M. Shahjamali, Z. Wang, Z. Yin, Y. Yang, F. Y. C. Boey, J. Barber, S. C. J. Loo and C. Xue, *CrystEngComm*, 2012, **14**, 7229.

285. H. Zhu, X. Chen, Z. Zheng, X. Ke, E. Jaatinen, J. Zhao, C. Guo, T. Xie and D. Wang, *Chem. Commun.*, 2009, 7524.
286. L. Kong, W. Chen, D. Ma, Y. Yang, S. Liu and S. Huang, *J. Mater. Chem.*, 2012, **22**, 719.
287. C. W. Yen, M. A. Mahmoud and M. A. El-Sayed, *J. Phys. Chem. A*, 2009, **113**, 4340.
288. R. Singh and B. Pal, *J. Mol. Catal. A*, 2013, **378**, 246.
289. N. Kumar, V. K. Komarala and V. Dutta, *Chem. Eng. J.*, 2014, **236**, 66.
290. H. Zhu, B. Yang, J. Xu, Z. Fu, M. Wen, T. Guo, S. Fu, J. Zuo and S. Zhang, *Appl. Catal. B*, 2009, **90**, 463.
291. T. Lv, L. Pan, X. Liu and Z. Sun, *Electrochim. Acta*, 2012, **83**, 216.
292. D. A. Birkholz, S. M. Stilson, H. S. Elliott and S. Ahuja, in *Comprehensive Water Quality and Purification*, ed. S. Ahuja, Elsevier, Waltham, MA, 2014, vol. 3, p. 212.
293. M. Farré, S. Pérez, L. Kantiani and D. Barceló, *Trend. Anal. Chem.*, 2008, **27**, 991.
294. N. Bolong, A. F. Ismail, M. R. Salim and T. Matsuura, *Desalination*, 2009, **239**, 229.
295. S.-i. Naya, T. Nikawa, K. Kimura and H. Tada, *ACS Catal.*, 2013, **3**, 903.
296. X. Z. Li, C. He, N. Graham and Y. Xiong, *J. Appl. Electrochem.*, 2005, **35**, 741.
297. C. Di Valentin, E. Finazzi, G. Pacchioni, A. Selloni, S. Livraghi, M. C. Paganini and E. Giamello, *Chem. Phys.*, 2007, **339**, 44.
298. C. Di Valentin and G. Pacchioni, *Catal. Today*, 2013, **206**, 12.
299. X. Wang and T.-T. Lim, *Appl. Catal. B*, 2010, **100**, 355.
300. B. Tian, C. Li, F. Gu and H. Jiang, *Catal. Commun.*, 2009, **10**, 925.
301. N. L. Panwar, S. C. Kaushik and S. Kothari, *Renew. Sust. Energy Rev.*, 2011, **15**, 1513.
302. D.-H. Lee, *Int. J. Hydrogen Energy*, 2012, **37**, 15726.
303. H. Bahruji, M. Bowker, P. R. Davies, L. S. Al-Mazroai, A. Dickinson, J. Greaves, D. James, L. Millard and F. Pedrono, *J. Photochem. Photobiol. A*, 2010, **216**, 115.
304. K. Maeda and K. Domen, *J. Phys. Chem. Lett.*, 2010, **1**, 2655.
305. A. Kudo and Y. Miseki, *Chem. Soc. Rev.*, 2009, **38**, 253.
306. M. Matsuoka, M. Kitano, M. Takeuchi, K. Tsujimaru, M. Anpo and J. M. Thomas, *Catal. Today*, 2007, **122**, 51.
307. M. Ni, M. K. H. Leung, D. Y. C. Leung and K. Sumathy, *Renew. Sust. Energy Rev.*, 2007, **11**, 401.
308. A. Fujishima and K. Honda, *Bull. Chem. Soc. Jpn.*, 1971, **44**, 1148.
309. A. Fujishima and K. Honda, *Nature*, 1972, **238**, 37.
310. Z. Zou and H. Arakawa, *J. Photochem. Photobiol. A*, 2003, **158**, 145.
311. D. Jing, L. Guo, L. Zhao, X. Zhang, H. Liu, M. Li, S. Shen, G. Liu, X. Hu, X. Zhang, K. Zhang, L. Ma and P. Guo, *Int. J. Hydrogen Energy*, 2010, **35**, 7087.
312. X. Chen, S. Shen, L. Guo and S. S. Mao, *Chem. Rev.*, 2010, **110**, 6503.
313. K. Maeda, *J. Photochem. Photobiol. C*, 2011, **12**, 237.

314. R. Abe, *J. Photochem. Photobiol. C*, 2010, **11**, 179.
315. A. Primo, T. Marino, A. Corma, R. Molinari and H. García, *J. Am. Chem. Soc.*, 2011, **133**, 6930.
316. P. Wang, B. Huang, Y. Dai and M.-H. Whangbo, *Phys. Chem. Chem. Phys.*, 2012, **14**, 9813.
317. W. Fan, Q. Zhang and Y. Wang, *Phys. Chem. Chem. Phys.*, 2013, **15**, 2632.
318. F. Gärtner, S. Losse, A. Boddien, M.-M. Pohl, S. Denurra, H. Junge and M. Beller, *ChemSusChem*, 2012, **5**, 530.
319. V. Subramanian, E. E. Wolf and P. V. Kamat, *J. Am. Chem. Soc.*, 2004, **126**, 4943.
320. M. Murdoch, G. I. N. Waterhouse, M. A. Nadeem, J. B. Metson, M. A. Keane, R. F. Howe, J. Llorca and H. Idriss, *Nat. Chem.*, 2011, **3**, 489.
321. J. A. Ortega Méndez, C. R. López, E. Pulido Melián, O. González Díaz, J. M. Doña Rodríguez, D. Fernández Hevia and M. Macías, *Appl. Catal. B*, 2013, **147**, 439.
322. G. R. Bamwenda, S. Tsubota, T. Nakamura and M. Haruta, *J. Photochem. Photobiol. A*, 1995, **89**, 177.
323. S. S. Rayalu, D. Jose, M. V. Joshi, P. A. Mangrulkar, K. Shrestha and K. Klabunde, *Appl. Catal. B*, 2013, **142–143**, 684.
324. D. Wang, Z. Ma, S. Dai, J. Liu, Z. Nie, M. H. Engelhard, Q. Huo, C. Wang and R. Kou, *J. Phys. Chem. C*, 2008, **112**, 13499.
325. G. L. Chiarello, E. Selli and L. Forni, *Appl. Catal. B*, 2008, **84**, 332.
326. G. L. Chiarello, L. Forni and E. Selli, *Catal. Today*, 2009, **144**, 69.
327. H.-Y. Lin and Y.-S. Chang, *Int. J. Hydrogen Energy*, 2010, **35**, 8463.
328. A. Iwase, H. Kato and A. Kudo, *Catal. Lett.*, 2006, **108**, 7.
329. A. Iwase, H. Kato and A. Kudo, *Appl. Catal. B*, 2013, **136–137**, 89.
330. J. Zhang, Y. Wang, J. Zhang, Z. Lin, F. Huang and J. Yu, *ACS Appl. Mater. Interfaces*, 2012, **5**, 1031.
331. P. Dhanasekaran and N. M. Gupta, *Mater. Res. Bull.*, 2012, **47**, 1217.
332. G. R. Bamwenda, S. Tsubota, T. Kobayashi and M. Haruta, *J. Photochem. Photobiol. A*, 1994, **77**, 59.
333. A. Naldoni, M. D'Arienzo, M. Altomare, M. Marelli, R. Scotti, F. Morazzoni, E. Selli and V. Dal Santo, *Appl. Catal. B*, 2013, **130–131**, 239.
334. M. A. Nadeem, M. Murdoch, G. I. N. Waterhouse, J. B. Metson, M. A. Keane, J. Llorca and H. Idriss, *J. Photochem. Photobiol. A*, 2010, **216**, 250.
335. V. Jovic, W.-T. Chen, D. Sun-Waterhouse, M. G. Blackford, H. Idriss and G. I. N. Waterhouse, *J. Catal.*, 2013, **305**, 307.
336. M. Bowker, L. Millard, J. Greaves, D. James and J. Soares, *Gold Bull.*, 2004, **37**, 170.
337. M. Bowker, P. Davies and L. S. Al-Mazroai, *Catal. Lett.*, 2009, **128**, 253.
338. M. Bowker, C. Morton, J. Kennedy, H. Bahruji, J. Greves, W. Jones, P. R. Davies, C. Brookes, P. P. Wells and N. Dimitratos, *J. Catal.*, 2014, **310**, 431.
339. P. Gomathisankar, D. Yamamoto, H. Katsumata, T. Suzuki and S. Kaneco, *Int. J. Hydrogen Energy*, 2013, **38**, 5517.

340. H. Yuzawa, T. Yoshida and H. Yoshida, *Appl. Catal. B*, 2012, **115–116**, 294.
341. O. Rosseler, M. V. Shankar, M. K.-L. Du, L. Schmidlin, N. Keller and V. Keller, *J. Catal.*, 2010, **269**, 179.
342. M. Ilie, B. Cojocaru, V. I. Parvulescu and H. García, *Int. J. Hydrogen Energy*, 2011, **36**, 15509.
343. P. Shen, S. Zhao, D. Su, Y. Li and A. Orlov, *Appl. Catal. B*, 2012, **126**, 153.
344. Z. W. Seh, S. Liu, M. Low, S.-Y. Zhang, Z. Liu, A. Mlayah and M.-Y. Han, *Adv. Mater.*, 2012, **24**, 2310.
345. T. Puangpetch, S. Chavadej and T. Sreethawong, *Energy Convers. Manage.*, 2011, **52**, 2256.
346. O. Legrini, E. Oliveros and A. M. Braun, *Chem. Rev.*, 1993, **95**, 671.
347. A. L. Linsebigler, G. Q. Lu and J. T. Yates, *Chem. Rev.*, 1995, **95**, 735.
348. A. Mills and S. LeHunte, *J. Photochem. Photobiol. A*, 1995, **108**, 171.
349. A. Corma and H. García, *J. Catal.*, 2013, **308**, 168.
350. F. B. Li, X. Z. Li, C. H. Ao, M. F. Hou and S. C. Lee, *Appl. Catal. B*, 2004, **54**, 275.
351. J. Hernández-Fernández, A. Aguilar-Elguezabal, S. Castillo, B. Ceron-Ceron, R. D. Arizabalo and M. Moran-Pineda, *Catal. Today*, 2009, **148**, 115.
352. M. A. Debeila, N. J. Coville, M. S. Scurrell, G. R. Hearne and M. J. Witcomb, *J. Phys. Chem. B*, 2004, **108**, 18254.
353. J. Zhang, T. Asuyama, M. Minagawa, K. Kinugaqa, H. Yamashita, M. Matsuoka and M. Anpo, *J. Catal.*, 2002, **198**, 1.
354. S. Zhao, G. Ramakrishnan, D. Su, R. Rieger, A. Koller and A. Orlov, *Appl. Catal. B*, 2011, **104**, 239.
355. C. Pan and L. Dong, *J. Macromol. Sci. B*, 2009, **48**, 919.
356. A. Mattson and L. Österlund, *J. Phys. Chem. C*, 2010, **114**, 14121.
357. T. Mallat and A. Baiker, *Annu. Rev. Chem. Biomol. Eng.*, 2012, **3**, 11.
358. A. Abad, C. Almela, A. Corma and H. Garcia, *Chem. Commun.*, 2006, 3178.
359. B. Zhu, M. Lazar, B. G. Trewyn and R. J. Angelici, *J. Catal.*, 2008, **260**, 1.
360. M. Lazar and R. J. Angelici, *J. Am. Chem. Soc.*, 2006, **128**, 10613.
361. A. Grirrane, A. Corma and H. García, *Science*, 2008, **322**, 1661.
362. M.-H. So, Y. Liu, C.-M. Ho, K.-Y. Lam and C.-M. Che, *ChemCatChem*, 2010, **3**, 386.
363. E. Kowalska, R. Abe and B. Ohtani, *Chem. Commun.*, 2009, 241.
364. S.-I. Naya, M. Teranishi, K. Kimura and H. Tada, *Chem. Commun.*, 2011, **47**, 3230.
365. S.-I. Naya, M. Teranishi, T. Isobe and H. Tada, *Chem. Commun.*, 2010, **46**, 815.
366. X. Ke, X. Zhang, J. Zhao, S. Sarina, J. Barry and H. Zhu, *Green Chem.*, 2013, **15**, 236–244.
367. Y. Zhang, X. Cui, F. Shi and Y. Deng, *Chem. Rev.*, 2011, **112**, 2467.
368. J. Zhao, Z. Zheng, S. Bottle, A. Chou, S. Sarina and H. Zhu, *Chem. Commun.*, 2013, **49**, 2676.

CHAPTER 8

Advances in Catalytic Reactions by Gold-based Catalysts Through the Radical Chain Mechanism

XUE-JING YANG, LIKE OUYANG, JING XU AND YI-FAN HAN*

East China University of Science and Technology, State Key Laboratory of Chemical Engineering, 130 Meilong Road, Shanghai, 200237, PR China
*Email: yifanhan@ecust.edu.cn.

8.1 Introduction

Gold (Au) has long been considered as an inert element for catalysis, because it cannot react with most chemical reagents and bulk gold surfaces chemisorb few molecules readily. Since the discovery of the low-temperature catalytic oxidation reaction using nano-gold catalysts by Haruta and co-workers in the 1980s,[1] gold catalysis has attracted wide interest.[2,3] Supported gold particles in nanosize scale have proven active and selective in other reactions such as hydrogenation,[4] selective oxidation[5–7] and various synthetic reactions for fine chemicals.[8]

As a model reaction, CO oxidation is ideal for the study of the relationship between catalytic performance and structure.[9,10] Thousands of papers have reported CO oxidation on Au catalysts from different aspects,[10] such as size effect and electronic effect of gold, and the types, crystal phases, sizes, and shape effect of supports, *etc.*[11,12] Several mechanistic studies of CO oxidation

RSC Catalysis Series No. 18
Heterogeneous Gold Catalysts and Catalysis
Edited by Zhen Ma and Sheng Dai

Published by the Royal Society of Chemistry, www.rsc.org

on Au-based catalysts have revealed that the active sites are the gold–support interfaces. In those processes, a clear explanation is that CO oxidation does not need the active oxygen atom and that the adducts formed by CO and molecular oxygen are the key intermediates for CO_2 formation. However, there are still some problems to be solved, such as the geometric and electronic state of the active oxygen species.

It is noted that the active oxygen species play a key role not only in CO oxidation but also in some other oxidation reactions. For example, atomic oxygen radicals abound in high-temperature reactions such as combustion and explosion.[13] Hydroxyl ($OH^•$) and peroxyl radicals ($OOH^•$) can be easily detected in low-temperature reactions over Au-based catalysts. For Fenton's reaction, the $OH^•$ from H_2O_2 catalytic decomposition or photocatalytic process is the main active oxygen species responsible for the degradation of organic pollutants.[14] Compared with traditional iron Fenton catalysts, gold catalysts have much higher $OH^•$ selectivity for H_2O_2 decomposition. Recently, Hutchings and co-workers[15] even found the existence of $OH^•$ and methyl radicals ($CH_3^•$) in the selective oxidation of methane using Pd-Au catalysts. This study put forward a new strategy for the transformation of persistent hydrocarbons. It proved that methane, the most stable hydrocarbon compound, can be selectively transformed into methanol under mild conditions using very active oxygen species, $OH^•$. Similarly, selective oxidation of other alkanes, alkenes, alcohols and aldehydes using oxygen species (superoxy, peroxy and hydroperoxy radicals) has also been reported. These studies are systematically summarized in this chapter.

Based on the oxidative capabilities of different oxygen species, it is possible to obtain a desired activity and selectivity for the oxidative transformation of organic compounds under mild conditions provided the corresponding active oxygen species for each oxidation reaction can be technically controlled. In fact, as well as active oxygen species, alkyl and alkoxy radical, hydrogen radicals and an Au–H adduct have also been detected in Au-based selective oxidation and hydrogenation reactions. However, there are few studies on Au-based catalysts.

To investigate the nature of these reactions, electron paramagnetic resonance (EPR) spectroscopy techniques with femtosecond resolution and radical trap techniques have been developed. Using these techniques, part of radicals, such as $OH^•$, can be semi-quantified. However, it is still difficult to fully obtain the signals of radicals or adsorbed intermediates due to their very short lifetime and the complexity of the reaction conditions which are easily influenced by other factors such as catalyst structure, gas phase or solvent properties, temperature and so on. How to reveal the effects of each factor and their interactions depends strongly on the development of new characterization techniques and experimental methods.

Several excellent reviews and book chapters have described the development of gold catalysis from different aspects, such as model gold catalysis,[9,10] design of structured gold catalysts,[3,6,16] and the transformation of chemicals on gold catalysts.[2,7,8] Unlike the published literature, this chapter focuses on the recent development of radicals involved in catalytic reactions on heterogeneous gold-based catalysts. This chapter is composed of four parts: (1) selective oxidation of alkanes; (2) selective oxidation of alkenes; (3) selective oxidation of alcohols and aldehydes; and (4) Fenton-like reactions and photocatalytic oxidation.

However, it must be noted that only parts of reactions are robustly identified to follow the radical mechanism. Reasonable explanation is still rare for both qualitative and quantitative description of these reactions. Frankly, the concept of 'radicals' is ambiguous or even not mentioned in some of the papers we cite here. However, we still classified these works as a 'radical process'. At least, the involvement of radicals in the process cannot be ignored.

8.2 Selective Oxidation of Alkanes

8.2.1 Selective Oxidation of Methane

Methane is the primary component of natural gas and shale gas. The abundance of methane makes it a potential raw material for the synthesis of bulk chemicals.[17,18] Except for the small amount of methane consumed through combustion and hydrogen production by the water gas shift reaction, the utilization of methane is relatively limited. In fact, the oxidation of hydrocarbons, particularly that of methane, has been studied for over a century. Dating back to 1940s, German scientists developed two processes for the direct oxidation of methane, using NO or O_3 as oxidant, respectively. The former process had been applied in practice at Copsa Mica, Romania.[19] Because direct oxidation of methane does not require energy input, highly efficient and selective conversion of methane to required chemicals by one step would be the best way to utilize the vast natural gas reserves. However, the robust C–H bond (bond energy as high as 439 kJ mol^{-1}) means that this process remains very challenging.

In 1994, Sokolovskii *et al.*[20] discovered the effect of reaction mixture quenching which happened in the so-called 'post-catalytic' zone after the catalytic bed. The generation of H_2O_2 was assumed to explain this phenomenon. Thus, quenching of the oxygen-containing radical at this point was thought to hamper the yield of C_{2+} product. Furthermore, they demonstrated that, in a non-catalytic system, the apparent activation energy of methane oxidation by H_2O_2 decreased from 208 kJ mol^{-1} to 65–95 kJ mol^{-1} in the same H_2O_2-free reaction system.[21]

As a widely used oxidant, H_2O_2 was also an excellent oxidative reagent for the partial oxidation of methane for oxidative coupling and aromatization. Over a supported Au catalyst (1%Au/5%La_2O_3/CaO), remarkable quantities of benzene were detected as the product of oxidative coupling of methane in the presence of H_2O_2.[22] The yields of C_{2+} over this catalyst with the addition of various oxidants as well as H_2O were investigated in comparative studies. Enhancement effects on methane conversion and C_{2+} product selectivity were observed in the presence of H_2O_2. The maximum yield of C_{2+} products was 27%, including benzene (7% in yield) at 700 °C.

The generation of $OH^{\bullet}$ by the thermal decomposition of H_2O_2 was thought to be responsible for the enhancement [eqn (8.1)]. $OH^{\bullet}$ reacted directly with methane and promoted the generation of methyl radical [eqn (8.2)]. In addition, the chain-branching process was also initiated by $OH^{\bullet}$ [eqn (8.3) to eqn (8.5)].[23] However, the conversion of methane using pure O_2 as oxidant was relatively low (8.7%) over an Au(1wt%)/MgO catalyst. The yield of C_{2+} products was only 0.17%.[24] For this process, the lattice oxygen was thought to be responsible for the oxidation, which was generated at the two-dimensional (2D) raft interface of gold and MgO.

$$H_2O_2 \rightarrow 2OH^{\bullet} \tag{8.1}$$

$$OH^{\bullet} + CH_4 \rightarrow CH_3^{\bullet} + H_2O \tag{8.2}$$

$$OH^{\bullet} + H_2 \rightarrow H_2O + H^{\bullet} \tag{8.3}$$

$$H^{\bullet} + O_2 \rightarrow OH^{\bullet} + O \tag{8.4}$$

$$O + H_2 \rightarrow OH^{\bullet} + H^{\bullet} \tag{8.5}$$

When H_2SeO_4 served as the oxidant, activation of the C–H bond in methane could take place due to the redox reaction between Au^{I}/Au^{III}.[25,26] H_2SeO_4 was demonstrated to be the dissolve reagent of Au^0 as well as the complex ligand of $Au^{I/III}$ ions in this system. The conversion of CH_4 reached 28% with a selectivity of 94% to CH_3OSO_3H at 180 °C and 2.7 MPa (CH_4) pressure. The reaction mechanism was predicted using the B3LYP/ LACVP^{**++} module. Au^{I} and Au^{III} species are both viable active sites, being responsible for electrophilic C–H activation and oxidative functionalization (Figure 8.1).

In the circumstance of an ionic liquid, a green chemical process for methane oxidation was proposed by Wang and co-workers.[27] This system consisted of Au nanoparticles supported on SiO_2 as the catalyst, [Bmin]Cl as the solvent, trifluoroacetic acid and trifluoroacetic anhydride as the acidic reagents, and $K_2S_2O_8$ as the oxidant. The conversion of methane was 24.9% with a selectivity of 71.5% to methanol at 90 °C, 2.0 MPa of CH_4. On the basis of electrophilic substitution mechanism, the authors suggested the existence of a metallic intermediate (M^{n+}–CH_3), which was formed *via* the attack by metal ions on methane. This intermediate then bound to the nucleophilic reagent (Nu) to form a stable species, CH_3–Nu. Simultaneously, the metal ion was reduced and then oxidized by $K_2S_2O_8$ to complete the catalytic cycle (Figure 8.2). The acidic reagents inhibited the over-oxidation to increase the selectivity.

Levchenco and co-workers[28,29] proposed that the Au^- containing protein from the auphilic bacteria *Micrococcus luteus* could act as a mimicking enzymatic catalyst for methane conversion at room temperature and ambient

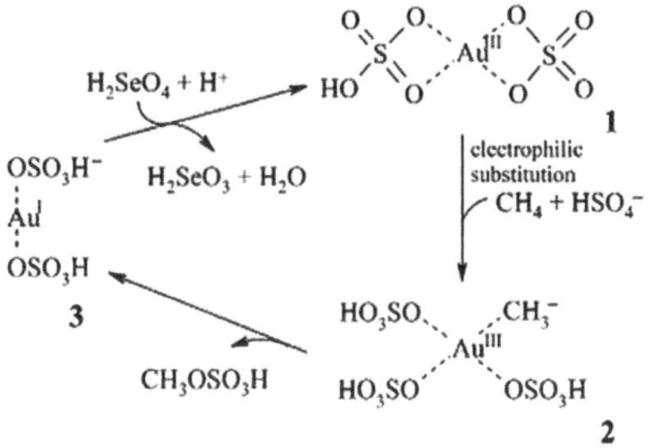

Figure 8.1 Possible pathways for gold-catalyzed methane oxidation. Reprinted with permission from ref. 26. Copyright 2004 Wiley.

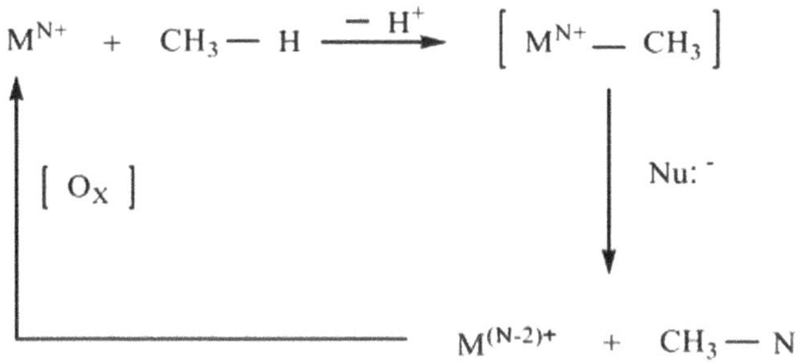

Figure 8.2 Mechanism of electrophilic substitution proposed by Wang and co-workers.
Reprinted with permission from ref. 27. Copyright Elsevier 2011.

pressure. A gold–rutin complex model was introduced and studied to explain the conversion process of methane. In the first step, oxidative addition occurred between methane and the gold complex accompanied by hydrolysis of the C–H and formation of the C–O bond. The generated Au^I could be oxidized to Au^{III} by Fe^{III} with cooperation from dioxygen, and the catalytic cycle completed (Figure 8.3).[30] The presence of nicotinamide adenine dinucleotide (NADH) acted as the intermediates in the oxidation of Au^I. The turnover number (TON) of gold for methane conversion in this system was up to *ca.* 0.6 h^{-1}. *In vivo*, the gold–protein complex perhaps helps the bacteria to survive when usual sources of carbon are limited.

The interaction between Au and H_2O_2 was proven to be an alternative method for methane oxidation. The ionic Au^{III} ($HAuCl_4$) has been recognized as the major species for this reaction in aqueous solution using H_2O_2 as the oxidant.[31] The TOF of C_1 oxygenates was up to over 10 h^{-1} with a selectivity of 57%. Compared with $OsCl_3$, the stability of this catalyst was relatively low due to the formation of Au nanoparticles. This process was demonstrated to be one involving radicals. Based on the results, the possibility of methane activation by H_2 and O_2 over a supported Au catalyst was predicted.[26]

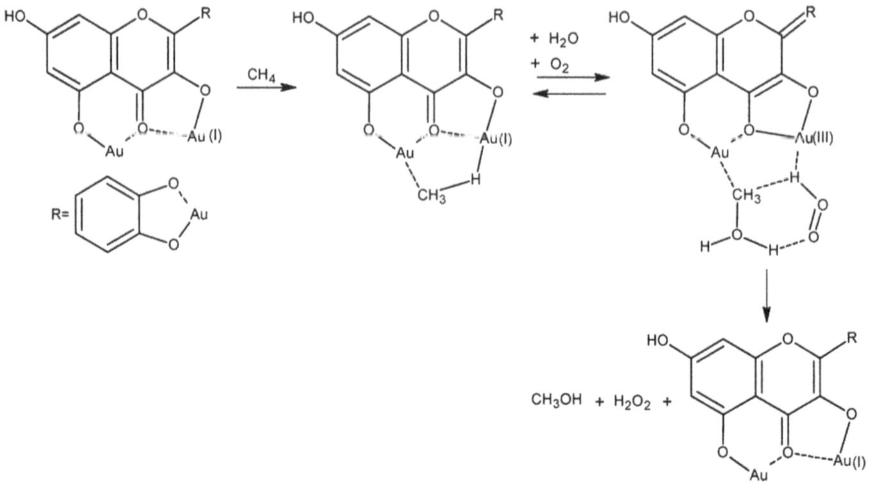

Figure 8.3 Activation of methane and oxygen by Au complex from *Micrococcus luteus* under scarce energy conditions.
Reprinted with permission from ref. 30. Copyright 2012 Royal Society of Chemistry.

Recently, Hutchings and co-workers[32] reported partial oxidation of methane into oxygenates over a Pd-Au/TiO_2 alloy catalyst with the assistance of H_2O_2. At 90 °C and 3.05 MPa pressure of CH_4, the TOF reached the maximum (25.72 h^{-1}) within 30 min over 1 wt% Au-Pd/TiO_2 prepared by the sol immobilization method. A methanol selectivity of 19% was obtained, while a selectivity of 88.4% to oxygenates (including CO_2) was also reported.

Methane conversion and methanol selectivity increased simultaneously with an increase in metal loading amount. The maximum selectivity to methanol was 49%. Furthermore, the selectivity of methanol was increased to 93% using *in situ* generated H_2O_2 from H_2 and O_2 at 2 °C. They also demonstrated that higher CH_4 pressure and initial concentration of H_2O_2 favored the production of oxygenate. The support materials, metal composition and surface treatment procedure influenced the catalytic activity significantly.[32] The reaction mechanism was further verified by ^1H nuclear magnetic resonance (NMR) analysis of the products using $^{13}CH_4$ or $^{13}CH_3OOH$ as starting reagents. The methyl hydroperoxide was demonstrated to be the primary reaction product, which was subsequently transformed to methanol and CO_2 in the presence of catalyst. The existence of $CH_3^\bullet$ and $OH^\bullet$ radicals was clearly evidenced by 5,5-dimethyl-1-pyrroline *N*-oxide (DMPO) trapped EPR (Figure 8.4). In contrast to CuFe/ZSM-5 system, the signal of $CH_3^\bullet$ was observed in the Pd-Au/TiO_2 system. The formation of $CH_3^\bullet$ is the result of a complicated interaction between different radicals. Thus, the existence of other radicals such as $OOH^\bullet$ or $CH_3OO^\bullet$ may also be possible.

Some progress was also achieved for methane activation over gold-based catalysts with input of extra energy. For instance, Marina and Mogensen[33]

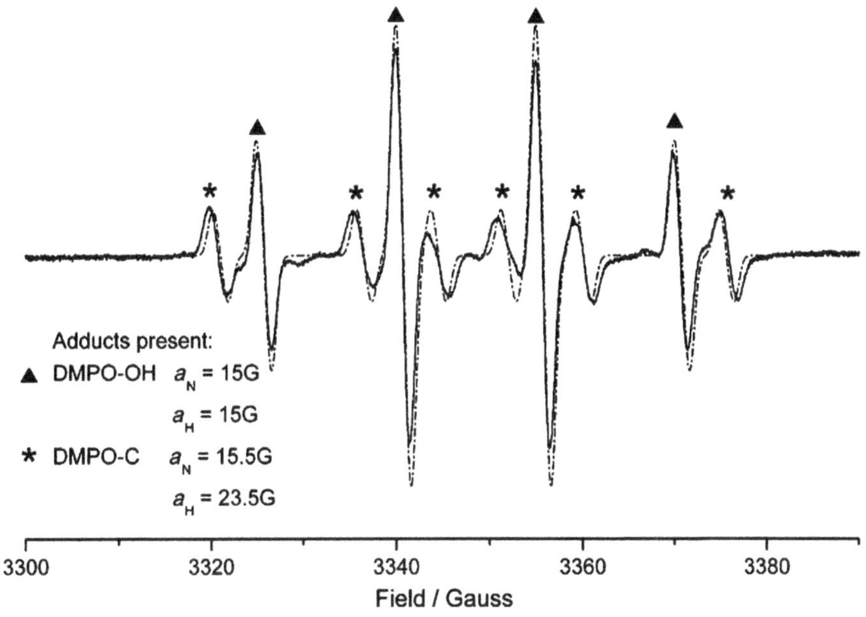

Figure 8.4 EPR spectrum showing radical species detected during the reaction of methane and H_2O_2 over a 5 wt%Au-Pd/TiO_2 (IW) catalyst with DMPO added to the reaction mixture as the radical trapping agent. Reprinted with permission from ref. 32. Copyright 2013 Wiley.

successfully set up a solid oxide fuel cell (SOFC) system consisting of porous Au and $Ce_{0.6}Gd_{0.4}O_{1.8}$ for direct methane electrochemical oxidation (Figure 8.5). The resistance to cracking on the Au-based electrode was one of the main advantages of this system. When thermal plasma was employed, methane could be selectively oxidized into formic acid with H_2 and O_2 over a core-shell Au nanoparticle film at ~140 °C and ambient pressure.[34] Dodecanethiol-stabilized Au nanoparticles were used as the precursor, which were self-assembled on the surface of mica (001) or gypsum (010) facet in toluene solutions. To avoid aggregation of the Au nanoparticles, plasma instead of thermal treatment was selected to remove the organic ligands. In the temperature range 140–200 °C, methane was selectively transformed into HCOOH (a selectivity of 92–97%) with a high rate, which was nearly two orders of magnitude higher than previously reported data. Llorca suggested some active oxidative species, *i.e.* peroxo or hydroperoxo, were involved in this system. However, only indirect evidence was obtained to verify this point.

Although the attempt at selective oxidation of methane by molecular oxygen over a supported Au catalyst at low temperature (~250 °C) failed,[35] with the combination of a reflection time-of-flight mass spectrometer (TOF-MS) and a laser ablation/supersonic expansion cluster source, several oxygen-rich clusters with metal cations were demonstrated to be crucial for the C–H activation of methane by the generation of $O^{\bullet -}$.[36–38]

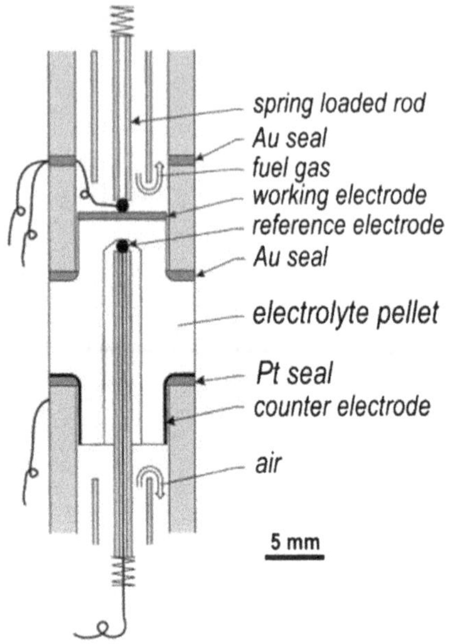

Figure 8.5 Sketch of the three-electrode cell set-up proposed by Marina and Mogensen.
Reprinted with permission from ref. 33. Copyright 1999 Elsevier.

He and co-workers[39] found that the Au-doped atomic metal cluster ($AuNbO_3^+$) showed great potential for the multiple activation of the C–H bond with high selectivity. The introduction of Au could effectively tune the local charge within the clusters during the reaction for C–H bond activation. Au acts as the electron acceptor for the storage of valence electron between Au and Nb in $AuNbO_3^+$. From this point of view, the design of supported gold catalysts may be easy in the future.

8.2.2 Selective Oxidation of Chain Alkanes

It is also very difficult to activate light alkanes, namely ethane, propane and butane. The transformation of these light alkanes into useful chemicals has been an attractive and challenging research field. Several techniques have been developed to convert alkanes, including non-catalytic thermal cracking of alkanes and catalytic processes, such as oxidative dehydrogenation, oxidative cracking, and selective oxidation to corresponding the alkenes and oxygenates. Metal oxides, zeolites, and supported catalysts have been applied using oxygen, nitrogen oxide, hydrogen peroxide, *etc.* as oxidants under different reaction systems. However, research on selective oxidation of light alkanes based on Au catalysts is very limited. In 2001, Ueda *et al.*[40] reported that a Au/TiO_2catalyst brought about a good ethylene and acetaldehyde yield for ethane oxidation at 475 °C.

The Oyama group[41] found that Au nanoparticles supported on microporous titanosilicate (TS-1) were highly selective (95%) towards the formation of acetone and isopropanol from propane, O_2 and H_2 under mild conditions (170 °C and 0.3 MPa). They proposed that the reaction follows the pathway: (1) H_2O_2 is formed from H_2 and O_2 on Au, Figure 8.6(a); (2) H_2O_2 is transferred to Ti centers to form Ti-hydroperoxo species, Figure 8.6(b); (3) Ti-hydroperoxo species react with propane to form 2-propoxy intermediate species; and (4) 2-propoxy species form acetone and 2-propanol.[42] *In situ* ultraviolet–visible (UV-vis) spectra gave clear evidence for step (2); Figure 8.7 shows that the hydrated hydroxyl species $Ti(OSi)_3\text{-}OH \cdot (H_2O)_2$ initially reacts (decrease in 270 nm band) with H_2O_2 to form hydroperoxide species. The whole process is summarized in Figure 8.8.

Using a kinetic study, the Oyama group subsequently found that the rate of oxygenate formation could be well described by a power-law model [eqn (8.6)] and the apparent activation energies for the formation of H_2O, acetone, and CO_2 were 42, 38, and 40 kJ · mol^{-1}, respectively.[42]

$$r_{oxyg} = k_{oxyg}(H_2)^{0.74}(O_2)^{0.36}(C_3H_8)^{0.29} \tag{8.6}$$

Moreover, propane epoxidation by sequential propane dehydrogenation and propylene epoxidation steps using a two-catalyst bed mixture was studied in the presence of a mixture of H_2 and O_2 (Figure 8.9). A propane conversion of 2% and overall propylene oxide (PO) and propylene selectivities of 8 and 57%, respectively, were observed for this system. *In situ* X-ray absorption near-edge structure (XANES) and UV-vis spectroscopies showed the formation of an Au–O complex which may react with H_2 to form hydroperoxide species. 2-Propoxy was proposed as the intermediate on the basis of production of acetone and 2-propanol on Au/TS-1, and of propylene and acetone on Au/TiO_2 during propane oxidation.[43]

Recently, Kung and co-workers[44,45] reported selective oxidation of propane to acetone by O_2 and H_2 at 220 °C over a new series of Au nano-catalysts decorated with metal oxides. Figure 8.10 shows a ligating molecule with controllable and quantified amine densities; Ti groups could be readily synthesized using polymethylhydrosiloxane (PMHS) functionalized with

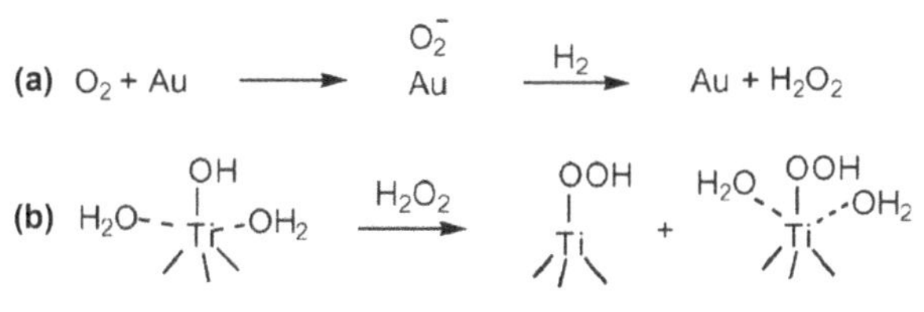

Figure 8.6 Sequence of steps in oxygen activation: (a) H_2O_2 formation on Au; (b) hydroperoxide formation on Ti centers.
Reprinted with permission from ref. 41. Copyright 2008 Royal Society of Chemistry.

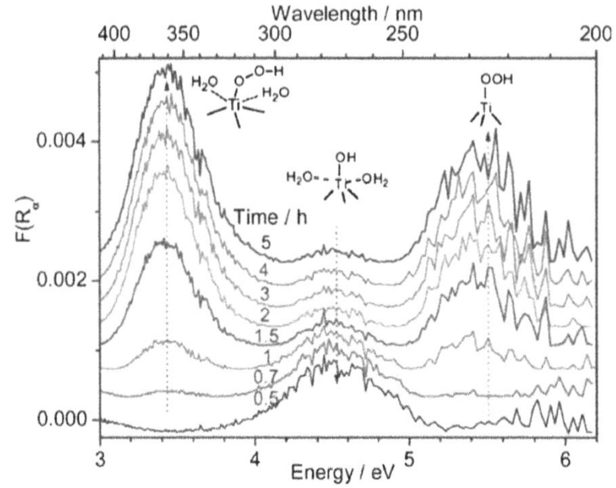

Figure 8.7 *In situ* UV-vis spectra for Au–Ba/TS-1 (Ti : Si = 3 : 100) before reaction under argon.
Reprinted with permission from ref. 41. Copyright 2008 Royal Society of Chemistry.

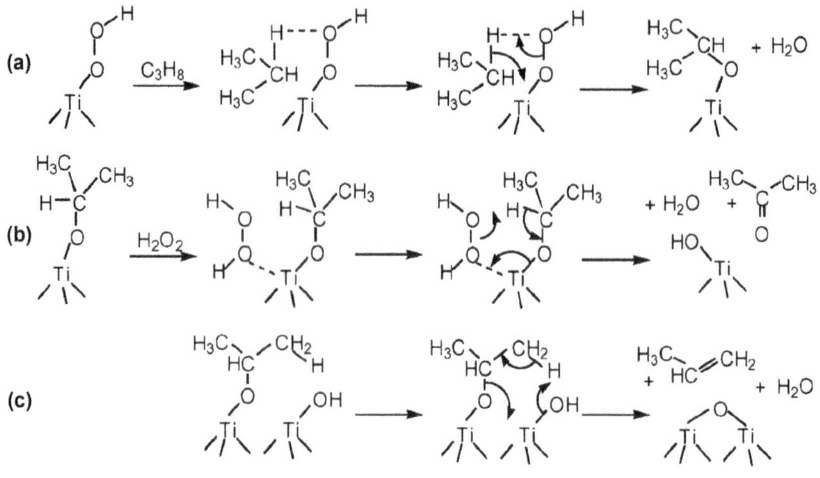

Figure 8.8 Possible sequence of steps during propane selective oxidation on Au-supported Ti-containing catalysts: (a) formation of the isopropoxide intermediate; (b) oxidation of the isopropoxide species to acetone; and (c) dehydration of the isopropoxide species to propylene.
Reprinted with permission from ref. 42. Copyright 2008 Elsevier.

amine and Ti acetylacetonate groups (Ti-N-PMHS). After deposition of Au and thermal treatments, Au nanoparticles were covered with thin patches of SiO_2 containing Ti with different amount of Au-TiO_x interface (Au-Ti/SiO_2). Compared with Au/SiO_2 and Ti/Au/SiO_2 (Ti was introduced as a second step), Au-Ti/SiO_2 has a higher TOF for propane and the activity increased with

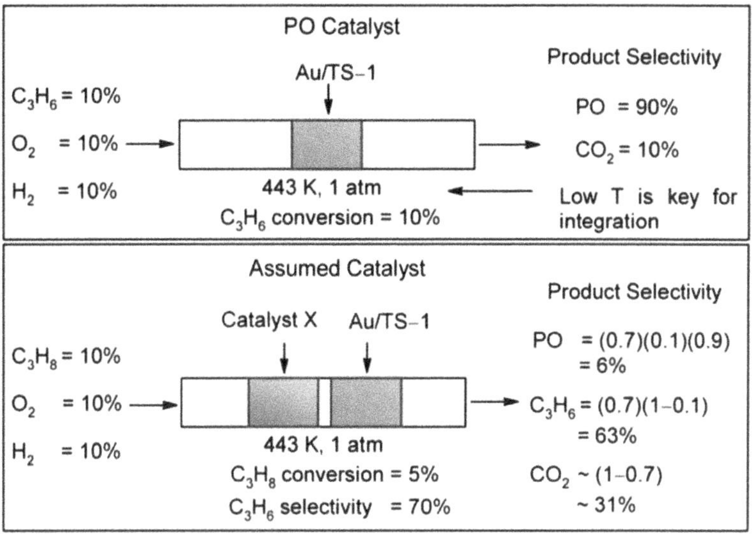

Figure 8.9 Hypothetical propane to propylene oxide process.
Reprinted with permission from ref. 43. Copyright 2008 Elsevier.

increasing Ti density. The selectivity to acetone for all catalysts was near 70%. Therefore, the Au-TiO_x interface was proposed as the active site. Although it was possible that hydrogen peroxide formed on Au could diffuse to a Ti site where propane oxidation to acetone occurs, hydrogen peroxide would decompose on passing through the silica patches to the Ti site, which was the reason for the low activity of Ti/Au/SiO_2.

Gas-phase catalytic cracking (GCC) and catalytic oxidative cracking (COC) are also used for propane conversion. Both processes are believed to follow a free radical mechanism, where activated oxygen species abstract H from hydrocarbons to form radicals ($2C_3H_8 + O_2 \rightarrow 3C_2H_4 + 2H_2O$). The Au/$Ce_{0.5}Zr_{0.5}O_2$ catalyst for this reaction was studied by Narasimharao and Ali.[46] Sulfation treatment of the support was found to be good for activity due to the formation of highly acidic and easily reducible interactive species between the Au particles and the $Ce_{0.5}Zr_{0.5}O_2$ surface on the more highly defective surfaces of the sulfated materials.

Barison *et al.*[47] prepared nanosized Au/$La_{1-x}Sr_xMnO_3$ (Au/LSM) and Au/$La_{1-x}Sr_xCrO_3$ (Au/LSC) composites using a modified nitrate–citrate Pechini method and studied their catalytic performance for steam reforming (SR), partial oxidation (POX) and autothermal reforming (ATR) of propane. Detailed results are shown in Table 8.1.

In SR, propane conversion increased from around 4% at 600 °C to 86% at 800 °C for Au/LSM catalysts, slightly higher than that over LSM catalysts. Variation in the Au amount had almost no influence on LSC catalysts for propane conversion, while higher ethane selectivity was detected. For POX, all catalysts attained nearly 100% conversion at 600 °C, and higher selectivity

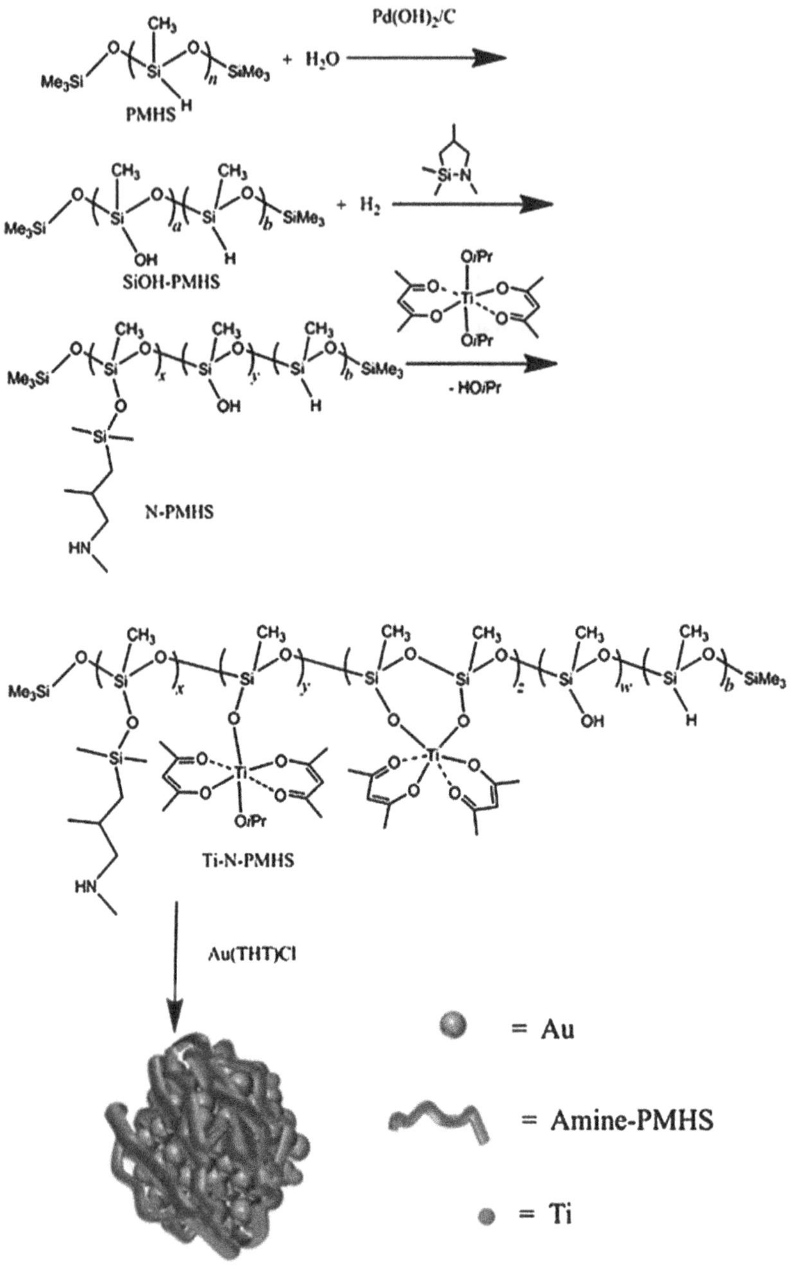

Figure 8.10 Synthesis of Au–Ti complexes.
Reprinted with permission from ref. 44. Copyright 2012 Royal Society of Chemistry.

Table 8.1 Conversion (%) and product composition (%) for propane SR, POX and ATR on 1 wt% and 5 wt% Au/LSM and on 1 wt% and 5 wt% Au/LSC. Reprinted with permission from ref. 47. Copyright 2007 Elsevier.

	T/ °C	SR 1%Au/ LSM	5%Au/ LSM	1%Au/ LSC	5%Au/ LSC	POX 1%Au/ LSM	5%Au/ LSM	ATR 1%Au/ LSM	5%Au/ LSM	1%Au/ LSC	5%Au/ LSC
Conversion/%	–	3.66	9.33	8.00	6.58	98.11	99.74	64.74	82.23	44.04	32.47
	700	13.66	23.95	16.27	13.79	98.39	99.58	76.31	86.67	83.68	79.79
	800	70.73	85.76	68.37	66.65	98.73	99.67	92.87	99.51	95.09	93.83
C_3H_8	600	87.49	76.74	80.04	83.07	0.53	0.07	17.36	6.51	23.44	33.18
	700	61.42	50.87	63.93	67.83	0.48	0.10	9.87	3.94	5.84	8.37
	800	11.04	5.25	14.90	15.85	0.38	0.08	2.00	0.08	1.64	2.02
H_2	600	2.18	2.67	0.69	1.16	26.27	30.04	25.27	16.24	2.52	1.11
	700	17.39	12.84	6.89	7.55	25.18	29.18	33.38	29.69	11.43	8.34
	800	32.52	26.59	17.52	17.70	24.08	30.58	36.03	52.79	16.62	18.18
CO	600	0.00	0.01	0.00	0.00	26.99	38.90	10.05	22.98	10.86	3.70
	700	0.38	0.18	0.00	0.02	26.69	38.21	7.71	16.31	28.99	31.49
	800	2.08	2.30	0.00	0.17	25.26	38.77	6.99	23.55	37.93	38.59
CO_2	600	0.06	0.10	0.03	0.06	3.50	4.20	16.34	21.55	10.35	25.47
	700	0.38	0.88	0.02	0.06	2.2	3.21	15.14	25.10	7.05	9.37
	800	3.66	4.45	0.04	0.03	2.82	3.40	16.30	14.98	5.46	5.87
C_2H_4	600	0.63	2.12	0.57	0.75	11.41	4.24	14.38	13.28	9.80	5.37
	700	8.16	10.99	7.19	7.01	13.27	4.20	14.91	10.55	15.07	19.28
	800	3.66	26.23	28.45	28.00	14.72	4.23	16.00	0.19	15.56	15.54
CH_4	600	0.64	1.47	0.03	0.44	20.14	13.72	13.03	11.33	4.84	2.26
	700	8.10	8.16	5.32	5.35	20.00	11.48	15.12	10.30	12.83	11.30
	800	23.84	23.38	22.59	21.93	22.31	11.96	18.09	6.47	16.18	10.09
O_2	600	0.00	0.00	0.00	0.00	0.00	0.01	0.08	0.06	27.58	22.29
	700	0.00	0.00	0.00	0.00	0.00	0.02	0.07	0.03	8.63	7.00
	800	0.00	0.00	0.00	0.00	0.00	0.01	0.04	0.02	1.82	6.79
Others	600	9.00	16.89	18.37	14.52	11.16	8.82	3.48	8.05	10.61	6.62
	700	4.46	16.08	16.65	12.18	12.11	13.69	3.80	4.08	10.16	4.85
	800	2.84	11.80	16.50	16.32	10.43	10.97	4.55	1.92	4.79	2.92

to H_2 and CO was detected for 5.0Au/LSM. For ATR, the propane conversions over Au/LSM were lower than those obtained from the POX process only at 800 °C. Low hydrogen selectivity was attained for ATR, analogous to SR, and high percentages of byproducts such as CO_2 and O_2 were detected.

$$C_4H_{10}+\frac{1}{2}O_2 \rightarrow 2C_2H_4+H_2O \quad (8.7)$$

Although there are a great number of studies on the oxidative cracking of *n*-butane [eqn (8.7)],[48–50] reports of Au catalysts appeared only recently. Hardacre and co-workers[51] reported the direct activation of *n*-butane into light alkenes with oxygen using supported Au catalysts (Table 8.2). In comparison with the bare supports, both conversion and selectivity decreased over supported Au catalysts, except for La_2O_3. Changes in the overall selectivity and the alkene speciation were both observed for Au/La_2O_3 with a rate of 0.29 mmol · min^{-1} g^{-1} and a selectivity to alkenes of 31.0% at 600 °C and a rate of 0.96 mmol · min^{-1} g^{-1} and a selectivity to alkenes of 47.7% at 650 °C. These results could be attributed to the excellent capability of radical generation for La_2O_3 and the efficient radical generation promoted by interaction between the cationic Au and La_2O_3 support (Figure 8.11). This provided active sites for the high selectivity instead of combustion compared with many other catalysts reported for this reaction. Moreover, the addition of Au to La_2O_3 led to a stabilization of oxycarbonate species.

Cavani and co-workers[52] studied the effect of Au doping for vanadyl pyrophosphate [$(VO)_2P_2O_7$] (VPP) on selective oxidation of *n*-butane to maleic anhydride (MA). Figure 8.12(a) shows the results of conversion of *n*-butane and Figure 8.12(b) the selectivity to MA with temperature. It was found that Au limited the over-oxidation of the catalyst surface that is

Table 8.2 Catalytic performance for the oxidative cracking of *n*-butane of the supports and 0.2 wt% gold-modified supports at 650 °C compared with BiOCl. Reprinted with permission from ref. 51. Copyright 2011 Royal Society of Chemistry.

		Selectivity/ %			Yield/ %		
Catalyst	Rate/mmol · min^{-1}g^{-1}	Alkene	Alkane	CO_X	C_2H_4	C_3H_6	C_4H_8
La_2O_3	0.84	28.8	3.2	23.8	1.7	1.7	-
Au/La_2O_3	0.96	47.7	1.4	33.7	3.5	1.3	1.7
$CeZrO_4$	0.86	17.9	2.2	69.5	1.0	0.3	-
Au/$CeZrO_4$	0.51	4.0	1.1	69.5	0.2	0.1	0.0
ZrO_2	0.38	16.3	3.8	24.5	0.4	0.5	-
Au/ZrO_2	0.40	7.0	1.3	12.2	0.2	0.2	0.0
TiO_2	0.51	17.3	2.1	26.3	1.0	0.3	-
Au/TiO_2	0.45	11.6	1.5	38.7	0.5	0.2	0.1
Nb_2O_5	0.25	20.5	5.3	27.2	0.2	0.2	-
Au/Nb_2O_5	0.20	7.0	2.6	18.2	0.1	0.1	0.0
BiOCl	0.27	21.1	4.6	18.1	0.3	0.4	0.0

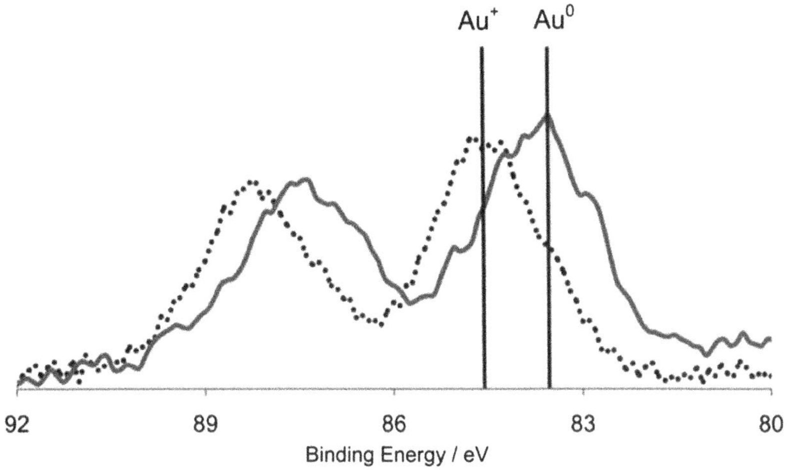

Figure 8.11 Comparison of Au_{4f} photoelectron spectra for 0.2 wt% Au/La_2O_3 (···) and 0.2 wt% Au/Nb_2O_5 (—) catalysts following calcination at 550 °C. Reprinted with permission from ref. 51. Copyright 2010 Wiley.

typically observed with stoichiometric VPP, and also hindered the development of bulk unselective $VOPO_4$ compounds due to the enhancement of the reducibility of V^{5+} in the presence of Au. From the X-ray photoemission spectroscopy (XPS) results, surface Au species in a positive oxidation state could be observed.

Compared with light alkanes, there are even fewer studies on the selective oxidation of long chain linear alkane using a Au-based catalyst. Hutchings and co-workers[53] studied the low-temperature aerobic oxidation of decane using an oxygen-free radical initiator, namely azobisisobutyronitrile (AIBN) and cumene hydroperoxide (CHP). Tables 8.3 and 8.4 show the formation of a range of oxygenated products using AIBN and CHP, respectively. It was found that the combination of AIBN and Au/nano CeO_2 does not enhance the conversion in the more challenging oxidation of *n*-decane compared with AIBN alone and just slightly increases alcohol selectivity.

Some of the peroxides produced by AIBN were considered to be selectively reduced to alcohols over the CeO_2 nanocatalysts (Figure 8.13). The combination of CHP and Au/nanoCeO_2 was found to significantly reduce decane conversion compared with the reaction with CHP alone (Table 8.4). This reduction of CHP by Au/nano CeO_2 may prevent the formation of decane peroxides, leading to the lower conversion of decane (Figure 8.14). The main decomposition products of CHP alone were acetophenone (the ketone product) and 2-phenyl-2-propanol (the alcohol product), in approximately a 50 : 50 ratio. However, for the combination of CHP and Au/nanoCeO_2, the radical initiator was found to decompose almost entirely into 2-phenyl-2-propanol. Moreover, AIBN was found to be more active than CHP under the conditions studied.

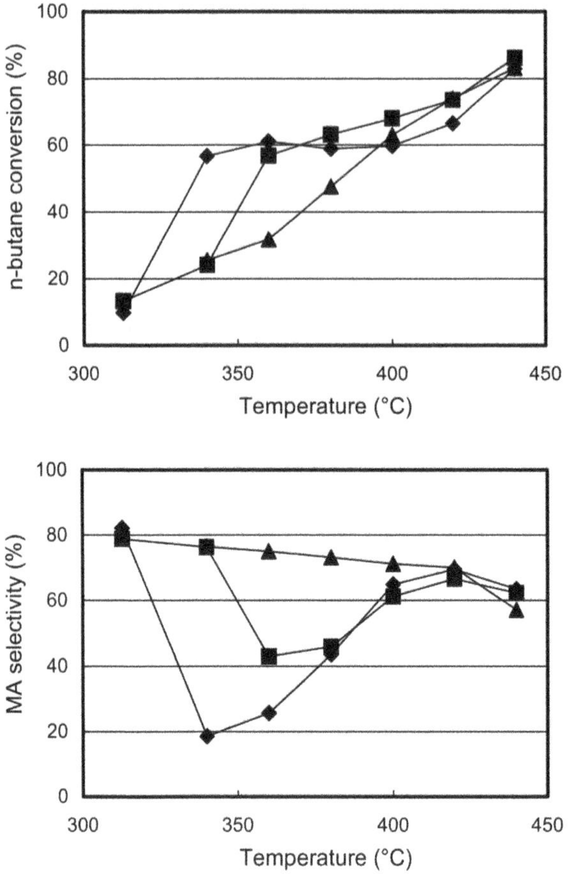

Figure 8.12 Conversion of n-butane (top) and selectivity to MA (bottom) in function of the reaction temperature. "Equilibrated" catalysts: VPO (◆), VPO1Au (■) and VPO3Au (▲).
Reprinted with permission from ref. 52 Copyright 2011 Elsevier.

8.2.3 Selective Oxidation of Cyclic Alkanes

The products of cyclohexane partial oxidation, cyclohexanone (Cy=O) and cyclohexanol (Cy–OH, the mixture with Cy=O and Cy–OH is referred to as KA-oil) are in great demand as the industrial precursors of ε-caprolactam and adipic acid. Both products are important for the manufacture of nylon-6 and nylon-6,6 fibers. In practical production, cyclohexane partial oxidation is carried out under aqueous aerobic conditions using cobalt naphthenate or cobalt acetylacetonate catalysts at 140–160 °C and 1–1.5 MPa. In order to maintain the high selectivity toward KA-oil (70–85%), the conversion was restricted to 4%.[54]

This homogeneous catalytic process took place according to a radical chain mechanism. The chain initiation was thought to proceed through the homolysis cleavage of cyclohexyl peroxide (CyOOH). The Haber–Weiss cycle (Figure 8.15) was triggered by a one-electron transfer catalyzed over transition metals or their complex (Co^{2+}, Mn^{2+} or Cr^{2+}), with the formation of oxygen-centered radicals ($CyO^{\bullet}$) as a consequence.[55] The $CyO^{\bullet}$ radical further reacted with CyH solvent to form cyclohexyl radicals ($Cy^{\bullet}$), which rapidly attacked the molecular oxygen, leading to the formation of cyclohexyl peroxy radical ($CyOO^{\bullet}$). As the main chain propagator, $CyOO^{\bullet}$ maintained the continuous formation of new radicals. The final production of CyOH or Cy=O was responsible for the chain termination.

The aerobic oxidation of cycloalkanes over transition metal complex catalysts has also been widely studied, such as Cr/SiO_2 or MCM-41,[56,57] and Ti, Cu, Co, Mn-containing zeolite.[58,59] The pioneering study of cycloalkanes conversion over Au catalysts can be dated back to 2004. Suo and co-workers reported that the supported Au catalysts, Au/ZSM-5[60] and Au/MCM-41[61] with a low loading amount (less than 1%), can oxidize cyclohexane with oxygen at ~150 °C and 1.0 MPa O_2 in N_2. The highest TON of cyclohexane reached 49 059 over an Au(0.04%)/TiO_2(0.24%)/SiO_2 catalyst.[62] A mesoporous silica (SBA-15) supported hybrid Au catalyst was employed for the cyclohexane oxidation by Zhu *et al.*[63]

Very recently, based on a similar strategy of catalyst design, Au nanoparticles were highly dispersed on modified SBA-15 *via* a one-pot synthesis method (Figure 8.16).[64] The conversion and the product distribution varied with solvent. Acetone, ethanol, acetic acid, benzonitrile, ethylbenzene, and acetonitrile were evaluated under the same reaction conditions. With acetonitrile, the catalyst, Au(0.2%)/M-SiO_2, exhibited the best performance [turnover frequency (TOF) = 21 097 h^{-1}].

Recently, the dependence of the reactivity for cyclohexane oxidation on particle size was systematically demonstrated.

Tsukuta and co-workers[65] successfully deposited Au clusters, Au_n ($n = 10$, 18, 25, 39) with an atomically controlled size on hydroxyapatite (HAp). A volcano-shaped curve was obtained when correlating between the turnover frequency of cyclohexane and the number of Au atoms (Figure 8.17). The optimal size of Au_n clusters for the aerobic oxidation was determined to correspond to $n = 39$–85. The introduction of Pd could improve the conversion of cyclohexane and tune the selectivity of cyclohexone. Pd upon alloying with gold to form bimetallic Pd-Au catalysts, which have a rich electron density, is favorable for O_2 adsorption. Subsequent formation of $O_2^{\bullet -}$ would be feasible for the oxidation of cyclohexane to produce cyclohexyl hydroperoxide (CHHP), which could be easily verified by Fourier transform infrared (FTIR) and gas chromatography–mass spectrometry (GC-MS). The surface composition had a strong effect on the stability of CHHP, leading to a different distribution of products (Figure 8.18).[66] All the activity results were summarized in Table 8.5.

With the addition of *t*-butyl hydroperoxide (TBHP) as the radical initiator, Hutchings and co-workers[67,68] investigated the possibility of cyclohexane oxidation by molecular oxygen at temperatures below 100 °C. Within 17 h, the

Table 8.3 Product breakdown for the oxidation of *n*-decane, with and without AIBN, with and without catalyst, and with and without addition of PPh_3 (for selected reactions). Reprinted with permission from ref. 53. Copyright 2011 Elsevier.

				Selectivity[c]/ %			
Catalyst/ initiator	Conditions	–	Conversion/%	C_{10} 5/4-one	C_{10} 3-one	C_{10} 2-one	C_{10} 4-ol
AIBN[a]	70 °C, 20 h, air, reflux	–	0.26	15.0	7.4	9.2	6.8
		+ PPh_3	0.27	11.7	5.8	6.1	18.7
AIBN[b]	90 °C, 2 h, 1.2 MPa O_2	–	0.49	20.1	10.0	12.7	4.6
		+ PPh_3	0.45	19.6	9.6	11.8	9.3
Au/ nanoCeO_X + AIBN[b]	90 °C, 2 h, 1.2 MPa O_2	–	0.45	14.0	6.6	8.8	6.6
		+ PPh_3	0.47	11.2	5.9	7.4	13.6
NanoCeO_X + AIBN[b]	90°C, 2 h, 1.2 MPa O_2	–	0.36	17.4	8.9	13.1	11.4
		+ PPh_3	0.40	14.0	6.4	10.2	18.7
Blank	90 °C, 20 h, 1.2 MPa O_2	–	0.028	21.9	12.2	21.2	6.0
AIBN	90 °C, 20 h, 1.2 MPa O_2	–	1.13	27.2	17.9	19.3	3.0
		+ PPh_3	1.47	23.7	15.1	16.3	5.4
Au/ nanoCeO_X + AIBN[b]	90 °C, 20 h, 1.2 MPa O_2	–	0.87	12.5	7.8	10.9	10.3
		+ PPh_3	0.88	11.6	8.1	9.7	16.0
NanoCeO_2	90 °C, 20 h, 1.2 MPa O_2	–	0.11	11.1	5.5	11.1	8.7
NanoCeO_2 + AIBN	90 °C, 20 h, 1.2 MPa O_2	–	0.78	5.3	8.0	11.0	10.3
		+ PPh_3	0.79	11.2	7.8	9.3	13.2
Aldrich CeO_2 + AIBN	90 °C, 20 h, 1.2 MPa O_2	–	1.13	20.0	10.2	15.3	7.4
TiO_2 + AIBN	90 °C, 20 h, 1.2 MPa O_2	–	0.58	15.3	7.6	11.7	12.8
ZSM-5 + AIBN	90 °C, 20 h, 1.2 MPa O_2	–	0.91	25.6	13.2	18.0	5.3

Note: Decrease in conversion and selectivity to others after addition of PPh_3 considered to be from dilution of products to below the GC detection limits and possible degradation.
[a]10.0 g decane, 50 mg AIBN.
[b]7.69 g decane, 30 mg AIBN.
[c]C_{10}-one = decanone, C_{10}-ol = decanol and acid = carboxylic acid.

C_{10} 3-ol	C_{10} 2-ol	C_{10} 1-ol	C_{10} acid	C_9 acid	C_8 acid	C_7 acid	C_5 acid	C_4 acid	C_3 acid	C_2 acid	C_{10} diones	Others	C balance/%
3.6	4.9	0.3	0	0.4	0.1	1.2	1.2	0	0	0.3	8.4	41.2	93
10.0	14.6	0.9	0	0.9	0.3	1.6	0	0	0	0	6.6	23.0	86
2.7	4.6	0.1	0	0.3	0	1.3	1.4	0.3	0	0.6	11.7	29.6	83
5.3	8.4	0	0	0.5	0	1.9	0.4	0	0	0.2	11.2	21.7	77
3.8	6.3	0	0	0.7	0	0.8	1.3	0.3	0	1.2	10.7	38.9	94
7.8	11.3	0	0.1	0.9	0.1	2.8	0.3	0.1	0.2	0.7	12.2	25.4	81
5.5	7.8	0.9	0.4	0	0	1.6	1.1	0.5	0	0	9.6	21.8	105
10.0	14.3	1.1	1.4	0	0	0.7	0	0	0	0	4.0	19.1	114
2.6	4.6	0	0	0	0	0	0	0	0	0	0	31.5	95
1.7	2.8	1.1	0.8	0	0.3	2.0	3.7	3.5	2.8	0	7.0	6.9	89
3.1	4.9	0.7	1.0	0	0.3	1.0	2.4	2.1	3.2	0	8.7	12.1	99
6.1	8.8	0.9	0.5	0.2	0.6	1.3	5.3	6.4	3.4	4.3	11.4	9.5	79
8.7	11.5	1.3	0.8	0.3	0.3	1.4	4.1	3.6	3.8	0.5	10.7	7.6	89
4.1	4.9	2.7	1.7	1.0	0	0.9	5.9	8.1	8.7	0	11.3	14.3	107
5.9	9.1	1.0	0.4	0.4	0.4	1.0	4.2	4.6	4.3	7.1	11.0	16.0	70
7.3	10.2	1.0	0.5	0.3	0.2	2.2	3.0	3.1	2.5	3.2	10.3	14.6	78
4.1	5.9	0.8	0.9	0.1	0.4	3.2	5.2	3.9	2.5	0	10.6	9.5	107
7.0	10.0	1.1	0.8	0	0.2	1.5	1.4	2.3	0	0	10.3	18.0	107
2.9	4.3	0.5	1.9	0	0.5	2.2	2.7	3.4	2.0	0	8.6	8.9	106

Table 8.4 Oxidation of decane using CHP with and without Au/nanoCeO_2 catalyst.[a] Reprinted with permission from ref. 53. Copyright 2011 Elsevier.

Catalyst/ initiator	Conversion/ %	Selectivity/% C_{10} ketones	C_{10} internal alcohols	C_{10} terminal alcohols	C_1–C_9 acids	others
CHP	0.15	24.8	6.1	0.2	2.5	66.4
CHP + Au/ nanoCeO_2	0.036	76.5	0.5	0	6.5	16.5

[a]Decane (7.69 g), CHP (30 mg), Au/nanoCeO_2 catalyst (100 mg), 2 h, 90 °C, 1.2 MPa O_2.

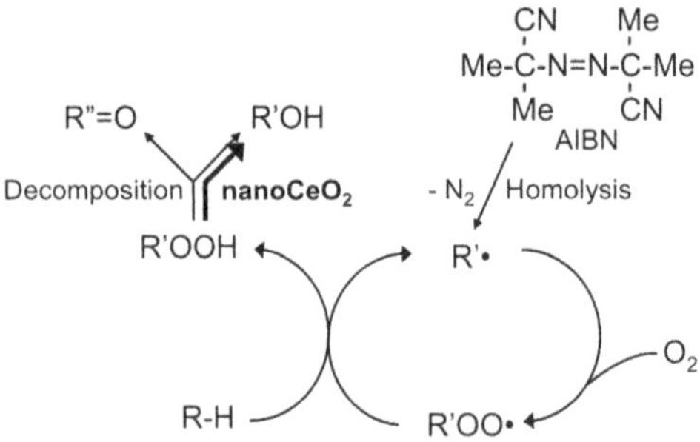

Figure 8.13 Proposed reaction scheme for the oxidation of decane (R–H) with AIBN. $R' = C_{10}H_{21}$ or $Me_2C(CN)$, $R'' = C_{10}H_{20}$. Reprinted with permission from ref. 53. Copyright 2011 Elsevier.

conversion of cyclohexane was 6.0% with selectivity towards cyclohexanone and cyclohexanol, 11.0% and 6.2%, respectively, at 70°C and ambient pressure. The dosing of 1,4-difluorobenzene as additive and the modification of the catalyst surface composition by bismuth can enhance the selectivity of KA-oil. Other prompt protocols, such as the application of H_2O_2 as initiator[69] and the assistance of photoradiation (Figure 8.19),[70] were also reported to increase the yield of KA-oil as well as decreasing the reaction temperature.

From the analysis of the products and a systemic comparison between heterogeneous gold and homogenous cobalt, Hereijers and Weckhuysen[71] argued that surface of the Au-based catalyst interacted strongly with the propagating radical intermediates, products or other byproducts. Through the fast selective addition trapping of DMPO, spin trapping X-band EPR spectroscopy was used to demonstrate the reaction mechanism. The intermediates with or without the loading of Au/MgO were compared, indicating that the amount of alkoxyl radicals increased with the dosing of Au/MgO (Figure 8.20). The catalytic homolysis of the O–O bond in TBHP took place on the surface of Au/MgO.[72]

Recently, the possibility of a one-pot synthesis of adipic acid was investigated by Alshammari *et al.*[73] Using a nano-Au/TiO_2 catalyst, low purity adipic acid with a selectivity of 26% was produced at 150 °C and 1.0 MPa O_2.

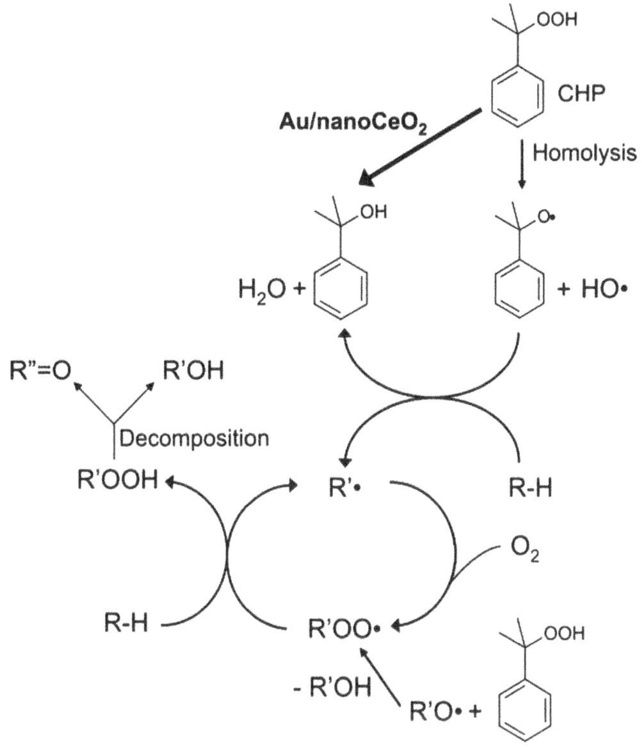

Figure 8.14 Proposed reaction scheme for the oxidation of decane (R–H) with CHP. $R' = C_{10}H_{21}$ or $PhC(Me)_2$, $R'' = C_{10}H_{20}$.
Reprinted with permission from ref. 53. Copyright 2011 Elsevier.

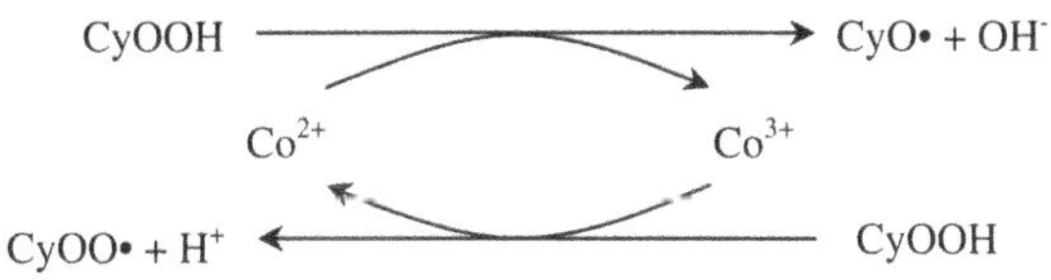

Figure 8.15 Haber–Weiss cycle of the cobalt-catalyzed decomposition of cyclohexyl hydroxyperoxide.

8.2.4 Selective Oxidation of Toluene

As one of the most widely used alkyl aromatic chemicals, toluene is a basic commodity in organic chemistry. The selective conversion of toluene by molecular oxygen is a promising alternative process for the production of benzyl alcohol, benzaldehyde, benzoic acid, and benzoate. These products are of crucial importance in the manufacture of various high value-added chemicals such as solvents, dyes, pharmaceuticals, fragrance and plasticizers. However, the selective activation of primary C–H bonds in toluene is extremely tricky. In industry, benzaldehyde is synthesized by the

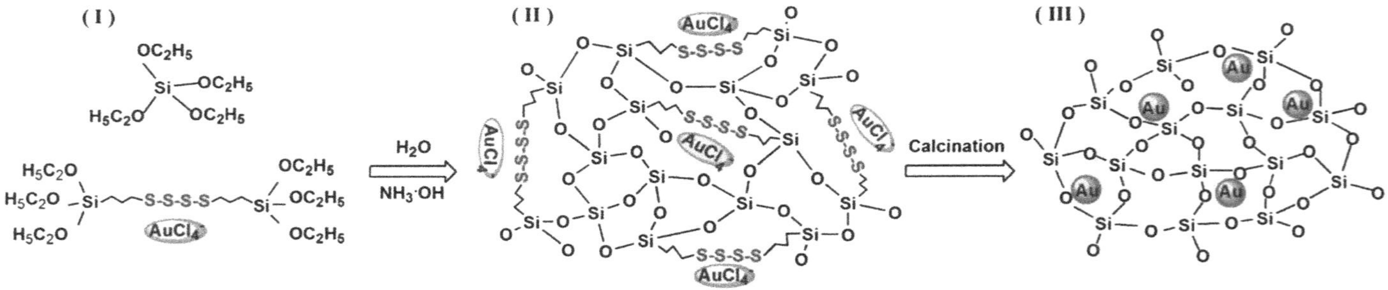

Figure 8.16 Illustration of the one-pot synthesis process of Au/M-SiO_2.
Reprinted with permission from ref. 64. Copyright 2013 Royal Society of Chemistry.

Table 8.5 Summary of cyclohexane oxidation activity over various supported gold catalysts.

Catalyst	Conversion/ %)	Selectivity/ %				TON or TOF/ h^{-1}	Reaction conditions	Ref.
		Cy=O	Cy–OH	CyOOH	Others			
Au(0.51%)/ZSM-5	7.0	36.0	55.0	–	9.0	3092	150 °C, O_2 1 MPa, Cy 20 ml, catalyst 5 mg, 3 h	46
Au(0.13%)/MCM-41	12.0	37.0	60.0		3.0	11 214	150°C, O_2 1 MPa, Cy 20 ml, catalyst 5 mg, 6 h	47
Au(0.2%)/Al_2O_3	12.6	52.6	32.1	3.8	11.5	16 136	150 °C, O_2 1 MPa, Cy 20 ml, catalyst 5 mg, 3 h	48
Au(0.04%)/TiO_2(0.24%)/ SiO_2	8.7	51.5	37.3	1.7	11.2	49 059	150 °C, O_2 1 MPa, Cy 20 ml, catalyst 5 mg, 3 h	48
$Au_{0.79}Si_{13.9}AlO_x$	6.4	54.8	21.8	21.0	–	3013	150 °C, O_2 1 MPa, Cy 20 ml, catalyst 5 mg, 2 h	59
Au(2.09%)/IL-SBA-15	28.0	57.0	37.0	–	6.0	1628[a]	150 °C, O_2 1 MPa, Cy 20 ml, catalyst 5 mg, 6 h	49
Au(0.2%)/M-SiO_2	17.3	56.8	30.3	2.4	10.5	21 097	150 °C, O_2 1 MPa, Cy 20 ml, catalyst 5 mg, in 20 ml acetone, 2 h	43
Au(0.5%)/Graphite	1.0	48.1	43.5	–	8.4	–	70 °C, ambient pressure, Cy 10 ml, catalyst 120 mg, TBHP 80 mg, 1,4-difluorobenzene, 120 mg, 17h	42
Au_{39}(0.2%)/HAp	14.9	50.0	49.0	–	1.0	18 500	150 °C, O_2 1MPa, Cy 10 ml, catalyst 100 mg, 4 h, TBHP 10 mg	50
Au(1.0%)/MIL-10	16.2	18.2	81.8	–	–		150 °C, O_2 1.0 MPa, Cy 10 ml, catalyst 50 mg, 4 h	51
Au(0.52%)Pd(0.48%)/ MIL-10	28.4	16.1	78.3	–	–			
Au(0.8%)/CNT	1.9	57.9	42.1	–	–	171	Room temp., ambient pressure, Cy 0.54 ml, catalyst 100 mg, H_2O_2(30%) 1.03 ml, acetonitrile 3.0 ml	53
Au/CQDs	63.3	99.9	–	–	–	–	Room temp., ambient pressure, Cy 50 ml, catalyst 25 mg, H_2O_2 (30%) 143 ml	55

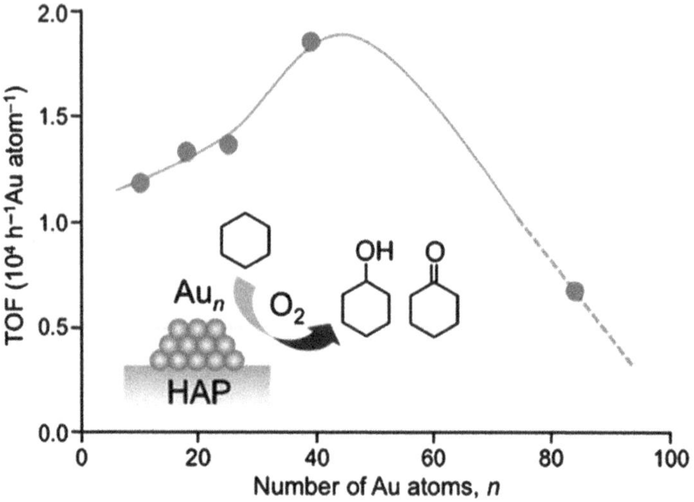

Figure 8.17 TOF values as a function of the cluster size *n*.
Reprinted with permission from ref. 65. Copyright 2011American Chemical Society.

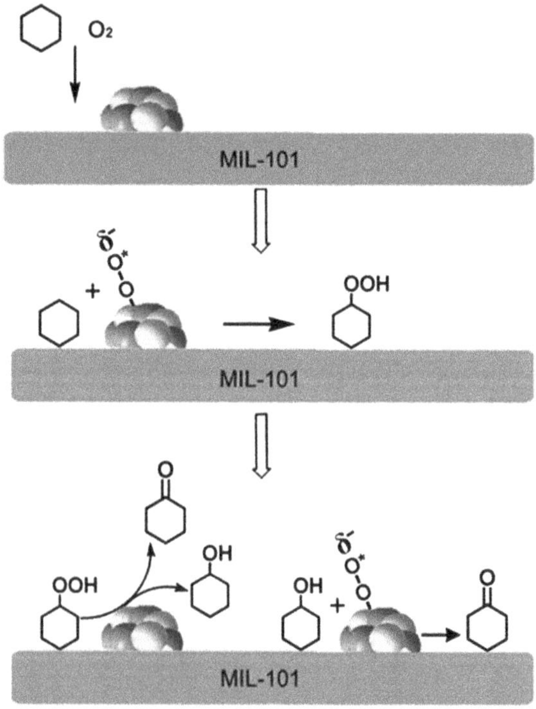

Figure 8.18 Schematic representation of the mechanism for cyclohexane aerobic oxidation over Au-Pd/MIL-101.
Reprinted with permission from ref. 66. Copyright 2013 American Chemical Society.

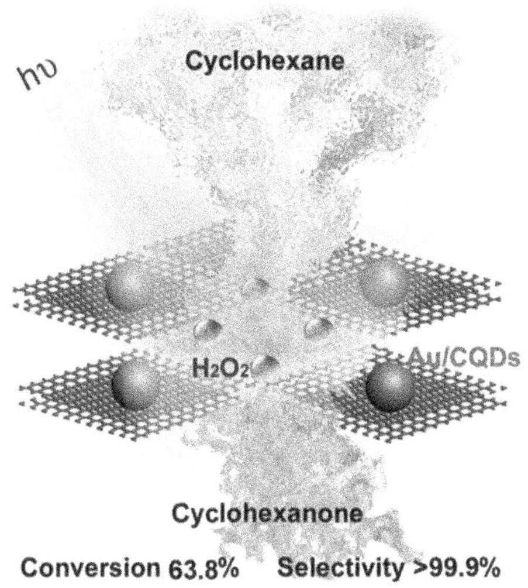

Figure 8.19 Gold/carbon quantum dot (Au/CQD) composites as a photocatalyst for selective oxidation of cyclohexane in the presence of H_2O_2 under visible light.
Reprinted with permission from ref. 55. Copyright 2014 American Chemical Society.

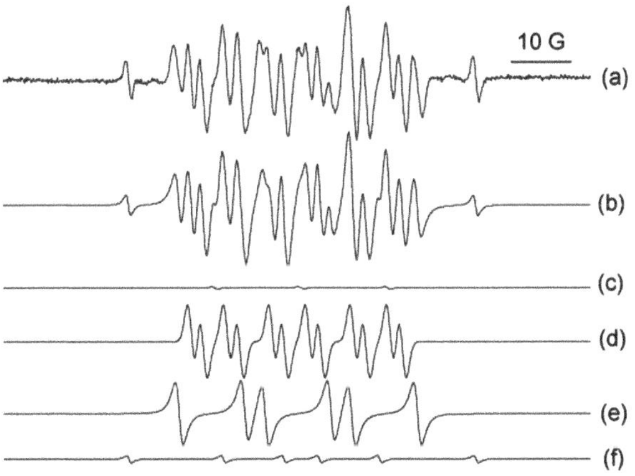

Figure 8.20 Deconvoluted EPR spectra of the DMPO spin adducts obtained during cyclohexane autoxidation at room temperature in the presence of CHHP: (a) experimental spectrum; (b) simulated spectrum; (c) di-*tert*-butyl-nitroxide derivative, (d) DMPO–O–C_6H_{11} spin adduct; (e) DMPO–OO–C_6H_{11} adduct; and (f) carbon-centered adduct which is possibly a DMPO–C(OH)R_2 species.
Reprinted with permission from ref. 72. Copyright 2012 Royal Society of Chemistry.

chlorination and saponification of toluene, and benzoic acid is produced *via* homogenous catalytic process using transition metal catalysts (Co, Cu or Cr) and acetic acid as solvent at 165 °C. In order to obtain a high yield of benzoic acid, conversion has to be limited to less than 15%.

An innovative work was presented by Hutchings and co-workers[74] in 2011. A high yield (conversion of nearly 100%, selectivity of 90%) of benzyl benzoate was obtained under mild conditions (150 °C, 0.1 MPa O_2) over a bimetallic Au-Pd catalyst. The TON was as high as 3300. As indicated by a control test, in which 2-, 3-, and 4-methoxytoluene and 2-, 3-, and 4-nitrotoluene were anaerobically oxidized over a PdAu/C catalyst, an electron-deficient intermediate was deduced to be involved in the reaction (Figure 8.21). A plausible reaction mechanism was then proposed to explain the excellent selectivity towards benzyl benzoate; consecutive oxidation proceeded in this system to yield benzyl alcohol and benzaldehyde. The further esterification led to the high selectivity towards benzyl benzoate. The relationship between the reactivity and the composition of active metals as well as the support effect was systematically investigated. The subtle differences in the micro-morphology of the interface between bimetallic nanoparticle and the support material (TiO_2 P25 or carbon G60) may greatly influence the reaction activity. Au nanoparticles with low coordination number are easy to stabilize on carbon compared with TiO_2. It was

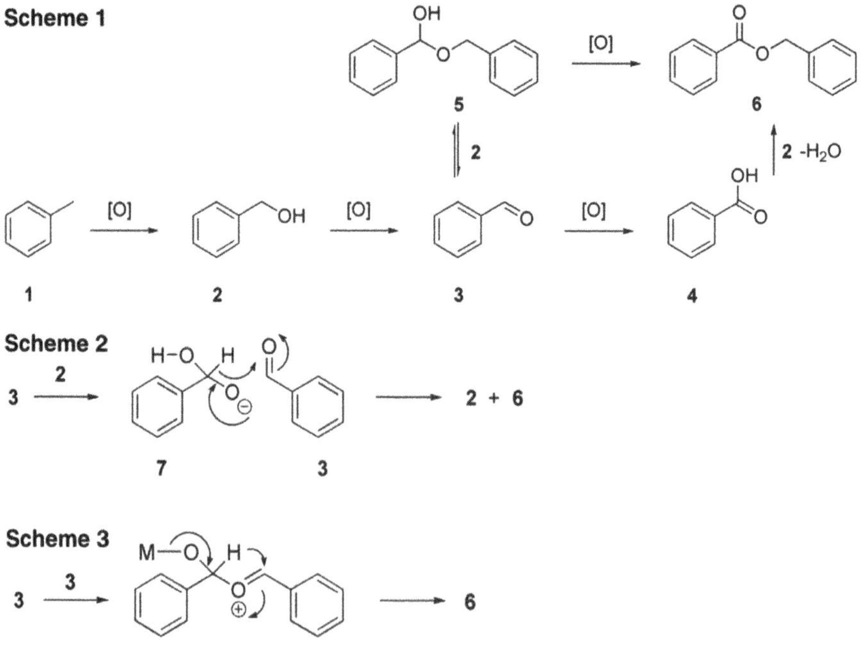

Figure 8.21 Reaction scheme of toluene oxidation over Pd-Au/C catalyst. Reprinted with permission from ref. 74. Copyright 2011 AAAS.

emphasized that a surface hydroperoxy intermediate was assumed to participate in the reaction cycle on the basis of the results.

In further work by Hutchings and co-workers[15] with introduction of TBHP as the principal oxygen donor, the reaction condition became much milder (80 °C, ambient in air). The catalysts were analyzed by X-band EPR and signals from Pd^+ ($g_1 \approx 2.800$, $g_2 \approx 2.130$ and $g_3 \approx 2.080$) and Ti^{3+} ($g_\perp = 1.978$, $g_\parallel \approx 1.962$) were observed for the fresh catalyst (Figure 8.22, line a). However, after interacting with TBHP, the majority of Pd^+ signal disappeared and a new set of signals from Ti^{4+}-O_2^- ($g_1 = 2.003$, $g_2 = 2.004$ and $g_3 = 2.024$) were detected. After the subsequent addition of toluene, only a signal located at $g = 2.004$ was remained (Figure 8.22, lines b and c). These observations demonstrated that the surface bond in hydroperoxy species radicals played a direct role in the activation of toluene.

Xiao and co-workers[75] recently reported toluene oxidation over a series of Au/MnO_x catalysts. α-, δ-, and γ- MnO_2 were used as the support materials and the highest TOF of toluene, 62.7 h^{-1}, was obtained over γ- MnO_2 supported Au catalysts. A plausible reaction mechanism is shown in Figure 8.23.

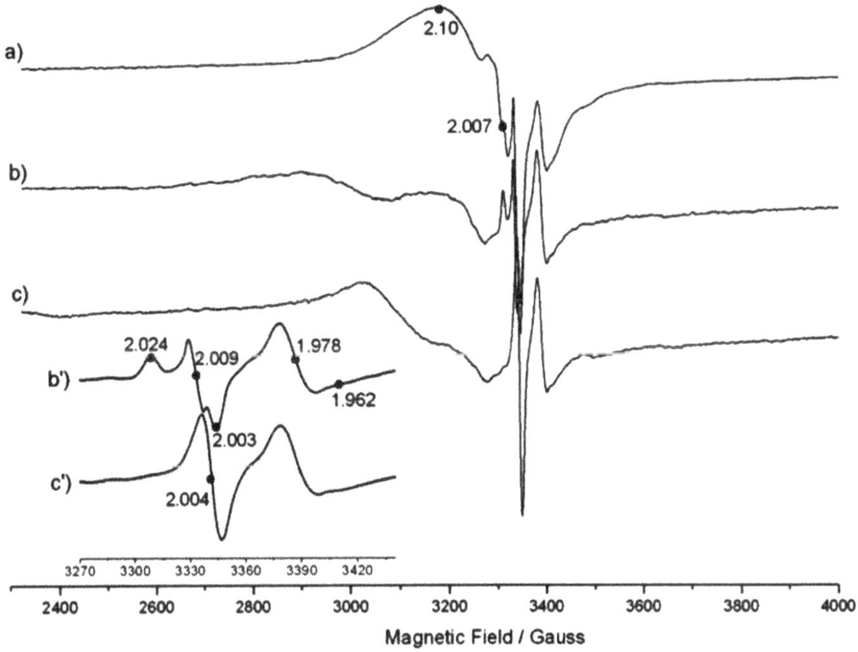

Figure 8.22 X-band CW-EPR spectra (−196 °C) of Au-Pd/TiO_2 (line a), Au-Pd/TiO_2 after treatment with TBHP (line b), and Au-Pd/TiO_2 after treatment with TBHP + toluene (line c). The expanded views of the center field regions of lines b) and c) are shown in the inset, and labeled as b′ and c′, respectively.
Reprinted with permission from ref. 15. Copyright 2012 Wiley.

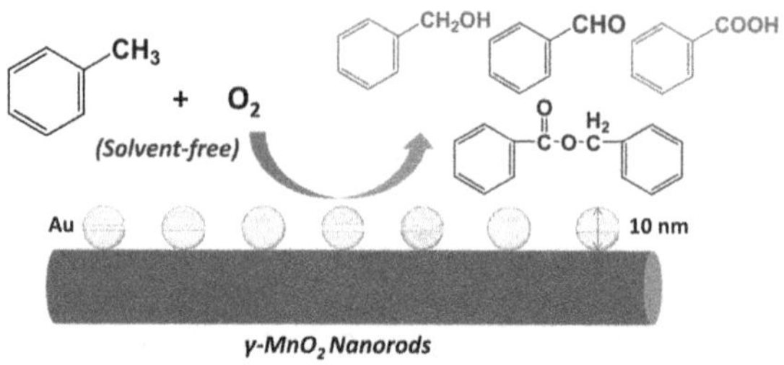

Figure 8.23 The plausible reaction mechanism proposed by Xiao and co-workers. Reprinted with permission from ref. 75. Copyright 2013 Editorial Office of Chinese Journal of Catalysis.

8.3 Selective Oxidation of Alkenes

The selective oxidation of alkenes is much easier than that of alkanes due to the existence of a π bond in the former. The epoxidation of ethane over Au nanoparticles when hydrogen is used to activate the molecular oxygen was discovered by Haruta and co-workers.[76,77] It was thought to be one of the greatest achievements of Au catalysis. However, in practical production, the epoxidation of ethene is still carried out over a silver catalyst using molecular oxygen as oxidant, while the higher alkenes are generally epoxidized by H_2O_2 over a TS-1 catalyst.

8.3.1 Propene Epoxidation

Propene oxide (PO) is a high value-added chemical for the manufacture of polyurethane (PU), polyethers, diols, and other chemical reagents. Using a similar process as ethene epoxidation, which is a heterogeneous oxidation promoted by Ag/Al_2O_3, an attempt at direct oxidation of propene was by no means successful (Figure 8.24). Attack of the acidic allylic hydrogen leads to the combustion of propene. PO manufacture is still based on the chlorohydrin process or indirect hydroperoxidation.

In 1998, Haruta and co-workers[78] found that Au/TiO_2 showed high selectivity to propene epoxidation using O_2 and H_2 as reaction agents. Active oxygen species (*i.e.* peroxo or hydroperoxy species) formed from H_2 and O_2 could selectively interact with the double bond rather than attacking the weak bonded allylic hydrogen. On Au/TiO_2 catalysts, the particle size of Au, properties of TiO_2, and preparation methods were considered to be crucial for both the activity and selectivity of propene epoxides.

Stangland *et al.*[79] studied the effect of the degree of connectivity of Ti on propene epoxidation and found that titanium isolation was beneficial for propene conversion and PO selectivity. Au/TS-1 showed the best catalytic performance and stability compared with other Au catalysts supported on different titanium-containing materials. Hydroperoxy species would decompose to $OH^{\bullet}$ radical in the presence of TiO_2 with extended or octahedral

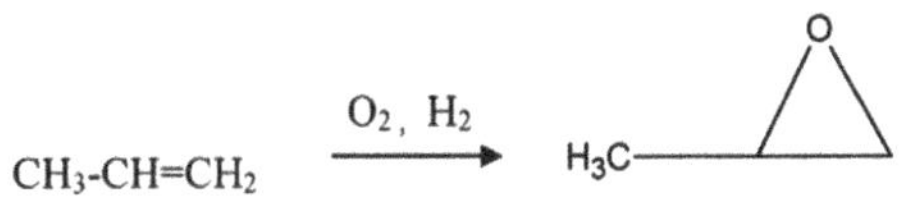

Figure 8.24 Scheme of direct epoxidation of propylene by H_2 and O_2.

coordinated structures, which then catalyzed the nonselective oxidation of C_3H_6, or the further oxidation of PO to ethanol and CO_2.

Weckhuysen and co-workers[80] studied the adsorption of propene on Au nanoparticles using H_2O_2 and XANES spectra measured *in situ*; propene was proved to be adsorbed on Au *via* π-bonding. Because the adsorption of propene on Au supported on an inert support SiO_2 was observed, the formation of bidentate propoxy species was considered *via* the adsorption of propene on Au, followed by a spillover to TiO_2. However, whether the bidentate propoxy species is a reaction intermediate for the formation of propene oxide or the species responsible for catalyst deactivation is still not clear. Density functional theory (DFT) calculation results showed that on Au (111) surface, the selective oxidation of propene could not happen, even when pre-covered by atomic oxygen and the formation of oxametallacycle intermediates on Au was thermodynamically favored but kinetically inhibited. However, stripping of the hydrogen from the C–H bond leading to an adsorbed allyl and $OH^•$ radicals was by far more probable than the formation of oxametallacycle intermediates, which is the necessary step towards the partial oxidation of propene.[81]

8.3.2 Selective Oxidation of Other Aliphatic Alkenes

Kung and co-workers[82] observed that Au/TiO_2 and Au/TS-1 co-catalysts could catalyze 1-butene epoxidation using molecular oxygen in aqueous solution even without a peroxy initiator but with carbon monoxide as a sacrificial reductant. Moreover, Hutchings and co-workers[83] found that Au supported on graphite was a promising catalyst for the oxidation of 1-hexene with air and the addition of TBHP as peroxy initiator. Based on the experimental results, a possible reaction pathway was proposed which involved several kinds of alkoxy radicals and peroxy radical species (Figure 8.25). The key intermediate, an allylic radical, was formed through hydrogen abstraction by an oxygen-derived mono-radical.

Corma and co-workers[84] studied the epoxidation of 1-octene with molecular oxygen over an Au/CeO_2 + Ti-MCM-41-silylated catalytic system in the presence of a hydrocarbon and with AIBN as a promoter. Organo-gold species formed from AIBN could promote the formation of hydrocarbon hydroperoxides with alkene conversion of 9% (40% of the maximum conversion attainable) and peroxide selectivity of 90% when working with cumene as a sacrificial hydrocarbon (Figure 8.26). Indeed, Corma and coworkers observed that gold chlorides as well as Au nanoparticles were able to trap radicals, forming well-defined organo-gold compounds that were stable at room temperature in the presence of O_2. However, they would decompose

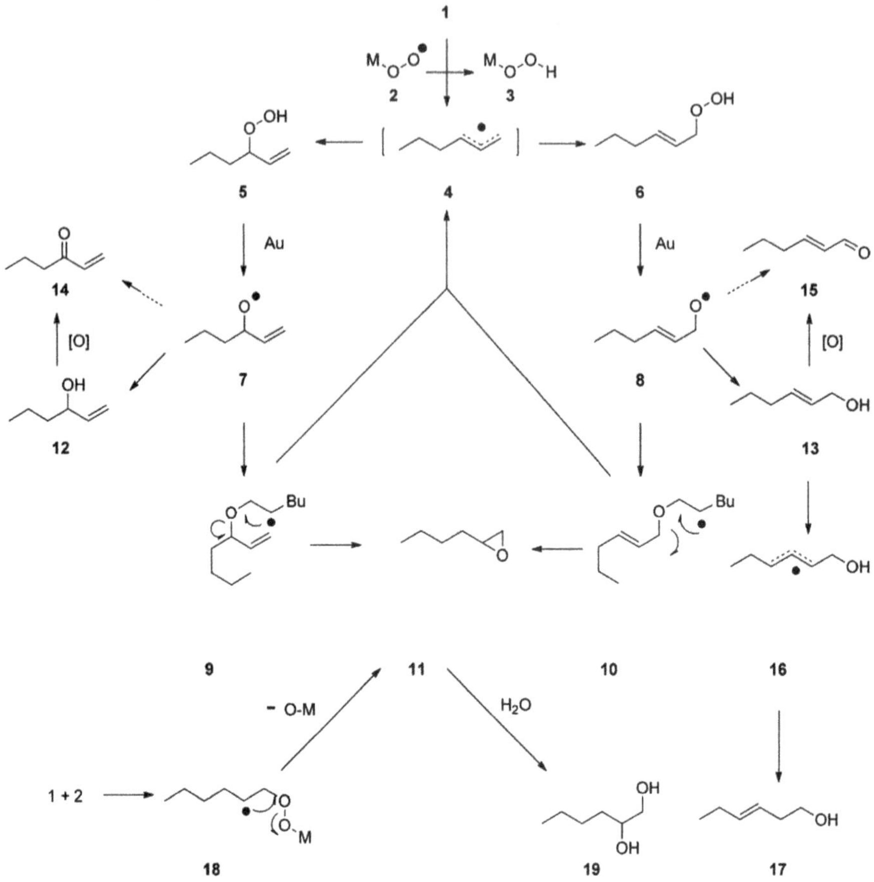

Figure 8.25 Possible pathways that account for the observed products from 1-hexene oxidation reported by Hutchings and co-workers. Reprinted with permission from ref. 83. Copyright 2012 Wiley.

at higher temperature or in the presence of acids, generating potentially reactive surface species.[85,86]

8.3.3 Selective Oxidation of Cycloalkenes

A major breakthrough was the discovery by Hutchings and co-workers in 2005 of the tunable activity of Au-based catalysts for the epoxidation of cycloalkenes (Figure 8.27).[87] The adjustment of solvent polarity, radical initiator, and the surface composition of the Au/C catalyst can influence the selectivity of cyclohexane oxidation with molecular oxygen. When triggered by TBHP at 80 °C, the distribution of the C_6 products was determined by the solvent. The highest selectivity of cyclohexene oxide (~50%) was reached with 1,2,3,5-tetramethylbenzene as solvent. However, the selectivities towards 2-cyclohexen-1-ol and 2-cyclohexen-1-one were 35.1% and 25.1%, respectively, using toluene as solvent. The addition of bismuth further

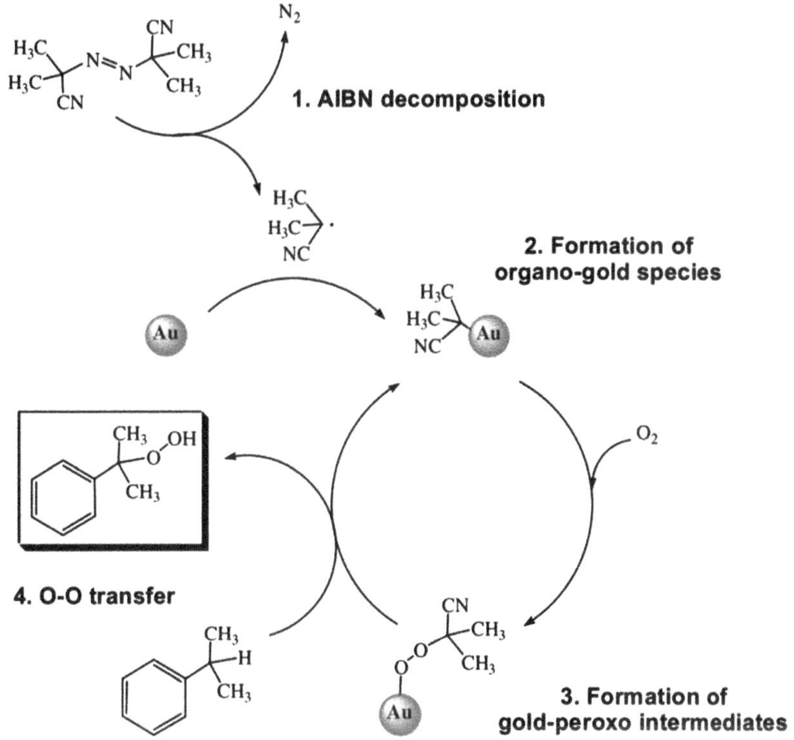

Figure 8.26 Proposed mechanism for the main initiation route of hydroperoxide generation reported by Corma and co-workers.
Reprinted with permission from ref. 84. Copyright 2009 Elsevier.

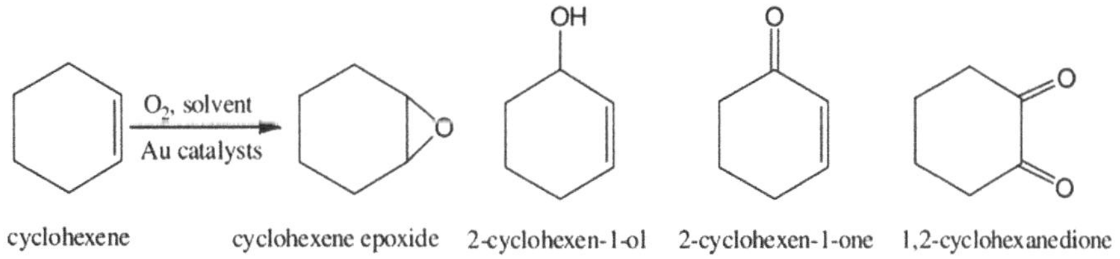

Figure 8.27 Reaction scheme of aerobic oxidation of cyclohexene.

increased the selectivity of C_6 production to 97.9%. Moreover, the tunable properties were also observed in the oxidation of *cis*-cyclooctene.[87–89] In their research, the authors excluded the existence of the oxygen-centered radical and inferred the existence of a weak interaction between the solvent and the surface of Au/C catalyst. The interaction was of crucial importance for electron transfer and the Au nanoparticles acted as the active chain carrier during the reaction.[87,89]

Hutchings and co-workers further investigated the aerobic oxidation of cyclic alkenes over supported Pd-Au alloy catalysts. The distribution of products was little affected by the Au to Pd ratio and the conversion of cycloalkene oscillated with the variation of this ratio. However, further investigation of oxidation of the crotyl alcohol (*trans*-buten-2-en-1-ol) indicated that the metal composition greatly affected the selectivity of 3-buten-1-ol (Figure 8.28). They demonstrated that the oxidation to crotonaldehyde was the preferred pathway with a high concentration of Au. In the presence of Pd, however, the dominant pathway was perhaps isomerization *via* a Wacker-like oxidation.[90]

By a consequent grafting method (Figure 8.29), a mesoporous SBA-15 material functionalized by (s)-(−)-2-pyrrolidinone-5-carboxylic acid (Py) was successfully prepared by Xiao and co-workers.[91] Gold nanoparticles could be highly dispersed on the surface of this material, which showed a superior reactivity towards the oxidation of cyclohexene. Even without the extra addition of a radical initiator, the selectivity to 2-cyclohexen-1-ol was as high as 99.5% with a TOF of 12 000 h^{-1} in toluene solution at 100 °C. The addition of radical initiator could further increase the conversion and the adjustment of solvent could tune the catalytic performance significantly.

Based on DFT calculation of the grafted Au_{55} cluster model, the origin of the high activity can be reasonably explained as follows. The attaching of Py alters the density of states (DOS) considerably due to the strong overlap between the frontier orbitals of the Py molecule and the frontier states of the Au clusters on the surface. In this system, Py served as an electron-rich functional group on the surface of Au nanoparticles and induced more electronic modification of the entire Au nanoparticle, which caused the charge redistribution of

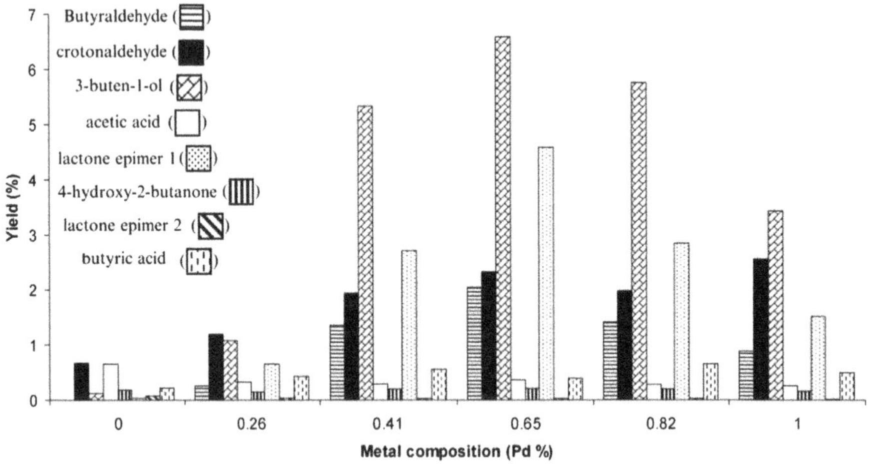

Figure 8.28 Effect of metal composition on the product yield of crotyl alcohol conversion. Reaction conditions: catalyst (0.12 g), crotyl alcohol (10 mL, 0.12 mol), temperature 80 °C, TBHP (0.01 mL, 1.03 $\times 10^{-4}$ mol), 24 h, atmospheric pressure.
Reprinted with permission from ref. 90. Copyright 2011 Royal Society of Chemistry.

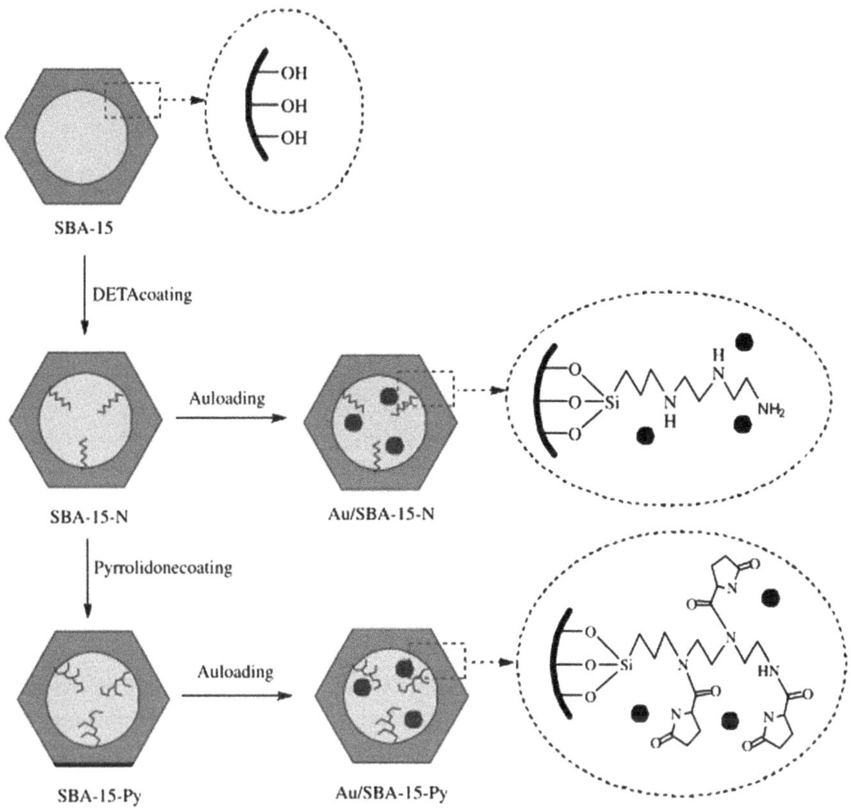

Figure 8.29 Procedures for preparation of Au/SBA-15-N and Au/SBA-15-Py catalysts.
Reprinted with permission from ref. 91. Copyright 2012 Elsevier.

Au/Py-SBA-15. The negative charge property of Au nanoparticles was also indicated by XPS characterization. Moreover, the thin film assembling from the silicon quantum dots prepared by the electrochemical method was also used as the support for Au nanoparticles.[92] Without the addition of TBHP, 65.0% conversion based on *cis*-cyclooctene was achieved with a 78.3% selectivity for 2-hydroxycyclooctanone. A silicon-modified Au catalyst with an Au-loading amount of 62 wt% was reported by Lee and co-workers[93] (Figure 8.30). Specifically, the *cis*-cyclooctene conversion of 38.3% and the selectivity toward epoxy-cyclooctene with 90% were achieved simultaneously.

The application of so-called 'active support' materials were also studied in the oxidation of cyclohexene. Zhu and co-workers[94] reported that Au nanoparticles deposited on the surface of La-modified δ-MnO_2 with a 2×2 tunnel structure (OMS-2) exhibited an excellent performance for the aerobic oxidation of cyclohexane. The very high selectivity (~86%) to the C_6 product was obtained even for a high conversion of 48%, which is comparable with a Cr/MCM-41 catalyst. From detailed analysis of the product distribution, they inferred

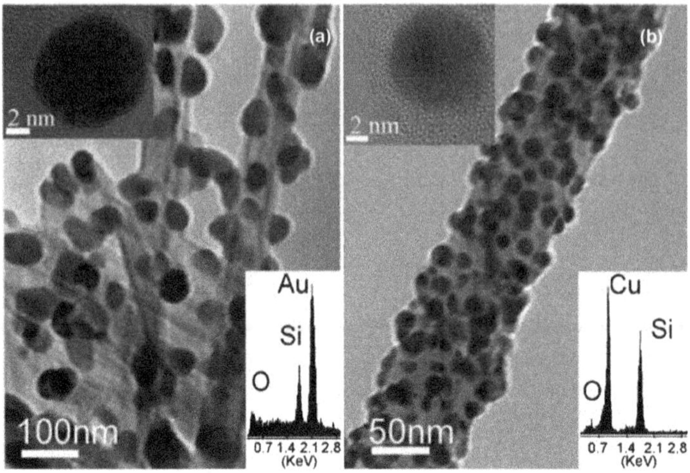

Figure 8.30 Transmission electron microscopy (TEM) images of (a) Au/SiNWs and (b) Cu/SiNWs. High resolution TEM image is shown in the upper inset and the energy dispersive X-ray (EDX) spectrum in the lower inset. Reprinted with permission from ref. 93. Copyright 2009 Royal Society of Chemistry.

that a cyclohexene peroxide intermediate was initially formed by the oxidative adduction. On the surface of the Au catalyst, a radical chain mechanism proceeded (Figure 8.31). The distribution of different oxidation products could be tuned easily for this process. A similar phenomenon was observed when a halloysite nanotube (HNT) was used as the support material.[95]

In addition, the influence of other factors such as the oxidation state of Au, kind of radical initiator, and ring size of the cyclic alkenes were also studied. Bachir *et al.*[172] stated that Au^0 was non-selective for all the oxidation products. However, $Au^{\delta+}$ species could enhance the production of 2-cyclohexene-1-one and 2-cyclohexene-1-ol with high selectivity. Compared with other radical initiators, including cumene hydroperoxide (CHP), di-*t*-butyl peroxide (DTBP), and AIBN, the O–O bond homolysis energy of TBHP was much more suitable for the oxidation of cyclic alkene. The relatively long lifetime of the radicals generated by TBHP was also a key factor in the high conversion of cyclooctene and the selectivities of the C_6 products.[85,89] Moreover, the experimental and DFT calculation results indicated that the ring size of the cyclic alkene would greatly influence the selectivity towards epoxides, which could be attributed to the different energy reflected on the ring closure barrier to form the epoxide. For a small alkene ring (C_5 compared with C_8), an additional ring strain may cause the extra barrier for ring closure.[96]

Recently, the radical chain mechanisms that occurred during the oxidation of cyclic alkenes were demonstrated by Hutchings and co-workers.[97] By treating the byproduct of cyclooctene oxidation with triphenylphosphine (PPh_3), they demonstrated the existence of cyclooctene peroxide. With controllable introduction of corresponding radical scavengers, a complete

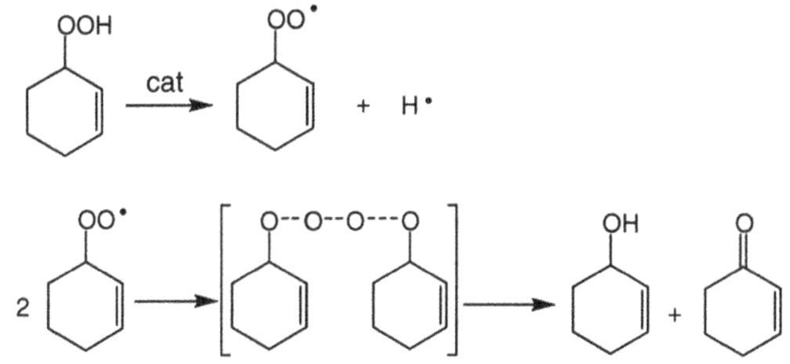

Figure 8.31 Radical chain sequence mechanism proposed by Zhu and co-workers for the explanation of cyclohexene oxidation over Au/(La-)K-OMS-2. Reprinted with permission from ref. 94. Copyright 2010 Elsevier.

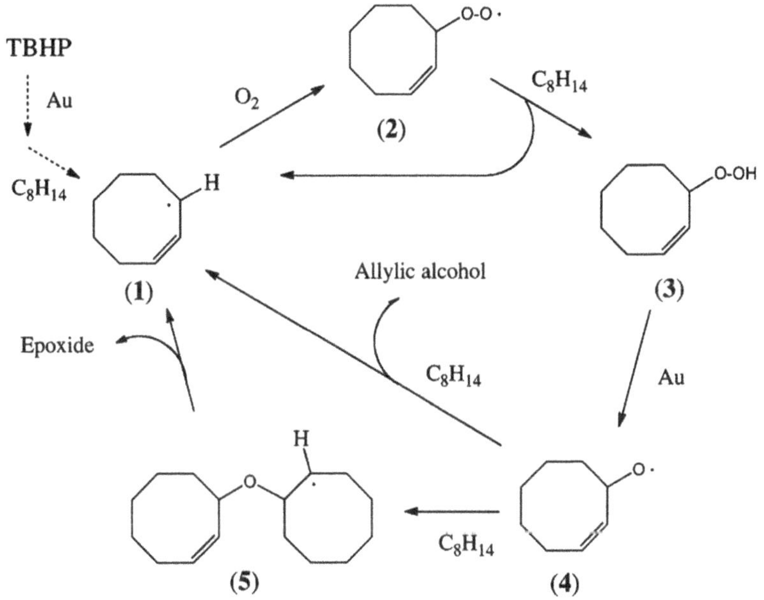

Figure 8.32 Proposed mechanism for cyclooctene oxidation using TBHP as radical initiator over supported gold catalysts.
Reprinted with permission from ref. 97. Copyright 2011 Royal Society of Chemistry.

reaction cycle involving several radical steps was proposed (Figure 8.32). First, TBHP decomposes into a *t*-BuOO• or *t*-BuO• radical. The generated radicals then participate in a hydrogen abstract reaction with cyclooctene to form an allylic radical. In the cycle, cyclooctene hydroperoxide was generated *via* the allylic radical adducted with oxygen. Subsequently, cyclooctene hydroperoxide was saturated on the surface of the Au catalyst and the further oxidation of cyclic octane occurred. The cyclooctenyloxy radical was assumed to exist and played a key role in product distribution.

8.3.4 Epoxidation of Styrene

Styrene epoxide is a versatile and useful intermediate in the organic synthesis of fine chemicals and pharmaceuticals. The epoxidation of styrene is traditionally brought about using stoichiometric amounts of per acids as oxidant.[98] For the sake of reaction safety, alternative processes based on the so-called 'Herrmann epoxidation system', which took methyl-trioxorhenium as the catalyst and H_2O_2 as the oxidant, were widely studied.[99,100] Furthermore, heterogeneous catalysts such as TS-1 or Ti-SiO_2 were investigated to enhance the facility of separation.[101,102] Besides its great fascination for inorganic synthesis, styrene can also be used as a probe molecule for the evaluation of molecular oxygen activation.

Choudhary and co-workers have presented a series of studies on the epoxidation of styrene over supported Au catalysts. The particle size of Au nanoparticles was finely controlled by a homogeneous deposition–precipitation method, in which urea was applied as the precipitation agent. Identified by XPS, both metallic and cationic Au species existed. When anhydrous TBHP (in benzene) was introduced at 82 °C, the conversion of styrene over Au/TiO_2 was nearly 60% with a selectivity of 53% toward styrene.[103] With a similar method, Au nanoparticles were deposited onto the surface of various group IIIA metal oxide, including γ-Al_2O_3, Ga_2O_3, In_2O_3, and Tl_2O_3.[104] It was found that the conversion of styrene and selectivity toward styrene oxide for these supported Au catalysts followed the order of $Al_2O_3 < Ga_2O_3 < In_2O_3 < Tl_2O_3$. The amount of loading of the Au nanoparticles resulted in a re-hydroxylation of the oxide support. During this oxidation–reduction process, the adsorbed Au^{3+} ion interacted with the metal oxides at the liquid–solid interface; this interaction force was determined by the standard electrode potential of the group IIIA metal. Consequently, cationic Au was stabilized on In_2O_3 much more easily than that on Al_2O_3, leading to an $Au_{4f,5/2}$ XPS peak shift of 1.8 eV on Au(6.81 wt%)/In_2O_3. Thus the correlation between the reactivity and the existence of the positive Au was established. In contrast, upon chlorination, positive Au was greatly removed and the epoxidation reaction was strongly hampered.

A similar reaction behavior could be observed when alkaline earth metal oxides or rare earth metal oxides were used as the supports.[105] Due to differences in the isoelectric point of each material, the deposition method should be switched. On the surface of alkaline or alkaline rare earth oxides, precipitation was often carried out with NaOH or Na_2CO_3. The size of Au nanoparticles could be controlled into a range of 5.7–17.9 nm on the surface of MgO, CaO, SrO, and BaO.[105,106] The conversion was about 44.6% with a selectivity of styrene oxide up to 36.1% over Au/MgO. By controlling the calcination temperature, the micro-morphology and surface charge of Au nanoparticles on CaO could be finely tuned to improve catalytic performance. When the calcination temperature was 500 °C, the selectivity of styrene epoxides was the highest with 78.3%.[107]

Some rare earth oxides and transition metal oxides, such as Yb_2O_3, Tb_2O_3, Eu_2O_3, Sm_2O_3, NiO, CuO, and U_3O_8, were also studied by the same group.[108–110] Using a layered double hydroxide compound (LDH), Zhang *et al.*[111] reported the tunable properties of supported Au catalysts for the oxidation of styrene by TBHP. A preferential deposition of gold nanoparticles with a narrow size distribution was formed on the lateral (10–10) facet of LDH platelets. The highest selectivity was up to 89.1% in this system. *Via* a ligand assistant method, an Au_{25} cluster was finely synthesized and deposited on the surface of HAp, which exhibits the best catalytic performance up to now. In toluene solution, styrene was completely converted with a selectivity of styrene oxide up to 92%.[112]

Besides the interaction with support, the naked Au nanoparticles with selected size can also promote the activation of oxygen.[112,113] In 2008, Lambert and co-workers[113] reported that an Au_{55} cluster with the size of 0.5 nm could activate molecular oxygen for the oxidation of styrene. The electronic property of this Au nanoparticle with a 'magic size' was positively charged as revealed by XPS. The adsorbed atomic O from O_2 dissociation over Au_{55} was thought to be responsible for the oxidation of styrene in the toluene solution system.

Moreover, the tunable porous structure of Al_2O_3 and SiO_2 can also enhance the reactivity of styrene epoxidation. For instance, compared with Au supported on commercial γ-Al_2O_3 with microporous architecture, the increase in the conversion of styrene ($\sim$20.2%) was achieved over Au/mesoporous γ-Al_2O_3 (64.5% *vs* 84.3%).[114] The Au nanoparticles encapsulated by SiO_2 with a core-shell structure were designed by Das and Asefa.[115] Furthermore, the introduction of organic functional groups anchored or hydrided into the surface of SiO_2 was demonstrated to be an efficient strategy to improve the yield of styrene oxide (Figure 8.33).[116–118] In addition, grafting of ionic liquid onto the surface of Au nanoparticles could modify the performance of the catalysts and extend the reaction into the solvent-free system.[119]

8.3.5 Oxidation of *trans*-Stilbene

Caps and co-workers have studied the selective oxidation of alkene using *trans*-stilbene as the probe molecule since 2007. First they evaluated catalysts from the World Gold Council.[120] They used methylcyclohexane as solvent and TBHP as the so-called 'initiator'. The selectivity toward *trans*-stilbene oxide was up to 88% over Au/TiO_2 with a conversion of *trans*-stilbene of about 95%. In the reaction cycle, the Ti species was assumed to trap the OH• and the gold propagated the radical species. The solvent effect was not only

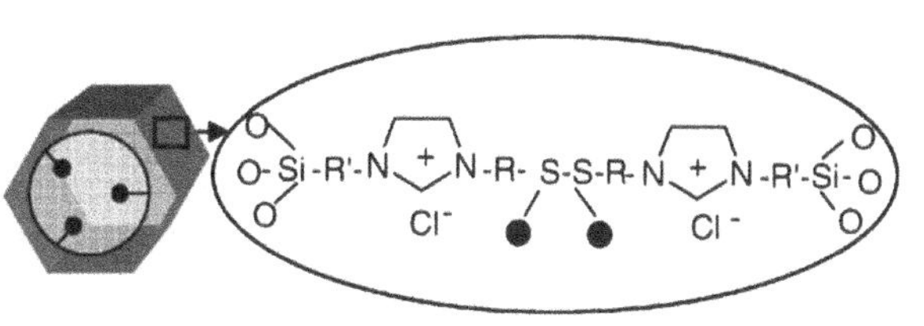

Figure 8.33 Proposed structural scheme of Au-PMO-SBA-15 catalyst presented by Yin *et al.*
Reprinted with permission from ref. 117. Copyright 2008 Elsevier.

reflected in the polarity but also in the solubility of oxygen, the complex capacity of gold, and the feasibility of H donation.[121] *Via* a controllable method for synthesis of Au nanoparticles, the influence of the surface properties of the support on the styrene reactivity was studied. The surface properties had little impact on the intrinsic activity of gold but more on the apparent reaction rates, which are a combination of reaction activity and the diffusion limitation. Additionally, the surface modification and the design of porous structure could also enhance the apparent reaction rate.[122,123] The SiO_2 Aerosil R972 was proposed as a good support for the epoxidation of stilbene. An Au(42%)/SiO_2, Aerosil R972, has great potential in practical applications with a TOF of 30 h^{-1} and an epoxidation rate of 0.066 $mol \cdot g_{Au}{}^{-1} \cdot h^{-1}$.[123] Moreover, *via* a mild hydrolysis of $Gd_4TiO(O^iPr)_{14}$, the surface wettability of the material was improved. The TOF of the as-prepared materials was up to 40 h^{-1}.[124] A similar strategy was applied in the preparation of In_2O_3. The TOF of Au was further increased to 69 h^{-1}.[125]

Over a Au/TiO_2 catalyst, the epoxidation reaction was strongly hampered when a scavenging agent, 2,6-di-*tert*-butyl-4-methylphenol was added. The free radical mechanism was evidenced by a systematic analysis of byproduct as well as the fragments from the oxidation of solvent. Further confirmation was given by triphenylphosphine (PPh_3) titration followed by ^{31}P NMR.[126] The presence of 1-methylcyclohexyl hydroperoxide was directly evidenced over Au. During this process, molecular oxygen was activated by the radical process and several steps were involved in the free radical reaction cycle. The initiation of the chain reaction was involved in the thermolysis of TBHP and the consequent production of methylcyclohexyl radical (steps 1, 2 in Figure 8.34). The methylcyclohexyl peroxy radical may act as the chain propagator (step 3 in Figure 8.34), which can be added to *trans*-stibene and trigger the hydrogen abstraction reaction from methylcyclohexane for the regeneration of methylcyclohexyl radical (steps 4 and 5 in Figure 8.34). Following unimolecular decomposition, the stilbene epoxides were produced (step 6 in Figure 8.34).

The octylsilane-stabilized crystalline gold nanoparticles were also highly active and selective for the aerobic oxidations of stilbene and cyclohexene in methylcyclohexane solution. Active species provided by the 'surface' polymerization of the oxidized ligands near the gold nanoparticles triggered oxygen activation in the liquid phase, according to the proposed radical mechanism.[127]

8.4 Selective Oxidation of Alcohols and Aldehydes

Selective oxidation of alcohols and aldehydes is an important process for the synthesis of chemicals. Catalytic conversion using molecular O_2 or H_2O_2 under mild conditions is considered a green process for the selective oxidation of alcohols and aldehydes. The mechanism of reaction and oxygen activation are still hot topics in fundamental study. Several key questions about this process are of major significance for further improvement of catalytic performance. How does the OH-bond activate alcohols and how

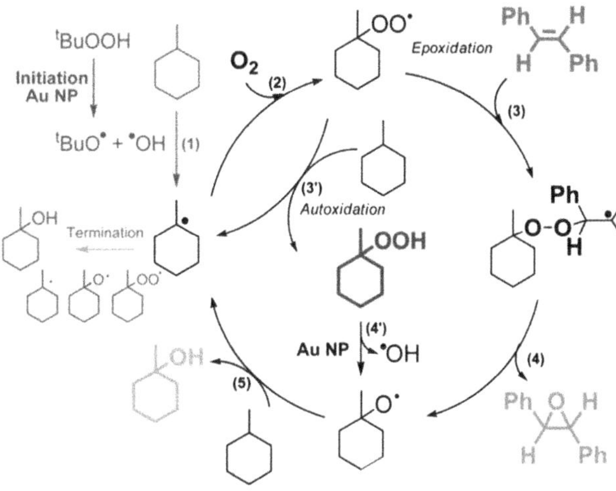

Figure 8.34 Postulated mechanism for the Au-catalyzed aerobic co-oxidation of methylcyclohexane and *trans*-stilbene, involving an autoxidation and an epoxidation pathway.
Reprinted with permission from ref. 121. Copyright 2013 Elsevier.

does the C–H bond activate both alcohol and aldehydes? What are real active oxygen species (*e.g.* •OH, •OOH, peroxo-type oxygen, and superoxide-type oxygen) as the terminal oxidants? Is there another radical species existing in the process (*e.g.* carbon-centered radicals or H radicals)? This chapter gives a brief summary of the selective oxidation of alcohols and aldehydes from the view of species (or radicals).

Using DFT calculation and isotopic labeling studies, Zope *et al.*[128] found that oxygen atoms originating from hydroxide ions instead of molecular oxygen were incorporated into the alcohol during the oxidation reaction (Figure 8.35). First, the surface-bound OH fascinated O–H activation of ethanol *via* proton transfer. The activation barrier for this pathway was 25 kJ mol^{-1}, much lower than that of direct ethanol dissociation with activation barriers of 204 and 116 kJ mol^{-1} over the Au(111) and Pt(111) surfaces, respectively (Figure 8.36). Meanwhile, the adsorbed -OH could activate the C–H bond and CH_3CHO intermediate formed from CH_3C_2O. Then, CH_3CHO reacted with surface-bound hydroxide and CH_3CHOOH formed. The subsequent C–H activation to CH_3COOH is very easy with activation barriers of only 10–20 kJ mol^{-1} over Pt and Au. In this process, the surface bound -OH originated from the reaction between H_2O and O_2, and then the interaction of parallel intermediate OOH with another H_2O molecule resulted in H_2O_2 and a new adsorbed OH. These two steps are irreversible. However, the regeneration of OH from the adsorption of OH^- by donating an electron and OH^- formation from OH- and an electron are reversible. On the basis of the reaction cycle, molecular oxygen participates in the catalytic cycle by regenerating hydroxide ions formed *via* the catalytic decomposition of a peroxide intermediate but not by dissociating to atomic oxygen.

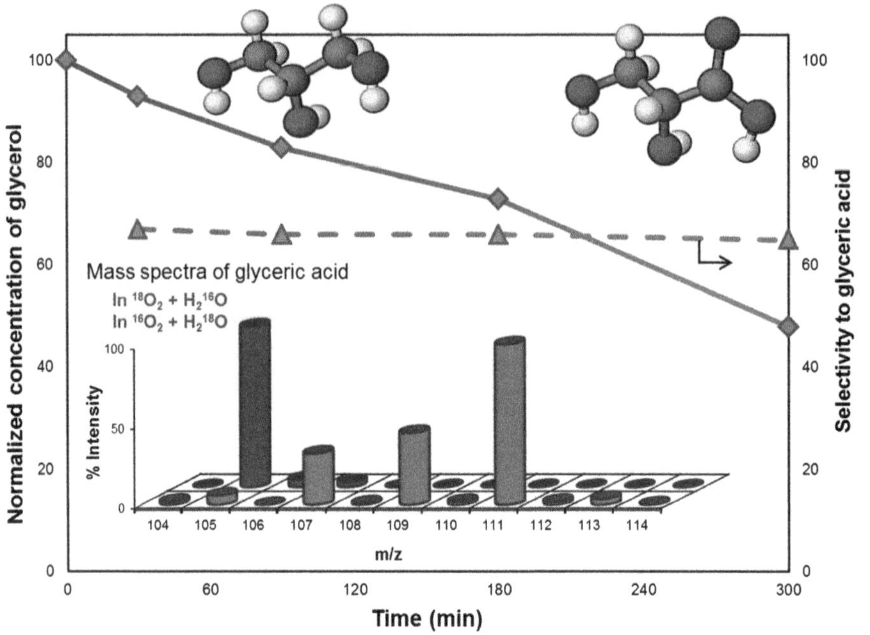

Figure 8.35 Reaction profile for the oxidation of glycerol over Au/C in liquid water. Reaction conditions were as follows: 0.3 M glycerol; NaOH: glycerol = 2.0 (mol : mol); glycerol: Au = 8000 : 1 (mol : mol); at a temperature of 60 °C; partial pressure of O_2 (P_{O2}) = 11 bar. (Inset at bottom) Liquid chromatography (LC) mass spectrum (electronegative ion mode) of glyceric acid (molecular weight 106) formed during glycerol oxidation over Au/C in $^{18}O_2 + H_2{}^{16}O$ or $^{16}O_2 + H_2{}^{18}O$. Glycerol oxidation in $^{16}O_2 + H_2{}^{18}O$ results in the incorporation of multiple ^{18}O atoms in the product glyceric acid. m/z, mass/charge ratio. Reprinted with permission from ref. 128. Copyright 2010 AAAS.

By using an integrated approach on the basis of DFT calculation and the periodic continuum solvation model, Shang and Liu[129] provided a complete reaction network of aerobic alcohol oxidation on Au nanoparticles. They considered that Au nanoparticles (~4 nm) are capable of activating O_2 at the solid–liquid interface by doubling the O_2 adsorption energy to ~0.4 eV. An external base (OH^-) is essential to break the hydroxyl bond of alcohol, and the α-C–H bond cleavage can be readily removed by O_2 and OOH *via* a H_2O_2 pathway instead of atomic O. At the initial stage, the pathway follows $H + O_2 \rightarrow OOH$. During the steady state, $O_2 + H_2O \leftrightarrow OOH + OH$ becomes the major channel to produce OOH, and Au particles are negatively charged. The similar system was extended to Au(111) bulk structure by Li *et al.*[130] The assistance effect of methanol or water on the formation of OOH intermediate for the activation of oxygen molecule was also investigated. Indeed, the water molecule promoted the hydrogen transfer reaction between alcohol•••O_2 and H_2O•••O_2, and lowered the energy barrier for OOH formation.

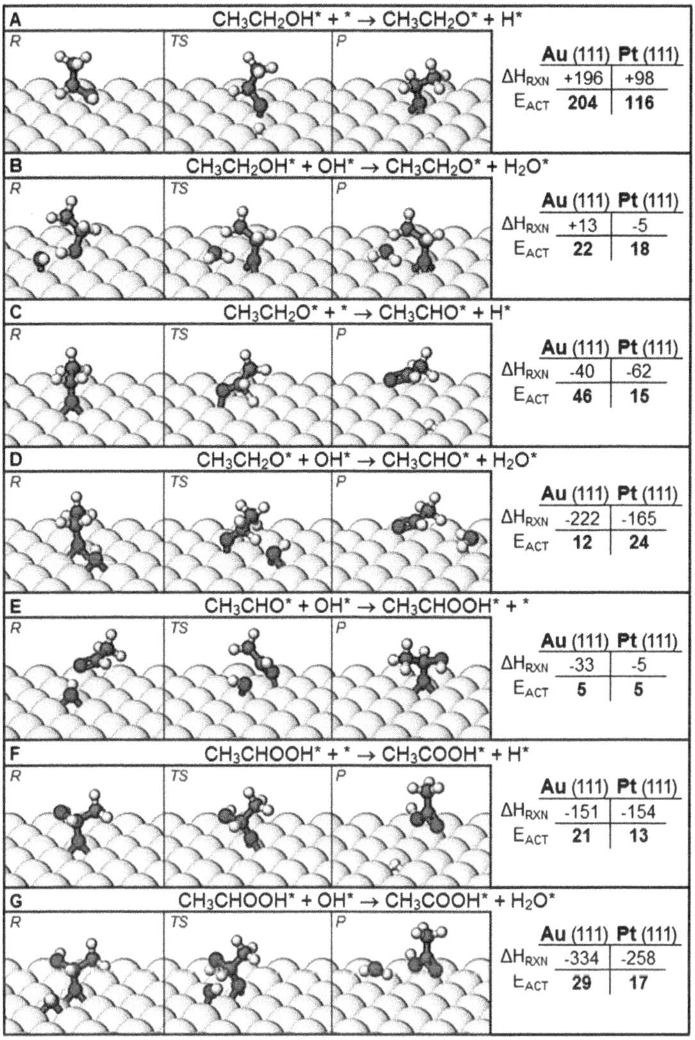

Figure 8.36 (a–g) Selected reaction energies and activation barriers (in kilojoules per mole) for the oxidation of ethanol to acetic acid on Au(111) and Pt(111) surfaces in the presence of liquid water. The reactant (R), transition state (TS), and product (P) on the surfaces were computed with density functional theory. The solution-phase water has been omitted for clarity.
Reprinted with permission from ref. 128. Copyright 2010 AAAS.

Hutchings and co-workers[131] studied the aerobic oxidation of benzyl alcohol over TS-1 supported Pd, Au and Pd-Au catalysts with and without H_2. Because the presence of H_2 enhanced the oxidation rates, the *in situ* generated OOH species from H_2 and O_2 is proposed as the terminal oxidant instead of O_2.

For aerobic oxidation of alcohol in a photocatalysis system, the reactive oxidative species is proposed as peroxo-type oxygen anions (O–O$^-$) but not superoxide-type oxygen anions (O_2^-). With the excitation of light, the photoactivated Au particle would transfer an electron to TiO_2; this electron is consumed by the reduction of O_2 to form O–O$^-$. By using ESR techniques, Yasuhiro and co-workers[132] observed an interesting phenomenon that the signal of O–O$^-$ from Au/P25 is much stronger than that of Au/TiO_2 (anatase and rutile) and P25 under the light (Figure 8.37). However, the Au/P25 catalyst treated with O_2 at 90 °C in the dark did not show the O–O$^-$ signal, which indicates that the light-to-heat conversion, even if it occurs on the photoactivated Au particles, did not promote O_2 reduction. They proposed that the

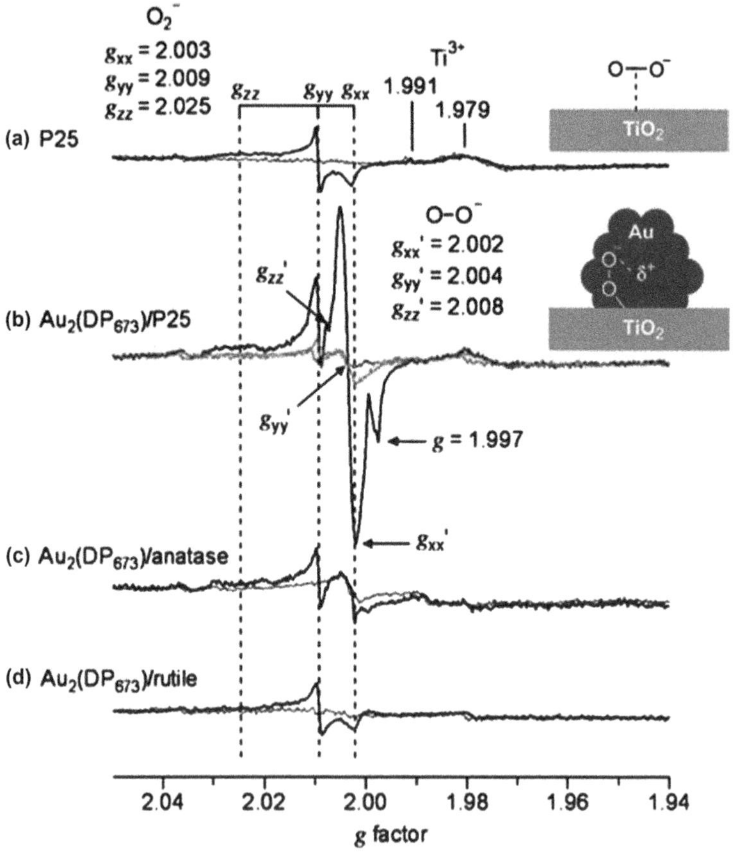

Figure 8.37 ESR spectra of the catalysts. The catalysts were treated with 20 Torr O_2 (gray) in the dark or (black) under visible-light irradiation at 25 °C or (green) in the dark at 90 °C. After evacuation, the samples were measured at −196 °C. The $g = 1.997$ signal (spectrum b) is assigned to e$^-$ at the lattice trapping site of TiO_2.
Reprinted with permission from ref. 132. Copyright 2010 American Chemical Society.

higher activity for Au/P25 was due to the existence of a joint active site consisting of Au/TiO_2. This joint active site facilitates smooth electron transfer from the photoactivated gold particles to TiO_2 and efficient O_2 reduction.

For the selective oxidation of *sec*-phenethyl and benzyl alcohols to acetophenone and benzaldehyde, Hallet-Tapley *et al.*[133] considered that the reactive species should be the more selective $^{\bullet}OOH$ instead of the more active $^{\bullet}OH$ through surface plasmon excitation of supported Au in the presence of H_2O_2. Under photoexcitation at 530 nm light, an electron-transfer from Au nanoparticle to H_2O_2 was took place. Then the OH radical and OH- were generated. The supported Au nanoparticle can be viewed as an integral participant in a nanoscale radical chain process. The propagated $OH^{\bullet}$ then continues to abstract additional hydrogen from another molecule of H_2O_2 to form the less reactive $OOH^{\bullet}$. A ketyl radical would be formed through the secondary H-atom abstraction by $OOH^{\bullet}$ and an electron transferred back to the Au nanoparticle. The supports at least act as an adsorber for the organic alcohol molecule and basic supported materials may be explained by stronger support–alcohol interactions.

In fact, the effect of the interaction between gold and supports is not negligible. It is of high complexity based on the properties of support, alcohol substrates, and reaction mechanism. Activation of the C–H bond is also the rate-determining step for aerobic oxidation of aldehydes compared with alcohol oxidation. Instead of considering active oxygen species, more attention is paid to the substrates.

Fristrup *et al.*[134] investigated Au-catalyzed aerobic oxidation of aldehydes using the Hammett studies and isotopic labeling experiments. A radical mechanism was followed for the substrates with electron-withdrawing substituents (X = Cl, CF_3, CN, and COOMe) because of the good fit with Creary's σ-values. The reaction could be abstraction of $H^{\bullet}$ with reactive species (*e.g.* Au–O–O) in a radical recombination (H^+ *vs.* $H^{\bullet}$). But for the substrates with an electron-donating substituent, the observed low reactivity is expected from several of the possible σ-values (σ, σ^+ and σ^-), indicating a non radical mechanism was followed. Conte *et al.* studied aerobic oxidation of aldehydes on polymer-encapsulated Au nanoparticles. EPR spectroscopy and spin-trapping experiments showed that aldehyde oxidation *via* a radical pathway by initiating formation of acyl radicals and C–H bond was activated by O_2 probably *via* $AuOO^{\bullet}$ species (Figures 8.38 and 8.39).[134]

Although both the activation of the O–H bond and C–H bond involve the abstraction of H, it is easy to ignore the behavior of H species. Indeed, for the oxidation of alcohols and aldehydes, especially in the absence of base, the behavior of the H species from the O–H and C–H bonds is important for understanding the whole oxidative process. Pt–H and Pd–H have proven key intermediates which can be oxidized by oxygen to yield water. For a long time, the existence of an Au-H species has only been proposed for homogenous catalysis because of the weak bonding strength of Au-H. $H^{\bullet}$ prepared by pre-dissociation of H_2 can bond to Au surfaces up to $-123\ ^{\circ}C$.

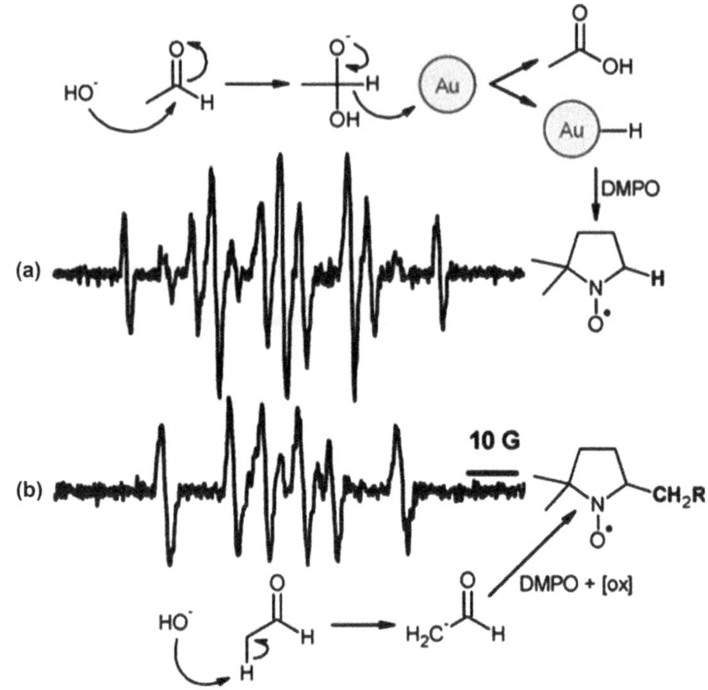

Figure 8.38 DMPO spin adducts formed in a reaction of acetaldehyde in basic media. (a) DMPO–H formation in the presence of PI-Au and (b) DMPO–CH_2R adduct formation in the absence of catalyst (R = COH). Reprinted with permission from ref. 134. Copyright 2010 Royal Society of Chemistry.

Recently, the shift of β-hydride from alcohol to Au by forming an Au–H intermediate was found as a key step of aerobic alcohol by supported gold catalysts.[135,136] Using EPR and spin trapping methods, Conte *et al.*[137] proved the existence of Au-H species over supported and unsupported Au catalysts (Figure 8.40). Additionally, the Au-catalyzed aerobic oxidation has a primary kinetic isotope effect on the activation of C–H bonds.[134,137] An identical DMPO-H spin adduct was detected for $(CH_3)_2CHOH$ and $(CH_3)_2CHOD$, and the perdeuterated isopropanol gave no hydrogen adduct (Figure 8.41). This results also proved that the detected H spin adduct is due to C–H cleavage rather than O–H bond cleavage.

8.5 Selective Oxidation Fenton-like Reactions and Photocatalytic Degradation

8.5.1 Fenton's and Fenton-like Reactions

Fenton's reaction encompasses reactions of hydrogen peroxide in the presence of iron to generate highly reactive species such as the $OH^•$ and possibly

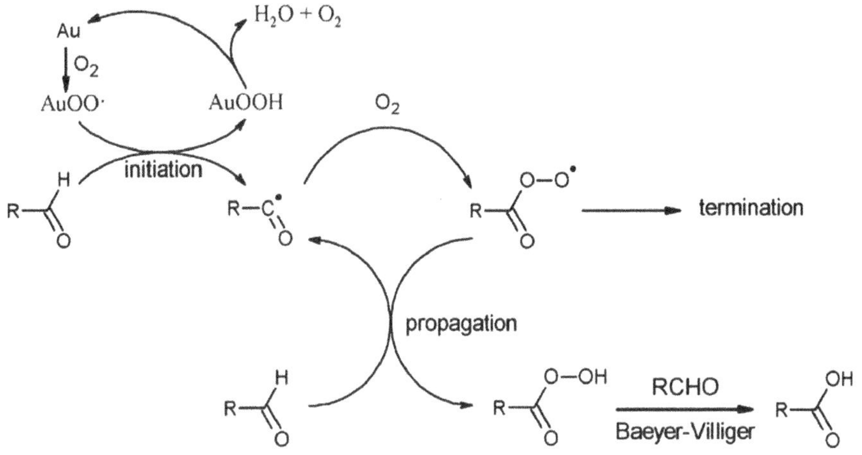

Figure 8.39 Proposed aldehyde oxidation pathway.
Reprinted with permission from ref. 134. Copyright 2010 Royal Society of Chemistry.

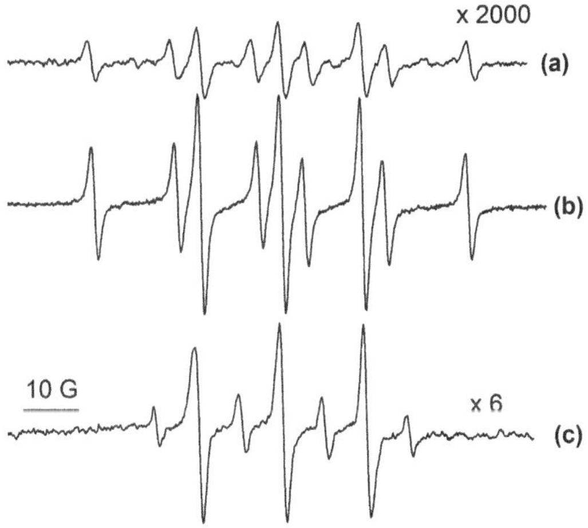

Figure 8.40 EPR spectra of H spin adducts formed during aerobic oxidation of secondary alcohols over supported and unsupported Au catalysts: (a) DMPO–H adduct for $AuPPh_3$ nanoparticles and 2-pentanol; (b) DMPO-H adduct for Au/CeO_2 and 3-octanol; and (c) PBN-H adduct for Au-PI and 1-phenylethanol.
Reprinted with permission from ref. 137. Copyright 2009 American Chemical Society.

others.[138] The British chemist, Henry J. Fenton, discovered the remarkable oxidation power of H_2O_2 in the presence of Fe^{II} salt in 1894.[139] Nowadays, Fenton's reaction has been widely applied in environmental, biological, and synthetic chemical industries.[140,141] Fenton's reaction is generally based on

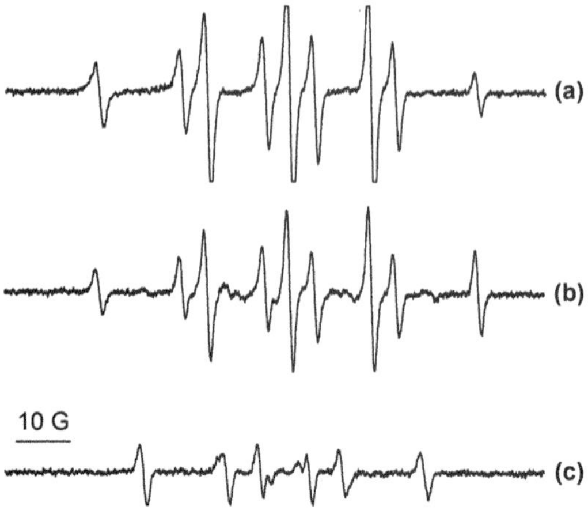

Figure 8.41 Spin adducts formed from DMPO in a reaction of Au/CeO_2 with (a) $(CH_3)_2CHOH$, (b) $(CH_3)_2CHOD$, and (c) $(CD_3)_2CDOD$.
Reprinted with permission from ref. 137. Copyright 2009 American Chemical Society.

the redox process of iron ions and the $OH^{\bullet}$ produced by H_2O_2 decomposition is thought to be responsible for the degradation of toxic chemicals in aqueous solution. The self-catalysis process simultaneously takes place with the hydrolysis of iron ions, which leads to the precipitation of homogeneous metal cations. Moreover, the generation capability of $OH^{\bullet}$ is related to the species of iron cations. For the ferrous cation, the pH usually needs to be adjusted to ~2.5. In order to enhance the stability of Fenton's reagents and to extend the pH to neutral, the development of heterogeneous Fenton-like system is of great importance.

Han *et al.*[142] developed Au-containing Fenton catalysts in 2008 for the first time, which was represented as a 'golden' chapter for Fenton's reaction.[143] The Au nanoparticles were highly dispersed on hydroxyapatite with an average size of 4.9 nm. The activation of H_2O_2 can be achieved at neutral even in alkaline pH solutions (pH ~9). In particular, the Au/HAp catalyst exhibited a superior stability to the conventional Fe-containing heterogeneous Fenton catalyst, Fe/ZSM-5. Using the modified ion-exchange method, Au nanoparticles of uniform size were successfully deposited on the surface of activated carbon. The generation of $OH^{\bullet}$ was verified by DMPO-trapped EPR and the origin for the activity was studied by oxygen temperature programmed desorption (O_2-TPD). The interaction between the dangling bond on carbon and the Au nanoparticles was thought to be responsible for $OH^{\bullet}$ production.[144] To increase the reaction rate for the total degradation of organic pollutants, Quintanilla *et al.*[145] reported several Au catalysts for the wet peroxide process; Au/TiO_2 was more efficient for the decomposition of H_2O_2 whereas Au/C was much more active for the degradation of phenol. Other oxidants, such as peroxomonosulfate (PMS) and peroxodisulfate (PDS) were also evaluated over Au/TiO_2 for the degradation of Acid Red 88.

Besides being interested in the stability of the catalyst and the pH range for its operation, Garcia and co-workers presented a so-called 'nano-jewel' catalyst for the Fenton degradation of phenol. The Au nanoparticles were grafted onto nanoparticulate diamond (DNP). DNP reached an extremely high efficiency for the generation of $OH^\bullet$ from H_2O_2 decay, which was as high as 79%.[146] Under irradiation from visible light, the Fenton activity of this catalyst is further enhanced;[147] in an experiment involving laser flash photolysis, photoinduced electron ejection from gold was observed (Figure 8.42). On the basis of electron transfer theory, Garcia and co-workers successfully demonstrated the mechanism of photochemical enhancement effect on Fenton's reaction. In combination with the *Vibrio fischeri* toxicity test, the optimized Fenton reaction was proposed for the pretreatment of phenol-containing waste water. The optimum conditions (pH = 4, 50 °C) was found for Fenton-biochemical treatment over the Au/DNP catalyst.[148,149] In this system, the anchoring of Au nanoparticles *via* the interaction with surface functional groups on DNP was crucial for improving catalytic performance. Thus the surface treatment of DNP was systemically studied.

The outer-shell of DNP is normally a sooty material without the treatment. When DNP was treated with Fenton's reagent, the surface hydroxyl group (OH-) was increased remarkably.[146,147,150] Light combustion following the hydrogen annealing was an alternative way.[151,152] By the direct current (DC) sputtering method, Au nanoparticles was deposited on the surface of silicon(111) wafer. With the combination of grazing incidence X-ray fluorescence (GIXRF) and theoretical calculation, Ramalho and co-workers proposed that the introduction of gold increased the defective structure on

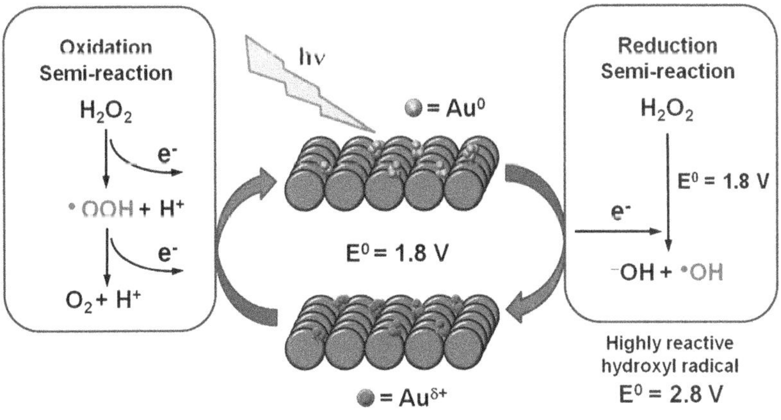

Figure 8.42 Proposed mechanism for the photochemical enhancement of Fenton catalysis.
Reprinted with permission from ref. 147. Copyright 2012 American Chemical Society.

silicon, thus enhancing the affinity for H_2O_2. The efficiency of photo peroxidation was consequently improved.[153,154]

8.5.2 Gold-based Photocatalytic Degradation of Organic Pollutants

On the photosensitizer surface, molecular oxygen can be directly activated over gold nanoparticles for the degradation of organic pollutants in water. Semiconductor oxides, such as TiO_2 and ZnO, were widely studied for this aim. The doping of gold shifted the Fermi level; in addition, the identity of the metal, metal loading, the ratio of cationic to metallic species, nanoparticle size and the coordination behavior affected the charge transfer and charge separation processes significantly.

Various techniques were proposed to enhance the photochemical properties of Au/TiO_2 catalysts. The pioneering work in this area by Kisch and co-workers[155] dates back to 1998. By the doping of noble metal chloride, including $HAuCl_4$, on the surface or into the bulk structure of amorphous TiO_2, the hybrid TiO_2 acted efficiently for the degradation of organic pollutants under irradiation with visible light. On the heterolytic cleavage of the metal–chloride bond under visible radiation, the metal nanoparticle was formed as the consequence of the precursor decomposition. Probed by 4-chlorophenol, the visible-active center was attributed to metal nanoparticles. The reduced nanoparticles or ions transfer the electrons to the conduction band of TiO_2, and get electrons from the pollutant to complete the catalytic cycle and regenerate the active sites. The containments were totally degraded in this process. For instance, methyl blue was degraded completely over a 0.5 wt% Au(0.5 wt%)/rutile-type TiO_2 catalyst.[156] The authors assumed a new phase $Au_xTi_{1-x}O_2$ was formed, which resulted in the changes in surface and semiconductor properties. From photoluminescence analysis, recombination of the electron–hole pair was shown to be slow, while the overcapacity electron transferred at a higher rate to an adsorbed oxygen molecule and generated superoxide radical ($O_2^{\bullet-}$) as consequence. A gold hybrid ZnO nanorod with large surface area was prepared by the wet chemical method, totally degrading rhodamine B solution with an initial concentration of 2.0×10^{-5} M within 300 min.[157]

Since the innovative work of Asahi *et al.*,[158] nitrogen-doped TiO_2 has received much attention. They found that the doping of nitrogen can generate an 'intra-band gap' state. The so-called 'intra-band gap' overlapped with the edge of the energy band and prevented charge recombination. Studies of this aspect are widely reviewed.[159]

By hydrolysis of titanium sulfate and ammonia followed by a Au deposition–precipitation method, Zhang and co-workers[160] prepared Au/N-TiO_2 for the degradation of methyl orange (MO) under visible light and ultraviolet (UV) light irradiation. The degradation ratio of MO was increased about 10% over Au(4.0wt%)/N-TiO_2 compared with commercial P25 under visible light irradiation. The surface species in the form of N-Ti-O or Ti-O-N were detected by XPS. Thus nitrogen was introduced into TiO_2 matrix in the form of substitutional and interstitial nitrogen atoms. As indicated by diffuse reflectance UV-vis spectra, the light absorption of Au/N-TiO_2 was extended into the visible region; results from photoluminescence analysis confirmed the inhibition effect of Au nanoparticles on electron–hole recombination. Similar systems were also evaluated for the degradation of methylene blue (MB) and 2,4-dichlorophenol (2,4-DCP), respectively.[161] A plausible mechanism was proposed as shown in Figure 8.43. Au loading and N doping exhibited a synergetic effect on enhancing the visible light photocatalytic activity of TiO_2.

The synergistic effect between gold and nitrogen was systematically studied by Sanz and co-workers from the aspects of theoretical calculation and practical experiment. Usually, the implanted nitrogen was stabilized through the electron transfer from the oxygen vacancy in TiO_2 toward the partially occupied N_{2p} band.[162] The introduction of an Au atom or ion donated extra electrons to the 2p orbitals of nitrogen by forming an Au/TiN_xO_{2-y} species, with a $SrTiO_3$ structure. In the optimized structural matrix, N was located at the substituted in-plane site in the first TiO_2 layer [Figure 8.44(a)], at the bridge site in the second layer [Figure 8.44(b)], and at the top of the adsorbed oxygen [Figure 8.44(c)] on the bridge or pseudo-hollow site [Figure 8.44(d)], respectively. When the Au atom was introduced, the preferred site was found to be with the Au atom on top of N with an adsorption energy of −2.55 eV [Figure 8.44(a)]. The analysis of the electronic structure indicated that the simultaneous presence of both empty N_{2p} states in the low band gap region and the (higher energy) partially filled Au_{6s} band allows for electron transfer between them. The electron density of Au (hybridized spd band) was pumped out and redirected toward the 2p orbitals of an N atom, which closes to its shell. Moreover, the implantation energy of N was higher

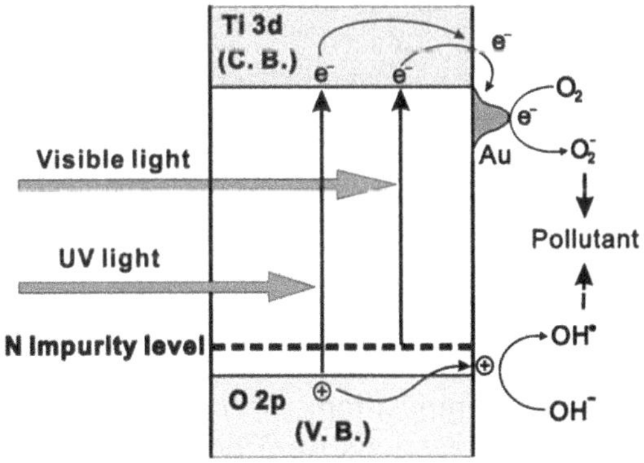

Figure 8.43 Proposed mechanism for phototocatalytic degradation of organic pollutant under UV and visible light irradiation over an Au/N–TiO_2 catalyst proposed by Tian *et al.* CB, conduction band; VB, valence band.
Reprinted with permission from ref. 161. Copyright 2009 Elsevier.

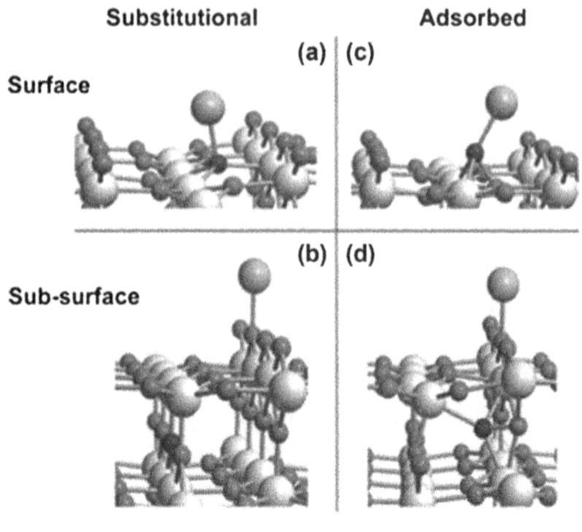

Figure 8.44 Optimized structures of Au atom deposited on N-doped $TiO_2(110)$ rutile.
Reprinted with permission from ref. 162. Copyright 2008 American Chemical Society.

when Au was pre-adsorbed on $TiO_2(110)$ surface (−2.50 eV *vs.* −0.12 eV). This indicated a synergistic effect between implanted N and deposited Au atoms, even though the presence of Au hampered the creation of an oxygen vacancy. With the doping of N atom into the Ti-O terminated Au/$SrTiO_3(001)$, the interaction between Au and N led to the appearance of new levels just above the valence band while below the conduction band of $SrTiO_3$. These extra levels could play a key role in photocatalytic reactions.

In a combination of the sol-gel method and hydrothermal treatment, Fe^{3+} ion was doped into TiO_2. Upon the deposition of Au nanoparticles, excellent performance was exhibited for the degradation of methyl orange. As indicated by XRD and EPR, Fe^{3+} replaced Ti^{4+} in the TiO_2 lattice, introducing a dopant energy level into the bandgap of TiO_2. Under UV light radiation, the excited electrons with a large amount of energy could be rapidly trapped by Au and the remaining holes were trapped by Fe^{3+}, which were then transferred to the surface to boot up the photocatalytic processes.[163,164]

The surface treatment of Au/TiO_2 by $HClO_4$ was investigated by Ye and coworkers.[165] The removal of surfactant prompted the interaction between Au and TiO_2, and facilitated the charge carriers or energy transfer in the plasmonic composite. Li *et al.* presented an Au/TiO_2 with mesoporous structure and tailored framework composition; the photoassisted phenol oxidation and chromium reduction reactions were both enhanced.[166] Three-dimensionally ordered macroporous TiO_2 was synthesized by Srinivasan and White.[167]

Upon the deposition of Au, the degradation rate of methyl blue was double that with commercial P25 catalyst. With consecutive solvothermal and hydrothermal treatments, Au nanoparticles were homogeneously

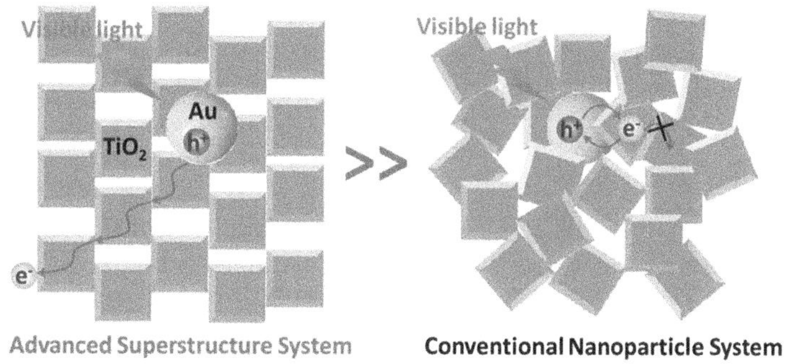

Figure 8.45 Plausible scheme of the electron transfer process of advanced superstructure and conventional nanoparticle system proposed by Bian *et al.*
Reprinted with permission from ref. 170. Copyright 2014 American Chemical Society.

encapsulated into the core-shell TiO_2 microsphere.[168] The design and modification of micro-geometry were demonstrated to be efficient for the improvement of quantum productivity. Liu and Sen[169] encapsulated the Au nanorod into the TiO_2 nanowire. The core-shell structure could be fabricated by varying the morphology of the seeding metal nanostructure or controlling selective TiO_2 growth on different surfaces of the metallic nano material. The separation of the electron–hole pair was enhanced and the quantum efficiency was improved as a consequence. Interestingly, the superstructure, which was assembled by the TiO_2 mesocrystals, was tunable simply by an impregnation method.[170] The electron ejected from the excited Au nanoparticles on the basal surfaces of mesoporous TiO_2 effectively migrated to the lateral TiO_2 surfaces through the ordered TiO_2 nano crystal networks. Consequently, the activity for organic compound degradation was significantly improved (Figure 8.45).

8.6 Summary and Outlook

The generation of oxygen-containing radicals, more generally known as reactive oxygen species (ROS), is an important process in nature, occurring from the human body to the nebulae of outer space.[171] Due to the weak capability for adsorption, the radical-related process is more facile over Au-based catalysts. Many cases have been introduced in the present review, which are the landmark for Au catalysis in the new area. Ideas from another vision could be inspired for the strategy for catalyst design and understanding of the reaction process. In another aspect, these studies have enriched the catalogue of radical reactions.

Until very recently, however, no theoretical studies focused on radical-related processes happening on the heterogeneous interface, especially on the interface of Au nanoparticles. No direct evidence has been proposed to demonstrate the effect of the Au catalyst in decomposition. Nowadays or in the future, the radical-related processes over Au catalysts still represent the place where fantasy meets ability in designing catalytic reactions for the processes of green energy and clean environment.

Acknowledgements

The authors are grateful to the support from the National Science Foundation (21176071, 21106041, 21273070), the Program for New Century Excellent Talents in university (NCET-12-0852), Science and Technology Commission of Shanghai Municipality (11JC1402700), Innovation Program of Shanghai Municipal Education Commission (11ZZ52, 12ZZ051), Shanghai Natural Science Foundation (11ZR1408400), Fundamental Research Funds for the Central Universities, and the Chinese Education Ministry 111 project (B08021).

References

1. M. Haruta, T. Kobayashi, H. Sano and N. Yamada, *Chem. Lett.*, 1987, **16**, 405.
2. B. K. Min and C. M. Friend, *Chem. Rev.*, 2007, **107**, 2709.
3. Z. Ma and S. Dai, *ACS Catal.*, 2011, **1**, 805.
4. M. Pan, A. J. Brush, Z. D. Pozun, H. C. Ham, W.-Y. Yu, G. Henkelman, G. S. Hwang and C. B. Mullins, *Chem. Soc. Rev.*, 2013, **42**, 5002.
5. C. Della Pina, E. Falletta, L. Prati and M. Rossi, *Chem. Soc. Rev.*, 2008, **37**, 2077.
6. M. Boronat, A. Leyva-Pérez and A. Corma, *Acc. Chem. Res.*, 2013.
7. M. Stratakis and H. Garcia, *Chem. Rev.*, 2012, **112**, 4469.
8. Y. Zhang, X. Cui, F. Shi and Y. Deng, *Chem. Rev.*, 2011, **112**, 2467.
9. R. Meyer, C. Lemire, S. K. Shaikhutdinov and H. J. Freund, *Gold Bull.*, 2004, **37**, 72.
10. J. Gong, *Chem. Rev.*, 2011, **112**, 2987.
11. M. S. Chen and D. W. Goodman, *Science*, 2004, **306**, 252.
12. I. X. Green, W. Tang, M. Neurock and J. T. Yates, *Science*, 2011, **333**, 736.
13. R. Atkinson, *Chem. Rev.*, 1986, **86**, 69.
14. A. Dhakshinamoorthy, S. Navalon, M. Alvaro and H. Garcia, *ChemSusChem*, 2012, **5**, 46.
15. M. I. bin Saiman, G. L. Brett, R. Tiruvalam, M. M. Forde, K. Sharples, A. Thetford, R. L. Jenkins, N. Dimitratos, J. A. Lopez-Sanchez, D. M. Murphy, D. Bethell, D. J. Willock, S. H. Taylor, D. W. Knight, C. J. Kiely and G. J. Hutchings, *Angew. Chem., Int. Ed.*, 2012, **51**, 5981.
16. M.-C. Daniel and D. Astruc, *Chem. Rev.*, 2004, **104**, 293.
17. R. Pitchai and K. Klier, *Catal. Rev.*, 1986, **28**, 13.
18. H. D. Gesser, N. R. Hunter and C. B. Prakash, *Chem. Rev.*, 1985, **85**, 235.
19. M. M. Holm and E. H. Reichl, *The Oxidation of Methane*, Germany Field Information Agency Technical Final Report No. 1085, Office of Military Government for Germany, United States, 1947.

20. V. Sokolovskii, C. Pearcey and N. Coville, *React. Kinet. Catal. Lett.*, 1994, **52**, 341.
21. I. Iskendirov, V. Sokolovskii and N. Coville, *J. Chem. Soc., Chem. Comm.*, 1994, 2033.
22. I. Eskendirov, N. J. Coville, A. Parmaliana and V. D. Sokolovskii, in *Studies in Surface Science and Catalysis*, Elsevier, 1997, pp. 301–306.
23. I. Eskendirov, N. J. Coville and V. D. Sokolovskii, *Catal. Lett.*, 1995, **35**, 33.
24. K. Blick, T. Mitrelias, J. J. Hargreaves, G. Hutchings, R. Joyner, C. Kiely and F. Wagner, *Catal. Lett.*, 1998, **50**, 211.
25. C. J. Jones, D. Taube, V. R. Ziatdinov, R. A. Periana, R. J. Nielsen, J. Oxgaard and W. A. Goddard, *Angew. Chem., Int. Ed.*, 2004, **43**, 4626.
26. D. E. De Vos and B. F. Sels, *Angew. Chem., Int. Ed.*, 2005, **44**, 30.
27. T. Li, S. Wang, C. Yu, Y. Ma, K. Li and L. Lin, *Appl. Catal. A: Gen.*, 2011, **398**, 150.
28. L. A. Levchenko, A. P. Sadkov, N. V. Lariontseva, E. M. Koldasheva, A. K. Shilova and A. E. Shilov, *Doklady Biochem. Biophy.*, 2001, **377**, 123.
29. L. A. Levchenko, A. P. Sadkov, N. V. Lariontseva, V. S. Kulikova, A. K. Shilova and A. E. Shilov, *Doklady Biochem. Biophy.*, 2004, **394**, 33.
30. C. D. Pina, E. Falletta and M. Rossi, *Chem. Soc. Rev.*, 2012, **41**, 350.
31. Q. Yuan, W. Deng, Q. Zhang and Y. Wang, *Adv. Synth. Catal.*, 2007, **349**, 1199.
32. M. H. Ab Rahim, M. M. Forde, R. L. Jenkins, C. Hammond, Q. He, N. Dimitratos, J. A. Lopez-Sanchez, A. F. Carley, S. H. Taylor, D. J. Willock, D. M. Murphy, C. J. Kiely and G. J. Hutchings, *Angew. Chem., Int. Ed.*, 2013, **52**, 1280.
33. O. A. Marina and M. Mogensen, *Appl. Catal. A: Gen.*, 1999, **189**, 117.
34. J. Llorca, A. Casanovas, M. Domínguez, I. Casanova, I. Angurell, M. Seco and O. Rossell, *J. Nanopart. Res.*, 2008, **10**, 537.
35. B. C. Hereijgers and B. Weckhuysen, *Catal. Lett.*, 2011, **141**, 1429.
36. Z.-C. Wang, T. Weiske, R. Kretschmer, M. Schlangen, M. Kaupp and H. Schwarz, *J. Am. Chem. Soc.*, 2011, **133**, 16930.
37. N. Dietl, R. F. Höckendorf, M. Schlangen, M. Lerch, M. K. Beyer and H. Schwarz, *Angew. Chem., Int. Ed.*, 2011, **50**, 1430.
38. J.-B. Ma, Z.-C. Wang, M. Schlangen, S.-G. He and H. Schwarz, *Angew. Chem., Int. Ed.*, 2012, **51**, 5991.
39. X.-N. Wu, X.-N. Li, X.-L. Ding and S.-G. He, *Angew. Chem., Int. Ed.*, 2013, **125**, 2504.
40. A. Ueda, Y. Yamada and T. Kobayashi, *Appl. Catal. A: Gen.*, 2001, **209**, 391.
41. J. J. Bravo-Suarez, K. K. Bando, T. Akita, T. Fujitani, T. J. Fuhrer and S. T. Oyama, *Chem. Commun.*, 2008, 3272.
42. J. J. Bravo-Suárez, K. K. Bando, T. Fujitani and S. T. Oyama, *J. Catal.*, 2008, **257**, 32.
43. J. J. Bravo-Suárez, K. K. Bando, J. Lu, T. Fujitani and S. T. Oyama, *J. Catal.*, 2008, **255**, 114.

44. N. A. Mashayekhi, Y. Y. Wu, M. C. Kung and H. H. Kung, *Chem. Commun.*, 2012, **48**, 10096.
45. Y. Y. Wu, N. A. Mashayekhi and H. H. Kung, *Catal. Sci. Technol.*, 2013, **3**, 2881.
46. K. Narasimharao and T. Ali, *Catal. Lett.*, 2013, **143**, 1074.
47. S. Barison, M. Battagliarin, S. Daolio, M. Fabrizio, E. Miorin, P. L. Antonucci, S. Candamano, V. Modafferi, E. M. Bauer, C. Bellitto and G. Righini, *Solid State Ionics*, 2007, **177**, 3473.
48. N. Kijima, K. Matano, M. Saito, T. Oikawa, T. Konishi, H. Yasuda, T. Sato and Y. Yoshimura, *Appl. Catal. A: Gen.*, 2001, **206**, 237.
49. M. M. Bhasin, J. H. McCain, B. V. Vora, T. Imai and P. R. Pujadó, *Appl. Catal. A: Gen.*, 2001, **221**, 397.
50. C. Larese, J. M. Campos-Martin, J. J. Calvino, G. Blanco, J. L. G. Fierro and Z. C. Kang, *J. Catal.*, 2002, **208**, 467.
51. J. Sá, M. Ace, J. J. Delgado, A. Goguet, C. Hardacre and K. Morgan, *ChemCatChem*, 2011, **3**, 394.
52. S. Luciani, F. Cavani, V. Dal Santo, N. Dimitratos, M. Rossi and C. L. Bianchi, *Catal. Today*, 2011, **169**, 200.
53. R. Lloyd, R. L. Jenkins, M. Piccinini, Q. He, C. J. Kiely, A. F. Carley, S. E. Golunski, D. Bethell, J. K. Bartley and G. J. Hutchings, *J. Catal.*, 2011, **283**, 161.
54. R. Sheldon, *Metal-Catalyzed Oxidations of Organic Compounds: Mechanistic Principles and Synthetic Methodology including Biochemical Processes*, Academic Press, New York and London, 1981.
55. C. A. Tolman, J. D. Druliner, M. J. Nappa and N. Herron, Alkane oxidation studies in Du Pont's central research department, in *Activation and Functionalization of Alkanes*, ed. C. L. Hill, Wiley-Interscience, New York, 1989, p. 136.
56. A. Sakthivel and P. Selvam, *J. Catal.*, 2002, **211**, 134.
57. E. Breynaert, I. Hermans, B. Lambie, G. Maes, J. Peeters, A. Maes and P. Jacobs, *Angew. Chem., Int. Ed.*, 2006, **118**, 7746.
58. B. Notari, Microporous Crystalline Titanium Silicates, *Advances in Catalysis Vol. 41*, Academic Press, New York, 1996.
59. E. Armengol, A. Corma, V. Fornés, H. García and J. Primo, *Appl. Catal. A: Gen.*, 1999, **181**, 305.
60. R. Zhao, D. Ji, G. Lv, G. Qian, L. Yan, X. Wang and J. Suo, *Chem. Commun.*, 2004, 904.
61. G. Lü, R. Zhao, G. Qian, Y. Qi, X. Wang and J. Suo, *Catal. Lett.*, 2004, **97**, 115.
62. L.-X. Xu, C.-H. He, M.-Q. Zhu, K.-J. Wu and Y.-L. Lai, *Catal. Lett.*, 2007, **118**, 248.
63. K. Zhu, J. Hu and R. Richards, *Catal. Lett.*, 2005, **100**, 195.
64. C. Wang, L. Chen and Z. Qi, *Catal. Sci. Technol.*, 2013, **3**, 1123.
65. Y. Liu, H. Tsunoyama, T. Akita, S. Xie and T. Tsukuda, *ACS Catal.*, 2010, **1**, 2.
66. J. Long, H. Liu, S. Wu, S. Liao and Y. Li, *ACS Catal.*, 2013, **3**, 647.

67. Y.-J. Xu, P. Landon, D. Enache, A. Carley, M. W. Roberts and G. Hutchings, *Catal. Lett.*, 2005, **101**, 175.
68. G. Hutchings, S. Carrettin, P. Landon, J. Edwards, D. Enache, D. Knight, Y.-J. Xu and A. Carley, *Top. Catal.*, 2006, **38**, 223.
69. S. A. C. Carabineiro, L. M. D. R. S. Martins, M. Avalos-Borja, J. G. Buijnsters, A. J. L. Pombeiro and J. L. Figueiredo, *Appl. Catal. A: Gen.*, 2013, **467**, 279.
70. R. Liu, H. Huang, H. Li, Y. Liu, J. Zhong, Y. Li, S. Zhang and Z. Kang, *ACS Catal.*, 2014, **4**, 328.
71. B. P. Hereijgers and B. M. Weckhuysen, *J. Catal.*, 2010, **270**, 16.
72. M. Conte, X. Liu, D. M. Murphy, K. Whiston and G. J. Hutchings, *Phys. Chem. Chem. Phys.*, 2012, **14**, 16279.
73. A. Alshammari, A. Koeckritz, V. N. Kalevaru, A. Bagabas and A. Martin, *ChemCatChem*, 2012, **4**, 1330.
74. L. Kesavan, R. Tiruvalam, M. H. A. Rahim, M. I. bin Saiman, D. I. Enache, R. L. Jenkins, N. Dimitratos, J. A. Lopez-Sanchez, S. H. Taylor, D. W. Knight, C. J. Kiely and G. J. Hutchings, *Science*, 2011, **331**, 195.
75. F. Jiang, X. Zhu, B. Fu, J. Huang and G. Xiao, *Chin. J. Catal.*, 2013, **34**, 1683.
76. A. K. Sinha, S. Seelan, S. Tsubota and M. Haruta, *Angew. Chem., Int. Ed.*, 2004, **43**, 1546.
77. A. K. Sinha, S. Seelan, S. Tsubota and M. Haruta, *Top. Catal.*, 2004, **29**, 95.
78. T. Hayashi, K. Tanaka and M. Haruta, *J. Catal.*, 1998, **178**, 566.
79. E. E. Stangland, B. Taylor, R. P. Andres and W. N. Delgass, *J. Phys. Chem. B*, 2005, **109**, 2321.
80. T. A. Nijhuis, E. Sacaliuc, A. M. Beale, A. M. J. van der Eerden, J. C. Schouten and B. M. Weckhuysen, *J. Catal.*, 2008, **258**, 256.
81. A. Roldan, D. Torres, J. M. Ricart and F. Illas, *J. Mol. Catal. A: Chem.*, 2009, **306**, 6.
82. J. Jiang, H. H. Kung, M. C. Kung and J. Ma, *Gold Bull.*, 2009, **42**, 280.
83. H. Alshammari, P. J. Miedziak, S. Bawaked, D. W. Knight and G. J. Hutchings, *ChemCatChem*, 2012, **4**, 1565.
84. C. Aprile, A. Corma, M. E. Domine, H. Garcia and C. Mitchell, *J. Catal.*, 2009, **264**, 44.
85. M. Álvaro, C. Aprile, A. Corma, B. Ferrer and H. García, *J. Catal.*, 2007, **245**, 249.
86. C. Aprile, M. Boronat, B. Ferrer, A. Corma and H. García, *J. Am. Chem. Soc.*, 2006, **128**, 8388.
87. M. D. Hughes, Y.-J. Xu, P. Jenkins, P. McMorn, P. Landon, D. I. Enache, A. F. Carley, G. A. Attard, G. J. Hutchings, F. King, E. H. Stitt, P. Johnston, K. Griffin and C. J. Kiely, *Nature*, 2005, **437**, 1132.
88. B. Li, P. He, G. Yi, H. Lin and Y. Yuan, *Catal. Lett.*, 2009, **133**, 33.
89. S. Bawaked, N. F. Dummer, N. Dimitratos, D. Bethell, Q. He, C. J. Kiely and G. J. Hutchings, *Green Chem.*, 2009, **11**, 1037.

90. S. Bawaked, Q. He, N. F. Dummer, A. F. Carley, D. W. Knight, D. Bethell, C. J. Kiely and G. J. Hutchings, *Catal. Sci. Technol.*, 2011, **1**, 747.
91. L. Wang, H. Wang, P. Hapala, L. Zhu, L. Ren, X. Meng, J. P. Lewis and F.-S. Xiao, *J. Catal.*, 2011, **281**, 30.
92. X. Han, H. Huang, H. Zhang, X. Zhang, H. Li, R. Liu, Y. Liu and Z. Kang, *Mater. Res. Bull.*, 2013, **48**, 3717.
93. C. H. A. Tsang, Y. Liu, Z. Kang, D. D. D. Ma, N.-B. Wong and S.-T. Lee, *Chem. Commun.*, 2009, 5829.
94. Z.-Y. Cai, M.-Q. Zhu, J. Chen, Y.-Y. Shen, J. Zhao, Y. Tang and X.-Z. Chen, *Catal. Comm.*, 2010, **12**, 197.
95. Z.-Y. Cai, M.-Q. Zhu, H. Dai, Y. Liu, J.-X. Mao, X.-Z. Chen and C.-H. He, *Adv. Chem. Eng. Sci.*, 2011, **1**, 15.
96. H. Alshammari, P. J. Miedziak, D. W. Knight, D. J. Willock and G. J. Hutchings, *Catal. Sci. Technol.*, 2013, **3**, 1531.
97. S. Bawaked, N. F. Dummer, D. Bethell, D. W. Knight and G. J. Hutchings, *Green Chem.*, 2011, **13**, 127.
98. D. Swern, *Organic Peroxides*, Wiley, New York and London, 1961.
99. C. Coperet, H. Adolfsson and K. Barry Sharpless, *Chem. Commun.*, 1997, 1565.
100. J. Rudolph, K. L. Reddy, J. P. Chiang and K. B. Sharpless, *J. Am. Chem. Soc.*, 1997, **119**, 6189.
101. S. B. Kumar, S. P. Mirajkar, G. C. G. Pais, P. Kumar and R. Kumar, *J. Catal.*, 1995, **156**, 163.
102. Q. Yang, S. Wang, J. Lu, G. Xiong, Z. Feng, Q. Xin and C. Li, *Appl. Catal. A: Gen.*, 2000, **194–195**, 507.
103. N. S. Patil, B. S. Uphade, P. Jana, R. S. Sonawane, S. K. Bhargava and V. R. Choudhary, *Catal. Lett.*, 2004, **94**, 89.
104. N. S. Patil, R. Jha, B. S. Uphade, S. K. Bhargava and V. R. Choudhary, *Appl. Catal. A: Gen.*, 2004, **275**, 87.
105. N. S. Patil, B. S. Uphade, P. Jana, S. K. Bharagava and V. R. Choudhary, *J. Catal.*, 2004, **223**, 236.
106. V. R. Choudhary, P. Jana and S. K. Bhargava, *Int. J. Nanopart.*, 2009, **2**, 11.
107. D. K. Dumbre, V. R. Choudhary, N. S. Patil, B. S. Uphade and S. K. Bhargava, *J. Colloid Interf. Sci.*, 2014, **415**, 111.
108. N. S. Patil, B. S. Uphade, D. G. McCulloh, S. K. Bhargava and V. R. Choudhary, *Catal. Comm.*, 2004, **5**, 681.
109. N. S. Patil, B. S. Uphade, P. G. Jana, S. K. Bhargava and V. R. Choudhary, *Chem. Lett.*, 2004, **33**, 400.
110. V. R. Choudhary, D. K. Dumbre, N. S. Patil, B. S. Uphade and S. K. Bhargava, *J. Catal.*, 2013, **300**, 217.
111. F. Zhang, X. Zhao, C. Feng, B. Li, T. Chen, W. Lu, X. Lei and S. Xu, *ACS Catal.*, 2011, **1**, 232.
112. Y. Liu, H. Tsunoyama, T. Akita and T. Tsukuda, *Chem. Commun.*, 2010, **46**, 550.

113. M. Turner, V. B. Golovko, O. P. Vaughan, P. Abdulkin, A. Berenguer-Murcia, M. S. Tikhov, B. F. Johnson and R. M. Lambert, *Nature*, 2008, **454**, 981.
114. D. Yin, L. Qin, J. Liu, C. Li and Y. Jin, *J. Mol. Catal. A: Chem.*, 2005, **240**, 40.
115. S. Das and T. Asefa, *Top. Catal*, 2012, **55**, 587.
116. J. Liu, F. Wang, S. Qi, Z. Gu and G. Wu, *New J. Chem.*, 2013, **37**, 769.
117. Y. Jin, D. Zhuang, N. Yu, H. Zhao, Y. Ding, L. Qin, J. Liu, D. Yin, H. Qiu, Z. Fu and D. Yin, *Micro. Meso. Mater.*, 2009, **126**, 159.
118. Y. Jin, P. Wang, D. Yin, J. Liu, H. Qiu and N. Yu, *Micro. Meso. Mater.*, 2008, **111**, 569.
119. L. Luo, N. Yu, R. Tan, Y. Jin, D. Yin and D. Yin, *Catal. Lett.*, 2009, **130**, 489.
120. P. Lignier, F. Morfin, L. Piccolo, J.-L. Rousset and V. Caps, *Catal. Today*, 2007, **122**, 284.
121. P. Lignier, S. Mangematin, F. Morfin, J.-L. Rousset and V. Caps, *Catal. Today*, 2008, **138**, 50.
122. F. Kerdi, V. Caps and A. Tuel, *Micro. Meso. Mater.*, 2011, **140**, 89.
123. K. Guillois, L. Burel, A. Tuel and V. Caps, *Appl. Catal. A: Gen.*, 2012, **415–416**, 1.
124. V. Mendez, K. Guillois, S. Daniele, A. Tuel and V. Caps, *Dalton Trans.*, 2010, **39**, 8457.
125. S. Mishra, V. Mendez, E. Jeanneau, V. Caps and S. Daniele, *Eur. J. Inorg. Chem.*, 2013, **2013**, 500.
126. K. Guillois, S. Mangematin, A. Tuel and V. Caps, *Catal. Today*, 2013, **203**, 111.
127. M. Boualleg, K. Guillois, B. Istria, L. Burel, L. Veyre, J.-M. Basset, C. Thieuleux and V. Caps, *Chem. Commun.*, 2010, **46**, 5361.
128. B. N. Zope, D. D. Hibbitts, M. Neurock and R. J. Davis, *Science*, 2010, **330**, 74.
129. C. Shang and Z.-P. Liu, *J. Am. Chem. Soc.*, 2011, **133**, 9938.
130. C.-R. Chang, X.-F. Yang, B. Long and J. Li, *ACS Catal.*, 2013, **3**, 1693.
131. I. Moreno, N. F. Dummer, J. K. Edwards, M. Alhumaimess, M. Sankar, R. Sanz, P. Pizarro, D. P. Serrano and G. J. Hutchings, *Catal. Sci. Technol.*, 2013, **3**, 2425.
132. D. Tsukamoto, Y. Shiraishi, Y. Sugano, S. Ichikawa, S. Tanaka and T. Hirai, *J. Am. Chem. Soc.*, 2012, **134**, 6309.
133. G. L. Hallett-Tapley, M. J. Silvero, C. J. Bueno-Alejo, M. González-Béjar, C. D. McTiernan, M. Grenier, J. C. Netto-Ferreira and J. C. Scaiano, *J. Phys. Chem. C*, 2013, **117**, 12279.
134. P. Fristrup, L. Bahn Johansen and C. Hviid Christensen, *Chem. Commun.*, 2008, 2750.
135. A. Abad, A. Corma and H. García, *Chem. Euro. J*, 2008, **14**, 212.
136. P. Fristrup, L. Johansen and C. Christensen, *Catal. Lett.*, 2008, **120**, 184.
137. M. Conte, H. Miyamura, S. Kobayashi and V. Chechik, *J. Am. Chem. Soc.*, 2009, **131**, 7189.

138. J. J. Pignatello, E. Oliveros and A. MacKay, *Crit. Rev. Environ. Sci. Technol.*, 2006, **36**, 1.
139. H. J. H. Fenton, *J. Chem. Soc.,Trans.*, 1894, **65**, 899.
140. P. Bautista, A. F. Mohedano, J. A. Casas, J. A. Zazo and J. J. Rodriguez, *J. Chem. Tech. Biotech.*, 2008, **83**, 1323.
141. S. Venny, Gan and H. K. Ng, *Chem. Eng. J*, 2012, **213**, 295.
142. Y.-F. Han, N. Phonthammachai, K. Ramesh, Z. Zhong and T. White, *Envrion. Sci. Technol.*, 2007, **42**, 908.
143. G. Z. Chen, *Angew. Chem., Int. Ed.*, 2010, **49**, 5413.
144. X. Yang, P.-F. Tian, C. Zhang, Y.-Q. Deng, J. Xu, J. Gong and Y.-F. Han, *Appl. Catal. B: Environ.*, 2013, **134–135**, 145.
145. A. Quintanilla, S. García-Rodríguez, C. M. Domínguez, S. Blasco, J. A. Casas and J. J. Rodriguez, *Appl. Catal. B: Environ.*, 2012, **111–112**, 81.
146. S. Navalon, R. Martin, M. Alvaro and H. Garcia, *Angew. Chem., Int. Ed.*, 2010, **49**, 8403.
147. S. Navalon, M. de Miguel, R. Martin, M. Alvaro and H. Garcia, *J. Am. Chem. Soc.*, 2011, **133**, 2218.
148. R. Martín, S. Navalon, M. Alvaro and H. Garcia, *Appl. Catal. B: Environ.*, 2011, **103**, 246.
149. S. Navalon, R. Martin, M. Alvaro and H. Garcia, *ChemSusChem*, 2011, **4**, 650.
150. R. Martin, S. Navalon, J. J. Delgado, J. J. Calvino, M. Alvaro and H. Garcia, *Chem. Eur. J.*, 2011, **17**, 9494.
151. D. Sempere, S. Navalon, M. Dančíková, M. Alvaro and H. Garcia, *Appl. Catal. B: Environ.*, 2013, **142–143**, 259.
152. S. Navalon, D. Sempere, M. Alvaro and H. Garcia, *ACS Appl. Mater. Interf.*, 2013, **5**, 7160.
153. T. Ramalho, H. P. Carvalho, A. L. Batista, C. Pérez and A. Gobbi, *J. Mater. Sci.*, 2009, **44**, 1029.
154. H. W. P. de Carvalho, A. P. L. Batista, T. C. Ramalho, C. A. Pérez and A. L. Gobbi, *Sepctochimica Acta. A: Mol. Biomol. Spec.*, 2009, **74**(292).
155. H. Kisch, L. Zang, C. Lange, W. F. Maier, C. Antonius and D. Meissner, *Angew. Chem., Int. Ed.*, 1998, **37**, 3034.
156. X. Z. Li and F. B. Li, *Envrion. Sci. Technol*, 2001, **35**, 2381.
157. Q. Wang, B. Geng and S. Wang, *Envrion. Sci. Technol*, 2009, **43**, 8968.
158. R. Asahi, T. Morikawa, T. Ohwaki, K. Aoki and Y. Taga, *Science*, 2001, **293**, 269.
159. J. Zhang, Y. Wu, M. Xing, S. A. K. Leghari and S. Sajjad, *Energy Environ. Sci.*, 2010, **3**, 715.
160. Y. Wu, H. Liu, J. Zhang and F. Chen, *J. Phys. Chem. C*, 2009, **113**, 14689.
161. B. Tian, C. Li, F. Gu and H. Jiang, *Catal. Comm.*, 2009, **10**, 925.
162. J. Graciani, L. J. Alvarez, J. A. Rodriguez and J. F. Sanz, *J. Phys. Chem. C*, 2008, **112**, 2624.
163. L. Xiao, J. Zhang, Y. Cong, B. Tian, F. Chen and M. Anpo, *Catal. Lett.*, 2006, **111**, 207.

164. Y. Wu, J. Zhang, L. Xiao and F. Chen, *Appl. Catal. B: Environ.*, 2009, **88**, 525.
165. L. Liu, S. Ouyang and J. Ye, *Angew. Chem., Int. Ed.*, 2013, **125**, 6821.
166. H. Li, Z. Bian, J. Zhu, Y. Huo, H. Li and Y. Lu, *J. Am. Chem. Soc.*, 2007, **129**, 4538.
167. M. Srinivasan and T. White, *Envrion. Sci. Technol.*, 2007, **41**, 4405.
168. Z. Bian, J. Zhu, F. Cao, Y. Lu and H. Li, *Chem. Commun.*, 2009, 3789.
169. R. Liu and A. Sen, *J. Am. Chem. Soc.*, 2012, **134**, 17505.
170. Z. Bian, T. Tachikawa, P. Zhang, M. Fujitsuka and T. Majima, *J. Am. Chem. Soc.*, 2013, **136**, 458.
171. G. Herzberg, *The Spectra and Structures of Simple Free Radicals: An Introduction to Molecular Spectroscopy*, Dover Publications, New York, 2003.
172. N. Ameur, S. Bedrane, R. Bachir and A. Choukchou-Braham, *J. Mol. Catal. A: Chem.*, 2013, **1**, 374–375.

CHAPTER 9

Heterogeneous Gold Catalysts for Selective Oxidation Reactions

POERNOMO GUNAWAN,[a] RONG XU[b] AND ZIYI ZHONG*[a,b]

[a] Institute of Chemical and Engineering Sciences, A*Star, 1 Pesek Road, Jurong Island, Singapore 627833; [b] School of Chemical & Biomedical Engineering, Nanyang Technological University, 62 Nanyang Drive, Singapore 637459
*Email: Zhong_ziyi@ices.a-star.edu.sg

9.1 Introduction

Since the pioneering work by the groups of Bond, Haruta, and Hutchings on olefin hydrogenation,[1] low-temperature CO oxidation[2] and ethyne hydrochlorination to vinyl chloride[3] over supported gold catalysts, there has been a rapid growth in the number of publications on gold (Au) catalysis, which completely overturns the previous impression that gold was inert for catalysis. Although gold is one of the noble and coinage metals, and is widely used in decorative items such as jewelry, its chemistry is not well known. Au belongs to group 11 (IB) in the periodic table and can exist in the oxidation states of (−I), (0), (I), (II), (III), (IV), and (V). However, only Au(0), (I) and (III) are stable in aqueous solution. Among them, Au(0) is the most stable. Au(0) does not oxidize or burn in air, even with heating, and is inert to strong alkalis and acids.[4] However, when the particle size is small enough, usually below 5 nm, Au nanoparticles are found to be surprisingly active in catalysis. The structure and surface chemistry of Au-based catalysts have been well

RSC Catalysis Series No. 18
Heterogeneous Gold Catalysts and Catalysis
Edited by Zhen Ma and Sheng Dai

Published by the Royal Society of Chemistry, www.rsc.org

reviewed by Gong[5] and Meyer *et al.*,[6] although some issues still need to be clarified.

Au-based catalysts are capable of catalyzing numerous reactions homogeneously and/or heterogeneously. These reactions include reduction and oxidation reactions, acidic site catalyzed isomerization of epoxides into allylic alcohols, benzylation of aromatics, rearrangements, C–C coupling reactions, cycloadditions, and the water gas shift reaction (WGSR),[7,8] which can be made possible because of various Au speciation as well as the rich coordination and organometallic chemistry of Au.[7] For example, cationic Au species in supported catalysts stabilized by the support can act as soft Lewis acids or participate in reactions involving Au redox catalysis.[5,8,9] It is anticipated that the use of Au catalysts in chemical synthesis will have a rapid growth, not only changing reaction selectivities towards specific target products, but also generating some unprecedented reactions.[10,11]

In general, supported Au catalysts exhibit the following features: size-dependent catalytic performance; milder catalytic reaction conditions as compared with many other catalysts; high tolerance to air and moisture as well as strong leaching-resistance in aqueous solutions; and high chemoselectivity towards specific chemical products. Therefore, Au catalysts are to be a new generation of green catalysts.[12,13]

Selective oxidation reactions are employed to produce about a quarter of monomers and chemical intermediates in modern industry.[14] In recent years, they have been applied to convert biomass or biomass-derived products to various value-added chemicals. In this chapter, we present and discuss the applications of supported Au catalysts in several important selective oxidation reactions, particularly focusing on the literature results published in recent years. These reactions include: selective oxidation of hydrocarbons, *e.g.* epoxidation of propylene to propylene oxide; and selective oxidation of sugars, alcohols, glycerol, and other polyols in the liquid and/or gas phase. Besides these often mentioned catalytic activity issues, we also address the optimization of selectivity and the use of cheap oxidants such as air and oxygen. In addition, we examine several fundamentally important aspects, such as the preparation methods of Au catalysts, reaction mechanisms, and the activation of molecular oxygen (O_2), a crucial step in oxidation reactions. Some theoretical calculations and modeling results, part of which are not available under the current experimental conditions, are mentioned as they shed light on various aspects including the structure of the active sites, activation of O_2, and the role of the supports. In this way, we hope to provide readers with a relatively full picture in this area, and inspire further research interests in Au catalysis.

9.2 Preparation Methods of Supported Gold Catalysts

9.2.1 Physics and Chemistry Involved in Various Gold Catalyst Preparation Methods

One of the features of heterogeneous Au catalysis is the dependence of catalytic performance on Au particle size. For example, in CO oxidation and the epoxidation of propylene to propylene oxide using Au catalysts, the Au particle size should usually be below 5 nm. Otherwise, low catalytic activity or byproducts will be obtained. To date, it remains a challenging task to prepare supported Au catalysts with a well-controlled particle size and morphology.

Briefly, the preparation methods for supported Au catalysts can be classified into physical and chemical methods. Chemical approaches are usually adopted in aqueous solution using relatively cheap and common Au precursors, such as tetrachloroauric acid ($HAuCl_4$) and $AuCl_3$. However, in some other preparations, more expensive Au precursors such as dimethyl gold acetylacetonate[15] and phosphine gold complex, $Au_3(Ph_2pz)_3$[16] are also employed. Since there are several excellent books and review papers that summarize the preparation procedures of supported Au catalysts,[13,17,18] here we focus our discussion on the wet chemical methods in aqueous solution using $HAuCl_4$ as the precursor. These methods include impregnation, co-precipitation, deposition–precipitation, grafting, and reduction–deposition. Below, we mainly address the physics and chemistry involved during the catalyst preparation, and possible measures to obtain good control of Au particle size.

To deposit Au particles onto a catalyst support with good control of particle size, there need to be proper interactions between the Au species and the catalyst support, as well as among the Au species. Here, 'Au species' refer to any of the Au precursor molecules and their derived intermediates, clusters or colloids, *etc.* In general, the collective effect of these interactions should drive and anchor the Au species onto the support surface while avoiding their agglomeration in liquid phase or on the support surface. To achieve this, a subtle balance should be reached among these interactions, which can be generated and finely tuned *via* various chemical and physical measures. These interactions include covalent bonds, hydrogen bonds, van der Waals force, electrostatic interactions, *etc.*, whose relative strength should have relevance to the adopted experimental conditions and nature of the components. For example, when the support surface is functionalized with SH-terminal groups that bind strongly to the Au species, the chemical bonding between the SH group and Au species becomes the predominant force. In general, supported Au catalysts are usually prepared *via* the following steps:

(1) Generation of necessary Au species and support surfaces
(2) Adsorption/interaction of Au species with surface active sites
(3) Subsequent transformation of surface Au species to the desired active Au particles/oxidation states.

For example, in ion exchange reaction of Au species with the support surface, Au species first access the support surface *via* electrostatic attraction, followed by ion exchange between the Au species and the support surface, and surface polymerization reaction of the deposited Au species.

Brown *et al.*[19] summarized the possible factors that may influence the binding strength of an aqueous metal ion to a metal oxide surface as follows:

(1) The intrinsic nature of the surface, *i.e.* the surface charge as a function of pH, the double-layer properties, and the number and nature of reactive surface sites, including surface defects;
(2) The pH and ionic strength of the solution, the total concentration of the aqueous metal ion of interest, and the metal ion-to-solid ratio;
(3) The coordination chemistry of the aqueous metal ion and the possible changes in coordination chemistry during sorption, including cation coordination number, valence state, and hydrolysis;
(4) The mode of adsorbate bonding to the surface, *e.g.* inner sphere, outer sphere, or as a surface precipitate;
(5) The possibility that the adsorbent can act as a reductant or an oxidant for aqueous metal ions that exist in different oxidation states;
(6) The effect of other complexants in solution such as organic species and CO_3^{2-}, which can cause either desorption of the metal ion or enhanced adsorption;
(7) The presence and reactivity of other adsorbates (*e.g.* PO_4^{3-} or SO_4^{2-}), which can compete with a metal ion for reactive surface sites;
(8) The intrinsic nature of the support and Au species surface, *e.g.* the reactivity and density of the surface sites and the Au ion-to-solid ratio.

To achieve a high Au loading on the catalyst support surface while maintaining a high dispersion, the above mentioned factors should be finely considered and tuned. Table 9.1 lists some of the common approaches to vary the interactions between Au species–support surface and Au–Au species.

Electrostatic interaction between the Au species and support surface is very common; it depends on the surface charging status of both the Au

Table 9.1 Interactions between Au species and support surface, and their application to various catalyst preparation methods.

Interaction	Measures to Vary the Interaction	Application to Preparation Methods
Electrostatic interaction	Changing solution pH values is a common approach to tuning the electrostatic interaction. Functionalization of the support surface with specific groups is also able to tune charging status/ isoelectric point (IEP) values of the supports.	DP method CP method Ion adsorption method Colloid-based method
Chemical/ hydrogen bonds	Creating conditions for formation of chemical/ hydrogen bonds between Au species and support active sites: ion exchange and coordination reactions.	Grafting method Colloid-based method *etc.*
Alien forces	Sonication, magnetic forces, heating, *etc.*	Sonication-based method, *etc.*

Table 9.2 IEP values of some common catalyst supports.[19,23–26]

Support Type	IEP Value	Support Type	IEP Value
MgO	12.4	CaO	12.3
α-Al_2O_3	5–9.4	TiO_2 (rutile)	5.8
γ-Al_2O_3	6–9.5	CeO_2	6.7–8.6
CeO_2 (hydrous)	6.7	TiO_2	6.0–6.9
α-Fe_2O_3	8.5	Fe_3O_4	6.6
α-FeO(OH)	9.0–9.7	α-SiO_2 (quartz)	2.9
SiO_2 (SBA-15)	4.2	Stöber SiO_2	2.1
WO_3	0.2–0.3	V_2O_5	1–2
ZrO_2	4–11	SWNT	3.5
Zeolite-β	3.6	Montmorillonite	3.4
Zeolite Y	12.2	Activated carbon	5–11

Note: the IEP values of materials not only depend on their composition, but also on their specific products or brand name, because various brands may have had different treatments thus possessing different surface groups. This is why some materials have a wide range of IEP values. More IEP values can be found in ref. 23.

species and the support, and can be adjusted by varying the solution pH value. The early pioneering work on solid–liquid interfacial electrical double layer (EDL) as well as triple layer model (TML), which was investigated by Gouy,[20] Chapman,[21] Stern[22] and some others, is still useful to explain many experimental phenomena.[19] The metal oxide surfaces are often covered with OH groups, which can be either protonated into OH_2^+, deprotonated into O^-, or in the neutral state of OH in an aqueous solution depending on the solution pH value. An important concept is the iso-electric point (IEP) or the point of zero charge (PZC). When the solution pH is at, below or above this point, the metal oxide or support surface becomes neutral, positively or negatively charged, respectively. Therefore, by knowing the IEP value of the support, solution pH and the structure of Au species, it is easy to determine whether the Au species and the support surface are electrostatically attractive or repulsive, or have no electrostatic interaction. Table 9.2 lists the IEP values of some common metal oxides and catalyst supports.[19,23–25]

The IEP values can be adjusted by modifying the support surface with certain functional groups. For example, McPhail *et al.* modified single-walled carbon nanotubes (SWNTs) with carboxylic acid, maleic anhydride and nitroso groups to change the IEP value of high pressure CO "HiPco" SWNTs (single walled carbon nanotubes that are produced from gas-phase decomposition of CO at high pressure) from 3.5 to 1.2, 2.0, and 7.5 respectively.[26] Therefore, any changes in surface structure may vary the IEP value of a material, providing ample room for exploring the 'adsorption and deposition' chemistry of active particles on catalyst supports.

$HAuCl_4$ is widely used as the Au precursor in the preparation of supported Au catalysts because it is commercially available and relatively of low price. Figure 9.1 shows the hydrolysis of Au species in $HAuCl_4$ solution at different pH.[27] Similar results can also be found in a study by Pitchon and co-workers.[28]

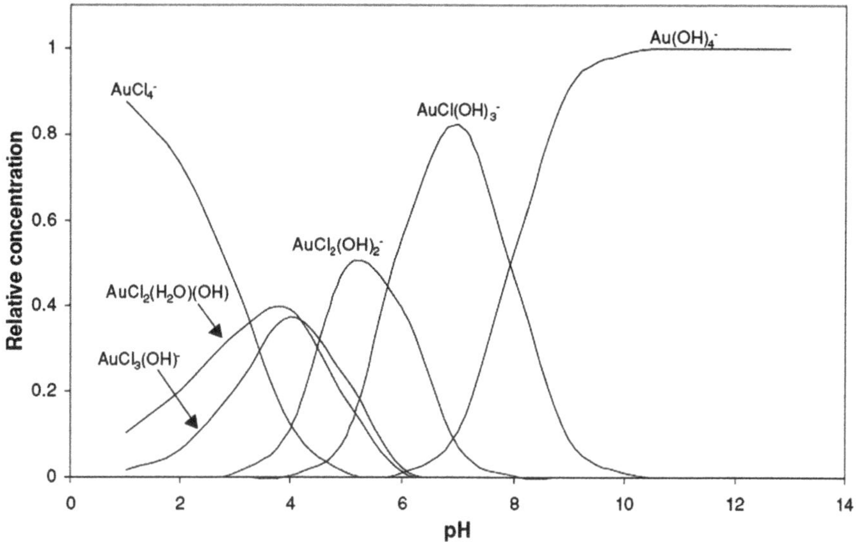

Figure 9.1 Calculated equilibrium concentration of gold complexes in 2.5×10^{-3} M $HAuCl_4$ solution.
Reproduced with permission from ref. 27.

9.2.2 Impregnation Method

The impregnation (IMP) method is widely used in catalyst preparation because it is very simple and low cost, and often leads to the formation of robust catalysts.[29] A catalyst support can be simply added into an aqueous solution of $HAuCl_4$ or $AuCl_3$ followed by drying of the suspension. Afterwards, the dried catalyst is calcined at a high temperature to obtain metallic Au particles. The liquid phase volume can be adjusted to be equal to or greater than the support's pore volume. The former is called incipient wetness impregnation, while the latter is conventional impregnation method. However, this method often provides poor control on particle size as a wide Au particle size distribution ranging from 2 to 30 nm was observed in our experiments. In addition, the obtained catalysts contain a lot of residual chloride, which is a poison to the supported Au catalysts, and causes aggregation of Au particles. Moreover, residual chloride suppresses the reducibility of Au, causing a significant amount of oxidized Au after calcination.[30,31]

A number of efforts have been made to improve IMP-derived Au catalysts. According to Figure 9.1, the Au species will have less Cl^- ions at higher pH values. Therefore, increasing the pH value of the solution makes it possible to reduce the residual chloride content in the prepared Au catalysts. This procedure is often used in the deposition–precipitation[31] and ion exchange methods,[28] which are discussed below. In preparing Au/γ-Al_2O_3 catalysts, Xu *et al.*[32] found that base treatment of the adsorbed Au species on the support is able to lower the chloride content. Tsai *et al.*[33] synthesized Au/TiO_2, Au/γ-Al_2O_3 and Au/C catalysts by the IMP method, and found that reducing the catalysts in H_2 gas at high temperature improved the catalytic activity of Au/TiO_2 and Au/C. In addition, a base (NH_4OH) treatment was able to enhance the activity of Au/γ-Al_2O_3 and Au/TiO_2. Recently Hutchings' group modified the IMP method in their preparation of Au–Pd/TiO_2 catalysts.[34] They used an excess of hydrogen chloride to prepare $PdCl_2$ solution, as the Pd precursor tends to form insoluble salts in water. Reduction in H_2 was carried out at 400 °C after drying of the samples to remove chloride ions.

The improved IMP method allows the reproducibility of catalysts of high activity in the direct synthesis of hydrogen peroxide from hydrogen and oxygen ($H_2 + O_2 \rightarrow H_2O_2$) and in the solvent-free aerobic oxidation of benzyl alcohol. The merits of these modified impregnation catalysts compared with those made by conventional impregnation include a smaller average particle size, an optimized random alloy composition, and an improved compositional uniformity from particle to particle. The above improvements are explicable because the reduction in H_2 releases HCl gas, thus removing chloride, while the high pH impregnation and base treatment convert the gold chloride into gold hydroxide.

Alternatively, chloride-free precursors such as the $Au(CH_3)_2(C_5H_7O_2)$ complex can be used to avoid the problem,[35] but they are usually much more expensive than $HAuCl_4$. Because of the above obvious drawbacks, the IMP method has not been sufficiently investigated in supported Au catalyst preparation. There is probably still room for further improvement for this method.

9.2.3 Co-precipitation and Deposition–Precipitation Methods

9.2.3.1 Supports with IEP value above 5

The co-precipitation (CP) and deposition–precipitation (DP) methods are commonly used for the preparation of supported Au catalysts. In particular, the latter method usually yields Au nanoparticles of 2–3 nm. In the CP method, $HAuCl_4$ and a metal nitrate as the precursor for the support are hydrolyzed and precipitated in alkali, *e.g.* Na_2CO_3 and NaOH,[2,36,37] whereas in the DP method, the metal nitrate is replaced with a metal oxide, a metal hydroxide or a metal salt. Both methods are conducted at high pH values (above 7) using NaOH or Na_2CO_3 as the precipitating agent and at around 70 °C or room temperature. They are usually applicable to metal oxides with high IEP values above 5.

In both methods, the primary factor is the solution's pH value, which should be close to or a little higher than the IEP value of the support (usually above 7). Kung and co-workers investigated the adsorption of $HAuCl_4$ on Al_2O_3 at various pH values using extended X-ray absorption fine structure (EXAFS) spectroscopy.[38] It was found that the Au–Cl coordination number was 4 at pH <4, while at pH >4, the $AuCl_4^-$ complex started to hydrolyze and the Au–Cl coordination number decreased with pH value. At *ca.* pH 7, the Au–Cl coordination number became almost zero. As determined by inductively coupled plasma (ICP), the maximum amount of adsorbed Au species on Al_2O_3 is obtained between pH 5 and 6 (near the IEP value). The samples prepared at high pH values (above 7) had much lower chloride contents than those prepared at low pH values. Similar Au uptake was observed on TiO_2 by Moreau and Bond.[39]

However, it can be a complex process to achieve an optimum Au loading as it involves various hydrolyzed intermediates and different interactions between the Au species and the support surface, as well as among the Au species. These interactions include electrostatic interaction, hydrolysis and precipitation reactions, ligand exchange, and polymerization of hydrolyzed Au species on the support surface. The electrostatic repulsion between the negatively charged OH-containing Au species and the metal hydroxide (in CP method) or metal oxide surface (in DP method) will become very weak when the pH value is at the IEP value or a little higher. By using TiO_2 as a support, the OH-containing Au species can react with the surface OH groups as follows:[18]

$$\mathrm{TiOH} + \mathrm{AuCl(OH)_3}^- \rightarrow [(\mathrm{TiO})\mathrm{Au(OH)_2Cl}]^- + \mathrm{H_2O} \qquad (9.1)$$

The main difference between the two methods is that, with the CP method, most of the Au particles are buried in the metal hydroxide matrix, thus only a portion of Au particles are exposed to reactant molecules. On the other hand, with DP method, most of Au particles are located on the surface of the support. However, both methods have low Au loading efficiency, usually below 30%, which means the majority of Au is washed off during the catalyst preparation.[40]

Louis and co-workers modified the conventional DP method by using urea as the precipitating agent.[41,42] The reaction was conducted at around 80 °C for 4 h. The Au particles obtained are small, with an average size of *ca.* 2 nm. High Au loading up to 8 wt% can be reached with an Au loading efficiency of almost 100%. In the early stage of the reaction, the adsorption of Au anions takes place at low pH (pH 3). Upon heating, urea slowly decomposes and the solution pH gradually increases to 7. With the increase in the pH value, the adsorbed Au species start to hydrolyze and polymerize on the support surface, and are further converted to gold hydroxide and metallic gold particles. The generated NH_4OH not only promotes the above conversions, but also reduce the chloride content in the catalyst.[41]

9.2.3.2 *Supports with IEP below 5*

Some catalyst supports with low IEP values, *e.g.* silica (IEP ~ 2), are negatively charged at the normal pH value used for DP method (pH above 7). Thus the conventional DP method is not applicable to these supports because of the strong electrostatic repulsion between the Au anions and the negatively charged support surface. Dai's[43] group developed a unique DP method to deposit very small and uniform Au nanoparticles onto silica supports using the cationic Au ethylenediamine complex $[Au(en)_2]Cl_3$, where (en) = ethylenediamine, as the precursor. Initially, the $[Au(en)_2]^{3+}$ cations are adsorbed on the support surface *via* electrostatic attraction, followed by a series of ion exchange reactions:[43]

$$[Au(en)_2]^{3+} + 2Cl^- + M^{+-}OSi{\equiv} \rightarrow M^+(aq) + [Au(en)_2Cl_2]^{+-}OSi{\equiv} \quad (9.2)$$

$$[Au(en)(d\text{-}en)]^{2+} + Cl^- + M^{+-}OSi{\equiv} \rightarrow M^+(aq) + [Au(en)(d\text{-}en)Cl]^{+-}OSi{\equiv} \quad (9.3)$$

Note: M^+ is a charged single cation, and [d-en] is deprotonated ethylenediamine ligand ($NH_2CH_2CH_2NH^-$).

The ion exchange reactions are necessary to maintain a high dispersion of Au nanoparticles on the silica support. The as-formed small Au particles are thus anchored on the surface and have low mobility for further aggregation.

Cumaranatunge and Delgass treated low IEP and surface hydrophobic titanium silicalite-1 (TS-1) support with NH_4NO_3 before deposition of Au by the DP method.[44] Such a treatment increased the Au loading fourfold compared with the untreated sample. A similar phenomenon was observed in Au deposition on the same support after addition of cobalt by the CP method[45] and alkaline earth metal salts by the DP method.[46] The variation of charging status or IEP values *via* this kind of surface modification is effective in overcoming the weak or even repulsive interaction between the Au species and the supports. However, a systematic investigation in this area is still lacking.

9.2.4 Anion and Cation Adsorption Methods

The above CP and DP methods work at high pH values (>7) and are not applicable for low-pH preparations. For example, TiO_2 has an IEP of around 6. Below pH 6, the TiO_2 surface is positively charged (protonated Ti-OH_2^+). On the other hand, $HAuCl_4$ exists in the form of anion complexes in aqueous solution, which provides the possibility for electrostatic attraction between the Au anions and the positively charged titania surface. Louis and co-workers[41] prepared an Au/TiO_2 catalyst at pH around 2 and observed fast adsorption of Au anions. However, at low pH values, $HAuCl_4$ still does not hydrolyze completely and the Au anionic species has a high content of Cl^-. As a result, the amount of chloride is much higher in these samples than those prepared by the DP-NaOH method, which is detrimental to the catalyst. This method yields a maximum Au loading of about 1.5 wt%. On the contrary, when the solution pH is above IEP, the TiO_2 surface becomes negatively charged. The authors[41] used $[Au(en)_2]Cl_3$ to adsorb on TiO_2 and obtained a slightly higher Au loading (2%) by the cation adsorption method. Kiwi-Minsker and co-workers applied the same method to prepare Au/ACF (activated carbon fibers) and Au/FeO_x/ACF catalysts.[47]

It should be noted that all these preparation procedures involve several interactions and intermediates. For example, in the anion adsorption method, besides the electrostatic attraction, the subsequent surface reactions also occur:[41]

$$TiOH + AuCl_3(OH)^- \leftrightarrow TiOAuCl_2 + H_2O + Cl^- \quad (9.4)$$

In the initial access of Au anions to the support surface, the electrostatic interaction plays the major role; once the Au anions are close enough to the support surface, surface ion exchange or coordination reactions take place and become the dominant interaction. Otherwise, it is difficult to explain why the DP method often leads to large Au particles on supports with low IEP values such as silica and carbon.

Finally, it should be pointed out that the supported Au catalysts derived from the anion and cation adsorption methods are not active before calcination, which is quite different from the CP and DP derived catalysts using $HAuCl_4$ as the precursor.

9.2.5 Grafting Method: Generating Chemical Bonds to Immobilize Gold Species on Catalyst Supports

In the case of strong electrostatic repulsion between the Au species and the catalyst support, and/or weak interaction during the ion exchange or coordinating reaction of the Au species on the support surface, it is very difficult to obtain highly dispersed Au nanoparticles on catalyst supports. The typical examples are silica and carbon, both of which have low IEP values (around 2) and limited surface OH groups.

One strategy is to create chemical bonding between the Au species and the catalyst support through surface functionalization. Some ligands have two terminal functional groups that can bind to silica/carbon and Au species, *e.g.* 3-aminopropyltriethoxyilane (APTES) for modification of silica surface,[48] and 3-chloropropylamine-hydrochloride (CPAHCL) for the carbon surface.[49] The grafting method is also applicable to clays[50] and molecular sieves. Okumura and co-workers[51,52] deposited Au nanoparticles on acidic supports, such as aluminosilicates, activated carbon and silica by a gas phase grafting method. Mou and co-workers[53] confined small Au particles in MCM-41 by employing propylamine as a surface modifier. However, after calcination, some Au particles were still agglomerated due to the weak confinement exerted by the thin wall of MCM-41. Later they used APTES as the surface modifier and deposited Au on SBA-15,[54] which has a thicker wall compared to MCM-41, and obtained a very uniform size distribution of Au nanoparticles within the channels of Al-SBA-15. The Au nanoparticle size distribution in the as-prepared catalyst was 1.5–2 nm, and after reduction in H_2 at 600 °C, the average size became *ca.* 2.7 nm.

Although the grafting method usually leads to the formation of relatively highly dispersed supported Au catalysts, it is necessary to note that a thin layer of silica, originated from silyl molecules in the modifier, is introduced at the interface between the Au particles and the catalyst support. This often results in decrease of the catalytic activity.[55] Nevertheless, high catalytic activities for Au-supported on silica catalysts were reported by Budroni and Corma,[56] and Dai and co-workers.[43]

9.2.6 Reduction–Deposition Methods

In all the preceding methods, Au ions are adsorbed and anchored onto the catalyst support surfaces, and further converted into Au nanoparticles after a series of surface reactions and/or heat treatment. An alternative approach is to form Au nanoclusters/nanoparticles *via* reduction in the aqueous solution to form Au colloids prior to deposition on the support surface. The *in situ* reduction of Au can be done independently or in the presence of supports. The former is also called the sol immobilization method.

In these procedures, reducing agents like $NaBH_4$, and NH_2NH_2 (hydrazine), or photoreduction are often employed, and capping agents such as thiols, amine, organic acids and polymers, *e.g.* polyvinylpyrrolidone (PVP) and polyvinylalcohol (PVA), are used to prevent the aggregation of the Au colloids. A general principle for selection of capping agents is that they should be able to stabilize the Au nanoparticles to give the desired small particle size, as well as being easily removed at relatively low temperatures without any obvious effect on the Au particle size. For example, thiols are very useful in preparing small Au particles, but they are difficult to remove, and thus not often employed in the preparation of Au catalysts. The Au nanoparticles can be either directly deposited onto the catalyst support once they are formed, or post-deposited in a separate step. The latter is often called as colloid-based method.

Grunwaldt *et al.*[57,58] first prepared Au colloids by employing tetrakis(hydroxymethyl)-phosphonium chloride (THPC) as the capping agent and subsequently adsorbed them on TiO_2 and ZrO_2 supports. Later Porta *et al.*[59] and Mirescu and Prüße[60] prepared Au/C and Au/TiO_2 catalysts with PVA and poly(diallyldimethylammonium bromide) (PDDA) as the protective agents in a one-pot reaction. Haruta and co-workers successfully obtained a supported Au catalyst on polymethyl methacrylate (PMMA) *via* reduction of $Au(en)_2Cl_3$ by $NaBH_4$, with an average Au particle size of around 6.9 nm.[61] Prati's group prepared an Au catalyst supported on a commercial anionic resin (Dowex M-43) *via* reduction of $HAuCl_4$ by $NaBH_4$ in the presence of THPC.[62] Li's group prepared Au colloids with average sizes of 2, 5, 7, 13, and 16 nm and $LaVO_4$ nanocrystals with a well-defined shape (square plates with a size of 40 nm×40 nm×10 nm) separately in the presence of oleic acid, then immobilized the Au colloids on the $LaVO_4$ nanocrystals.[63] The Au/$LaVO_4$ catalyst exhibited high activity in catalytic CO oxidation at low temperatures. This method provides a new way to fabricate supported Au catalysts with well-defined local structures.

Photoreduction can replace the conventional reducing agents to produce Au colloids. Chan and Barteau[64] obtained 1–2 nm Au and Ag nanoparticles on commercial TiO_2 after irradiation of $AgNO_3$ and $HAuCl_4$ solution under light at 380–450 nm. Yuan and co-workers[65] made use of the photoreduction of aromatic ketone to generate ketyl radicals that readily reduced Au and Ag ions to colloids.

One of the issues requiring attention is the deposition efficiency of Au colloids on the support surface. Such data are missing in most of the literature reports. However, according to some published data, the efficiency is not high (lower than 50%), which means that the majority of the Au colloid is washed away during the catalyst preparation.

Sonication is an effective tool to facilitate the deposition of metal colloids onto support surfaces. Gedanken's group[66,67] carried out systematic work on the sonochemical synthesis of supported catalysts. During the sonication process, together with the formation, growth and implosive collapse of bubbles in a liquid medium,[68] extremely high transient temperature (≥5000 K), pressure (≥20 MPa) and very high cooling rates ($\geq 10^{10}$ K s^{-1}) can be reached,[69] which can decompose some of the volatile compounds used in the sonochemical synthesis. Meanwhile, the collapse of the bubbles can generate numerous micro-jets in the liquid medium near the solid surface, which can reach velocities of hundreds of meters per second.[70] These micro-jets are able to propel the small particles to the solid surface, thus depositing or inserting the small particles into the support pores.

In the sonication-assisted synthesis reported by Gedanken *et al.*,[66,67] a long duration (above 1 h) is usually applied. However, further experimental results showed that an extended sonication can cause the aggregation of Au nanoparticles in solution.[71] Zhong and co-workers[72–75] conducted a series of experiments on the aggregation behaviors of Au colloids, and observed that L-lysine, an amino acid possessing two terminal amine (NH_2) groups, could act as a capping agent to the as-formed Au colloids after reduction by $NaBH_4$. At pH values below 6, the Au colloids were severely aggregated; while at pH above 6, linear Au colloid assembly was obtained. Under sonication and at pH > 6, isolated Au colloids were observed despite a short lifetime. Further experiments showed that, when the catalyst support and L-lysine were premixed with an aqueous solution of $HAuCl_4$ followed with sonication and subsequent addition of the reducing agent ($NaBH_4$), the formed isolated Au colloids could be quickly deposited on the support with high efficiency during the sonication. This method is able to deposit noble metals such as Au and Pt onto a number of catalyst supports with IEP above 5. The obtained noble metal particle size is between 2 and 5 nm, with a short sonication time of 20–60 seconds, much shorter than those in other Au catalyst preparation methods. In addition, the final pH value should be above 6 to prevent the aggregation of Au colloids in liquid phase.

In short, in wet chemical approaches, we should consider not only various interactions between the Au species and the support surface, and among the Au colloids, but also the subsequent reactions of Au species on the support surface. In order to obtain a highly dispersed Au catalyst, a delicate balance among these various interactions should be reached.

9.3 Nature of the Active Sites in Supported Gold Catalysts and the Activation of Molecular Oxygen

For various oxidation reactions over supported Au catalysts, their active sites and reaction pathways may be different. However, there should be some common features for these reactions, including activation of molecular oxygen (O_2), and the effect of the Au oxidation state and particle size on the catalytic property. Because CO oxidation is often conducted under mild conditions (*e.g.* at low temperatures and atmospheric pressure), and has simple reactant and product molecules, it often serves as a model reaction to other relevant oxidation reactions, and is the one most investigated. Much more information regarding CO oxidation is available in the literature than on other oxidation reactions. In this section, we mainly present information about the catalyst active sites, reaction mechanism, and molecular oxygen activation obtained in CO oxidation. In the subsequent sections, we discuss some other selective oxidation reactions in turn, and present their specific reaction mechanism (if any). It should be emphasized that, even though the CO oxidation is the simplest oxidation reaction, it is not yet fully understood, and various results and views can be found in the literature.

9.3.1 Effect of Gold Particle Size and Oxidation States

In general, for a catalytically active Au catalyst, the Au particle size should be less than 5 nm. Haruta and co-workers[37] prepared Au catalysts supported on TiO_2, α-Fe_2O_3 and Co_3O_4, and found that when the Au particle size was less than 4 nm, the turnover frequency (TOF) for CO oxidation per surface Au atom increased sharply. Later, they further confirmed that in Au/TiO_2, the optimized Au particle size should be *ca.* 3 nm.[76] Goodman and co-workers[77–80] performed a series of scanning tunneling microscopy (STM), scanning tunneling spectroscopy (STS) and CO oxidation kinetics studies on Au vapor-deposited onto a TiO_2(110) single crystal and a TiO_2(001) thin film. The catalytic activity was found to be a function of the Au cluster size and the maximum catalytic activity for these clusters was found to coincide with the metal-to-nonmetal transition occurring in Au clusters of approximately 3.0 nm in diameter. This cluster diameter also coincided with a cluster growth transition from the nucleation of flat, two-dimensional clusters to their agglomeration into hemispherical, three-dimensional structures, as measured by STM. Over dried Au/FeOOH catalysts, Hutchings and co-workers[81] reported that the high catalytic activity for CO oxidation was correlated with the presence of bilayer Au clusters $\sim$0.5 nm in diameter, which contained only $\sim$10 Au atoms. Later we explain that, in the epoxidation of propylene to propylene oxide on supported catalysts, some sub-nanometer Au particles are believed to be the active Au species.

Regarding the oxidation state of active Au species, there are different views in the literature. Louis and co-workers[82] prepared Au/TiO_2 catalysts by DP methods (DP-NaOH and DP-urea) and found that, after preparation, the Au species were in the oxidic state III. They began to transform into metallic gold under air at ~100 °C for DP-NaOH and ~150 °C for DP-urea. At 200 °C, metallic Au species were formed for both catalysts. The particle size increased from 1.5 to 3.5 nm after calcination, and the catalytic activity for CO oxidation increased with the increasing fraction of metallic gold, which reached its maximum after calcination at 200 °C for both types of the samples. Chen and Goodman[83] observed the formation of non-metallic Au species (negatively charged Au) in Au/TiO_2 catalyst. Toshima and coworkers[84] prepared catalytically active Au–Pd catalysts that contained abundant top (vertex or corner) Au atoms. The genesis of the catalysts with high activity for aerobic glucose oxidation was ascribed to the presence of negatively charged top Au atoms. Lin *et al.*[85] prepared negatively charged Au particles on MgO thin films by two different approaches and characterized them with respect to their CO binding behavior. On the ultra-thin films, the ad-atoms were charged up due to electron transfer from the support, and the charging on thicker films was realized by electron donation from defects. In both cases, the excess electrons gave rise to unusual CO adsorption properties. On the planar Au islands on thin MgO films, CO was exclusively bound to the low coordinated Au atoms along the island perimeter. The adsorption was accompanied by a redistribution of the cluster electrons away from the boundary toward the interior. On the thicker films, the Au charging gave rise to an increasing red shift of the CO stretch mode with decreasing particle size, reflecting the gradual filling of the antibonding CO $2\pi^*$ orbital.

Catalysts containing cationic Au species were prepared for CO oxidation by Guzman and Gates,[86] and several other groups, using an Au complex as the Au precursor. The spectroscopic evidence showed that both cationic and zerovalent Au clusters were present in working CO oxidation catalysts.[86,87] The relative amounts of Au(I) and Au(0) depended on the composition of the reacting atmosphere and the catalytic sites that incorporated Au(I). During the reaction, CO played a dual role as a reactant and a reducing agent that converted Au(I) into Au(0), thereby diminishing the catalytic activity.[88] It was also reported by Gates' group that dimethyl gold acetylacetonate, $Au(CH_3)_2(acac)$, bonded to a MgO powder support lacked measurable catalytic activity for CO oxidation at room temperature.[89] However, as the temperature was increased to greater than 100 °C, while keeping CO and O_2 flowing at atmospheric pressure, removal of methyl ligands from the Au complex was observed by infrared (IR) and EXAFS spectroscopies, and accordingly, the sample became active for CO oxidation catalysis. EXAFS characterization of the sample immediately after the activation indicated that the Au species had aggregated into clusters consisting, on average, of approximately 4–6 Au atoms each, which were inferred to be the catalytically active species. The X-ray absorption near-edge structure (XANES) data indicated that Au in the activated catalyst was reduced to the metallic state. Obviously, the catalytic functions of cationic Au are not only evidenced in the homogeneous Au complexes in homogeneous catalytic reactions, but also in their supported forms (supported Au complexes without and with calcination). However, it would be dangerous to ascribe all the catalytic properties of gold to cationic species, as there is also evidence linking the catalytic activity to small Au clusters.[90] Ding and co-workers[91,92] reported that nanoporous gold also exhibited high catalytic activity toward CO oxidation in the absence of any catalyst supports. The detailed mechanism is still not clear and should be further studied.

It should be noted that spectroscopic techniques such as EXAFS, IR, XANES, X-ray photon spectroscopy (XPS), and Mössbauer spectroscopy have their limitations as they provide only average structural information. Thus, when mixtures of Au(I) and Au(III) are present in a catalyst, spectroscopic results cannot easily resolve whether they are present in the same species or in different ones. Again, the structural complexity of typical supported gold catalysts complicates the interpretation of experiments and motivates research with structurally well-defined supported catalysts.[90]

9.3.2 Location of the Catalyst Active Sites

According to many literature results, the active sites for CO oxidation should be located mainly at the interfaces between the active Au particles and the catalyst support particles, although it is still difficult to exclude the possibility that Au particles play their roles independently.[91,92] Haruta and co-workers first pointed out the importance of the Au/oxide perimeter interface in CO oxidation[93] and proposed that oxygen molecules were activated at the perimeter interface, assuming the generation of oxygen vacancies.[94] Widmann and Behm reported that the active oxygen species for CO oxidation was highly stable atomic oxygen species or surface lattice oxygen located at the Au/TiO_2 perimeter interface, which was easily removed in a CO atmosphere.[95] Fujitani and Nakamura believed that oxygen molecules were activated at the perimeter interface in the low-temperature range (<47 °C), while it took place mainly on the low-coordinated Au surface at the higher temperature range.[96] Zhong *et al.*[97] conducted *in situ* measurement of thermogravimetry (TG) coupled with on-line Fourier transform infrared spectroscopy (TG-FTIR) on Au/α-Fe_2O_3 catalyst and α-Fe_2O_3 in a CO–He gas mixture (Figure 9.2). According to thermogravimetric analysis, there were three distinct stages of CO interaction with 3 wt% Au/α-Fe_2O_3 catalyst but only two stages for the α-Fe_2O_3 support. At low temperatures (<100 °C), only the Au catalyst had a rapid weight loss. This observation confirmed that CO could react with highly active surface lattice oxygen or adsorbed oxygen species (associated with O-vacancy), which were associated with and promoted by the Au nanoparticles. Around 300 °C, both the catalyst and the support experienced the reduction of α-Fe_2O_3 to Fe_3O_4, while above 400 °C, further reduction to FeO and Fe metal took place. Au played no role in the kinetics of Fe_3O_4 formation because lattice O mobility was the rate-limiting step. At higher temperature where Fe_3O_4 was further reduced to FeO and Fe^0, the initially formed metallic Fe^0 nuclei could decompose CO molecules and release O species, which together with the lattice oxygen could react with CO

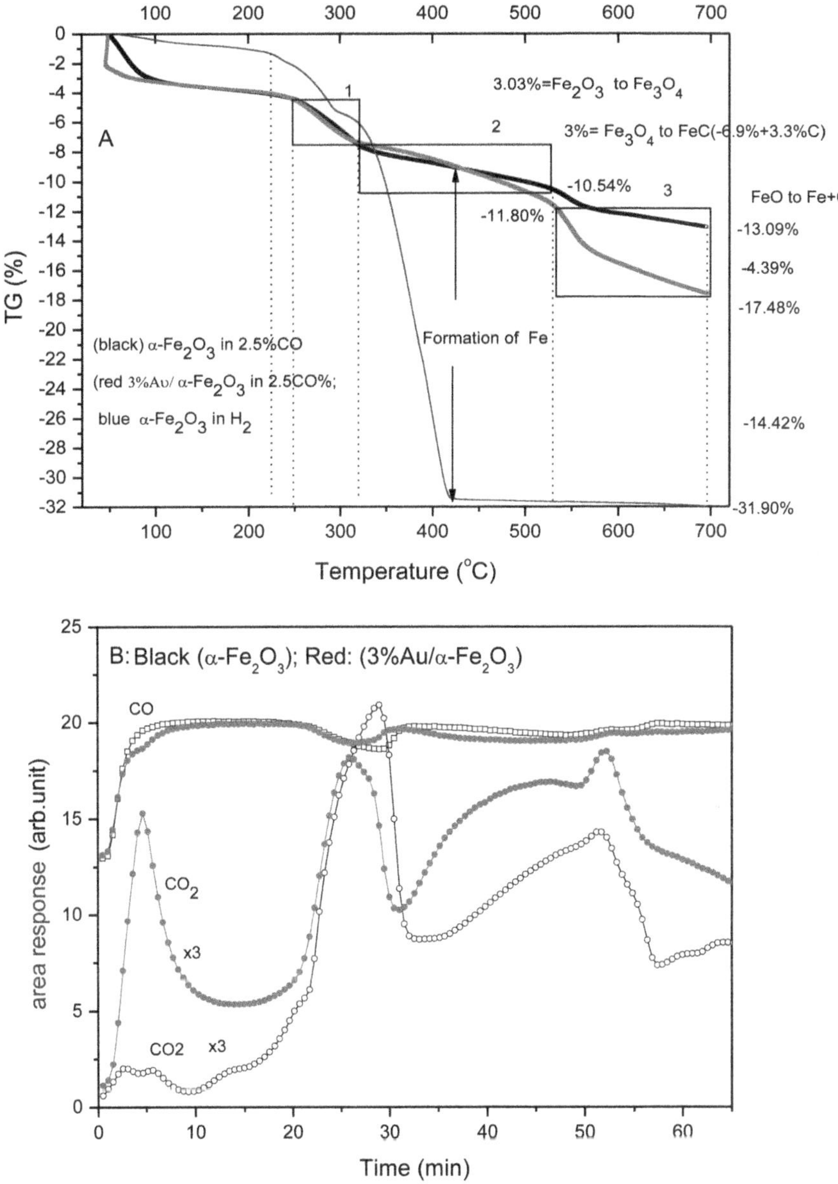

Figure 9.2 Recorded (a) TG and (b) FTIR profiles of α-Fe_2O_3 (black line) and 3% Au/α-Fe_2O_3 (red line) in 2.5% CO/He. The TG reduction profile of α-Fe_2O_3 (blue line) in pure H_2 shown for comparison in (a). Samples (20 mg) were exposed to 2.5% CO (or pure H_2) at 30 mL min^{-1}, and then the temperature was ramped to 700 °C at a rate of 10 °C min^{-1}. Reproduced with permission from ref. 97.

molecules. These findings confirm that the combination of Au and α-Fe_2O_3 is necessary to activate molecular oxygen at low temperatures (<100 °C), but that it does not play a role in the activation of the bulk lattice oxygen. Probably these active sites are located at the interface between Au and α-Fe_2O_3, and the O-vacancies are also involved in the synergistic action with Au.

Similar conclusions are deduced in the case of monodispersed and thiolate-protected $Au_{25}(SR)_{18}$ (where R = CH_2CH_2Ph) nanoclusters deposited on TiO_2, CeO_2, and Fe_2O_3, respectively, for CO oxidation in a fixed-bed reactor.[98] Among the three catalysts, $Au_{25}(SR)_{18}/CeO_2$ was the most active. O_2 pretreatment of $Au_{25}(SR)_{18}/CeO_2$ at 150 °C for 1.5 h significantly enhanced the catalytic activity, which was still well below the thiolate desorption temperature (~200 °C), indicating that the CO oxidation reaction was catalyzed by the intact $Au_{25}(SR)_{18}/CeO_2$. However, further increasing the O_2 pretreatment temperature to 250 °C (above the thiolate desorption temperature) did not lead to any further increase in activity, subverting the common concept that surface thiolates should be removed when activating the Au catalyst. The authors concluded that this relatively low pretreatment temperature could generate perimeter sites between $Au_{25}(SR)_{18}$ and the CeO_2 support, thus activating the catalyst. Maeda *et al.*[99] conducted *in situ* electrical conductance measurements on Au/TiO_2 catalysts under catalytic CO oxidation with O_2. After introduction of the reaction mixture of CO and O_2, a sharp increase in electrical conductance of the Au/TiO_2 catalyst was observed. Since TiO_2 is an n-type semiconductor, the increase indicates the generation of oxygen vacancies during CO oxidation. Also, the generation of the oxygen vacancies and the conduction of the electrons did not change in the whole temperature range investigated; this differs from assumption by Fujitani and Nakamura that activation of the oxygen molecules takes place at the perimeter interface at low temperature but on the Au surface at high temperatures.[96] The amount of easily removable oxygen in the CO atmosphere was estimated to correspond to 41% of surface lattice oxygen around the perimeter interface between Au nanoparticles and TiO_2. These results support the mechanism proposed by Haruta and co-workers[93,94] that activation of O_2 molecules takes place at the oxygen vacancies produced at the perimeter interface, which are probably generated in the whole temperature range. With respect to the CO oxidation reaction, they play a much more important role in the low temperature range but not in the higher temperature range.

Recently, Kang *et al.*[100] fabricated highly ordered, stable and well-defined Au–FeO_x binary superlattices (BNSLs) by self-assembling Au and FeO_x nanocrystals. By adjusting the relative size and concentration of the constituent nanocrystals, the superlattices with AB, AB_2, AB_6, and AB_{13} structures (here, FeO_x is A, Au is B), analogous to the atomic crystals of CuAu, AlB_2, CaB_6, and $NaZn_{13}$ respectively, were obtained. Since each nanocrystal had a precisely controlled size and morphology, this eliminated any uncertainties arising from polydispersity of the catalyst or support. In the study, the size of Au nanocrystals was 6 nm to ensure the individual Au nanocrystal would not contribute significantly to the overall activity unless in contact with FeO_x (because Au nanocrystals themselves are almost inactive if their size is larger than 5 nm). Furthermore, the structure of BNSLs (*e.g.* the stoichiometry of Au to Fe) was carefully tuned to rigorously control the number of contacts between Au and FeO_x nanocrystals. For example, in the AB structure, each Au nanocrystal makes contact with four neighboring FeO_x nanocrystals, and vice versa. In the AB_2 monolayer BNSL, the number of contacts is six per FeO_x nanocrystal and three per Au nanocrystal, while in the AB_{13} structure, there are 24 contacts per FeO_x and two per Au. The fabricated BNSLs were used for CO oxidation and their specific activities were measured. Because the AB_2 BNSL has the largest domain size, the rates obtained on AB_2 Au–FeO_x were averaged to provide a relative rate per Au–FeO_x contact and this reference was further used to predict the expected activity of AB and AB_{13} based on their respective number of Au–FeO_x contacts. Interestingly, the experimental rates of CO oxidation on AB and AB_{13} exhibit a linear correlation from the rate per contact calculated for the AB_2 structure. Given that the Au and FeO_x show negligible activity under the same reaction conditions, this linear correlation between the overall activities and the density of Au–FeO_x contacts verifies that the Au–FeO_x contacts are where the active sites are located for CO oxidation. Moreover, the authors observed that changing the size of FeO_x nanocrystals in the BNSLs changed the activity per FeO_x nanocrystal, but it did not significantly change the activity per Au nanocrystal, thus providing further confirmation of the above conclusion.

Green *et al.*[101] observed dual catalytic sites at the perimeter of Au particles (2–8 nm diameter, and with a most probable diameter of 3 nm) supported on TiO_2 during CO oxidation. In the case of TiO_2, they identified chemisorbed CO species with absorbance peaks at 2179 and 2168 cm^{-1} using FTIR, due to CO absorbed on titanium cation sites such as Ti^{4+} (denoted as CO/TiO_2). In the case of Au/TiO_2, two additional features centered at 2102 and 2341 cm^{-1} were identified, assigned to CO chemisorbed on metallic Au (CO/Au) and chemisorbed CO_2 on TiO_2 sites (CO_2/TiO_2), respectively. Interestingly, after exposure to 1.0 Torr of O_2 (g) at −153 °C, CO oxidation began immediately on the CO-saturated Au/TiO_2 surface, accompanied by the gradual disappearance of CO absorbance on TiO_2 sites and the growth of the CO_2 feature on TiO_2 sites. The main participant in the reaction was CO/TiO_2, which depleted completely. The contribution from CO/Au was minor, accounting for only about 12% of the CO/Au coverage occurring in over 120 min. Meanwhile, little CO oxidation was detected on the pure TiO_2 control sample, indicating the importance of the Au/TiO_2 interface. Infrared-kinetic measurements indicated that O–O bond scission was activated by the formation of a CO–O_2 complex at dual Ti–Au sites at the Au/TiO_2 interface. Density functional theory (DFT) calculation, which provided the activation barriers for the formation and bond scission of the CO–O_2 complex, confirmed this model. The observation of sequential delivery and reaction of CO first from TiO_2 sites and then from Au sites indicates that reaction occurs at the three-phase boundary between the Au particle, the support, and the gas phase. However, it should be pointed out that most of the literature results show that CO molecules adsorbed on Au nanoparticles participate in the reaction.

Recently, Yoshida *et al.*[102] used a newly developed aberration-corrected environmental transmission electron microscopy (TEM) technique and observed that the adsorbed CO molecules caused the (100) facets of an Au nanoparticle to reconstruct during CO oxidation at room temperature. The CO molecules adsorbed at the on-top sites of Au atoms in the reconstructed surface, and the energetic favorability of this reconstructed structure was confirmed by *ab initio* calculations and image simulations.

9.3.3 Activation of O_2

9.3.3.1 Activation of O_2 on Cationic and Metallic Au

As mentioned above, the catalytic oxidation of CO most probably occurs at the three-phase boundary region as a request of synergistic action among Au nanoparticles, catalyst support, and gas reactant molecules. The reaction involves a series of steps such as CO adsorption, O_2 adsorption and activation. Among them, activation of O_2 is the most critical step in catalytic oxidation reactions, particularly in CO oxidation.

It is well known that bulk gold is poor at activating O_2 to form the active oxygen species required for an oxidation proceeding *via* the Mars–van Krevelen mechanism.[101] However, many experimental results show that supported nanosized Au particles (below 5 nm) are highly active for CO oxidation. However, unlike CO adsorption, which is often characterized by diffuse reflectance infrared Fourier transform spectroscopy (DRIFTS), O_2 adsorption and activation are quite difficult to characterize by conventional spectroscopic techniques. For example, the IR spectroscopic technique fails to provide useful information on the active oxygen species while the XPS technique often works at conditions quite different from the practical reaction conditions. Instead, in experimental and theoretical studies, size-selected gas phase clusters are often selected and serve as well-defined and controllable models for mechanistic understanding of nanogold catalysis at the molecular level.

In the remainder of this section, we present some theoretical calculations and experimental results to elaborate the interaction of Au and O_2. Several points from the theoretical calculations can be summarized as follows:

(1) The reactivity towards O_2 is dominated by Au clusters possessing an odd number of electrons, *i.e.*, the odd-numbered clusters.
(2) For most small cluster sizes, O_2 is energetically favored to be dissociatively bound, although the barriers to dissociation are expected to

be large (in the order of 1 eV and greater) and thus may not be observed experimentally owing to kinetic trapping.
(3) In the cases where O_2 adsorption is observed, charge transfer to O_2 occurs with an increase of the O–O distance, suggesting activation toward a superoxo-like state.
(4) The extent of activation (for molecular adsorption) is predicted to be less for the neutral clusters than the anionic ones.[103–106]

Theoretical calculations also predict that cationic Au clusters are not able to adsorb or activate O_2 effectively.[107] DFT studies showed that the interaction of neutral Au clusters with O_2 is rather weak, and insufficient to activate molecular oxygen.[105,108] For example, Ding *et al.*[105] showed that there is no O_2 chemisorption in Au_nO_2 ($n = 2, 4, 6$) neutral complexes, and their O–O stretching frequencies should be close to or almost the same as that in free O_2 (1580 cm^{-1}). *Ab initio* calculations by Metiu and co-workers showed that O_2 interacts only weakly with either Au_3^- or Au_3.[109] The conclusion for the neutral Au clusters is found not to be completely consistent with the experimental results. However, experimental study on the neutral Au clusters towards O_2 is difficult because there is a lack of a direct experimental probe for the uncharged species.

Van Bokhoven *et al.*[110] probed the interaction of Au/Al_2O_3 catalyst with O_2 using *in situ* XANES technique and confirmed the charge transfer from small Au particles to oxygen when the catalyst was exposed to oxygen, along with partial depletion of the Au *d* band. It is apparent that the small Au particles on a non-reducible oxide support reacted with O_2 to form partially oxidized gold: $Au^0 + O_2 \rightarrow Au_yO_x$. After introduction of CO, the partially oxidized gold was readily reduced by CO to yield CO_2, thus closing a catalytic cycle: $Au_yO_x + CO \rightarrow Au^0 + CO_2$. After the complete reduction, the adsorbed CO was observed on the surface. Clearly, the reduction is much faster than the reoxidation, and the rate-limiting step is the activation of oxygen on the gold surface.[111] In a mixed O_2/CO gas environment, fully reduced gold is expected,[112] probably with adsorbed CO on its surface. Van Bokhoven *et al.*[110] believed that partially oxidized gold was formed upon reaction with oxygen but only momentary under the catalytic conditions.

Woodham *et al.*[113] investigated the interaction of Au clusters with O_2 using anion photoelectron spectroscopy (PES) and infrared multiple photon dissociation (IR-MPD). Mass spectrum obtained by single-photon ionization (at 7.9 eV) of a cluster complex distribution showed the addition of O_2 at -100 °C to the odd-sized (open-shell) clusters Au_7, Au_9, Au_{11}, and Au_{21}, in agreement with the model of the donation of single, unpaired electron into the highest occupied molecular orbital (HOMO) of the oxygen model.[114] Contrary to this, however, there were also mass peaks corresponding to the reaction of even-sized (closed-shell) clusters Au_4, Au_{10}, and Au_{12}, as well as a noticeable absence of reaction for the odd-sized clusters Au_5 and Au_{13}–Au_{19}. The IR-MPD spectra recorded from the mass channels of $Au_n(O_2)_m$ ($n = 4$, 7, 9, 11, and 21) between 1000 and 1700 cm^{-1} showed two distinct motifs, one at 1060 cm^{-1} observed only for the odd-sized clusters, and the other at *ca.* 1500 cm^{-1} observed only for $Au_7(O_2)$ and $Au_4(O_2)$, which are attributed to the O–O stretch vibrations of a superoxo (O_2^-) and a slightly activated O_2 species [*cf.* ν(O–O) = 1556 cm^{-1} for free O_2], respectively. All the bands correspond to around 70% depletion, with the exception of $Au_7(O_2)$, which splits roughly 60 : 40 between the two modes. The observation of a superoxo stretch for all the odd-sized clusters is in agreement with theoretical predictions. It is also interesting to note that the observed stretching frequencies are similar to those of the anions, suggesting a similar degree of activation by the neutral clusters, contrary to the theoretical predictions. The DFT calculations suggest that the charge transfer induces significant rearrangement of the Au cluster such that it more closely resembles the structure of the cation. Such a binding mechanism highlights the importance of structural flexibility for oxygen activation by small Au particles.[113]

A comparative study on the adsorption of the O_2 molecule on pure Au_{n+1}^+ and doped MAu_n^+ cationic Au clusters, where $n = 3$–7 and M = Ti, Fe, was carried out by Torres *et al.* by means of first principles calculations based on norm-conserving pseudo-potentials and numerical basis sets.[115] The adsorption energy on cationic Au clusters is similar to that on neutral Au clusters but smaller than that on anionic Au clusters. For pure Au_{n+1}^+ clusters, the O_2 molecules adsorb preferably on top of the low coordinated Au atoms, except for Au_5^+ and Au_8^+ clusters, which prefer bridge adsorption sites. The geometry of Au_n^+ is preserved after O_2 adsorption. The adsorption energy on top of Au_n^+ is smaller than 0.5 eV, the limit for experimental observation. Instead, the adsorption energy in bridge positions of Au_n^+ is larger than 0.5 eV and becomes higher for $n = 8$ than for $n = 5$. Interestingly, the doped MAu_n^+ clusters are good candidates for co-adsorption of CO and O_2 and subsequent CO oxidation because CO prefers to bind the less coordinated apex atoms of Au cluster cations, whereas O_2 prefers to bind the impurity metal site, which is always highly coordinated in the doped MAu_n^+ cluster.

A recent DFT calculation[116] showed that, for O_2 dissociation on an Au(111) surface substituted with transition metal atoms (Cu, Ni, Pd, Pt, Ir, Rh), the doping of Au surfaces with any of the six transition metal atoms has an important effect in stabilizing the oxygen molecule adsorption, and reduces the calculated activation energy barrier for dissociation of the molecular oxygen significantly. Interestingly, doping with nickel atoms coincides with large rate constants for the reaction of O_2 dissociation at moderate temperatures ($T = 25$ or 75 °C). Probably doping Au with a transition metal is a practical way to improve the catalytic performance of supported Au catalysts for oxidation reactions.

9.3.3.2 Activation of O_2 on Anionic Au

Early studies showed that the reaction of Au_n^- with O_2 *via* molecular O_2 addition is the primary reaction pathway, but only for even-sized clusters, whereas the odd-sized clusters are inert toward O_2.[114] The even–odd alternation correlates well with a similar trend in the electron affinities of Au_n,[117] suggesting that electron transfer from Au_n^- to O_2 should be the primary reaction mechanism.[118] PES results showed that the even-sized $Au_nO_2^-$ clusters are indeed molecularly chemisorbed complexes through the observation of O–O vibrational structures, with estimated vibrational spacings ranging from ~ 180 meV (1450 cm^{-1}) for $n = 2$ and 6 to 152 meV (1230 cm^{-1}) for $n = 4$.[119,120] There have been numerous theoretical investigations on the reactivity of Au clusters with O_2.

Bernhardt and co-workers[121] performed experiments in which mass-selected Au_2^- ions were exposed to a $CO/O_2/He$ gas mixture inside a radio frequency (RF) octopole ion trap at a defined reaction temperature. Temperature-dependent RF ion trap mass spectrometry and first-principle simulations revealed the detailed reaction mechanism of the catalytic gas-phase oxidation of CO by free Au_2^- ions in the presence of O_2. A metastable intermediate with a mass of $Au_2CO_3^-$ was observed at low temperatures, which may have two alternative structures: digold carbonate or peroxyformate. Both structures are characterized by low activation barriers for the formation of CO_2.

Wang and co-workers systematically investigated the interactions between O_2 and small Au cluster anions Au_n^- ($n = 1$–7) using PES.[122] Higher resolution photoelectron spectra showed the formation of molecularly chemisorbed even-sized $Au_nO_2^-$ ($n = 2$, 4, 6) complexes; well-resolved vibrational structures due to O–O stretching were observed and could be readily distinguished from the Au-derived PES bands. Physisorbed $Au_n^-(O_2)$ complexes for $n = 1$, 3, 5, 7 were observed for the first time, providing direct evidence for the inertness of the closed-shell odd-sized Au_n^- clusters toward O_2. Neutral even-sized Au_n clusters are closed-shell and are expected to be inert toward O_2; this is not consistent with the reduced O–O vibrational frequencies observed in the photoelectron spectra relative to free O_2. Probably the photo-detachment transitions can only access excited states of the neutral even-sized Au_nO_2 complexes.

Later, the same authors[123] reported the spectroscopic evidence of two modes of O_2 activation by the small even-sized Au_n^- clusters: superoxo and peroxo chemisorption (Figure 9.3). Photoelectron spectroscopy of $O_2Au_8^-$ revealed two distinct isomers, which can be converted from one to the other depending on the reaction time. *Ab initio* calculations showed that there are two close-lying molecular O_2^- chemisorbed isomers for $O_2Au_8^-$; the lower energy isomer involves a peroxo-type binding of O_2 onto Au_8^-, while the superoxo chemisorption is a slightly higher energy isomer. The computed detachment transitions of the superoxo and peroxo species are in good agreement with the experimental observation. There is a superoxo to peroxo chemisorption transition of O_2 on Au clusters at Au_8^-; $O_2Au_n^-$ ($n = 2$, 4, 6) involves superoxo binding and $n = 10$, 12, 14, 18 involves peroxo binding, whereas the superoxo binding re-emerges at $n = 20$ due to the high symmetry tetrahedral structure of Au_{20}, which has a very low electron affinity. Hence,

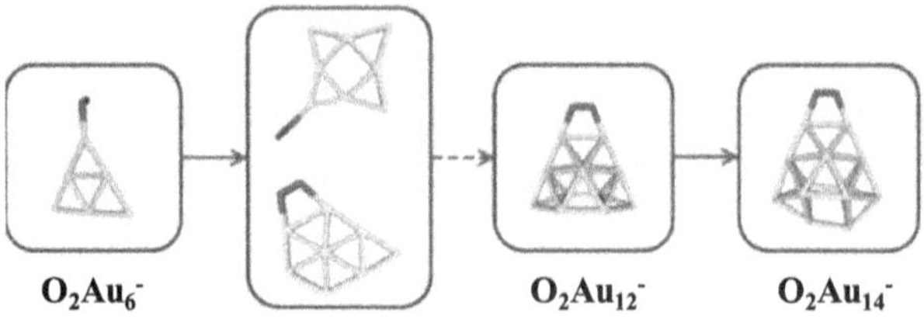

Figure 9.3 O_2 activation by size-selected gold clusters: Transition from Superoxo to peroxo chemisorption.
Reproduced with permission from ref. 123.

the two-dimensional (2D) Au_8^- is the smallest anionic gold nanoparticle that prefers peroxo binding with O_2. At Au_{12}^-, although both 2D and three-dimensional (3D) isomers coexist in the cluster beam, the 3D isomer prefers the peroxo binding with O_2.

9.3.4 Role of Oxygen Vacancies and Moisture/Surface OH Groups in CO Oxidation

The detailed role of O vacancies is discussed in a separate chapter in this book. Here we briefly discuss some experimental results related to the roles of O vacancies and moisture/surface OH groups in CO oxidation.

Besenbacher and co-workers[124] investigated the rutile $TiO_2(110)$ surface using scanning tunneling microscopy (STM) and found a new pathway for oxygen diffusion. When titania was reduced by annealing *in vacuo*, O vacancies were formed, taking the form of missing O atoms in the bridging O rows at the surface. With adsorption of O_2 molecules on the $TiO_2(110)$ surface at low temperatures, the O_2 molecules resided on top of the Ti atoms that constituted the troughs along the [001] direction in between the protruding O rows. A fast-scanning Aarhus STM was used to follow the dynamics of individual O_2 molecules on the $TiO_2(110)$ surface. From a time-lapsed sequential STM study, diffusion of O_2 molecules only along the [001] direction was observed, and evidence for a charge transfer-induced diffusion mechanism for O_2 molecules adsorbed on a rutile $TiO_2(110)$ surface was identified.

The O_2 hopping rate depends on the number of surface donors (oxygen vacancies), which determines the density of conduction band electrons. Landman and co-workers investigated Au/MgO model catalysts for CO oxidation.[125,126] The gold octamers (Au_8) bound to the F centers of the magnesia surface are the smallest known gold heterogeneous catalysts that can oxidize CO to CO_2 at temperatures as low as -133 °C. The same cluster bound to MgO surface without O vacancies is catalytically inactive for CO combustion.

It was found that charging the supported clusters plays a key role in promoting their chemical activity. Infrared measurements of the stretch vibration of CO adsorbed on mass-selected Au_8 soft-landed on MgO(001) with co-adsorbed O_2 showed a red shift on an F-center–rich surface with respect to the perfect surface, consistent with quantum *ab initio* calculations that predicted a red shift of the C–O vibration should arise *via* electron back-donation to the CO antibonding orbital. This partial electron transfer from the F centers to the adsorbed cluster complex correlates with frequency shifts of the intramolecular vibration of the adsorbed CO. In addition, these sites serve to strongly anchor the deposited clusters, thereby inhibiting their coalescence into larger inert ones.

Corma's group[127] studied the CeO_2 and Au/CeO_2 samples in O_2 and a CO/O_2 gas mixture, respectively, using IR and Raman spectroscopy. Raman bands at 1123, 966, 870, and 831 cm^{-1} were observed at the beginning (in O_2). These bands are assigned to η^1 superoxide species, peroxide adspecies at the one-electron defect site, nonplanar bridging peroxide species, and η^2 peroxide species, respectively. After CO oxidation, the intensity of the Raman band at 1123 cm^{-1} was decreased and the band at 966 cm^{-1} disappeared, indicating the involvement of η^1 superoxide species and peroxide adspecies at the one-electron defect site in the oxidation of CO. Also, the nonplanar bridging peroxide species, η^2 peroxide species, molecular oxygen, or lattice oxygen did not participate in the direct oxidation of CO. These results suggest that cationic gold was stabilized on nanoparticulated CeO_2 by creating Ce^{3+} and O vacancy sites, where peroxide species could be eventually formed for CO oxidation.

CO oxidation was studied over Au/γ-Al_2O_3 catalyst by Kung and co-workers.[87] They observed that the initial rapid loss of activity could be prevented if either hydrogen or water vapor was present in the reaction mixture. Otherwise, it could be recovered by exposure of the deactivated catalyst to either hydrogen or water vapor at room temperature. Thermal treatment above 100 °C in a dry atmosphere also deactivated the catalyst. These results suggest that the hydroxyl group is associated with the active site. Daté *et al.*[52] investigated the effect of moisture in the reactant gas on the low-temperature oxidation of CO over Au/TiO_2, Au/Al_2O_3, and Au/SiO_2. The moisture concentration ranged from about 0.1 to 6000 parts per million (ppm) H_2O. They found that the moisture enhanced the catalytic activities by no less than two orders of magnitude and that the effect of moisture depended on the type of metal oxide. It was believed that moisture plays dual roles in the reaction : activation of oxygen and decomposition of carbonate.

It is somewhat puzzling that the moisture concentration required for a high CO oxidation activity is different on various metal oxide supports. Zheng *et al.*[128] prepared four kinds of TiO_2 nanostructures with different phase compositions: NTM, a metastable phase between a hydrogen titanate and a TiO_2(B) phase (a monoclinic polymorph of titanium dioxide); NTMA, a thin anatase-layer covered metastable phase; NTB, a mixed TiO_2(B) and anatase phase; and NTA, an anatase phase. After deposition with nanosized Au particles, the four catalysts were used for three catalytic oxidation reactions: photocatalytic formaldehyde (HCHO) oxidation; CO oxidation; and photocatalytic oxidation of the dye sulforhodamine B (SRB) in aqueous solution (in the last reaction only four TiO_2 nanostructure structures were used). The order of the catalytic activity in the three reactions is the same as that of the surface OH regeneration ability of the catalyst supports: NTA > NTBA > NTMA > NTM. This observation reveals the common nature of these three catalytic oxidation reactions: the surface OH groups of the catalyst supports contribute to the three types of catalytic oxidation reactions, particularly to the adsorption and activation of the O_2 molecules. However, different supports have different abilities to 'breathe in' the moisture and convert them into surface OH groups, probably originating from the difference in their surface structures. In addition, XPS measurements identified one O 1s peak at 534.9 eV on Au/TiO_2 (NTA) after exposure to CO, revealing the strong interaction between the coexisting CO and OH species on this catalyst and the abundance of OH groups on the NTA support. This peak is assigned to the hydroxycarbonyl-like species. The authors proposed a model that showed that the adsorption of H_2O molecules facilitates the adsorption and activation of O_2.

Recently Gao and Zeng[129] reported a DFT study on the O_2 dissociation on small-sized anionic gold nanoclusters Au_n^- ($n = 1$–6) with and without the involvement of a water molecule. The results showed that the activation barriers from the molecular adsorption state of O_2 to dissociative adsorption can be significantly lowered with the involvement of an H_2O molecule. Once the O_2 dissociates on small-size Au clusters, atomic oxygen is readily available for subsequent oxidation reactions, such as CO oxidation, on the surface of Au clusters.

9.3.5 Reaction Models

Several reaction models have been proposed to explain the complex experimental phenomena. The proposed reaction pathways should answer the following questions:

(1) The importance and involvement of the support in the reaction
(2) The effect of oxidation state, size, and morphology of the Au particles
(3) The influence of moisture level in the reactant feed on the catalyst activity
(4) The activation, reaction and compensation of molecular O_2.

The models proposed by Bond and Thompson[111] and Kung *et al.*[130,131] contain cationic Au species as well as metallic gold (Au^0). In the Kung model, O_2 is adsorbed dissociatively to the Au^0 sites of the cluster while CO adsorbs at the cationic Au sites forming hydroxycarbonyl species with the OH^- counter anion. The surface atomic oxygen is then used to oxidize the hydroxycarbonyl to bicarbonate, which then decomposes to CO_2, restoring the surface OH^- anion. The Bond–Thompson model invokes O_2 adsorption at the oxide lattice defects near the Au particle/oxide interface. These mechanisms take into account the main components of the catalysts prepared by wet chemistry that are thought to be important for CO oxidation.[107]

Behm and co-workers[95,132] performed multipulse measurements for CO oxidation on a temporal analysis of products (TAP) reactor between 80 and 400 °C on Au/TiO_2, and determined the amount of stable adsorbed oxygen that could be reversibly deposited and removed by reaction under these conditions [oxygen storage capacity (OSC)]. They first treated the catalyst by *in situ* calcination in 10% O_2/N_2 at 400 °C to prepare a well-defined and fully oxidized catalyst. Afterwards, the microreactor was evacuated and exposed alternately to sequences of CO/Ar pulses and O_2/Ar pulses (1 : 1, 10^{16} molecules per pulse each) to determine the amount of stable adsorbed oxygen. They observed that CO_2 was produced solely during the CO/Ar pulses over the oxidized catalyst and not during the subsequent O_2/Ar pulses, indicating that CO was reversibly adsorbed under these conditions and desorbed instantaneously after the CO pulse, except during the first cycle. It was observed that the amount of oxygen deposited by O_2 pulses was equal to the amount of oxygen consumed during subsequent CO pulses. Also, the amount of oxygen reversibly deposited and reactively removed depended strongly on the temperature and increased by a factor of 4.5 between 80 and 400 °C, due to the increase in thermally activated oxygen and oxygen vacancies at high temperatures. Despite this increase, the amount of removable surface oxygen was rather small, and even at 400 °C, it accounted for just 3.3% of the entire surface oxygen content calculated from the overall surface area of the TiO_2 support, implying the active oxygen was mainly located at the perimeter or interface between the Au nanoparticles and the support. It was also found that the pulse-deposited oxygen species was highly stable even at temperatures as high as 400 °C. It should be noted that the CO pulse was introduced with a time delay one hour after the O_2 pulse. During that period, there was no measurable desorption of the active oxygen species. Moreover, there was no measurable oxygen removal or deposition on the pure support even at 400 °C. Obviously, the roles of the Au nanoparticles are to adsorb and accumulate CO on the surface, as well as to activate surface active oxygen species at the Au–TiO_2 perimeter sites. The active oxygen is formed by adsorption, activation, and dissociation of molecular oxygen and is mainly located at the perimeter sites on the catalyst surface. These observations are similar to the observations in the coupled TG-FTIR measurement for CO oxidation on Au/α-Fe_2O_3 catalyst.[97] However, more investigations to distinguish the adsorbed oxygen and surface lattice oxygen at different temperature regions should still be considered.

Takeda and co-workers conducted a systematic environmental transmission electron microscopy (ETEM) investigation on Au/TiO_2 catalysts under reaction conditions.[133] This study revealed detailed structural information on the Au nanoparticles and perimeter interface with a TiO_2 support within the spatial resolution available in ETEM under the reaction conditions. However, the role of surface OH is not included even in two new models proposed by the Behm and Takeda groups, respectively.

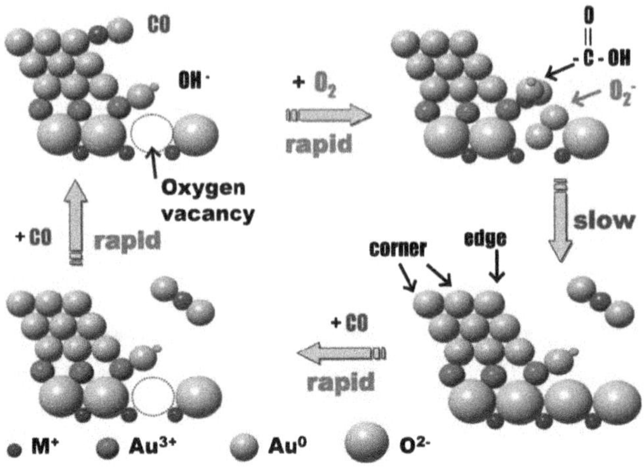

Figure 9.4 Proposed CO oxidation reaction mechanism for CO oxidation over supported gold catalysts at room temperature.
Reproduced with permission from ref. 94.

Recently, Haruta proposed a more comprehensive reaction mechanism (Figure 9.4) of CO oxidation at room temperature based on the catalytic performances of various supported Au catalysts, and on kinetic and spectroscopic investigations.[94] In this mechanism, CO molecules are adsorbed on the corner or edge sites on Au nanoparticles, and O_2 on the O vacancy sites. The two adsorbed species react to form hydroxycarbonyl-like intermediate at the periphery, which is the rate-determining step. At temperatures above 60 °C, oxygen molecules can be activated on the Au nanoparticle surfaces, which is different from the case at room temperature.[134]

In short, the knowledge obtained on CO oxidation should be applicable to other oxidation reactions on supported Au catalysts. With changes in reactants and reaction conditions, however, the active sites and reaction pathways may also change.

9.4 Synthesis of Propylene Oxide from Propylene

9.4.1 Current Industrial Processes for Propylene Oxide Production

Propylene oxide (PO) is a valuable propylene derivative widely used for the production of urethane polyether polyols, propylene glycol, glycol ethers, allyl alcohols, and isopropanolamines.[135] Annual PO output worldwide in 2009 was *ca.* 8 million tons and its market size was still expanding.[136] Currently there are several industrial processes for the manufacture of PO *via* the epoxidation of propylene: (1) the chlorohydrin process (CHPO); (2) the Halcon process using various organic hydroperoxide oxidants; and (3) the hydrogen peroxide (HPPO) process. The Halcon process can be further classified into the *tert*-butylalcohol (PO/TBA) process, the styrene monomer (PO/SM) process, and the cumyl hydroperoxide process (as it was invented by Sumitomo Chemical Company, hereafter it is called as the Sumitomo process), depending on the kind of organic hydroperoxide used.[13,136,137]

The CHPO process was first operated in the 1930s, and is a mild temperature reaction in aqueous solution. In this process, propylene reacts with chlorine to yield α- and β-chlorohydyrine (9 : 1), which further reacts with an alkali [$Ca(OH)_2$ or NaOH] to yield PO as the main product,[138] with other co-products and byproducts such as alkali salts, 1,2-dichloropropane, and dichloropropanols. The CHPO process consumes a large amount of water and yields almost the same amount of corrosive waste water containing alkali salts and organic wastes.[139] Developed in the 1970s, the PO/TBA and PO/SM processes first produce *tert*-butyl hydroperoxide (TBHP) and ethylbenzene hydroperoxide *via* oxidation of isobutene and ethylbenzene, respectively, which further react with propylene to yield PO, with production of main co-products of *tert*-butanol (TBA) and styrene monomer (SM). The catalyst used for the PO/SM process is Mo-based homogeneous catalyst[140] or Ti-supported silica catalyst.[141,142] For these two technologies, the main concern is the production of a large amount of co-products, which lowers their economic viability.[136] The Sumitomo process, which was invented by Sumitomo Chemical Company in 2003,[143] involves the oxidation of cumene to cumyl hydroperoxide (CHP) without any catalyst. CHP further reacts with propylene to produce PO and α,α-dimethyl benzyl alcohol (CMA) over supported Ti-silica catalyst; CMA is dehydrated and converted to cumene again by a dehydrogenation reaction on metal catalysts. This process recycles cumene and allows safe processing of CHP with a high concentration (75–85%). As a result, the reactor dimensions can be significantly reduced to lower the investment cost. However, in the epoxidation of propylene in liquid phase at 60 °C, a high propylene/CHP molar ratio is maintained at 10 : 1,[138] which leads to a low propylene conversion. Therefore, a large amount of propylene has to be purified and recycled, which eventually increases the cost of the whole process.

The above processes either involve multiple reactions, producing large amounts of byproducts and co-products, or require recycling of the excess chemical reactant. BASF-Dow and Degussa-Headwaters then commercialized the HPPO process[136,144] and have built plants in Europe and Asia in the past few years. In this process, aqueous H_2O_2 solution reacts with propylene in methanol solvent over TS-1 catalyst to produce PO. A high yield (95 mol%) of PO can be obtained by this method and the major byproducts are 1,2-propanediol, 1-methoxy-2-propanol, and 2-methoxyl-1-propanol.

The HPPO process uses the mild and green oxidizing agent, H_2O_2. However, H_2O_2 is highly explosive, making its transportation and storage difficult, and limiting its large-scale application. In addition, the separation of propylene oxide from the excess reactants and other products remains a problem.[145,146] Therefore, efforts have been made to integrate the production of propylene oxide with that of H_2O_2. For example, Meiers *et al.*

prepared 1 wt% Pd-0.02 wt% Pt/TS-1 catalyst and obtained a maximum PO yield of 11.7% and a PO selectivity of 46% in a pressured reactor filled with propylene, H_2, O_2, N_2, and methanol at 43 °C.[147] However, other problems remain. For instance, the reaction temperatures for the H_2O_2 synthesis (<20 °C) and for the epoxidation reaction (40 °C) are different, and thus different catalyst systems should be used for them; the noble metal catalysts, which are often used for H_2O_2 synthesis, may catalyze the hydrogenation reaction of propylene. From the literature results, the selectivity of PO based on H_2 is still low.[136] In addition, this reaction is a complex three-phase (gas, liquid, and solid) system, which imposes challenges on reactor design and control. In short, this approach still suffers from the low reaction yield for propylene oxide, difficulty in separation and purification of PO,[148] and low H_2 efficiency.[149]

9.4.2 Direct Gas-phase Production of Propylene Oxide on Supported Gold Catalysts

A single-step, direct catalytic oxidation of propylene to PO in the gas phase, analogous to the commercial ethylene oxide (EO) process, has long been desired.[150] Although Ag/α-Al_2O_3 catalyst is good at catalyzing ethylene epoxidation with molecular O_2, it is quite poor in the epoxidation of propylene in the same way.[151] One reason is that the generated O^- radicals in the reaction preferentially attack the allylic C–H bond, the weakest bond in propylene, leading to the formation of acrolein.[139] There are two gas-phase epoxidation routes of propylene to PO over supported Au catalysts: (1) the reaction conducted in the presence of H_2 and O_2; and (2) direct epoxidation with molecular O_2.

9.4.2.1 Gas-Phase Epoxidation of Propylene in the Presence of H_2 and O_2 on Supported Gold Catalysts

Table 9.3 lists some representative catalyst systems for PO synthesis in the presence of H_2 and O_2 on supported Au catalysts, from which the readers can find information on the catalyst preparation and reaction conditions. Haruta and co-workers[152] first found that PO with high selectivity (99%) could be synthesized on 1 wt% Au/TiO_2 catalyst directly with a feed gas composition of $C_3H_6/O_2/H_2/Ar = 10/10/10/70$ at low temperature of 30–120 °C and atmospheric pressure. However, the propylene conversion was quite low (1.1%) with PO formation rate of only 12 $g_{PO}\ kg_{cat}^{-1}\ h^{-1}$, much lower than the minimum requirement of *ca.* 100 $g_{PO}\ kg_{cat}^{-1}\ h^{-1}$ for a viable commercial process; propylene conversion of 10%, PO selectivity of 90%, and H_2 efficiency of 50%. The H_2 efficiency is defined as mole of PO produced per mole of H_2 consumed×100.[153] In addition, this catalyst deactivated fast.[154] Since titanium silicalites, particularly TS-1, are active in the HPPO process, various Ti-silica and titanosilicate materials have therefore been developed as

Table 9.3 Several representative supported Au catalysts for the epoxidation of propylene to PO.

Catalyst	Preparation Method and Conditions	Mean Au Size/nm	Reaction Conditions (Volumic Composition of the Reactant Gases and Reaction Temperature)	PO Formation rate/g_{PO} kg^{-1}_{Cat} h^{-1})	Other Information	Ref.
1 wt% Au/TiO_2	DP method, pH = 7	2.4	$C_3H_6/O_2/H_2/Ar$ = 10/10/10/70; GHSV: 4000 ml g^{-1}_{cat} h^{-1}; Temp: 50 °C	12	PO selectivity above 99%	152
0.3–0.4 wt% Au/mesoporous Ti–SiO_2	DP method, pH = 6.5	3.0	$C_3H_6/O_2/H_2/Ar$ = 10/10/10/70; GHSV: 4000 ml g^{-1}_{cat} h^{-1}; Temp: 160 °C	~80	Catalyst was silylated with methoxytrimethylsilane	156
0.05 wt% Au/TS-1 (molar ratio of Si/Ti = 36)	DP method, pH = 9	≤2	$C_3H_6/O_2/H_2/He$ = 10/10/10/70; GHSV: 7000 ml g^{-1}_{cat} h^{-1}; Temp: 200 °C	116		157
0.25 wt% Au/TS-1 (molar ratio of Si/Ti = 48)	Solid grinding, using dimethyl Au(III) acetylacetonate as precursor	≤2	$C_3H_6/O_2/H_2/Ar$ = 10/10/10/70; GHSV: 8000 ml g^{-1}_{cat} h^{-1}; Temp: 200 °C.	137	TS-1 was treated with NaOH before Au deposition	158
0.12–0.14 wt% Au/TS-1 (Si/Ti = ~100)	DP method,	≤1	$C_3H_6/O_2/H_2/N_2$ = 10/10/10/70, GHSV = 14 000 ml g^{-1}_{cat} h^{-1}; Temp: 200 °C	~160	Optimized preparation conditions: pH = 7.3, mixing for 9.5 h, reacted at RT	159
0.05 wt% Au/S-1/TS-1, S-1 is silicate-1, which was coated on TS-1	DP method	7–7.5	$C_3H_6/O_2/H_2/N_2$ = 10/10/10/70, GHSV = 14 000 ml $g^{-1}_{TS\text{-}1}$ h^{-1}	~240[a]		163

[a]The PO formation rate is normalized to TS-1. GHSV, gas hourly space velocity.

catalyst supports.[154,155] After intensive efforts to improve the catalyst activity, the PO formation rate has been significantly increased (Table 9.3),[156–159] generating new hope for the commercialization of this process.

There are a number of factors that influence the catalytic performances of the supported Au catalysts in the gas-phase epoxidation of propylene to PO, such as Au particle size, the Si to Ti ratio of the titanium silicalite (for example, TS-1 (100), in which the Si to Ti molar ratio is 100), nature of the supports and their surface hydrophilic/hydrophobic properties, *etc.* Some representative literature results will also be given to show the influences of these factors.

9.4.2.1.1 Influence of Au Particle Size, Location and Oxidation States. The Au particle size and their oxidation states are crucial to the catalytic performance. Qi *et al.* selected four types of catalyst supports [anatase TiO_2 microsphere, MCM-41, HPG (a hybrid porous organic/inorganic Ti–SiO_2 material), and Ti–TUD] and deposited Au particles by the DP-NaOH method, DP-NH_4OH method, DP-urea method and solid grinding (SG) method, respectively.[160] The supported Au catalysts were tested for PO production in a mixture of H_2, O_2, and propylene. They found that Au nanoparticles were deposited in the proximity of tetrahedral Ti species. The presence of alkali was also found to promote the epoxidation reaction, which predominantly took place on Au particles in the size range 2.0–5.0 nm. Meanwhile, the hydrogenation of propylene prevailed on the Au nanoparticles larger than 5.0 nm with or without alkali, and on Au clusters smaller than 2.0 nm in the absence of alkali, irrespective of the location of Au clusters deposited either on Ti or Si sites. A certain amount of alkali could switch hydrogenation to epoxidation over Au clusters located on Ti^{4+} sites.

Huang *et al.* pre-treated TS-1 with alkaline-treated TS-1-Na1 and deposited Au particles by the DP and SG methods on the support, respectively.[161] They found that, with the SG method, Au was selectively deposited on the exterior surface of alkaline-treated TS-1 as clusters (1–2 nm) and partly as nanoparticles (>2.0 nm), while with the DP method, Au was deposited on the exterior surfaces of TS-1 and alkaline-treated TS-1, as clusters (1–2 nm) and partly as nanoparticles (>2.0 nm), and inside the microporous channels as tiny clusters (*ca.* 0.55 nm). In propylene epoxidation with an O_2 and H_2 mixture, the PO formation rate over 0.23 wt% Au/TS-1-Na1 (DP) with a mean diameter of 1.7 nm was 74 $g_{PO}\ kg_{cat}^{-1}\ h^{-1}$, lower than that of 137 $g_{PO}\ kg_{cat}^{-1}\ h^{-1}$ over 0.20 wt% Au/TS-1-Na1 (SG) with a mean diameter of 1.8 nm. Since the main difference between the two catalysts was the location of Au clusters, it is believed that the Au clusters (1.0–2.0 nm) over the exterior surfaces are more active for PO synthesis than tiny Au clusters distributed inside the microporous channels.

Deglass and co-workers prepared Au/TS-1(Si/Ti = 100) catalysts with low Au loadings by the DP method at pH 7 and used the catalysts for both CO oxidation in O_2 and for PO formation.[162] The experimental conditions were the same as that used by Lee *et al.*,[159,163] as given in Table 9.3. They found that, when the Au loading was below 0.05 wt%, the PO formation rate increased approximately linearly with Au loading, suggesting that the PO formation rate initially increased with the number of active sites. Once the Au loading reached 0.07 wt%, the PO rate reached a plateau. As determined by TEM, the Au particle size was <2 nm when the Au loading was 0.07 wt%. The 0.07 wt% Au/TS-1 (100) catalyst was subsequently calcined at different temperatures to deliberately increase the Au particle size. The CO_2 formation rate increased with the Au particle size. Therefore, Delgass and co-workers concluded that sintering of the Au particles and/or increasing the Au loading of Au/TS-1 caused an increase in the specific CO oxidation rate, but a decrease of the propylene epoxidation rate in O_2 and H_2 (Figure 9.5). Probably only the very tiny Au particles have good activity for PO synthesis.

Gaudet *et al.*[164] prepared Au/TS-1 catalysts and leached them with sodium cyanide solutions. The low concentration of sodium cyanide resulted in preferential removal of small Au particles, while a high concentration of sodium cyanide led to dissolution of Au and its re-precipitation as Au^+ cyanide. X-ray absorption spectroscopy demonstrated that the catalysts, active for propylene oxide production in the presence of hydrogen and oxygen mixtures, had supported Au(III) oxide nanoparticles 3 nm in size. Under the

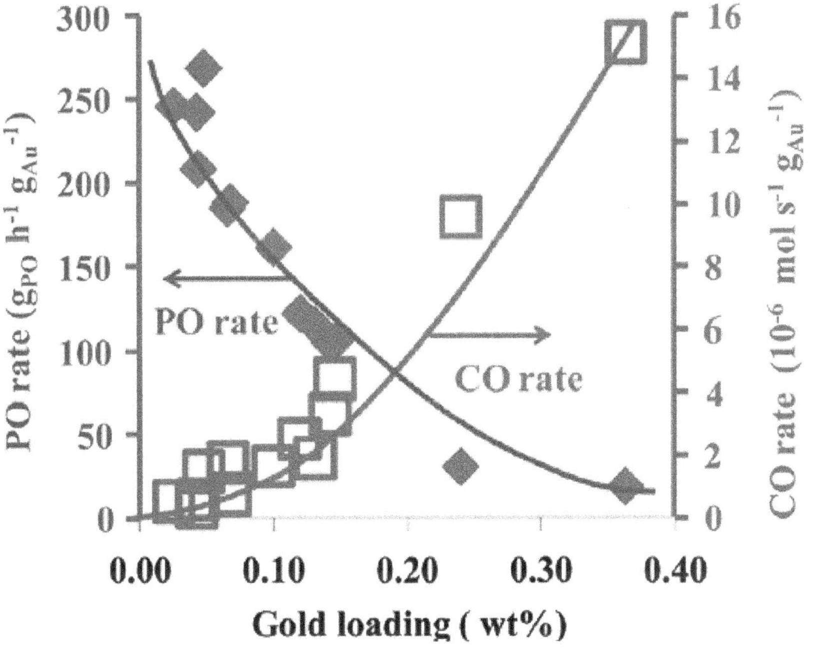

Figure 9.5 PO and CO formation rates on Au/TS-1 catalysts against Au loadings. Reproduced with permission from ref. 162.

reaction conditions, they were reduced to metallic Au. Meanwhile, the samples treated with strong solutions of sodium cyanide resulted in large supported Au(I) cyanide particles (9–11 nm). These particles did not produce propylene oxide but, surprisingly, showed high selectivity toward propylene hydrogenation. Therefore, it is not the large Au particles but the small ones in the metallic form that are active for PO synthesis.

It is generally believed that Au nanoparticles should be located on or nearby the isolated Ti^{4+} in titanium silicalites. However, different observations were obtained by Wu's group.[165] They synthesized TS-1 crystals and modified their surface with poly(diallyldimethylammonium) chloride to make their external surface positively charged, followed by coating of the TS-1 surface with a shell of mesoporous silica. After depositing Au particles in the silica shell by the DP method, the obtained Au/TS-1@meso-SiO_2 catalyst was used for PO synthesis; a steady PO selectivity as high as 93–96% was observed, independent of its Si/Ti ratio. Moreover, this catalyst was stable and gave no sign of deactivation within 60 h. However, we still cannot exclude the possibility that some small Au clusters below 1 nm were inserted into the nanopores of TS-1 and/or located at the interface of the silica and TS-1, which might be involved in the PO formation.

Delgass and co-workers[163] coated TS-1 seeds with silicalite-1 (S-1) and prepared a core-shell material of S-1/TS-1, possessing a well-defined MFI structure. They then deposited Au to obtain Au/S-1/TS-1 catalysts. Surprisingly, they observed a PO formation rate of $\sim$240 g_{PO} $kg_{TS\text{-}1}{}^{-1}$ h^{-1} at 200 °C on 0.05 wt% Au/S-1/TS-1 catalyst, higher than that on the 0.05 wt% Au/TS-1 catalysts with a Si to Ti molar ratio from 65 to 100 (60–140 g_{PO} $kg_{TS\text{-}1}{}^{-1}$ h^{-1}). The H_2O_2 molecules were probably not formed on Au/S-1 then transferred to TS-1 sites in the core, as they were very short-lived at the reaction temperature of 200 °C, but formed on the Au clusters located mainly inside the TS-1 nanoporous channels which acted as the dominant active sites for the PO reaction. It should be noted that, when the Au loading is very low, and particularly the formed tiny particles are located at the deep channels, it is very difficult to characterize them by TEM. However, these core-shell structures probably represent a new kind of catalyst promising for high catalytic activity and stability for PO synthesis.

9.4.2.1.2 Influence of Various Catalyst Supports, Preparation Methods, Pre- and Post-Treatments, *etc.* It is generally accepted that the catalyst supports should contain Ti because, in epoxidation of propylene to PO, Ti–OOH species is the active intermediate (the reaction mechanism is further discussed below). The Ti species in the support matrix should be isolated tetrahedral Ti sites. In the bulk, the Ti atoms are coordinated by four Si–O ligands (tetrapodal Ti site), but on the surface they are likely to be tripodally held to Si–O in order to be sterically accessible.[166] The deposited Au particles should be anchored to these Ti sites. The existence of these isolated Ti species means the absence of any TiO_2 crystal phases (octahedral sites) such as anatase. The identification of the isolated Ti species is performed with ultraviolet–visible (UV-vis) spectroscopy, which should not show an anatase absorbance peak at ~330 nm, and/or Ti-edge XANES analysis, which should show a pre-edge characteristic peak of tetrahedral Ti sites.[167,168]

So far, the catalyst supports for PO synthesis are often selected from Ti-containing silicates, particularly TS-1, which has a stable crystal structure and widely used in supported Au catalysts. This situation may change if a new type of catalyst support is discovered. Besides TS-1, many other Ti-containing mesoporous materials are employed as the catalyst support, including Ti-MCM-41,[169] Ti-MCM-48,[170] Ti-SBA-15,[171] Ti-TUD,[156] 3D mesoporous titanosilicate,[172,173] and mixed SiO_2-TiO_2 oxides.[174]

Haruta and co-workers[156,173] reported the preparation and catalytic performance of 0.3–0.4 wt% Au/3D mesoporous titanosilicate. The catalyst support had a Ti content of 1–12 mol%, a surface area of 838–896 $m^2\ g^{-1}$, pore size of 5–10 nm, and porosity of 1.0–1.4 ml g^{-1}. At 160 °C, $C_3H_6/O_2/H_2/Ar = 10/10/10/70$ and the gas hourly space velocity (GHSV) = 4000 mL $h^{-1}\ g_{cat}^{-1}$, the propylene conversion was 7%, PO selectivity >90%, and H_2 efficiency about 40%. However, these catalyst supports are not very stable under the reaction conditions, limiting their industrial application. Deglass and co-workers[175] prepared TS-1 with mesoporous-scale defects by using carbon pearl as the hard template. At 200 °C, $C_3H_6/O_2/H_2/He = 10/10/10/70$ and GHSV = 7000 mL $h^{-1}\ g_{cat}^{-1}$, the propylene conversion was 9.7%, PO selectivity >87% (corresponding to a PO rate of 132 $g_{PO}\ kg_{Cat}^{-1}\ h^{-1}$), and H_2 efficiency about 40%.

The most popular method for Au deposition on the titanium-containing silicates is the DP method. Delgass and co-workers[159] optimized the DP parameters to prepare ~0.1 wt% Au/TS-1 (Si/Ti = 100). In brief, $HAuCl_4$, H_2O, TS-1, and Na_2CO_3 were mixed and the end pH of the slurry was adjusted to ~7.3. The mixing time at room temperature was 9.5 h, after which the temperature was raised to 70 °C and maintained for 1.5 h. Thereafter, the mixture was separated by centrifugation, washed, and dried. The observed PO formation rate was *ca.* 160 $g_{PO}\ kg_{cat}^{-1}\ h^{-1}$. The pH value should be above 7. At low pH values, the Au species in the solution still contains Cl^- ligand, which is detrimental to catalytic performance. The Si to Ti molar ratio is also an important factor that influences catalytic performance. In most reports, the Si to Ti molar ratio is between 30 and 150, preferentially at around 100. If the Si to Ti ratio is too low (Ti content is high), extra framework Ti or anatase nanophase will lead to deactivation of Au/TS-1 catalyst by accelerating PO oligomerization, which subsequently blocks the active sites.[176,177] If the Si to Ti ratio is too high (>500), the existence of surface Si–OH groups will cause degradation of PO to propanal.[164] In an ideal DP preparation, the Au particles will be selectively adsorbed on the isolated tetrahedral Ti^{4+} sites, and thus the Au loading will be closely correlated to the Ti content. One of the main drawbacks of the DP method is the very low Au capture or loading efficiency, usually below 3%, which means that most of the Au species are washed away during the preparation.

Haruta's group used the SG method by employing dimethyl Au(III) acetylacetonate $[(CH_3)_2Au(acac)]$ as the precursor. A recent study by the same group[161] showed that, by the SG method, most of Au was selectively deposited on the exterior TS-1 surface as clusters (1.0–2.0 nm) and partly as Au nanoparticles with a size of above 2.0 nm, while the DP method led to the deposition of Au particles both on the exterior surface and inside the microporous channels ($\sim$0.55 nm). Compared with the very low Au capture/loading efficiency of the DP method, the SG method has a very high Au loading efficiency of almost 100%. However, the SG method relies heavily on the availability of cheap and chlorine-free Au precursor, which is still a major drawback to this method.

Recently, Li's group[178] reduced $HAuCl_4$ with *Cacumen platycladi* (bioreduction) to generate an Au sol (colloid), followed by addition of TS-1 support. The prepared 1 wt% Au/TS-1 had an Au particle size distribution between 2.2 and 4.4 nm, and the catalyst exhibited good catalytic performance and stability. By adding an ionic liquid, 1-butyl-3-methylimidazolium tetrafluoroborate, to the Au sol when adding TS-1, the IEP value of TS-1 (Si/Ti = 35) was tuned to 8 from 2.6. As a result, the Au colloid could be fully adsorbed on the TS-1 support. On the 1.0 wt% Au/TS-1 (Si/Ti = 35) catalyst, a PO formation rate of 164 g_{PO}. kg_{cat}^{-1} h^{-1} at 300 °C was obtained.[47]

It seems that the DP method using $[Au(en)_2]Cl_3$[43,53] as the Au precursor is not widely used to prepare Au/Ti-silicate catalysts. With this method, the Au particles are likely to adsorb on both the isolated Ti^{4+} sites and Si^{4+} sites. However, it deserves further exploration to better understand its influence on catalytic performance.

Various treatments before and after catalysts synthesis have been used to improve the catalysts and their catalytic properties. These treatments can be roughly classified as pre-treatment, post-treatment, and adding gas-phase promoter during the reaction. Cumaranatunge and Delgass pre-treated TS-1 support with NH_4NO_3 before deposition of Au, which increased the Au loading significantly.[44] Pre-treatment of the TS-1 with cobalt nitrate,[45] alkaline earth metal salts (Mg, Ba, Sr salts),[46] and indium nitrate,[179] respectively, was observed to enhance the interaction of Au species with the support and to increase the Au loading, thus increasing the number of active sites. In addition, the improvement in catalytic performance should also be related to the change of the surface acidic property, as strong Brönsted acidic sites can cause degradation of PO. This can be seen more clearly in a series of Au/TS-1 catalysts prepared by the SG method,[158] in which TS-1 was treated with NaOH, LiOH, KOH, and CsOH, respectively, before Au deposition. The Au nanoparticles larger than 2 nm had a PO formation rate of 11 g_{PO} kg_{cat}^{-1} h^{-1} at 300 °C. After being treated with NaOH, the same catalyst of 0.25 wt% Au/TS-1 (Si/Ti = 48) had a PO rate of 137 g_{PO} kg_{cat}^{-1} h^{-1}, and H_2 efficiency of 47%. Dow filed a patent application and claimed that addition of halides such as NaCl and CsCl could increase the H_2 efficiency and PO yield.[180] However, the presence of halide ions may sacrifice the stability of the catalysts.

Silylation of titanium silicate is another effective way to make the catalyst support surface more hydrophobic, which could facilitate fast desorption of PO molecules from the support surface and thus prevent their degradation. There are mainly two ways to silylate titanium-substituted molecular sieves: (1) by post-synthesis modification using different silylating agents;[181,182] and (2) by direct organic functionalization during Ti-silicate synthesis using monoalkylsilanes or dialkylsilanes.[183,184] Prasad *et al.* treated Ti-TUD-1 with hexamethyldisilazane (HMDS) at 200 °C, and confirmed the silylation of silanol (Si–OH) and titanol groups (Ti–OH) on the mesoporous silica by FTIR spectroscopy, hydrophobicity, TGA, and ammonia-temperature programmed desorption (NH_3-TPD) measurements.[185] Some other examples can be found in the work of Haruta's group.[156,173]

Haruta and co-workers also co-fed a trace amount (10–20 ppm) of gaseous trimethylamine together with propylene, O_2, and H_2 over Au nanoparticles deposited on 3D mesoporous titanosilicates. A steady propylene conversion up to 8.5% and a PO selectivity of 91% under an hourly space velocity of 4000 mL h^{-1} $g_{cat}{}^{-1}$ at atmospheric pressure, which is close to that of ethylene oxide in the current industrial process, was achieved for more than 250 min.[153]

There remains much work to be done for catalyst improvement, particularly through the characterization and clarification of catalyst structures and their changes after various modifications.

9.4.2.1.3 Possible Reaction Mechanism. There are several possible reaction pathways in PO synthesis with H_2 and O_2, which are listed in Table 9.4. The hydrogen oxidation reaction can take place at room temperature, while propylene combustion usually occurs above 250 °C. Qi *et al.* found that the hydrogenation of propylene and H_2 oxidation are competitive reactions, whereas the epoxidation of propylene and H_2 oxidation are parallel reactions.[160] Various measures have been taken to promote the major epoxidation reaction and to limit the side reactions,[137] including varying the Au particle, catalyst treatment by adding alkaline salt/oxides, *etc.*[160,161]

Table 9.4 Possible reactions in epoxidation of propylene to PO.[160]

No.	Reaction	Chemical Equation	Conclusive Points
1	Propylene epoxidation	$C_3H_6 + H_2 + O_2 \rightarrow C_3H_6O + H_2O$	Major reaction
2	Propylene hydrogenation	$C_3H_6 + H_2 \rightarrow C_3H_8$	Major side reaction
3	Hydrogen oxidation	$H_2 + \frac{1}{2} O_2 \rightarrow H_2O$	Started at low temperatures
4	Propylene combustion	$C_3H_6 + \frac{9}{2} O_2 \rightarrow 3\ CO_2 + 3\ H_2O$	Side reaction at high temperatures
5	Propylene oxide degradation	$C_3H_6O + H_2O \rightarrow$ propanal, propanol, *etc.*	Successive side reaction

Note: adapted from ref. 160.

It is a little complex to explain the role of H_2 in this reaction. First of all, there is a requirement to cleave the O–O bond in the reaction to yield O atoms. According to the literature, direct dissociation of molecular oxygen (O_2) into oxygen radicals is energy-intensive, which requires 497 kJ mol^{-1}, even larger than the bonding energy (431 kJ mol^{-1}) of the C–H bond in methane.[139,186] However, the presence of H_2 or CO molecules could reduce the activation energy to about *ca.* 5.9 kJ mol^{-1}.[187] Instead of generating very active oxygen radicals, the presence of H_2 leads to the formation of a much milder oxidizing agent, H_2O_2, which binds to Ti to form Ti-hydroperoxide species in the subsequent step.

How CO promotes the reaction is still not very clear. Kung and co-workers[188] conducted an experiment by bubbling a gas mixture of CO, O_2 (isotope-labeled), propylene, and helium into a suspension of Au/TiO_2 and TS-1 in methanol and H_2O, and found the O in PO was exclusively from O_2, not from H_2O or methanol. The formation of H_2O_2 from H_2O and O_2 is highly thermodynamically unfavorable:

$$H_2O_{(liq.)} + \frac{1}{2}O_2 \rightarrow H_2O_{2(liq.)},\ \Delta G^0 = 116.8\ \text{kJ mol}^{-1} \tag{9.5}$$

After being coupled with CO oxidation reaction, it becomes thermodynamically favorable:

$$H_2O_{(liq.)} + CO + O_2 \rightarrow H_2O_{2(liq.)} + CO_2,\ \Delta G^0 = -140.4\ \text{kJ mol}^{-1} \tag{9.6}$$

In the gas-phase synthesis of PO at $C_3H_6/H_2/O_2/He = 10/10/10/70$ on Au/Ti–SiO_2 catalyst, Nijhuis and co-workers[189] found that trace amounts (1 vol.%) of CO could completely switch off the propylene hydrogenation reaction (Reaction 2 in Table 9.4) while suppressing the hydrogen oxidation (Reaction 3 in Table 9.4). Probably CO molecules are competitively adsorbed on Au where hydrogen adatoms are located, thus abating the propylene hydrogenation and reducing the hydrogen combustion with the activated oxygen, as the hydrogen oxidation occurs exclusively on Au nanoparticles. It was also proven experimentally that propylene molecules are able to adsorb on Au nanoparticles *via* π-bonding.[190] A kinetic study of propylene epoxidation with H_2 and O_2 over Au/Ti–SiO_2 showed that the formation rate of PO is mostly dependent on the hydrogen concentration, in which the formation of an active peroxo species on the Au nanoparticles is the rate-determining step. Water formation and propylene epoxidation are strongly correlated to each other. Water is formed either through the active peroxo-intermediate responsible for epoxidation, or from direct hydrogen oxidation without involving this active intermediate. The active peroxo-intermediate in epoxidation is competitively consumed in hydrogenation and epoxidation reactions.[191]

Compared with the above reaction pathways, there is less debate on the formation of peroxo-intermediate during the reaction. Goodman and co-workers[192] investigated the vapor phase H_2–O_2 reaction on Au/TiO_2 catalyst using vibrational spectroscopy by inelastic neutron scattering (INS), and first obtained direct spectroscopic evidence for the formation of a hydroperoxo species (–OOH) during the reaction. Oyama and co-workers[193] studied the formation of Ti-hydroperoxo species during the gas-phase epoxidation of

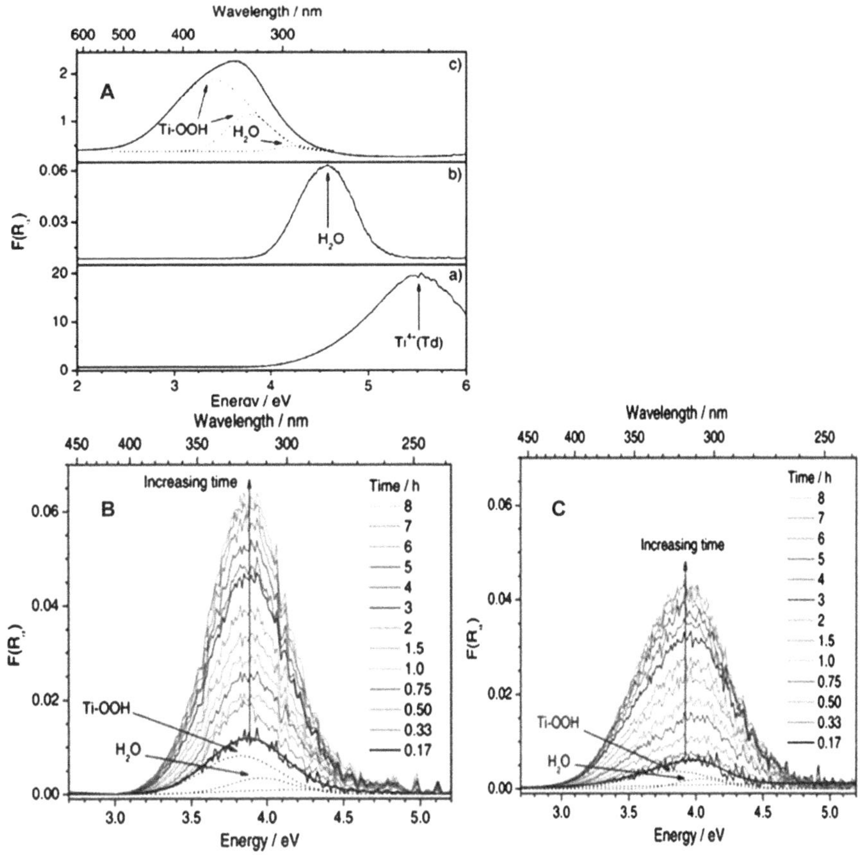

Figure 9.6 UV-vis spectra of: (a) Ti-TUD (i) fresh sample; (ii) exposed to 3.5% H_2O in Ar at 150 °C for 8 h, and (iii) contacted with 35% H_2O_2. TiOOH Species was Detected in (a)(iii); (b) Measured UV-vis Spectra of Au–Ba/Ti–TUD Exposed to $H_2/O_2/Ar = 1/1/8$ at 150 °C; and (c) the measured UV-vis spectra of Au–Ba/Ti–TUD exposed to pyopylene/$H_2/O_2/Ar = 1/1/1/8$ at 150 °C.
Reproduced with permission from ref. 193.

propylene with H_2 and O_2 over an Au–Ba/Ti–SiO_2 (Ti–TUD) catalyst by employing *in situ* spectroscopy technologies such as UV-vis diffuse reflectance spectroscopy, Ti K-edge XANES, and FTIR. Figure 9.6(a)(iii) shows the evolution of TiOOH species after contacting Ti–TUD with H_2O_2, which is not seen for the fresh sample [Figure 9.6(a)(i)] and after its exposure to H_2O [Figure 9.6(a)(ii)]. Figure 9.6(b) and 9.6(c) show the UV-vis spectra of Au–Ba/Ti–TUD exposed to $H_2/O_2/Ar = 1/1/8$ and propylene/$H_2/O_2/Ar = 1/1/1/8$ at 150 °C, respectively. In Figure 9.6(b), it is observed that the TiOOH species is generated, while in Figure 9.6(c), a fraction of the TiOOH species is consumed (reacted with propylene) as evidenced by the decrease in TiOOH peak intensity. Also, *in situ* FTIR detected the bidentate propoxy species during the reaction, which could result in PO decomposition on acidic Ti sites,

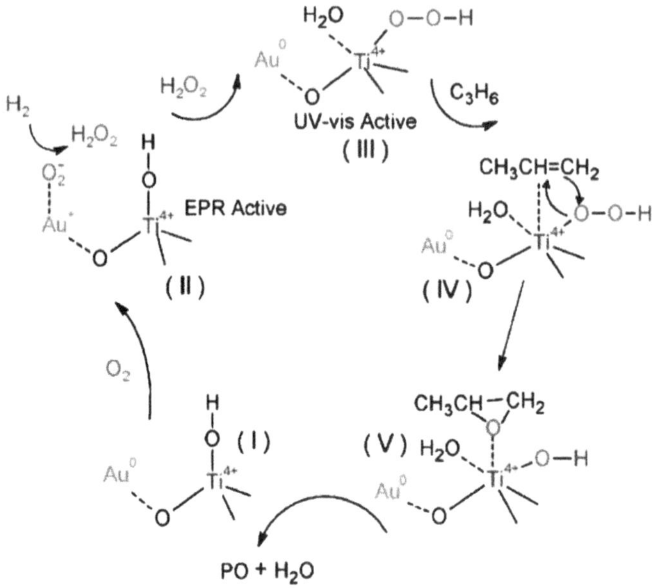

Figure 9.7 Possible reaction mechanism of propylene epoxidation to PO on Au/Titanosilicates.
Reproduced with permission from ref. 193.

indicating the catalyst deactivation is due to the formation of bidentate besides the Au particles agglomeration.

Based on the above experimental results, a reaction mechanism is proposed and summarized in Figure 9.7.[193] It is proposed that the PO synthesis involves four reaction steps:

(1) Formation of H_2O_2 from hydrogen and oxygen on Au nanoparticles
(2) Formation of Ti-hydroperoxo or peroxo species from H_2O_2 on tetrahedral Ti centers
(3) Reaction of propylene with the Ti-hydroperoxide species to form PO
(4) Decomposition of H_2O_2 to water

However, it should be noted that the detailed structure of Au nanoparticles and the interface between Au and support, as well as the location of Au, are still not very clear. More importantly, for industrial application, the catalyst stability, activity and H_2 efficiency would need to be improved significantly.

9.4.2.2 *Gas-phase Epoxidation of Propylene Using Molecular O_2 (without H_2) as Oxidant on Supported Catalysts*

The above epoxidation process with H_2 still suffers from low H_2 efficiency and short catalyst lifetime. The ideal process is the direct epoxidation of propylene to PO using the cheap molecular O_2 or air as the sole oxidant under atmospheric pressure, which is a one-step process as follows:

$$C_3H_6+\frac{1}{2}O_2\rightarrow C_3H_6O \tag{9.7}$$

The catalysts used for this process mainly include Ag-, Cu-, and Au-based catalysts, as well as some other catalysts (Ti-containing silica, mixed metal oxides). Some representative experiments are summarized in Table 9.5, and the results are discussed below.

9.4.2.2.1 Ag-based Catalysts. Silver (Ag) has been successfully used in the epoxidation of ethylene with high selectivity.[194,195] However, its use for direct propylene epoxidation by molecular oxygen often results in substantial formation of carbon dioxide with poor activity and selectivity towards PO.[196] Lei *et al.* prepared unpromoted, size-selected Ag_3 clusters and $\sim$3.5 nm Ag nanoparticles on alumina supports, and found they could actively catalyze the PO formation reaction with O_2 at low temperatures yielding only a negligible amount of carbon dioxide.[197] The formation rate of PO molecules per surface silver atom is $\sim 1\ s^{-1}$ at 110 °C for both Ag_3 and the silver aggregates (3.5 nm), much higher than that of other reported silver catalysts, *e.g.* the 50–660 nm unpromoted silver particles at 220 °C, of which the turnover rate is $10^{-2}\ s^{-1}$, and the Ag–Ni catalyst (3 : 1 weight ratio), which is only $5\times10^{-3}\ s^{-1}$.[198] Lambert and co-workers performed a series of surface chemistry studies over Ag and Cu single crystal surfaces with pre-adsorbed oxygen species, and found that the Cu surfaces were more selective than Ag surfaces for the epoxidation of alkenes with allylic C–H bonds.[199,200]

9.4.2.2.2 Cu-based Catalysts. Lambert and co-workers prepared Cu/SiO_2 catalysts and observed a low activity.[201] Li and co-workers prepared Na-$VCe_{0.5}Cu_{0.2}$-NaCl(20 wt%) catalyst[202] and K-promoted Cu/SiO_2 catalyst[203] for the direct oxidation of propylene with molecular oxygen, and obtained very low PO formation rates with the former but a very high PO formation rate with the latter. In the former, Cu^0 was believed to be the active phase, while in the latter, the isolated Cu species and small CuO_x clusters with low valence state (Cu^0 or Cu^{I}) modified by potassium acetate were proposed to be the main active sites for C_3H_6 epoxidation. Wang and co-workers[204] observed a similar phenomenon on the $K\text{-}Cu/SiO_2$ catalyst but believed that only Cu^+ was active, as the addition of K suppressed the Lewis acidic sites. Other promoters include Ru,[205] V,[206] and Cs.[207] Senkan and co-workers[208] developed a $RuO_2\text{-}CuO_x\text{-}NaCl/SiO_2$ catalyst, which also exhibited high PO selectivity (40–50%) and propylene conversion (10–20%). However, in this work, the propylene feed concentration (1%) was quite low compared with that used in some other studies. For comparison, we have roughly evaluated the PO formation rate for these catalysts and listed them in Table 9.5. Most of the catalysts exhibit much lower PO formation rates than those in Table 9.3. The kinetic studies

Table 9.5 Gas-phase epoxidation of propylene to PO using molecular O_2 as the oxidizing agent.

Catalyst	Preparation Method	Mean Cu or Au size/nm	Reaction Conditions (Volumic Composition of the Reactant Gases and Reaction Temperature)	PO Formation rate/g_{PO} kg^{-1}_{cat} h^{-1}	Other Information	Ref.
1 wt% Cu/SiO_2	Microemulsion	5	$C_3H_6/O_2/$ He = 5/5/90; total flow rate = 50 ml min^{-1}; 0.1 g catalyst, 225 °C	0.8	Active phase: Cu^0	201
Na-$VCe_{0.5}Cu_{0.5}$-NaCl (20 wt%)	Sol-gel method	–	C_3H_6/O_2 = 10/1; SV = 1.5×10^4 h^{-1}, 250 °C	9.7	Active phase: Cu^0	202
K^+-9 wt% Cu/SiO_2	DP method,	2.9 (CuO)	$C_3H_6/O_2/N_2$ = 1/1/15; GHSV: 30 000 h^{-1}, Temp: below 227 °C	145	Active phase: Cu^0 and Cu^+	203
K^+-5 wt% Cu/SiO_2	Sol-gel method	–	Reaction conditions: W = 0.20 g, F = 60 ml min^{-1}, $P(C_3H_6) = P(O_2)$ = 50.7 kPa, Temp: 275 °C	12^b	K^+ reduced CuO size and surface acidity Active phase: Cu^+	204
Cs^+-5 wt% Cu/SiO_2	Sol-gel method	–	Reaction conditions: W = 0.20 g, F = 60 ml min^{-1}, $P(C_3H_6)$ = 2.53 kPa, $P(O_2)$ = 98.8 kPa, Temp: 250 °C	28^b	Cs^+ reduced CuO size and surface acidity Active phase: Cu^+	207
RuO_2-CuO_x-NaCl/SiO_2	Co-impregnation	10×30 nm (Cu nanorods)	$C_3H_6/O_2/He$ = 1/1-7/92-98/; GHSV: 5000–40 000 h^{-1}; Temp: 240–270 °C	21 (at 265 °C[c])	Stable within 7 h (testing time)	208
0.19 wt% Au/TS-1-K	SG method	<2.0	$C_3H_6/O_2/H_2O/Ar$ = 10/10/2/78; GHSV: 4000 ml.g^{-1}_{cat}.h^{-1}; Temp: 200 °C	4.[b]	Ti–OOH species detected	187
K-0.2 wt% Au/TS-1-KH	SG method	1.6–1.8	$C_3H_6/O_2/H_2O/Ar$ = 10/10/2/78; GHSV: 4000 ml.g^{-1}_{cat}.h^{-1}; Temp: 200 °C	8.8^b	Sel_{PO} : *ca.* 37% $Conv_{C3H6}$: 2.7%	210
Au_{6-10}/Al_2O_3 (amorphous)	Atomic layer deposition (ALD)	Cluster	Gas[a] $C_3H_6/O_2/$ = 2:1. Max TOF at 170 °C: 0.45 molecule $(site)^{-1}$ s^{-1}	–	Adding H_2 or H_2O improved the stability	211
Ti-Al-HMS (Ti/Al/Si = 4/1/100)			Catalyst amount: 1 g; quartz sand: 2 g; C_3H_6: 3.3 mL min^{-1}; 10% O_2/Ar: 16.7 ml min^{-1}; Temp: 250 °C	68^b	Required the simultaneous existence of Ti and Al in the framework of HMS	212

[a]Sel_{PO} = PO selectivity; $Conv._{C3H6}$ = conversion of propylene.
[b]PO formation rate was estimated from the highest catalytic activity in the literature.
[c]Assuming the catalyst density is 1 g ml^{-1} in specific rate calculation.

showed that PO was formed as a primary product in parallel with acrolein on CuO_x/SiO_2 catalyst.[207] In the absence of Cs^+, PO readily underwent isomerization to allyl alcohol followed by oxidation to acrolein. With the addition of Cs^+, the isomerization reaction of PO and allylic oxidation were suppressed, thus increasing PO selectivity (see Table 9.5).

9.4.2.2.3 Au-based Catalysts. Huang *et al.*[187] reported that, although Au/TS-1 prepared with Au clusters <2.0 nm exhibited high stability (for more than 24 h), high catalytic activity (C_3H_6 conversion about 8.8%) and propylene oxide selectivity (about 82%) in C_3H_6 epoxidation with an $O_2 + H_2$ mixture, the 0.19 wt% Au/TS-1-K1 catalyst (TS-1-K1 represents alkaline-treated TS-1, where K stands for KOH and 1 stands for 1.0 h) with a mean Au diameter of 1.8 nm showed a low PO production rate in the absence of H_2. However, after the addition of H_2O vapor to the feed gas, the PO formation rate on the same catalyst was markedly enhanced. In contrast, Au/TiO_2 and Au/Al_2O_3 did not show any activity for PO production even in the presence of H_2O vapor in the feed gas. Similar to the case of propylene epoxidation in the presence of H_2 and O_2 (see Figure 9.6), *in situ* UV-vis diffuse reflectance spectroscopy measurement confirmed the formation of active Ti–OOH on Au/TS-1-K1 in the O_2 and H_2O gas mixture and its consumption after introduction of propylene in the feed gas. A similar experiment on TS-1-K1 (without Au clusters) showed the absence of any Ti–OOH species, suggesting that Au clusters are indispensable for the formation of hydroperoxides (–OOH species). As indicated by a theoretical calculation,[209] hydroperoxides could be produced directly from $O_2 + H_2O$ over Au clusters. The proposed reaction route is as follows. First, O_2 reacts with H_2O over Au cluster surfaces to produce –OOH species; then this –OOH species is reversibly transferred from the Au cluster surfaces to the neighboring Ti sites to form Ti–OOH species, which are responsible for subsequent C_3H_6 epoxidation. It is likely that during the reaction $O^•$ radicals would be produced and further used to oxidize C_3H_6 to PO, acrolein, and CO_2. Accordingly, PO selectivity should be improved by controlling the reactivity of $O^•$ radicals over the surfaces of Au clusters.[187]

Huang *et al.* further carried out a series of treatments to the TS-1 support.[210] First TS-1 was treated with aqueous KOH solution to introduce surface defects. The KOH-treated TS-1 (TS-1K) was then further treated with aqueous HNO_3 solution to remove residual potassium, followed with deposition of 1.6 nm Au clusters by the SG method to form Au/TS-1KH. Finally, Au/TS-1KH was modified by impregnating potassium salts, alkali hydroxides, or alkaline earth metal acetates. It was found that, in propylene epoxidation with molecular O_2 and H_2O, Au/TS-1KH promoted by basic potassium salts (CH_3COO^-, $CO_3{}^{2-}$ and $PO_4{}^{3-}$) or hydroxides of Na, K, Rb, and Cs could produce PO with a selectivity of 12–45%, while Au/TS-1KH combined with neutral potassium salts (KCl, KNO_3, and K_2SO_4), LiOH or alkaline earth metal acetates could not produce PO. Combining all the above experimental results, it could be concluded that direct propylene epoxidation with molecular O_2 over Au catalysts requires the following conditions:

(1) Au exists as clusters smaller than 2.0 nm.
(2) basic salts or hydroxides of alkali (except Li) are added as promoters.
(3) H_2O is introduced in the feed gas.
(4) Ti sites are indispensable and should exist as isolated ones.

However, Lee *et al.*[211] reported in 2009 that the deposited Au_{6-10} clusters (denoted as Au_n) on amorphous alumina films also exhibited high activity and selectivity for direct propene epoxidation in the presence of O_2 and H_2O. The *ab initio* DFT calculations comparing Au/TiO_2 and Au_n/Al_2O_3 confirmed the active characteristics of the Au_n/Al_2O_3 interface. These recent results prompt further development on non-Ti containing supports for PO synthesis.

9.4.2.2.4 Other Catalyst Systems. Liu *et al.*[212] synthesized titanium and aluminium containing hexagonal mesoporous silica (Ti–Al-HMS) for the epoxidation of propylene with molecular O_2. In a fixed bed catalytic system, Ti–Al-HMS (Ti/Al/Si molar ratio = 4/1/100) showed 47.8% propylene conversion and 30.6% selectivity at 250 °C, which corresponds to *ca.* 68 g_{PO} kg_{cat}^{-1} h^{-1}. The simultaneous existence of Ti and Al in the framework of HMS seems to be important for improving the PO yield, as both Ti-HMS and Al-HMS are not active. An epoxide-like intermediate forming on Ti and Al sites as well as the adsorbed O_2 was proposed to elucidate the synergistic effect of Ti and Al in Ti–Al-HMS. However, this catalyst is not stable, as evidenced by a continuous decrease in propylene conversion due to formation of surface carbonaceous deposits. Adding a little amount of hydrogen in the feed gas could reduce the carbonaceous formation and the catalyst deactivation, but it could not completely suppress the latter. The addition of KNO_3 could improve the selectivity for PO, and the $Ca(NO_3)_2$ additive could improve the propylene conversion over Ti–Al-HMS.

Besides developing new catalyst supports, bimetallic or multimetallic catalysts should be promising for propylene epoxidation. High-throughput experiments showed that Ag–Cu/SiO_2 and Mn–Cu/SiO_2 exhibited good catalytic performances.[213] The other two examples are reported by Li *et al.*[202] and Suebsai *et al.*,[208] showing that Na-$VCe_{0.5}Cu_{0.2}$-NaCl and RuO_2-CuO_x-NaCl/SiO_2, respectively, are promising. However, a systematic study is still lacking in this area.

9.4.2.3 *Gas-phase Epoxidation of Propylene using N_2O as the Oxidant on Supported Catalysts*

Duma and Hönicke[214] first found that the decomposition of nitrous oxide (N_2O) could be coupled with the epoxidation of propylene reaction. This is because the release of an oxygen atom from N_2O is more favorable thermodynamically than that from the oxygen molecule:

$$N_2O \rightarrow N_2 + \frac{1}{2}O_2,\ \Delta r H^0{}_{298} = 167.6\ \text{kJ mol}^{-1} \tag{9.8}$$

$$\frac{1}{2}O_2 \rightarrow O,\ \Delta r H^0{}_{298} = 249.2\ \text{kJ mol}^{-1} \tag{9.9}$$

The common catalysts are iron oxide supported on silica (FeO_x/SiO_2). The optimal iron loading is reported to be in the range 0.01–0.1 wt% and the iron oxide particle dimension is smaller than 2 nm. The reaction occurs through the non-dissociative adsorption of N_2O at the active centers of the catalysts, followed by the reaction between N_2O and propylene. The addition of sodium suppresses the side reactions, thus improving the selectivity toward propylene oxide. Ananieva and Reitzmann[215] modified the FeO_x/SiO_2 catalyst with caesium and increased the PO formation rate 4–6 times. They also observed that the isomerization of propylene oxide to other C_3-oxygenates, such as propanal, acetone, and allyl alcohol, as well as PO dimerisation were the main side reactions. Acid–base properties and silanol groups on the catalyst surface are responsible for these reactions. Jacobs and co-workers[216] systematically screened many catalysts:

(1) Sodium acetate promoted active metal oxide (Zn, Mn, Co, Ru, Cu, Ni, Au, Rh, Fe, Cr, Sm, Eu, La, Nd, and Ag)/silica catalysts
(2) FeO_x supported on various supports (ZSM-5, Al_2O_3, TiO_2, $CaCO_3$, $BaCO_3$, and amorphous SiO_2/Al_2O_3 with various Si to Al ratios)
(3) FeO_x/silica promoted by various alkali (Li, Na, K, Rb, and Cs) and alkali earth (Mg and Ba) acetate salts, sulfates, chlorides, and hydroxides.

They observed that $Fe/Rb_2SO_4/SiO_2$ catalyst showed a PO selectivity of 85–90%. Interestingly, the sodium acetate-promoted iron oxide on silica exhibited superior PO formation yields with N_2O, but no epoxidation selectivity with O_2 as oxidant. In the latter case, mainly CO_2 was formed, along with small amounts of acrolein and acetaldehyde.

However, there are not many examples using supported Au catalysts for this reaction. Medina and co-workers[217] found that 4 wt% Au_1Cu_3/TiO_2 catalyst (prior to the catalytic test, Au–Cu alloy was formed after reduction in H_2 at 400 °C) was superior to the respective Au/TiO_2 and Cu/TiO_2 monometallic catalysts for the reaction. One of the advantages of the Au-containing catalyst is the much lower reaction temperature needed for the reaction than for the other catalysts (see Table 9.6). Probably Au-containing bi-, tri-metallic catalysts or multiple component catalysts should be explored for this reaction.

Table 9.6 Gas-phase epoxidation of propylene to PO using N_2O as the oxidizing agent.

Catalyst	Preparation Method	Mean Fe or Au size/nm	Reaction Conditions (Volumic Composition of the Reactant Gases and Reaction Temperature)	PO formation rate/ g_{PO} kg_{cat}^{-1} h^{-1}	Other Information	Ref.
FeO_x(0.01–0.1) wt% Cs/SiO_2	Impregnation	<2	$C_3H_6/N_2O/He = 1/15/84$; GHSV = 4000 ml h^{-1} $g^{-1}{}_{cat}$; 380–450 °C	4.7[a]	PO selectivity *ca.* 50%; propylene conversion *ca.* 10% at 375 °C	214
FeO_x(0.03–0.1) wt% Cs/SiO_2	Impregnation	3.0	$C_3H_6/N_2O/He = 1/15/84$; GHSV = 10 l g^{-1} cat h^{-1} (gcat: active mass of catalyst); 375 °C	20–30[a]	PO selectivity *ca.* 75%; propylene conversion *ca.* 10% at 375 °C	215
$Fe/Rb_2\ SO_4/SiO_2$, Fe: 0.15 wt%, Rb/Fe = 12	Impregnation method	10–15	$C_3H_6/N_2O/He = 1/5/30$; GHSV = 4500 ml h^{-1} g_{cat}^{-1}; 410 °C	14–30[a]	PO selectivity *ca.* 85–90%; propylene conversion *ca.* 4.3% at 300 °C	216
4% Au_1Cu_3/TiO_2	Impregnation method	10–15	$C_3H_6/N_2O/Ar = 10/10/80$; GHSV = 9000 h^{-1}; 300 °C	24[b]	Cu/Au ratio of 3/1 showed the best catalytic behavior	217

[a]specific PO formation rate was evaluated from the catalytic activity data.
[b]Assuming the catalyst density is 1 g ml^{-1} in the specific rate calculation.

9.5 Selective Oxidation of Other Hydrocarbons

9.5.1 Selective Oxidation of Aliphatic Alkanes

In the preceding section, we discussed the epoxidation of propylene to propylene oxide. The selective oxidation of other hydrocarbons, in particular light alkanes (*e.g.* methane, ethane, and propane) to olefins, alcohols, and epoxides is also industrially important for the manufacture of many other fine and bulk chemicals.

In these reactions, there are mainly two technical issues that remain challenging. One is the activation of the C–H bond in saturated hydrocarbons.[218] The commonly used catalysts to carry out such processes are platinum group metals as they provide empty d-orbitals for C–H activation, typically the rate-determining step.[219] The second one is the selectivity, as the products are, in general, more reactive than the reactants, which is thermodynamically likely to lead to over oxidation to CO_x species. Much effort has been made to lower combustion activity of the catalysts, *e.g.*, by introducing an Au component in the catalysts, despite the fact that the latter often has lower activity for the initial C–H bond activation.[220] In this section, besides addressing the basic problems met in these reactions, we also focus on the catalyst development aspect, particularly the part related to Au catalysis.

9.5.1.1 Selective Oxidation of Methane

As the main component in natural gas, methane is an important raw material in C_1 chemistry. Selective oxidation of methane to methanol is one of the options that may provide a good feedstock for synthesis of valuable hydrocarbons,[221] as well as a solution to storage and transportation problems for natural gas. Unfortunately, due to high stability of methane, the conventional activation of the C–H bond (439 kJ mol^{-1}) often requires high temperature and pressure.[222] To prevent over-oxidation of methanol, an oxidation reaction in concentrated sulfuric acid was developed to produce methylbisulfate, followed with hydrolysis to form methanol. Periana and coworkers successfully used Pt(II) bipyrimidine complexes for methane oxidation and obtained 90% methane conversion with 81% selectivity (turnover number ~300) towards methylbisulfate at 3.4 MPa and 220 °C.[223] However, the reaction requires a large amount of concentrated sulfuric acid, which incurs additional cost due to corrosion and waste water treatment. Recently, Palkovits *et al.* developed a solid Pt(II) poly(benzimidazole) complex catalyst and tested it at 4.0 MPa, 215 °C in sulfuric acid. A high activity of ~550 turnover number and >75% methylbisulfate selectivity was achieved.[224] Although the bisulfate route is promising, direct methane oxidation using molecular oxygen or peroxides is still preferred as it is more environmentally friendly.

In recent years, supported Au catalysts have been explored for the partial oxidation of methane. Li *et al.* reported methane oxidation to methanol over supported Au/SiO_2 in [Bmim]Cl ionic liquid using trifluoroacetic acid as the acidic reagent and $K_2S_2O_8$ as the oxidant.[225] The reaction was conducted at 2.0 MPa of methane and 90 °C for 20 h. Under this reaction condition, a maximum 24.9% methane conversion and 71.5% methanol selectivity were achieved. In this reaction, the methanol product is present in the form of methyl trifluoroacetate. Over the heterogeneous catalytic system of Au/SiO_2 in ionic liquid, the results showed that the ionic liquid played a dual role: improving the dispersion and stability of Au/SiO_2 nanoparticles; and promoting the reactivity of gold, which rendered high methanol selectivity.

However, there are also different conclusions from the literature. Hereijgers and Weckhuysen prepared various Au catalysts (0.8–1.3 wt%) supported on alumina, silica, SBA-15, titania, zirconia, and ceria for methane oxidation.[222] The average Au nanoparticles were about 4 nm on ZrO_2, SBA-15, SiO_2 and TiO_2, and about 30–200 nm on ZSM-5. At 400 °C, all the catalysts exhibited very low methane conversion (<1%) with CO_2 as the sole carbon-containing product. Similar results were also obtained by Walther *et al.*,[226,227] who observed complete combustion of methane to CO_2 at low temperature (30–250 °C) irrespective of the type of supports. Based on the obtained results, they concluded that supported Au catalysts promote total combustion of methane, instead of selective oxidation of methane to methanol.

The oxidative coupling of methane to ethane using 1 wt% Au-5 wt% La_2O_3/CaO catalyst was also studied by Eskendirov *et al.* in the presence of hydrogen peroxide as an initiator.[228] A relatively high methane conversion of about 35% and *ca.* 49% ethane selectivity was achieved at 700 °C. In contrast, Blick *et al.* observed that the introduction of Au (0.04–15 wt%) onto MgO support by impregnation decreased the methane conversion and C_2 hydrocarbon selectivity at 750 °C, and instead promoted the formation of CO_x as the main oxidation products.[229] It was argued that the impregnation of Au poisoned the active sites of methane coupling oxidation on MgO by occupying the stepped surface planes, which usually contain reactive O^{2-} vacancies.

From these contradicting results, it seems that the application of supported Au catalysts to these reactions is still in the infant stage, and a more systematic study is needed to clarify the complex experimental phenomena.

9.5.1.2 Selective Oxidation of C_3 and C_4 Hydrocarbons

Partial oxidation of *n*-butane to lighter alkenes using oxygen was investigated by Sá *et al.*[220] over 0.2 wt% Au/La_2O_3 catalyst at 650 °C. After 48 h reaction, the average *n*-butane conversion was stable at 38.1% while the total alkene selectivity was 62.5%, yielding a mixture containing ethene, propene, 1-butene, and 2-methylpropene. The reason for the increase in the catalyst performance on 0.2 wt% Au/La_2O_3 is believed due to the ability of La_2O_3 to stabilize cationic Au, as the XPS results showed that Au/La_2O_3 retained much of the cationic gold species (Au^+). In addition, La_2O_3 is also known to be an excellent radical generator, which is further promoted by the interaction of cationic gold and the oxide support. The active site is believed to be a combination of cationic gold and La_2O_3, which activates *n*-butane at low temperature but does not provide a site for efficient combustion of the hydrocarbon, thus leading to low CO_x formation and favoring selective oxidation of the hydrocarbon. The addition of Au also appears to stabilize lanthanum oxycarbonate species that may play a role in enhancing the performance of the Au-promoted oxides catalysts.

Due to many useful derivatives from propane, in particular propylene, acetone, and isopropanol, direct selective oxidation of propane is very attractive. Acetone and isopropanol are major solvents used in many applications and common reagents in organic synthesis. About 6.7 million tons of acetone were produced worldwide in 2010,[230] while the production of isopropanol was about 2.7 million tons in 2012.[231] Commercially, acetone is mainly produced via the cumene process which involves benzene and propylene as reagents. Isopropanol, however, is produced by indirect hydration of propylene in the presence of sulfuric acid as the catalyst.

In the early work by Kalvachev *et al.*, it was reported that supported Au/Ti-MCM-41 (1.2 wt% Au, Ti/Si = 2.8/100) could selectively catalyze propane and isobutane to form acetone and *tert*-butyl alcohol, respectively.[169] At the reaction temperature of 120 °C, propane and isobutane conversions were 0.3% and 2.2%, respectively, with selectivities towards acetone and *tert*-butyl alcohol about 48% and 85%, respectively.

Direct oxidation of propane to acetone and, in lesser amounts, isopropanol, could be carried out under mild conditions (170 °C and 0.3 MPa) in the presence of O_2 and H_2 on Au–Ba/TS-1.[232] The catalyst was prepared by the DP method at pH 9 using Na_2CO_3. Ba was added to improve Au capture by the support. With an Au loading of about 0.1 wt% ($d_p = 4$ nm) and Si to Ti ratio of 100 : 3, the highest propane conversion of 6.5% with total oxygenate selectivity of 95% were obtained, which consisted of 90% acetone and 5% isopropanol. The total oxygenate production rate was 33 g kg_{cat}^{-1} h^{-1} with a TOF of 284.4 h^{-1} based on the exposed Au. These results are comparable with or higher than those obtained with other catalyst systems. Previously, oxidation reaction using H_2O_2 and TS-1 as catalyst was reported by Clerici in a batch system.[233] Using 0.3 M H_2O_2 solution in methanol, the reaction was carried out at 55 °C with stirring. The product distribution comprised 66.2% isopropanol and 33.8% acetone, and the total production rate was 79 g kg_{cat}^{-1} h^{-1}. Yamanaka *et al.* investigated propane oxidation in an H_2–O_2 cell using Pd-black as catalyst at 25 °C; acetone, acetic acid, and CO_2 were the main products, with the oxygenate production rate of only 6 g $kg_{cathode}^{-1}$ h^{-1}.[234] Espro *et al.* studied selective propane oxidation using a multiphase catalytic hollow fiber membrane reactor with an Fe^{2+}–H_2O_2 Fenton system.[235] Different oxygenates, such as *n*-propanol, isopropanol, propionaldehyde, and acetone, were obtained at mild reaction conditions (70–120 °C, 0.14 MPa). The total oxygenate selectivity was 96%, but the production rate was only 5 g $kg_{membrane}^{-1}$ h^{-1}.

Later, Bravo-Suárez and co-workers demonstrated that Au/TiO_2[236] effectively catalyzed the oxidation of propane to propylene, while Au/TS-1[232,237] converted propane to acetone and isopropanol. A detailed mechanistic study on Au/TS-1 was performed using an *in situ* UV-vis spectrophotometer and XANES spectra.[232,237] Similar to the PO synthesis mechanism discussed in the previous section, they concluded that the molecular oxygen is first adsorbed on Au nanoparticles, likely in the form of O^{2-}. The activated oxygen further reacts with H_2 to form H_2O_2. The as-formed H_2O_2 is subsequently transferred to Ti centers on TS-1 and reacted with the surface hydrated hydroxyl groups forming hydroperoxide (Ti–OOH) species. Recalling that TS-1 support mainly produces acetone and isopropanol, while TiO_2 support yields propylene, the authors proposed that upon contact with Ti–OOH species, propane is activated on TS-1 forming isopropoxy intermediates. On TS-1 support, propane molecule is subsequently attacked by the terminal OH groups of the hydroperoxide, followed by electron transfer from the C–H bond to the proximal peroxide oxygen (most electropositive) to form acetone. However, the dehydration of the isopropoxide moiety with surface OH groups on TiO_2 will produce propylene. The different selectivities of the TS-1 and TiO_2 supports could be due to difference in their acid–base properties.[238] Temperature-programmed desorption (TPD) of CO_2 and NH_3 from TS-1 and TiO_2 supports, as well as their corresponding supported Au catalysts clearly showed that Au/TS-1 primarily had a small numbers of weak acid and base sites, while Au/TiO_2 catalyst had stronger acid and base sites in higher concentrations. Therefore, under the oxidizing conditions in the study, acetone formation is expected to proceed by reaction of an isopropoxide intermediate with H_2O_2. However, propylene formation likely occurs by dehydration of isopropoxy species.

So far, the low catalytic activity remains a challenge that needs to be overcome for future selective oxidation of aliphatic alkanes. Hence more investigations are required to further explore the feasible catalysts for this reaction.

9.5.2 Selective Oxidation of Cycloalkanes

Selective oxidation of cyclohexane to cyclohexanol and cyclohexanone mixture [ketone-alcohol (KA) oil] (Figure 9.8) is an important reaction in the polymer industry as it is the first step in the commercial manufacture of the widely used nylon-6 and nylon-6,6. The current commercial oxidation reaction is typically carried out at 150–160 °C in a high pressure tank (1–2 MPa) using cobalt naphthenate as the catalyst. However, to prevent total oxidation of the products, the reaction is only run at around 4% conversion with 70–85% selectivity toward cyclohexanol and cyclohexanone.[239] This process is therefore deemed very inefficient. With increasing global production of nylon from 22.4 million tons in 2000 to 24 million tons in 2011,[240] as well as greater demand for the precursors, a much more efficient catalytic oxidation of cyclohexane needs to be developed. Benign oxidizing agents, such as

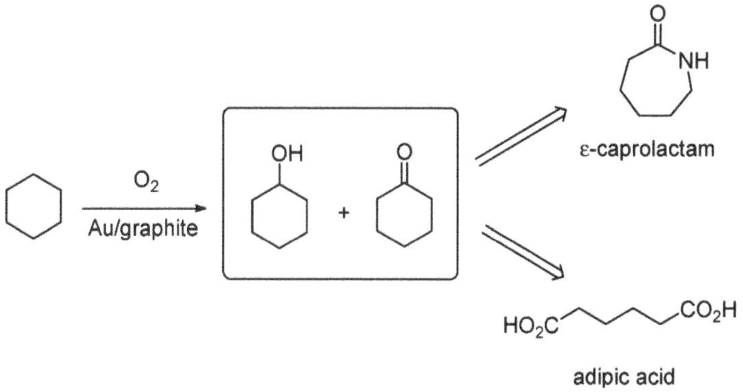

Figure 9.8 Selective oxidation of cyclohexane to caprolactam and adipic acid using cyclohexanone–cyclohexanol mixture as precursors.
Reproduced with permission from Ref. 414.

molecular oxygen or hydrogen peroxide, are also preferred to provide green oxidation chemistry.

Thomas and co-workers prepared aluminophosphate molecular sieves substituted with four-coordinate metal ions (Mn^{III}, Fe^{III}, Co^{III}) for cyclohexane oxidation using air as oxidant. Among the three catalysts, Fe-AlPO-5 was exceptionally good, giving a 19.8% conversion and a turnover number (TON) of 346 after a 24 h reaction at 130 °C and 1.5 MPa air pressure. The obtained products comprised 21.7% cyclohexanol, 32.5% cyclohexanone, and 32.3% adipic acid.[241] Other notable catalysts such as TS-1,[242] Ti-MCM-41,[243] and Ti-aluminophosphates redox molecular sieves[244] were also tested with H_2O_2 or other organic peroxides, such as TBHP. However, most of them exhibited low cyclohexane conversions (<20%).

A much more efficient and stable catalyst based on vanadium phosphorus oxide (VPO) catalyst was developed by Pillai and Sahle-Demessie for cyclohexane oxidation under mild reaction conditions using H_2O_2 as oxidant. A high TON of 400 was achieved at the optimum V to P ratio of 1 : 1. At 65 °C, it gave the highest conversion of 91% with products comprising 36% cyclohexanol and 64% cyclohexanone, using acetonitrile as a solvent.[245] The solvent used in the reaction, which determines the polarity of the medium in the reaction system, plays an important role in the activity of catalysts. When acetonitrile was replaced with methanol, 1,4-dioxane, or dichloromethane, the overall catalytic activity became markedly lower.[245]

From a safety and environmental viewpoint, however, substitution of peroxides with molecular oxygen, as well as the use of solvent-free systems are preferred. In view of this, Au-based catalysts have been developed as potential candidates and their catalytic performances are summarized in Table 9.7. The use of Au catalysts in the direct oxidation of cyclohexane was initiated by Zhao and co-workers.[246,247] Using molecular oxygen as the oxidant and in the absence of any solvent, a very high TOF of 3096 h^{-1} was

Table 9.7 Catalytic activity of different Au-based catalysts for selective oxidation of cyclohexane.

Support	Au Loading/ wt%	Catalyst Synthesis Method	T/°C	t/h	Solvent	Conversion/%	Product Selectivity/%		TOF/h^{-1}	Ref.
							Cyclohexanol	Cyclohexanone		
ZSM-5	0.50	HT[b]	150	3	–	6.5	27.0	73.0	3096	246
ZSM-5	1.30	HT	150	3	–	16.0	25.0	67.0	2242	246
MCM-41	0.13	Self-assembly[c]	150	6	–	12.0	37.0	60.0	11 214	247
MCM-41	0.76	Self-assembly	150	6	–	17.0	16.0	76.0	1907	247
SBA-15	4.53	*In situ* reduction[d]	150	6		32.0	32.0	62.0	858	248
Graphite[a]	1.00	IMP[e]	70	17	–	3.7	14.4	8.7	33	249
TiO_2/SiO_2	0.05	DP[f]	150	3	–	8.4	54.3	37.4	40 133	250
Al_2O_3	0.60	IE[g]	150	3	–	11.5	49.2	37.5	4508	251
TiO_2[h]	1.60	IMP	150	4	Acetonitrile	16.4	21.6% adipic acid		–	252

[a]Using *tert*-butyl hydroperoxide as oxidant.
[b]Hydrothermal method.
[c]One-pot synthesis of Au in MCM-41.
[d]*In situ* reduction of Au by silanol groups on SBA-15.
[e]Impregnation method.
[f]Deposition-precipitation method.
[g]Anion exchange method.
[h]the catalyst produced adipic acid instead of cyclohexanol and cyclohexanone.

achieved on 0.51 wt% Au/ZSM-5 at 150 °C and 1.0 MPa O_2. With 18.5 mmol cyclohexane, 5 mg catalyst, 6.5% conversion and *ca.* 90% selectivity were obtained after 3 h of reaction. The product distribution was 27% cyclohexanol and 73% cyclohexanone. When the Au loading was increased to 1.3 wt%, the overall conversion increased to 16% while maintaining a high selectivity of 92%. However, the TOF was decreased to 2242 h^{-1}. When MCM-41 was used as the support, high conversion and TOF of up to 17% and 3091 h^{-1}, respectively, were achieved at the Au loading of 0.76 wt%.[247] The obtained products consisted of 16% cyclohexanol and 76% cyclohexanone. In both cases, the catalysts were quite stable upon recycling, without any marked decrease in activity although the product selectivity shifted towards the formation of cyclohexanol.

Au/SBA-15 was tested by Zhu *et al.*[248] In order to improve the immobilization of Au nanoparticles on the support, SBA-15 was modified with amine, thiol, or ionic liquid prior to Au deposition. The resulting catalysts exhibited different Au content and particle size, which affected the activity of the catalysts. The catalytic reaction was carried out at conditions similar to those used by Zhao *et al.*[246] (18.5 mmol cyclohexane, 5 mg catalyst, T = 150 °C, $pO_2 = 1.5$ MPa, and reaction time 6 h). The highest conversion of 32% with TON of 858 was achieved on the thiol-modified catalyst. The total product selectivity was 94%, which consisted of 32% cyclohexanol and 62% cyclohexanone. As the Au content (4.53%) is higher than that in the catalyst prepared by Zhao *et al.* (0.51%), it is reasonable to observe a higher conversion.

Xu *et al.* used supported Au catalysts on graphite for aerobic oxidation reaction under mild conditions (70 °C and 0.3 MPa O_2).[249] Au/C catalysts with different Au content (0.25–1 wt%) were prepared *via* formaldehyde reduction. For comparison, Pd/C and Pt/C catalysts with a similar metal content were also prepared and tested. The catalytic test was carried out by adding TBHP as an initiator and the results showed a limited activity with the conversion less than 8%. Au/C catalysts exhibit very similar catalytic behavior to that of the supported Pt or Pd catalysts; the selectivity observed is solely a function of the conversion, which in turn is a function of the reaction time. High selectivity can only be achieved at very low conversions and within a short period, after which the selectivity declines steadily as the reaction progresses.

In more recent publications, Au catalysts supported on Ti-doped silica[250] and alumina[251] were investigated. Amorphous silica was modified by doping with TiO_2 *via* a surface sol-gel process and used as the support for the Au catalyst.[250] The obtained Au catalysts were tested in the selective cyclohexane oxidation using molecular oxygen as the oxidant. With Au loading as low as 0.05 wt% and TiO_2 doping of 0.26 wt%, 8.4% cyclohexane conversion could be achieved with 91.7% selectivity of cyclohexanone (37.4%) and cyclohexanol (54.3%) at 150 °C, 1.5 MPa O_2, and a 3 h reaction time, similar to the conditions used by Zhao *et al.*[246] and Zhu *et al.*[248] A high TOF was achieved at 40 133 h^{-1}. Moreover, the catalytic activity and selectivity were well retained after four recycling oxidation reactions, showing a good stability of the Au catalyst supported on titania-doped silica.

Using a direct ion exchange method, Au nanoparticles were deposited on γ-Al_2O_3, which also exhibited high catalytic performance.[251] During the synthesis, the loading of Au on the support was proceeded by a direct anionic exchange of $[Au(OH)_nCl_{(4-n)}]^-$ ($n = 0$–4) ions with other anions chemisorbed on the surface of Al_2O_3, making Au highly dispersed at the molecular scale.[251] In this work, a conversion of 11.5% as well as a TOF of 4508 h^{-1} was obtained over 0.6 wt% Au/Al_2O_3 under the same reaction conditions. Furthermore, a surprisingly high TOF up to 16 136 h^{-1} was attained over 0.2 wt% Au/Al_2O_3. It is well-established that the activation of molecular oxygen is a rate-determining step in aerobic cyclohexane oxidation, and the small Au particles of around 5 nm are effective in activating molecular oxygen under the reaction conditions. Therefore, the high activities of these supported Au catalysts should be relevant to the nanosized effect of gold.[250]

Recently, Alshammari *et al.* reported direct oxidation of cyclohexane to adipic acid using Au nanoparticles supported on CaO, MgO, ZrO_2, TiO_2, and Al_2O_3.[252] The activity test was performed in a Parr reactor with acetonitrile as the solvent and TBHP as the initiator. The reaction was conducted at 150 °C and 1 MPa O_2 for 4 h. Among the oxide supports, TiO_2 and Al_2O_3 yielded small Au nanoparticles with a narrow particle size distribution (1–5 nm), while the remaining supports exhibited a wide distribution of Au from 1 to 10 nm. The nature of the support has significant effects on the conversion of cyclohexane as well as the products selectivity. As the TiO_2 supported Au nanoparticle exhibited superior activity to all other catalysts, it suggests that a smaller size of Au particle is an important parameter. The cyclohexane conversion and adipic acid selectivity obtained over Au/TiO_2 were 16.4% and *ca.* 21.6%, respectively, without a noticeable change in the catalyst stability. In contrast, Au/MgO and Au/CaO catalysts showed the poorest performance, yielding large amounts of CO and CO_2 (up to 35% selectivity).

9.5.3 Selective Oxidation of Aliphatic and Aromatic Alkenes

Due to their versatility, epoxides, in particular, are highly desired products from the selective oxidation of alkenes. Epoxidation of lower alkenes, such as propene and 1-hexene, are usually performed over microporous TS-1 using a stoichiometric amount of hydrogen peroxide as the oxidant.[253] Higher alkenes, such as cyclohexene or *cis*-stilbene can be selectively oxidized over Ti-MCM-41 with TBHP as the primary oxidant,[254,255] while the epoxidation of 1-octene[256] and monoterpenes[257] has been reported using metal-substituted polyoxometalates (POM) and excess H_2O_2.

In 2005, selective oxidation of different types of alkenes were comprehensively studied by Hughes *et al.*[258] over supported Au/graphite (Johnson Matthey) catalysts in the presence of oxygen and peroxide as an initiator. With 1.0 wt% Au loading, the catalyst was found to be very effective in catalyzing different substrates such as cyclohexene, styrene, *cis*-cyclooctene, and *cis*-stilbene. A range of solvents was used and the product distribution was found to be strongly dependent on the type of solvent used. Table 9.8 summarizes the catalytic oxidation of different substrates.

Selective oxidation of cyclohexene primarily produced cyclohexene oxide when 1,2,3,5-tetramethyl benzene and TBHP was used as the solvent and initiator, respectively. The highest selectivity was obtained at 80 °C after 24 h reaction, at which 29.7% of cyclohexene conversion was achieved with 50.2% of cyclohexene oxide and 26.3% of 2-cyclohexe-1-one as the products. Modification of the catalyst with bismuth markedly increased the selectivity of cyclohexene-1-one, despite lowering the overall conversion. Over Bi-modified 0.5% Au/C, the cyclohexene conversion was 24.2% and the selectivity of the combined C_6 products was more than 97%; the products comprised 49.6% of cyclohexene oxide, 44.2% of 2-cyclohexene-1-one, and 4.1% of 2-cyclohen-1-ol. An XPS study revealed that the reactive layer of Bi was on the catalyst surface.

When *cis*-cyclooctene was used as the substrate, a very high selectivity towards C_8 products was obtained (>97%) in 1,2,3,5-tetramethyl benzene at 27.8% of conversion. Cyclooctene oxide was the major product with 94% of selectivity. High selectivity to epoxide could be achieved even with a cyclooctene to TBHP molar ratio of 300. In the absence of solvent and under the same reaction condition (80 °C, 24 h), the overall conversion was 7.9% with a 95% C_8 products selectivity, of which the majority consisted of cyclooctene oxide (81.2%). To confirm the general application of the catalyst, the oxidation reaction was also conducted for styrene and *cis*-stilbene. Under the same reaction conditions, the oxidation of styrene in 1,2,3,5-tetramethyl benzene resulted in 90% of total C_8 products selectivity at 17.3% conversion, and benzaldehyde was the major product. The oxidation of *cis*-stilbene to stilbene oxides was also successfully performed by using *p*-xylene as solvent. An overall conversion of 11% and stilbene oxides selectivity 91% were achieved, with a *cis*- to *trans*- isomer ratio of 0.74.

The observed high selectivity toward epoxide demonstrates direct oxidation of the C=C bond by molecular oxygen with the assistance of the initiator.[258] As the reaction yields more *trans*-stilbene, it strongly suggests that the two C–O bonds in the epoxide are not formed simultaneously. The formation of the *trans*- isomer provides evidence for the rotation about the C–C bond following the formation of the initial C–O bond. It indicates the adsorption of *cis*-stilbene molecules onto Au catalyst surface, which creates sterical constraints, thus decreasing both the symmetry and accessibility of the molecule and forcing the reaction to proceed in a specific way.[259] This is supported by the results obtained from the oxidation of *trans*-stilbene, from which there is no *cis*-epoxide observed.[260] It is also considered that the active intermediates that lead to the formation of epoxide are closely associated with Au surface. The role of solvent in the oxidation reaction was thought to control the rates of the consecutive oxidation steps as well as on the formation of the intermediates *via* electron transfer to and from Au nanoparticles.

Table 9.8 Selective oxidation of alkenes over 1 wt% Au/C catalyst.[258]

Substrate	Solvent	Conversion/%	Product Selectivity/%		
			Cyclohexene oxide	2-Cyclohexene-1-one	2-Cyclohen-1-ol
Cyclohexene	Water	100	0	0	0
	1,4-dimethylbenzene	53.5	0	12	0
	1,2,3,5-tetramethylbenzene	29.7	56.2	26.3	0
Substrate	Solvent	Conversion/%	Product selectivity/%		
			Cyclooctene oxide	2-Cyclooctene-1-one	2-Cycloocten-1-ol
cis-Cyclooctene	1,2,3,5-tetramethylbenzene	27.8	94	0	3.7
	–	7.9	81.2	4.1	9.3
Substrate	Solvent	Conversion/%	Product selectivity/%		
			Styrene oxide	Acetophenone	Benzaldehyde
Styrene	1,4-dimethylbenzene	34.2	13.5	10.5	37.7
	1,2,3,5-tetramethylbenzene	17.3	28.9	15	46.8
Substrate	Solvent	Conversion/%	Product selectivity/%		
			cis-Stilbene oxide	*trans*-Stilbene oxide	*cis*- : ***trans***-ratio
cis-Stilbene	*p*-xylene	11.1	38.7	52.3	0.74
	isopropylbenzene	47.6	7.1	73.9	0.1

Reaction conditions: T = 80 °C; t = 24 h. Initiator: *tert*-butyl hydroperoxide.

By using azobis(isobutyronitrile) (AIBN) as the initiator, Alvaro *et al.* further studied the radical trapping capability of Au and the relationship with its catalytic activity in the oxidation of cyclohexene.[261] The aerobic oxidation of cyclohexene was performed at 60 °C using 1.3% Au/CeO_2 supported catalyst. Without the addition of AIBN, only trace amounts of oxidation products were observed. However, by adding a small quantity of AIBN (1%) into the feed, a significant increase in cyclohexene conversion was observed. The products included cyclohexene epoxide (20%), 2-cyclohexenol (15%), and some other minor products. Control experiments with Au and without AIBN, or with AIBN and without Au showed negligible conversion under the working conditions. Hence, radical initiators most likely promote the catalytic activity of Au in hydrocarbon aerobic oxidation by the intermediacy of organogold species arising from radical trapping by Au nanoparticles. The presence of positive gold atoms in gold nanoparticles, such as Au(III) or Au(I), as observed with XPS, would likely form R–Au organogold species on the surface of the gold catalyst. These organogold compounds can promote aerobic oxidation without the need for additional amounts of radical initiator through a reversible capping/decapping of gold atoms in radical chain mechanisms.

Caps and co-workers outlined the free radical chain mechanism of gold-catalyzed *trans*-stilbene by using commercial Au/TiO_2, Au/Fe_2O_3, and Au/C.[259,260] The correlation between the solvent effect and catalytic behavior was studied in detail and a reaction mechanism was proposed. The oxidation reaction was performed in the presence of TBHP as oxidant, and was held in air at atmospheric pressure for 24 h. With 1 wt% Au/C as the catalyst, the reaction in methyl cyclohexane exhibited the highest yield (42%) of *trans*-stilbene oxide compared with other solvents such as toluene, cyclohexane, and acetonitrile. The activities of the catalysts varied in the following order: 1.5 wt% Au/TiO_2 > 5 wt% Au/Fe_2O_3 > 1 wt% Au/C. Overall, the Au/TiO_2 catalyst exhibited the highest conversion (95%) and epoxide selectivity (77%), compared with 81% and 53%, respectively, for Au/C. When only a limited amount of peroxide was used, Au/TiO_2 still remained superior to the remaining catalysts, although the initial reaction rate decreased dramatically (60% conversion). In a detailed mechanistic study, it was evidenced that free radical chain reactions occurred during the oxidation reaction, being strongly dependent on the amount of the initiator. The presence of free radicals further activates oxygen molecules closely associated with the nature of the solvent. An analysis with gas chromatography–mass spectrometry (GC-MS) revealed that a fraction of methylcyclohexane was oxidized. These pieces of evidence indicate that co-oxidation of the solvent and *trans*-stilbene takes place during the reaction. Both Au and titania seem to take part in this free radical reaction mechanism sustained by molecular oxygen, possibly by trapping the unselective hydroxyl radicals, catalyzing the active radical formation, and stabilizing the intermediate resulting from the addition of the methyl cyclohexyl peroxy radical onto *trans*-stilbene. The proposed reaction mechanism is summarized in Figure 9.9.

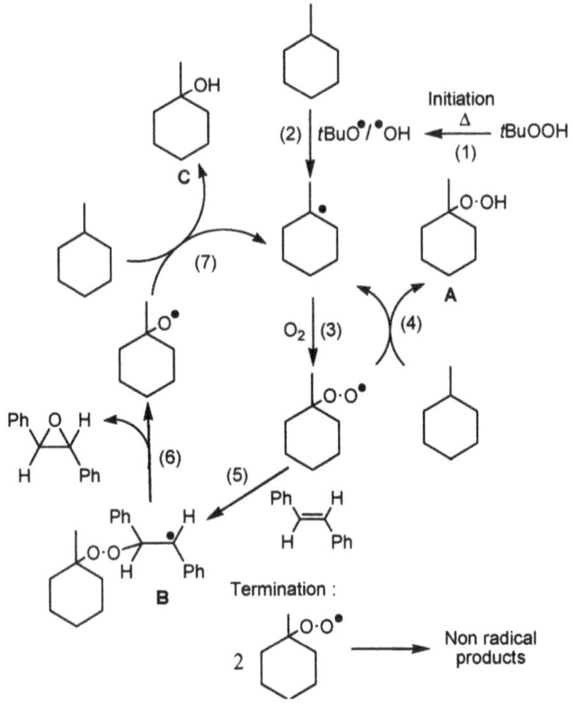

Figure 9.9 Proposed reaction mechanism of *trans*-stilbene selective oxidation in methylcyclohexane in the presence of *tert*-butyl hydroperoxide over Au/TiO_2 catalyst.
Reproduced with permission from ref. 259.

The application of various supported Au catalysts for selective oxidation of styrene using TBHP has been reported in several papers and the catalytic activity is presented in Table 9.9. Patil *et al.* deposited Au nanoparticles on various metal oxides by the homogeneous DP method using urea as the precipitation agent.[262–264] This method yields catalysts with higher Au loading and better Au dispersion compared with the conventional DP method using NaOH, thus leading to higher styrene conversion. The catalytic study was carried out in a stirred batch reactor at atmospheric pressure and temperature of 80 °C in the presence of benzene as the solvent. The highest yield of styrene oxide (35%) was obtained over Au/Tl_2O_3 catalyst, mostly attributed to the high basicity and redox property of Tl_2O_3. However, the specific activity per unit mol of Au was lower than that of the other catalysts, in particular Au/CaO, which with Au loading of about half of that of Au/Tl_2O_3 (4.7%), resulted in a styrene oxide yield of 32%. The high Au loading and low surface area of the latter led to the formation of a large Au nanoparticle size (24.4 nm) compared with that of Au/CaO (5.7 nm).

A detailed study on the effect of surface basicity on the Au loading and catalytic activity was conducted by Yin *et al.* through synthesis and

Table 9.9 Catalytic activity of different Au-based catalysts for styrene selective oxidation.

Support	Au Loading/ wt%	Catalyst Synthesis Method[b]	T/°C	t/h	Solvent	Conversion/%	Products Selectivity/% Syrene oxide	Benzaldehyde	Phenyl Acetaldehyde	Ref.
TiO_2	6	HDP[c]	82	3	Benzene	60	51	14	18	262
MgO	7.5	HDP	80	3	Benzene	62.6	54.3	10.8	16.8	263
CaO	4.7	HDP	80	3	Benzene	53.6	60.2	6.3	19.2	263
SrO	5.1	HDP	80	3	Benzene	53	44.8	11.7	26	263
BaO	5.3	HDP	80	3	Benzene	55.9	53.5	10.2	17.6	263
Al_2O_3	6.4	HDP	82	3	Benzene	45	27	23	12	264
Ga_2O_3	6.4	HDP	82	3	Benzene	48	29	20	20	264
In_2O_3	6.8	HDP	82	3	Benzene	50	44	11	22	264
Tl_2O_3	8	HDP	82	3	Benzene	64	55	8	21	264
γ-Al_2O_3	2	HDP	82	12	Benzene	64.5	67.8	25.6	3.3	265
Meso-Al_2O_3	2	HDP	82	12	Benzene	84.3	69	23	3.6	265
PMO-SBA-15[a]	2	Post-syn[d]	82	12	Benzene	94.8	75	16.6	2.5	266
HAP	0.5	Deposition[e]	80	12	Toluene	100	92			267
Silica	4	One-pot[f]	80	48	ACN-DMF[g]	98.5	82.8	15.3	1.9	268

[a]Periodic mesoporous organosilica SBA-15.
[b]*Tert*-butyl hydroperoxide was used as oxidant in all reactions.
[c]Homonegeous deposition-precipitation using urea.
[d]HAuCl4 was added into suspension of PMO-SBA-15.
[e]Ligand-protected Au nanoclusters (Au25) were deposited on HAP.
[f]Au precursor was added during the synthesis of silica.
[g]Solvent mixture of acetonitrile and dimethylformamide.

modification of mesoporous alumina.[265] Au nanoparticles were subsequently deposited by the homogeneous DP method using urea. In particular, the mesoporous alumina prepared *via* an electrostatic S^+T^- pathway, using cetyltrimethylammonium bromide (CTABr) as a structure-directing agent, had a large surface area and pore volume of 302 $m^2\ g^{-1}$ and 0.78 mL g^{-1}, respectively. A CO_2-TPD study showed that the mesoporous alumina contained a large number of surface basic sites, advantageous for the formation and stabilization of uniform Au nanoparticles. With a 2 wt% Au loading, the supported Au catalyst achieved 84% of conversion with 69% of styrene oxide selectivity after 12 h of the reaction. The used catalyst was recycled for eight times and no considerable reduction in catalytic activity was observed, making it a promising epoxidation catalyst.

Several recent publications have reported even higher catalytic activity,[266–268] showing complete conversion and a very high selectivity of 92% styrene oxide after a 12 h reaction over small Au_{25} clusters supported on hydroxyapatite (0.5 wt% Au_{25}/HAP).[267] With high monodispersity and small Au nanoparticles (1.4 nm), the catalyst was highly active and remained stable upon recycling reactions. The high selectivity of Au_{25}/HAP can be explained on the basis of the extremely slow decomposition of styrene oxide by Au_{25} clusters. However, given the complex preparation method and difficulty in controlling the cluster size, a more straightforward method is preferred.

Liu *et al.* recently reported the high activity of Au/SiO_2 catalyst prepared in an one-pot synthesis.[268] The effects of different parameters on the catalyst activity were investigated such as the solvent mixture, peroxide concentration, and reaction time. The highest conversion of 98% and 82% styrene oxide selectivity were achieved after 48 h reaction in the mixture of acetonitrile–dimethylformamide (ACN-DMF) as the solvent. Uniform Au nanoparticles with a mean diameter of about 6.4 nm were highly dispersed on the SiO_2 surface. The Au-SiO_2 catalyst had a large surface area of 1019 $m^2\ g^{-1}$, making a high catalytic activity possible for styrene epoxidation.

Overall, despite giving high product selectivity, the application of supported Au catalysts for selective oxidation of hydrocarbons is still limited by low conversion due to the stable C–H bond. Further study of tailored catalyst preparation and optimum reaction conditions is required. In addition, the development of a solvent-free system should be considered, as the use of benzene as the common solvent is harmful towards health and environment.

9.6 Selective Oxidation of Alcohols

Selective oxidation of alcohols to their corresponding aldehydes, ketones, and/or carboxylic acids covers a wide range of organic synthetic reactions, which are not only used for synthesis of many organic products, but are also important in fundamental research. The heterogeneous catalysts used for these reactions have evolved from traditional supported Pt- and Pd-based catalysts to Au monometallic and polymetallic systems. In general, the new generation catalysts should be more efficient in facilitating the conversion of the reactants to the targeted products, to make the whole catalytic systems greener and safer, *e.g.* by the use of molecular oxygen as the oxidant and more environmentally benign solvents or solvent-free systems. Extensive recent reviews are available for further reading and have covered selective oxidation of various alcohols and their derivatives over different types of noble metal catalysts.[12,269–272] In this section, selective oxidation of different types of alcohols is discussed, particularly catalyst development, together with possible reaction mechanisms. Other important polyols, like glycerol and sugar alcohols, are covered separately in subsequent sections.

9.6.1 Selective Oxidation of Monoalcohols over Conventional Noble Metal Catalysts

Alcohols can be obtained in large quantities from natural and renewable resources, and have been used to produce numerous chemical derivatives and products. Traditionally, the industrial oxidation of alcohols was performed by a chemical method using dichromate and permanganate.[273,274] Sequential oxidation of alcohols produces various carbonylic and carboxylic acids, which are particularly useful for the production of flavoring agents, fragrances, pharmaceutical, and agrochemical compounds. A variety of supported noble metal catalysts, such as ruthenium,[275–280] palladium,[281–287] and platinum[288–293] are highly active in oxidizing various types of alcohols, and the results have been well documented in the literature. Table 9.10 summarizes the oxidation of alcohols over various supported Ru, Pd, and Pt catalysts.

Ruthenium supported on HAP has been reported effective for oxidation of alcohols. The incorporation of Ru species into HAP structure results in the formation of a monomeric Ru cation in the form of a phosphate complex on the Ru-HAP surface. Kaneda and co-workers prepared 17 wt% Ru-HAP and investigated its activity at 80 °C under O_2 atmosphere in toluene using different alcohols as reactants, such as benzyl alcohol, 1-octanol, 2-octanol, and cinnamyl alcohol.[275] The catalyst was highly reactive and the corresponding carbonyl compounds were successfully produced with high conversion and selectivity. For example, benzaldehyde with 98% yield was obtained within 3 h, while 99% cinnamaldehyde was obtained within only 1 h. Upon recycling, the Ru-HAP catalyst was able to maintain its high catalytic activity and selectivity; *ca.* 93% of benzaldehyde yield was obtained during the three recycling reactions. Despite the high Ru content, inductively coupled plasma (ICP) analysis confirmed there was no obvious Ru leaching during the reaction, even during recycles.

Yamaguchi and Mizuno prepared 1.4 wt% Ru/Al_2O_3 catalyst for solvent-free oxidation of alcohols with molecular oxygen.[276] At 83 °C and 1 atm oxygen, very high conversion and selectivity (>99%) were achieved for the oxidation of various alcohols. Moreover, Ru/Al_2O_3 was also used to catalyze

Table 9.10 Selective oxidation of alcohols over supported Ru, Pd, and Pt catalysts.

Alcohol type	Catalyst	T/°C	Solvent	Yield[a]/%	Time/h	Ref.
Benzyl alcohol	Ru/HAP	80	Toluene	>99	3	275
1-Octanol	Ru/HAP	60	Toluene	89	16	275
2-Octanol	Ru/HAP	80	Toluene	92	6	275
Benzyl alcohol	Ru/Al_2O_3	83	Trifluorotoluene	>99	1	276
2-Octanol	Ru/Al_2O_3	83	Trifluorotoluene	>90	2	276
Benzyl alcohol	RuO_2-FAU	80	Toluene	>99	1.5	415
1-Heptanol	RuO_2-FAU	80	Toluene	>92	20	415
Cyclohexanol	RuO_2-FAU	80	Toluene	>16	4	415
Cinnamyl alcohol	Ru/C	70	Toluene	79	–	279
2-Decanol	Ru/C	90	Toluene	61	–	279
Cinnamyl alcohol	$Ru(OH)_x/TiO_2$	80	Toluene	>98	7	280
Benzyl alcohol	$Ru(OH)_x/TiO_2$	80	Toluene	>98	1	280
Cinnamyl alcohol	$Pt–Bi/Al_2O_3$	40	Water	94	4.8	298
Cinnamyl alcohol	Pt–Bi/C	60	Ethanol	51	–	296
Cyclohexanol	Pt/C	150	Water	50	2	416
2-Butene-1-ol	Pd/SiO_2	80	Supercritical CO_2	97	–	283
Cyclooctanol	Pd/SiO_2	80	Supercritical CO_2	45	–	283
Benzyl alcohol	Pd/OMC[b]	100	Supercritical CO_2	>72	13	417
Cinnamyl alcohol	Pd/OMC	80	Supercritical CO_2	>81	13	417
Benzyl alcohol	Pd/HAP	90	Trifluorotoluene	99	13	282
Cinnamyl alcohol	Pd/HAP	90	Toluene	87	13	282

[a]Yield of the corresponding aldehydes/ketones.
[b]Ordered mesoporous carbon.

the oxidation of alcohols containing nitrogen or sulfur atoms and successfully produced the corresponding aldehydes in high yields (>99%). Alternatively, a commercial Ru/Carbon (10 wt% Ru) catalyst was tested using toluene as solvent and 1 atm O_2 as oxidant at 50–90 °C. A variety of primary and secondary benzylic, allylic, and aliphatic alcohols were oxidized to the corresponding carbonyl compounds with relatively good yields, albeit lower than those obtained by the aforementioned supported Ru catalysts. The authors proposed the following reaction pathway for aerobic alcohol oxidation over the Ru catalysts: upon contact with alcohols, Ru alcoholate species was first formed by ligand exchange between alcohol and Ru–OH, which subsequently underwent β-hydride elimination to produce the corresponding carbonylic compound and a Ru hydride species. The Ru hydride species was further re-oxidized with molecular oxygen to close the catalytic cycle.[275,280]

Since Döbereiner[294] discovered the activity of Pt in catalyzing liquid-phase alcohol oxidation with molecular oxygen, there have been major developments in using Pt group metals (particularly Pt and Pd) as a heterogeneous catalytic system. The most commonly used catalysts are Pt or Pd supported on carbon[287,288,292,293,295,296] or Al_2O_3.[281,285,297,298] The reactive metal centers of these catalysts are able to activate oxygen and C–H bonds, but during the oxidation process, deactivation by oxygen often occurs,[295] due to over--oxidation of the active sites. It was discovered earlier that the rate of alcohol oxidation is much higher on a reduced metal surface than on the oxidized surface.[299] As a consequence, the reactor must be operated under an oxygen transport limited region, so that the oxygen supply during the reaction is lower than the alcohol dehydrogenation.[271] This phenomenon prompted the development of bimetallic catalysts by adding other metals as promoters to suppress the over-oxidation reaction, in particular Bi and Pb.[296–298,300]

A series of alumina supported Pt-Bi catalysts was prepared by Baiker and co-workers by deposition of Bi onto 5 wt% Pt/Al_2O_3, and used for oxidation of cinnamyl alcohol to cinnamaldehyde.[298] Under the same reaction conditions (40–80 °C), the initial reaction rate improved by a factor of more than 26 and cinnamaldehyde yield increased from 9% to 94–96% on the Bi-promoted catalyst (Bi : Pt = 0.5). By increasing the Bi to Pt surface ratio, higher Bi surface coverage was achieved, which suppressed hydrogen adsorption on Pt and therefore improved the catalyst activity and selectivity to cinnamaldehyde. The authors proposed that the main role of Bi as promoter is *via* geometric blocking of Pt sites, acting as new active centers, as well as preventing over-oxidation of cinnamaldehyde.

However, as these metals are less noble than Pt and Pd, they are hence more susceptible to leaching. Moreover, oxidative conditions and chelating agents will lead to excessive promoter leaching, detrimental for industrial applications. In addition, metal particle growth due to sintering and dissolution–re-deposition could result in irreversible deactivation. Site coverage by side products is also a major factor to catalyst deactivation.[295] Typical side reactions are the aldol condensation and oligomerization of the carbonyl compound product, as well as the decomposition of alcohols mainly *via* the carbonyl compound formed *in situ*, resulting in adsorbed CO and carbonaceous species, C_xH_y.

The proposed mechanism for the aerobic alcohol oxidation over heterogeneous catalysts can be described using three models.[270] First, the classical oxidative–dehydrogenation mechanism suggests that the adsorbed alcohol dehydrogenates in two elementary steps.[272,301] The alcohol O–H bond is first dissociated upon adsorption on the catalyst surface, resulting in the formation of alkoxide and hydrogen, as observed by IR spectroscopy.[302] Due to the strong electronegative oxygen atom, the β-C–H (β-hydride) bond of the adsorbed alkoxide becomes weaker, causing the elimination of hydrogen atom and leaving behind a carbonyl species.[303] This dehydrogenation step is usually catalyzed by alkalis.[301] Subsequently, the adsorbed dissociated oxygen molecules oxidize the evolving hydrogen to form water molecules, which shifts the equilibrium towards the carbonyl formation. This accelerates the liberation of carbonyl compounds from the catalyst surface. In this model, oxygen plays an inferior role as it does not interact directly with the adsorbed reactant. In the second model, it is assumed that the rate-determining step involves direct interaction of the adsorbed O_2 with the adsorbed reactant or the dehydrogenated intermediate.[271,289,290] This assumption is supported by both X-ray absorption study[304] and electrochemical study,[295,298,305] which indicate partial oxygen coverage on the metal surface during the reaction. The third approach adopts the classical oxidative–dehydrogenation mechanism, but presumes that the adsorbed oxygen suppresses catalyst deactivation by either oxidative removal of the degradation products[281] or prevention of the decomposition pathway.[306] Keresszegi *et al.* observed that the reaction rate acceleration achieved by introducing oxygen is not an indication of the direct involvement of adsorbed oxidizing species in the alcohol dehydrogenation mechanism, but is rather due to removal of strongly adsorbed species and an increase in the number of active metallic sites available for alcohol dehydrogenation.[281]

Despite the high activity and product yield obtained by using Pt-group catalysts, over-oxidation and fast deactivation of the catalysts remain a problem. Therefore, novel catalysts that are highly active and selective need to be developed.

9.6.2 Selective Oxidation of Alcohols by Gold Catalysts

The works by Rossi, Prati, and co-workers have demonstrated effective catalytic properties exhibited by supported Au catalysts for the selective oxidation of a wide range of alcohols, including polyols.[307–316] In addition, several extensive reviews on the subject have been published.[10,11,317,318]

9.6.2.1 Selective Oxidation of Ethanol

Annual global production of ethanol increased from 6.4 million kL in 2000 to about 34 million kL in 2008,[319] largely due to the rise of bio-based ethanol production. Selective oxidation of ethanol is expected to produce acetaldehyde, a valuable chemical commercially derived from ethylene using a corrosive $PdCl_2$–$CuCl_2$ catalyst. Haruta and co-workers explored the gas-phase oxidation of ethanol over Au catalysts supported on 23 types of metal oxides.[319] A careful screening and detailed study of the effect of supports on product distribution was carried out in a continuous fixed bed reactor at a space velocity of 20 000 mL g_{cat}^{-1} h^{-1} and 100–280 °C. The gaseous reactant was fed with a molecular composition of $C_2H_5OH/O_2/N_2 = 1/3/126$, containing a stoichiometric amount of molecular oxygen for complete oxidation of ethanol. In order to obtain finely dispersed Au nanoparticles (<10 nm), the deposition of Au nanoparticles on metal oxides was conducted by the CP, DP, or SG method depending on the surface properties of the support.

Table 9.11 presents the catalytic performances over Au catalysts supported on various oxide supports at the maximum yield of acetaldehyde.[319] It is observed that metal oxide supports define the product distribution. Based on the selectivity of acetaldehyde and the formation of acetic acid and CO_2, the extent of the oxidation reaction can be classified into three groups—mild, deep, and complete oxidation, depending on the characteristics of metal oxides. Mild oxidation is achieved by Au nanoparticles deposited on

Table 9.11 Product selectivity of gas-phase oxidation of ethanol over Au supported on metal Oxides.[319]

Group	Support Type	Deposition Method	Au Loading/ %	T/°C	Yield/%		
					ACTAL	ACTAC	CO_2
Mild oxidation	MoO_3	SG	1.0	240	94	0	0
	La_2O_3	CP	6.0	260	81	0	0
	Bi_2O_3	CP	4.3	180	68	0	0
	SrO	CP	9.1	280	67	0	0
	Al_2O_3	DP	1.0	220	65	0	24
	TiO_2	DP	1.0	260	62	1	22
	SnO_2	DP	1.0	280	64	2	0
	SiO_2	DP	1.0	240	57	8	0
	Y_2O_3	CP	8.4	240	57	0	9
	MgO	CP	20.6	280	34	0	0
	Nb_2O_5	SG	1.0	260	30	0	0
	BaO	CP	6.3	280	27	0	0
	WO_3	SG	1.0	180	18	0	0
Deep oxidation	ZnO	CP	11.3	180	74	15	0
	In_2O_3	CP	6.9	140	45	14	0
	V_2O_5	SG	1.0	200	67	3	0
Complete oxidation	MnO_2	DP	1.0	100	10	0	0
	CeO_2	DP	1.0	140	49	0	14
	CuO	CP	11.5	160	80	4	5
	Co_3O_4	CP	11.4	140	38	0	0
	NiO	CP	12.2	160	21	0	0
	ZrO_2	DP	1.0	180	38	12	0
	Fe_2O_3	CP	11.5	200	35	0	14

ACTAL, acetaldehyde; ACTAC, acetic acid.

acidic, basic, or neutral metal oxides, which can selectively oxidize ethanol to form acetaldehyde with high selectivity. A maximum acetaldehyde yield of 94% was achieved at 240 °C by using MoO_3, a highly acidic oxide, as support. On the other hand, the deposition of Au on La_2O_3, which is weakly basic, obtained a maximum yield of 81% at 260 °C without the formation of acetic acid and CO_2. Similarly, other basic metal oxides (SrO, Al_2O_3, TiO_2) showed similar catalytic properties.

However, deposition of Au on n-type semiconductors like ZnO, In_2O_3, and V_2O_5 results in the formation of acetic acid and even CO_2 at higher temperatures, in addition to acetaldehyde. Hence, a lower temperature is required to maintain high acetaldehyde selectivity. The best support for deep oxidation of ethanol is ZnO. At 180 °C, Au/ZnO catalyst had the highest acetaldehyde yield of 74%, producing acetic acid as the byproduct with 15% yield. At higher temperatures, acetaldehyde selectivity was decreased while acetic acid selectivity was increased, together with the formation of CO_2. Complete oxidation of ethanol primarily produces CO_2 and H_2O. It occurs over Au supported on p-type semiconductor supports such as MnO_2, CeO_2, CuO, Co_3O_4, NiO, ZrO_2, and Fe_2O_3 at low temperature (below 200 °C).

Figure 9.10 illustrates the mechanism of ethanol oxidation over Au catalysts.[319] Initially, the adsorption of ethanol on the oxide support leads to the formation of surface ethoxide groups, formed by the reaction of ethanol molecules with hydroxyl groups or metal cations of (a) basic metal oxide supports [Figure 9.10(a)], acidic metal oxide supports [Figure 9.10(b)], and n-type semiconductor metal oxide supports [Figure 9.10(c)]. Subsequently, the adsorbed ethoxide species is oxidized by breaking the C–H bond to form acetaldehyde. The deposited Au nanoparticles play an important role in activating molecular oxygen to O_2^- or O^-, depending on the temperature,

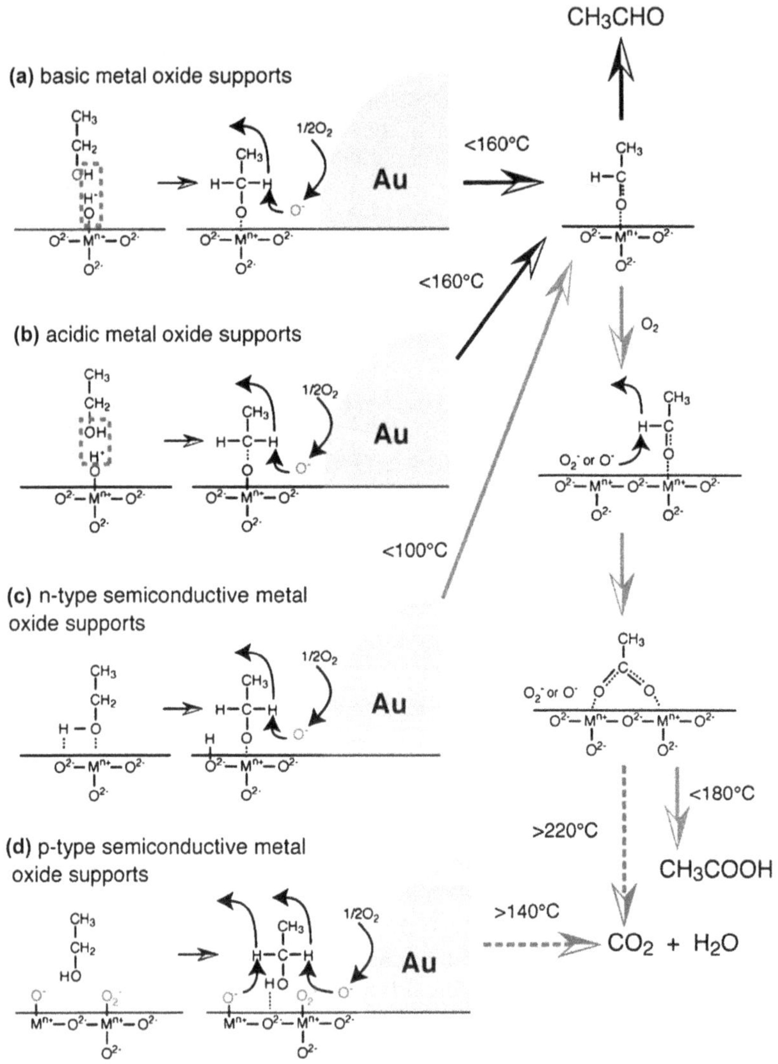

Figure 9.10 Probable routes for the formation of surface ethoxide from ethanol and its transformation to acetaldehyde, acetic acid, and CO_2 over supported gold catalysts.
Reproduced with permission from ref. 319.

which might be formed on the edges and corners of Au nanoparticles as well as at the perimeter interface between Au nanoparticles and metal oxide supports. In the case of n-type semiconductor metal oxide supports, *e.g.* ZnO, successive oxidation to acetic acid can be initiated with the active O_2^- or O^- species, which forms carboxylates over the support surfaces. In the case of p-type semiconductor metal oxide supports [Figure 9.10(d)], characterized by an excess of surface oxygen species, the oxygen density on the oxide surface enables simultaneous oxidation of ethanol, acetaldehyde, acetic acid into CO_2 and H_2O.[319]

In contrast to gas-phase oxidation, aerobic liquid-phase oxidation of ethanol is reported to mainly produce acetic acid instead of acetaldehyde. Christensen *et al.* first reported the aerobic oxidation of aqueous phase ethanol over Au catalysts to produce acetic acid.[320] The reaction was conducted at 100–200 °C with pO_2 of 2.5–5.0 MPa using Au supported on $MgAl_2O_4$. The catalyst was prepared by the DP method. For comparison, the corresponding Pd and Pt catalysts were also prepared using the same support *via* the incipient wetness impregnation method. The resulting catalysts contained 1 wt% of metal. $MgAl_2O_4$ was chosen as the support material due to its stability at high water pressure and its inertness in redox processes. Therefore, there should be a very weak synergistic effect with the support, and the observed activity can be attributed solely to the metal nanoparticles. At 180 °C and $pO_2 = 0.6$ MPa, Au/$MgAl_2O_4$ exhibited 97% conversion, giving 83% yield of acetic acid within 4 h of reaction. Under the same conditions, the Pd catalyst exhibited 93% conversion and 60% yield, while the Pt catalyst yielded only 16% acetic acid at 82% conversion. TEM analysis showed that, in the 1 wt% Au/$MgAl_2O_4$ catalyst, the Au particle size was about 3–6 nm; this remained almost the same after the reaction, and thus sintering of the catalyst could be ruled out. Later the same authors compared ethanol oxidation to acetic acid and ethyl acetate over 1 wt% Au/$MgAl_2O_4$ and commercial 1.5 wt% Au/TiO_2 catalysts under the same reaction conditions.[321] These two catalysts exhibited similar results, yielding ~95% of acetic acid at 99% ethanol conversion. The initial Au particle size was 3–6 nm in the two catalysts and there was no coke formation after the reaction. The kinetic study revealed that the rate-limiting step is ethanol dehydrogenation to an 'CH_3CH_2O' intermediate that subsequently undergoes either C–C cleavage forming CO_2, or transformation to acetaldehyde before forming acetic acid. However, ethyl acetate was formed at higher ethanol concentrations. The maximum selectivity toward ethyl acetate was obtained at ethanol concentrations between 80 and 100%, indicating that the water content had a limiting effect.

The above proposed mechanism was further clarified by Aghaei and Berger.[322] By using powdered commercial 1 wt% Au/TiO_2, high selectivity to acetic acid was obtained. In this catalyst, the Au nanoparticles were finely dispersed with an average diameter of about 2.6 nm. The reaction conditions (*e.g.* temperature, pressure, initial ethanol concentration, catalyst concentration, and reaction time) were varied systematically and a set of experimental data were obtained and could be fitted adequately into the Langmuir–Hinshelwood–Hougen–Watson, kinetic model based on catalytic surface reactions. Hydrogen abstraction by surface hydroxyl species was assumed to be the rate-determining step. The model also adequately described the decreasing conversion at higher ethanol concentration and also the selectivity towards the products acetaldehyde, acetic acid, ethyl acetate, and CO_2.

9.6.2.2 *Selective Oxidation of Octanol*

The oxidation of octanol is often difficult because of its low reactivity. However, it can be used as a probe molecule to determine the catalyst effectiveness toward oxidation of alkanols.[12] Enache *et al.* studied the oxidation of 1-octanol in the absence of solvent over various supported Au catalysts, as listed in Table 9.12.[316] The oxidation reaction was carried out at 100 °C and pO_2 of 2 atm for 3 h in the absence of solvent. Only Au/Fe_2O_3 and Au/CeO_2 were active, albeit with low conversion, while the remaining catalysts were practically inactive. However, the former was also found to be less selective compared with the latter as it also produced ester and octyl octanoate.

It was reported earlier that the combination of small Au nanoparticles (2–5 nm) and nanocrystalline CeO_2 (~5 nm) yielded high catalytic activity and stability for oxidation of CO.[323] The resulted CO oxidation activity on Au supported on nanocrystalline CeO_2 was two orders of magnitude higher than that of Au deposited on a regular ceria support at 5 °C. In view of this, Abad *et al.* prepared this Au–CeO_2 nanoconjugate by precipitation of Au(III) salt on ceria nanoparticles at pH 10.[324,325] The as-prepared catalyst was tested for oxidation of various alcohols, including 3-octanol, in the absence of base and solvent. At 80 °C and 1 atm of oxygen partial pressure, it gave a 3-octanol conversion of 97% with 3-octanone selectivity of more than 99% within 3.5 h. The catalyst also exhibited a high TOF (420 h^{-1}) compared with that of the conventional Au/CeO_2 (150 h^{-1}) as well as Au/C catalysts (~20 h^{-1}). The authors proposed a reaction mechanism.[324] XPS analysis showed the presence of Au^{3+}, Au^{+}, and Au^{0}, indicating a strong interaction of Au nanoparticles with CeO_2 nanosupport. The latter stabilized the positive oxidation

Table 9.12 Solvent-free oxidation of 1-Octanol over Au catalysts.[316]

Sample Name	Preparation Method	Alcohol Type	Conversion/%	1-Octanal Selectivity/%
0.7 wt% Au/SiO_2	Co-precipitation	1-octanol	0	–
5 wt% Au/Fe2O_3	Co-precipitation	1-octanol	1.5	55.9
2 wt% Au/CeO_2	Co-precipitation	1-octanol	0.57	100
2 wt% Au/TiO_2	Impregnation	1-octanol	0	–
2 wt% Au/C	Impregnation	1-octanol	0	–

Reaction conditions: 40 ml alcohol, 0.2 g catalyst, 1500 rpm, 100 °C, pO_2 = 0.2 MPa, 3 h reaction.

states of Au by creating Ce^{3+} species and oxygen-deficient sites in ceria. The adsorption of alcohol molecules on the Lewis acid sites of Au/CeO_2 could cause the formation of metal alkoxide, succeeded by the formation of metal hydrides (Ce–H and Au–H) and the release of ketone. With the presence of oxygen in the system, as well as the coordination to the oxygen-deficient sites of ceria, formation of cerium-coordinated superoxide (Ce-OO*) species occurred. These superoxide species evolved into cerium hydroperoxide by hydrogen abstraction from Au–H (step 5), followed by the formation of the initial Au^+ species, after reduction of Ce^{4+}. Without the addition of Au, this step would be rendered impossible and lead to a depletion of Ce^{3+}.

A similar study was conducted by Villa *et al.* using a supported Au catalyst (~3.6 nm) on NiO nanocrystals (~5–6 nm).[326] The nanostructured catalyst exhibited higher activity in oxidizing aliphatic alcohols than Au supported on commercial NiO (~40 nm). In the base-free reaction system, the former converted 25% 1-octanol to give 40% 1-octanal selectivity, with octanoic acid and octanoate ester as the byproducts. Higher conversion (51%) was obtained when the secondary alcohol, 2-octanol, was oxidized to 2-octanone with 99% selectivity. In contrast, Au supported on commercial NiO was inactive with only <1% conversion for both alcohols. The effect of support basicity and the interaction of Au particles with the nanosized NiO are discussed in the next section.

Bimetallic Au–Pd and Au–Pt supported on carbon were prepared by Prati *et al.* by sol immobilization, which gave 1 wt.% of total metal loading.[315] The reactivity against the oxidation of octanol was investigated in the absence and presence of alkali, and their catalytic behavior was compared to the corresponding monometallic catalysts. Table 9.13 summarizes the catalytic data for 1-octanol oxidation. The materials characterization showed that both were constituted of a single alloy phase. The oxidation of 1-octanol was carried out at 50 °C, 0.3 MPa O_2, an octanol to metal ratio of 1000, and 8 h of

Table 9.13 Catalytic results of 1-Octanol oxidation in the absence and presence of base.[315]

Catalyst	Conversion/%	1-Octanal Selectivity/%	TOF/h^{-1}
In the absence of base			
1% Au/C	0	0	0
1% Pd/C	2	70	8
1% Pt/C	2	>99	10
1% Au–Pd/C	17	70	15
1% Au–Pt/C	20	75	22
Catalyst	**Conversion/%**	**Octanoate selectivity/%**	**TOF/h^{-1}**
In the presence of base (NaOH/octanol = 4)			
1% Au/C	23	96	58
1% Pd/C	2	97	8
1% Pt/C	4	90	10
1% Au–Pd/C	93	98	413
1% Au–Pt/C	62	78	210

reaction time in the absence of base. Despite poor reactivity, the bimetallic Pt-Au/C catalyst performed better than its Pd–Au/C counterpart in terms of catalytic activity and product selectivity. It can be observed in Table 9.13 that, in the absence of base, monometallic Au/C catalysts are inactive while Pt and Pd catalysts exhibit much poorer conversion and selectivity compared with the respective bimetallic catalysts. Pt–Au/C is more active than Pd–Au/C as the former exhibits higher conversion and selectivity. However, the addition of Au on Pt/C lowers the selectivity from >99% to 75%. Synergistic effects are normally explained by electronic and geometric modification occurring in metal particles, but the results obtained here indicate otherwise. The authors suggested that a possible mismatch between electronic/geometric effects obtained by adding Au to Pt and the octanol-like skeleton molecule requirements could occur, thus limiting the overall effect on the catalytic activity.[315]

As alcohol dehydrogenation requires alkaline condition, the activity of all catalysts in general increases in the presence of NaOH (four times for octanol), as presented in Table 9.13. The highest conversion (93%) and selectivity (98%) are obtained over 1% Au–Pd/C catalyst. It should be noted that, under this basic condition, octanoate is the main product. The formation of sodium salt prevents the irreversible adsorption of organic acid, thus avoiding deactivation. In contrast, in the absence of alkali, Au–Pt/C catalyst is more active than Au–Pd/C. These results confirm that the overall catalytic effect observed when Au is alloyed to Pd or Pt not only depends on the support and the substrate, but also on the specific reaction conditions. The final activity of the catalyst is the result of the matching of all the effects involved.

9.6.2.3 Selective Oxidation of Benzyl Alcohol

Due to its reactivity, benzyl alcohol has been intensively studied as a probe molecule and a large number of reports are available on its selective oxidation reactions. Like other alcohols, the oxidation of benzyl alcohol produces benzaldehyde as intermediate and benzoic acid as the final product. Benzaldehyde has been mainly used in the food industry as a flavoring additive that confers almond flavor, as well as in the pharmaceutical and perfume industries. However, benzoic acid is widely used as a preservative due to its antimicrobial property. Although Pd and Pt have been commonly used as catalysts, supported Au and its bimetallic alloys are also proven to be highly effective. The effects of supports, type of base, solvent, and Au particle size have been thoroughly investigated.

Solvent-free oxidation of benzyl alcohol was studied by Enache *et al.* over various supported Au catalysts.[316] Carbon, SiO_2, CeO_2, Fe_2O_3, and TiO_2 were used as a support and the deposition of Au was carried out by different methods, such as the CP, DP and IMP methods using $HAuCl_4$ solution. The oxidation reaction was carried out at 100 °C and pO_2 of 2 atm for 3 h in the absence of solvent. Table 9.14 presents the catalytic performance of different

Table 9.14 Solvent-free oxidation of benzyl alcohol over Au catalysts.[316]

Sample Name	Preparation Method	Alcohol Type	Conversion/ %	Benzaldehyde Selectivity/%
0.7 wt% Au/SiO_2	co-precipitation	Benzyl alcohol	1.9	100
5 wt% Au/Fe_2O_3	co-precipitation	Benzyl alcohol	7.1	87.6
2 wt% Au/CeO_2	co-precipitation	Benzyl alcohol	3.4	100
2 wt% Au/TiO_2	impregnation	Benzyl alcohol	0.65	100
2 wt% Au/C	impregnation	Benzyl alcohol	2.3	90.4

Reaction conditions: 40 ml alcohol, 0.2 g catalyst, 1500 rpm, 100 °C, $pO_2 = 2$ bar, 3 hr reaction.

catalysts. With SiO_2, CeO_2, and TiO_2 supports, selective oxidation to benzaldehyde took place over 3 h of reaction and no byproduct was obtained. The best results were observed using 2 wt% Au/CeO_2, with 3.4% conversion and 100% selectivity. In contrast, higher conversion (7.1%) was achieved over 5 wt% Au/Fe_2O_3, albeit the selectivity (87.6%) was lower due to the formation of ester as byproduct. However, the latter contains more Au. It is considered that the ester byproduct formation was due to the acid sites on the support materials, as verified by TPD using NH_3, which revealed strong acidic sites on Au/Fe_2O_3.

The doping of CeO_2 support with alkaline earth ions, such as Ba^{2+}, Mg^{2+}, and Ca^{2+}, was found to improve the catalytic activity of Au/CeO_2.[327] With a M^{2+} to Ce^{4+} molar ratio of 4:100, the alkaline earth cations were incorporated into CeO_2 lattice *via* a sol-gel procedure with a triethanolamine–water mixture as the solvent. It was thought that the introduction of dopants would change the acid–base and redox properties of CeO_2, which would eventually influence the catalyst performance. Au was deposited *via* the DP method at pH 7 and the reaction was conducted at 120 °C, 50 ml min^{-1} O_2, for 3 h, in the presence of toluene as a solvent. The aerobic catalytic results showed that 0.03 wt% Au/Ba-CeO_2 catalyst had the highest activity, albeit with a very small Au content, followed with Au/Ca-CeO_2, Au/Mg-CeO_2, and Au/CeO_2 in descending order. A maximum benzyl alcohol conversion of 36.8% was achieved with 100% benzaldehyde selectivity and TOF of 3221 h^{-1}. As observed with TEM, the Au nanoparticles were well dispersed over Ba-CeO_2 supports with a small particle size (3–4 nm), compared with others (4 6 nm). Hydrogen temperature-programmed reaction (H_2-TPR) analysis showed that the reducibility of CeO_2 was markedly enhanced by Ba^{2+}, while upon deposition of Au, the reduction temperature was lowered by 400 °C. Probably the Au nanoparticles were located at the oxygen vacancies and acted as sites for oxygen adsorption, which successively spilled over to the Ba-CeO_2 surface, making it highly active for alcohol oxidation. A large number of oxide vacancies along with a mixed valence Ce^{3+}/Ce^{4+} species were identified by XPS analysis, especially for that doped with Ba^{2+}. The strong gold–support interaction resulted in the shift of Ce 3d core electron peaks, which were observed when both the Au nanoparticles and dopant cations were present. Simultaneously, the lattice oxygen atoms were more labile in the case of Au/Ba-CeO_2, making it more active in the benzyl alcohol oxidation reaction. Since the presence of Au increases the activity of the doped CeO_2, Au nanoparticles most probably constitute the dominant active sites involved in the complete catalytic cycle while the acid–base and redox sites on the CeO_2 play complementary roles.[327]

Villa *et al.* examined the effect of the basicity of the supports on the catalytic activity without using any alkali. Au catalyst supported on nanosized NiO (n-NiO) was prepared *via* sol immobilization and was found active by an order of magnitude higher than that of the commercial micrometer NiO.[326] The catalytic test was carried out in cyclohexane, $pO_2 = 3$ atm, and 80 °C. With an average Au particle size of 3.6 nm, 1 wt% Au/n-NiO exhibited 55% conversion with 66% selectivity toward benzaldehyde within 6 h of reaction, compared with 6% conversion and 75% benzaldehyde selectivity over Au/NiO. At a higher temperature of 100 °C and longer reaction time (10 h), an impressive 99% conversion could be achieved with 64% selectivity. The high activity of Au/n-NiO is likely due to both the high basicity of the support (0.21 mmol g^{-1} for nanosized NiO *vs.* 0.12 mmol g^{-1} for commercial NiO) and the synergistic effect between the metal component and the support. The aldehyde selectivity was inversely proportional to the basicity of the support, as the increased basicity promoted the sequential oxidation to carboxylic acids. However, the basicity of the support is not the sole factor for the catalytic activity. Although MgO was more basic (basic site density = 0.42 mmol · g^{-1}) than n-NiO, it was a much less active support for Au (6% conversion *vs.* 55%, respectively). Investigation by XPS showed that there was a shift in the Au-binding energy attributable to changes in the initial and final state electronic structures because of the reduced Au–Au coordination numbers. Thus, it was concluded that an electronic interaction was established between Au and the n-NiO surface, which contributed to the catalytic activity.

In order to further increase the catalytic activity, addition of base, such as NaOH, is necessary. Zhou *et al.* deposited Au nanoparticles on NaY zeolite by the DP method and tested the catalysts for oxidation of benzyl alcohol in alkaline aqueous solution.[328] In the absence of NaOH, the main product was benzaldehyde. Upon addition of NaOH, benzaldehyde selectivity decreased and more benzoic acid was produced. The yield of the obtained benzoic acid was dependent on the catalyst synthesis parameters, such as pH, deposition temperature, and the initial $HAuCl_4$ precursor concentration. The optimum synthesis yields Au loading of 1.9 wt% with an average Au diameter of 6.1 nm. A maximum yield of 94.6% benzoic acid was obtained at 80 °C, 0.1 MPa O_2, and NaOH/benzyl alcohol = 2. To examine the reusability of Au/NaY catalysts, four recycling runs were carried out; there was no obvious loss in catalytic activity and benzoic acid yield, indicating stable characteristics of the catalyst.

Ni *et al.* reported a highly efficient Au/TiO_2 (1 wt% Au) catalyst using 5% H_2O_2 solution as the oxidant. A complete conversion of benzyl alcohol was obtained at 90 °C to produce benzoic acid with high selectivity (85%) and benzaldehyde as the byproduct.[329] The primary role of H_2O_2 is to facilitate the abstraction of the α-hydrogen as a hydrogen scavenger, which leads to the formation of a *meta*-alcoholate species during the initial step of the reaction. The Au-alcoholate complex may be further attacked by H_2O_2 or the Au-hydridoperoxide complex is generated from the reaction of Au^0 with H_2O_2, to form the final carbonyl compound *via* the β-elimination pathway. Hence, H_2O_2 is able to significantly increase the activity of alcohol oxidation in contrast to the normal aerobic process in which dioxygen or air is used as the oxidant.

The effect of Au particle size on the selective oxidation of alcohols was studied by Boronat *et al.* on a series of catalyst having Au atoms with different coordination number.[330] At first, supported Au/MgO catalysts with different Au loading (0.4–10 wt%) were prepared by the impregnation method and kinetic experiments were performed at different concentrations of benzyl alcohol. The experimental data showed a strong correlation between the reactivity on the catalyst surface and the Au coordination number, further confirmed by the periodic DFT calculation. A similar phenomenon was also encountered by Chen *et al.*,[331] who noticed an optimum ratio of edge and corner sites to terrace sites in the oxidation of benzyl alcohol over Pd nanoparticles. This clear trend means that not all active sites are equivalent and that real Au catalysts might have heterogeneous surfaces with Au atoms in different coordination states showing different reactivities towards alcohol dehydrogenation. Hence, as supported by DFT calculation, the alcohol dehydrogenation on Au nanoparticles is considered a structure-sensitive reaction. In addition, the oxidation of benzyl alcohol exhibited the highest rates (18 000 h^{-1}) over Au catalysts in the basic environment. Although the oxidation of alcohol to aldehyde and further to carboxylic acid is greatly accelerated by the presence of base, the sequential oxidation of aldehyde and further to acid may be inhibited by careful selection of the solvent and catalyst support. The incorporation of acid and/or basic sites on catalyst support can influence the product selectivity of the overall reaction, as discussed earlier.

Bimetallic Au–Pd and Au–Pt nanoalloys supported on activated carbon were prepared by Dimitratos *et al.*[332] They found that the former drastically improved catalytic activity while the latter showed a negative effect compared with the respective Pd and Pt monometallic catalysts. Without any added base, at 60 °C and 1.5 atm O_2 gas, 0.73 wt% Au-0.27 wt% Pd/C gave the highest conversion of 32% and 94% benzaldehyde selectivity within 3 h reaction in the presence of toluene as a solvent. When water was used as the solvent, a remarkably high conversion was achieved (96%) while maintaining high selectivity (94%). In addition, the overall oxidation rate was also higher in water (TOF = 160 h^{-1}) than in toluene (TOF = 54 h^{-1}). A similar trend was also observed for cinnamyl alcohol oxidation, which is discussed below.

9.6.2.4 *Selective Oxidation of Cinnamyl Alcohol*

Selective oxidation of cinnamyl alcohol to produce cinnamaldehyde involves a complex network reactions such as hydrogenation, hydrogenolysis, and decarbonylation, depending on the reaction conditions and metal catalyst used.[281] The commonly used catalysts are Ru-, Pd-, and Pt-based, in the presence of toluene as a solvent. However, the introduction of Au into Pd- and Pt- catalysts enhances the catalyst activity, as reported by Dimitratos *et al.*[332] As cinnamyl alcohol is similar to benzyl alcohol in electronic structure, the same catalyst 0.73 wt% Au-0.27 wt% Pd/C gave the highest conversion at 72% conversion and 85% cinnamaldehyde selectivity when toluene was used as the solvent. Higher conversion (95%) was obtained upon replacing toluene with water while maintaining high selectivity (83%). It has been speculated that water could facilitate hydrogen abstraction due to its weak basicity, hence promoting catalyst activity. With TOF values of 237 h^{-1} and 180 h^{-1} in water and toluene system, respectively, the activity of the bimetallic catalyst was almost threefold as much as that of Pd and fivefold as much as that of Pt, although they exhibited higher selectivity, whereas Au monometallic catalyst was found inactive.

Relatively high conversion (80%) was also achieved by supported Au/nanosized NiO, which has high basicity. The selectivity of cinnamaldehyde was found 64% with 12% cinnamic acid as byproduct. The reaction was carried out in the absence of base, at 100 °C and 3 atm O_2.[326]

The investigation of cinnamyl alcohol oxidation highlights the importance of solvent choice in alcohol oxidation with regard to product selectivity and catalyst activity. As with benzyl alcohol oxidation, the product selectivity is greatly influenced by the solvent as well as the nature of the catalyst support. As in other alcohol oxidations, the Au catalyst features the highest oxidation rate at modest temperature, <100 °C, though the solvent and pH of the solution must be carefully monitored.[12]

9.6.2.5 *Oxidation of Allyl Alcohol and Other Unsaturated Alcohols*

The oxidation of allyl alcohol to 3-hydroxypropionic acid was explored using Au/Carbon (X40S) catalyst with an Au loading of 0.3%.[311] For comparison, the activity of commercial 1.5 wt% Au/TiO_2 was also investigated. The reaction was carried out at 25–80 °C, pO_2 of 0.3 MPa, alcohol to metal molar ratio of 4000, and in the presence of different amounts of base (NaOH/alcohol = 1–3). A high selectivity of 79% at full conversion was obtained by 0.3 wt% Au/C at 50 °C and excess NaOH (3 : 1). Higher temperature led to lower selectivity (74%) and increase in byproducts such as acrylic acid and glyceric acid. In contrast, the commercial Au/TiO_2 was inferior, yielding only 53% 3-hydroxypropionic acid at 97% conversion.

Abad *et al.* prepared small Au (2–5 nm) nanoparticles supported on small CeO_2 nanocrystals (~5 nm) as the support for oxidation of different types of alcohols.[324] The oxidation reaction of 1-octen-3-ol was performed in a magnetically stirred, glass batch reactor in the absence of solvent and base at 50 °C, with 1 atm of oxygen partial pressure. Within 3.5 h, 80% conversion with >99% selectivity towards 1-octen-3-one was achieved.

Table 9.15 Catalytic results for oxidation of 2-Octene-1-ol in the absence of base.[315]

Catalyst	Conversion/%	Selectivity/%			TOF/h^{-1}
		2-Octenal	3-Octene-1-ol	Octanal	
1% Au/C	0	0	0	0	0
1% Pd/C	3	0	0	0	24
1% Pt/C	1	0	0	0	5
1% Pd–Au/C	97	20	58	20	552
1% Pt–Au/C	8	10	90	0	39

Reaction conditions: substrate/metal = 1000, pO_2 = 3 atm, T = 50 °C, reaction time = 2 h.

Using bimetallic Au–Pd/C and Au–Pt/C catalysts as discussed earlier, Prati *et al.* also tested the catalysts for oxidation of 2-octene-1-ol in the absence of base.[315] The results are presented in Table 9.15 which shows that all the monometallic catalysts exhibited very poor activity. The highest activity was achieved by 1 wt% Au-Pd/C, with 97% conversion, while 1 wt-% Au–Pt/C only exhibited 62% conversion. The comparison between the catalytic results of bimetallic catalyst is interesting. Despite being active, only 1% Au–Pd/C catalyst could produce octanal, albeit a small amount. The other major byproduct is 3-octene-1-ol, the isomer of 2-octene-1-ol. The reaction pathway is more complex as the presence of a double bond in the allylic position makes isomerization and hydrogen transfer reactions possible, thus making 3-octen-1-ol the main product in all the cases, in particular, when 1 wt% Pt–Au–C catalyst is used. However, the activity of 1 wt% Pt–Au/C was considerably lower than that obtained with 1 wt% Pd–Au/C catalyst, even if a positive synergistic effect is demonstrated in both cases.

In short, selective oxidation of various alcohols on supported Au catalysts depends on several factors. The catalyst support, solvent, reaction conditions, and catalyst preparation methods all impact on catalytic performance. To overcome the inherent drawbacks of Au catalysts, bimetallic catalyst systems are probably a good choice. As demonstrated by Prati *et al.*[315] and Dimitratos *et al.*,[332] bimetallic Au catalysts are superior to their monometallic counterparts. Therefore, to make the process greener, solvent-free and base-free systems are preferred and should be given more attention.

9.7 Selective Oxidation of Polyols

9.7.1 Selective Oxidation of Glycerol

Glycerol is a simple polyol that has many applications, ranging from food additives, drugs/pharmaceuticals, cosmetics, personal care products, to explosives. It is typically obtained as a byproduct from soap manufacture, fatty acid and fatty ester production, as well as microbial fermentation. Chemically, it can be synthesized from propylene *via* the epichlorohydrin process.[333] Despite a slower growth of biodiesel production worldwide recently, there is still a large surplus in global supply of glycerol, with a projection of about 4 billion gallons of crude glycerol by 2016,[334] in which, *ca.* 10 wt% is from biodiesel production.[335] Hence, it becomes increasingly important to utilize glycerol as an alternative feedstock for producing various value-added chemicals.

As a highly functionalized compound, a large number of chemicals and intermediates can be derived from the oxidation of glycerol, as shown in Figure 9.11.[336] For example, the oxidation of the primary hydroxyl group yields glyceraldehydes, glyceric acid, and ultimately tartronic acid. Oxidation of the secondary hydroxyl group produces dihydoxyacetone (DHA), a common chemical widely used as a skin coloring agent. The oxidation of all three hydroxyl groups produces mesoxalic acid, an important intermediate for malonic acid and oxalic acid. Chemically, the stoichiometric oxidation of glycerol can be performed by mineral acids, such as potassium permanganate, which is able to produce tartronic acid, while nitric acid primarily yields glyceric acid. However, these mineral acids are toxic and harmful to the environment.[336] In addition, due to similarly reactive hydroxyl groups, selective oxidation of glycerol remains a great challenge. To overcome this problem, catalytic oxidation of glycerol over various noble metal catalysts has been investigated, and the results are reviewed in several papers.[337–339]

Palladium and platinum catalysts were widely used for glycerol oxidation.[340–345] By controlling the reaction conditions, particularly the solution pH, different oxidation products can be obtained. Under acidic pH

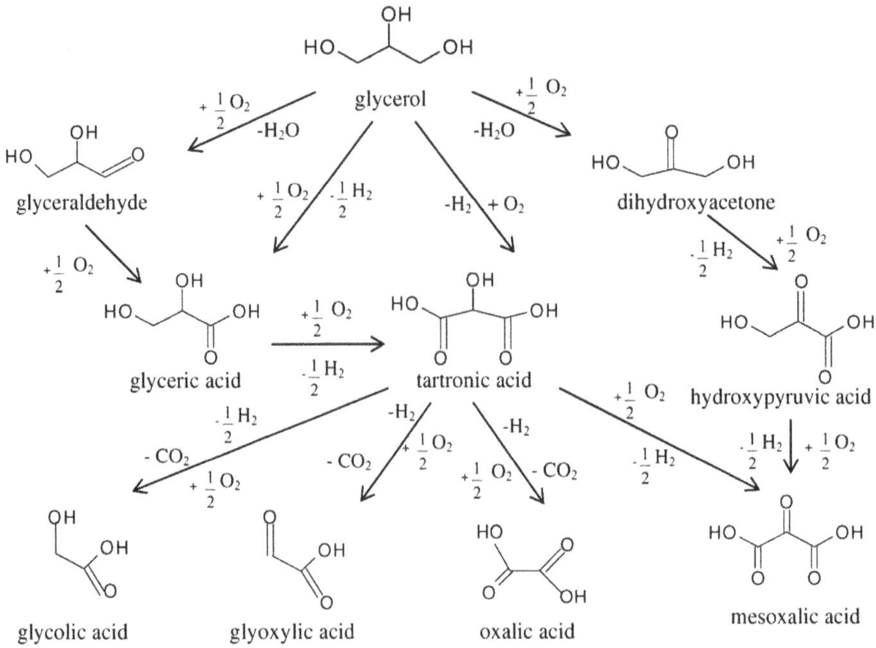

Figure 9.11 Derived chemicals from the selective oxidation of glycerol. Reproduced with permission from ref. 336.

(2–4), the oxidation of secondary hydroxyl group takes place to produce DHA; while under basic pH (>10), the primary hydroxyl group is oxidized to produce glyceric acid or tartronic acid. An early work by Miya and Matsuda reported the selective oxidation of the primary hydroxyl group of glycerol to produce glyceric acid and mesoxalic acid over supported Pd catalyst on charcoal.[346]

9.7.1.1 *Selective Oxidation of Glycerol to Produce DHA*

In the early 1990s, Kimura *et al.*[340] first studied the catalytic synthesis of DHA, which is commercially produced by low productivity fermentation using *Acetobacter xylinum.*[347] Different noble metal catalysts such as Pt, Pd, Ru, and Re were supported on charcoal with a metal loading of 5 wt%. The oxidation reaction was carried out at atmospheric pressure, 50 °C and acidic pH ($\sim$2). The supported Pt/C catalyst had a weak activity in producing DHA with a maximum yield of 4% at 37% glycerol conversion, while the other metals exhibited much lower activity. The catalytic activity was further optimized by adding different metals such as Bi, Te, Pb, Sb, or Se as promoter. The incorporation of Bi on Pt catalyst with a Bi/Pt ratio of 0.2 (1% Bi-5% Pt/C) increased the DHA yield and selectivity up to 20% and 80%, respectively. An XPS study revealed that Bi was concentrated on the Pt(111) surface with a surface coverage of 0.33. It was also proposed that the Bi submonolayer on Pt(111) could act as site blocker, thus controlling the glycerol orientation favorable for the oxidation of the secondary hydroxyl group to produce DHA. Higher glycerol conversion of up to 80% could be achieved by running the reaction in a trickle bed reactor using wood-based granular charcoal as a support.[341] With the same catalyst, the oxidation of 50% aqueous solution of glycerol at 50 °C, an oxygen to glycerol molar ratio of 2, and a liquid hourly space velocity (LHSV) of 0.06 h^{-1} produced DHA with a selectivity of *ca.* 80% at the conversion of 80%. Moreover, long-term catalyst performance of more than 1000 h was achieved, which is an important feature for commercial application.

Using supported Pd and Pt catalysts, Gallezot *et al.* carried out liquid-phase glycerol oxidation with air at different pH values.[342] The reaction was conducted at atmospheric pressure, 60 °C, and various pH values adjusted by adding 30% NaOH solution. The supported Pd and Pt catalysts with 5 wt% metal loading were prepared by impregnation of active charcoal (CECA 50S) with H_2PdCl_4 and H_2PtCl_6 as precursors, respectively, followed with reduction by formaldehyde. Under alkaline condition (pH 11), 90–100% glycerol conversion was achieved on 5 wt% Pd/C, yielding glyceric acid as the main product with a maximum yield of 70%. At pH 7, however, 5 wt% Pt/C produced glyceric acid with 55% yield at 90% glycerol conversion and 12% yield of DHA. The initial reaction rate on the Pt catalyst was six times that on the Pd catalyst, because the former has a higher redox potential and thus more resistance to oxygen poisoning compared with the latter. Further addition of Bi into the catalyst was proven to suppress the formation of glyceric acid and to promote the production of DHA, as reported elsewhere. Co-impregnation of Pt and Bi *via* reduction by formaldehyde was found to give higher yield of DHA at pH 2. At a Bi to Pt molar ratio of 0.13, the maximum DHA yield was 34% at 70% conversion, while at higher Bi to Pt ratio of 0.37, 37% yield of DHA was obtained at 75% conversion. The Bi atom has two roles in the oxidation reaction: (1) to minimize over-oxidation by strongly binding with the carboxylic acid molecules thus preventing their adsorption on Pt surface; and (2) to promote the secondary hydroxyl oxidation of glycerol by forming a surface complex.[343]

Despite extensive investigations, the product selectivity and yield obtained by Pt and Pd catalysts are still far from satisfactory. In addition, strong deactivation of the catalyst during the reaction remains challenging. This deactivation not only lowers the activity but also the selectivity to DHA.[348] Hence, alternative catalysts should be developed. One of the options is to develop Au-based catalysts. Unlike Pt, which is susceptible to oxygen poisoning, Au has a higher resistance to it, allowing a higher oxygen partial pressure during the reaction. Under alkaline conditions, the distribution of products can differ depending on the type of support used, but generally carbon-supported catalysts show the highest activities, besides their high stability in acid–base conditions.[349,350]

Demirel *et al.* first reported the use of carbon-supported Au catalysts.[351] A 1 wt% Au/C catalyst was prepared by the Au-sol method using tetrakis-(hydroxylmethyl)-phosphonium chloride (THPC) as the reducing agent, which generated an Au particle size smaller than 10 nm. The oxidation reaction was performed in a semi-batch reactor at 1 atm, 60 °C, and pH 12. A maximum DHA selectivity of 26% could be achieved at 50% glycerol conversion, together with glyceric acid as the major byproduct (44% selectivity). The addition of Pt could further increase glycerol conversion and DHA selectivity. Catalysts with 1 wt% Au and 0.5 wt% Pt supported on activated carbon and carbon black 'Black Pearls 2000' (BP) were prepared, and the glycerol conversion rate was found to be significantly increased. Probably the promotional effect of Pt is related to the formation of Au–Pd alloys and the catalytic activity is dependent on the molar fraction of Pt. The highest activity was achieved when the Au to Pd molar ratio ranged from 0.2 to 0.4, at which 36% DHA could be obtained at 50% conversion of glycerol. X-ray diffraction (XRD) analysis showed the formation of an alloy with $Au_{0.8}Pd_{0.2}$ composition.

9.7.1.2 *Selective Oxidation of Glycerol to Glyceric Acid*

A list of supported catalysts used for selective oxidation glycerol and the product distribution is presented in Table 9.16. Detailed investigations of the Au catalysis for glycerol oxidation to produce glyceric acid were conducted by Carrettin and co-workers.[344,352,353] A series of Au catalysts supported on activated carbon (AC) or graphite were prepared with different Au loadings of 0.25, 0.5, and 1 wt%, employing formaldehyde as the

Table 9.16 Glyceric acid selectivity and products distribution obtained by various supported catalysts.

			Selectivity/%					Reaction Conditions				Au	
Support	Synthesis Method	Conversion/ %	GA	GLYAL	GLYCA	TA	OXA	T/K	pO_2/ atm	NaOH/ gly	Gly/Metal Ratio	Diameter/ nm	Ref.
1%Au/Graphite	Reduction with	54.0	100.0	0.0	n/c	0.0	0.0	333	3	1	538	5–50	353
	formaldehyde	91.0	92.0	0.0	n/c	6.0	0.0	333	6	1	214		
5%Pd/Carbon	Reduction with	57.2	67.0	0.0	n/c	12.2	0.0	333	3	1	500	2–5	344
5%Pd/Graphite	formaldehyde	53.0	50.0	7.5	n/c	7.5	0.0	333	3	1	500	n/c	
5%/Pt/Carbon	Reduction with	56.0	69.0	30.0	n/c	0.0	0.0	333	3	2	500	1.5–3	344
	formaldehyde	88.0	54.5	0.0	n/c	23.0	0.0	333	3	1	100		
1%Au/Carbon X40S	Citrate-protected sol	50.0	95.0	n/c	n/c	n/c	n/c	303	3	1	500	5–20	354
	immobilization	90.0	92.0	n/c	n/c	n/c	n/c	303	3	1	500		
1%Au/Carbon	Sol immobilization	50.0	64.5	n/c	12.2	9.2	9.5	323	3	4	500	n/c	358
	with BH_4^-	90.0	51.8	n/c	12.9	9.6	23.2	323	3	4	500		
1%Pd/Carbon	Sol immobilization	50.0	80.6	n/c	2.7	14.4	0.0	323	3	4	500	n/c	358
	with BH_4^-	90.0	37.9	n/c	23.9	29.6	7.5						
1%Pt/Carbon	Sol immobilization	50.0	42.4	n/c	31.1	5.9	8.2	323	3	4	500	n/c	358
	with BH_4^-												
0.65%Au-0.35%Pd/	Sol immobilization	50.0	77.2	n/c	4.5	17.6	0.0	323	3	4	500	2.6	358
Carbon	with BH_4^-	90.0	57.7	n/c	11.9	29.2	0.0						
0.49%Au-0.51%Pt/	Sol immobilization	50.0	72.8	n/c	22.7	2.9	0.6	323	3	4	500	n/c	358
Carbon	with BH_4^-	90.0	50.4	n/c	40.3	6.4	1.3						
1%Au/TiO_2	DP, reduced with	50.0	55.0	n/c	28.0	17.0	0.0	323	3	4	500	2–5	355
	BH_4^-	90.0	62.0	n/c	26.0	12.0	0.0						
1%Au/TiO_2	DP, calcination	50.0	81.0	n/c	13.0	4.0	0.0	323	3	4	500	5	355
2.5%Au-2.5%Pd/	Wet impregnation	50.0	78.7	n/c	6.8	6.5	1.7	33	10	2	500	2–10	360
TiO_2		90.0	73.6	n/c	6.8	10.2	2.8	3				80	
2.5%Au-2.5%Pd/	Wet impregnation	50.0	75.6	n/c	9	5.7	1.4	33	10	2	500	2–10	360
Carbon		90.0	72.3	n/c	6.1	7	1.8	3				35–80	
0.5%Au-0.5%Pd/ TiO_2	Sol immobilization	100.0	52.7	n/c	10.9	21.2	7.3	333	10	2	2000	2–6	360
0.5%Au-0.5%Pd/ Carbon	Sol immobilization	100.0	44.1	n/c	7.8	42.7	4.9	333	10	2	2000	2–10	360

GA, glyceric acid; GLYAL, glyceraldehydes; GLYCA, glyconic acid; TA, tartronic acid; OXA: oxalic acid.

reducing agent. The oxidation reaction was carried out in a stirred autoclave and pressurized with O_2 gas ($pO_2 = 3$–6 atm) at 60 °C. NaOH solution was added into the solution to maintain an alkaline pH. It was figured out that a basic condition is essential to obtain highly selective oxidation. In the absence of alkali, no glycerol conversion was observed. It was also found that the selectivity towards glyceric acid was dependent proportionally on the molar ratio of NaOH to glycerol as well as on the Au loading in the catalyst.[352] A highly alkaline solution and high Au loading resulted in high selectivity of glyceric acid. The highest yield of glyceric acid (83.7%) was achieved using 1 wt% Au/C with an equimolar amount of NaOH and glycerol (NaOH/glycerol = 1). The glycerol conversion was 91% and glyceric acid selectivity was 92%. It is proposed that the presence of NaOH facilitates the initial dehydrogenation *via* H-abstraction from the primary hydroxyl group of glycerol. TEM analysis revealed that the active 1 wt% Au/graphite catalyst comprised widely distributed Au nanoparticles with diameters ranging from 5 to 50 nm, averaging around 25 nm, whereas an inactive Au/graphite catalyst displayed larger particle diameters (>50 nm) with a narrow size distribution. Compared with the Pd and Pt catalysts, in general, the Au catalyst showed superior catalytic performance. Under the same reaction conditions, 5 wt% Pd/C exhibited a maximum glycerol conversion of 57% with a 67% selectivity towards glyceric acid, while on 5 wt% Pt/C, the corresponding conversion was 88% with a selectivity of 55%.[344]

Employing the same synthesis method used by Carrettin *et al.*,[352] Porta *et al.* prepared 1 wt% Au/C catalyst using graphite (Lonza HSAG300) and activated carbon (Camel X40S) as the support,[354] and optimized glycerate selectivity by fine tuning the catalyst preparation and reaction conditions. Using similar reaction conditions of Carrettin *et al.* ($pO_2 = 3$ atm, $T = 60$ °C, NaOH/glycerol = 1, glycerol/Au = 500), Au on activated carbon was found to be more active than Au on graphite. After finely tuning the particle dimension and dispersion by modifying the preparation procedures on both supports, two types of catalysts with different characteristics were obtained. The catalyst with well-dispersed Au nanoparticles of 4–5 nm was found not able to maintain its initial high selectivity, and the selectivity at 50% conversion was higher than that at 90% conversion. Interestingly, the catalyst with larger Au particles (>20 nm) was able to maintain its initial selectivity throughout the course of the reaction. The former was known to have a good dispersion, thus exhibiting high activity. However, it could subsequently lead to the oxidation of glycerates to glycolic acid or tartronic acid, thus lowering the glyceric acid selectivity.

This conclusion is further supported by the results obtained by Demirel-Gülen *et al.*,[336] who prepared carbon-supported catalysts with different Au particle sizes (2.7–42 nm) and Au loadings (0.5–10 wt%). It was observed that glycerol conversion was inversely proportional to the Au particle size. As the Au particle size decreased from 42 nm to the optimum size of 3.7 nm, the conversion increased from 84.7% to 100%. Meanwhile, the selectivity towards glyceric acid at 30% conversion increased from 49% to 75% within

3 h of the reaction. However, smaller Au nanoparticles (2.7 nm) caused a decrease in the selectivity to 40% while conversion remained at 100%.

The influence of the amount of NaOH was studied by varying the molar ratio of NaOH to glycerol. Initially, there was barely any reaction going on in the absence of NaOH. As the molar ratio was increased from 1. 2 to 4, the reaction rate increased accordingly.[336] However, according to Porta and Prati, there was little influence on initial selectivity from an increase on the base amount,[354] although at a prolonged time. It should be noted that high base concentration (NaOH/glycerol = 4) promotes oxidation of glyceric acid to tartronic acid, hence decreasing glyceric acid selectivity.[336] The results clearly confirm that the presence of the base has an important role in the dehydrogenation step of glycerol, as reported elsewhere for the oxidation of alcohols.[272] The conversion and selectivity are also dependent on the reaction temperature. Increasing the temperature leads to an increase in conversion but a decrease in selectivity as more glyceric acid is further oxidized to tartronic acid. Therefore, the overall selectivity of the reaction is dependent on the combination of several factors, such as initial selectivity of the catalyst, base-catalyzed interconversion, and stability of the products. It was observed that under basic conditions, the initial interconversion between the glyceraldehydes and DHA occurred very rapidly. As the oxidation of the aldehydic group is faster than that of the hydroxyl group, the subsequent rapid oxidation of the glyceraldehydes favors glyceric acid production. Reaction conditions, such as temperature and base concentration, play an important role in determining product stability; hence fine control of these parameters will yield high selectivity. By optimizing the reaction conditions ($pO_2 = 3$ atm, T = 30 °C, NaOH/glycerol = 4, glycerol/Au = 500) and the catalyst preparation, Porta and Prati obtained 92% selectivity to sodium glycerate at the full conversion.[354]

Besides the carbon supports, few have reported the use of metal oxides as a support for glycerol oxidation. Dimitratos *et al.* prepared 1 wt% Au/TiO_2 and studied the effect of the preparation conditions on the catalytic performance.[355] The catalytic reaction was carried out at 50 °C with a pO_2 of 3 atm, glycerol to metal ratio of 500, and a NaOH to glycerol ratio of 4. Different preparation methods, such as sol immobilization and deposition–precipitation, as well as different Au reduction methods (calcination, reduction with $NaBH_4$) were used to prepare the catalysts. The freshly prepared catalyst without calcination by the DP method (1 wt% Au/TiO_2-DP) mainly produced Au(III) species, which was not active. Calcination of 1 wt% Au/TiO_2-DP at 450 °C generated Au particles with an average size of 5.2 nm, which comprised 80% Au^0 and 20% Au^{III}, was active for glycerol oxidation with a TOF of 113 h^{-1}. The selectivity toward glyceric acid at 50% conversion was 81%. The highest catalytic activity with a TOF of 721 h^{-1} was achieved by chemically reduced catalyst with $NaBH_4$, which contained Au particles ranging from 2 to 5 nm, and metallic Au (Au^0) only. On this catalyst, however, the glyceric acid selectivity was only 55% at 50% conversion, with glycolic acid and tartronic acid as byproducts. Further calcination of the catalyst in air was detrimental as it promoted the agglomeration of Au to form large particles around 100 nm. XPS data also revealed the presence of two oxidation states for Au (83% of Au^0 at 83.7 eV and 17% of Au^{III} at 85.2 eV).

These results hence demonstrate that progressive increase in particle size will lead to a significant decrease in the activity but an increase in glyceric acid selectivity. This is in agreement with the results obtained by Porta and co-workers[354,356] and Demirel-Gülen and co-workers,[336] although carbon was used as support in their cases. From the XPS analysis, probably Au^0 is the active phase in the liquid-phase oxidation of glycerol, whereas the presence of Au^{III} species is found to be detrimental. To study the effect of different support on the activity and selectivity of the catalyst, 1 wt% Au on activated carbon (1% Au/AC) was also prepared. With the same synthesis method, the effect of the support should be reflected mainly in the activity difference of the catalyst. It was found that 1% Au/AC had much higher activity (TOF 1090 h^{-1}) than 1% Au/TiO_2 (TOF 178 h^{-1}), whereas the selectivity appeared to be determined particularly by the metal nanoparticles itself.

A report by Villa *et al.* showed that, compared with those supported on TiO_2 and MgO, Au supported on n-NiO was highly active for the oxidation of alcohols in the absence of base.[326] Recently, mixtures of $NiO_{(1-x)}$-$TiO_{2(x)}$ with different ratios were used as supports for the pre-formed Au nanoparticles and the catalytic activity in the oxidation of glycerol and ethane-1,2-diol was examined.[357] For glycerol oxidation, the initial supported 1 wt% Au/n-NiO was very active with a TOF value of 1418 h^{-1}, while that of 1 wt% Au/TiO_2 was lower (339 h^{-1}). However, under the same reaction conditions (50 °C, 3 atm O_2, NaOH/glycerol = 4), the former exhibited a lower selectivity of 55% towards glyceric acid compared with 75% at 90% glycerol conversion of the latter. Glycolic acid, tartronic acid, and lactic acid were the major byproducts. The addition of TiO_2 (10 wt%) to NiO was observed to increase the glyceric acid selectivity to 70% at the same conversion. Further addition of TiO_2 up to 90 wt% continuously increased the glyceric acid selectivity in the following order: NiO_{90}–$TiO_{2\ 10}$ (70%) < NiO_{50}–$TiO_{2\ 50}$ (72%) < NiO_{10}–$TiO_{2\ 90}$ (75%). These results indicate that the interface between NiO and TiO_2 plays an important role in affecting catalytic activity. High resolution TEM images showed that the Au nanoparticles preferentially deposited on the NiO phase, although in all samples, the Au particles had a similar size of about 3–4 nm. Therefore, the activity modification is likely attributed to a changed role of the support. Through FTIR and XPS analysis, it was considered that when in contact with NiO, TiO_2 changed the interactions between the NiO and the Au particles, hence changing the overall catalytic activity.[357]

To further improve the catalytic activity, a bimetallic catalyst system containing either Au–Pt or Au–Pd was prepared and compared with the corresponding monometallic Au catalysts.[351,358–360] While monometallic supported Au catalysts have been reported to be effective for the oxidation of glycerol, the bimetallic Au–Pd and Au–Pt catalysts show enhanced catalytic activity and selectivity. In addition, the oxidation product distribution can be better controlled by using these bimetallic catalysts.[358] Au–Pd or Au–Pt was reduced by $NaBH_4$ and deposited on a carbon support in the presence of polyvinyl alcohol as a stabilizing agent. In the presence of NaOH (molar ratio of NaOH to glycerol = 4), pO_2 of 3 atm, and reaction temperatures of 30 and 50 °C, the oxidation reaction yielded glyceric acid together with glycolic acid and tartronic acid as the main byproducts. Similar to the results obtained by using monometallic Au catalyst, the higher temperature (50 °C) produced glyceric acid with lower selectivity. It was observed that 0.61 wt% Au-0.23 wt% Pd/C (Au_6Pd_4/C) exhibited the highest glyceric acid selectivity of 69.3% at 90% conversion at 30 °C. Under the same catalyst preparation and reaction conditions, however, Au–Pt/C exhibited lower glyceric acid selectivity of ~50% at 90% conversion and 30 °C. The alloyed Au–Pd bimetallic catalyst also showed higher activity (TOF = 1077.4 h^{-1}) than the monometallic 1% Au/C (TOF = 779 h^{-1}) and 1% Pd/C (TOF = 514 h^{-1}) catalysts, indicating a synergistic effect of Au–Pd. After reaching 100% conversion, the reaction was kept on-stream at 50 °C for 5 h. Glyceric acid selectivity was quite constant with only a small degradation observed at the expense of tartronic acid. In contrast, a marked decrease in glyceric acid selectivity was observed over Au–Pt/C. It was concluded that glyceric acid was not practically over-oxidized by Au–Pd catalysts, whereas on Au–Pt catalysts, consecutive oxidation to tartronic acid proceeded. XPS analysis showed that the surface atomic ratio of Au to Pd was 0.9 for the obtained Au–Pd catalyst, with an average Au–Pd alloy particle size of 2.5 nm.

Dimitratos *et al.* prepared bimetallic supported Au–Pd catalysts by the impregnation and sol immobilization methods.[360] The catalysts of 2.5 wt% Au-2.5 wt% Pd/C and 2.5 wt% Au-2.5 wt% Pd/TiO_2 prepared by the impregnation method were tested at 60 °C, 0.1 MPa O_2, glycerol/metal = 500, and NaOH/glycerol = 2 for 4 h. The latter catalyst showed a significantly higher activity (TOF 662 h^{-1}) in a short reaction time compared with the former (TOF 381 h^{-1}), indicating that TiO_2 is a better support for Au–Pd alloy than carbon, in contrast to the results obtained by the monometallic catalysts. Within 1 h, the latter reached 95% conversion while the former only reached 45% conversion. TiO_2-supported catalysts are also more selective to glyceric acid while carbon-supported catalysts tend to produce more glycolic acid and formic acid. This could be due to more acidic nature of the carbon support. However, the glyceric acid selectivity of the former steadily declined and conversion to byproducts, in particular, tartronic acid and oxalic acid, increased with reaction time. On the other hand, carbon-supported catalysts showed constant glyceric acid selectivity.

Despite having lower metal loadings, the catalysts prepared by the sol immobilization method are more reactive than those prepared by the impregnation method. Very high TOFs (3300–4000 h^{-1}) were achieved on both 1 wt% (Au–Pd)/TiO_2 and 1 wt% (Au–Pd)/C. At 100% conversion, the glyceric selectivity was 47–52%, lower than that obtained on catalysts prepared by the wet impregnation method. Again, this is due to smaller particle size obtained by the sol immobilization method, which induced the sequential oxidation of glyceric acid. Observation by TEM and XPS analysis revealed that, on TiO_2, core-shell Au–Pd nanoparticles were formed with an Au-rich core and Pd-rich shell, while a homogeneous Au–Pd alloy was formed on carbon.

So far, various catalysts have been examined mainly in a batch autoclave reactor. Heterogeneous catalytic liquid phase oxidation reaction is, however, a complex process, as it involves gas–liquid–solid contact. Therefore, mass transfer across different phases may vary with the configuration of the reactor, which will eventually affect the catalytic performance. Pollington *et al.*[361] studied a monolithic reactor and down-flow slurry bubble column with Au/C as catalyst, while Zope and Davis ran a continuous fixed bed reactor with Au/TiO_2 as a catalyst.[362] In the former case, both monolith and down-flow slurry bubble column designs led to an enhancement in the reaction rate over autoclave batch studies by an order of magnitude on the Au/carbon coated monolith, and two orders of magnitude on the mesoscale structured down-flow slurry bubble column reactor.[361] A change in the product distribution was observed between the two reactors. The monolith and batch data showed very high selectivity to glyceric acid, whereas the thin channel slurry bubble column yielded approximately equal quantities of DHA and glyceric acid under similar conditions. Since the same batch of catalyst was used for both systems, the difference in product selectivity cannot be attributed to a particle size effect, but rather to the enhanced interaction between bubbles and particles from multiphase gas–liquid–solid reacting flows in thin channels or capillaries.

Zope and Davis also found that the product distribution during glycerol oxidation is influenced by the gas–liquid–solid contacting profile in different reactor configurations.[362] Using the similar reaction conditions of concentration, pressure, temperature, and pH, both tartronic acid and oxalic acid were formed in very small quantities over the course of the reaction in the batch study (<5% selectivity). However, in a continuous flow fixed-bed reactor, a higher selectivity towards secondary oxidation products like tartronic and oxalic acid was observed, even at low conversions. Thus, the selectivity of Au catalysts for primary oxidation is a strong function of reactor configuration.

Despite intensive studies, it is still very challenging to achieve a high activity and selectivity toward glyceric acid while maintaining a good stability by preventing further oxidation. Although prior works on glycerol oxidation have reported high activities for supported Au catalysts, the product distribution can vary widely.[362] Some research groups have observed very high selectivity (>90%) to glyceric acid, while others have obtained substantially lower selectivity (50–70%) due to the production of other byproducts, such as tartronic acid and glycolic acid. Therefore, to obtain the desired product with high selectivity and yield, it is still necessary to further fine tune the catalysts and to carefully optimize the reactor design and reaction conditions.

9.7.2 Selective Oxidation of Alkanediols

In spite of the abundant literature results available on the selective oxidation of alcohols, in general there are fewer studies on the oxidation of lower polyols like C_2 and C_3 diols. With regard to ethane-1,2-diol oxidation, a few studies indicate a limited application of platinum[363] and palladium catalysts due to the over-oxidation that normally occurs with C–C bond cleavage, leading to the formation of CO_2 *via* HCOOH formation.[364] Also in the case of the catalytic oxidation of propane-1,2-diol, there is a lack of systematic work in the literature concerning the selective oxidation of the hydroxyl groups.

Prati and Rossi first reported the catalytic oxidation of liquid-phase diols such as ethane-1,2-diol and propane-1,2-diol over Au catalysts supported on different substrates, such as Al_2O_3, SiO_2, Fe_2O_3, CeO_2, and carbon.[349,364] From the initial catalyst screening, it was found that 1 wt% Au/C catalyst was highly selective to produce glycolic acid and lactic acid, respectively. Under a mild reaction condition (70–90 °C, $pO_2 = 0.3$ MPa, diol/metal = 1000, NaOH/diol = 1), 1 wt% Au/C prepared by the DP method could selectively oxidize ethane-1,2-diol to produce glycolic acid with 94% conversion and 90% selectivity, while the same catalyst oxidized propane-1,2-diol to form lactic acid with 78% conversion and 100% selectivity within 1 h. These results showed the superior performance of Au catalysts compared with their Pd and Pt counterparts. In addition, Au catalysts also exhibited good resistance and stability without any obvious deactivation and metal leaching being observed.

Biella *et al.* reported the use of 1 wt% Au/C catalyst to selectively oxidize phenylethane-1,2-diol to mandelic acid, an important pharmaceutical intermediate.[365] Different types of activated carbon, depending on the treatment, were used as supports.[365] The reaction was carried out at 70 °C and 3 atm O_2 with or without the addition of NaOH. In the presence of a base (NaOH/diol = 2), 100% conversion of phenylethane-1,2-diol could be achieved. Besides mandelate, benzoate was also formed in large amounts. The highest yield of mandelate (65%) was obtained using Au supported on MV 125 carbon with a large surface area (1095 m^2 g^{-1}) and pore diameter (0.42 nm). Further optimization by increasing the amount of base (NaOH/diol – 4) and reaction temperature (90 °C) yielded a high selectivity of mandelic acid (83%).

As discussed in the preceding section, a mixed oxide composite of NiO and TiO_2 was also very active in oxidizing ethane-1,2-diol to produce glycolic acid.[357] Overall, high selectivity of 99% at 90% conversion could be achieved by depositing Au nanoparticles on $NiO_{(1-x)}$-$TiO_{2(x)}$ composites (x >0.5), with only a little amount of oxalic acid as the byproduct. The reaction was conducted at 50 °C, 3 atm O_2 and NaOH/diol of 4.

Recently, Hutchings and co-workers also reported selective oxidation of propane-1,2-diol to lactic acid using monometallic Au/C as well as bimetallic Au–Pt and Au–Pd nanoalloy catalysts prepared by the sol immobilization method.[366] The Au alloyed with Pt is much more active than Au alloyed with

Pd under the reported conditions (30–60 °C, $pO_2 = 0.3$ MPa, NaOH/diol = 1–2). Moreover, the reaction could be carried out under much milder conditions than those reported previously,[349] with a significant reduction in the catalyst amount. In addition, air could also be used as the oxidant. The enhanced activity of the supported Au–Pt catalysts compared with the Au–Pd catalysts is probably related to the smaller mean particle size of the Au–Pt catalyst, which leads to more active sites on the surface.

Although many Au-based catalysts can be effective for selective oxidation of alcohols, it may not be the case for polyols. More detailed and in-depth studies on reaction mechanism and the optimization of catalyst–catalyst configuration are required, as it is believed that the oxidation of different hydroxyl groups requires specific orientation and unique interaction with the catalysts. The physicochemical properties of the catalysts should be tailored to better direct the reaction pathways.

9.8 Selective Oxidation of Sugars

Development of novel processes for the transformation of renewable biomass into value-added chemicals is of great importance for a sustainable chemical industry. It is estimated that the global annual production of biomass amounts to *ca.* 170 billion tons (in 2007), of which only a small portion (3–4%) is utilized for food and non-food purposes.[367] Carbohydrates are by far the most abundant renewable resources on Earth and represent the major portion of the annual biomass generation. With the recent increase in petroleum prices as well as the depletion of natural oil, production of chemicals from biomass or biomass-derived platform chemicals, *e.g.* glucose and glycerol, *via* a catalytic approach is attracting more and more attention. For instance, the transformation of biomass to chemicals and selective oxidation of sugars are comprehensively reviewed respectively by Corma *et al.*[367] and Kusema and Murzin.[368] In this section, we mainly review the selective oxidation of sugars by supported Au catalysts.

In general, the selective oxidation of sugars mainly produces aldonic acids, which find many applications in foods, cosmetics, detergents, and pharmaceuticals. For example, gluconic acid and its sodium salts are excellent chelating agents for calcium and other calcareous deposits to prevent scale formation on surfaces. On the other hand, calcium gluconate is used for treating calcium deficiency and is an antidote for the treatment of burns due to hydrogen fluoride.[369]

The transformation of sugars is usually carried out *via* fermentation and chemical routes. With a wide range of microorganisms, the fermentation process remains an important route. However, it has several drawbacks. It usually requires multiple stages, as the microbial cells must be first cultivated. The feedstock contamination also poses a major issue, as well as the difficult product recovery.[367,370] Furthermore, substrate and product inhibition will also limit the overall yield, as the culture cells would be poisoned beyond the maximum allowable concentration.[369,370] On the other hand, chemical oxidation of sugars involves hazardous compounds, such as bromine or chlorine in highly acidic or alkaline solution.[371] In addition, it is less efficient as excess reagents are usually used, which result in a large amount of waste, especially in the form of halide salts. This requires further treatment and eventually increases the production cost. Heterogeneous catalytic reaction in aqueous solution can offer great advantages in solving the aforementioned issues. It allows easy separation, re-use of the catalysts and continuous reaction, and minimizes waste and the generation of hazardous materials.

A pioneer work by Wieland showed that in the presence of oxygen, D-glucose was partially oxidized over pure palladium black within 80 h.[372] Subsequently Heyns and co-workers investigated the use of carbon-supported and pure platinum catalysts for the oxidation of different types of sugars, namely sorbose, D-glucose, D-galactose, L-arabinose, D-mannose, and D-xylose to produce their respective aldonic acids in alkaline solution.[373–375] In this method, the aldoses are selectively oxidized at room temperature, while further oxidation occurs at higher temperatures.

Various studies have since dealt with oxidation of mono- and disaccharides on various catalytic systems, primarily based on Pt and Pd.[376–383] Different types of supports have been used to immobilize the noble metal nanoparticles, such as carbon, Al_2O_3, SiO_2, TiO_2, and zeolites. In liquid-phase oxidation reactions, carbon seems to be the support of choice compared with other oxide materials, hypothetically due to the high density of phenolic groups on the surface, which could interact closely with the deposited metal particles, as well as the high stability in the reaction media.[384]

9.8.1 Selective Oxidation of Monosaccharides

With an estimated production of about 5 million metric tons per year (2001),[385] D-glucose is the most produced monosaccharide in the world, mainly from the fermentation of starch, cellulose, and sucrose in industry. With its abundance and simple structure, D-glucose can serve as a readily available raw material for the manufacture of various industrial chemicals. Among many important products derived from D-glucose, gluconic acid is the main product from the selective oxidation of D-glucose. It is currently produced by enzymatic fermentation of D-glucose derived from *Aspergillus niger*,[369] with an estimated annual global production of 60,000 tons.[385] Besides this route, heterogeneous catalytic oxidation with oxygen or air shows great potential as it can be carried out under mild conditions. Figure 9.12 shows various chemicals that can be derived from glucose.

In the early work, direct oxidation of D-glucose on carbon-supported Pd or Pt nanoparticles was found to be competitive with high selectivity and yield. Similar to that of alcohols, the oxidation of sugar can be described as an oxidative-dehydrogenation reaction.[272,377] In general, the oxidation reactions are performed in dilute aqueous solution in a batch reactor, under mild conditions at atmospheric pressure and a temperature ranging from 20

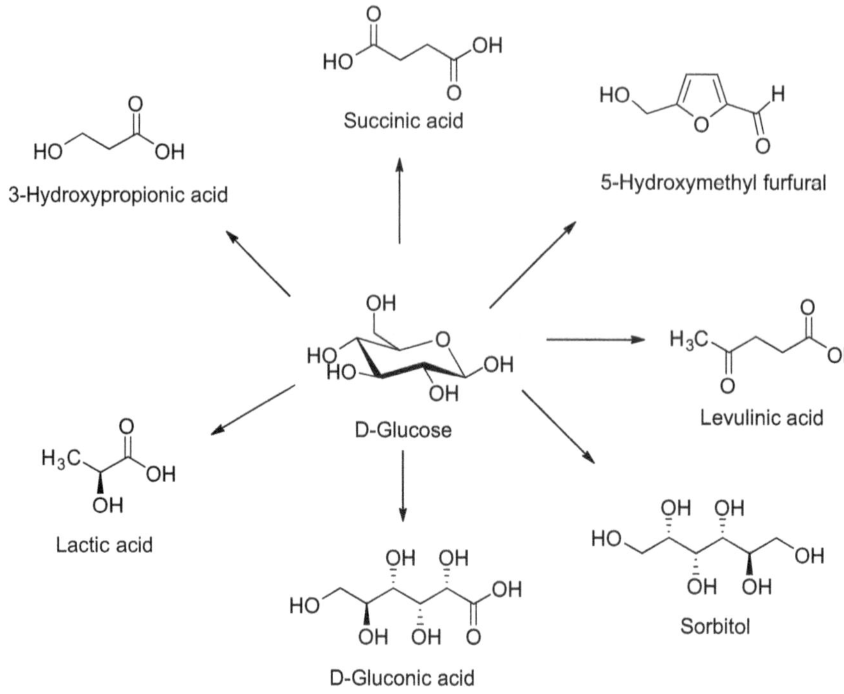

Figure 9.12 Various chemicals derived from D-glucose.

to 80 °C. Air or molecular oxygen is bubbled into the solution in the presence of catalyst. The reactions are usually performed at basic pH (pH 7–10).

Besides the catalyst itself, there are several factors that influence the reaction selectivity and yield, including reaction temperature, pH value, and oxygen concentration. For example, side reactions of glucose are favored at high temperature (>70°C) and high pH (>10). Under these conditions, glucose may isomerize to form fructose, which can further react to form mannose, as depicted in Figure 9.13, while in the presence of oxygen, glucose is selectively oxidized to gluconic acid.[386]

The influence of solution pH and oxygen concentration on the performance of Pd- and Pt- based catalysts over a prolonged time is related to catalyst poisoning. There are two possible reasons for the catalysts poisoning:

1. The oxidation of the active site of noble metal nanoparticles due to chemisorption of oxygen
2. Strong adsorption of the product on the catalyst surface at low pH.

In the alkaline environment, the carboxylate anions are easily desorbed from the catalyst surface, while at neutral/acidic pH, the carboxylic acid remains strongly adsorbed, hence promoting over-oxidation reactions, which lead to degradation of products and catalyst poisoning.

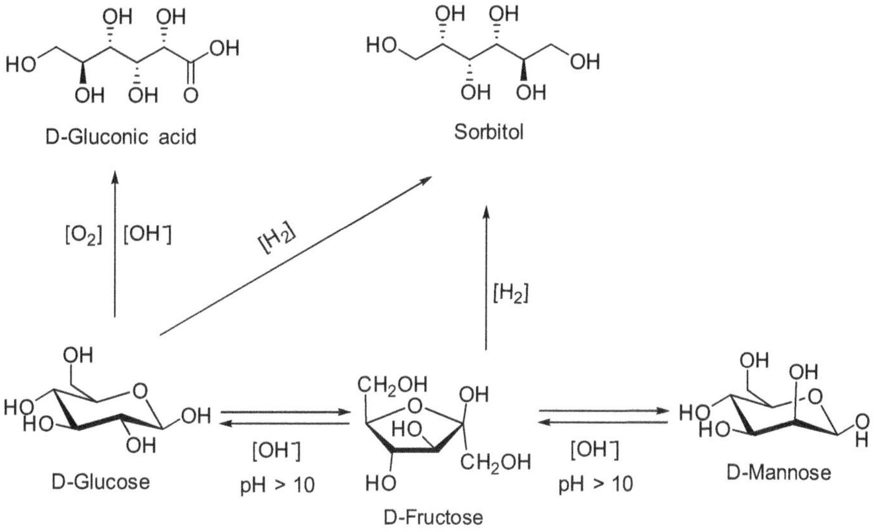

Figure 9.13 Possible reactions of D-glucose in alkaline solution. Reproduced with permission from ref. 386.

The promotion of the Pd/C and Pt/C catalysts with heavy metals, such as Bi and Pb, is found to be effective in increasing gluconic acid selectivity. Despevroux *et al.* reported that on 4% Pd-1% Pt-5% Bi (wt%) supported on carbon, a selectivity of 98% toward gluconic acid could be reached,[387] while Besson *et al.* obtained as high as 99.3% selectivity and 100% conversion within 130 min at 40 °C and pH 9.[388] The loading of Pd was 4.7% with Bi/Pd of 0.2 on activated charcoal. The promotional effect of Bi is believed due to several reasons, such as its ability to stabilize the metallic state of the noble metal nanoparticles by slowing the oxidation of the metal,[287] the creation of new active centers,[297,298,305] and the formation of a Bi-glucose complex on the catalyst surface, which could facilitate the dehydrogenation reaction.[379] However, due to the poisonous property of Bi and Pb, as well as substantial leaching into the solution under the reaction conditions, the further use of both metals is not favored.

9.8.1.1 *Selective Oxidation of D-glucose*

The transformation of glucose over Au catalysts has been extensively studied. They exhibit a unique feature as they are active either in the form of supported metal or colloidal nanoparticles (sol). The catalyst preparation methods,[389–391] reaction conditions, reaction mechanisms over supported Au/carbon catalysts[386] and colloidal Au,[392] the effects of supports and Au nanoparticle size,[393] as well as the stability and durability of the catalysts[394,395] have also been investigated.

Biella *et al.* reported that high conversion in selective oxidation of glucose could be achieved on 1 wt% Au/C at pH 7–9 with a high selectivity similar to that obtained with Pd–Bi/C catalyst (>99%).[396] The supported Au catalyst was prepared by the sol immobilization method using $NaBH_4$ and PVA as the reducing and protective agent, respectively. The supported Au catalyst exhibited higher activity, with a TOF value of $\sim$3000 h^{-1} at pH 9.5 compared with that of Pd–Bi catalyst ($\sim$500 h^{-1}). Another advantage of using an Au catalyst is its ability to retain high selectivity upon recycled reactions. Interestingly, the catalyst remained active even at low pH of 2.5, which could suppress the isomerization reaction of glucose to fructose.[396] The reaction was carried out under mild conditions at 50–100 °C with O_2 pressure of 1–3 atm. The correlation of the catalytic activity and the carbon surface properties was also studied by applying different pre-treatments on various carbon supports.[384] With the same Au loading and particle size, it was observed that a specific Au–carbon interaction exists, which can be related to the density of phenolic groups on the carbon surface.

Meanwhile, Comotti *et al.* observed that unsupported Au nanoparticles with a mean diameter of 3.6 nm dispersed in aqueous solution were also very active in converting 21% of glucose within 200 sec. With an initial TOF of 50 120 h^{-1}, the activity is comparable with that of enzymatic reactions.[397] In contrast, colloidal nanoparticles of a similar size of other noble metals, such as Cu, Ag, Pd, and Pt, were hardly active. However, due to the coagulation of Au nanoparticles upon prolonged reaction ($\sim$400 sec), the suspension became inactive. Using a colloidal gold suspension, Beltrame *et al.* calculated the glucose oxidation reaction kinetics and found that the experimental data fitted well with the Eley–Rideal mechanism, where the adsorbed glucose molecules interact with the dissolved oxygen.[392] In agreement with Comotti *et al.*,[397] the specific activity of colloidal gold catalysis was found quite similar to the enzymatic one.

On the other hand, Önal *et al.* studied the effect of different Au particle size (3–6 nm) and carbon supports (black pearls and Vulcan) on catalytic activity, and obtained a kinetic model[386] which showed the dependence of the catalytic activity in the aqueous phase on the specific Au surface area and, hence, Au particle diameter. Such dependence can be related to the adsorption behavior of the reactant on different crystal sites. As the metal particle size decreases, the relative proportions of surface atoms on corners and edges increase; this is where the adsorption/hydrogenation of D-glucose, probably *via* the carbonyl group, occurs. The authors proposed that the oxidation reaction over Au/C could be described by a Langmuir–Hinshelwood type model. This is because, in the presence of support, oxygen molecules are not in direct contact with the adsorbed reactant, but rather with the surface hydrides formed immediately after glucose dehydrogenation.

The whole reaction mechanism involves three elementary reactions: adsorption of D-glucose; surface reaction of D-glucose; and desorption of gluconic acid. The initial adsorption of glucose on the catalyst surface is followed with the dehydrogenation of glucose. The product, D-gluconic acid, is then immediately desorbed from the surface while the hydrides bound to the catalyst surface are oxidized by the dissociatively adsorbed oxygen. Under vigorous stirring, the adsorption of glucose can occur very rapidly and should not be the rate-limiting step. As the reaction is carried out at alkaline pH, the product desorption from the catalyst surface should also be faster than the surface reaction. Therefore, the dehydrogenation of glucose should be the rate-determining step.

To increase the catalytic activity of Au-based catalysts, especially at the acidic pH, a bimetallic system was investigated by combining Au with other noble metals, such as Pd, Pt, and Rh, supported on carbon by the sol immobilization method. With the total metal loading of 1 wt%, the experiments were conducted at both uncontrolled and controlled pH (9.5).[398] The former reaction was carried out at 3 atm of O_2 and temperature of 70–90 °C, while the latter was at atmospheric pressure and 50 °C. Comparing both reactions, supported Au–Pt/C had overall higher activity. In the absence of alkali (low pH), Au–Pt/C (Au/Pt = 2 : 1) showed a strong synergistic effect with a catalytic activity much higher than that of the monometallic catalyst, which gave TOF value of 924 h^{-1} compared with 51 and 60 h^{-1} for Au and Pt monometallic catalysts, respectively. Interestingly, at a controlled pH of 9.5, the bimetallic Au–Pt catalyst showed very similar activity to that of monometallic Au catalyst, with TOF value of ~17 000 h^{-1}. The different promoting effect of Pt and Pd under acidic and alkaline solution could be related to different reaction mechanisms. The presence of alkali plays a strategic role, which favors the slow dehydrogenation step of glucose producing gluconate and H_2O_2. As such, no important effects of Pt and Pd metals on Au should be expected. However, no mechanistic studies are yet available in the case of glucose oxidation in acidic conditions for understanding the strong synergetic effect of platinum.[398]

Recently, Zhang and Toshima prepared a colloidal dispersion of multimetallic nanoparticles coated with PVP as a protective agent.[399] The catalytic results at 60 °C and pH 9.5 showed that Au–Ag, Au–Pt and Au–Pd bimetallic nanoparticles with average diameter less than 2 nm (~1.4–1.6 nm) exhibited higher catalytic activity compared with the monometallic Au nanoparticles. The highest activity was obtained on $Au_{80}Ag_{20}$ nanoparticles (Ag_{core}–Au_{shell} structure) with a specific activity of 16 890 $mol_{glucose}\ mol_{metal}^{-1}\ h^{-1}$, about three times higher than that of the Au nanoparticles. The introduction of Pt into the Au–Ag bimetallic system further increased the activity and selectivity. With the same particle size, the trimetallic $Au_{0.7}Pt_{0.2}Ag_{0.1}$ alloy exhibited specific activity of about five times higher than that of the Au nanoparticles (20 090 $mol_{glucose}\ mol_{metal}^{-1}\ h^{-1}$) during a 2 h testing. It is believed that good dispersion of nanoparticles, small particle size, and the electronic charge transfer effect among different noble metal atoms contribute to the high activity of the above multimetallic systems.

Besides using carbon as support, Au nanoparticles on metal oxides have also been actively studied. Baatz and co-workers prepared Au/Al_2O_3 catalysts by the DP method and incipient wetness impregnation. The oxidation reactions were conducted at atmospheric pressure, pH 9 and 40 °C. It was found that 0.3 wt% Au/Al_2O_3 with an Au particle size of 1–2 nm was very active in oxidizing glucose, yielding more than 98% gluconate selectivity, and being able to maintain a high activity of about 500–600 $mmol_{glucose}$ g_{Au}^{-1} min^{-1} over 20 batches of reactions.[390,391] Another highly active and stable catalyst of Au/TiO_2 (0.45 wt% Au) was reported by Mirescu and co-workers.[382,394] The catalyst was stable upon 17 recycles without losing activity (~200 $mmol_{glucose}$ g_{Au}^{-1} min^{-1}). It was prepared *via* DP at pH 6.5. The catalytic reaction was carried out at pH 11, 40 °C and atmospheric pressure of O_2. A full conversion was achieved after a 2 h reaction with 98% gluconic acid selectivity.

Haruta and co-workers prepared supported Au on different oxides, such as ZrO_2, Al_2O_3, TiO_2, and CeO_2, by the SG and DP methods.[393] The size of Au nanoparticles and the interaction of Au and oxide supports depend strongly on the preparation method. The SG method works efficiently for preparing small Au nanoclusters on ZrO_2 and Al_2O_3, while the DP method is more suitable for TiO_2 and CeO_2. Unlike CO oxidation, the catalytic activity of glucose oxidation was more dependent on the Au particle size than on the nature of the supports, indicating that the liquid-phase reaction likely occurs on the Au surface rather than at the Au–support perimeter. At the reaction condition of 1 atm, 50 °C and pH 9, the gluconic acid selectivity obtained by all of the catalysts was always very high (above 98%). However, the highest activity was achieved by Au/ZrO_2 with an Au diameter of about 3 nm and TOF value of 201 600 h^{-1} at 50 °C and pH 9.5, the highest value for glucose oxidation reaction over Au catalysts so far.

9.8.1.2 Selective Oxidation of L-arabinose

Besides glucose, supported Au catalysts are also active for selective oxidation of other monosaccharides, such as oxidation of L-arabinose (a pentose) to arabinonic acid. Using 0.45 wt% Au/TiO_2 catalyst, Mirescu and Prüße obtained at least 90% conversion of L-arabinose, high selectivity towards arabinonic acid (>99.5%), and a specific activity of 1440 $mmol_{arabinose}$ g_{Au}^{-1} h^{-1}.[382] Under the same condition (40 °C, pH 9, 1 atm O_2), Au catalysts demonstrated higher activity and selectivity compared with supported Pt and Pd catalysts.

Murzin and co-workers investigated the aqueous-phase oxidation of L-arabinonic acid in a shaker reactor with semi-batch mode.[400] The solution pH was varied in the range of 6–9 while maintaining the temperature at 50–70 °C. Commercial Au and Pd catalysts supported on carbon and different oxides, such as Al_2O_3, TiO_2, ZnO, were used for the reaction. High activity and selectivity to arabinonic acid were achieved on the Au catalysts while the Pd catalysts exhibited lower activity and selectivity due to the generation of more byproducts. With an original L-arabinose concentration of 0.1 mol L^{-1}, 99% of conversion and selectivity were achieved within 200 min over 1 wt% Au/Al_2O_3 and 1 wt% Au/ZnO (supplied by Mintek) at an optimum temperature of 60 °C and pH 8. Both catalysts had a uniform and narrow distribution of Au nanoparticles with an average size of 2.3 and 2.8 nm, and similar Au dispersion of 46 and 42%, respectively. In contrast, a 1 wt% Au/C catalyst (supplied by Johnson Matthey) showed much lower activity at the reaction temperature of 60 °C (259 h^{-1}) compared with that of Au/Al_2O_3 (828 h^{-1}) and Au/ZnO (1030 h^{-1}), due to the broad particle size distribution of Au nanoparticles in the Au/C catalyst, ranging from 3 to 80 nm. The effects of reaction parameters, such as pH, temperature, oxygen flow rate, were also studied. The kinetic data revealed that the oxidation rates increased with increase of the reaction temperature and oxygen flow rates, while alkaline conditions (up to pH 9) were favorable for high initial rates, catalyst activity, and product selectivity.[400]

Further investigation showed that bimetallic Au–Pd catalysts exhibited higher activity and selectivity than their corresponding monometallic catalysts, a phenomenon observed in the oxidation of other alcohols and saccharides as described elsewhere. Smolentseva *et al.*[401] prepared 4 wt% Au supported on Al_2O_3 and CeO_2, respectively, by the DP method with subsequent chemisorption of 1 wt% Pd; the effect of a different reduction treatment in formaldehyde, hydrogen, and oxygen was also studied. In general, bimetallic Au–Pd catalyst on CeO_2 is more active than that on Al_2O_3. The highest TOF value (990 h^{-1}) and selectivity (90%) were obtained when the catalyst was reduced in formaldehyde because it led to the formation of smaller Au particles (~1.7 nm). Also, XPS analysis confirmed a strong interaction between Au and Pd with reducible CeO_2, which resulted in the higher activity too.[401]

The structure sensitivity of arabinose oxidation over Au/Al_2O_3 was revealed by Kusema and co-workers,[402] who prepared the catalysts by direct ion exchange (DIE) and DP-urea methods. The particle size variation was achieved by tuning the initial concentration of the precursors, the washing procedure, and the calcination temperature. The DP-urea method generated Au nanoparticles with a mean diameter of 2.3 nm, whereas those derived from the DIE method had a mean diameter of 1–1.3 nm. According to the mechanism proposed by Claus for the oxidation of glucose,[386] L-arabinose is assumed to be adsorbed on the low-coordinated Au atoms at the corner, edge and steps *via* the carbonyl group. The size of the low coordinated atoms has an optimum value of around 2 nm.[403] Below this size, the particles are monolayers and all the atoms are in contact with the support, whereas above 2 nm, the number of these sites decreases. Then, it can be assumed that the catalysts prepared by the DP-urea method with an average diameter of 2.3 nm had the maximum number of actives sites for the adsorption of L-arabinose. The catalyst with a higher number of low-coordinated atoms exhibited better catalytic performance, which could also be associated with the fact that these sites might be responsible for the O_2 activation[404] as well as the H-abstraction.[405] Therefore, the volcano-type relationship between the activity and the Au particle size can be explained by the population of the active sites for the adsorption of L-arabinose; dehydrogenation step, and oxygen activation.

9.8.1.3 Selective Oxidation of D-galactose

The versatility of 0.45 wt% Au/TiO_2 catalyst prepared by Mirescu and Prüße for various types of sugars was demonstrated by high selectivity of galactonic acid, which was produced from the selective oxidation of D-galactose. At 40 °C, pH 9 and 1 atm O_2, a specific activity of 2040 $mmol_{galactonic}$ acid g_{Au}^{-1} h^{-1} was achieved with >99.5% selectivity.[382] The Au catalysts showed higher activity and selectivity than the supported Pt and Pd catalysts.

Murzin and co-workers also reported that 100% conversion in the selective oxidation of galactose over Al/Al_2O_3 was achieved within 200 min with a high selectivity of up to 99% at pH 10 and an optimum temperature of 60 °C. The catalytic properties of the Au nanoparticles correlates to the Au cluster size, with a mean particle size of 2.6 nm exhibiting the highest activity.[406]

Through the comparative study on the oxidation of different monosaccharides, it appears that overall, Au-based catalysts are superior to Pd and Pt catalysts. In addition, it is also revealed that Au displays higher activity over the oxidation of glucose compared with other monosaccharides. Nonetheless, independent of the investigated sugar substrate, Au catalysts in all the cases showed the highest catalytic activity and a permanently high selectivity towards the aldonic acid.[368]

9.8.2 Selective Oxidation of Disaccharides

9.8.2.1 Selective Oxidation of Lactose

Lactose, a milk sugar, is one of the widely produced disaccharides. It consists of glucose and galactose monomers bound together *via* a β-1,4-glycosidic linkage. Several million tons are produced annually as a byproduct of the dairy industry. However, only about 400,000 tons are further processed in cheese while the remaining is disposed of and likely to cause environmental issues by increasing the chemical oxygen demand (COD) and biological oxygen demand (BOD) in aquatic ecosystems.[381] Lactose finds its application primarily in the food and pharmaceutical industries as a stabilizer, capsule filler, and tablet ingredient. Due to its excess availability, chemical transformation of lactose to other value-added products is highly desired. For example, hydrogenation of lactose will produce lactitol, a sugar alcohol, while the isomerization reaction will produce lactulose. Selective oxidation of lactose produces lactobionic acid, an important industrial product primarily used as a preservative agent of human organs during transplantation and hand-protecting cream. Figure 9.14 shows the reaction network of the oxidation and hydrogenation of lactose. Biochemical oxidation of lactose can be performed by aerobic microorganisms, in particular, the genus *Pseudomonas*.[407] However, difficulties in separating the microorganism, and the control of byproducts and waste water lead to another avenue that involves heterogeneous catalysis.

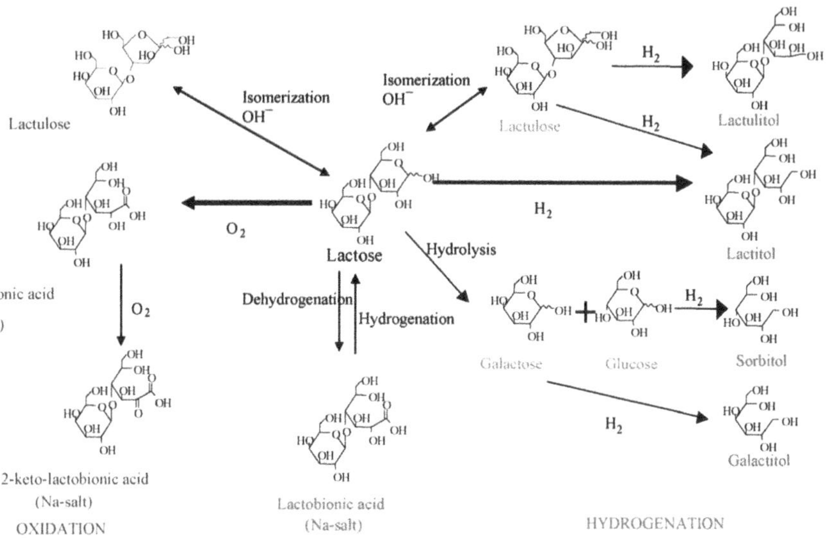

Figure 9.14 Various products derived from oxidation and hydrogenation of D-lactose.
Reproduced with permission from ref. 381.

An early work by Hendriks *et al.* presented several catalyst systems for the oxidation of lactose to different products. With a stirred tank reactor, 5 wt% Pd/C could obtain 100% selectivity toward lactobionic acid when promoted with Bi (Bi/Pd = 0.5–0.67). A total lactose conversion of 95% was achieved at pH 7–10 and temperature 60 °C.[408]

Murzin and co-workers studied the oxidation of lactose in a shaker reactor and on various catalysts, such as Pd/C, Au/Al_2O_3, and Au/CeO_2. The oxidation of lactose occurs in two consecutive steps: the oxidation of cyclic hemiacetal lactose to lactobionic acid, followed by further oxidation of the alcohol groups to keto groups forming 2-keto-lactobionic acid. However, the formation of the latter is undesirable as it decreases the lactobionic acid selectivity over prolonged time. The optimum temperature was found to be 50–70 °C at pH 6–8. In addition, a small fraction of lactose was also isomerized and hydrogenated to form lactulose and lactitol, respectively, as the isomerization reaction is highly favored in strong basic media.[381] Similar to the oxidation of glucose, lactose oxidation accelerates as pH increases due to easy desorption of lactobionate salt. Among the tested catalysts, 2 wt% Au/CeO_2 was observed to suppress the formation of 2-keto-lactobionic acid, while it was significantly formed when lactose conversion was above 60% over 5 wt% Pd/C catalyst. Hence, it was concluded that Au catalysts are more selective towards lactobionic acid, while Pd catalysts can be of importance when the aim is the production of 2-keto-lactobionic acid.

As the catalytic activity of Au depends strongly on the particle size and the types of supports, various supported Au catalysts have been studied.[409] Different metal oxides, such as Al_2O_3, TiO_2, ZnO, ZrO_2, SiO_2, Fe_2O_3, and Fe_3O_4, were used as supports; 2 wt% of Au was loaded onto the respective supports by the DP method. The catalytic activity was compared by measuring the time reached for lactose conversion of 20 and 50% at 60 °C and pH 8. With an oxygen flow rate of 2.5 ml min^{-1} and catalyst amount of 50 mg, the most active catalyst was 2 wt% Au/Al_2O_3. With lactobionic acid selectivity of about 95%, almost zero-order dependence of lactose concentration was observed. On the other hand, 2 wt% Au/Fe_3O_4 showed a substantial initial activity but deactivated after 20% conversion was achieved. Several reasons may contribute to the high activity of 2%Au/Al_2O_3, such as very small Au particle size in the range of 1–2 nm compared with the other tested catalysts, as well as the large surface area of Al_2O_3 support, which enhances dispersion of Au nanoparticles. XPS analysis showed that gold was in the metallic Au^0 state.

Recently, a supported Au catalyst on mesoporous silica (m-SiO_2) was reported to be highly active and selective towards lactobionic acid. A simple one-pot synthesis of Au/m-SiO_2 was employed by using bis-[3-(triethoxysilyl) propyl] tetrasulfide (BTSPT), a silane coupling agent, to immobilize gold.[410] The catalyst with 0.7 wt% Au content was found to be optimum for the highest activity. Full conversion and 100% selectivity could be achieved within 100 min of the reaction at 1 atm of air, 65 °C and pH 9 and a catalyst to lactose ratio of 0.2. The catalyst had the smallest Au particle size of about 7 nm, and these Au nanoparticles seemed to be located in the mesoporous framework and on the surface of silica. XPS analysis revealed that most Au was present as metallic Au^0 with a small fraction of oxidized Au^+.

Further catalyst modification by introducing ceria (CeO_2) into SBA-15 matrix could further enhance the catalytic activity.[411] A catalyst with the same Au loading (0.7 wt%) was prepared by the DP method using Au-ethylenediamine complex as the precursor. Under the same reaction conditions, Au/SBA-15-CeO_2 (Ce/Si = 0.1) showed an initial rate of 0.16 g acid L^{-1} min^{-1}, higher than that of Au/m-SiO_2 (0.12 g acid L^{-1} min^{-1}), which was reported earlier.[410] The catalytic activity was dependent on the Ce to Si molar ratio. A maximum activity was achieved at a ratio of 0.2. In these catalysts, the Au particle size was mostly invariant with the Ce to Si molar ratio, ranging from 4–5 nm. Meanwhile, the XPS study indicated the coexistence of metallic and oxidized species on the catalyst, with relative abundances of $Au^0 > Au^+ > Au^{3+}$, whereas the XPS spectra of the Ce 3d core level suggested the presence of both Ce^{3+} and Ce^{4+} oxidation states in the 0.7 wt% Au/SBA-15-CeO_2 catalyst. Compared with Au/m-SiO_2, the introduction of CeO_2 resulted in a larger fraction of oxidized Au^+ and Au^{3+}.

The high activity and selectivity of the 0.7 wt% Au/SBA-15-CeO_2 catalyst could be due to a synergy of the following factors: the small size (4–5 nm) of the well-dispersed Au nanoparticles; the presence of CeO_2 that can stabilize the dispersion of Au particles and enhance the mobility of surface oxygen species; and the redox properties of Ce that might contribute to the reaction mechanism.[411] Timofeeva *et al.* found that the coordination and the agglomeration states of the Ce atoms as well as the density and the content of silanol groups on the surface of the Ce-SBA-15 are important in determining the catalytic activity of Ce-modified SBA-15 supports.[412]

Despite the extensive work on Au catalysis, there are several controversial views regarding the nature of the active species responsible for the oxidation reactions. Moreover, since most of the studies deal with CO oxidation because it is the most frequently adopted catalytic reaction to evaluate the performance of Au catalysts, in the case of the carbohydrate oxidation over Au catalysts, discussion of the role of metallic and oxidized Au species in the reaction mechanisms is ongoing. Therefore, in order to better understand the reaction mechanism as well as to improve the catalytic performance, more detailed investigations on the specific Au active sites are necessary.

9.8.2.2 Selective Oxidation of Cellobiose

Cellobiose, consisting of two D-glucose monomers with a β-1,4-glycosidic bond, is one of the hydrolysis products of cellulose. So far, there has been only a little effort in developing the catalytic conversion of cellobiose into chemicals. A report by Tan *et al.* showed that Au nanoparticles supported on carbon nanotubes (CNTs) are a potential catalyst for direct oxidation of cellobiose to produce gluconic acid.[413] In the absence of Au, CNTs alone was able to convert cellobiose into glucose with 27% conversion and 80% glucose selectivity without the formation of gluconic acid; this could be due to the acidic groups on CNTs. At the optimum temperature of 145 °C and 0.015 M cellobiose concentration, when the loading of Au was increased up to 0.5%, glucose selectivity became lower while gluconic acid selectivity was raised up to 84%. In addition, cellobiose conversion was also significantly enhanced up to 81%, giving 68% overall yield of gluconic acid. The role of Au nanoparticles accounts for the selective oxidation of glucose, and intermediate and cellobiose conversion. The acidity of the catalyst also improves gluconic acid selectivity.

9.9 Conclusions and Future Perspectives

In this chapter, we have reviewed and discussed the possible approaches to the preparation of supported Au catalysts, various factors that influence the dispersion of Au and the Au particle size, nature and structure of active sites, and O_2 activation on the supported Au catalysts. Subsequently, we have summarized the progress achieved in selective oxidation of hydrocarbons (including the industrially important epoxidation of propylene to propylene oxide), alcohols, polyols and sugars on supported Au catalysts, and compared the pros and cons of the Au catalysts. In most of the cases, supported Au catalysts exhibit promising catalytic performance under relatively mild reaction conditions. In addition, Au displays higher resistance to leaching and oxygen poisoning than other metal elements such as Pt. From the extensive literature presented, there is no doubt that Au catalysis offers ample opportunities for both fundamental and applied research, and that Au catalysts in heterogeneous and homogeneous forms will become prominent players in the green and sustainable chemistry, probably in this century.

However, it is clear that the detailed structure of the active sites, interactions between active Au species and catalyst supports, and the reaction mechanism of the selective oxidation reactions are still not completely understood. More experimental and theoretical investigations should be carried out to obtain deeper insights into these aspects. For example, more precisely designed and fabricated Au catalysts, extensive *in situ* characterizations of the catalysts and kinetic studies will be of great importance.

The relatively mild reaction conditions exhibited by supported Au catalysts probably have their advantages and disadvantages. They will be beneficial in saving energy and cost as well as changing the reaction pathways, *e.g.* reducing the possibility for total oxidation. However, the strong adsorption of both reactant and intermediate molecules on the catalyst surface may lead to catalyst poisoning. Hence, tuning the catalyst surface's acidic/basic property, hydrophilic/hydrophobic property and composition may alleviate the problem.

The size-dependent catalytic performance for supported Au catalysts, despite some exceptions, indeed imposes big challenges to the preparation and stability of the Au catalysts. Sintering and agglomeration of the small particles with increase in aging/reaction time and temperature is a common phenomenon. It becomes worse due to the fact that gold has a relatively low boiling point compared with other noble metals, and furthermore, it is significantly lowered with a decrease in Au particle size. Probably the formation of core-shell structures,[166] alloying the Au element with some other metal elements, the formation of bimetallic or even multi-metallic catalysts, or the formation of some hybrid nanostructures will provide practical solutions to this problem. For practical applications, the stability of Au catalysts will be the crucial factor that will determine the fate of the catalysts. As indicated in this chapter, these measures are also effective in affecting catalytic performances such as the product selectivity and distribution for a number of reactions.

For three-phase (containing gas and liquid reactants and solid catalysts) oxidation reactions, aside from catalyst development, the engineering aspects of the reactor should be given more attention. Efficient contact between the reactant molecules and the catalyst surface will significantly improve catalytic performance.

In view of the above aspects, it is expected that application of heterogeneous Au catalysts in selective oxidation for organic transformations will undergo rapid growth in the near future, and it will be an active research front in chemistry.

Acknowledgements

The authors are grateful to ICES and NTU for supporting this work. Z.Z. thanks Professor M. Haruta of Tokyo Metropolitan University and Professor Lai-Sheng Wang of Brown University, Rhode Island, USA, for their kind help.

References

1. G. C. Bond, P. A. Sermon, G. Webb, D. A. Buchanan and P. B. Wells, *J. Chem. Soc., Chem. Commun.*, 1973, 444b.
2. M. Haruta, T. Kobayashi, H. Sano and N. Yamada, *Chem. Lett.*, 1987, **16**, 405.
3. G. J. Hutchings, *J. Catal.*, 1985, **96**, 292.
4. A. S. Thakor, J. Jokerst, C. Zavaleta, T. F. Massoud and S. S. Gambhir, *Nano Lett.*, 2011, **11**, 4029.
5. J. Gong, *Chem. Rev.*, 2011, **112**, 2987.
6. R. Meyer, C. Lemire, S. K. Shaikhutdinov and H. J. Freund, *Gold Bull.*, 2004, **37**, 72.
7. A. S. K. Hashmi and G. J. Hutchings, *Angew. Chem., Int. Ed.*, 2006, **45**, 7896.
8. M. Stratakis and H. Garcia, *Chem. Rev.*, 2012, **112**, 4469.
9. M. Haruta, *Catal. Today*, 1997, **36**, 153.
10. A. Corma and H. Garcia, *Chem. Soc. Rev.*, 2008, **37**, 2096.
11. C. D. Pina, E. Falletta and M. Rossi, *Chem. Soc. Rev.*, 2012, **41**, 350.
12. S. E. Davis, M. S. Ide and R. J. Davis, *Green Chem.*, 2013, **15**, 17.
13. T. Takei, T. Akita, I. Nakamura, T. Fujitani, M. Okumura, K. Okazaki, J. Huang, T. Ishida and M. Haruta, *Adv. Catal.*, 2012, **55**, 1.
14. G. Centi and M. Misono, *Catal. Today*, 1998, **41**, 287.
15. T. Ishida, M. Nagaoka, T. Akita and M. Haruta, *Chem. Eur. J.*, 2008, **14**, 8456.
16. Z. Yan, S. Chinta, A. Mohamed, J. Fackler, Jr. and D. W. Goodman, *Catal. Lett.*, 2006, **111**, 15.
17. G. C. Bond, C. Louis and D. T. Thompson, *Catalysis by Gold*, Imperial College Press, London, 2006.
18. C. Louis and O. Pluchery, in *Chemical Preparation of Gold Nanoparticles on Surfaces*, ed. C. Louis and O. Pluchery, Imperial College Press, London, 2012, p. 139–169.
19. G. E. Brown, V. E. Henrich, W. H. Casey, D. L. Clark, C. Eggleston, A. Felmy, D. W. Goodman, M. Grätzel, G. Maciel, M. I. McCarthy, K. H. Nealson, D. A. Sverjensky, M. F. Toney and J. M. Zachara, *Chem. Rev.*, 1998, **99**, 77.
20. G. Gouy, *J. Phys.*, 1910, **9**, 457.
21. D. L. Chapman, *Philos. Mag.*, 1913, **25**, 475.
22. O. Stern, *Z. Elektrochem.*, 1924, **30**, 508.
23. M. Kosmulski, *J. Colloid Interface Sci.*, 2009, **337**, 439.

24. C. O. Ijagbemi, M.-H. Baek and D.-S. Kim, *J. Hazard. Mater.*, 2009, **166**, 538.
25. T. Kuzniatsova, Y. Kim, K. Shqau, P. K. Dutta and H. Verweij, *Micropor. Mesopor. Mater.*, 2007, **103**, 102.
26. M. R. McPhail, J. A. Sells, Z. He and C. C. Chusuei, *J. Phys. Chem. C*, 2009, **113**, 14102.
27. S.-J. Lee and A. Gavriilidis, *J. Catal.*, 2002, **206**, 305.
28. S. Ivanova, C. Petit and V. Pitchon, *Appl. Catal. A*, 2004, **267**, 191.
29. G. C. Bond and D. T. Thompson, *Catal. Rev. Sci. Eng.*, 1999, **41**, 319.
30. M. Haruta, *CATTECH*, 2002, **6**, 102.
31. H. S. Oh, J. H. Yang, C. K. Costello, Y. M. Wang, S. R. Bare, H. H. Kung and M. C. Kung, *J. Catal.*, 2002, **210**, 375.
32. Q. Xu, K. C. Kharas and A. K. Datye, *Catal. Lett.*, 2003, **85**, 229.
33. H.-Y. Tsai, Y.-D. Lin, W.-T. Fu and S. Lin, *Gold Bull.*, 2007, **40**, 184.
34. M. Sankar, Q. He, M. Morad, J. Pritchard, S. J. Freakley, J. K. Edwards, S. H. Taylor, D. J. Morgan, A. F. Carley, D. W. Knight, C. J. Kiely and G. J. Hutchings, *ACS Nano*, 2012, **6**, 6600.
35. J. C. Fierro-Gonzalez and B. C. Gates, *J. Phys. Chem. B*, 2005, **109**, 7275.
36. M. Haruta, N. Yamada, T. Kobayashi and S. Iijima, *J. Catal.*, 1989, **115**, 301.
37. M. Haruta, S. Tsubota, T. Kobayashi, H. Kageyama, M. J. Genet and B. Delmon, *J. Catal.*, 1993, **144**, 175.
38. J. H. Yang, J. D. Henao, C. Costello, M. C. Kung, H. H. Kung, J. T. Miller, A. J. Kropf, J. G. Kim, J. R. Regalbuto, M. T. Bore, H. N. Pham, A. K. Datye, J. D. Laeger and K. Kharas, *Appl. Catal. A*, 2005, **291**, 73.
39. F. Moreau and G. C. Bond, *Catal. Today*, 2007, **122**, 260.
40. M. Khoudiakov, M. C. Gupta and S. Deevi, *Appl. Catal. A*, 2005, **291**, 151.
41. R. Zanella, S. Giorgio, C. R. Henry and C. Louis, *J. Phys. Chem. B*, 2002, **106**, 7634.
42. R. Zanella, L. Delannoy and C. Louis, *Appl. Catal. A*, 2005, **291**, 62.
43. H. Zhu, C. Liang, W. Yan, S. H. Overbury and S. Dai, *J. Phys. Chem. B*, 2006, **110**, 10842.
44. L. Cumaranatunge and W. N. Delgass, *J. Catal.*, 2005, **232**, 38.
45. C. Mennemann and P. Claus, *Catal. Lett.*, 2010, **134**, 31.
46. J. Lu, X. Zhang, J. J. Bravo-Suárez, T. Fujitani and S. T. Oyama, *Catal. Today*, 2009, **147**, 186.
47. D. A. Bulushev, I. Yuranov, E. I. Suvorova, P. A. Buffat and L. Kiwi-Minsker, *J. Catal.*, 2004, **224**, 8.
48. A. Simon, T. Cohen-Bouhacina, M. C. Porté, J. P. Aimé and C. Baquey, *J. Colloid Interface Sci.*, 2002, **251**, 278.
49. M. L. Gray, Y. Soong, K. J. Champagne, J. Baltrus, R. W. Stevens Jr, P. Toochinda and S. S. C. Chuang, *Sep. Purif. Technol.*, 2004, **35**, 31.
50. H. He, J. Duchet, J. Galy and J.-F. Gerard, *J. Colloid Interface Sci.*, 2005, **288**, 171.
51. M. Okumura, S. Tsubota and M. Haruta, *J. Mol. Catal. A*, 2003, **199**, 73.

52. M. Daté, M. Okumura, S. Tsubota and M. Haruta, *Angew. Chem., Int. Ed.*, 2004, **43**, 2129.
53. Y.-S. Chi, H.-P. Lin and C.-Y. Mou, *Appl. Catal. A*, 2005, **284**, 199.
54. C.-W. Chiang, A. Wang, B.-Z. Wan and C.-Y. Mou, *J. Phys. Chem. B*, 2005, **109**, 18042.
55. Y.-H. Liu, H.-P. Lin and C.-Y. Mou, *Langmuir*, 2004, **20**, 3231.
56. G. Budroni and A. Corma, *Angew. Chem., Int. Ed.*, 2006, **45**, 3328.
57. J.-D. Grunwaldt, M. Maciejewski, O. S. Becker, P. Fabrizioli and A. Baiker, *J. Catal.*, 1999, **186**, 458.
58. J.-D. Grunwaldt, C. Kiener, C. Wögerbauer and A. Baiker, *J. Catal.*, 1999, **181**, 223.
59. F. Porta, L. Prati, M. Rossi and G. Scari, *J. Catal.*, 2002, **211**, 464.
60. A. Mirescu and U. Prüße, *Catal. Commun.*, 2006, **7**, 11.
61. K. Kuroda, T. Ishida and M. Haruta, *J. Mol. Catal A*, 2009, **298**, 7.
62. A. Villa, C. E. Chan-Thaw and L. Prati, *Appl. Catal. B*, 2010, **96**, 541.
63. J. Liu, W. Chen, X. Liu, K. Zhou and Y. Li, *Nano Res.*, 2008, **1**, 46.
64. S. C. Chan and M. A. Barteau, *Langmuir*, 2005, **21**, 5588.
65. J. C. Scaiano, P. Billone, C. M. Gonzalez, L. Maretti, M. L. Marin, K. L. McGilvray and N. Yuan, *Pure Appl. Chem.*, 2009, **81**, 635.
66. A. Gedanken, X. Tang, Y. Wang, N. Perkas, Y. Koltypin, M. V. Landau, L. Vradman and M. Herskowitz, *Langmuir*, 2001, **7**, 4546.
67. S. Bhattacharyya, A. Gabashvili, N. Perkas and A. Gedanken, *J. Phys. Chem. C*, 2007, **111**, 11161.
68. K. S. Suslick, in *Ultrasound: Its Chemical, Physical, and Biological Effects*, ed. K. S. Suslick, VCH, Weinheim, 1988, p. 1–6.
69. K. S. Suslick, S.-B. Choe, A. A. Cichowlas and M. W. Grinstaff, *Nature*, 1991, **353**, 414.
70. K. S. Suslick and G. J. Price, *Ann. Rev. Mater. Sci.*, 1999, **29**, 295.
71. Z. Zhong, F. Chen, A. S. Subramanian, J. Lin, J. Highfield and A. Gedanken, *J. Mater. Chem.*, 2006, **16**, 489.
72. Z. Zhong, J. Luo, T. P. Ang, J. Highfield, J. Lin and A. Gedanken, *J. Phys. Chem. B*, 2004, **108**, 18119.
73. Z. Zhong, S. Patskovskyy, P. Bouvrette, J. H. T. Luong and A. Gedanken, *J. Phys. Chem. B*, 2004, **108**, 4046.
74. Z. Zhong, J. Lin, S. P. Teh, J. Teo and F. M. Dautzenberg, *Adv. Funct. Mater.*, 2007, **17**, 1402.
75. Z. Zhong, J. Teo, M. Lin and J. Ho, *Top. Catal.*, 2008, **49**, 216.
76. G. R. Bamwenda, S. Tsubota, T. Nakamura and M. Haruta, *Catal. Lett.*, 1997, **44**, 83.
77. X. Lai, T. P. S. Clair, M. Valden and D. W. Goodman, *Prog. Surf. Sci.*, 1998, **59**, 25.
78. M. Valden, X. Lai and D. W. Goodman, *Science*, 1998, **281**, 1647.
79. M. Valden, S. Pak, X. Lai and D. W. Goodman, *Catal. Lett.*, 1998, **56**, 7.
80. M. Valden and D. W. Goodman, *Israel J. Chem.*, 1998, **38**, 285.
81. A. A. Herzing, C. J. Kiely, A. F. Carley, P. Landon and G. J. Hutchings, *Science*, 2008, **321**, 1331.

82. R. Zanella, S. Giorgio, C.-H. Shin, C. R. Henry and C. Louis, *J. Catal.*, 2004, **222**, 357.
83. M. S. Chen and D. W. Goodman, *Science*, 2004, **306**, 252.
84. H. Zhang, T. Watanabe, M. Okumura, M. Haruta and N. Toshima, *Nat. Mater.*, 2012, **11**, 49.
85. X. Lin, B. Yang, H.-M. Benia, P. Myrach, M. Yulikov, A. Aumer, M. A. Brown, M. Sterrer, O. Bondarchuk, E. Kieseritzky, J. Rocker, T. Risse, H.-J. Gao, N. Nilius and H.-J. Freund, *J. Am. Chem. Soc.*, 2010, **132**, 7745.
86. J. Guzman and B. C. Gates, *J. Phys. Chem. B*, 2002, **106**, 7659.
87. C. K. Costello, M. C. Kung, H. S. Oh, Y. Wang and H. H. Kung, *Appl. Catal. A*, 2002, **232**, 159.
88. J. Guzman and B. C. Gates, *J. Am. Chem. Soc.*, 2004, **126**, 2672.
89. Y. Hao and B. C. Gates, *J. Catal.*, 2009, **263**, 83.
90. J. C. Fierro-Gonzalez and B. C. Gates, *Chem. Soc. Rev.*, 2008, **37**, 2127.
91. C. Xu, J. Su, X. Xu, P. Liu, H. Zhao, F. Tian and Y. Ding, *J. Am. Chem. Soc.*, 2007, **129**, 42.
92. C. Xu, X. Xu, J. Su and Y. Ding, *J. Catal.*, 2007, **252**, 243.
93. S. Tsubota, D. A. H. Cunningham, Y. Bando and M. Haruta, *Stud. Surf. Sci. Catal.*, 1995, **91**, 227.
94. M. Haruta, *Faraday Discuss.*, 2011, **152**, 11.
95. D. Widmann and R. J. Behm, *Angew. Chem., Int. Ed.*, 2011, **50**, 10241.
96. T. Fujitani and I. Nakamura, *Angew. Chem., Int. Ed.*, 2011, **50**, 10144.
97. Z. Zhong, J. Highfield, M. Lin, J. Teo and Y.-f. Han, *Langmuir*, 2008, **24**, 8576.
98. X. Nie, H. Qian, Q. Ge, H. Xu and R. Jin, *ACS Nano*, 2012, **6**, 6014.
99. Y. Maeda, Y. Iizuka and M. Kohyama, *J. Am. Chem. Soc.*, 2012, **135**, 906.
100. Y. Kang, X. Ye, J. Chen, L. Qi, R. E. Diaz, V. Doan-Nguyen, G. Xing, C. R. Kagan, J. Li, R. J. Gorte, E. A. Stach and C. B. Murray, *J. Am. Chem. Soc.*, 2013, **135**, 1499.
101. I. X. Green, W. Tang, M. Neurock and J. T. Yates, Jr., *Science*, 2011, **333**, 736.
102. H. Yoshida, Y. Kuwauchi, J. R. Jinschek, K. Sun, S. Tanaka, M. Kohyama, S. Shimada, M. Haruta and S. Takeda, *Science*, 2012, **336**, 317.
103. G. Mills, M. S. Gordon and H. Metiu, *Chem. Phys. Lett.*, 2002, **359**, 493.
104. A. Franceschetti, S. J. Pennycook and S. T. Pantelides, *Chem. Phys. Lett.*, 2003, **374**, 471.
105. X. Ding, Z. Li, J. Yang, J. G. Hou and Q. Zhu, *J. Chem. Phys.*, 2004, **120**, 9594.
106. E. M. Fernández, P. Ordejón and L. C. Balbás, *Chem. Phys. Lett.*, 2005, **408**, 252.
107. R. Coquet, K. L. Howard and D. J. Willock, *Chem. Soc. Rev.*, 2008, **37**, 2046.
108. X. Ding, B. Dai, J. Yang, J. G. Hou and Q. Zhu, *J. Chem. Phys.*, 2004, **121**, 621.

109. S. A. Varganov, R. M. Olson, M. S. Gordon and H. Metiu, *J. Chem. Phys.*, 2003, **119**, 2531.
110. J. A. van Bokhoven, C. Louis, J. T. Miller, M. Tromp, O. V. Safonova and P. Glatzel, *Angew. Chem., Int. Ed.*, 2006, **45**, 4651.
111. G. Bond and D. Thompson, *Gold Bull.*, 2000, **33**, 41.
112. C. K. Costello, J. Guzman, J. H. Yang, Y. M. Wang, M. C. Kung, B. C. Gates and H. H. Kung, *J. Phys. Chem. B*, 2004, **108**, 12529.
113. A. P. Woodham, G. Meijer and A. Fielicke, *J. Am. Chem. Soc.*, 2013, **135**, 1727.
114. D. M. Cox, R. Brickman, K. Creegan and A. Kaldor, *Z. Phys. D*, 1991, **19**, 353.
115. M. B. Torres, E. M. Fernández and L. C Balbås, *J. Phys. Chem. A*, 2008, **112**, 6678.
116. J. L. C. Fajín, M. N. D. S. Cordeiro and J. R. B. Gomes, *Appl. Catal. A*, 2013, **458**, 90.
117. K. J. Taylor, C. L. Pettiette-Hall, O. Cheshnosky and R. E. Smalley, *J. Chem. Phys.*, 1992, **96**, 3319.
118. B. E. Salisbury, W. T. Wallace and R. L. Whetten, *Chem. Phys.*, 2000, **262**, 131.
119. Y. D. Kim, M. Fischer and G. Ganteför, *Chem. Phys. Lett.*, 2003, **377**, 170.
120. Q. Sun, P. Jena, Y. D. Kim, M. Fischer and G. Gantefor, *J. Chem. Phys.*, 2004, **120**, 6510.
121. L. D. Socaciu, J. Hagen, T. M. Bernhardt, L. Wöste, U. Heiz, H. Häkkinen and U. Landman, *J. Am. Chem. Soc.*, 2003, **125**, 10437.
122. W. Huang, H.-J. Zhai and L.-S. Wang, *J. Am. Chem. Soc.*, 2010, **132**, 4344.
123. R. Pal, L.-M. Wang, Y. Pei, L.-S. Wang and X. C. Zeng, *J. Am. Chem. Soc.*, 2012, **134**, 9438.
124. E. Wahlström, E. K. Vestergaard, R. Schaub, A. Rønnau, M. Vestergaard, E. Lægsgaard, I. Stensgaard and F. Besenbacher, *Science*, 2004, **303**, 511.
125. A. Sanchez, S. Abbet, U. Heiz, W. D. Schneider, H. Häkkinen, R. N. Barnett and U. Landman, *J. Phys. Chem. A*, 1999, **103**, 9573.
126. B. Yoon, H. Häkkinen, U. Landman, A. S. Wörz, J.-M. Antonietti, S. Abbet, K. Judai and U. Heiz, *Science*, 2005, **307**, 403.
127. J. Guzman, S. Carrettin and A. Corma, *J. Am. Chem. Soc.*, 2005, **127**, 3286.
128. Z. Zheng, J. Teo, X. Chen, H. Liu, Y. Yuan, E. R. Waclawik, Z. Zhong and H. Zhu, *Chem. Eur. J.*, 2010, **16**, 1202.
129. Y. Gao and X. C. Zeng, *ACS Catal.*, 2012, **2**, 2614.
130. H. H. Kung, M. C. Kung and C. K. Costello, *J. Catal.*, 2003, **216**, 425.
131. M. C. Kung, R. J. Davis and H. H. Kung, *J. Phys. Chem. C*, 2007, **111**, 11767.
132. M. Kotobuki, R. Leppelt, D. A. Hansgen, D. Widmann and R. J. Behm, *J. Catal.*, 2009, **264**, 67.
133. Y. Kuwauchi, H. Yoshida, T. Akita, M. Haruta and S. Takeda, *Angew. Chem., Int. Ed.*, 2012, **51**, 7729.

134. B. K. Min and C. M. Friend, *Chem. Rev.*, 2007, **107**, 2709.
135. D. L. Trent, *Propylene Oxide*, Kirk-Othmer Encyclopedia of Chemical Technology, John Wiley & Sons, New York, 2000.
136. F. Cavani and J. H. Teles, *ChemSusChem*, 2009, **2**, 508.
137. C. Qi, *Gold Bull.*, 2008, **41**, 224.
138. M. Ishino, presented at the 16th Saudi Arabia-Japan Joint Symposium, Dhahran, Saudi Arabia, 2006.
139. J. Huang and M. Haruta, *Res. Chem. Intermed.*, 2012, **38**, 1.
140. J. Kollar, *US Pat.*, 3350422, 1967.
141. W. G. Reman, A. H. Joustra, W. Drent and W. D. Bruijn, *Eur. Pat.*, EP0345856 B1, 1989.
142. J. K. F. Buijink, J. J. M. van Vlaanderen, M. Crocker and F. G. M. Niele, *Catal. Today*, 2004, **93–95**, 199.
143. M. Ishino and J. Tsuji, *Jap. Pat.*, EP1437350 B1, 2008.
144. B. S. Lane and K. Burgess, *Chem. Rev.*, 2003, **103**, 2457.
145. H. Abekawa and T. Seo, *US Pat.*, 20090270641 A1, 2009.
146. M. Mizuno and H. Narahara, *World Pat.*, WO2009008493 A2, 2009.
147. R. Meiers, U. Dingerdissen and W. F. Hölderich, *J. Catal.*, 1998, **176**, 376.
148. K. P. Keating, E. T. Marquis, J. R. Sanderson and W. A. Smith, *US Pat.*, 4691034 A, 1986.
149. D. Barton, R. Bowman, S. Dhingra, L. Ito, R. Schroden, D. Trent, K. Watson and H. Weiner, *World Pat.*, WO 2008/063880 A1, 2008.
150. J. P. Dever, K. F. George, W. C. Hoffman and H. Soo, *Ethylene Oxide*, John Wiley & Sons, New York, 1995.
151. E. A. Carter and W. A. Goddard III, *J. Catal.*, 1988, **112**, 80.
152. T. Hayashi, K. Tanaka and M. Haruta, *J. Catal.*, 1998, **178**, 566.
153. B. Chowdhury, J. J. Bravo-Suárez, M. Daté, S. Tsubota and M. Haruta, *Angew. Chem., Int. Ed.*, 2006, **45**, 412.
154. E. E. Stangland, K. B. Stavens, R. P. Andres and W. N. Delgass, *J. Catal.*, 2000, **191**, 332.
155. T. A. Nijhuis, B. J. Huizinga, M. Makkee and J. A. Moulijn, *Ind. Eng. Chem. Res.*, 1999, **38**, 884.
156. A. K. Sinha, S. Seelan, S. Tsubota and M. Haruta, *Angew. Chem., Int. Ed.*, 2004, **43**, 1546.
157. B. Taylor, J. Lauterbach and W. N. Delgass, *Appl. Catal. A*, 2005, **291**, 188.
158. J. Huang, T. Takei, T. Akita, H. Ohashi and M. Haruta, *Appl. Catal. B*, 2010, **95**, 430.
159. W.-S. Lee, M. Cem Akatay, E. A. Stach, F. H. Ribeiro and W. N. Delgass, *J. Catal.*, 2012, **287**, 178.
160. C. Qi, J. Huang, S. Bao, H. Su, T. Akita and M. Haruta, *J. Catal.*, 2011, **281**, 12.
161. J. Huang, E. Lima, T. Akita, A. Guzmán, C. Qi, T. Takei and M. Haruta, *J. Catal.*, 2011, **278**, 8.
162. W.-S. Lee, R. Zhang, M. C. Akatay, C. D. Baertsch, E. A. Stach, F. H. Ribeiro and W. N. Delgass, *ACS Catal.*, 2011, **1**, 1327.

163. W.-S. Lee, L.-C. Lai, M. Cem Akatay, E. A. Stach, F. H. Ribeiro and W. N. Delgass, *J. Catal.*, 2012, **296**, 31.
164. J. Gaudet, K. K. Bando, Z. Song, T. Fujitani, W. Zhang, D. S. Su and S. T. Oyama, *J. Catal.*, 2011, **280**, 40.
165. L. Xu, Y. Ren, H. Wu, Y. Liu, Z. Wang, Y. Zhang, J. Xu, H. Peng and P. Wu, *J. Mater. Chem.*, 2011, **21**, 10852.
166. C. M. Barker, D. Gleeson, N. Kaltsoyannis, C. R. A. Catlow, G. Sankar and J. M. Thomas, *Phys. Chem. Chem. Phys.*, 2002, **4**, 1228.
167. J. M. Thomas and G. Sankar, *Acc. Chem. Res.*, 2001, **34**, 571.
168. F. Farges, G. E. Brown, Jr. and J. J. Rehr, *Phys. Rev. B*, 1997, **56**, 1809.
169. Y. A. Kalvachev, T. Hayashi, S. Tsubota and M. Haruta, *J. Catal.*, 1999, **186**, 228.
170. B. S. Uphade, T. Akita, T. Nakamura and M. Haruta, *J. Catal.*, 2002, **209**, 331.
171. E. Sacaliuc, A. M. Beale, B. M. Weckhuysen and T. A. Nijhuis, *J. Catal.*, 2007, **248**, 235.
172. J. Lu, X. Zhang, J. J. Bravo-Suárez, S. Tsubota, J. Gaudet and S. T. Oyama, *Catal. Today*, 2007, **123**, 189.
173. A. K. Sinha, S. Seelan, M. Okumura, T. Akita, S. Tsubota and M. Haruta, *J. Phys. Chem. B*, 2005, **109**, 3956.
174. Q. Jin, J. Bao, H. Sakiyama and N. Tsubaki, *Res. Chem. Intermed.*, 2011, **37**, 177.
175. B. Taylor, J. Lauterbach and W. N. Delgass, *Catal. Today*, 2007, **123**, 50.
176. G. Mul, A. Zwijnenburg, B. van der Linden, M. Makkee and J. A. Moulijn, *J. Catal.*, 2001, **201**, 128.
177. N. Yap, R. P. Andres and W. N. Delgass, *J. Catal.*, 2004, **226**, 156.
178. G. Zhan, M. Du, J. Huang and Q. Li, *Catal. Commun.*, 2011, **12**, 830.
179. J.-Q. Lu, N. Li, X.-R. Pan, C. Zhang and M.-F. Luo, *Catal. Commun.*, 2012, **28**, 179.
180. R. G. Bowman, R. C. Schroden, G. R. Meima, K. J. Watson and D. G. Barton, *World Pat.*, WO 2009/061623 A2, 2009.
181. A. Corma, M. Domine, J. A. Gaona, J. L. Jorda, M. T. Navarro, F. Rey, J. Perez-Pariente, J. Tsuji, B. McCulloch and L. T. Nemeth, *Chem. Commun.*, 1998, 2211.
182. T. Tatsumi, K. A. Koyano and N. Igarashi, *Chem. Commun.*, 1998, 325.
183. A. Bhaumik and T. Tatsumi, *J. Catal.*, 2000, **189**, 31.
184. A. Bhaumik and T. Tatsumi, *Catal. Lett.*, 2000, **66**, 181.
185. M. R. Prasad, M. S. Hamdy, G. Mul, E. Bouwman and E. Drent, *J. Catal.*, 2008, **260**, 288.
186. M. Haruta, *Nature*, 2005, **437**, 1098.
187. J. Huang, T. Akita, J. Faye, T. Fujitani, T. Takei and M. Haruta, *Angew. Chem., Int. Ed.*, 2009, **48**, 7862.
188. J. Jiang, S. M. Oxford, B. Fu, M. C. Kung, H. H. Kung and J. Ma, *Chem. Commun.*, 2010, **46**, 3791.
189. J. Chen, S. J. A. Halin, D. M. Perez Ferrandez, J. C. Schouten and T. A. Nijhuis, *J. Catal.*, 2012, **285**, 324.

190. T. A. Nijhuis, T. Visser and B. M. Weckhuysen, *Angew. Chem., Int. Ed.*, 2005, **44**, 1115.
191. T. A. Nijhuis, E. Sacaliuc, A. M. Beale, A. M. J. van der Eerden, J. C. Schouten and B. M. Weckhuysen, *J. Catal.*, 2008, **258**, 256.
192. C. Sivadinarayana, T. V. Choudhary, L. L. Daemen, J. Eckert and D. W. Goodman, *J. Am. Chem. Soc.*, 2003, **126**, 38.
193. J. J. Bravo-Suárez, K. K. Bando, J. Lu, M. Haruta, T. Fujitani and T. Oyama, *J. Phys. Chem. C*, 2008, **112**, 1115.
194. R. M. Lambert, F. J. Williams, R. L. Cropley and A. Palermo, *J. Mol. Catal. A*, 2005, **228**, 27.
195. J. G. Serafin, A. C. Liu and S. R. Seyedmonir, *J. Mol. Catal. A*, 1998, **131**, 157.
196. M. Luo, J. Lu and C. Li, *Catal. Lett.*, 2003, **86**, 43.
197. Y. Lei, F. Mehmood, S. Lee, J. Greeley, B. Lee, S. Seifert, R. E. Winans, J. W. Elam, R. J. Meyer, P. C. Redfern, D. Teschner, R. Schlogl, M. J. Pellin, L. A. Curtiss and S. Vajda, *Science*, 2010, **9**, 224.
198. A. Takahashi, N. Hamakawa, I. Nakamura and T. Fujitani, *Appl. Catal. A*, 2005, **294**, 34.
199. R. L. Cropley, F. J. Williams, A. J. Urquhart, O. P. H. Vaughan, M. S. Tikhov and R. M. Lambert, *J. Am. Chem. Soc.*, 2005, **127**, 6069.
200. R. L. Cropley, F. J. Williams, O. P. H. Vaughan, A. J. Urquhart, M. S. Tikhov and R. M. Lambert, *Surf. Sci.*, 2005, **578**, L85.
201. O. P. H. Vaughan, G. Kyriakou, N. Macleod, M. Tikhov and R. M. Lambert, *J. Catal.*, 2005, **236**, 401.
202. J. Lu, M. Luo, H. Lei, X. Bao and C. Li, *J. Catal.*, 2002, **211**, 552.
203. W. Su, S. Wang, P. Ying, Z. Feng and C. Li, *J. Catal.*, 2009, **268**, 165.
204. W. Zhu, Q. Zhang and Y. Wang, *J. Phys. Chem. C*, 2008, **112**, 7731.
205. W. Long, Q. Zhai, J. He, Q. Zhang, W. Deng and Y. Wang, *ChemPlusChem*, 2012, **77**, 27.
206. L. Yang, J. He, Q. Zhang and Y. Wang, *J. Catal.*, 2010, **276**, 76.
207. J. He, Q. Zhai, Q. Zhang, W. Deng and Y. Wang, *J. Catal.*, 2013, **299**, 53.
208. A. Seubsai, M. Kahn and S. Senkan, *ChemCatChem*, 2011, **3**, 174.
209. A. Bongiorno and U. Landman, *Phys. Rev. Lett.*, 2005, **95**, 106102.
210. J. Huang, T. Takei, H. Ohashi and M. Haruta, *Appl. Catal. A*, 2012, **115**, 435–436.
211. S. Lee, L. M. Molina, M. J. López, J. A. Alonso, B. Hammer, B. Lee, S. Seifert, R. E. Winans, J. W. Elam, M. J. Pellin and S. Vajda, *Angew. Chem., Int. Ed.*, 2009, **48**, 1467.
212. Y. Liu, K. Murata, M. Inaba and N. Mimura, *Appl. Catal. A*, 2006, **309**, 91.
213. I. Onal, D. Düzenli, A. Seubsai, M. Kahn, E. Seker and S. Senkan, *Top. Catal.*, 2010, **53**, 92.
214. V. Duma and D. Hönicke, *J. Catal.*, 2000, **191**, 93.
215. E. Ananieva and A. Reitzmann, *Chem. Eng. Sci.*, 2004, **59**, 5509.
216. B. Moens, H. De Winne, S. Corthals, H. Poelman, R. De Gryse, V. Meynen, P. Cool, B. F. Sels and P. A. Jacobs, *J. Catal.*, 2007, **247**, 86.

217. R. J. Chimentão, F. Medina, J. L. G. Fierro, J. Llorca, J. E. Sueiras, Y. Cesteros and P. Salagre, *J. Mol. Catal. A*, 2007, **274**, 159.
218. J. M. Thomas, *Nature*, 1985, **314**, 669.
219. V. M. Akhmedov and S. H. Al-Khowaiter, *Catal. Rev.*, 2002, **44**, 455.
220. J. Sá, M. Ace, J. J. Delgado, A. Goguet, C. Hardacre and K. Morgan, *ChemCatChem*, 2011, **3**, 394.
221. F. J. Keil, *Micropor. Mesopor. Mater.*, 1999, **29**, 49.
222. B. P. C. Hereijgers and B. M. Weckhuysen, *Catal. Lett.*, 2011, **141**, 1429.
223. R. A. Periana, D. J. Taube, S. Gamble, H. Taube, T. Satoh and H. Fujii, *Science*, 1998, **280**, 560.
224. R. Palkovits, C. C. Malotki, M. Baumgarten, K. Mullen, C. Baltes, M. Antonietti, P. Kuhn, J. Weber, A. Thomas and F. Schüth, *ChemSusChem*, 2010, **3**, 277.
225. T. Li, S. J. Wang, C. S. Yu, Y. C. Ma, K. L. Li and L. W. Lin, *Appl. Catal. A*, 2011, **398**, 150.
226. G. Walther, L. Cervera-Gontard, U. J. Quaade and S. Horch, *Gold Bull.*, 2009, **42**, 13.
227. G. Walther, G. Jones, S. Jensen, U. J. Quaade and S. Horch, *Catal. Today*, 2009, **142**, 24.
228. I. Eskendirov, N. J. Coville and V. D. Sokolovskii, *Catal. Lett.*, 1995, **35**, 33.
229. K. Blick, T. D. Mitrelias, J. S. J. Hargreaves, G. J. Hutchings, R. W. Joyner, C. J. Kiely and F. E. Wagner, *Catal. Lett.*, 1998, **50**, 211.
230. IHS Chemical, *Acetone*, http://www.ihs.com/products/chemical/planning/ceh/acetone.aspx, 2010.
231. P. Dutia, *Isopropyl alcohol: A techno-commercial profile*, www.chemicalweekly.com, 2012.
232. J. J. Bravo-Suárez, K. K. Bando, T. Akita, T. Fujitani, T. J. Fuhrer and S. T. Oyama, *Chem. Commun.*, 2008, 3272.
233. M. G. Clerici, *Appl. Catal.*, 1991, **68**, 249.
234. I. Yamanaka, S. Hasegawa and K. Otsuka, *Appl. Catal. A*, 2002, **226**, 305.
235. C. Espro, F. Arena, F. Tasselli, A. Regina, E. Drioli and A. Parmaliana, *Catal. Today*, 2006, **118**, 253.
236. J. J. Bravo-Suárez, K. K. Bando, J. Lu, T. Fujitani and S. T. Oyama, *J. Catal.*, 2008, **255**, 114.
237. J. J. Bravo-Suárez, K. K. Bando, T. Fujitani and S. T. Oyama, *J. Catal.*, 2008, **257**, 32.
238. J. Cunningham, B. K. Hodnett, M. Ilyas, J. Tobin, E. L. Leahy and J. L. G. Fierro, *Faraday Discuss. Chem. Soc.*, 1981, **72**, 283.
239. K. U. Ingold, *Aldrichim. Acta*, 1989, **22**, 69.
240. Chemicals & Petrochemicals Manufacturer's Association of India, NFY – Nylon Filament Yarn, http://cpmaindia.com/nfy_about.php, 2012.
241. R. Raja, G. Sankar and J. M. Thomas, *J. Am. Chem. Soc.*, 1999, **121**, 11926.
242. E. V. Spinace, H. O. Pastore and U. Schuchardt, *J. Catal.*, 1995, **157**, 631.
243. W. A. Carvalho, P. B. Varaldo, M. Wallau and U. Schuchardt, *Zeolites*, 1997, **18**, 408.

244. M. H. Zahedi-Niaki, M. P. Kapoor and S. Kaliaguine, *J. Catal.*, 1998, **177**, 231.
245. U. R. Pillai and E. Sahle-Demessie, *Chem. Commun.*, 2002, 2142.
246. R. Zhao, D. Ji, G. Lv, G. Qian, L. Yan, X. Wang and J. Suo, *Chem. Commun.*, 2004, 904.
247. G. Lü, R. Zhao, G. Qian, Y. Qi, X. Wang and J. Suo, *Catal. Lett.*, 2004, **97**, 115.
248. K. Zhu, J. Hu and R. Richards, *Catal. Lett.*, 2005, **100**, 195.
249. Y.-J. Xu, P. Landon, D. Enache, A. F. Carley, M. W. Roberts and G. J. Hutchings, *Catal. Lett.*, 2005, **101**, 175.
250. L.-X. Xu, C.-H. He, M.-Q. Zhu, K.-J. Wu and Y.-L. Lai, *Catal. Lett.*, 2007, **118**, 248.
251. L.-X. Xu, C.-H. He, M.-Q. Zhu and S. Fang, *Catal. Lett.*, 2007, **114**, 202.
252. A. Alshammari, A. Kockritz, V. N. Kalevaru, A. Bagabas and A. Martin, *Appl. Petrochem. Res.*, 2012, **2**, 61.
253. R. A. Sheldon, M. Wallau, I. W. C. E. Arends and U. Schuchardt, *Acc. Chem. Res.*, 1998, **31**, 485.
254. K. U. Ingold, D. W. Snelgrove, P. A. MacFaul, R. D. Oldroyd and J. M. Thomas, *Catal. Lett.*, 1997, **48**, 21.
255. R. D. Oldroyd, J. M. Thomas, T. Maschmeyer, P. A. MacFaul, D. W. Snelgrove, K. U. Ingold and D. D. M. Wayner, *Angew. Chem., Int. Ed.*, 1996, **35**, 2787.
256. A. Lambert, P. Plucinski and I. V. Kozhevnikov, *Chem. Commun.*, 2003, 714.
257. I. C. M. S. Santos, M. M. Q. Simões, M. M. M. S. Pereira, R. R. L. Martins, M. G. P. M. S. Neves, J. A. S. Cavaleiro and A. M. V. Cavaleiro, *J. Mol. Catal. A*, 2003, **195**, 253.
258. M. D. Hughes, Y.-J. Xu, P. Jenkins, P. McMorn, P. Landon, D. I. Enache, A. F. Carley, G. A. Attard, G. J. Hutchings, F. King, E. H. Stitt, P. Johnston, K. Griffin and C. J. Kiely, *Nature*, 2005, **437**, 1132.
259. P. Lignier, F. Morfin, S. Mangematin, L. Massin, J.-L. Rousset and V. Caps, *Chem. Commun.*, 2007, 186.
260. P. Lignier, F. Morfin, L. Piccolo, J.-L. Rousset and V. Caps, *Catal. Today*, 2007, **122**, 284.
261. M. Alvaro, C. Aprile, A. Corma, B. Ferrer and H. Garcia, *J. Catal.*, 2007, **245**, 249.
262. N. S. Patil, B. S. Uphade, P. Jana, R. S. Sonawane, S. K. Bhargava and V. R. Choudhary, *Catal. Lett.*, 2004, **94**, 89.
263. N. S. Patil, B. S. Uphade, P. Jana, S. K. Bharagava and V. R. Choudhary, *J. Catal.*, 2004, **223**, 236.
264. N. S. Patil, R. Jha, B. S. Uphade, S. K. Bhargava and V. R. Choudhary, *Appl. Catal. A*, 2004, **275**, 87.
265. D. Yin, L. Qin, J. Liu, C. Li and Y. Jin, *J. Mol. Catal. A*, 2005, **240**, 40.
266. Y. Jin, P. Wang, D. Yin, J. Liu, H. Qiu and N. Yu, *Micropor. Mesopor. Mater.*, 2008, **111**, 569.

267. Y. Liu, H. Tsunoyama, T. Akita and T. Tsukuda, *Chem. Commun.*, 2010, **46**, 550.
268. J. Liu, F. Wang, S. Qi, Z. Gu and G. Wu, *New J. Chem.*, 2013, **37**, 769.
269. N. Dimitratos, J. A. Lopez-Sanchez and G. J. Hutchings, *Chem. Sci.*, 2012, **3**, 20.
270. T. Mallat and A. Baiker, *Chem. Rev.*, 2004, **104**, 3037.
271. J. H. J. Kluytmans, A. P. Markusse, B. F. M. Kuster, G. B. Marin and J. C. Schouten, *Catal. Today*, 2000, **57**, 143.
272. M. Besson and P. Gallezot, *Catal. Today*, 2000, **57**, 127.
273. M. Hudlucky, *Oxidations in Organic Chemistry*, American Chemical Society, Washington DC, 1990.
274. R. A. Sheldon and J. K. Kochi, *Metal Catalyzed Oxidations of Organic Compound*, Academic Press, New York, 1981.
275. K. Yamaguchi, K. Mori, T. Mizugaki, K. Ebitani and K. Kaneda, *J. Am. Chem. Soc.*, 2000, **122**, 7144.
276. K. Yamaguchi and N. Mizuno, *Angew. Chem., Int. Ed.*, 2002, **41**, 4538.
277. Z. Opre, J. D. Grunwaldt, M. Maciejewski, D. Ferri, T. Mallat and A. Baiker, *J. Catal.*, 2005, **230**, 406.
278. Z. Opre, D. Ferri, F. Krumeich, T. Mallat and A. Baiker, *J. Catal.*, 2006, **241**, 287.
279. S. Mori, M. Takubo, K. Makida, T. Yanase, S. Aoyagi, T. Maegawa, Y. Monguchi and H. Sajiki, *Chem. Commun.*, 2009, 5159.
280. K. Yamaguchi, J. W. Kim, J. He and N. Mizuno, *J. Catal.*, 2009, **268**, 343.
281. C. Keresszegi, T. Bürgi, T. Mallat and A. Baiker, *J. Catal.*, 2002, **211**, 244.
282. K. Mori, T. Hara, T. Mizugaki, K. Ebitani and K. Kaneda, *J. Am. Chem. Soc.*, 2004, **126**, 10657.
283. Z. Hou, N. Theyssen and W. Leitner, *Green Chem.*, 2007, **9**, 127.
284. Z. Hou, N. Theyssen, A. Brinkmann, K. V. Klementiev, W. Grünert, M. Bühl, W. Schmidt, B. Spliethoff, B. Tesche, C. Weidenthaler and W. Leitner, *J. Catal.*, 2008, **258**, 315.
285. D. Ferri, C. Mondelli, F. Krumeich and A. Baiker, *J. Phys. Chem. B*, 2006, **110**, 22982.
286. J.-D. Grunwaldt, M. Caravati and A. Baiker, *J. Phys. Chem. B*, 2006, **110**, 25586.
287. M. Hronec, Z. Cvengrošová and J. Kizlink, *J. Mol. Catal.*, 1993, **83**, 75.
288. H. E. Van Dam, L. J. Wisse and H. Van Bekkum, *Appl. Catal.*, 1990, **61**, 187.
289. T. Mallat and A. Baiker, *Catal. Today*, 1994, **19**, 247.
290. J. A. A. van den Tillaart, B. F. M. Kuster and G. B. Marin, *Appl. Catal. A*, 1994, **120**, 127.
291. A.-B. Crozon, M. Besson and P. Gallezot, *New J. Chem.*, 1998, **22**, 269.
292. Y. H. Ng, S. Ikeda, T. Harada, Y. Morita and M. Matsumura, *Chem. Commun.*, 2008, 3181.
293. M. Hronec, Z. Cvengrosova, J. Tuleja and J. Ilavsky, *Stud. Surf. Sci. Catal.*, 1990, **55**, 169.
294. J. W. Dobereiner, *Ann.*, 1845, **53**, 145.

295. J. H. Vleeming, F. A. d. Bruijn, B. F. M. Kuster and G. B. Marin, *Stud. Surf. Sci. Catal.*, 1994, **88**, 467.
296. A. F. Lee, J. J. Gee and H. J. Theyers, *Green Chem.*, 2000, **2**, 279.
297. T. Mallat, Z. Bodnar, C. Brönnimann and A. Baiker, *Stud. Surf. Sci. Catal.*, 1994, **88**, 385.
298. T. Mallat, Z. Bodnar, P. Hug and A. Baiker, *J. Catal.*, 1995, **153**, 131.
299. J. M. H. Dirkx and H. S. van der Baan, *J. Catal.*, 1981, **67**, 14.
300. P. Gallezot, *Catal. Today*, 1997, **37**, 405.
301. P. Vinke, H. E. van Dam and H. van Bekkum, *Stud. Surf. Sci. Catal.*, 1990, **55**, 147.
302. M. K. Weldon and C. M. Friend, *Chem. Rev.*, 1996, **96**, 1391.
303. K. Yamaguchi and N. Mizuno, *Chem. Eur. J.*, 2003, **9**, 4353.
304. J.-D. Grunwaldt, M. Caravati, M. Ramin and A. Baiker, *Catal. Lett.*, 2003, **90**, 221.
305. T. Mallat, Z. Bodnar, A. Baiker, O. Greis, H. Strubig and A. Reller, *J. Catal.*, 1993, **142**, 237.
306. X. Xu and C. M. Friend, *Surf. Sci.*, 1992, **260**, 14.
307. C. Bianchi, F. Porta, L. Prati and M. Rossi, *Top. Catal.*, 2000, **13**, 231.
308. C. L. Bianchi, S. Biella, A. Gervasini, L. Prati and M. Rossi, *Catal. Lett.*, 2003, **85**, 91.
309. S. Biella, G. L. Castiglioni, C. Fumagalli, L. Prati and M. Rossi, *Catal. Today*, 2002, **72**, 43.
310. S. Biella and M. Rossi, *Chem. Commun.*, 2003, 378.
311. C. d. Pina, E. Falletta and M. Rossi, *ChemSusChem*, 2009, **2**, 57.
312. F. Porta, L. Prati, M. Rossi, S. Coluccia and G. Martra, *Catal. Today*, 2000, **61**, 165.
313. L. Prati and G. Martra, *Gold Bull.*, 1999, **32**, 96.
314. L. Prati and F. Porta, *Appl. Catal. A*, 2005, **291**, 199.
315. L. Prati, A. Villa, C. Campione and P. Spontoni, *Top. Catal.*, 2007, **22**, 319.
316. D. I. Enache, D. W. Knight and G. J. Hutchings, *Catal. Lett.*, 2005, **103**, 43.
317. C. Della Pina, E. Falletta, L. Prati and M. Rossi, *Chem. Soc. Rev.*, 2008, **37**, 2077.
318. A. S. K. Hashmi and G. J. Hutchings, *Angew. Chem., Int. Ed.*, 2006, **45**, 7896.
319. T. Takei, N. Iguchi and M. Haruta, *New J. Chem.*, 2011, **35**, 2227.
320. C. H. Christensen, B. Jorgensen, J. Rass-Hansen, K. Egeblad, R. Madsen, S. K. Klitgaard, S. M. Hansen, M. R. Hansen, H. C. Andersen and A. Riisager, *Angew. Chem., Int. Ed.*, 2006, **45**, 4648.
321. B. Jorgensen, S. E. Christiansen, M. L. D. Thomsen and C. H. Christensen, *J. Catal.*, 2007, **251**, 332.
322. P. Aghaei and R. J. Berger, *Appl. Catal. B*, 2013, **195**, 132–133.
323. S. Carrettin, P. Concepción, A. Corma, J. M. López Nieto and V. F. Puntes, *Angew. Chem., Int. Ed.*, 2004, **43**, 2538.
324. A. Abad, P. Concepción, A. Corma and H. García, *Angew. Chem., Int. Ed.*, 2005, **44**, 4066.

325. A. Abad, C. Almela, A. Corma and H. Garcia, *Tetrahedron*, 2006, **62**, 6666.
326. A. Villa, C. E. Chan-Thaw, G. M. Veith, K. L. More, D. Ferri and L. Prati, *ChemCatChem*, 2011, **3**, 1612.
327. C. Santra, S. Rahman, S. Bojja, O. O. James, D. Sen, S. Maity, A. K. Mohanty, S. Mazumder and B. Chowdhury, *Catal. Sci. Technol.*, 2013, **3**, 360.
328. L. Zhou, W. Yu, L. Wu, Z. Liu, H. Chen, X. Yang, Y. Sun and J. Xu, *Appl. Catal. A*, 2013, **451**, 137.
329. J. Ni, W.-J. Yu, L. He, H. Sun, Y. Cao, H.-Y. He and K.-N. Fan, *Green Chem.*, 2009, **11**, 756.
330. M. Boronat, A. Corma, F. Illas, J. Radilla, T. Rodenas and M. J. Sabater, *J. Catal.*, 2011, **278**, 50.
331. J. Chen, Q. Zhang, Y. Wang and H. Wan, *Adv. Syn.Catal.*, 2008, **350**, 453.
332. N. Dimitratos, A. Villa, D. Wang, F. Porta, D. Su and L. Prati, *J. Catal.*, 2006, **244**, 113.
333. D. M. Newitt and P. S. Mene, *J. Chem. Soc.*, 1946, 97.
334. X. H. Fan, R. Burton and Y. Zhou, *Open Fuel Energy Sci. J*, 2010, **3**, 17.
335. F. Yang, M. A. Hanna and R. Sun, *Biotechnol. Biofuels*, 2012, **5**, 13.
336. S. Demirel-Gülen, M. Lucas and P. Claus, *Catal. Today*, 2005, **102–103**, 166.
337. M. Li, C. Zhou, J. N. Beltramini, W. Yu and Y. Fan, *Prog. Chem.*, 2008, **20**, 1474.
338. A. Brandner, K. Lehnert, A. Bienholz, M. Lucas and P. Claus, *Top. Catal.*, 2009, **52**, 278.
339. B. Katryniok, H. Kimura, E. Skrzynska, J.-S. Girardon, P. Fongarland, M. Capron, R. Ducoulombier, N. Mimura, S. Paul and F. Dumeignil, *Green Chem.*, 2011, **13**, 1960.
340. H. Kimura, K. Tsuto, T. Wakisaka, Y. Kazumi and Y. Inaya, *Appl. Catal. A*, 1993, **96**, 217.
341. H. Kimura, *Appl. Catal. A*, 1993, **105**, 147.
342. R. Garcia, M. Besson and P. Gallezot, *Appl. Catal. A*, 1995, **127**, 165.
343. P. Fordham, R. Garcia, M. Besson and P. Gallezot, *Stud. Surf. Sci. Catal.*, 1996, **101**, 161.
344. S. Carrettin, P. McMorn, P. Johnston, K. Griffin, C. J. Kiely and G. J. Hutchings, *Phys. Chem. Chem. Phys.*, 2003, **5**, 1329.
345. W. Hu, D. Knight, B. Lowry and A. Varma, *Ind. Eng. Chem. Res.*, 2010, **49**, 10876.
346. H. Miya and M. Matsuda, *Jap. Pat.*, Laid Open No. 116415, 1977.
347. K. Nabe, N. Izuo, S. Yamada and I. Chibata, *Appl. Environ. Microbiol.*, 1979, **38**, 1056.
348. N. Worz, A. Brandner and P. Claus, *J. Phys. Chem. C*, 2009, **114**, 1164.
349. L. Prati and M. Rossi, *J. Catal.*, 1998, **176**, 552.
350. S. Gil, M. Marchena, C. M. Fernández, L. Sánchez-Silva, A. Romero and J. L. Valverde, *Appl. Catal. A*, 2013, **450**, 189.
351. S. Demirel, K. Lehnert, M. Lucas and P. Claus, *Appl. Catal. B*, 2007, **70**, 637.

352. S. Carrettin, P. McMorn, P. Johnston, K. Griffin and G. J. Hutchings, *Chem. Commun.*, 2002, 696.
353. S. Carrettin, P. McMorn, P. Johnston, K. Griffin, C. Kiely, G. Attard and G. Hutchings, *Top. Catal.*, 2004, **27**, 131.
354. F. Porta and L. Prati, *J. Catal.*, 2004, **224**, 397.
355. N. Dimitratos, A. Villa, C. L. Bianchi, L. Prati and M. Makkee, *Appl. Catal. A*, 2006, **311**, 185.
356. N. Dimitratos, J. Lopez-Sanchez, D. Lennon, F. Porta, L. Prati and A. Villa, *Catal. Lett.*, 2006, **108**, 147.
357. A. Villa, G. M. Veith, D. Ferri, A. Weidenkaff, K. A. Perry, S. Campisi and L. Prati, *Catal. Sci. Technol.*, 2013, **3**, 394.
358. C. L. Bianchi, P. Canton, N. Dimitratos, F. Porta and L. Prati, *Catal. Today*, 2005, **102–103**, 203.
359. D. Wang, A. Villa, F. Porta, D. Su and L. Prati, *Chem. Commun.*, 2006, 1956.
360. N. Dimitratos, J. A. Lopez-Sanchez, J. M. Anthonykutty, G. Brett, A. F. Carley, R. C. Tiruvalam, A. A. Herzing, C. J. Kiely, D. W. Knight and G. J. Hutchings, *Phys. Chem. Chem. Phys.*, 2009, **11**, 4952.
361. S. D. Pollington, D. I. Enache, P. Landon, S. Meenakshisundaram, N. Dimitratos, A. Wagland, G. J. Hutchings and E. H. Stitt, *Catal. Today*, 2009, **145**, 169.
362. B. Zope and R. Davis, *Top. Catal.*, 2009, **52**, 269.
363. H. H. C. M. Pinxt, B. F. M. Kuster and G. B. Marin, *Appl. Catal. A*, 2000, **191**, 45.
364. L. Prati and M. Rossi, *Stud. Surf. Sci. Catal.*, 1997, **110**, 509.
365. S. Biella, L. Prati and M. Rossi, *Inorg. Chim. Acta*, 2003, **349**, 253.
366. Y. Ryabenkova, Q. He, P. J. Miedziak, N. F. Dummer, S. H. Taylor, A. F. Carley, D. J. Morgan, N. Dimitratos, D. J. Willock, D. Bethell, D. W. Knight, D. Chadwick, C. J. Kiely and G. J. Hutchings, *Catal. Today*, 2013, **203**, 139.
367. A. Corma, S. Iborra and A. Velty, *Chem. Rev.*, 2007, **107**, 2411.
368. B. T. Kusema and D. Y. Murzin, *Catal. Sci. Technol.*, 2013, **3**, 297.
369. S. Ramachandran, P. Fontanille, A. Pandey and C. Larroche, *Food Technol. Biotechnol.*, 2006, **44**, 185.
370. T. Miyake and Y. Sato, *US Pat.*, 3862005, 1975.
371. H. G. J. d. Wilt, *Ind. Eng. Chem. Prod. Res. Dev.*, 1972, **11**, 370.
372. H. Wieland, *Ber. Dtsch. Chem. Ges.*, 1912, **45**, 2606.
373. K. Heyns, *Justus Liebigs Ann. Chem.*, 1947, **558**, 177.
374. K. Heyns and R. Heinemann, *Justus Liebigs Ann. Chem.*, 1947, **558**, 187.
375. K. Heyns and O. Stöckel, *Justus Liebigs Ann. Chem.*, 1947, **558**, 192.
376. G. de Wit, J. J. de Vlieger, A. C. Kock-van Dalen, R. Heus, R. Laroy, A. J. van Hengstum, A. P. G. Kieboom and H. van Bekkum, *Carbohyd. Res.*, 1981, **91**, 125.
377. P. Vinke, D. de Wit, A. T. J. W. de Goede and H. van Bekkum, *Stud. Surf. Sci. Catal.*, 1992, **72**, 1.
378. A. Abbadi and H. van Bekkum, *J. Mol. Catal. A*, 1995, **97**, 111.

379. M. Wenkin, P. Ruiz, B. Delmon and M. Devillers, *J. Mol. Catal. A*, 2002, **180**, 141.
380. S. Karski, I. Witońska and J. Gołuchowska, *J. Mol. Cat. A*, 2006, **245**, 225.
381. J. Kusisto, A. V. Tokarev, E. V. Murzina, M. U. Roslund, J.-P. Mikkola, D. Y. Murzin and T. Salmi, *Catal. Today*, 2007, **121**, 92.
382. A. Mirescu and U. Prüße, *Appl. Catal. B*, 2007, **70**, 644.
383. A. V. Tokarev, E. V. Murzina, P. K. Seelam, N. Kumar and D. Y. Murzin, *Micropor. Mesopor. Mater.*, 2008, **113**, 122.
384. C. Bianchi, S. Biella, A. Gervasini, L. Prati and M. Rossi, *Catal. Lett.*, 2003, **85**, 91.
385. F. W. Lichtenthaler, *Acc. Chem. Res.*, 2002, **35**, 728.
386. Y. Önal, S. Schimpf and P. Claus, *J. Catal.*, 2004, **223**, 122.
387. B. M. Despevroux, K. Deller and E. Peldszus, *Stud. Surf. Sci. Catal.*, 1990, **55**, 159.
388. M. Besson, F. Lahmer, P. Gallezot, P. Fuertes and G. Fleche, *J. Catal.*, 1995, **152**, 116.
389. C. Baatz, N. Thielecke and U. Prüße, *Appl. Catal. B*, 2007, **70**, 653.
390. C. Baatz and U. Prüße, *J. Catal.*, 2007, **249**, 34.
391. C. Baatz, N. Decker and U. Prüße, *J. Catal.*, 2008, **258**, 165.
392. P. Beltrame, M. Comotti, C. D. Pina and M. Rossi, *Appl. Catal. A*, 2006, **297**, 1.
393. T. Ishida, N. Kinoshita, H. Okatsu, T. Akita, T. Takei and M. Haruta, *Angew. Chem., Int. Ed.*, 2008, **47**, 9265.
394. A. Mirescu, H. Berndt, A. Martin and U. Prüße, *Appl. Catal. A*, 2007, **317**, 204.
395. N. Thielecke, K.-D. Vorlop and U. Prüße, *Catal. Today*, 2007, **122**, 266.
396. S. Biella, L. Prati and M. Rossi, *J. Catal.*, 2002, **206**, 242.
397. M. Comotti, C. Della Pina, R. Matarrese and M. Rossi, *Angew. Chem., Int. Ed.*, 2004, **43**, 5812.
398. M. Comotti, C. D. Pina and M. Rossi, *J. Mol. Catal. A*, 2006, **251**, 89.
399. H. Zhang and N. Toshima, *Catal. Sci. Technol.*, 2013, **3**, 268.
400. B. T. Kusema, B. C. Campo, P. Maki-Arvela, T. Salmi and D. Y. Murzin, *Appl. Catal. A*, 2010, **386**, 101.
401. E. Smolentseva, B. T. Kusema, S. Beloshapkin, M. Estrada, E. Vargas, D. Y. Murzin, F. Castillon, S. Fuentes and A. Simakov, *Appl. Catal. A*, 2011, **392**, 69.
402. O. A. Simakova, B. T. Kusema, B. C. Campo, A.-R. Leino, K. Kordas, V. Pitchon, P. Maki-Arvela and D. Y. Murzin, *J. Phys. Chem. C*, 2011, **115**, 1036.
403. M. Mavrikakis, P. Stoltze and J. K. Norskov, *Catal. Lett.*, 2000, **64**, 101.
404. N. Lopez, T. V. W. Janssens, B. S. Clausen, Y. Xu, M. Mavrikakis, T. Bligaard and J. K. Norskov, *J. Catal.*, 2004, **223**, 232.
405. Y. Guan and E. J. M. Hensen, *Appl. Catal. A*, 2009, **361**, 49.
406. B. T. Kusema, B. C. Campo, O. A. Simakova, A.-R. Leino, K. Kordas, P. Maki-Arvela, T. Salmi and D. Y. Murzin, *ChemCatChem*, 2011, **3**, 1789.

407. F. H. Stodola and L. B. Lockwood, *J. Biol. Chem.*, 1947, **171**, 213.
408. H. E. J. Hendriks, B. F. M. Kuster and G. B. Marin, *Carbohyd. Res.*, 1990, **204**, 121.
409. E. V. Murzina, A. V. Tokarev, K. Kordas, H. Karhu, J.-P. Mikkola and D. Y. Murzin, *Catal. Today*, 2008, **131**, 385.
410. L.-F. Gutierrez, S. Hamoudi and K. Belkacemi, *Appl. Catal. A*, 2011, **402**, 94.
411. L.-F. Gutierrez, S. Hamoudi and K. Belkacemi, *Appl. Catal. A*, 2012, **425**, 213.
412. M. N. Timofeeva, S. H. Jhung, Y. K. Hwang, D. K. Kim, V. N. Panchenko, M. S. Melgunov, Y. A. Chesalov and J. S. Chang, *Appl. Catal. A*, 2007, **317**, 1.
413. X. Tan, W. Deng, M. Liu, Q. Zhang and Y. Wang, *Chem. Commun.*, 2009, 7179.
414. A. Corma and H. Garcia, *Chem. Soc. Rev.*, 2008, **37**, 2096.
415. B.-Z. Zhan, M. A. White, T.-K. Sham, J. A. Pincock, R. J. Doucet, K. V. R. Rao, K. N. Robertson and T. S. Cameron, *J. Am. Chem. Soc.*, 2003, **125**, 2195.
416. J. C. Béziat, M. Besson and P. Gallezot, *Appl. Catal. A*, 1996, **135**, L7.
417. A.-H. Lu, W.-C. Li, Z. Hou and F. Schüth, *Chem. Commun.*, 2007, 1038.

CHAPTER 10

Selective Oxidation of Biomass-Derived Secondary Alcohols

OLGA A. SIMAKOVA[a,b] AND DMITRY YU. MURZIN*[b]

[a] Georgia Institute of Technology, School of Chemical and Biomolecular Engineering, 311 Ferst Drive NW, Atlanta, GA, 30332-0100, USA;
[b] Laboratory of Industrial Chemistry and Reaction Engineering, Process Chemistry Centre, Åbo Akademi University, FI-20500, Åbo/Turku, Finland
*Email: dmurzin@abo.fi

10.1 Oxidation of Biomass-Derived Compounds

The discovery of gold's catalytic activity by Haruta and co-workers[1] led to a growth of research interest in this area. The origin of gold's catalytic activity is still a matter of debate. Activation treatment of gold led to the formation of partially oxidized gold ($Au^{\delta+}$) with unoccupied outer *d*-orbitals, similar to the electronic structure of Pt.[2] The catalytic activity of gold catalysts is strongly dependent on the gold particle size.[3] It was reported that the major effect on the activity of gold catalysts is related to the presence of low-coordinated gold atoms,[4,5] which are abundant in the smallest gold nanoparticles. The most reactive atoms are located at the edges and corners of the particles, which were suggested to facilitate the adsorption of reactants. The number of low-coordinated corners and edges sites decreases with increasing Au particle size.[5] Furthermore, other effects, such as interactions with the support,[6,7] charge transfer,[8] and strain[9] may also contribute to the catalytic activity of gold particles, though to a lesser extent.

The main advantage of gold-based catalysts is environmentally-friendly manufacturing of organic chemicals employing molecular oxygen or air as

RSC Catalysis Series No. 18
Heterogeneous Gold Catalysts and Catalysis
Edited by Zhen Ma and Sheng Dai

Published by the Royal Society of Chemistry, www.rsc.org

oxidizing agents and relatively mild reaction conditions. Moreover, in contrast to typically applied platinum and palladium catalysts, gold is stable under an oxidative atmosphere. Since the discovery of gold's catalytic activity in low-temperature CO oxidation,[1] the remarkable activity of nanosized gold catalysts in oxygen-atom transfer reactions was further shown in numerous reactions of the selective oxidation of a variety of organic molecules such as alcohols and aldehydes,[10,11] alkanes,[12] olefins,[13,14] and polyfunctional alcohols.[15–17] Activation of molecular oxygen on gold particles is a crucial step in the oxidation reactions. Low-temperature activation of dioxygen over gold was studied by Salisbury *et al.*[18] The results suggested that dioxygen is adsorbed on gold clusters, forming a superoxide (O_2^-), which is the activated form of oxygen.

The unique properties of nanosized gold to activate oxygen as well as high stability in an oxidative atmosphere make gold an excellent catalyst in oxidizing a number of biomass-derived compounds. Utilization of biorenewable feedstock is essential for the sustainable development of the chemical industry.[19–21] The main components of wood biomass such as cellulose (40–50%), lignin (16–33%), hemicelluloses (15–30%), and extractives (1–10%)[22] are rich sources of platform chemicals. Therefore sugars produced by cellulose and hemicelluloses hydrolysis are available in large amounts and can be selectively oxidized over gold into value-added products,[23–35] such as aldonic acids valuable for food industry cosmetics and medicine.[36] Moreover, gold catalysts demonstrated outstanding long-term stability in the oxidation of sugars.[37]

Selective oxidation of sugar-derived 5-hydroxymethyl-2-furfural (HMF) into 2,5-furandicarboxylic acid (FDCA), a possible replacement monomer for terephthalic acid utilized in the production of polyethylene terephthalate (PET), was successfully performed over gold catalysts.[38–40] As an example, the yield of FDCA was more than 99% after 8 h over Au/CeO_2,[40] while catalysts based on transition metals can be deactivated due to over-oxidation of the metal surface. The reaction occurs *via* subsequent oxidation of hydroxyl and carbonyl groups.[41]

Gold-catalyzed selective oxidation of glycerol has been intensively studied.[17,42–47] Traditionally applied catalysts based on platinum and palladium suffer from deactivation by over-oxidation of their surface, while gold catalysts were found to be more stable and moreover possess higher selectivity. Reaction selectivity can be adjusted by applying certain reaction conditions and catalyst selection.[48–63]

Selective aerobic oxidation of bioethanol/ethanol to acetic acid and ethyl ester over gold has been investigated previously. It was reported that 90% yield of acetic acid can be reached at high ethanol conversion over gold.[64,65] Gold catalysts allowed the use of less severe reaction conditions compared with those needed for other catalysts active in the gas phase such as supported V_2O_5 (175–200 °C, 270 kPa), Nb-Mo-V-Ox (237 °C, 1600 kPa).[66–68] Gold catalysts can be applied in both liquid[69–72] and gas phase[73] oxidation. The kinetics of acetic acid formation was investigated and were consistent with the following proposal regarding the reaction mechanism. As a first step, ethanol is adsorbed on the gold surface to form an activated intermediate species, which then is oxidized/dehydrogenated to the aldehyde. Consequently, aldehyde is oxidized to acetic acid, while CH_3CH_2O can also undergo C–C bond cleavage to produce CO_2.[64] Efforts were made to understand the reaction mechanism revealing the important role of the catalyst support[74] and reaction conditions.[75]

The high activity and selectivity of gold catalysts in the secondary alcohols oxidation has been reported elsewhere.[76] However, the oxidation of biomass-derived secondary alcohols over gold has been investigated to a lesser extent primarily due to the lower availability of the substrate compared with sugars and primary alcohols. Selective oxidation of the lignin, hydroxymatairesinol (HMR), over gold represents an example of selective oxidation of biomass-derived secondary alcohols.

Lignans are a group of natural phenolic compounds found in different parts of plants. Coniferous trees are a rich source of different lignans. For example, Norway spruce (*Picea abies*) knots contain large quantities of lignans, among which HMR[77–79] is the most abundant. HMR, constituting 65–85 wt% of the total lignans, appears in unconjugated free form,[80] simplifying the isolation process. Extracted from wood biomass, HMR is a mixture of two diastereomers, namely (7*R*,8*R*,8′*R*)-(−)-7-allo-hydroxymatairesinol (HMR 1 or RRR-HMR) and (7*S*,8*R*,8′*R*)-(−)-7-allo-hydroxymatairesinol (HMR 2 or SRR-HMR); HMR 1 and HMR 2 can be isomerized into each other (Scheme 10.1). The ratio between HMR 1 and HMR 2 depends on the feedstock.

Lignans have been shown to provide benefits for human health due to their anticarcinogenic and antioxidative effects.[81–83] Investigation of the properties of lignans revealed that oxomatairesinol (oxoMAT) exhibits the highest superoxide scavenging activity. Lignan oxoMAT and its derivatives, due to their antioxidative activity[84] and ultraviolet (UV) protection properties,[85] are valuable for cosmetic and pharmaceutical uses (skincare and haircare products), as well as color-retention agents for the textile industry. While oxoMAT cannot be extracted directly in considerable amounts from wood biomass, it can be synthesized by selective oxidation of HMR (see Scheme 10.1).

The formation of oxoMAT from HMR was performed using light irradiation,[86] and a homogeneous catalyst.[87] Pd/C catalysts have been studied for the synthesis of oxoMAT in anaerobic conditions using organic solvents.[88] This was the first attempt to perform the oxidation reaction *via* heterogeneous catalysis, which is more attractive for industrial application. However, oxoMAT was not the only product in this reaction. Application of heterogeneous gold catalysts was demonstrated for the first time by Murzin and co-workers,[89] when the selectivity of gold-catalyzed oxidation was 100%. Moreover, conversion of HMR over gold catalysts was higher than over palladium ones.

This chapter provides an overview of the selective oxidation of biomass-derived lignan as an example of the oxidation of secondary alcohols over gold

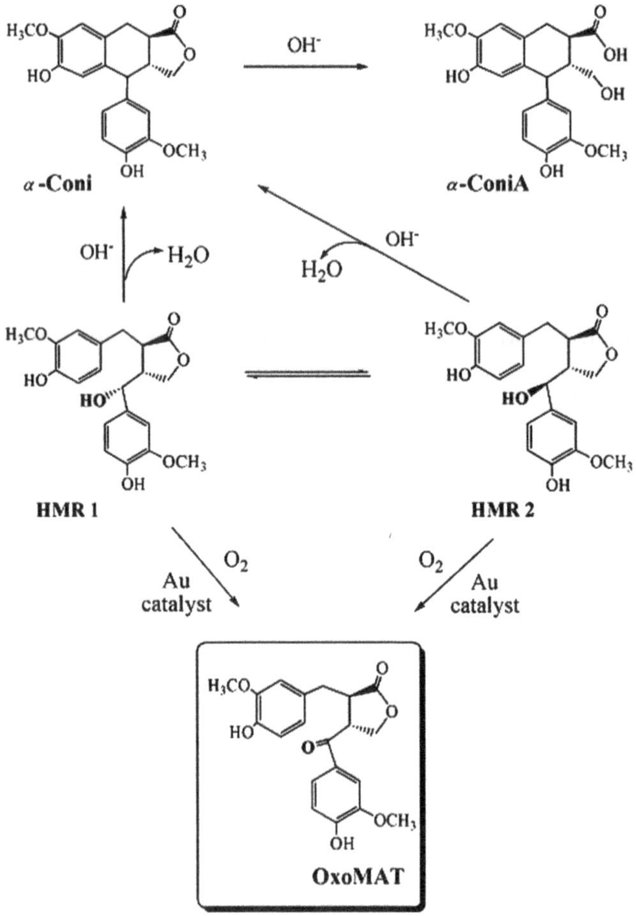

Scheme 10.1 Reaction pathway of the transformation of the lignan hydroxymatairesinol (HMR) over gold catalysts under aerobic conditions. Coni, conidendrin; ConiA, conidendric acid; OxoMAT, oxomatairesinol.

catalysts. The focus is on describing the influence of the reaction conditions, catalyst deactivation, reaction mechanism, and reaction kinetics.

10.2 Selective Oxidation of the Lignan Hydroxymatairesinol

10.2.1 Influence of Reaction Conditions

The lignan HMR is poorly soluble in water and hydrocarbons, but well soluble in alcohols. Therefore the selective oxidation reaction has been performed in different alcohols and their mixtures with water. In order to avoid competitive oxidation of HMR and alcohols over the gold catalyst, the reaction has also been carried out using 1,4-dioxane and tetrahydrofuran (THF) and their mixtures with water as a solvent.[89]

Applied gold catalysts were found not to be active in the synthesis of oxoMAT when neat alcohols were used as solvents. However, utilization of alcohol–water mixtures resulted in oxoMAT formation due to HMR oxidation (Table 10.1).[89] These changes were attributed to the activity of gold catalysts in oxidizing the solvent. The products of the oxidation of alcohols such as aldehydes, ketones and carboxylic acid most likely poison gold catalysts due to the strong adsorption of carboxylic and carbonyl groups. Gold catalysts became active in the oxoMAT synthesis when an alcohol–water mixture was used as a result of minor alcohol oxidation. Moreover, catalytic activity was enhanced with a decrease in the alcohol fraction of the alcohol–water mixture.

Table 10.1 Effect of solvent on the activity and selectivity in HMR oxidative dehydrogenation. Reprinted from ref. 89 with permission from Elsevier.

Solvent	Selectivity towards OxoMAT/ %	Conversion after 4 h / %	Activity/ mol $l^{-1} \times$ sec $\times g_{cat} \times 10^5$	pH of HMR solution
Alcohols				
Ethanol	–	0	n.a[a]	5
Propan-2-ol	–	0	n.a	5
Butan-2-ol	–	0	n.a	5
Pentan-2-ol	–	0	n.a	5
Ethanol in water, vol.%				
2	0	30	n.a	5
Propan-2-ol in water, vol.%				
2	100	70	6	5
5	100	70	6	5
10	100	53	3	5
20	100	52	3	5
50	100	52	2	5
80	100	30	2	5
Butan-2-ol, vol.%				
2	0	30	n.a	5
Pentan-2-ol, vol.%				
2	35[b]	65	9	5
Cyclohexanol, vol.%				
2	75[c]	24	0.2	5
Acetone, vol.%				
2	83[c]	15	0.1	5
1,4-Dioxane, vol.%				
2	100	44	1.3	5
80	100	15	0.6	6
100	–	0	n.a	6
Tetrahydrofuran (THF), vol.%				
2	55	21	0.5	5
80	–	0	n.a	5.5
100	–	0	n.a	6

[a]n.a = not active in oxoMAT formation. The main byproduct was lignan.
[b]Conidendric acid (ConiA).
[c]Hydroxymatairesinol (HMR 1).

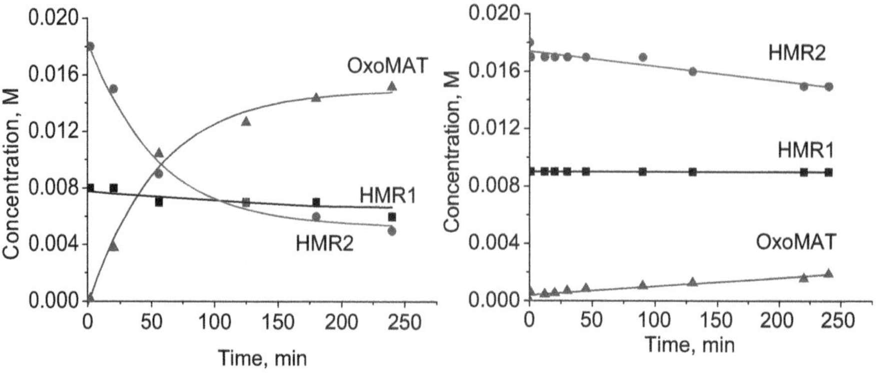

Figure 10.1 Oxidation of HMR in 2 vol.% 2-propanol in water (250 mg of 2 wt% Au/Al_2O_3-DIE catalyst at 70°C under the flow of synthetic air). Reprinted from ref. 89 with permission from Elsevier.

The highest activity and selectivity were obtained using 2 and 5 vol.% 2-propanol in water as a solvent. The role of water could be related to three phenomena: (1) suppression of the possible deactivation described above due to organic solvent interactions with the catalyst surface; (2) activation of adsorbed oxygen by adsorbed water due to the formation of hydroperoxyl-like intermediate species ($^{\bullet}OOH$) followed by their decomposition and release of $O^{\bullet}$ and H_2O; and (3) the hydrolysis of the metal oxide support, followed by the formation of the surface OH groups, enhancing the adsorption of HMR on the catalyst surface in a geometry suitable for further dehydrogenation.[90] It is worth noting that the activity of Pd catalysts towards oxoMAT formation decreases in the presence of water. Therefore, synthesis of oxoMAT over Pd catalysts requires utilization of larger amounts of organic solvents compared with the same reaction over gold.

The presence of oxygen was found to be essential for oxoMAT formation. The stability of gold catalysts in an oxidative atmosphere as a key advantage over transition metals is well known. In contrast to previously applied Pd catalysts, gold demonstrated higher activity in the presence of oxygen than under the same reaction conditions in its absence.[89] Moreover, gold catalysts demonstrated a complete selectivity to the desired oxoMAT compared with a selectivity of 70% obtained over Pd catalysts (Figure 10.1). It has also been reported that gold catalysts have a catalytic activity in alcohol oxidation in the absence of oxygen. However, the presence of oxygen leads to the removal of adsorbed hydrogen atoms by co-adsorbed oxygen, thus increasing the reaction rate.[91]

10.2.2 Effect of the Catalyst Support and Active Phase

The unusually high catalytic activity of gold is often attributed to the metal–support interface, which is strongly dependent on the nature of the support.

From this point of view ceria was reported as being one of the most efficient metal oxide supports due to its oxygen storage/release capacity.[11,92] Furthermore, it is possible to improve the redox behavior of ceria by incorporation of other elements such as Zr(IV) cations in its lattice.[93] At the same time, it was shown that gold catalysts supported on zeolites were highly active in the aerobic selective oxidation of alcohols.[94] Moreover, doping of Au supported on zeolites by other metals such as Fe, Ni, and Cu changes the oxidation state of gold, which in turn results in the increase of the activity of gold catalysts in CO oxidation.[95–97]

Oxidation of lignan HMR to oxoMAT was investigated using various gold catalysts such as Au/C, SiO_2, TiO_2, Al_2O_3, MgO, CeO_2, ZrO_2, La_2O_3 as well as ceria–alumina–zirconia mixed oxides and zeolites.[89,98] The tested gold catalysts showed different activities towards the oxidation/dehydrogenation of HMR (Figure 10.2). In the case of Au/MgO catalysts, HMR was transformed into other lignans: conidendrin (Coni) and conidendric acid (ConiA) (see Scheme 10.2) by HMR dehydration, followed by cyclisation. This behavior is related to the basicity of magnesia, as ConiA is normally formed with increasing solution pH (Scheme 10.2). Since Au particles are located inside the zeolite channels where bulky HMR molecules cannot penetrate (HMR size is 5–20 Å depending on the conformer), zeolite-supported gold catalysts were shown to be inactive in HMR oxidation. The activity and selectivity of ceria–alumina–zirconia mixed oxides were lower than those obtained over Au on alumina catalysts (Table 10.2). Among the tested catalysts, the most active catalyst was gold supported on alumina. Moreover, the activity of alumina-supported catalysts increased with increasing support pore size.[98]

The catalytic performance of the studied catalysts correlates with the acidity of the supports and, therefore, the presence of Lewis acidity on the support surface was suggested to promote oxoMAT formation.[89] Since

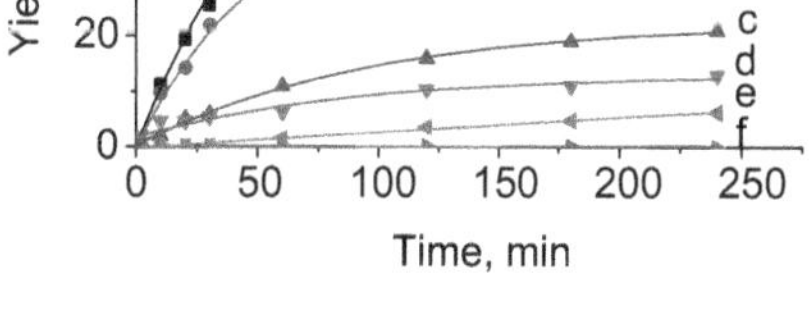

Figure 10.2 The oxoMAT yield as a function of time at 70 °C using 2 vol.% 2-propanol in water as a solvent in air flow over 250 mg catalyst: (a) Au/Al_2O_3 (prepared by direct ionic exchange); (b) Au/Al_2O_3 (prepared by deposition–precipitation with urea); (c) Au/TiO_2; (d) Au/SiO_2; (e) Au/C; and (f) Au/MgO.
Reprinted from ref. 89 with permission from Elsevier.

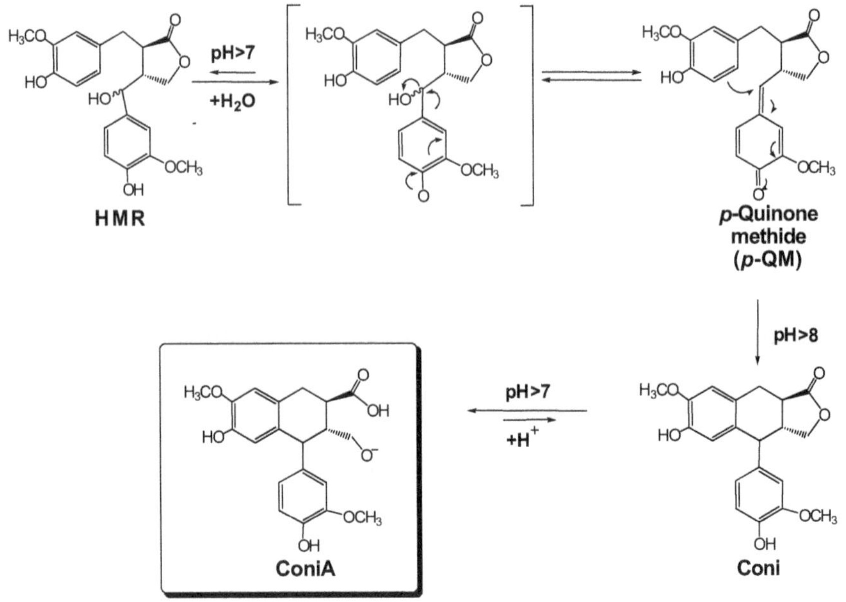

Scheme 10.2 Transformations of the lignan hydroxymatairesinol (HMR) to the lignans conidendrin (Coni) and conidendric acid (ConiA) in alkaline solution (pH 7–10).

the reaction solvent is mainly water, the alumina surface is hydroxylyzed, and water is adsorbed on the Lewis acid sites, forming Al–OH groups.[99] Functional groups on the support surface were suggested to coordinate the structure of the HMR complex in the adsorption mode, making it more favorable for consequent dehydrogenation (Figure 10.4). Since the number of surface hydroxyl groups is proportional to the amount of Lewis acid sites, the catalyst activity was found to rise with an increasing number of Lewis acid sites on the support surface. Similar to previously reported work,[100] HMR could be adsorbed on Al–OH or Al^{3+} surface groups by oxygen bonded with α-H, while proton elimination occurred on the surface of the gold particles or oxygen adsorbed on gold clusters.[101]

In order to investigate the influence of bimetallic interactions and support properties, gold, gold–palladium, and palladium catalysts were tested in the selective oxidation of lignan HMR. It was demonstrated that supports applied *per se* were not active in the oxoMAT synthesis, while supported gold catalysts showed different activities in the HMR transformation reaction. The major difference in catalytic behavior was the selectivity towards various products. Bimetallic Au-Pd catalyst had similar or slightly lower activity than monometallic Au catalyst (Table 10.2). The effect of the support on selectivity was dominating in the case of magnesia, zirconia, and lanthana. Utilization of Au and Au-Pd catalysts on ceria resulted in the formation of oxoMAT and Coni with the same yields. Au and Au-Pd/alumina catalysts led to the formation of oxoMAT only (Figure 10.3). In general, if the support activity is the

Table 10.2 Product yields of HMR transformation obtained over Au, Au-Pd, Pd catalysts, and corresponding supports at 70°C, under the synthetic air flow, after 4 h. Reprinted from ref. 98 with permission from Elsevier.

Catalyst	TOF/s^{-1}	Yield/ %[a]				
		OxoMAT	HMR 1	Coni	ConiA	Others
MgO	0	0	0	0	100	0
Pd/MgO	0	0	0	0	100	0
Au/MgO	0	0	0	0	100	0
Au-PdMgO	0	0	0	0	100	0
La_2O_3	0	0	20	46	0	0
Pd/La_2O_3	0	0	22	19	3	0
Au/La_2O_3	0.007	3	25	21	2	0
Au-Pd/La_2O_3	–	4	22	21	1	0
ZrO_2	0	0	0	1	0	55
Pd/ZrO_2	0	0	0	2	1	65[b]
Au/ZrO_2	0.002	1	12	19	2	0
Au-Pd/ZrO_2	–	1	0	0	2	61[b]
CeO_2	0	0	4	16	0	0
Pd/CeO_2	0	0	2	3	2	0
Au/CeO_2	0.035	20	0	11	0	0
Au-Pd/CeO_2	–	20	0	12	0	0
Al_2O_3-I	0	0	11	24	0	0
Pd/Al_2O_3-I	0	0	5	3	1	0
Au/Al_2O_3-I	0.117	74	0	0	0	0
Au-Pd/Al_2O_3-I	–	68	0	0	0	0

[a]Yields of products were obtained after 4 h.
[b]Main product was lignan dimer of unknown structure.

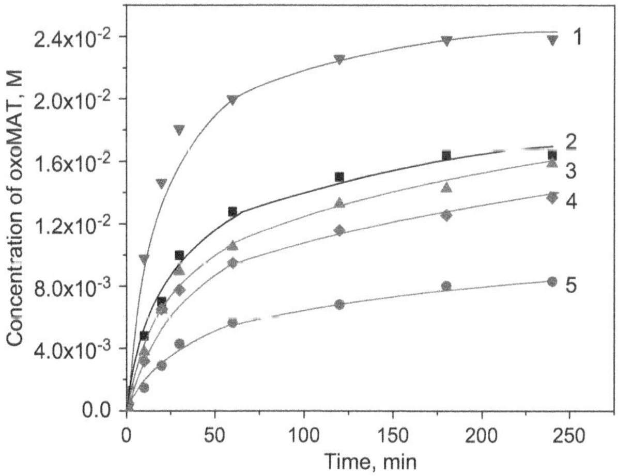

Figure 10.3 The dependence of the oxoMAT concentration on time over 3 wt% Au catalysts supported on oxides prepared by sol-gel: 1 = Au/Al_2O_3-II; 2 = Au/Ce(5)Zr(5)-Al, 3 = Au/Ce(10)-Al; 4 = Au/Ce(15)Zr(15)-Al; 5 = Au/Ce(30)-Al. Conditions: m_{cat} = 0.2 mg, under synthetic air flow, at 70 °C. Reprinted from ref. 98 with permission from Elsevier.

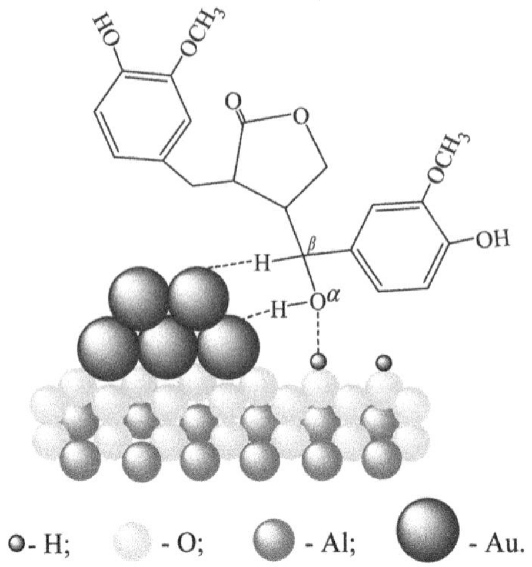

Figure 10.4 Proposed scheme of HMR adsorption on Au/Al_2O_3. Reprinted from ref. 98 with permission from Elsevier.

dominating factor, side reactions of the substrate, *i.e.* isomerization HMR 2 to HMR 1, and formation of Coni and ConiA, dominated. Otherwise, selectivity is shifted towards the oxidation of HMR.

10.2.3 Catalyst Deactivation and Regeneration

Long-term stability of the catalyst is essential for industrial applications. Gold is known to be stable in an oxidizing atmosphere. The reasons for catalyst deactivation could be poisoning by reagent and solvent impurities, catalyst leaching, gold particles sintering and catalyst surface reconstruction, or strong adsorption of the main and side products, which can form oligomeric or polymeric structures. Since deactivation of gold catalysts during the HMR oxidation was observed, the reasons for decreasing catalyst activity and possibilities to regenerate the spent catalyst were studied.[102]

The biomass-derived lignan HMR contains inorganic impurities, although no poisoning elements, such as Cl, S, P, *etc.*, were detected. Organic impurities of HMR are the lignans Coni and ConiA, which can be adsorbed on the catalyst and reduce its activity. The influence of the product oxoMAT was also investigated. It was discovered that catalyst activity decreased during product formation. Moreover, those experiments involving the addition of fresh catalyst in the presence of the product demonstrated an increase in catalyst activity in each subsequent experiment (Figure 10.5). The presence of adsorbed organic species was confirmed by thermo-programmed oxidation, where their combustion was observed at increasing temperature. Therefore, the adsorption of oxoMAT could be a reason for catalyst deactivation, even if its impact is rather moderate. The effects of potential changes in the catalyst surface area, Au particle size and the metal loading

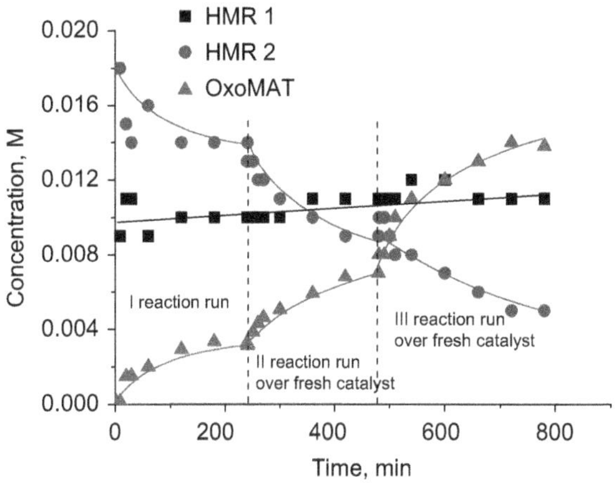

Figure 10.5 Selective oxidation of the lignan HMR with addition of fresh 2 wt% Au/Al_2O_3, 70 °C.
Reprinted from ref. 102 (Figure 3, Copyright © 2012, Springer Science + Business Media, LLC). With kind permission from Springer Science + Business Media.

have also been elucidated; no catalyst structure changes and leaching of the active phase were observed.[102]

Since the main reason for catalyst deactivation was considered to be the adsorption of organic compounds, attempts to regenerate the catalyst by extraction and subsequent calcination were made. Calcination was found to be the most efficient method (Figure 10.6) to restore at least the initial catalyst activity, although the calcination temperature should not exceed 600 °C to avoid catalyst sintering and support changes.

10.2.4 Reaction Kinetics

The selective oxidation of HMR to oxoMAT is in fact selective oxidation of secondary alcohols into the corresponding ketones over gold catalysts, as described in many papers.[103–107] However, the reaction kinetics is challenging to investigate as HMR is a mixture of two diastereomers, HMR 1 and HMR 2, which can undergo both reactions, *i.e.* oxidation to oxoMAT and isomerization to each other (Scheme 10.1).

Studies revealed reaction kinetic regularities.[108] The reaction pathway of HMR oxidative dehydrogenation was described according to Scheme 10.3, where oxoMAT is formed due to oxidation of both HMR isomers. The proposed kinetic model and the equations of the batch reactor were examined using MODEST software;[109] the code was compiled and used for parameter estimation of the data obtained in the experimental work.

The observed difference in the reactivity of diastereomers HMR 1 and HMR 2 was attributed to the different potential energy of the intermediates formed from each isomer, as well as steric hindrances. The apparent reaction

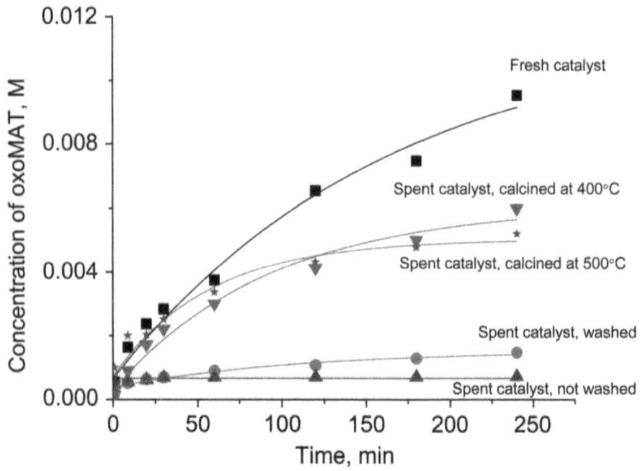

Figure 10.6 The oxoMAT concentration profile during the selective oxidation of HMR over 2 wt% Au/Al_2O_3: fresh, spent/calcined, spent/washed and spent/non-washed catalysts.
Reprinted from ref. 102 (Figure 4, Copyright © 2012, Springer Science + Business Media, LLC). With kind permission from Springer Science + Business Media.

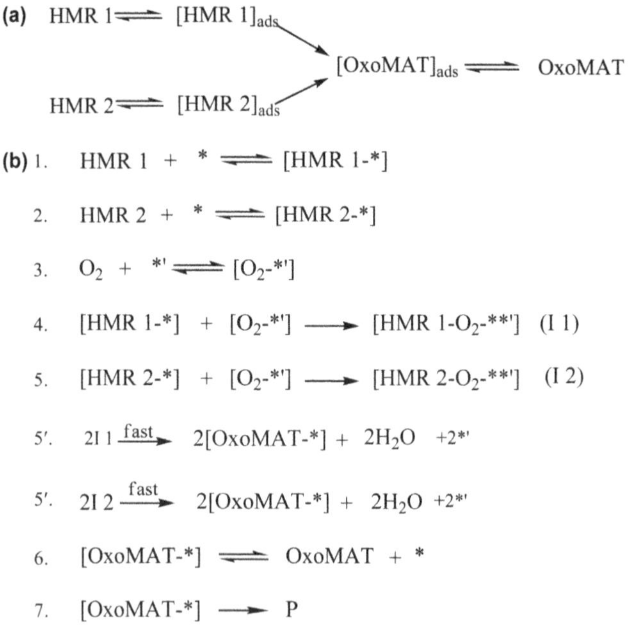

Scheme 10.3 (a) Reaction pathway and (b) reaction steps of the selective aerobic oxidation of HMR to oxoMAT over Au catalysts (where * and *′ are different types of sites).
Reprinted from ref. 108 with permission from Elsevier.

order in reagents was found to be 0.42 ± 0.05 and 0.29 ± 0.07 for the HMR 1 and HMR 2 isomers, respectively. Calculations gave a close to first-order dependence with respect to the catalyst concentration (0.80 ± 0.08). The slight deviation from first-order behavior was suggested to be related to the catalyst deactivation demonstrated.[102] The apparent energy of activation calculated from the slope of the Arrhenius plot was found to be 43 ± 4.3 kJ mol^{-1} for HMR oxidation. The proposed kinetic model correlated well with the experimental data.

The reaction rate was found to be independent of the oxygen partial pressure in the range of 5 to 49 vol.% of oxygen amount, thus following zero-order behavior. The apparent zero-order with respect to oxygen can be explained by the following assumptions, conceptually similar to the hydrogenation case.[110] Oxygen and HMR molecules do not compete for the adsorption sites and the adsorbed oxygen is in equilibrium with oxygen dissolved in the liquid phase. Thus, the reaction rate does not depend on the concentration of adsorbed oxygen.

In general, during selective oxidation reactions over gold catalysts, activated oxygen can interact with water, forming oxidizing active species.[111,112] Moreover it can act as a Brønsted basic site, enhancing O–H bond cleavage[113] and β-H removal, resulting in the formation of the corresponding aldehydes/ketones,[91] and can also scavenge the electrons deposited on gold during the reaction.[114]

10.2.5 Structure Sensitivity

The activity of gold catalysts is strongly dependent on the particle size.[3,5] The structure sensitivity of HMR selective oxidation has been studied.[115] The reaction was performed over gold catalysts supported on alumina with different Au particle sizes (1.9–10 nm). The relationship of the catalyst activity with gold cluster size has a volcano-shape dependence with a maximum at 3.2 nm (Figure 10.7). The proposed mechanism assumes noncompetitive adsorption of HMR and oxygen on different catalyst sites, and can be essentially simplified if coverage of the product is neglected. The dependence of the reaction rate on Au cluster size is thus:

$$rate = \frac{k_3' e^{\alpha \chi_A / d_{cluster}} K_1 C_A}{(1 + K_1 e^{\chi_A / d_{cluster}} C_A) d_{cluster}} \tag{10.1}$$

As can be seen from Figure 10.7, the model correctly captures the dependence of the reaction rate on the cluster size, and is consistent with the global kinetic model.[108]

10.2.6 Reaction Mechanism

The selective oxidation of HMR to oxoMAT is challenging due to the different reactivity of the HMR diastereomers. The epimerization occurs through the

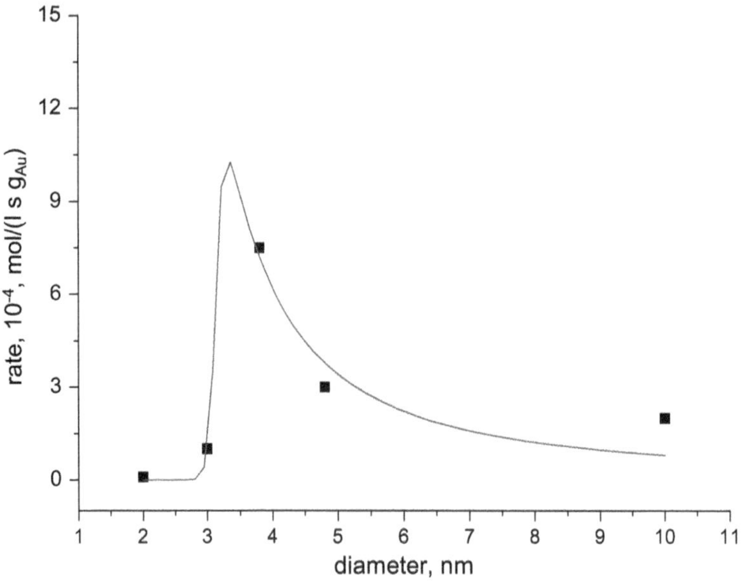

Figure 10.7 Dependence of the initial rate in the selective oxidation of the lignan HMR on the Au particle size of Au/Al_2O_3.

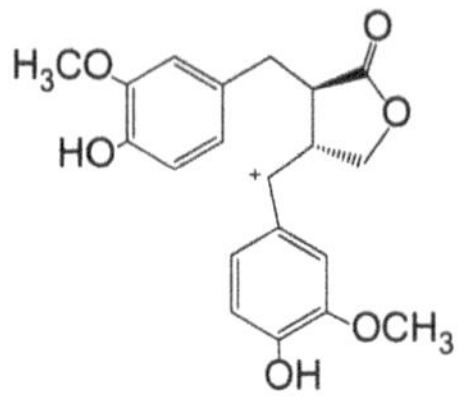

Figure 10.8 HMR 1 and HMR 2 epimerization intermediate: carbenium ionic species.

transfer of one proton from the acidic support surface to the alcohol hydroxyl group of both HMR 1 and HMR 2, resulting in release of a water molecule; this leaves the same carbenium ionic species on the catalytic surface, irrespectively of the starting HMR epimer (Figure 10.8). This carbenium species can be easily rehydrated and deprotonated to either HMR 1 or HMR 2.

As already mentioned (Section 10.2.1) the lignan HMR can be transformed to oxoMAT under both anaerobic and aerobic conditions,[89] although the reaction rate in the presence of oxygen is approximately five times higher (see Figure 10.1). Therefore, the HMR transformation over gold includes dehydrogenation steps, while the presence of oxygen enhances the oxoMAT synthesis. The difference in the reaction mechanism was investigated by quantum chemical calculations.[116] All the calculations were performed using the Gaussian 03 package. The hybrid Becke's three parameters exchange–correlation functional B3LYP was used. The Los Alamos LANL2 pseudopotential and the corresponding valence basis set were used for Au atoms, while the 6-31G(d,p) basis set was employed for O, C, and H atoms. Such a level of theory was proven to provide reliable structural findings for gold clusters. The starting structure for the Au_{28} cluster was obtained using a fragment extracted from one Au *fcc* bulk structure (space group Fm3m, a = 4.0782 Å). It was subjected to full geometry optimization.

Since an alumina support is not active in oxoMAT formation, the catalytic activity of Au catalysts in the HMR dehydration to oxoMAT can be straightforwardly associated with the presence of gold particles. The results of the calculations confirmed that the HMR transformation over gold includes the dehydrogenation route, not requiring the presence of oxygen.

The state of the activated oxygen on the surface of Au particles is still a matter of debate. There are many studies confirming the involvement of oxygen in the formation of two kinds of surface species, superoxo or peroxo, as well as atomic oxygen, active in oxidation. In any case, the participation of activated oxygen in proton removal is generally accepted and described elsewhere.[91] In ref. 116 only the activation of oxygen by the Au cluster was considered and dissociative chemisorption, which would certainly require an energy barrier, was not taken into account. The relaxed geometry of the O_2/Au_{28} aggregate has an O_2 molecule in the bridge position, which was used throughout the calculations. The optimized geometries of the HMR 1 and HMR 2 diastereomers interacting with the O_2/Au_{28} system were obtained. Geometrical and electronic parameters pointed out a pronounced structural difference (Figure 10.9).

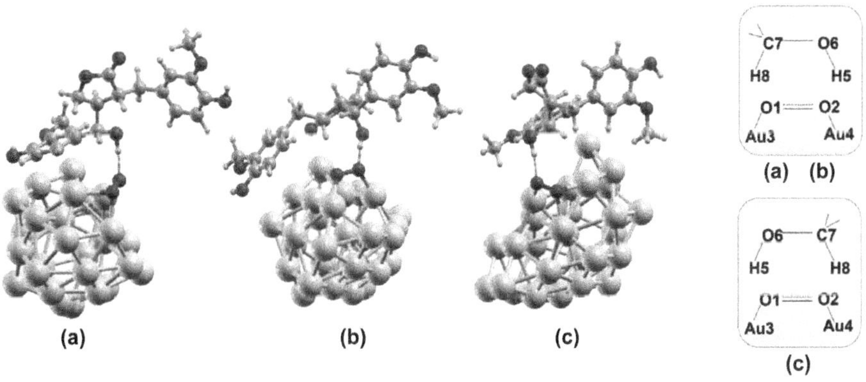

Figure 10.9 Optimized geometries of the HMR epimers on O_2/Au_{28}: (a) and (b) systems represent the most stable HMR 2 (SRR-HMR) and HMR 1 (RRR-HMR) interacting conformer, respectively; (c) is another HMR 1 (RRR-HMR) interacting conformer, which is *ca.* 8 kJ mol^{-1} less stable with respect of (b). The insets on the right report the numbering characterizing the (a) and (b) (top) and (c) systems (bottom). Reprinted from ref. 116. Copyright © 2013 WILEY-VCH Verlag GmbH & Co. KGaA, Weinheim.

One of the differences is the O2–H5 distance, which is considerably shorter in the HMR 2 + O_2/Au_{28} (1.686 Å) than in the HMR 1 + O_2/Au_{28} aggregate (1.747 Å). It was found that the most stable conformation of HMR 2 has the right geometry to interact efficiently with the activated oxygen molecule adsorbed on the gold cluster, while the HMR 1 molecule has presumably to adopt a low population conformation. Even when HMR 1 interacts with the right geometry, small steric effects give raise to an energy barrier for the first hydrogen migration which is ca. 30 kJ mol^{-1} higher than the one calculated for the corresponding step in the case of the HMR 2 (SRR-HMR) diastereomer. Therefore, theoretical calculations on the dehydrogenation mechanism were presented for the case of more reactive HMR 2 epimer as a model.

The whole dehydrogenation energetic profile is presented in Figure 10.10. The first phase in the process is characterized by a mechanism in which the O_2 molecule approaches the reactant giving favorable specific interactions, and the orientation to promote the C–H hydrogen removal and the subsequent facile O–H bond cleavage. This directly produces an oxoMAT molecule and a characteristic surface dihydroxide system, $(OH)_2/Au_{28}$. In the second reaction phase, desorption of the oxoMAT molecule is assumed and thus a second SRR-HMR molecule could interact with the two hydroxyl groups on the Au_{28} surface. Therefore, also in this case the presence of oxygen, as hydroxyl group, positively affects the C–H8 bond cleavage by adjusting the orientation of the reactant with the help of a hydrogen bond. Final products are constituted by one $(H_2O)_2/Au_{28}$ moiety and one more molecule of oxoMAT. The overall reaction is highly exothermic, with an energy difference between reactants and products being −300 kJ mol^{-1} and

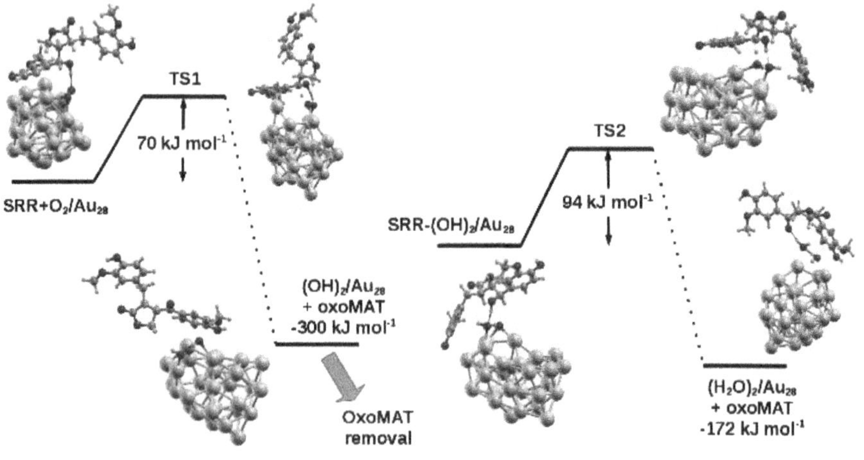

Figure 10.10 HMR 2 (SRR-HMR) dehydrogenation energetic profile under aerobic conditions.
Reprinted from ref. 116. Copyright © 2013 WILEY-VCH Verlag GmbH & Co. KGaA, Weinheim.

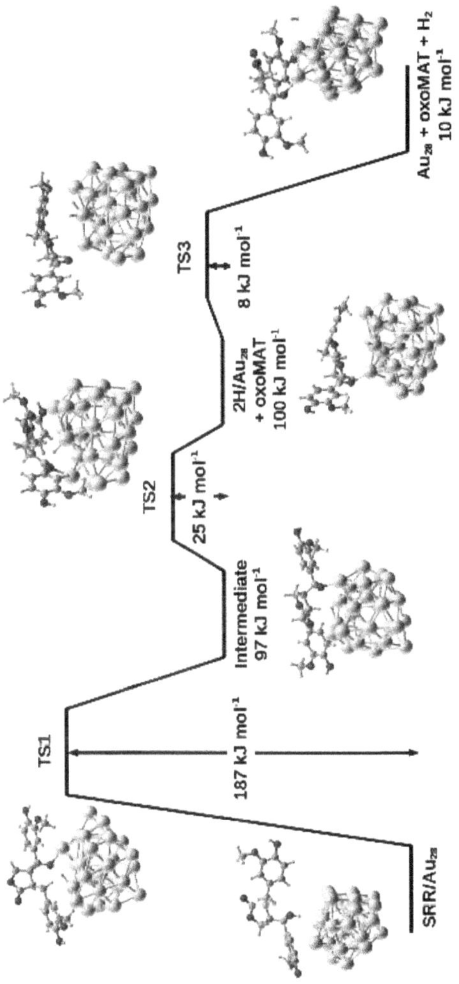

Figure 10.11 HMR 2 (SRR-HMR) dehydrogenation energetic profile under anaerobic conditions. Reprinted from ref. 116. Copyright © 2013 WILEY-VCH Verlag GmbH & Co. KGaA, Weinheim.

−173 kJ mol^{-1} for the stages leading to the formation of the first and the second oxoMAT molecules, respectively.

Under anaerobic conditions, the dehydrogenation of HMR 2 (SRR-HMR) isomer catalyzed by the Au_{28} cluster occurs *via* a mechanism different to the one when activated oxygen is co-adsorbed on the gold surface. The first step, which leaves a first hydrogen atom on the cluster surface, now involves deprotonation of the hydroxyl group and is followed by an alcoholate intermediate formation, *i.e.* the deprotonated HMR 2 (SRR-HMR) species of Figure 10.11. In the second step, the C–H hydrogen is also transferred to the gold cluster, with subsequent oxoMAT formation. Finally, one hydrogen molecule desorbs from the cluster. The activation energies for the first and second step are in the order of 187 and 25 kJ mol^{-1}, respectively. The whole reaction was found to be slightly endothermic.

Calculations underlined the role of the activated oxygen in the oxidation of HMR catalyzed by gold. It was demonstrated that in the absence of activated oxygen available on the cluster, the dehydrogenation of HMR should occur with much slower kinetics involving a much higher energy barrier. Furthermore, under aerobic conditions, the entire reaction pathway produces two molecules of oxoMAT instead of one. The dehydrogenation and oxidative dehydrogenation are found to be energetically different, with the latter reaction pathway being preferred.

10.3 Conclusions

Gold is a highly active and selective catalyst in various reactions of biomass transformation. Since gold cannot be deactivated by molecular oxygen, reactions of the selective oxidation over gold catalysts were studied extensively. Gold was showed to be active and highly selective for the oxidation of biomass-derived alcohols. Selective oxidation of the lignan hydroxymatairesinol (HMR) described in this chapter as an example of the oxidation reaction of secondary alcohols demonstrated complete selectivity to the desired product, despite the complexity of the substrate. The observed outstanding performance of gold catalysts can be applied in the synthesis of fine chemicals where high selectivity is essential.

References

1. M. Haruta, T. Kobayashi, H. Sano and N. Yamada, *Chem. Lett.*, 1987, **2**, 405.
2. Z. Hao, L. An, H. Wang and T. Hu, *React. Kinet. Catal. Lett.*, 2000, **70**, 153.
3. G. C. Bond and D. T. Thompson, *Catal. Rev. Sci. Eng.*, 1999, **41**, 319.
4. N. Lopez, T. V. P. Janssens, B. S. Claussen, Y. Xu, M. Mavrikakis, T. Bligaard and J. K. Nørskov, *J. Catal.*, 2004, **223**, 232.
5. B. Hvolbæk, T. V. W. Janssens, B. S. Claussen, H. Falsig, C. H. Christensen and J. K. Nørskov, *Nanotoday*, 2007, **2**, 14.

6. F. Boccuzzi, A. Chiorino, S. Tsubota and M. Haruta, *J. Phys. Chem. B*, 1996, **100**, 3625.
7. S. Miscino, S Scire, C. Crisafulli, A. M. Visco and S. Galvagno, *Catal. Lett.*, 1997, **44**, 273.
8. A. Sanchez, S. Abbet, U. Heiz, W.-D. Schneider, H. Hakkinen, R. N. Barnett and U. Landman, *J. Phys. Chem. A*, 1999, **103**, 9573.
9. Y. Xu and M. Mavrikakis, *J. Phys. Chem. B*, 2003, **107**, 690.
10. A. Abad, P. Conception, A. Corma and H. Garcia, *Angew. Chem., Int. Ed.*, 2005, **44**, 4066.
11. A. Corma and M. E. Domine, *Chem. Commun.*, 2005, **32**, 4042.
12. Mobil Oil Corp., *US Pat.*, 374 2079, 1973.
13. M. D. Hughes, Y.-J. Xu, P. Jenkins, P. McMorn, P. Landon, D. I. Enache, A. F. Carley, G. A. Attard, G. J. Hutchings, F. King, E. H. Stitt, P. Johnstone, K. Griffin and C. J. Kiely, *Nature*, 2005, **437**, 1132.
14. R. Zhao, D. Ji, G. Lv, G. Qian, L. Yan, X.-L. Wang and J. Suo, *Catal. Lett.*, 2004, **97**, 115.
15. M. Comotti, C. Della Pina, R. Matarrese and M. Rossi, *Angew. Chem., Int. Ed.*, 2004, **43**, 5812.
16. F. Porta and L. Prati, *J. Catal.*, 2004, **224**, 397.
17. S. Carrettin, P. McMorn, P. Johnstone, K. Griffin and G. J. Hutchings, *Chem. Commun.*, 2002, 696.
18. B. E. Salisbury, W. T. Wallance and R. L. Whetten, *Chem. Phys.*, 2000, **262**, 131.
19. Report of the United Nations Conference on Environment and Development, Rio de Janeiro, 3–14 June 1992, http://www.un.org/esa/dsd/agenda21/Agenda%2021.pdf.
20. American Chemical Society, American Insitute of Chemical Engineers, Chemical Manufacturers Association, Council for Chemical Research and Synthetic Organic Chemical Manufacturers Association, *Technology Vision 2020, The US Chemical Industry*, American Chemical Society, Washington DC, 1996; http://energy.gov/sites/prod/files/2013/11/f4/chem_vision.pdf.
21. The European Chemical Industry Council (CEFIC), www.cefic.org.
22. P. Mäki-Arvela, B. Holmbom, T. Salmi and D. Yu. Murzin, *Catal. Rev.*, 2007, **49**, 197.
23. A. V. Tokarev, E. V. Murzina, J.-P. Mikkola, J. Kuusisto, L. M. Kustov and D. Yu. Murzin, *Chem. Eng. J.*, 2007, **134**, 153.
24. E. V. Murzina, A. V. Tokarev, K. Kordas, H. Karhu, J.-P. Mikkola and D. Yu. Murzin, *Catal. Today*, 2008, **131**, 385.
25. D. Yu. Murzin, E. V. Murzina, A. V. Tokarev and J.-P. Mikkola, *Russ. J. Electrochem.*, 2009, **45**, 1017.
26. A. Mirescu and U. Prüβe, *Appl. Catal. B*, 2007, **70**, 644.
27. B. Kusema, B. C. Campo, P. Maki-Arvela, T. Salmi and D. Yu. Murzin, *Appl. Catal. A*, 2010, **386**, 101.
28. A. Corma, S. Ibora and A. Velty, *Chem. Rev.*, 2007, **107**, 2411.
29. C. Baatz and U. Prüβe, *Catal. Today*, 2007, **122**, 325.

30. U. Prüße, M. Herrmann, C. Baatz and N. Decker, *Appl. Catal. A*, 2011, **406**, 89.
31. T. Ishida, S. Okamoto, R. Makiyama and M. Haruta, *Appl. Catal. A*, 2009, **353**, 243.
32. T. Benkó, A. Beck, O. Geszti, R. Katona, A. Tungler, K. Frey, L. Guczi and Z. Schay, *Appl. Catal. A*, 2010, **388**, 31.
33. O. A. Simakova, B. T. Kusema, B. C. Campo, A.-R. Leino, K. Kordas, V. Pitchon, P. Mäki-Arvela and D. Yu. Murzin, *J. Phys. Chem. C*, 2011, **115**, 1036.
34. B. T. Kusema, B. C. Campo, O. A. Simakova, A.-R. Leino, K. Kordas, P. Mäki-Arvela and D. Yu. Murzin, *ChemCatChem*, 2011, **3**, 1789.
35. L.-F. Gutierrez, S. Hamoudi and K. Belkacemi, *Appl. Catal. A*, 2011, **402**, 94.
36. J. N. Cheda, G. W. Huber and J. A. Dumesic, *Angew. Chem., Int. Ed.*, 2007, **46**, 7164.
37. N. Thielecke, M. Aytemir and U. Prüße, *Catal. Today*, 2007, **121**, 115.
38. E. Taarning, I. S. Nielsen, K. Egeblad, R. Madsen and C. H. Christensen, *ChemSusChem*, 2008, **1**, 75.
39. Y. Y. Gorbanev, S. K. Klitgaard, J. M. Woodley, C. H. Christensen and A. Riisager, *ChemSusChem*, 2009, **2**, 672.
40. O. Casanova, S. Iborra and A. Corma, *ChemSusChem*, 2009, **2**, 1138.
41. S. E. Davis, L. R. Houk, E. C. Tamargo, A. K. Datye and R. J. Davis, *Catal. Today*, 2011, **160**, 55.
42. S. Carrettin, P. McMorn, P. Johnston, K. Griffin, C. J. Kiely and G. J. Hutchings, *Phys. Chem. Chem. Phys.*, 2003, **5**, 1329.
43. L. Prati and M. Rossi, *J. Catal.*, 1998, **176**, 552.
44. C. Bianchi, F. Porta, L. Prati and M. Rossi, *Top. Catal.*, 2000, **13**, 231.
45. S. Biella, L. Prati and M. Rossi, *Inorg. Chem. Acta*, 2003, **349**, 253.
46. C. Bianchi, S. Biella, A. Gervasini, L. Prati and M. Rossi, *Catal. Lett.*, 2003, **85**, 91.
47. F. Porta and L. Prati, *J. Catal.*, 2004, **224**, 397.
48. S. Demirel-Gülen, M. Lucasv and P. Claus, *Catal. Today*, 2005, **102–103**, 166.
49. L. Guczi, *Catal. Today*, 2005, **53**, 53.
50. M. Chen, D. Kumar, C. W. Yi and D. W. Goodman, *Science*, 2005, **310**, 291.
51. C. L. Bianchi, P. Canton, N. Dimitratos, F. Porta and L. Prati, *Catal. Today*, 2005, **102–103**, 203.
52. D. Wang, A. Villa, F. Porta, D. Su and L. Prati, *Chem. Commun.*, 2006, 1956.
53. N. Dimitratos, C. Messi, F. Porta, L. Prati and A. Villa, *J. Mol. Catal. A*, 2006, **256**, 21.
54. W. C. Ketchie, M. Murayama and R. J. Davis, *Top. Catal.*, 2007, **44**, 307.
55. W. C. Ketchie, Y.-L. Fang, M. S. Wong, M. Murayama and R. J. Davis, *J. Catal.*, 2007, **250**, 94.
56. S. Demirel, K. Lehnert, M. Lucas and P. Claus, *Appl. Catal. B*, 2007, **70**, 637.

57. L. Prati, A. Villa, F. Porta, Di Mang and D. Su, *Catal. Today*, 2007, **122**, 386.
58. W. C. Ketchie, M. Murayama and R. J. Davis, *J. Catal.*, 2007, **250**, 264.
59. N. Dimitratos, A. Villa and L. Prati, *Catal. Lett.*, 2009, **133**, 334.
60. S. D. Pollington, D. I. Enache, P. Landon, S. Meenakshisundaram, N. Dimitratos, A. Wagland, G. J. Hutchings and E. H. Stitt, *Catal. Today*, 2009, **145**, 169.
61. B. N. Zope, D. D. Hibbitts, M. Neurock and R. J. Davis, *Science*, 2010, **330**, 74.
62. I. Sobczak, K. Jagodzinska and M. Ziolek, *Catal. Today*, 2010, **158**, 121.
63. E. G. Rodrigues, S. A. C. Carabineiro, J. J. Delgado, X. Chen, M. F. R. Pereira and J. J. M. Órfão, *J. Catal.*, 2012, **285**, 83.
64. C. H. Christensen, B. Jørgensen, J. Rass-Hansen, K. Egeblad, R. Madsen, S. K. Klitgaard, S. M. Hansen, M. R. Hansen, H. C. Andersen and A. Riisager, *Angew. Chem., Int. Ed.*, 2006, **45**, 4648.
65. B. Jørgensen, S. E. Christiansen, M. L. D. Thomsen and C. H. Christensen, *J. Catal.*, 2007, **251**, 332.
66. X. Li and E. Iglesia, *Chem. Eur. J.*, 2007, **13**, 9324.
67. T. Mallat and A. Baiker, *Chem. Rev.*, 2004, **104**, 3037.
68. B. Jørgensen, S. B. Kristensen, A. J. Kunov-Kruse, R. Fehrmann, C. H. Christensen and A. Riisager, *Top. Catal.*, 2009, **52**, 253.
69. C. Della Pina, E. Falletta, L. Prati and M. Rossi, *Chem. Soc. Rev.*, 2008, **37**, 2077.
70. A. Corma and H. Garcia, *Chem. Soc. Rev.*, 2008, **37**, 2096.
71. A. S. K. Hashmi, *Chem. Rev.*, 2007, **107**, 3180.
72. A. S. K. Hashmi and G. J. Hutchings, *Angew. Chem., Int. Ed.*, 2006, **45**, 7896.
73. O. A. Simakova, V. I. Sobolev, K. Yu. Koltunov, B. Campo, A.-R. Leino, K. Kordas and D. Yu. Murzin, *ChemCatChem*, 2010, **2**, 1535.
74. V. I. Sobolev, O. A. Simakova and K. Y. Koltunov, *ChemCatChem*, 2011, **3**, 1422.
75. V. I. Sobolev, K. Yu. Koltunov, O. A. Simakova, A.-R. Leino and D. Yu. Murzin, *Appl. Catal. A*, 2012, **433–434**, 88.
76. D. Saio, T. Amaya and T. Hirao, *Adv. Synth. Catal.*, 2010, **352**, 2177.
77. R. Ekman, *Holzforschung*, 1976, **30**, 79.
78. S. Willför, P. Eklund, R. Sjöholm, M. Reunanen, R. Sillanpää, S. von Schoultz, J. Hemming, L. Nisula and B. Holmbom, *Holzforschung*, 2005, **59**, 413.
79. P. C. Eklund, S. Willför, A. Smeds, B. R. Holmbom and R. E. Sjöholm, *J. Nat. Prod.*, 2004, **67**, 927.
80. S. Willför, J. Hemming, M. Reunanen, C. Eckerman and B. Holmbom, *Holzforschung*, 2003, **57**, 27.
81. S. M. Willför, M. O. Ahotupa, J. E. Hemming, M. H. T. Reunanen, P. C. Eklund, R. E. Sjöholm, C. S. E. Eckerman, S. P. Pohjamo and B. R. Holmbom, *J. Agr. Food Chem.*, 2003, **51**, 7600.

82. S. Yamaguchi, T. Sugahara, Y. Nakashima, A. Okada, K. Akiyama, T. Kishida, M. Maruyama and T. Masuda, *Biosci. Biotechnol. Biochem.*, 2006, **70**, 1934.
83. P. C. Eklund, O. K. Långvik, J. P. Wärnå, T. O. Salmi, S. M. Willför and R. E. Sjöholm, *Org. Biomol. Chem.*, 2005, **3**, 3336.
84. H. Korte, M. Unkila, M. Hiilovaara-Teijo, V.-M. Lehtola, M. Ahotupa, *World Pat.*, WO/2004/000304 A1, 2003.
85. T. Ahlnäs, *World Pat.*, WO 2008/020112 A1, 2008.
86. F. Kawamura, M. Miyachi and S. Kaway, *J. Wood Sci.*, 1998, **44**, 47.
87. P. Eklund, A. Lindholm, J.-P. Mikkola, A. Smeds, R. Lehtilä and R. Sjöholm, *Org. Lett.*, 2003, **5**, 491.
88. H. Markus, A. J. Plomp, T. Sandberg, V. Nieminen, J. H. Bitter and D. Yu. Murzin, *J. Mol. Catal. A.*, 2007, **274**, 42.
89. O. A. Simakova, E. V. Murzina, P. Mäki-Arvela, A.-R. Leino, B. C. Campo, K. Kordas, S. Willför, T. Salmi and D. Yu. Murzin, *J. Catal.*, 2011, **282**, 54.
90. H. Liang, H. Gao, Q. Kong and Z. Chen, *J. Chem. Eng. Data*, 2006, **51**, 119.
91. Y. Guan and E. J. M. Hensen, *Appl. Catal. A*, 2009, **361**, 49.
92. A. Abad, C. Amela, A. Corma and H. Garcia, *Chem. Commun.*, 2006, **30**, 3178.
93. M. Daturi, E. Finocchio, C. Binet, J. C. Lavalley, F. Fally, V. Perrichon, H. Vidal, N. Hickey and J. Kaspar, *J. Phys. Chem. B*, 2000, **104**, 9186.
94. M. Zahmakiran and S. Ozkar, *Mater. Chem. Phys.*, 2010, **121**, 359.
95. E. Smolentseva, N. Bogdanchikova, A. Simakov, A. Pestryakov, I. Tuzovskaya, M. Avalos, M. H. Farias, J. A. Diaz and V. Gurin, *Surf. Sci.*, 2006, **600**, 4256.
96. E. Smolentseva, N. Bogdanchikova, A. Simakov, A. Pestryakov, M. Avalos, M. H. Farias, A. Tompos and V. Gurin, *J. Nanosci. Nanotechnol.*, 2007, **7**, 1882.
97. A. Simakov, I. Tuzovskaya, A. Pestryakov, N. Bogdanchikova, V. Gurin, M. Avalos and M. H. Farias, *Appl. Catal. A*, 2007, **331**, 121.
98. O. A. Simakova, E. Smolentseva, M. Estrada, E. V. Murzina, S. Beloshapkin, S. M. Willför, A. V. Simakov and D. Yu. Murzin, *J. Catal.*, 2012, **291**, 95.
99. M. Digne, P. Sautet, P. Raybaud, P. Euzen and H. Toulhoat, *J. Catal.*, 2004, **226**, 54.
100. Z. Martinez-Ramirez, S. A. Jimenez-Lam and J. C. Fierro-Gonzalez, *J. Mol. Catal. A*, 2011, **344**, 47.
101. M. Boronat, A. Corma, F. Illas, J. Radilla, T. Rodenas and M. J. Sabater, *J. Catal.*, 2011, **278**, 50.
102. O. A. Simakova, E. V. Murzina, A.-R. Leino, P. Mäki-Arvela, S. Willför and D. Yu. Murzin, *Catal. Lett.*, 2012, **142**, 1011.
103. A. Abad, C. Almela, A. Corma and H. Garcia, *Tetrahedron*, 2006, **62**, 6666.

104. H. Li, B. Guan, W. Wang, D. Xing, Z. Fang, X. Wan, L. Yang and Z. Shi, *Tetrahedron*, 2007, **63**, 8430.
105. L.-C. Wang, L. He, Q. Liu, Y.-M. Liu, M. Chen, Y. Cao, H.-Y. He and K.-N. Fan, *Appl. Catal. A*, 2008, **344**, 150.
106. X. Wang, H. Kawanami, S. E. Dapurkar, N. S. Venkataramanan, M. Chatterjee, T. Yokoyama and Y. Ikushima, *Appl. Catal. A*, 2008, **349**, 86.
107. M. Kidwai and S. Bhardwaj, *Appl. Catal. A*, 2010, **387**, 1.
108. O. A. Simakova, E. V. Murzina, J. Warna and D. Yu. Murzin, *J. Mol. Catal. A*, 2014, **388–389**, 154.
109. H. Haario, *MODEST 6.0 – User's Guide*, Profmath Oy, Helsinki, 2001.
110. D. Yu. Murzin and N. V. Kul'kova, *Catal. Today*, 1995, **24**, 35.
111. X. Yang, X. Wang, C. Liang, W. Su, C. Wang, Z. Feng, C. Li and J. Qiu, *Catal. Commun.*, 2008, **9**, 2278.
112. B. N. Zope, D. D. Hibbitts, M. Neurock and R. J. Davis, *Science*, 2010, **330**, 74.
113. B. K. Min and C. M. Friend, *Chem. Rev.*, 2007, **107**, 2709.
114. S. E. Davis, B. N. Zope and R. J. Davis, *Green Chem.*, 2012, **14**, 143.
115. O. A. Simakova, E. V. Murzina, D. Yu. Murzin, *Comptes Rendus Chimie*, DOI: 10.1016/j.crci.2014.01.002.
116. A. Prestianni, F. Ferrante, O. A. Simakova, D. Duca and D. Yu. Murzin, *Eur. J. Chem*, 2013, **19**, 4577.

CHAPTER 11

Harnessing the Selective Catalytic Action of Supported Gold in Hydrogenation Applications

XIAODONG WANG, MAOSHUAI LI AND MARK A. KEANE*

Chemical Engineering, School of Engineering & Physical Sciences, Heriot-Watt University, Edinburgh, EH14 4AS, Scotland
*Email: M.A.Keane@hw.ac.uk

11.1 Introduction: Gold as a Hydrogenation Catalyst

Gold finds diverse uses that cut across the automotive, electronics, medicine, nanotechnology, space and engineering sectors.[1–3] Catalysis by gold is now a priority growth area in terms of technological applications.[4] This has resulted in a number of review articles that have considered different aspects of gold catalysis.[5–19] Bond and co-workers have provided an overview of the status of related research in 1999 and 2006,[5,6] focusing on physicochemical properties, synthesis, characterisation and applications, notably in CO oxidation and the water gas shift reaction. In a more recent article, Bond highlighted the reactivity of gold at the nanoscale in activating small molecules.[20] A number of reviews have examined the catalytic action of gold in selective oxidation and we can flag critical assessments by Hutchings, Edwards and co-workers (catalyst preparation and characterisation),[7] Chen and Goodman (use of Au/TiO_2 as model catalyst),[8] Rossi and co-workers

RSC Catalysis Series No. 18
Heterogeneous Gold Catalysts and Catalysis
Edited by Zhen Ma and Sheng Dai

Published by the Royal Society of Chemistry, www.rsc.org

(selective oxidation of organic compounds),[9] Gong (surface chemistry)[10] and Li and Jin (action of gold nanoclusters).[19] Beyond oxidation, we note comprehensive treatments of organic/total synthesis *via* homogeneous and heterogeneous Au catalysis by Corma and Garcia[11] and Hashmi and Rudolph.[12] Gold catalysis has also been the subject of theoretical and modelling/simulation work, which has been evaluated by Willock and co-workers[13] (application of quantum chemistry) Pyykkö,[14] and Soriano and Marco-Contelles[21] (computational methods).

Gold utilisation in hydrogenation has been driven by the enhanced selectivity exhibited in the conversion of functionalised nitroarenes, unsaturated aldehydes and ketones, acetylene, propyne and 1,3-butadiene.[4,15,16,22,23] We can highlight reviews by Cárdenas-Lizana and Keane (hydrogenation of nitro-compounds),[15] McEwan *et al.*[16] and Claus[17] (hydrogen–gold interactions), Mitsudome and Kaneda[22] (liquid phase hydrogenation) and Mullins and co-workers (model experiments and theoretical calculations).[18] Although bulk gold is inactive, when supported at the nanoscale (<10 nm) on a carrier, gold displays significant hydrogenation activity. Nevertheless, the associated rates are significantly lower than those delivered by conventional transition metal (*e.g.* Pt, Pd and Ni) catalysts, a response that has been attributed to less effective H_2 activation/dissociation.[17,23] Dissociative adsorption is dependent on gold site characteristics, notably particle size and electronic character, that are, in turn, influenced by metal–support interactions. Hydrogen chemisorption is favoured on low coordinated corner and edge sites and is enhanced with decreasing gold particle size,[17,23] as confirmed by theoretical calculations.[24,25] Moreover, H_2 activation is an activated process where uptake is increased at higher temperatures (298–523 K).[23,26–29]

To date, a wide range of supported gold catalysts have been investigated for use in hydrogenation. We provide a compilation of relevant literature since 2009 in Table 11.1, wherein we identify the catalyst and reactants that have been studied.[26,27,30–101] The majority of this work was conducted in batch liquid phase with fewer studies of continuous gas phase operation.[26,27,32,39,40,42,45]

Taking an overview of the publications given in Table 11.1, the consensus[26,38,50–52,82] is that smaller gold particles exhibit intrinsically higher activity. The gold–support interface and Au electronic structure play crucial roles in determining catalyst performance.[25,39,50,62,89] Enhanced activity while maintaining elevated hydrogenation chemoselectivity has been achieved by employing noble metals (*e.g.* Pd,[95,102] Pt[41]) as promoters. Catalyst deactivation has been linked to Au sintering,[103] carbon deposition[46,103] and poisoning by chlorine or sulfur.[39,104]

While high hydrogenation selectivity is established, the low associated reaction rates still represent a decided drawback. Moreover, there is significant scope to extend the range of functionalised reactants to fully exploit the catalytic action of gold. Work conducted in this laboratory, which forms the basis for this chapter, has been directed at correlating the supported

Table 11.1 Compilation of the literature (since 2009) dealing with hydrogenation reactions over supported Au catalysts.

Reactant	Catalyst
Nitrobenzene	Au/TiO_2,[30–33] Au/Al_2O_3,[30] Au/SiO_2,[31,34] Au/ZrO_2,[35] Au/CeO_2[36]
1-Methoxy-*2*-nitrobenzene	Au/Fe$(OH)_x$[37]
p-Chloronitrobenzene	Au/Al_2O_3[27,38–40], Au/$Ce_{0.62}Zr_{0.38}O_2$,[39] Au/CeO_2,[38] Au/TiO_2,[32,38,41] Au/β-Mo_2C,[40] Au/β-Mo_2C/Al_2O_3,[40] Au/Mo_2N,[42,43] Au/SiO_2,[34,44] Au/Fe_2O_3,[38,45] Au/Fe_3O_4[45]
o-Chloronitrobenzene	Au/TiO_2,[46] Au/Fe$(OH)_x$[37]
m-Chloronitrobenzene	Au/SiO_2,[34] Au/Fe_2O_3,[45] Au/Fe_3O_4[45]
m-Dinitrobenzene	Au/Al_2O_3,[47–50] Au/TiO_2,[50,51] Au/β-Mo_2C,[40] Au/β-Mo_2C/Al_2O_3,[40] Au/SiO_2,[34] Au/Fe_2O_3,[50] Au/CeO_2[50]
1,3,5-Trinitrobenzene	Au/Al_2O_3[47]
p-Nitrostyrene	Au/Al_2O_3,[52] Au/SiO_2,[52] Au/MgO,[52] Au/C,[52] Au/montmorillonite[52]
p-Nitroacetophenone	Au/Fe$(OH)_x$,[37] Au/Fe_2O_3[37]
p-Nitrobenzaldehyde	Au/Fe$(OH)_x$,[37] Au/TiO_2[53]
p-Bromonitrobenzene	Au/TiO_2,[32] Au/Fe$(OH)_x$[37]
p-Nitroaniline	Au/TiO_2[32]
p-Nitrotoluene	Au/TiO_2,[32] Au/SiO_2[34]
p-Nitrophenol	Au/TiO_2,[32] Au/Fe_3O_4[54]
p-Nitroanisole	Au/TiO_2[32]
Nitrocyclohexane	Au/Al_2O_3,[26,55] Au/SiO_2,[55] Au/MgO,[55] Au/TiO_2,[26,55] Au/CeO_2,[26] Au/ZrO_2[26]
Citral	Au/TiO_2,[56–58] Au/Nb_2O_5,[56] Au/Al_2O_3[59]
Phthalic anhydride	Au/TiO_2,[60] Au/FeO_x-TiO_2[61]
Cinnamaldehyde	Au/SiO_2,[44,62,63] Au/TiO_2,[63,64] Au/Mg_xAlO-hydrotalcite,[65] Au/CeO_2,[66] Au/CNT,[67] Au/ZnO,[68] Au/Al_2O_3,[69] Au/Fe_2O_3,[69] Au/C[70]
Crotonaldehyde	Au/TiO_2,[71,72] Au/Mg_2AlO-hydrotalcite,[73] Au/CeO_2,[74,75] Au/Al_2O_3,[69] Au/Fe_2O_3,[69] Au/SiO_2[76]
Benzalacetone	Au/SiO_2[62]
Benzaldehyde	Au/Al_2O_3,[77,78] Au/CeO_2,[75,78] Au/TiO_2,[78] Au/Fe_2O_3,[78] Au/C[78]
Acrolein	Au/ZnO[79]
Phenylacetylene	Au/Al_2O_3[80–82]
Acetylene	Au/ZnO,[83] Au/ZrO_2,[83] Au/SiO_2,[84–86] Au/Al_2O_3,[83,87] Au/FeO_x,[88] Au/TiO_2[83,85]
Propylene	Au/TS-1[89]
1-Heptyne	Au/TiO_2[90]
1,3-Butadiene	Au/ZrO_2,[91] Au/CeO_2,[92] Au/CNT,[67] Au/SiO_2,[93] Au/Al_2O_3,[94,95] Au/TiO_2[94]
2-Hydroxymethyl-5-furfural	Au/Al_2O_3[96]
Quinolines	Au/TiO_2[97]
Styrene	Au/CeO_2[75]
Biphenyl	Au/metallosilicates[98]
Carbon dioxide	Au/TiO_2,[99,100] Au/TiC[101]

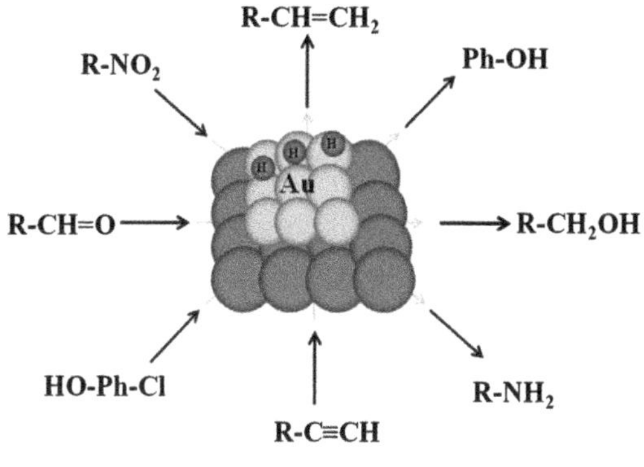

Figure 11.1 Examples of chemoselective hydrogenation (and hydrogenolysis) over supported Au catalysts discussed in the chapter.

Au surface structure with hydrogenation chemoselectivity in the gas phase conversion of a range of functional groups (R–NO_2 to R–NH_2, R–C≡CH to R–CH=CH_2, R–CH=O to R–CH_2OH), as illustrated in Figure 11.1. We examine each of these systems in turn, presenting key aspects of our catalysis tests that we couple with characterisation measurements and a critical analysis of the relevant literature. We address process variables (temperature, contact time and H_2 partial pressure), draw on reaction thermodynamics and kinetics, and consider overall process sustainability (mode of operation, solvent usage, H_2 utilisation and possible reaction coupling) in terms of practical application.

11.2 Selective Hydrogenation of Nitro-Compounds

Amines are extensively used as intermediates in the manufacture of fine chemicals, pharmaceuticals and agricultural products. Standard production routes based on Fe-promoted reduction in acid media (the Béchamp reaction) suffer from major drawbacks, notably low product yields and the generation of significant quantities of toxic Fe/FeO waste sludge.[105] Catalytic hydrogenation of nitro-compounds can serve as a viable alternative where selective –NO_2 reduction in functionalised nitro-reactants is challenging. In this section, we assess gold chemoselectivity in the generation of target amine products.

11.2.1 Hydrogenation of *p*-Chloronitrobenzene

The possible reaction pathways in the conversion of *p*-chloronitrobenzene (*p*-CNB) are identified in Figure 11.2. Aniline can result from –NO_2 reduction (step A) with subsequent hydrodechlorination (step D) or *via* sequential hydrodechlorination (to nitrobenzene, step B) and hydrogenation (step E). Hydrodeamination of *p*-chloroaniline (*p*-CAN) generates chlorobenzene

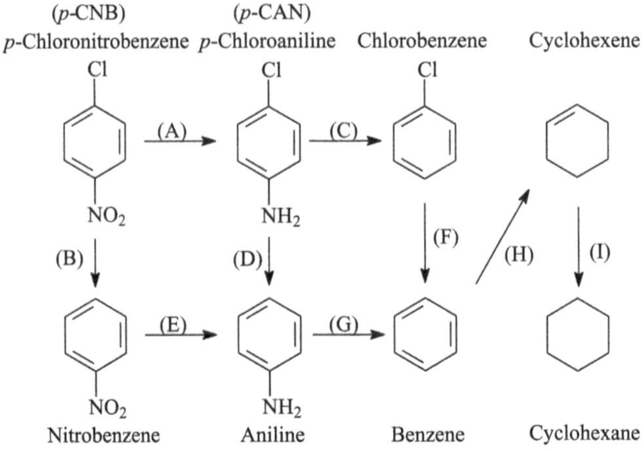

Figure 11.2 Reaction pathways in the hydrogenation of *p*-chloronitrobenzene (*p*-CNB).

(step C), while benzene can be formed *via* hydrodechlorination (step F) or hydrodeamination (step G). Subsequent benzene hydrogenation generates cyclohexene and cyclohexane (steps H and I).

Chemical thermodynamics provide a measure of the maximum conversion/selectivity under given reaction conditions and is an essential consideration in the design of a catalytic process. Thermodynamic calculations have established complete *p*-CNB conversion under equilibrium conditions at 373–473 K and inlet H_2/*p*-CNB from 4 (stoichiometric ratio for production of aniline) to 50 (H_2 far in excess).[27] Product distribution is invariant at H_2/*p*-CNB ≥ 10 where cyclohexane is the sole product, indicating that there is no thermodynamic barrier for full hydrogenation (with hydrogenolysis) of *p*-CNB.

Reaction selectivity (S %) under thermodynamic control (at 423 K) as a function of H_2/*p*-CNB can be assessed from the entries in Figure 11.3. *p*-CAN is the only product where H_2/*p*-CNB ≤ 3, which is the stoichiometric requirement for path A in Figure 11.2. Nitrobenzene is not formed at equilibrium even though H_2/*p*-CNB exceeds stoichiometry for path B. An increase in H_2/*p*-CNB limits *p*-CAN selectivity, favouring aniline (path D) with chlorobenzene (path C) as secondary product. The hydrogenolysis products pass through a selectivity maximum at H_2/*p*-CNB $= 4$ with exclusive formation of benzene at H_2/*p*-CNB $= 5$. Benzene hydrogenation is favoured at H_2/*p*-CNB >5 with sole formation of cyclohexane.

The catalytic response for Au/Al_2O_3 and Pd/Al_2O_3 is given in Table 11.2, in terms of conversion (X) and product selectivity (S) with varying temperature at a fixed H_2/*p*-CNB (=4). We should note that Pd/Al_2O_3 delivered an order of magnitude higher turnover frequency (TOF = 1260 h^{-1}) than Au/Al_2O_3 (137 h^{-1}) under the same reaction conditions.[27] Activity was enhanced at higher temperatures, but remained far below the complete conversion

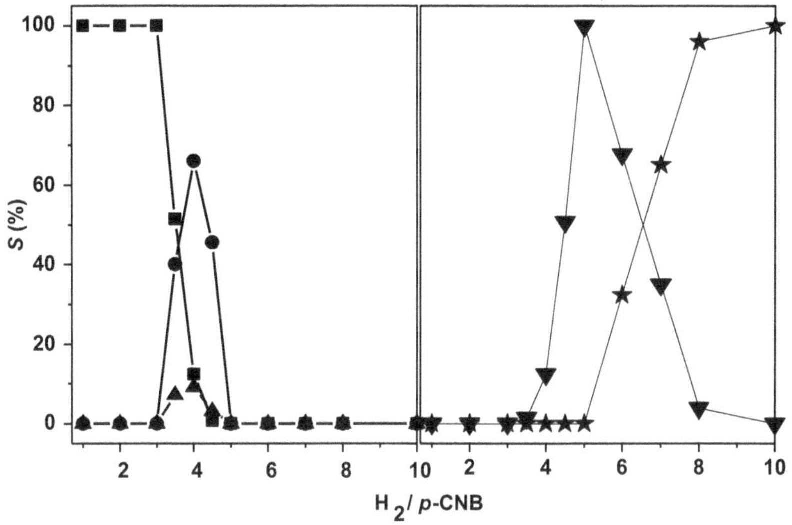

Figure 11.3 Thermodynamic analysis of product selectivity in the hydrogenation of *p*-CNB at equilibrium as a function of inlet H_2/*p*-CNB at 423 K: *p*-CAN (■); aniline (●); chlorobenzene (▲); benzene (▼); cyclohexane (★).

Table 11.2 Effect of reaction temperature on fractional conversion (*X*) and product selectivity (*S*) in the hydrogenation of *p*-chloronitrobenzene (*p*-CNB) over Au/Al_2O_3 and Pd/Al_2O_3: H_2/*p*-CNB = 4.

	Au/Al_2O_3		Pd/Al_2O_3	
Temperature (K)	*X*	Product (*S*/%)	*X*	Product(s) (*S*/%)
373	0.01	*p*-CAN (100)	0.02	Nitrobenzene (100)
423	0.12	*p*-CAN (100)	0.13	Nitrobenzene (82), Aniline (18)
473	0.18	*p*-CAN (100)	0.17	Nitrobenzene (74), Aniline (26)

achieved at the thermochemical equilibrium. Reaction over Au/Al_2O_3 generated *p*-CAN as sole product at each temperature. In contrast, Pd/Al_2O_3 exhibited hydrodechlorination exclusivity to nitrobenzene at 373 K where subsequent hydrogenation to aniline (path E in Figure 11.2) was promoted at higher temperatures. The predominance of *p*-CNB hydrodechlorination coupled with hydrogenation over Pd/Al_2O_3 suggests surface interaction that activates both C–Cl for hydrogenolytic attack and $-NO_2$ for hydrogen addition. Indeed, Pd/Al_2O_3 is established as an effective hydrodechlorination catalyst with significant chloroarene hydrodechlorination even at sub-ambient temperatures.[106] The hydrogenation of *p*-CNB over Au on a range of oxide supports (Al_2O_3, TiO_2, Fe_2O_3 and CeO_2) resulted in 100% selectivity to *p*-CAN.[38]

The dependence of specific *p*-CNB hydrogenation rate on Au particle size is presented in Figure 11.4(a), where Au size was determined by electron

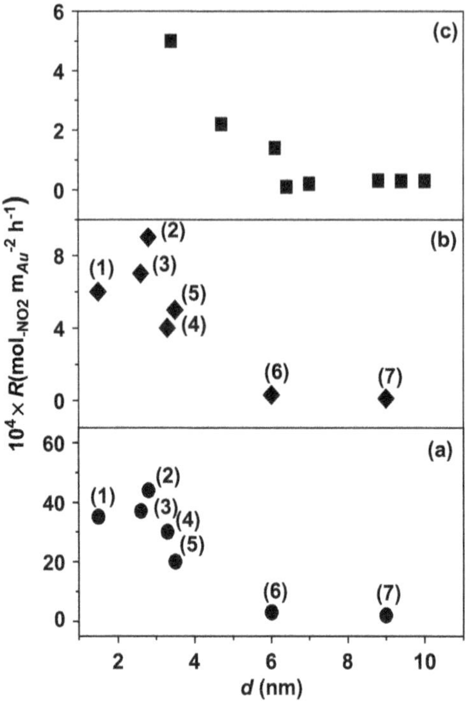

Figure 11.4 Dependence of selective $-NO_2$ hydrogenation rate on mean Au particle size (*d*) in the hydrogenation of (a) *p*-chloronitrobenzene (*p*-CNB) and (b) *m*-dinitrobenzene (*m*-DNB) at 423 K over (1) Au/CeO_2, (2) Au/Al_2O_3, (3) Au/Fe_2O_3 and (4) Au/TiO_2 prepared by deposition/precipitation, (5) Au/Fe_2O_3, (6) Au/TiO_2 and (7) Au/Al_2O_3 prepared by impregnation, and (c) hydrogenation of *m*-DNB at 473 K over Au/TiO_2 of varying rutile content.

microscopy (see Figure 11.5 for representative images). It can be seen that the rate increased with decreasing particle size (from 9 to 3 nm), which was independent of support. This can be correlated with greater availability of surface reactive hydrogen where dissociative H_2 chemisorption is favoured with decreasing Au particle size (see Section 11.1). The lower specific rate over particles <2 nm is attributed to a quantum size effect.[38]

The role of support redox character was considered by comparing the performance of Au on non-reducible Al_2O_3 with the reducible mixed oxide $Ce_{0.62}Zr_{0.38}O_2$ (CZ) as carrier.[39] An increase in activation temperature (from 473 to 673 K) resulted in Au sintering on Al_2O_3 (Table 11.3), as noted in the literature.[107] In contrast, Au size on CZ was insensitive to temperature, exhibiting a smaller mean (1.8–1.9 nm) than Au/Al_2O_3, suggesting stronger interactions with CZ that anchor the Au phase and minimise particle growth. Treatment of CZ in a reducing atmosphere generates oxygen vacancies[108,109] that are known to stabilise transition metal nanoparticles and inhibit sintering.[110]

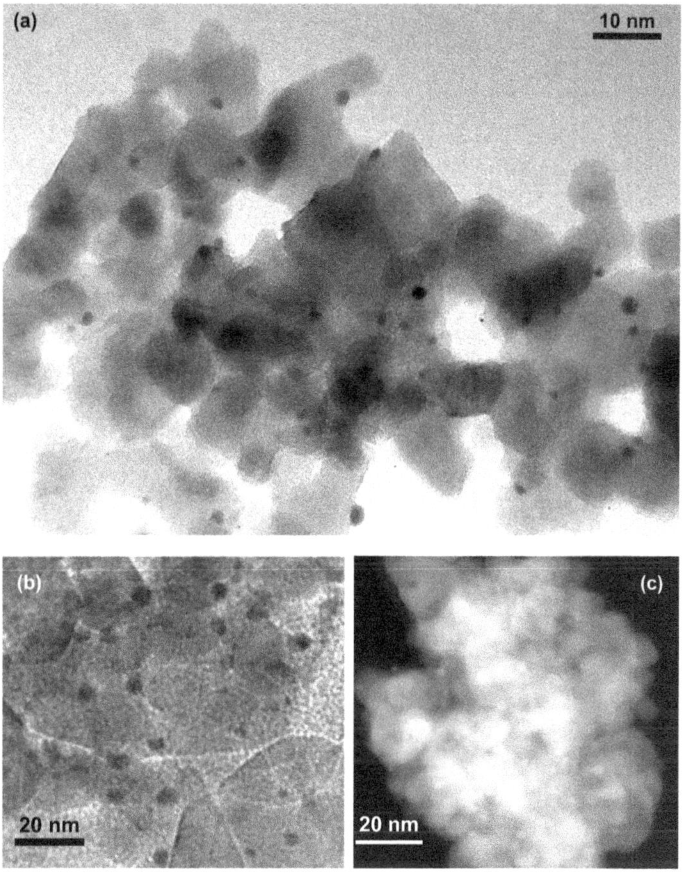

Figure 11.5 Representative TEM/STEM images of: (a) Au/Al_2O_3; (b) Au/TiO_2; and (c) Au/CeO_2.

Oxygen vacancy generation is demonstrated by the oxygen storage capacity (OSC) measurements presented in Table 11.3, with increasing Ce^{3+} content at higher temperature. The H_2 content in the reducing gas stream (from 5 to 100% v/v) had no influence on Au size or degree of CZ reduction. The *p*-CNB TOF delivered by Au/Al_2O_3 was unaffected by the activation gas composition (see 100 – 5 *vs.* 5 – 5 in Table 11.3), but catalyst performance was strongly influenced by reactant gas composition (100 – 100 *vs.* 100 – 5). We can associate this with the low capacity of Au to dissociate H_2, where higher H_2 content in the feed must serve to increase surface hydrogen with a consequent elevation of TOF. Reduction of Au/Al_2O_3 at 673 K, resulting in Au sintering, delivered an appreciably lower TOF with exclusive production of *p*-CAN. Reaction over Au/CZ was also accompanied by higher TOF with increased H_2 content in the feed, but the effect was not as marked as with Au/Al_2O_3. Reduction of Au/CZ in a H_2 rich stream or increasing temperature (at 5 – 5) generated lower TOF values. This can be ascribed to variations in

Table 11.3 Influence of support redox character [Au on Al_2O_3 *vs.* $Ce_{0.62}Zr_{0.38}O_2$ (CZ)] on *p*-chloronitrobenzene (*p*-CNB) hydrogenation performance at 393 K.

Catalyst	T^a/K	Au size/nm	H_2 % v/v activation-reaction	OSC^b/ Ce^{3+} %	S^c/%	S^c *p*-CAN	S^c Aniline
Au/Al_2O_3	473	5.7	100–100	–	315	100	–
	473	5.7	100–5	–	82	100	–
	473	5.7	5–5	–	80	100	–
	673	8.8	5–5	–	29	100	–
Au/CZ	393	1.8	100–100	12	34	84	16
	393	1.8	100–5	12	28	43	57
	393	1.8	5–5	12	37	23	77
	473	1.9	5–5	21	22	10	90
	673	1.9	5–5	33	14	10	90

[a]Activation temperature.
[b]Oxygen storage capacity.
[c]Product selectivity at 10% *p*-CNB conversion.

gold–support interaction (coincident with the formation of oxygen vacancies) that serve to modify adsorptive properties and limit activity of Au.[111] The higher TOF values recorded for Au/Al_2O_3 indicate a contribution due to the support by facilitating H_2 dissociation (to H^+/H^-) at the Au-Al_2O_3 interface[52,55] and accommodating spillover hydrogen.[27] In contrast to Au/Al_2O_3, Au/CZ generated *p*-CAN and aniline as products, where support oxygen vacancies facilitate C–Cl activation for hydrogenolytic attack. It has been established that, under reducing conditions, Cl is readily incorporated into ceria[112] and strong Cl interactions have been established for metals on reducible supports,[113–116] *i.e.* Rh/CeO_2, Pd/CeO_2 and Rh/CZ. X-ray photon spectroscopy (XPS) analysis demonstrated significant residual Cl associated with Au/CZ post reaction.[39] Increased selectivity to aniline over Au/CZ was observed for reaction in a H_2 lean feed and dechlorination was enhanced with increased support reduction (higher OSC, see Table 11.3).

11.2.2 Hydrogenation of *m*-Dinitrobenzene

Selectivity in di-nitro reactant hydrogenation represents an even greater catalytic challenge. Hydrogenation of *m*-dinitrobenzene (*m*-DNB) can generate *m*-nitroaniline (*m*-NAN) and *m*-phenylenediamine (*m*-PDM) as products of partial or complete –NO_2 reduction.[48–51] The relationship between specific activity and mean Au particle size [Figure 11.4(b)] matches the trends for the *p*-CNB reaction. The nature of the support does not impact significantly on hydrogenation rate, which is governed by Au size with enhanced intrinsic hydrogenation efficiency associated with smaller Au particles (9→3 nm). *m*-NAN was generated as the sole product over Au/TiO_2 and Au/Fe_2O_3, whereas Au/CeO_2 promoted exclusive formation of *m*-PDM and a mixture of both products was obtained over Au/Al_2O_3.[50]

The hydrogenation of nitroarenes proceeds *via* a nucleophilic mechanism, where a weak nucleophilic agent (hydrogen) attacks the activated $-NO_2$ function with the formation of a negatively charged intermediate.[117] The selectivity response can be attributed to differences in support acid–base Lewis properties, which influence Au electronic character and *m*-DNB adsorption/activation. Lewis basic sites on CeO_2 induce support → metal electron transfer to generate $Au^{\delta-}$ species,[118] whereas Lewis acid sites on Fe_2O_3 and TiO_2 generate supported $Au^{\delta+}$.[119,120] The occurrence of both $Au^{\delta+}$ and $Au^{\delta-}$ species is possible in the case of Au/Al_2O_3 due to the Lewis basic and acid character of the carrier.[121] *m*-DNB adsorption on Au/CeO_2 results in repulsion between π-delocalised electrons and the partially (negatively) charged metal sites leading to resonance structures with two positive localised charges on the ring where both $-NO_2$ groups are activated, leading to the formation of *m*-PDM *via* nucleophilic attack.[50]

Adsorption on Au/Fe_2O_3 and Au/TiO_2 can generate a resonance form with a single delocalised positive charge on the ring and only one $-NO_2$ group is activated to give *m*-NAN. The presence of negatively charged and electron-deficient Au particles in Au/Al_2O_3 accounts for the formation of both products. The effect of Au particle size on TiO_2 has been probed by varying catalyst reduction/activation temperature (603–873 K).[51] The incorporation of Au promoted a thermal allotropic transition (from anatase to rutile), but support composition did not influence $-NO_2$ reduction rate, which was controlled by Au particle size with increasing specific rates for particles <6 nm and rate invariance for particle sizes >6 nm [Figure 11.4(c)], where *m*-NAN formation was favoured on smaller Au particles.[51]

11.2.3 Hydrogenation of *p*-Nitrobenzonitrile

The hydrogenation of *p*-nitrobenzonitrile (*p*-NBN) to the target *p*-aminobenzonitrile (*p*-ABN) can generate up to 15 intermediates and products as established by Koprivova and Cerveny.[122] The overall reaction scheme involves disproportionation, reduction, condensation, hydrodeamination and hydrodecarbonation. A simplified pathway is presented in Figure 11.6 which applies to the published (liquid phase) catalytic studies.[122,123] Partially hydrogenated *p*-nitrosobenzonitrile and *p*-(cyanophenyl)-hydroxylamine[124] undergo condensation to 4,4′-dicyanoazobenzene.[122] The *p*-ABN formed can be hydrated to *p*-aminobenzylamide or hydrogenated to *p*-aminobenzylimine and *p*-aminobenzylamine with subsequent hydrogenolysis to *p*-aminotoluene (*p*-ATL).[122] The latter can also be generated *via* hydrogenolysis of 4,4′-diaminodibenzylamine as a condensation intermediate from *p*-aminobenzylimine and *p*-aminobenzylamine.[122,123] The catalytic action of Au/ZrO_2 and Au/Al_2O_3 in gas phase hydrogenation of *p*-NBN may be assessed against Pd/Al_2O_3 and Ni/Al_2O_3 from the entries in Table 11.4. Reaction over both Au catalysts, with equivalent Au size (7–8 nm), was fully selective to *p*-ABN.

Au/ZrO_2 delivered a higher TOF which results from the greater H_2 uptake (56 relative to 19 $\mu mol\ g_{Au}^{-1}$ on Au/Al_2O_3) under ambient conditions. A more

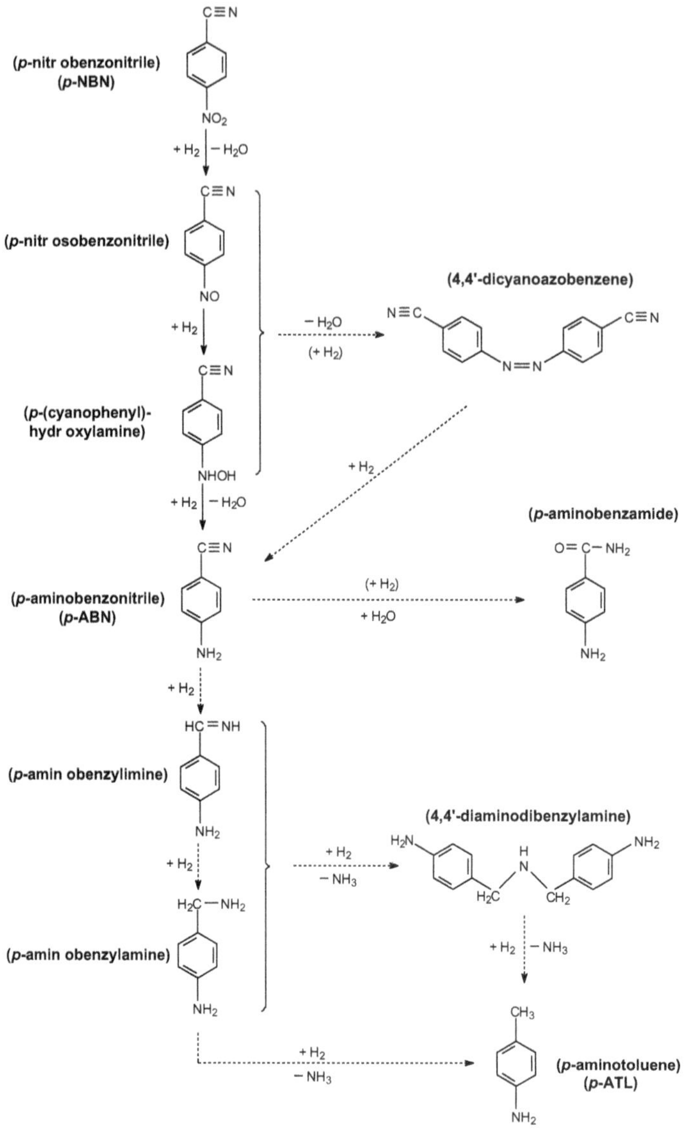

Figure 11.6 Reaction pathways in *p*-nitrobenzonitrile (*p*-NBN) hydrogenation.

effective $-NO_2$ activation on Au/ZrO_2 can be ascribed to differences in Au electronic character where increased electron density at Au sites on Al_2O_3 results in strong repulsive effects with respect to electron-rich $-NO_2$, leading to unfavourable nitro-group reduction. Both Au catalysts showed enhanced performance compared with Pd/Al_2O_3 and Ni/Al_2O_3 where the former only delivered selectivity to *p*-ABN at low conversions and Ni/Al_2O_3 was non-selective, promoting *p*-ATL as the sole product (Table 11.4).

Table 11.4 Catalytic performance of Au/ZrO_2, Au/Al_2O_3, Pd/Al_2O_3 and Ni/Al_2O_3 in terms of *p*-nitrobenzonitrile (*p*-NBN) turnover frequency (TOF) and *p*-aminobenzonitrile (*p*-ABN) selectivity ($S_{p\text{-}ABN}$) at partial ($X=0.1$) and complete ($X=1$) *p*-NBN conversion at 423 K.

Catalyst	TOF/h^{-1}	$S_{p\text{-}ABN}$/% at $X=0.1$	at $X=1$
Au/ZrO_2	267	100	100
Au/Al_2O_3	109	100	100
Pd/Al_2O_3[a]	1717	100	–[b]
Ni/Al_2O_3[a]	123	–[b]	–[b]

[a]X = 0.1 obtained using 10% v/v H_2 in He as carrier gas.
[b]*p*-aminotoluene (*p*-ATL) was the sole product.

11.2.4 Hydrogenation of Nitrocyclohexane

Standard cyclohexanone oxime production from cyclohexanone is limited by low yields (3–10%) of the starting ketone where the use of a hydroxylamine salt (usually the sulfate) generates significant quantities of ammonium sulfate waste.[125,126] The selective hydrogenation of nitrocyclohexane represents an alternative route to the oxime, which is commercially important in the production of nylon-6. The oxime generated from nitrocyclohexane hydrogenation can undergo further reduction to an amine (cyclohexylamine) and ketone (cyclohexanone), where the latter requires the participation of water [Figure 11.7(a)]. Condensation involving cyclohexylamine and cyclohexanone produces cyclohexyl-cyclohexylidene amine (CCA) with subsequent hydrogenation to dicyclohexylamine (DCA).[127] A higher nitrocyclohexane TOF in continuous gas phase operation was observed with decreasing Au size (from 7.0 to 4.3 nm) (Table 11.5). A similar dependence of activity on Au size has been reported for liquid phase nitrocyclohexane hydrogenation where an increase in size (from 2.5 to 6 nm) resulted in a drop in conversion from 100 to 4.5%.[55] Au/CeO_2 with the smallest mean Au size (3.0 nm) deviates from this trend and delivered the lowest TOF, which is consistent with the response shown in Figure 11.4(a) and Figure 11.4(b).

Thermodynamic analysis has revealed complete nitrocyclohexane conversion under reaction conditions with DCA as predominant product.[26] The results in Table 11.5 represent catalytic control where product selectivities suggest three catalyst groupings: (1) Au/Al_2O_3 generated oxime as major product; (2) Au/CeO_2 promoted near exclusive production of cyclohexanone; and (3) Au/TiO_2 and Au/ZrO_2 delivered cyclohexylamine and cyclohexanone as principal products (in varying proportions). These results find agreement with liquid phase work where Au/Al_2O_3 exhibited appreciably higher selectivity (83%) to the oxime compared with Au/TiO_2 (10%).[55] The relative surface concentration of oxime (to form the imine intermediate), reactive hydrogen and –OH (dissociated form of water) influence product selectivity, as illustrated in Figure 11.7(b). Surface reactive –OH results from reduction of nitrocyclohexane and/or oxime; the surface hydrogen reacts with imine to

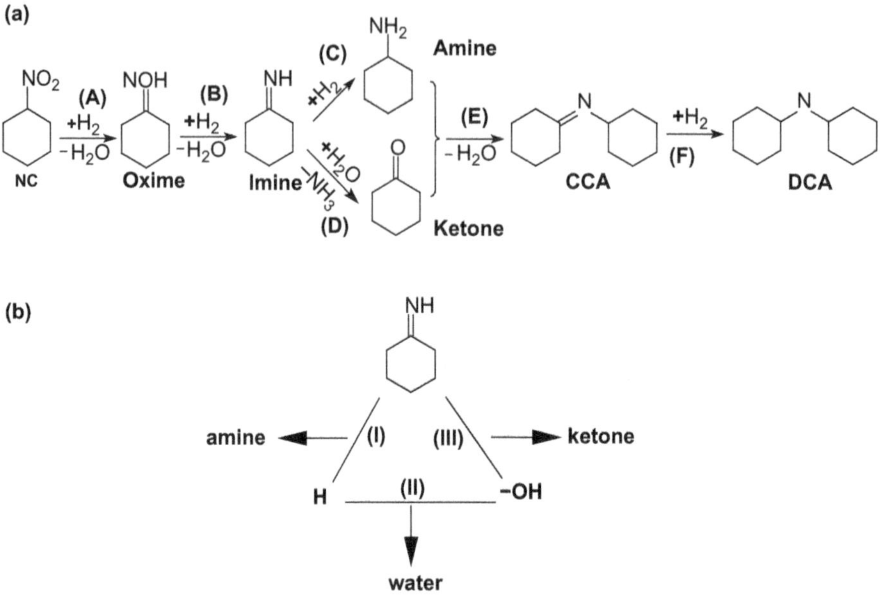

Figure 11.7 (a) Reaction pathway in nitrocyclohexane (NC) hydrogenation. (b) Scheme showing possible surface interactions between imine, reactive hydrogen (H) and –OH (dissociated form of water).

Table 11.5 Turnover frequency (TOF) and product selectivity (S at $X = 0.1$) in the hydrogenation of nitrocyclohexane over oxide (Al_2O_3, TiO_2, CeO_2 and ZrO_2) supported Au at 353 K.

			S^a/%				
Catalyst	Au size/nm	TOF/h^{-1}	Oxime	Amine	Ketone	CCA	DCA
Au/Al_2O_3	4.3	16.4	80	18	2	–	–
Au/TiO_2	4.7	9.1	6	34	40	–	20
Au/CeO_2	3.0	1.6	0	2	98	–	–
Au/ZrO_2	7.0	6.1	18	28	54	–	–

[a]CCA, cyclohexyl-cyclohexylidene amine; DCA, dicyclohexylamine.

give the amine product [step (I) in Figure 11.7(b)] or with –OH to form water [step (II) in Figure 11.7(b)], which desorbs from the catalyst, whereas ketone formation requires the combination of surface imine with -OH [step (III) in Figure 11.7(b)].[26] Preferential oxime production over Au/Al_2O_3 can be attributed to a higher rate for step (A) relative to step (B) in Figure 11.7(a), where the available surface hydrogen facilities consumption of reactive –OH (to form water that desorbs). In the case of Au/CeO_2, metal–support interaction(s) limit(s) hydrogen activation and inhibit(s) steps (I) and (II) in Figure 11.7(b) with the predominance of step (III), favouring the ketone. Both Au/TiO_2 and Au/ZrO_2 were non-selective with respect to the oxime with contributions from all surface interactions shown in Figure 11.7(b).

Condensation [step (E) in Figure 11.7(a)] only proceeded over Au/TiO_2, which is in line with the study of Shimizu *et al.* where the production of DCA and CCA was a feature of reaction over Au/TiO_2 at all conversions.[55]

11.2.5 Substituent Effects in Nitroarene Hydrogenation

Nitro group reduction exclusivity has been established in Section 11.2.1 for the conversion of *p*-CNB. Such ultra-selectivity extends to a range of *para*-substituted nitrobenzenes, as shown in Figure 11.8 for reaction over Au/TiO_2. The following activity sequence applies: *p*-DNB > *p*-CNB > nitrobenzene > *p*-nitrotoluene > *p*-nitroanisole > *p*-nitrophenol.

Hammett analysis can be employed to evaluate the effect of the *para*-substituent on reaction kinetics and to elucidate the reaction mechanism.[128] The rate constants for the substituted nitrobenzenes (k) are related to that obtained for the non-substituted nitrobenzene reference (k_0) according to $log[k/k_0] = \rho \times \sigma$. The ρ term (reaction constant) is an estimation of the charge development during the course of the reaction and provides a measure of the susceptibility of the system to substituent electronic effects.[129,130] In nucleophilic attack, the reaction rate is enhanced by electron-withdrawing substituents and $\rho > 0$.[130] The σ factor is an empirical parameter that is dependent on the substituent electron donating/acceptor character.[130] The fit of the experimental rate data to the Hammett relationship (Figure 11.8) generated a positive reaction constant ($\rho = 0.93$), consistent with a nucleophilic reaction mechanism. The extracted ρ value is comparable with that reported for liquid phase operation (MgFeO (0.690),[131] iron oxides (0.546)[132] and Pt/SiO_2–$AlPO_4$ (0.1–2.0),[133] suggesting a common reaction mechanism for these heterogeneous catalyst systems.

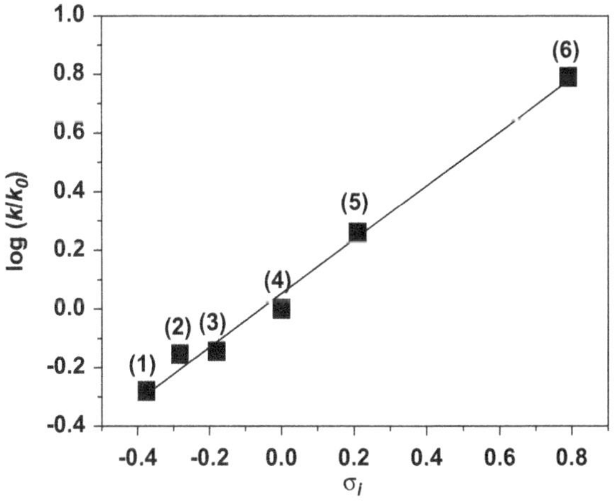

Figure 11.8 Hammett plot for selective $-NO_2$ group reduction in *para*-substituted nitroarenes [(1) *p*-nitrophenol, (2) *p*-nitroanisole, (3) *p*-nitrotoluene, (4) nitrobenzene, (5) *p*-chloronitrobenzene and (6) *p*-dinitrobenzene] over Au/TiO_2 at 473 K.

11.3 Selective Hydrogenation of Aromatic Carbonyl Compounds

The hydrogenation of the –CH=O function in the presence of other reducible groups (–CH=CH_2, –C≡CH and –NO_2) over supported Pd, Pt, Ni and Cu invariably results in side reactions, generating a mixture of unsaturated alcohols, saturated aldehdyes/ketones/alcohols and other byproducts.[134,135] Exclusive and continuous production of alcohols, used as flavours, fragrances and intermediates in the synthesis of pharmaceutical products, is of appreciable commercial importance.[136] There is evidence in the literature[16,17] for selective –CH=O reduction over nanosized Au particles with enhanced alcohol selectivity relative to standard hydrogenation catalysts. The structural features of gold catalysts, notably particle size, morphology and the nature of the gold–support interface have been shown to impact on hydrogenation efficiency.[30,62,73,74,91,137–141]

11.3.1 Hydrogenation of Benzaldehyde

Gold catalyst performance in benzaldehyde hydrogenation can be assessed against benchmark alumina-supported Pd and Ni in Table 11.6. Full selectivity to benzyl alcohol was achieved over Au but at an appreciably lower reactant turnover than that measured for Pd and Ni. Reaction over the latter generated toluene as the principal product *via* hydrogenolysis, as shown in Figure 11.9. The lower reaction rate over Au can again be linked to the limited capacity for H_2 dissociation/activation.

Selectivity in cinnamaldehyde hydrogenation has been attributed to nanoscale Au π back-bonding in –CH=O adsorption that enhanced cinnamyl alcohol formation.[142] Electron transfer from iron oxide as support was considered to facilitate –CH=O activation on charged Au particles with high selectivity to unsaturated alcohols in liquid phase hydrogenation of conjugated –CH=O.[139,143] Lenz *et al.*[144] ascribed preferential –CH=O

Table 11.6 Turnover frequency (TOF) and product selectivity (S) in the hydrogenation of benzaldehyde over oxide (Al_2O_3, ZrO_2, TiO_2, Fe_3O_4, Fe_2O_3 and CeO_2) supported Au and alumina supported Pd and Ni at 413 K.

Catalyst	Au size/nm	TOF/h^{-1}	Product(s) (S/%)
Au/Al_2O_3	4.3	86	Benzyl alcohol (100)
Au/ZrO_2	7.0	37	Benzyl alcohol (100)
Au/TiO_2	4.7	52	Benzyl alcohol (100)
Au/Fe_3O_4	7.6	–[a]	–
Au/Fe_2O_3	2.6	–[a]	–
Au/CeO_2	2.0	–[a]	–
Pd/Al_2O_3	3.0	130	Toluene (90), Benzyl alcohol (10)
Ni/Al_2O_3	30	155	Toluene (100)

[a]No detectable activity.

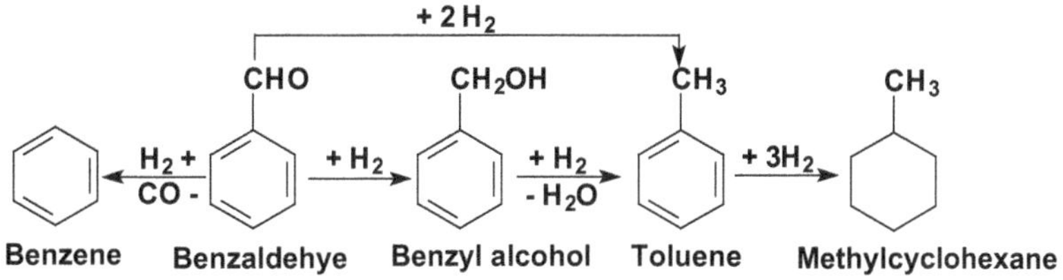

Figure 11.9 Reaction pathways in benzaldehyde hydrogenation.

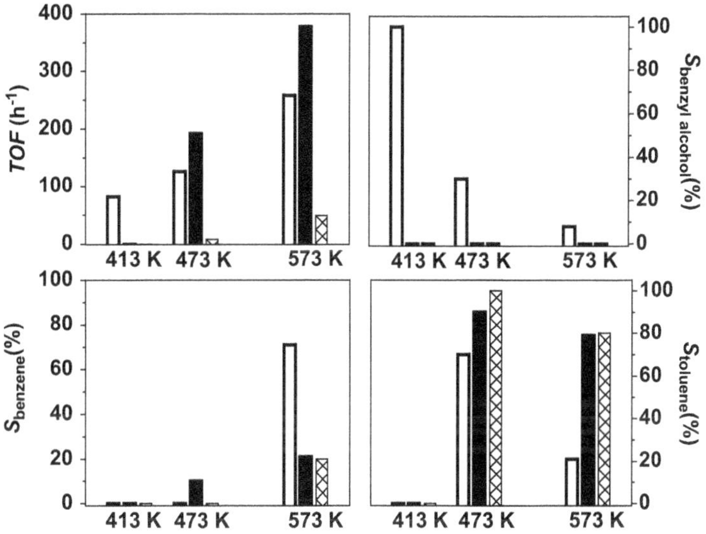

Figure 11.10 Effect of reaction temperature on benzaldehyde turnover frequency (TOF) and product (benzyl alcohol, toluene and benzene) selectivity (*S*) for reaction over Au/Al_2O_3 (open bars), Au/Fe_3O_4 (solid bars) and Au/CeO_2 (cross-hatched bars).

(over $-CH{=}CH_2$) reduction to differences in Au particle morphology. A higher benzaldehyde TOF is recorded in Table 11.6 with decreasing Au particle size (7 to 4 nm) with no detectable activity over Au < 3 nm, a trend that is consistent with $-NO_2$ reduction (see Sections 11.2.1, 11.2.2 and 11.2.4). An increase in acrolein TOF with increasing Au size (1–5 nm) has been reported for Au/ZrO_2 and Au/TiO_2.[145] In contrast, Milone and co-workers[146] could find no correlation of activity or selectivity with particle size for cinnamaldehyde hydrogenation over Au on iron oxide. Elevated reaction temperatures increased benzaldehyde consumption rate and altered product distribution (Figure 11.10). Reaction over Au/Al_2O_3 was accompanied by a shift in selectivity from benzyl alcohol as the sole product (at 413 K) to a progressive increase in hydrogenolysis (to toluene and benzene) from 473 to 573 K. Benzaldehyde conversion to benzyl alcohol was not observed over Au/Fe_3O_4 and Au/CeO_2 at any reaction temperature. Measurable conversion was

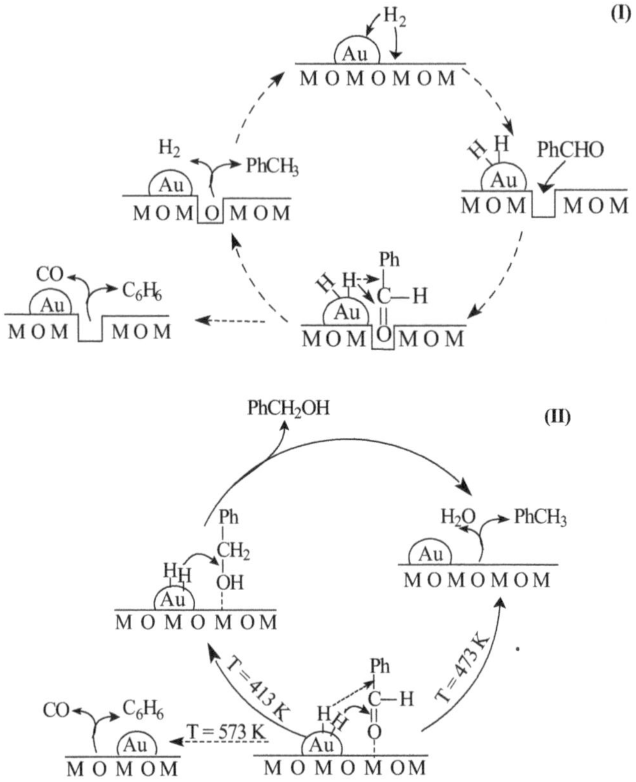

Figure 11.11 Reaction mechanisms for benzaldehyde hydrogenation over (I) Au on redox supports and (II) Au/Al_2O_3.

only recorded at $T > 413$ K with toluene as principal product and increased selectivity to benzene at 573 K, indicating a predominant hydrogenolytic character for Au on redox supports. Interaction with oxygen vacancies must stabilise the carbonyl function whereas at elevated temperature, cleavage of the C–O bond (by reactive H generated by dissociative adsorption on Au) generates toluene, as illustrated by the reaction scheme in Figure 11.11(I). In contrast, weaker interaction of the carbonyl function on Au/Al_2O_3 results in lower temperature (413 K) hydrogen addition to generate the alcohol [Figure 11.11(II)].

11.3.2 Hydrogenation of *p*-Nitrobenzaldehyde

Selectivity in the reduction of –NO_2 or –CH=O can be probed in the reaction of *p*-nitrobenzaldehyde. As –NO_2 is more reactive than –CH=O, the amine should be the preferred product.[147] Both aminobenzaldehyde (selective –NO_2 reduction) and nitrobenzyl alcohol (selective –CH=O reduction) are commercially important compounds.[148,149] This reaction has been studied over soluble metal (Pd, Ru and Co) complexes[148,150] and solid supported transition metal (Au, Ru, Ni, Pd and Pt) catalysts[149,151,152] in the liquid phase (P = 0.1–5 MPa, T = 303–423 K) with one reported gas phase application over MgO.[153]

Polymerisation of the aldehyde and alcohol products generates toxic azoxy compounds,[149] which is favoured by the longer contact times that apply in batch processes[154] and should be circumvented in continuous operation. Production of aminobenzyl alcohol has been reported for reaction over sol-gel entrapped Pd,[155] MgO[153] and Ru/Al_2O_3.[148] Taking an overview of the literature, aminobenzaldehyde has been the major product with selectivities up to 99%.[149,150,151,156,157] Preferential nitrobenzyl alcohol formation has been demonstrated in reactions involving Ru complex catalysts,[148] but the precise source of this selectivity or the means of controlling product distribution remains unresolved. Results generated in the gas phase continuous hydrogenation of *p*-nitrobenzaldehyde are given in Table 11.7, where smaller Au particle size again delivered higher TOF.

Reaction over Au/ZrO_2 and Au/TiO_2 generated *p*-aminobenzaldehyde as the sole product with no detectable hydrogenation or hydrogenolysis of the carbonyl group or evidence of any polymerisation or azoxy formation. Selectivity in nitro-group reduction requires effective polarisation of the N=O function with nucleophilic hydrogen attack and the formation of a negatively charged intermediate, as demonstrated by the Hammett correlation (Figure 11.8). In contrast, Au/Al_2O_3 (prepared by deposition–precipitation) generated *p*-nitrobenzyl alcohol in addition to *p*-aminobenzaldehyde as the major product. Carbonyl–surface interaction with the support can be ionic (*e.g.* on Mn_3O_4) or bridging (*e.g.* on Al_2O_3), the latter leading to the formation of a surface benzoate, which is an intermediate in alcohol formation. Fourier transform infrared (FTIR) measurements have established benzoate formation in benzaldehyde adsorption at Al^{3+} sites,[158] indicating that Lewis acidity is a critical catalyst property. Alumina-supported Au prepared by impregnation generated larger mean Au size with 100% selectivity to the alcohol. This can be attributed to increased Lewis acidity as a result of a higher residual Cl content associated with synthesis by impregnation (using $HAuCl_4$ as precursor).[102] Increased Au size is also a contributing factor where

Table 11.7 Turnover frequency (TOF) and product selectivity (S) in the hydrogenation of *p*-nitrobenzaldehyde over oxide (ZrO_2, TiO_2 and Al_2O_3) supported Au at 443 K.

Catalyst	Au size/nm	TOF/h^{-1}	Product (S/%)
Au/ZrO_2[a]	7.0	79	*p*-Aminobenzaldehyde (100)
Au/TiO_2[a]	4.7	108	*p*-Aminobenzaldehyde (100)
Au/Al_2O_3[a]	4.3	119	*p*-Nitrobenzyl alcohol (18)
			p-Aminobenzaldehyde (82)
Au/Al_2O_3[b]	7.8	79	*p*-Nitrobenzyl alcohol (100)

[a]Prepared by deposition–precipitation.
[b]Prepared by impregnation.

selectivity to allyl alcohol in the hydrogenation of acrolein[141,145] was associated with an increase in Au size (from 4 to 8 nm). Moreover, the increase in Au size reduces the number of low coordination Au sites, limiting the efficacy of the impregnated catalyst in $-NO_2$ reduction. These results demonstrate viability of controlling (*via* Au size and surface acidity) selective $-NO_2$ or $-CH{=}O$ reduction.

11.4 Selective -C≡CH Hydrogenation

Trace quantities of phenylacetylene can poison styrene polymerisation catalysts.[159,160] The phenylacetylene content must be less than 10 parts per million (ppm)[161] and selective hydrogenation (to styrene) is viewed as a viable process solution.[160,162] High styrene selectivities have been achieved in batch liquid phase over Pd catalysts.[163,164] Other transition metal (Ru,[165] Pt,[165] Rh[166] and Ni[167]) catalysts are active in this reaction with varying levels of selectivity. In terms of reaction pathway, it has been established (in liquid phase operation) that phenylacetylene undergoes consecutive reduction where styrene serves as an intermediate in the subsequent conversion to ethylbenzene (Figure 11.12).[168–170] Nikolaev and co-workers examined the gas phase hydrogenation of phenylacetylene/styrene mixtures over Au/Al_2O_3.[80–82] Selectivity to styrene was below 30% and activity was dependent on Au particle size (2.5–30 nm).[81,82]

The data given in Table 11.8 demonstrate exclusive phenylacetylene conversion to styrene over Au/Al_2O_3 at 353 K. An increase in reaction temperature (to 473 K) resulted in an order of magnitude increase in TOF with styrene as principal product and secondary formation of ethylbenzene. A shift in selectivity to ethylbenzene at higher temperatures has also been observed in the liquid phase over carbon-supported Pd and Pt.[161,171]

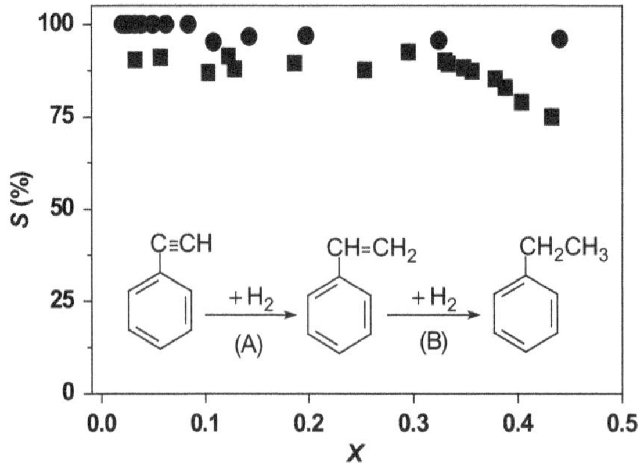

Figure 11.12 Variation of styrene selectivity (S) with phenylacetylene fractional conversion (X) over Au/Al_2O_3 at 353 K (●) and 473 K (■). Inset: Reaction pathways in the hydrogenation of phenylacetylene.

Table 11.8 Temperature dependence of turnover frequency (TOF) and associated product selectivity (S) at a common fractional conversion ($X = 0.1$) in the hydrogenation of phenylacetylene and styrene over Au/Al_2O_3.

	Phenylacetylene feed		Styrene feed	
T/K	TOF/h^{-1}	S/%	S^a/%	S/%^{a}
353	54	Styrene (100)	3	100
473	468	Styrene (90); Ethylbenzene (10)	32	100

^{a}Ethylbenzene was the sole product.

The extended selectivity/conversion response is shown in Figure 11.12. High styrene selectivity (>90%) was maintained at 353 K, but there was a clear decline at 473 K with increasing conversion, consistent with a stepwise mechanism [steps (A) and (B) in Figure 11.12]. Styrene and ethylbenzene were the only detected products with no observed dealkylation (to toluene) or condensation as reported for reaction over Ni and Ni-Pd catalysts.[167] We can note high alkene selectivities over Au/Al_2O_3 in the hydrogenation of acetylene,[87,172,173] propyne[174] and 1,3-butadiene.[175]

Use of styrene as reactant generated ethylbenzene as sole product, but with an appreciably lower TOF relative to the phenylacetylene feed (Table 11.8). This can account for the observed styrene selectivity from phenylacetylene where subsequent hydrogenation proceeded at a far lower rate. As both phenylacetylene and styrene act as Lewis bases, interaction with Lewis acid-base sites on Al_2O_3 ($Al^{\delta+}$–$O^{\delta-}$) leads to dissociative adsorption through the terminal carbon atom at $Al^{\delta+}$ with a proton bonded to the adjacent $O^{\delta-}$. Ivanov *et al.* have provided evidence from diffuse reflectance infrared Fourier transform spectroscopy (DRIFTS) analysis for this mode of bonding using acetylene as probe molecule.[176]

While there is no clear consensus regarding structure sensitivity in alkyne hydrogenation, there is evidence that the geometric and electronic properties of nanoscale Au influence phenylacetylene hydrogenation, where smaller Au size favours higher activity and styrene selectivity.[82] Nikolaev and Smirnov[81] have proposed preferential adsorption of phenylacetylene over styrene on Au edge and corner sites that are prevalent for smaller Au particles. Density functional theory (DFT) calculations have demonstrated stronger interaction of propyne (relative to propylene) with nanosized Au, resulting in high (80–90%) associated selectivity to propylene in the hydrogenation of propyne + propylene mixtures over Au/CeO_2.[174] Moreover, Jia *et al.* have recorded greater acetylene uptake than ethylene on Au/Al_2O_3, which was linked to enhanced (by a factor of 2000) acetylene hydrogenation rates.[172]

11.5 Use of Gold Catalysts in Hydrogenolysis

Monometallic Au catalysts have exhibited low activity in hydrogenolysis reactions. However, the incorporation of Au as a modifier has served to enhance performance and extend catalyst lifetime by limiting deactivation in the hydrogenolysis of methylcyclohexane over Ir–Au/Al_2O_3[177] and hydrodesulfurisation of thiophene over Au-Pd/SiO_2.[178] Gold has also served as an effective promoter of unsupported,[179] Al_2O_3[180] and carbon[181] supported Pd in hydrogen-mediated dehalogenation of trichloroethene[179] and dichlorodifluoromethane (CCl_2F_2).[181] Catalytic hydrodechlorination, hydrogen scission of C–Cl bonds, is an effective means of detoxifying chlorinated compounds, facilitating raw material recovery.[182] Oxide (SiO_2 and TiO_2) supported Au delivered low hydrodechlorination rates in the gas phase conversion of 2,4-dichlorophenol (2,4-DCP).[183] However, a combination of Au with Ni prepared by co-impregnation (on SiO_2 and TiO_2),[183,184] co-deposition/precipitation (on SiO_2 and TiO_2)[183] and reductive deposition of Au onto Ni (on Al_2O_3)[185] resulted in higher dechlorination rates relative to supported Ni. This was ascribed to a surface Au–Ni synergism where the Au component activated C–Cl bond(s), with subsequent attack from reactive hydrogen dissociated at Ni sites.

Dechlorination of chlorophenols is inhibited due to steric hindrance with respect to Cl positioned *ortho-* to the –OH substituent, as illustrated for 2,4-DCP in Figure 11.13 where 2-chlorophenol (2-CP) has been consistently isolated as the only partially dechlorinated product, regardless of the phase (gas[186–188] or liquid[189–191]), metal (Ni,[186,188] Pd[192–194]) or support (carbon,[195,196] Al_2O_3,[193,197] SiO_2[198]). However, Au supported on Fe_2O_3 and Fe_3O_4 has been shown to selectively cleave *ortho*-Cl, generating 4-CP as the principal product.[199,200] Hydrodechlorination proceeds *via* an electrophilic mechanism where hydrogen attacks the Cl bonded to the aromatic carbon bearing the greatest electron density to form a (cationic) intermediate that is stabilised by charge delocalisation in the aromatic ring.[201] The selective activation of *ortho*-Cl by Au on reducible iron oxide substrates is associated with the formation of electron deficient Au *via* interaction with surface oxygen vacancies, as demonstrated by XPS analysis.[199] Adsorption of 2,4-DCP on (positively) charged Au clusters *via* the –OH electron lone pair facilitates

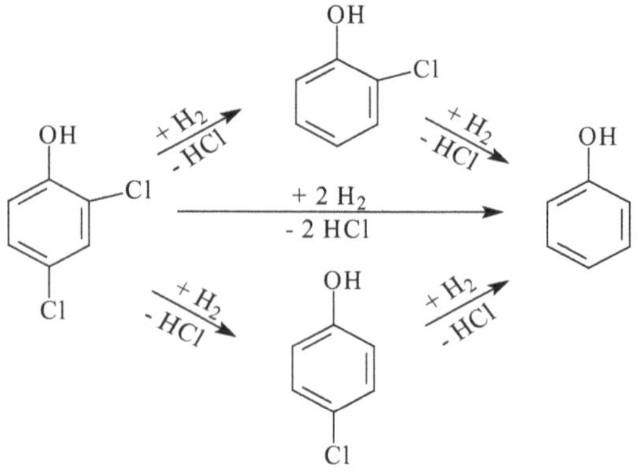

Figure 11.13 Possible reaction pathways in the hydrodechlorination of 2,4-dichlorophenol.

Table 11.9 Specific rate (R) and product [2-chlorophenol (2-CP), 4-chlorophenol (4-CP) and phenol (PhOH)] selectivity (S) in the hydrodechlorination of 2,4-dichlorophenol (2,4-DCP) at 423 K over Au/Fe_2O_3 and Au/Fe_3O_4 with associated mean Au size and H_2 chemisorption capacity.

Catalytic system	Au/Fe_2O_3	Au/Fe_3O_4
Au size/nm	2.0	7.6
H_2 uptake/μmol g_{Au}^{-1}	21	3
10^5 R/mol h^{-1} m_{Au}^{-2}	108	9.7
$S_{4\text{-CP}}$ ($S_{2\text{-CP}}$) (S_{PhOH})/%	92 (3) (5)	71 (24) (5)

polarisation (and activation) of the *ortho*- C–Cl bond as a result of inductive effects. This renders the *ortho*-Cl substituent more susceptible to electrophilic attack than the *para*- counterpart.

Preparation of Au/Fe_2O_3 by deposition–precipitation generated a narrower Au size distribution and smaller mean size than Au/Fe_3O_4 (Table 11.9). This is a consequence of differences in the point of zero charge of the supports, which result in stronger electrostatic Au precursor/support interactions in the case of Au/Fe_2O_3. The latter exhibited greater H_2 chemisorption capacity and delivered an order of magnitude higher specific hydrodechlorination rate relative to Au/Fe_3O_4.

11.6 Process Sustainability

11.6.1 Batch Liquid *vs.* Continuous Gas Phase Operation

Sustainability is a critical issue in the chemical industry, where the application of catalysis to reduce energy requirements and achieve high selectivity to target products is key to the application of 'green chemistry'.[202] A switch from batch to continuous processing has been highlighted as crucial for the sustainable manufacture of fine chemicals.[203] Adopting this imperative, we have examined the benefits of continuous gas phase operation relative to batch liquid conversion, taking *o*-chloronitrobenzene (*o*-CNB) hydrogenation as a model reaction. The feed was vaporised and carried through a fixed catalyst (supported Au) bed, which has the advantage of simultaneous reactant delivery and product removal where the process is operated at steady state. Exclusive formation of *o*-CAN was achieved over Au/TiO_2 in both continuous and batch hydrogenation. Gas phase reaction (at 423 K) delivered time invariant conversion for up to 140 h on-stream at 1 atm total pressure. A direct comparison with batch liquid operation is not possible as there was no measurable activity at 1 atm (and 423 K) and elevated pressure (to 13 atm) was required to deliver comparable reaction data. This represents greater energy demands and potential safety hazards at high H_2 pressure, which runs counter to green chemistry principles.[204] Any meaningful comparison should normalise *o*-CAN generated with respect to total process time to generate a measure of production capacity.[205]

Table 11.10 Comparison of batch (liquid) *vs.* continuous (gas) phase operation in the hydrogenation of *o*-chloronitrobenzene (*o*-CNB) over Au/TiO_2 at 423 K.

Operation mode	P/atm	R/h^{-1}	Downtime[a]/h, steps	Productivity/tonne mol_{Au}^{-1} $year^{-1}$	E-Factor[b]/ kg_{waste} $kg_{product}^{-1}$
Batch	13	10	10 $(t_1+t_2+t_3+t_4+t_5)$	1	8
Continuous	1	12	3 $(t_1+t_2+t_3)$	17	2[c]

[a]Time required for: reactor loading (t_1); catalyst activation (t_2); temperature/pressure stabilisation (t_3); unload/reload product/reactant (t_4); catalyst filtration-washing-reloading (t_5).
[b]Based on 90% recovery of ethanol (solvent).
[c]Unreacted H_2 is recycled.

The liquid phase reaction was run in time-sequential steps with appreciable periods between batches associated with reactor loading (t_1), catalyst activation (t_2), temperature/pressure stabilisation (t_3), extraction and recharge of product and new reactant (t_4) and catalyst filtration-washing-reloading (t_5) (Table 11.10). Any downtime ($t_1+t_2+t_3$) in gas phase operation is minimised due to the combined catalyst stability, continuous reactant supply/product removal and facile separation of catalyst from reactant/product.

Taking reaction conditions that delivered an equivalent rate of selective *o*-CNB hydrogenation in both modes, the downtime between batches was significantly greater than that required for the start-up of the continuous reactor, which in turn was negligible relative to total process time. The estimated annual production capacity (Table 11.10) is appreciably greater for continuous *o*-CAN production. The environmental or E-Factor (kg_{waste} $kg_{product}^{-1}$) introduced by Sheldon relates waste to product generation and has served to highlight sustainability gaps in chemical manufacture.[206] A move from batch to continuous operation was accompanied by a significant decrease in E-Factor to 2 (Table 11.10). Continuous Au promoted hydrogenation can remove an existing bottleneck in the production of amines 'on demand'.

11.6.2 Hydrogenation Rate Enhancement

Nanosized gold exhibits enhanced selectivity in the hydrogenation of unsaturated groups (–NO_2, –CH=O and –C≡CH), but low activity due to inefficiencies in H_2 chemisorption/dissociation. As H_2 activation is the limiting step, any additional source of reactive hydrogen can contribute to increased rate while retaining product exclusivity. Catalyst development focusing on (1) addition of a second metal as promoter and (2) use of a support capable of H_2 uptake, or process development in terms of (3) solvent as hydrogen donor or (4) coupling with catalytic dehydrogenation represent possible approaches to increasing the available hydrogen.

11.6.2.1 Use of Bimetallics (Pd-Au/Al_2O_3)

The incorporation of a second metal in bimetallic catalyst systems can induce surface structural, chemical and electronic modifications with a consequent change in activity/selectivity relative to monometallic systems.[207–209] It has been demonstrated in Section 11.2.1 that halo-nitroarene consumption rates delivered by supported Au fall far below those achieved with supported Pd, albeit the latter was non-selective and promoted unwanted dehalogenation. The greater H_2 chemisorption exhibited by supported Pd can be exploited as an additional source of reactive hydrogen in supported Pd-Au bimetallics.[106,210]

The incorporation of Pd with Au on Al_2O_3 by the deposition–precipitation resulted in increased H_2 uptake that was enhanced at higher Pd content (Table 11.11). Moreover, the presence of Pd served to significantly raise the hydrogenation rate with exclusive *p*-CAN formation at Au/Pd $\geq$20. A higher Pd content (Au/Pd = 8) resulted in hydrodechlorination activity to generate nitrobenzene (see Figure 11.2). The combination of Au/Al_2O_3 and Pd/Al_2O_3 as a physical mixture (Au/Pd = 20) delivered a higher rate and distinct product distribution (with the generation of aniline) relative to bimetallic Pd-Au/Al_2O_3 with an equivalent Au/Pd. This demonstrates that the catalytic contribution of Pd in Pd-Au/Al_2O_3 is not merely an additive effect but the result of surface Pd/Au synergism.

11.6.2.2 Support Effect

The support can play a significant role in H_2 chemisorption and storage.[211,212] The application of transition metal nitrides as hydrotreating catalysts is now established.[213] While the use of molybdenum nitride to promote hydrogenation has only been considered to a limited extent, there is evidence of significant catalytic hydrogenation activity associated with these materials.[48,49,214] This has been ascribed to a contraction of the *d* band and modification of electron density due to the incorporation of N interstitially in the metal lattice, which facilitates H_2 adsorption/activation.[215] Both

Table 11.11 Hydrogenation rate (*R*) for the conversion of *p*-chloronitrobenzene (*p*-CNB) with product selectivity (*S*) over alumina supported mono (Au) and bimetallic- (Pd-Au and Pd + Au) catalysts at 393 K.

Catalyst	Au/Pd	Metal size/nm	H_2 uptake/ $\mu mol\ g_{Au}^{-1}$	R/h^{-1}	Product(s)(*S*/%)
Au/Al_2O_3	–	2.9	33	12	*p*-CAN (100)
Pd-Au/Al_2O_3	88	1.9	62	24	*p*-CAN (100)
Pd-Au/Al_2O_3	20	2.0	68	39	*p*-CAN (100)
Pd-Au/Al_2O_3	8	2.7	75	92	*p*-CAN (80) Nitrobenzene (20)
Pd/Al_2O_3 + Au/Al_2O_3	20	–	–	77	Aniline (55) Nitrobenzene (45)

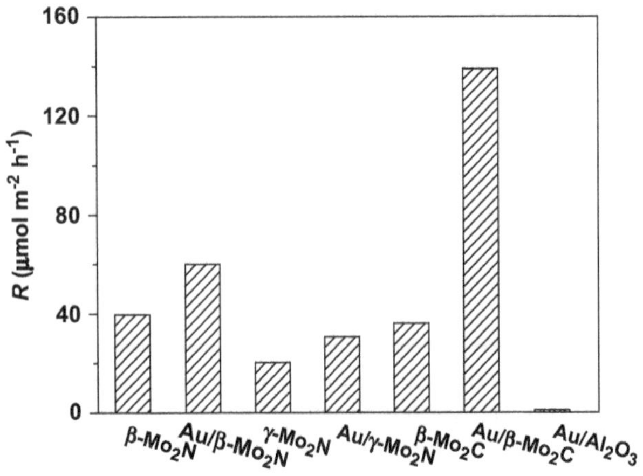

Figure 11.14 Specific rates in the hydrogenation of *p*-chloronitrobenzene (*p*-CNB) over β-Mo_2N, γ-Mo_2N, Mo_2C (with and without Au incorporation) and Au/Al_2O_3 at 493 K.

allotropic forms (cubic γ-Mo_2N and tetragonal β-Mo_2N) promoted the selective hydrogenation of *p*-CNB to *p*-CAN, as shown in Figure 11.14, with comparable activity over hexagonal close packed β-Mo_2C. Incorporation of Au on Mo_2N or Mo_2C served to increase hydrogenation rate, retaining 100% selectivity to *p*-CAN. Moreover, Au on Mo_2N or Mo_2C exhibited appreciably higher rates than Au/Al_2O_3. The nitride crystallographic phase had an influence on hydrogenation efficiency where tetragonal β-Mo_2N outperformed γ-Mo_2N, which can be linked to a greater number of nitrogen deficient sites that facilitate H_2 interaction.[42] The combination of Au with Mo nitride or carbide is a viable catalyst formulation to achieve improvements in selective nitroarene hydrogenation.

11.6.2.3 Water as Carrier

Solvent choice can influence activity and selectivity in heterogeneous catalysis, particularly in batch liquid phase reactions. Solvent effects in hydrogenation are generally associated with polarity, H_2 solubility and surface interactions.[216–218] Green chemistry principles highlight the use of innocuous (or non-toxic) solvents where water as a cleaner, inexpensive and benign polar carrier represents an important sustainability consideration.

Taking the gas phase hydrogenation of benzaldehyde over oxide (Al_2O_3, TiO_2 and ZrO_2) supported Au, use of water as carrier resulted in a dramatic increase in turnover relative to ethanol with full selectivity to the target benzyl alcohol (Figure 11.15). The promotional role of water in the liquid phase hydrogenation of crotonaldehyde was attributed to facilitated interaction of the hydrophilic –CH=O moiety with surface catalytic sites.[75] In Au promoted gas phase hydrogenation, the solvent can serve as an additional source of hydrogen.

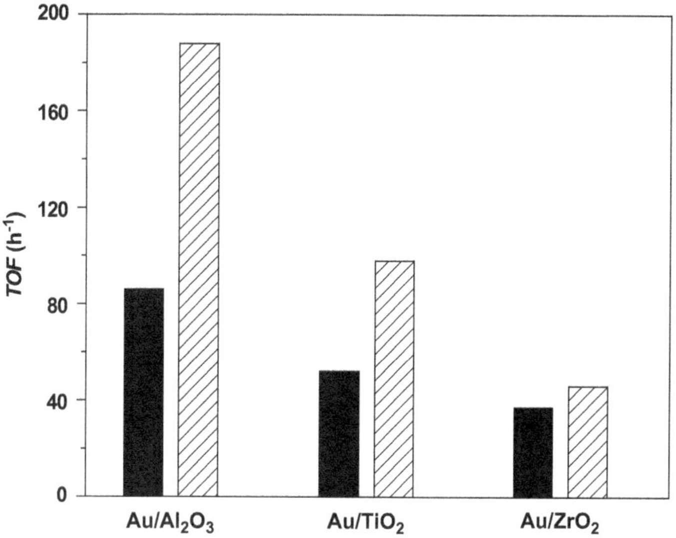

Figure 11.15 Turnover frequency (TOF) of benzaldehyde to benzyl alcohol for reaction over oxide supported Au using ethanol (solid bars) and water (hatched bars) as carrier at 413 K.

The generation of surface hydrogen (protons) from water dissociation is promoted by Lewis acid sites on Al_2O_3[219] and oxygen vacancies on TiO_2.[220] The abstracted protons can bond with two-coordinate O^{2-} sites on the support to form bridging hydroxyl groups. Buchanan and Webb[221] demonstrated that surface hydroxyl groups on Al_2O_3 act as a source of atomic 'hydrogen' and dehydroxylated Au/Al_2O_3 was inactive in butadiene hydrogenation. Moreover, activity of Au/ZrO_2 in 1,3-butadiene hydrogenation has been correlated with surface hydroxyl group density where dehydroxylated Au/ZrO_2 was inactive and activity was partially recovered by water treatment.[91] The observed (Figure 11.15) beneficial effect of water as carrier can be attributed to a surface dissociation that generates reactive hydrogen.

11.6.3 Hydrogen Utilisation

Over 95% of global H_2 production is fossil fuel based, notably methane steam reforming and coal gasification.[222] Hydrogen is not a naturally occurring feedstock and production, storage and transport represent sustainability constraints. A step change in Au hydrogenation catalysis must tackle H_2 utilisation. Conventional catalytic hydrogenation in fine chemical synthesis involves batch liquid phase operation at elevated pressure where recycle/reuse of unreacted H_2 is problematic, resulting in process inefficiency. An innovative approach is the exploitation of catalytic dehydrogenation as a source of reactive hydrogen for hydrogenation. There is evidence that dehydrogenation can be combined with hydrogenation without a supply of external H_2 (gas), notably the coupling of endothermic ethylbenzene dehydrogenation (to styrene) with exothermic nitrobenzene hydrogenation (to aniline) over supported Pt.[223]

Nagaraja *et al.*[224,225] have established coupling of cyclohexanol dehydrogenation with furfural hydrogenation over supported Cu. Furfural is a biomass (corncob and sugar cane bagasse) derived heterocyclic aldehyde that is attracting attention as a non-petroleum based feedstock.[226] Selective hydrogenation generates furfuryl alcohol, a high value chemical used to manufacture resins, rubbers and adhesives, and as a chemical building block for drug synthesis.[227] Selectivity is crucial as furan ring reduction/opening, –CH=O hydrogenolysis and decarbonylation generate a range of by-products (Figure 11.16). Selectivity is sensitive to the Au support where 2-methyl furan formation was observed for reaction over Au/Al_2O_3 and Au/TiO_2 and associated with elevated furfural consumption rates (Figure 11.17).

Reaction exclusivity to the target alcohol was achieved over Au/CeO_2 and Au/MgO. However, this was accompanied by appreciable unreacted hydrogen, expressed as mol H_2 required to produce 1 mol of the furfuryl alcohol in Table 11.12. Gold is inactive for dehydrogenation and a second catalytic metal is required in any coupled system. Results for the coupling of 2-butanol dehydrogenation over Cu/SiO_2 where the hydrogen released is used by Au (on CeO_2) in continuous furfural hydrogenation are presented in Table 11.12. Dehydrogenation of 2-butanol generated 2-butanone, which is widely used as solvent and in the production of plastics, coatings and films. The coupled reaction produced furfuryl alcohol and 2-butanone as sole products that can be readily separated (100 K difference in boiling points at 1 atm) by standard distillation. The reaction was conducted in N_2 as carrier where Au in effect 'borrows' hydrogen generated *in situ* by the Cu component, circumventing the use of compressed H_2 with important safety implications for large-scale production. The tabulated data demonstrate: significantly increased furfural TOF relative to the single component (Au)

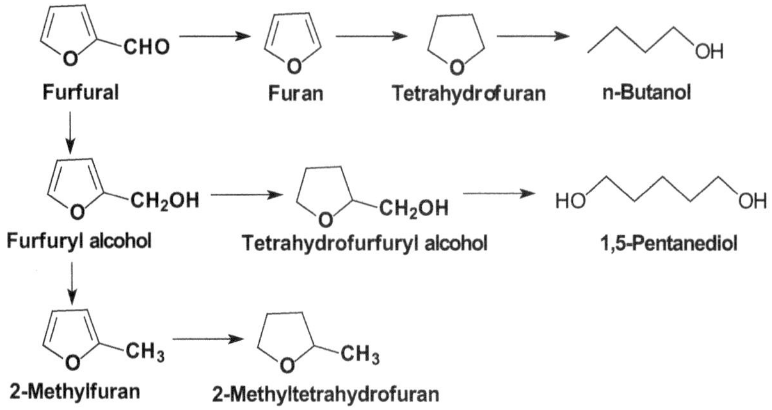

Figure 11.16 Reaction pathways in the hydrogenation of furfural.

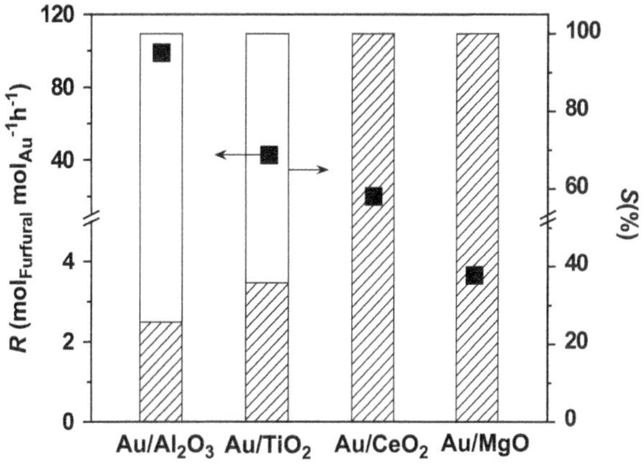

Figure 11.17 Furfural consumption rate (■) and product selectivity (*S*, hatched bars: furfuryl alcohol, open bars: 2-methylfuran) for reaction over oxide (Al_2O_3, TiO_2, CeO_2 and MgO) supported Au at 498 K.

Table 11.12 Furfural turnover frequency (TOF) and H_2 utilisation in direct furfural hydrogenation (over Au/CeO_2) and the coupled reaction with 2-butanol dehydrogenation (over $Cu/SiO_2 + Au/CeO_2$) at 498 K.

Catalyst	TOF/h^{-1}	H_2 utilisation/mol H_2/mol alcohol
Au/CeO_2	61	477
$Cu/SiO_2 + Au/CeO_2$	195	1

reaction with orders of magnitude enhanced H_2 utilisation in the coupled system where all the hydrogen generated was used to convert furfural to furfuryl alcohol (mol H_2/mol alcohol = 1). This represents a step change improvement in efficiency and closes the sustainability gap in terms of unreacted hydrogen in standard hydrogenation over Au.

11.7 Conclusions and a Look to the Future

Nanoscale supported Au offers clear advantages over standard transition metal (Ni and Pd) catalysts in the hydrogenation of unsaturated functions ($-NO_2$, $-CH{=}O$, $-C{\equiv}CH$) and hydrogenolytic C–Cl cleavage. Gold particle size is critical in determining the reaction rate with increasing reactant turnover as mean Au particle size is decreased (from 9 nm to 3 nm). Moreover, support properties, notably redox and acid/base characteristics, can be used to control or tune product distribution. Inefficiencies in H_2 activation by Au, which limit the rate can be counteracted by the incorporation of a second metal as promoter, application of a support capable of H_2 chemisorption, employment of a solvent/carrier as hydrogen donor or coupling catalytic dehydrogenation with the hydrogenation step.

At this juncture, the use of Au catalysts in oxidation and the water gas shift reaction is far more developed than hydrogenation applications, which are still at the formative stages. Effective exploitation of the unique selectivity of Au catalysis in hydrogen-mediated reactions at a commercial scale must continue to address H_2 utilisation in order to enhance efficiency. Further research is required, directed at facilitating H_2 activation or increasing the internal supply of active hydrogen *via* coupling methodologies.

Friend and co-workers have highlighted the importance of developing new theoretical models that extend DFT calculations in order to inform experimental measurements and elucidate reactant/surface dynamics.[228] The molecular level mechanistic information provided by surface science methodologies should serve as the basis for future catalyst development. This requires surface science measurements that bridge the pressure gap between the analysis of model surfaces and practical heterogeneous catalysts.

There is still considerable scope to extend the use of Au catalysis in new hydrogenation reactions. We can flag recent work that has demonstrated significant Au activity in H_2 mediated conversion of biomass-based feedstock to valuable chemicals, with immense potential significance in sustainable processing.[229,230] An important new finding is the establishment of high regioselective hydrogenation activity for unsupported 'nanoporous' Au.[231,232] This can be developed to gain greater insight into the intrinsic catalytic action of Au by removing support effects and to expand the repertoire of possible end uses. Hydrogenation by Au has largely been limited to batch liquid phase operation. An analysis of catalysts to promote hydrogenation in both (continuous) gas and (batch/continuous) liquid modes should be conducted in parallel with a full sustainability analysis that considers energy requirements and overall throughput in order to arrive at the best practicable option for the effective delivery of next generation fine chemicals to the marketplace.

Acknowledgements

We thank Dr F. Cárdenas-Lizana, Dr S. Gómez-Quero and Dr N. Perret for their contributions to this chapter.

References

1. C. Louis and O. Pluchery, *Gold Nanoparticles for Physics, Chemistry and Biology*, Imperial College Press, London, 2012.
2. D. A. Giljohann, D. S. Seferos, W. L. Daniel, M. D. Massich, P. C. Patel and C. A. Mirkin, *Angew. Chem., Int. Ed.*, 2010, **49**, 3280.
3. C. Corti, R. Holliday and D. Thompson, *Gold Bull.*, 2002, **35**, 111.
4. B. C. Gates, *Chem. Commun.*, 2013, **49**, 7876.
5. G. C. Bond and D. T. Thompson, *Catal. Rev.: Sci. Eng.*, 1999, **41**, 319.

6. G. C. Bond, C. Louis and D. T. Thompson, *Catalysis by Gold.*, Imperial College Press, London, 2006.
7. G. J. Hutchings, J. K. Edwards, L. J. Roy and J. P. Wilcoxon, *Front. Nanosci.*, 2012, **3**, 249.
8. M. Chen and D. W. Goodman, *Chem. Soc. Rev.*, 2008, **37**, 1860.
9. C. Della Pina, E. Falletta, L. Prati and M. Rossi, *Chem. Soc. Rev.*, 2008, **37**, 2077.
10. J. Gong, *Chem. Rev.*, 2012, **112**, 2987.
11. A. Corma and H. Garcia, *Chem. Soc. Rev.*, 2008, **37**, 2096.
12. A. S. K. Hashmi and M. Rudolph, *Chem. Soc. Rev.*, 2008, **37**, 1766.
13. R. Coquet, K. L. Howard and D. J. Willock, *Chem. Soc. Rev.*, 2008, **37**, 2046.
14. P. Pyykko, *Chem. Soc. Rev.*, 2008, **37**, 1967.
15. F. Cárdenas-Lizana and M. A. Keane, *J. Mater. Sci.*, 2013, **48**, 543.
16. L. McEwan, M. Julius, S. Roberts and J. Fletcher, *Gold Bull.*, 2010, **43**, 298.
17. P. Claus, *Appl. Catal. A: Gen.*, 2005, **291**, 222.
18. M. Pan, A. J. Brush, Z. D. Pozun, H. C. Ham, W.-Y. Yu, G. Henkelman, G. S. Hwang and C. B. Mullins, *Chem. Soc. Rev.*, 2013, **42**, 5002.
19. G. Li and R. Jin, *Acc. Chem. Res.*, 2013, **46**, 1749.
20. G. C. Bond, *Molecules.*, 2012, **17**, 1716.
21. E. Soriano and J. Marco-Contelles, *Top. Curr. Chem.*, 2011, **302**, 1.
22. T. Mitsudome and K. Kaneda, *Green Chem.*, 2013, **15**, 2636.
23. C. Kartusch and J. A. van Bokhoven, *Gold Bull.*, 2009, **42**, 343.
24. A. Corma, M. Boronat, S. González and F. Illas, *Chem. Commun.*, 2007, 3371.
25. M. Boronat, P. Concepción, A. Corma, S. González, F. Illas and P. Serna, *J. Am. Chem. Soc.*, 2007, **129**, 16230.
26. X. Wang, N. Perret and M. A. Keane, *Appl. Catal. A: Gen.*, 2013, **467**, 575.
27. X. Wang, N. Perret and M. A. Keane, *Chem. Eng. J.*, 2012, **210**, 103.
28. E. Bus and J. A. van Bokhoven, *Phys. Chem. Chem. Phys.*, 2007, **9**, 2894.
29. E. Bus, J. T. Miller and J. A. van Bokhoven, *J. Phys. Chem. B.*, 2005, **109**, 14581.
30. U. Hartfelder, C. Kartusch, M. Makosch, M. Rovezzi, J. Sa and J. A. van Bokhoven, *Catal. Sci. Technol.*, 2013, **3**, 454.
31. C. Torres, C. Campos, J. L. G. Fierro, M. Oportus and P. Reyes, *Catal. Lett.*, 2013, **143**, 763.
32. M. A. Keane and R. Larsson, *React. Kinet. Mech. Catal.*, 2012, **106**, 267.
33. G. Richner, J. A. van Bokhoven, Y. M. Neuhold, M. Makosch and K. Hungerbuhler, *Phys. Chem. Chem. Phys.*, 2011, **13**, 12463.
34. X. Y. Tan, Z. X. Zhang, Z. H. Xiao, Q. Xu, C. H. Liang and X. K. Wang, *Catal. Lett.*, 2012, **142**, 788.
35. S. Gómez, C. Torres, J. L. G. Fierro, C. R. Apesteguia and P. Reyes, *J. Chil. Chem. Soc.*, 2012, **57**, 1194.
36. C. Kartusch, M. Makosch, J. Sa, K. Hungerbuehler and J. A. van Bokhoven, *ChemCatChem.*, 2012, **4**, 236.

37. L. Q. Liu, B. T. Qiao, Z. J. Chen, J. Zhang and Y. Q. Deng, *Chem. Commun.*, 2009, 653.
38. F. Cárdenas-Lizana, S. Gómez-Quero, N. Perret and M. A. Keane, *Gold Bull.*, 2009, **42**, 124.
39. X. Wang, N. Perret, J. J. Delgado, G. Blanco, X. Chen, C. M. Olmos, S. Bernal and M. A. Keane, *J. Phys. Chem. C.*, 2013, **117**, 994.
40. N. Perret, X. Wang, L. Delannoy, C. Potvin, C. Louis and M. A. Keane, *J. Catal.*, 2012, **286**, 172.
41. D. P. He, X. D. Jiao, P. Jiang, J. Wang and B. Q. Xu, *Green Chem.*, 2012, **14**, 111.
42. N. Perret, F. Cárdenas-Lizana, D. Lamey, V. Laporte, L. Kiwi-Minsker and M. A. Keane, *Top. Catal.*, 2012, **55**, 955.
43. F. Cárdenas-Lizana, D. Lamey, N. Perret, S. Gómez-Quero, L. Kiwi-Minsker and M. A. Keane, *Catal. Commun.*, 2012, **21**, 46.
44. R.-Y. Zhong, X.-H. Yan, Z.-K. Gao, R.-J. Zhang and B.-Q. Xu, *Catal. Sci. Technol.*, 2013, **3**, 3013.
45. F. Cárdenas-Lizana, S. Gómez-Quero, L. Kiwi-Minsker and M. A. Keane, *Int. J. Nanotechnol.*, 2012, **9**, 92.
46. Y. F. Hao, R. X. Liu, X. C. Meng, H. Y. Cheng and F. Y. Zhao, *J. Mol. Catal. A: Chem.*, 2011, **335**, 183.
47. F. Cárdenas-Lizana, S. Gómez-Quero, G. Jacobs, Y. Y. Ji, B. H. Davis, L. Kiwi-Minsker and M. A. Keane, *J. Phys. Chem. C.*, 2012, **116**, 11166.
48. F. Cárdenas-Lizana, S. Gómez-Quero, C. J. Baddeley and M. A. Keane, *Appl. Catal. A: Gen.*, 2010, **387**, 155.
49. F. Cárdenas-Lizana, S. Gómez-Quero and M. Keane, *Catal. Lett.*, 2009, **127**, 25.
50. F. Cárdenas-Lizana, S. Gómez-Quero, N. Perret and M. A. Keane, *Catal. Sci. Technol.*, 2011, **1**, 652.
51. F. Cárdenas-Lizana, S. Gómez-Quero, H. Idriss and M. A. Keane, *J. Catal.*, 2009, **268**, 223.
52. K. Shimizu, Y. Miyamoto, T. Kawasaki, T. Tanji, Y. Tai and A. Satsuma, *J. Phys. Chem. C.*, 2009, **113**, 17803.
53. C. Kartusch, F. Krumeich, O. Safonova, U. Hartfelder, M. Makosch, J. Sá and J. A. van Bokhoven, *ACS Catal.*, 2012, **2**, 1394.
54. F. Zhang, N. Liu, P. Zhao, J. Sun, P. Wang, W. Ding, J. Liu, J. Jin and J. Ma, *Appl. Surf. Sci.*, 2012, **263**, 471.
55. K. Shimizu, T. Yamamoto, Y. Tai and A. Satsuma, *J. Mol. Catal. A: Chem.*, 2011, **345**, 54.
56. H. Rojas, J. J. Martinez, S. Mancipe, G. Borda and P. Reyes, *React. Kinet. Mech. Catal.*, 2012, **106**, 445.
57. L. N. Protasova, E. V. Rebrov, H. E. Skelton, A. E. H. Wheatley and J. C. Schouten, *Appl. Catal. A: Gen.*, 2011, **399**, 12.
58. R. X. Liu, Y. C. Yu, K. Yoshida, G. M. Li, H. X. Jiang, M. H. Zhang, F. Y. Zhao, S. Fujita and M. Arai, *J. Catal.*, 2010, **269**, 191.

59. C. Caballero, J. Valencia, M. Barrera and A. Gil, *Powder Technol.*, 2010, **203**, 412.
60. Y. X. Liu and T. F. Xing, *Asian J. Chem.*, 2013, **25**, 2833.
61. Y. X. Liu, T. F. Xing, Y. M. Luo, X. N. Li and W. Yan, *Chin. Chem. Lett.*, 2010, **21**, 1322.
62. H. Rojas, G. Diaz, J. J. Martinez, C. Castaneda, A. Gómez-Cortes and J. Arenas-Alatorre, *J. Mol. Catal. A: Chem.*, 2012, **363**, 122.
63. Y. X. Liu, Y. M. Luo and Z. J. Wei, *Asian J. Chem.*, 2013, **25**, 8617.
64. Y. X. Liu, L. F. Meng, Z. J. Wei and H. T. Shi, *Chin. J. Catal.*, 2011, **32**, 1269.
65. K. J. You, C. T. Chang, B. J. Liaw, C. T. Huang and Y. Z. Chen, *Appl. Catal. A: Gen.*, 2009, **361**, 65.
66. Z. M. Tian, X. Xiang, L. S. Xie and F. Li, *Ind. Eng. Chem. Res.*, 2013, **52**, 288.
67. X. Zhang, Y. C. Guo, Z. C. Zhang, J. S. Gao and C. M. Xu, *J. Catal.*, 2012, **292**, 213.
68. E. Castillejos, E. Gallegos-Suarez, B. Bachiller-Baeza, R. Bacsa, P. Serp, A. Guerrero-Ruiz and I. Rodríguez-Ramos, *Catal. Commun.*, 2012, **22**, 79.
69. J. Lenz, B. C. Campo, M. Alvarez and M. A. Volpe, *J. Catal.*, 2009, **267**, 50.
70. H. Gu, X. Xu, A.-A. Chen, P. Ao and X. Yan, *Catal. Commun.*, 2013, **41**, 65.
71. H. Yoshitake and N. Saito, *Micropor. Mesopor. Mater.*, 2013, **168**, 51.
72. G. Q. Xie, X. J. Liu, L. P. Tao, J. Q. Lu, M. F. Luo and X. N. Li, *Chin. J. Catal.*, 2009, **30**, 543.
73. H. Y. Chen, C. T. Chang, S. J. Chiang, B. J. Liaw and Y. Z. Chen, *Appl. Catal. A: Gen.*, 2010, **381**, 209.
74. B. Campo, G. Santori, C. Petit and M. Volpe, *Appl. Catal. A: Gen.*, 2009, **359**, 79.
75. M. M. Wang, L. He, Y. M. Liu, Y. Cao, H. Y. He and K. N. Fan, *Green Chem.*, 2011, **13**, 602.
76. X. Yang, A. Wang, X. Wang, T. Zhang, K. Han and J. Li, *J. Phys. Chem. C.*, 2009, **113**, 20918.
77. N. Perret, F. Cárdenas-Lizana and M. A. Keane, *Catal. Commun.*, 2011, **16**, 159.
78. L. He, J. Ni, L. C. Wang, F. J. Yu, Y. Cao, H. Y. He and K. N. Fan, *Chem. Eur. J.*, 2009, **15**, 11833.
79. Z. Li, W. P. Ding, G. J. Kang and Z. X. Chen, *Catal. Commun.*, 2012, **17**, 164.
80. S. A. Nikolaev, N. A. Permyakov, V. V. Smirnov, A. Y. Vasil'kov and S. N. Lanin, *Kinet. Catal.*, 2010, **51**, 288.
81. S. A. Nikolaev and V. V. Smirnov, *Catal. Today*, 2009, **147**, S336.
82. S. A. Nikolaev and V. V. Smirnov, *Gold Bull.*, 2009, **42**, 182.
83. S. A. Nikolaev, D. A. Pichugina and D. F. Mukhamedzyanova, *Gold Bull.*, 2012, **45**, 221.

84. X. Y. Liu, C. Y. Mou, S. Lee, Y. N. Li, J. Secrest and B. W. L. Jang, *J. Catal.*, 2012, **285**, 152.
85. A. Sarkany, *React. Kinet. Catal. Lett.*, 2009, **96**, 43.
86. X. Liu, Y. Li, J. W. Lee, C.-Y. Hong, C.-Y. Mou and B. W. L. Jang, *Appl. Catal. A: Gen.*, 2012, **439–440**, 8.
87. A. C. Gluhoi, J. W. Bakker and B. E. Nieuwenhuys, *Catal. Today*, 2010, **154**, 13.
88. A. Sarkany, Z. Schay, K. Frey, E. Szeles and I. Sajo, *Appl. Catal. A: Gen.*, 2010, **380**, 133.
89. J. Gaudet, K. K. Bando, Z. X. Song, T. Fujitani, W. Zhang, D. S. Su and S. T. Oyama, *J. Catal.*, 2011, **280**, 40.
90. B. Pongthawornsakun, S.-i. Fujita, M. Arai, O. Mekasuwandumrong and J. Panpranot, *Appl. Catal. A: Gen.*, 2013, **467**, 132.
91. X. Zhang, H. Shi and B.-Q. Xu, *J. Catal.*, 2011, **279**, 75.
92. Y. Guan and E. J. M. Hensen, *Phys. Chem. Chem. Phys.*, 2009, **11**, 9578.
93. X. F. Yang, A. Q. Wang, Y. L. Wang, T. Zhang and J. Li, *J. Phys. Chem. C.*, 2010, **114**, 3131.
94. A. Hugon, L. Delannoy and C. Louis, *Gold Bull.*, 2009, **42**, 310.
95. N. E. Kolli, L. Delannoy and C. Louis, *J. Catal.*, 2013, **297**, 79.
96. J. Ohyama, A. Esaki, Y. Yamamoto, S. Arai and A. Satsuma, *RSC Adv.*, 2013, 3, 1033–1036.
97. D. Ren, L. He, L. Yu, R. S. Ding, Y. M. Liu, Y. Cao, H. Y. He and K. N. Fan, *J. Am. Chem. Soc.*, 2012, **134**, 17592.
98. P. Castano, T. A. Zepeda, B. Pawelec, M. Makkee and J. L. G. Fierro, *J. Catal.*, 2009, **267**, 30.
99. M. Toth, J. Kiss, A. Oszko, G. Potari, B. Laszlo and A. Erdohelyi, *Top. Catal.*, 2012, **55**, 747.
100. D. Preti, C. Resta, S. Squarcialupi and G. Fachinetti, *Angew. Chem., Int. Ed.*, 2012, **50**, 12551.
101. J. A. Rodriguez, J. Evans, L. Feria, A. B. Vidal, P. Liu, K. Nakamura and F. Illas, *J. Catal.*, 2013, **307**, 162.
102. F. Cárdenas-Lizana, S. Gómez-Quero, A. Hugon, L. Delannoy, C. Louis and M. A. Keane, *J. Catal.*, 2009, **262**, 235.
103. T. V. Choudhary, C. Sivadinarayana, A. K. Datye, D. Kumar and D. W. Goodman, *Catal. Lett.*, 2003, **86**, 1.
104. S.-C. Qi, X.-Y. Wei, Z.-M. Zong and Y.-K. Wang, *RSC Adv.*, 2013, **3**, 14219.
105. C. Xiao, X. Wang, C. Lian, H. Liu, M. Liang and Y. Wang, *Curr. Org. Chem.*, 2012, **16**, 280.
106. G. Yuan and M. A. Keane, *Catal. Commun.*, 2003, **4**, 195.
107. Y. Zhou and J. Zhou, *J. Phys. Chem. Lett.*, 2010, **1**, 609.
108. S. E. Collins, J. M. Cíes, E. del Río, M. López-Haro, S. Trasobares, J. J. Calvino, J. M. Pintado and S. Bernal, *J. Phys. Chem. C.*, 2007, **111**, 14371.
109. H. Vidal, J. Kaspar, M. Pijolat, G. Colon, S. Bernal, A. Cordon, V. Perrichon and F. Fally, *Appl. Catal. B: Environ.*, 2000, **27**, 49.
110. C. T. Campbell and C. H. F. Peden, *Science*, 2005, **309**, 713.

111. J. M. Cíes, E. del Río, M. López-Haro, J. J. Delgado, G. Blanco, S. Collins, J. J. Calvino and S. Bernal, *Angew. Chem., Int. Ed.*, 2010, **49**, 9744.
112. S. Bernal, J. J. Calvino, G. A. Cifredo, J. M. Gatica, J. A. P. Omil, A. Laachir and V. Perrichon, *Stud. Surf. Sci. Catal.*, 1995, **96**, 419.
113. B. Aristizabal, C. A. Gonzalez, I. Barrio, M. Montes and C. M. de Correa, *J. Mol. Catal. A: Chem.*, 2004, **222**, 189.
114. L. Kepinski and J. Okal, *J. Catal.*, 2000, **192**, 48.
115. P. Fornasiero, N. Hickey, J. Kaspar, C. Dossi, D. Gava and M. Graziani, *J. Catal.*, 2000, **189**, 326.
116. F. Fajardie, J. F. Tempere, J. M. Manoli, G. Djega-Mariadassou and G. Blanchard, *J. Chem. Soc. Faraday Trans.*, 1998, **94**, 3727.
117. O. Exner and T. M. Krygowski, *Chem. Soc. Rev.*, 1996, **25**, 71.
118. A. Y. Stakheev and L. M. Kustov, *Appl. Catal. A: Gen.*, 1999, **188**, 3.
119. T. Okazawa, M. Fujiwara, T. Nishimura, T. Akita, M. Kohyama and Y. Kido, *Surf. Sci.*, 2006, **600**, 1331.
120. G. Busca, *Phys. Chem. Chem. Phys.*, 1999, **1**, 723.
121. M. Trueba and S. P. Trasatti, *Eur. J. Inorg. Chem.*, 2005, **2005**, 3393.
122. K. Koprivova and L. Cerveny, *Res. Chem. Intermed.*, 2008, **34**, 93.
123. K. Hata and K.-I. Watanabe, *Bull. Chem. Soc. Jpn.*, 1959, **32**, 861.
124. R. J. Rahaim and R. E. Maleczka, *Org. Lett.*, 2005, **7**, 5087.
125. H.-G. Liao, Y.-J. Xiao, H.-K. Zhang, P.-L. Liu, K.-Y. You, C. Wei and H. A. Luo, *Catal. Commun.*, 2012, **19**, 80.
126. K. Weissermel and H.-J. Arpe, *Industrial Organic Chemistry*, 4th edn, Wiley-VCH, Weinheim, 2003.
127. P. Serna, M. Lopez-Haro, J. J. Calvino and A. Corma, *J. Catal.*, 2009, **263**, 328.
128. F. Cárdenas-Lizana, Z. M. de Pedro, S. Gómez-Quero and M. A. Keane, *J. Mol. Catal. A: Chem.*, 2010, **326**, 48.
129. J. Shorter, *Chem. Lysty.*, 2000, 210.
130. C. D. Johnson, *The Hammett Equation.*, Cambridge University Press, Cambridge, 1973.
131. P. S. Kumbhar, J. Sánchez-Valente, J. M. M. Millet and F. Figueras, *J. Catal.*, 2000, **191**, 467.
132. M. Lauwiner, P. Rys and J. Wissmann, *Appl. Catal. A: Gen.*, 1998, **172**, 141.
133. M. A. A. Lopidana, V. B. Bolos, C. J. Sanchidrian, J. M. M. Rubio and F. R. Luque, *Bull. Chem. Soc. Jap.*, 1987, **60**, 3415.
134. P. Maki-Arvela, J. Hajek, T. Salmi and D. Y. Murzin, *Appl. Catal. A: Gen.*, 2005, **292**, 1.
135. H. Wei, C. Gomez, J. Liu, N. Guo, T. Wu, R. Lobo-Lapidus, C. L. Marshall, J. T. Miller and R. J. Meyer, *J. Catal.*, 2013, **298**, 18.
136. Y. Zhang, X. Cui, F. Shi and Y. Deng, *Chem. Rev.*, 2012, **112**, 2467.
137. Y. Zhu, Z. Wu, C. Gayathri, H. Qian, R. R. Gil and R. Jin, *J. Catal.*, 2010, **271**, 155.
138. P. G. N. Mertens, P. Vandezande, X. Ye, H. Poelman, I. F. J. Vankelecom and D. E. de Vos, *Appl. Catal. A: Gen.*, 2009, **355**, 176.

139. C. Milone, R. Ingoglia, L. Schipilliti, C. Crisafulli, G. Neri and S. Galvagno, *J. Catal.*, 2005, **236**, 80.
140. C. Mohr, H. Hofmeister, J. Radnik and P. Claus, *J. Am. Chem. Soc.*, 2003, **125**, 1905.
141. C. Mohr, H. Hofmeister and P. Claus, *J. Catal.*, 2003, **213**, 86.
142. E. Bus, R. Prins and J. A. van Bokhoven, *Catal. Commun.*, 2007, **8**, 1397.
143. C. Milone, C. Crisafulli, R. Ingoglia, L. Schipilliti and S. Galvagno, *Catal. Today*, 2007, **122**, 341.
144. J. Lenz, B. C. Campo, M. Alvarez and M. A. Volpe, *J. Catal.*, 2009, **267**, 50.
145. P. Claus, A. Bruckner, C. Mohr and H. Hofmeister, *J. Am. Chem. Soc.*, 2000, **122**, 11430.
146. C. Milone, R. Ingoglia, A. Pistone, G. Neri, F. Frusteri and S. Galvagno, *J. Catal.*, 2004, **222**, 348.
147. M. Arai, A. Obata and Y. Nishiyama, *React. Kinet. Catal. Lett.*, 1997, **61**, 275.
148. A. Indra, N. Maity, P. Maity, S. Bhaduri and G. K. Lahiri, *J. Catal.*, 2011, **284**, 176.
149. A. Corma, C. González-Arellano, M. Iglesias and F. Sánchez, *Appl. Catal. A: Gen.*, 2009, **356**, 99.
150. S. K. Mohapatra, S. U. Sonavane, R. V. Jayaram and P. Selvam, *Tetrahedron Lett.*, 2002, **43**, 8527.
151. L. Liu, B. Qiao, Y. Ma, J. Zhang and Y. Deng, *Dalton Trans.*, 2008, 2542.
152. A. Corma, P. Serna, P. Concepción and J. J. Calvino, *J. Am. Chem. Soc.*, 2008, **130**, 8748.
153. J. Kijeńiski, M. Gliński, R. Wiśniewski and S. Murghani, *Stud. Surf. Sci. Catal.*, 1991, **59**, 169.
154. S. K. Cao, S. L. Xu and S. G. Xu, *Polym. Advan. Technol.*, 1999, **10**, 43.
155. V. Pandarus, R. Ciriminna, F. Beland and M. Pagliaro, *Catal. Sci. Technol.*, 2011, **1**, 1616.
156. A. Corma and P. Serna, *Science*, 2006, **313**, 332.
157. S. G. Xu, X. L. Xi, J. Shi and S. K. Cao, *J. Mol. Catal. A: Chem.*, 2000, **160**, 287.
158. C. A. Koutstaal, P. Angevaare and V. Ponec, *J. Catal.*, 1993, **143**, 573.
159. F. Arena, G. Cum, R. Gallo and A. Parmaliana, *J. Mol. Catal. A: Chem.*, 1996, **110**, 235.
160. S. Domínguez-Domínguez, A. Berenguer-Murcia, D. Cazorla-Amorós and A. Linares-Solano, *J. Catal.*, 2006, **243**, 74.
161. C. Li, Z. Shao, M. Pang, C. T. Williams and C. Liang, *Catal. Today*, 2012, **186**, 69.
162. S. Domínguez-Domínguez, A. Berenguer-Murcia, B. K. Pradhan, A. Linares-Solano and D. Cazorla-Amorós, *J. Phys. Chem. C.*, 2008, **112**, 3827.
163. A. Bacchi, M. Carcelli, M. Costa, A. Leporati, E. Leporati, P. Pelagatti, C. Pelizzi and G. Pelizzi, *J. Organomet. Chem.*, 1997, **535**, 107.
164. S. Domínguez-Domínguez, A. Berenguer-Murcia, A. Linares-Solano and D. Cazorla-Amorós, *J. Catal.*, 2008, **257**, 87.

165. C. Li, Z. Shao, M. Pang, C. T. Williams, X. Zhang and C. Liang, *Ind. Eng. Chem. Res.*, 2012, **51**, 4934.
166. X.-Y. Quek, Y. Guan and E. J. M. Hensen, *Catal. Today*, 2012, **183**, 72.
167. M. D. Navalikhina, N. E. Kavalerskaya, E. S. Lokteva, A. A. Peristyi, E. V. Golubina and V. V. Lunin, *Russ. J. Phys. Chem. A.*, 2012, **86**, 1800.
168. F. Bautista, J. Campelo, A. Garcia, D. Luna, J. Marinas, R. Quiros and A. Romero, *Catal. Lett.*, 1998, **52**, 205.
169. T. Vergunst, F. Kapteijn and J. A. Moulijn, *Ind. Eng. Chem. Res.*, 2001, **40**, 2801.
170. W. Liu, C. Otero Arean, S. Bordiga, E. Groppo and A. Zecchina, *ChemCatChem*, 2011, **3**, 222.
171. S. D. Jackson and L. A. Shaw, *Appl. Catal. A: Gen.*, 1996, **134**, 91.
172. J. Jia, K. Haraki, J. N. Kondo, K. Domen and K. Tamaru, *J. Phys. Chem. B.*, 2000, **104**, 11153.
173. A. C. Gluhoi, J. W. Bakker and B. E. Nieuwenhuys, *Catal. Today*, 2010, **154**, 13.
174. Y. Segura, N. López and J. Pérez-Ramírez, *J. Catal.*, 2007, **247**, 383.
175. S. Schimpf, M. Lucas, C. Mohr, U. Rodemerck, A. Brückner, J. Radnik, H. Hofmeister and P. Claus, *Catal. Today*, 2002, **72**, 63.
176. A. V. Ivanov, A. E. Koklin, E. B. Uvarova and L. M. Kustov, *Phys. Chem. Chem. Phys.*, 2003, **5**, 4718.
177. R. J. Chimentao, G. R. Valenca, F. Medina and J. Perez-Ramirez, *Appl. Surf. Sci.*, 2007, **253**, 5888.
178. Z. H. Suo, C. Y. Ma, W. P. Liao, M. S. Jin and H. Y. Lv, *Fuel Process. Technol.*, 2011, **92**, 1549.
179. M. S. Wong, P. J. J. Alvarez, Y. L. Fang, N. Akcin, M. O. Nutt, J. T. Miller and K. N. Heck, *J. Chem. Technol. Biotechnol.*, 2009, **84**, 158.
180. M. O. Nutt, J. B. Hughes and M. S. Wong, *Environ. Sci. Technol.*, 2005, **39**, 1346.
181. M. Bonarowska, J. Pielaszek, V. A. Semikolenov and Z. Karpinski, *J. Catal.*, 2002, **209**, 528.
182. F. Alonso, I. P. Beletskaya and M. Yus, *Chem. Rev.*, 2002, **102**, 4009.
183. G. Yuan, C. Louis, L. Delannoy and M. A. Keane, *J. Catal.*, 2007, **247**, 256.
184. G. Yuan, J. L. Lopez, C. Louis, L. Delannoy and M. A. Keane, *Catal. Commun.*, 2005, **6**, 555.
185. M. A. Keane, S. Gómez-Quero, F. Cárdenas-Lizana and W. Q. Shen, *ChemCatChem*, 2009, **1**, 270.
186. G. Pina, C. Louis and M. A. Keane, *Phys. Chem. Chem. Phys.*, 2003, **5**, 1924.
187. E. J. Shin, A. Spiller, G. Tavoularis and M. A. Keane, *Phys. Chem. Chem. Phys.*, 1999, **1**, 3173.
188. E. J. Shin and M. A. Keane, *Chem. Eng. Sci.*, 1999, **54**, 1109.
189. L. Calvo, A. F. Mohedano, J. A. Casas, M. A. Gilarranz and J. J. Rodriguez, *Carbon*, 2004, **42**, 1377.

190. G. Yuan and M. A. Keane, *Appl. Catal. B: Environ.*, 2004, **52**, 301.
191. Y. Shao, Z. Y. Xu, H. Q. Wan, Y. Q. Wan, H. Chen, S. R. Zheng and D. Q. Zhu, *Catal. Commun.*, 2011, **12**, 1405.
192. I. Witonska, A. Krolak and S. Karski, *J. Mol. Catal. A: Chem.*, 2010, **331**, 21.
193. S. Gómez-Quero, F. Cárdenas-Lizana and M. A. Keane, *Chem. Eng. J.*, 2011, **166**, 1044.
194. S. Gómez-Quero, F. Cárdenas-Lizana and M. A. Keane, *Ind. Eng. Chem. Res.*, 2008, **47**, 6841.
195. G. S. Pozan and I. Boz, *J. Hazard. Mater.*, 2006, **136**, 917.
196. G. Yuan and M. A. Keane, *Chem. Eng. Sci.*, 2003, **58**, 257.
197. S. Gómez-Quero, F. Cárdenas-Lizana and M. A. Keane, *AIChE J.*, 2010, **56**, 756.
198. I. Witonska and A. Krolak, *Przem. Chem.*, 2011, **90**, 636.
199. S. Gómez-Quero, F. Cárdenas-Lizana and M. A. Keane, *Nanotechnology*, 2012, **23**, 294002.
200. S. Gómez-Quero, F. Cárdenas-Lizana and M. A. Keane, *J. Catal.*, 2013, **303**, 41.
201. Y. A. Serguchev and Y. V. Belokopytov, *Kinet. Catal.*, 2001, **42**, 174.
202. P. T. Anastas, M. M. Kirchhoff and T. C. Williamson, *Appl. Catal. A: Gen.*, 2001, **221**, 3.
203. C. Jiménez-González, P. Poechlauer, Q. B. Broxterman, B.-S. Yang, D. am Ende, J. Baird, C. Bertsch, R. E. Hannah, P. Dell'Orco, H. Noorman, S. Yee, R. Reintjens, A. Wells, V. Massonneau and J. Manley, *Org. Process. Res. Dev.*, 2011, **15**, 900.
204. P. T. Anastas and J. C. Warner, *Green Chemistry: Theory and Practice*, Oxford University Press, New York, 1998.
205. A. Goršek and P. Glavič, *Chem. Eng. Res. Des.*, 1997, **75**, 709.
206. R. A. Sheldon, *Pure Appl. Chem.*, 2000, **72**, 1233.
207. F. Gao and D. W. Goodman, *Chem. Soc. Rev.*, 2012, **41**, 8009.
208. M. Sankar, N. Dimitratos, P. J. Miedziak, P. P. Wells, C. J. Kiely and G. J. Hutchings, *Chem. Soc. Rev.*, 2012, **41**, 8099.
209. W. Yu, M. D. Porosoff and J. G. Chen, *Chem. Rev.*, 2012, **112**, 5780.
210. A. Borgschulte, R. J. Westerwaal, J. H. Rector, H. Schreuders, B. Dam and R. Griessen, *J. Catal.*, 2006, **239**, 263.
211. R. Prins, *Chem. Rev.*, 2012, **112**, 2714.
212. I. Nakamura, H. Mantoku, T. Furukawa, A. Takahashi and T. Fujitani, *Surf. Sci.*, 2012, **606**, 1581.
213. A. M. Alexander and J. S. J. Hargreaves, *Chem. Soc. Rev.*, 2010, **39**, 4388.
214. Z. X. Hao, Z. B. Wei, L. J. Wang, X. H. Li, C. Li, E. Z. Min and Q. Xin, *Appl. Catal. A: Gen.*, 2000, **192**, 81.
215. E. Furimsky, *Appl. Catal. A: Gen.*, 2003, **240**, 1.
216. U. K. Singh and M. A. Vannice, *Appl. Catal. A: Gen.*, 2001, **213**, 1.
217. N. M. Bertero, A. F. Trasarti, C. R. Apesteguia and A. J. Marchi, *Appl. Catal. A: Gen.*, 2011, **394**, 228.

218. B. S. Akpa, C. D'Agostino, L. F. Gladden, K. Hindle, H. Manyar, J. McGregor, R. Li, M. Neurock, N. Sinha, E. H. Stitt, D. Weber, J. A. Zeitler and D. W. Rooney, *J. Catal.*, 2012, **289**, 30.
219. K. C. Hass, W. F. Schneider, A. Curioni and W. Andreoni, *Science*, 1998, **282**, 265.
220. J. Blomquist, L. E. Walle, P. Uvdal, A. Borg and A. Sandell, *J. Phys. Chem. C*, 2008, **112**, 16616.
221. D. A. Buchanan and G. Webb, *J. Chem. Soc. Faraday Trans.*, 1975, **71**, 134.
222. E. Cetinkaya, I. Dincer and G. F. Naterer, *Int. J. Hydrogen Energy*, 2012, **37**, 2071.
223. M. E. E. Abashar, *Chem. Eng. Process.*, 2004, **43**, 1195.
224. B. M. Nagaraja, A. H. Padmasri, P. Seetharamulu, K. H. P. Reddy, B. D. Raju and K. S. R. Rao, *J. Mol. Catal. A: Chem.*, 2007, **278**, 29.
225. B. M. Nagaraja, A. H. Padmasri, B. D. Raju and K. S. R. Rao, *Int. J. Hydrogen Energy*, 2011, **36**, 3417.
226. Y. Yang, C. W. Hu and M. M. Abu-Omar, *ChemSusChem*, 2012, **5**, 405.
227. H. T. Wang and J. F. Yao, *Ind. Eng. Chem. Res.*, 2006, **45**, 6393.
228. T. A. Baker, X. Y. Liu and C. M. Friend, *Phys. Chem. Chem. Phys.*, 2011, **13**, 34.
229. J. Ohyama, A. Esaki, Y. Yamamoto, S. Arai and A. Satsuma, *RSC Adv.*, 2013, 3, 1033.
230. M. J. Climent, A. Corma, J. C. Hernandez, A. B. Hungria, S. Iborra and S. Martinez-Silvestre, *J. Catal.*, 2013, **292**, 118.
231. M. Yan, T. Jin, Y. Ishikawa, T. Minato, T. Fujita, L.-Y. Chen, M. Bao, N. Asao, M.-W. Chen and Y. Yamamoto, *J. Am. Chem. Soc.*, 2012, **134**, 17536.
232. M. Yan, T. Jin, Q. Chen, H. E. Ho, T. Fujita, L.-Y. Chen, M. Bao, M.-W. Chen, N. Asao and Y. Yamamoto, *Org. Lett.*, 2013, **15**, 1484.

CHAPTER 12

Metal–Support Interaction Effects on Gold Catalysts over Reducible Oxides

HONGJING WU[a,b] AND LEONARDA F. LIOTTA*[a]

[a] Institute for the Study of Nanostructured Materials (ISMN)-CNR, Via Ugo La Malfa 153, 90146, Palermo, Italy; [b] Department of Applied Physics, Northwestern Polytechnical University (NPU), 127 Youyi Xilu, 710072, Xi'an, PR China
*Email: liotta@pa.ismn.cnr.it

12.1 Introduction

It is well-known that, compared with homogeneous catalysis, heterogeneous catalysis has the advantages of repeatable use of catalysts, easy separation of catalysts from final products, *etc.*, which are well-described in standard textbooks of catalysis. Over the past decades, bi-/multi-component catalysts have also attracted increasing attention in several fields such as fine chemicals synthesis, electrochemical reactions, photocatalysis, energy production and environmental depollution.[1] These composite catalysts are generally composed of one or more catalytically active components and a functional support, such as easily reducible oxide.

In addition to the intrinsic physical and chemical properties of the constitutive components, heterogeneous catalytic performances are largely dependent on the catalyst nanostructures. For example, chemically processed nanomaterials, such as catalysts synthesized by the sol-gel method, usually show improved catalytic performance due to the high surface area

RSC Catalysis Series No. 18
Heterogeneous Gold Catalysts and Catalysis
Edited by Zhen Ma and Sheng Dai

Published by the Royal Society of Chemistry, www.rsc.org

and density of active sites on the surface of nanoparticles.[2,3] More recently, the possible synergetic catalytic effect in composite catalysts has been proposed by Shi.[1] The synergetic catalytic effect was defined as a certain kind of cooperation between different components and/or active sites in one catalyst, which resulted to an extent better than the arithmetic summation of those by corresponding individual components.[2–9] However, this kind of cooperation or interaction between varied components is generally complicated, and the underlying mechanism is not thoroughly addressed in the literature. Nevertheless, from a large amount of literature reports, such a cooperation/interaction effect indeed exists in some composite catalysts and this effect may play an important role in the heterogeneous catalytic reaction.

At the length scale of nanometres, however, the surface area to volume ratio increases significantly, so that surface/facet and interface effects become predominant and even modify significantly the macroscopic properties of the nanocatalysts to enhance their catalytic performance.[10] In surface science, model catalysts such as single-crystal surfaces of metal or metal oxide have been widely adopted by surface scientists to systematically study the effect of surface–interface structure on catalytic reactions. However, it has led to the large material gaps between model catalysis studies and real catalysis.[11–13] Right now, nanomaterials with well-defined surface/interface structures are ideal materials serving as novel model catalysts to bridge these gaps in heterogeneous catalysis.[14–18]

Interface effects in noble metal(s) reducible oxide(s) composite catalysts applied in heterogeneous catalysis can be broadly classified into several categories: (1) noble metal–noble metal interaction; (2) noble metal–reducible oxide interaction; (3) reducible oxide–reducible oxide interaction; and (4) interactions among noble metal(s) and reducible oxide(s).

Owing to the importance of surface/interface structures in the catalytic performance of noble metal–reducible oxide composite catalysts, over the past two decades considerable research attention has been focused on controlling the surface/interface structures of/between noble metal(s) and reducible oxide(s) so as to search for activity-determining factors during real reaction processes. In all cases, it has been well-documented that the size, composition, electronic structure, and facet structure of noble metal(s) and reducible oxide(s) are important factors in determining their catalytic performance.[10] In particular, creating new noble metal–noble metal or noble metal–reducible oxide interfaces by introducing the noble metal or reducible oxide leads to enhanced catalytic performance of the composite catalysts. For example, metal–oxide interfaces (*e.g.* Au-TiO_2, Au-Fe_3O_4, Pt-Fe_xO_y, *etc.*) have been found able to enhance catalysis in CO oxidation.[16,19–21] The Au-Pt interface could facilitate CO oxidation and also stabilize the Pt catalysts against dissolution under real reaction conditions.[22]

In this chapter, we first introduce briefly the syntheses of heterogeneous gold supported reducible oxide composite catalysts (Au-MO_x) prepared by chemical routes. Subsequently the emphasis is on the catalytic performance of various heterogeneous composite catalysts enhanced by the interaction effects between gold nanoparticles and reducible host supports, depending on the varied interfacial surfaces among them. Several kinds of possible synergetic/cooperation effects or interaction effects and the corresponding structural properties based on heterogeneous composite catalysts are proposed.

To understand the interaction effects more clearly, catalytic reactions catalyzed by composite catalysts both reported in the literature and the subject of our own investigations are listed as typical examples and discussed as to their specific behaviors and properties in terms of the interaction effects between the noble metal(s) and support(s). CO oxidation was used as a sensitive probe reaction in all the cases because it is easy to carry out. Finally, conclusions and perspectives are given to propose the challenges in design and control of synergetic/interaction effects in heterogeneous composite catalysts.

12.2 General Description of Chemical Routes for Synthesizing Heterogeneous Composite Catalysts

Gold nanocrystals such as nanoparticles (0D), nanowires (1D) and thin films containing nanophases (2D) can be synthesized by various chemical approaches. For a more recent overview in preparation of noble metals we suggest the reader refer to the literature available.[10] Traditional aqueous sol-gel chemistry has been employed for decades for the preparation of various oxides based on the formation of different intermediates and processes.[1] The basic concept of the chemical processes is the solid formation from a homogeneous solution or sol, by various approaches, such as conventional precipitation or coprecipitation, slow homogeneous precipitation, controlled hydrolysis, sol-gel transition, self-assembly on the molecule level, and crystallization.[1] For the syntheses of supports, the chemical sol-gel routes have become very attractive in catalysis because they can exhibit excellent and/or unique properties when the multi-interfacial oxide-based supports are prepared. In fact, very active components such as noble metal(s) supported on reducible oxide support(s) like TiO_2, CeO_2, Fe_2O_3, *etc.*, or their mixed composites, can exhibit strong synergetic/cooperation effects in heterogeneous catalysis because their supports can also play, to a certain extent, a positive role in the catalytic process.[1]

The general preparation method of supported gold heterogeneous catalysts is to deposit, graft or coat, both chemically and physically, active gold species or their precursors, onto or into the pre-synthesized reducible oxide supports, followed by post-treatment if necessary.[1] In this chapter, attention is given to the most commonly used preparation methods, *i.e.* impregnation (IMP), deposition–precipitation (DP), coprecipitation (CP), *etc.*, and to special deposition techniques such as solvated metal atom impregnation (SMAI) and solvated metal atom deposition (SMAD). The advantages of the common approaches include a very broad selection and combination of both supports and loaded noble metals, and the possible elaborate control over the structure and morphology of both noble metal(s) and support(s). However, they usually suffer from the drawbacks of multistep preparation, *i.e.* the dependence of the dispersivity of the active species on the processing route and the corresponding properties of the heterogeneous catalysts.

In impregnation the support is contacted with a solution of the metal precursor, and is then aged, dried, and calcined. The use of different conditions can lead to different interactions, of fundamental importance for catalytic applications. Conventional impregnation with chloroauric acid ($HAuCl_4$) has been shown to produce much less active gold catalysts than deposition–precipitation or coprecipitation. However, the simplicity of the methodology and the convenience of using chloroauric acid as the gold source make impregnation attractive for industrial scale-up purposes. Because of this, much research has been dedicated to improving the preparation method. It should be noted that this ion adsorption method strictly depends on the isoelectric point (IEP) of the supports. Oxides with an IEP around 7 (TiO_2, CeO_2, ZrO_2, Fe_2O_3) produced very active species, whereas acidic supports, such as SiO_2 or basic MgO usually appear less active. The interaction between the charge that the surface assumes in solution and the ionic gold species can be used fruitfully for preparing a supported Au catalyst.

Deposition–precipitation consists of the precipitation of a metal hydroxide or carbonate on the particles of a support *via* the reaction of a base with the precursor of the metal. It was the first efficient method reported to produce highly active gold supported nanoparticles due to the enhanced support–metal interaction. The optimum pH range for precipitation that also assures an efficient metal utilization (>90%) is primarily dictated by the IEP of the supports. However, this leads to the main constraint of this method, *i.e.* the inapplicability of acidic supports such as SiO_2 (IEP = 2). To overcome the limitations of deposition precipitation, SMAI and SMAD can be used, *i.e.* air-sensitive and thermally unstable bis(toluene)Au(0) complex solution or metallic gold used as the starting materials can be impregnated into the support or vaporized in a vacuum system by heating and then deposited on the support. However, the organometallic precursors or experimental facilities are expensive.

An alternative way to synthesize heterogeneous composite catalysts can be classified as the one-pot approach in which the desired species, mostly their precursors such as inorganic salts and/or organometallics are incorporated together in one system, as a homogeneous phase before extensive solid formation.[1] Such one-pot approaches include different routes such as coprecipitation and co-hydrolysis of different components, and instant forced solvent evaporation followed by post-treatments such as thermal decomposition.[3] For example, an aqueous solution of $HAuCl_4$ and a water soluble metal salt are poured in an alkaline solution. After precipitation the hydroxides or carbonates are filtered, washed and dried, and then calcined.

The one-pot synthesis has the valuable advantages of simplicity, component homogeneity, and the least waste of substances. Nevertheless, such a one-pot synthesis will inevitably suffer from difficulties in the facile and precise tuning of morphology, size distribution, and component dispersity.[1]

Ma and Dai recently reported that it is desirable to build up more complicated catalytic architectures by incorporating additional components (*e.g.* modifiers and additives).[23] They proposed three strategies to construct the heterogeneous composite gold catalysts: (1) pre-modification of the support; (2) co-addition of gold and inorganic component, and (3) post-modification of the composite gold catalyst. In particular, more complicated and sufficient interaction effects are formed between metal–support, metal–modifier, and modifier–support interfaces. As a result, new active sites may be created and the gold active center may be more stable at these relatively complicated multi-interfaces.[23]

12.3 Relationship between Interaction Effects at the Interfaces and CO Oxidation Activity

Heterogeneous catalysis first occurs with the reactants adsorbed to the surface/interface of a heterogeneous composite catalyst. Afterwards, the adsorbed reactant molecules usually find their own ways to interact with the catalytic surface/interface of the composite catalysts. Finally, the activated reactant molecules decompose into intermediate products and find paths to form products *via* the catalytic surface/interface of the composite catalysts. To enhance the catalytic activity, stability, and selectivity of the composite catalysts, it is therefore desirable to construct the proper surface/interface in the catalytic architecture by designing and controlling the composition, particle size, electronic effect, and structure of the composite catalyst. Table 12.1 summarizes the various Au-MO_x composite catalysts and their applications in CO oxidation to determine the structure–property correlation in heterogeneous catalytic architectures.

12.3.1 Au-TiO_2 Interaction

Semiconductive oxides have attracted considerable attention due to their wide application in heterogeneous catalysis. At present, TiO_2 has been mainly proven to be a promising photocatalyst due to its intriguing optical and electric properties, low cost, and easy of availability, and the fact that is eco-friendly and environmentally safe.[24–26] Moreover, considerable attention has been paid to the catalytic removal of CO, CH_4 and volatile organic compounds (VOCs) at low temperatures using TiO_2 as support because of its strong interaction with noble metals such as Au, Pt, Pd, *etc.* at the interface.[1]

Recently, Aït-Mansour and co-workers reported that Pt nanoclusters on rutile TiO_2 possessed a strong metal–support interaction (SMSI) between Pt and the reduced thin oxide layer of TiO_x from the support annealing.[27] As a

Table 12.1 Major synthetic approaches of selected composite catalysts and their catalytic performances.

Catalyst	Synthetic approach	Key factors	Ref.
TiO_2	Sputter annealing	SMSI	27
Au (Ag, Ir)-TiO_2	Deposition–precipitation (DP)	Shape, low-coordinated number, electronic state	28–34
Au (Ir)-TiO_2	Coprecipitation (CP)	SMSI	31
Au or Pt-TiO_2	Impregnation (IMP)	Electronic state	35,36
Au-TiO_2	Mechanical mixing	Perimeter interface	37
Au (Fe)-TiO_2	Colloidal deposition (CD)	SMSI	38
Pt-TiO_2	Arc plasma deposition (APD)	Electronic state	36,39
Au-TiO_2	Solvated metal atom impregnation (SMAI)	Particle size	40
Au-TiO_2-Ru(0001)	Molecule beam evaporation (MBE)	WMSI[a]	41
Au-TiO_2	Grafting	Electronic state	42
Au-CeO_2	DP	Electronic state, shape, SMSI	47–53
Au-CeO_2	CP	Electronic state	47,48
Ag or Pd-CeO_2	IMP	Electronic state, vacancy, SMSI	54,55
Au-CeO_2	Solvated metal atom deposition (SMAD)	Electronic state	47
Au-CeO_2	Solution–combustion	Electronic state, vacancy	55
Au-Fe_2O_3	DP	Particle size	59
Au-Fe_2O_3	CP	Particle size, electronic state	59–63
Au-Fe_2O_3	IMP	Particle size	60
Au-Fe_3O_4	CD	SMSI	58
Au-Fe_xO_y	Grafting	Particle size	64
Au-Fe_2O_3	Double impregnation (DIM)	Electronic state	65
Au-Fe_2O_3	Liquid phase reductive deposition (LPRD)	Electronic state	65
Au-Fe_2O_3	Ultrasonication	Electronic state	65
Au-MnO_x	DP	Electronic state	49,66–68
Au-MnO_x	CP	Electronic state, SMSI	69
Au-MnO_x	Direct thermal decomposition (DTD)	Electronic state, active oxygen, vacancy, SMSI	70,71
Au-$M(OH)_x$	Grafting	Particle size, electronic state	72,73
Au-ZnO	CD	SMSI, shape	74,80
Au-Co_3O_4	CD	SMSI, shape	75
Au-CuO	CP	Particle size	76
Au (Pd, Pt)-SnO_2	CP	SMSI	77
Au-Fe (La)PO_4	DP	Electronic state	78
Au-$MgFe_2O_4$	CD	Active oxygen	79

[a]Weak metal support interaction.

result, suboxide-encapsulated Pt clusters were found to be almost inactive in catalytic CO oxidation. Fortunately, they could transfer this inert SMSI state into a very active and thermally stable one for CO oxidation using the sputter annealing procedure, thus overcoming the SMSI limitation.[27] This suggests that the method of synthesis has a considerable or even decisive influence on the interaction effects between metals and supports, and that we can design and control these effects by choosing a suitable synthesis method.

Up to now, various synthesis methods have been developed to prepare a heterogeneous Au-TiO_2 composite catalyst. These have mainly included deposition–precipitation,[28–34] coprecipitation,[31] wetness impregnation,[35,36] mechanically mixing,[37] colloidal deposition,[38] arc plasma deposition (APD),[36,39] solvated metal atom impregnation,[40] sputter annealing,[27] molecule beam evaporation (MBE),[41] and grafting.[42] As can be imagined, the extent of interaction effects between metals and supports varies with the changing of synthesis methods. For example, a relatively weak interaction between the Au and the TiO_2 substrate was found in the literature in which the catalyst was prepared by MBE,[41] while another kind of strong SMSI was revealed when the catalyst was prepared by sputter annealing.[27] Louis and co-workers have systematically studied an Au-TiO_2 catalyst prepared by DP with NaOH and urea, and calcined at various temperatures between 100 and 400 °C, in CO oxidation.[29] They found that the catalytic activity increased with the percentage of metallic gold, being maximum after calcination at 200 °C for both types of samples. At higher calcination temperatures, the catalytic activity dropped (Figure 12.1).

The reduction in activity was mainly due to a change in the particle shape with the calcination temperature rather than to the increase in the particle size. At a low pre-treatment temperature [Figure 12.2(a)], the gold particles presented small facets with rounded parts, and their outer surface contained a large proportion of low-coordinated sites (steps and corners). However, they were truncated octahedra with smooth facets under a higher pre-treatment temperature [Figure 12.2(b)]. Figure 12.2(b) showed that the (111) plane of gold was parallel to the (110) plane of the rutile. By numerical analysis, the intensity profiles in the (111) lattice planes, parallel to the interface, have been recorded from the interface until the top of the particle. As seen in Figure 12.2(c), the gold particle was accommodated in the lattice of the support in the first three layers from the interface, which means a 10% expansion of the gold lattice. Then, it was relaxed to about 3% of the bulk value far away from the interface. This result indicates that the gold particles interact strongly with the TiO_2 support at the interface under higher calcination temperatures. However, this kind of SMSI is so strong that the gold particle shape is changed, which in turn leads to a decrease in the number of active low-coordinated sites for CO oxidation.[29]

In contrast, Haruta and co-workers elucidated the role of the contact structure between gold and metal oxide support in low-temperature CO oxidation by mechanically mixing a colloidal gold with TiO_2 powder under

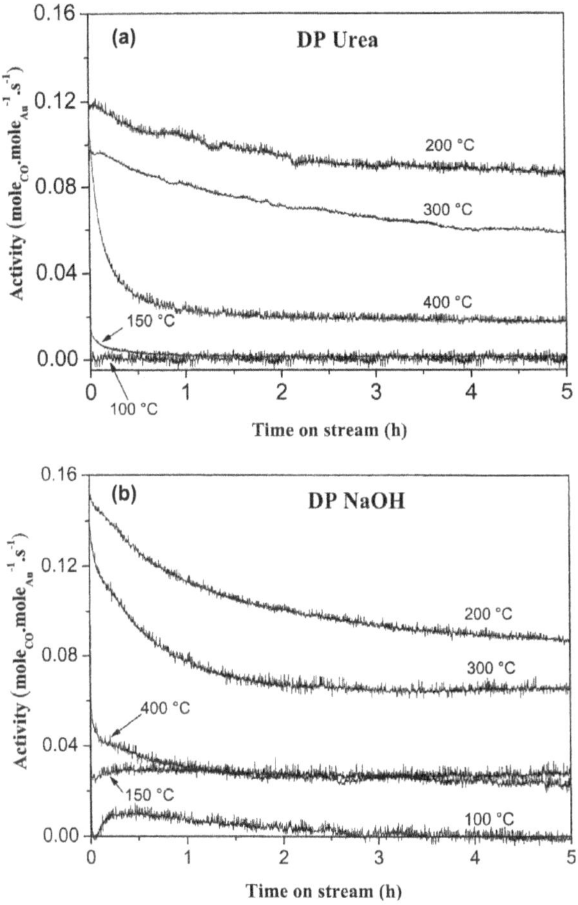

Figure 12.1 Activity in CO oxidation at 5 °C for the DP urea (a) and DP NaOH samples (b) *vs.* the time-on-stream for various calcination temperatures (100, 150, 200, 300, and 400 °C).
Reproduced with permission from ref. 29. Copyright 2004 Elsevier BV.

calcination at different temperatures.[37] The catalytic activity increased appreciably with an increase in calcination temperature up to 600 °C even though the gold particles grew to larger ones, reaching a level with almost the same turnover frequency (TOF) as that of Au-TiO_2 produced by the DP method. This means that the distance of the perimeter interface around gold particles could be related to the high catalytic activity (Figure 12.3). By increasing the gold particle size at high temperatures and changing the gold particle shape due to the SMSI effects, the distance of the perimeter interface increases and thus enhances CO oxidation.

Furthermore, some authors have also reported that different preparation methods would lead to the varied electronic state of Au or Pt over TiO_2. Gold

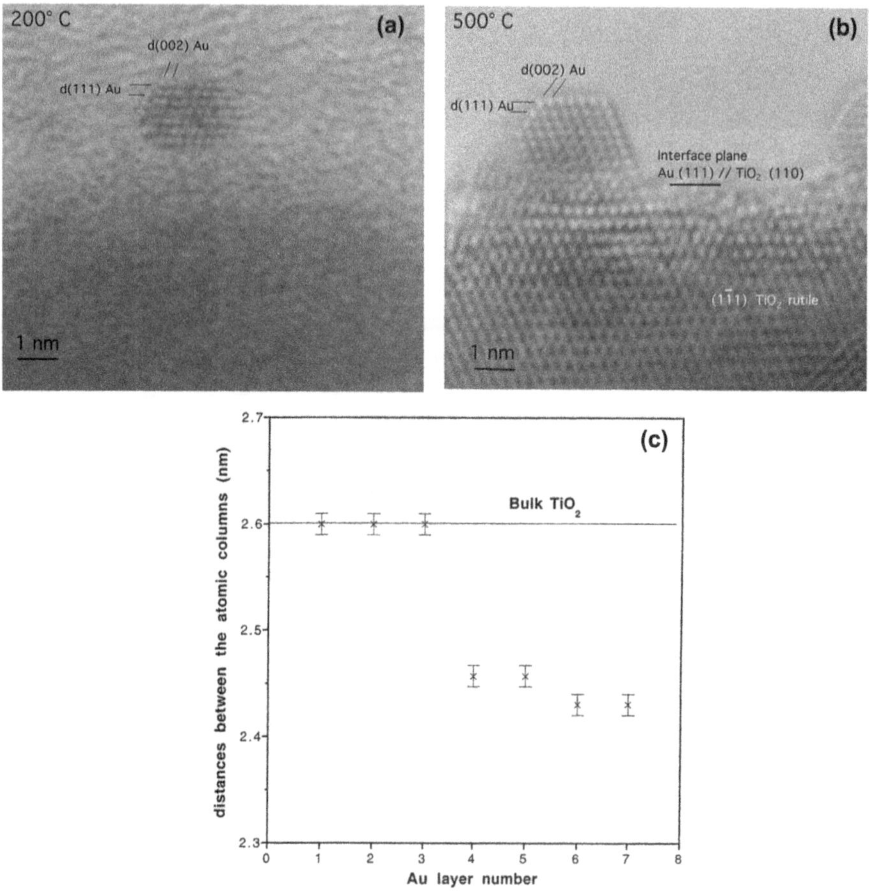

Figure 12.2 High-resolution transmission emission microscopy (TEM) images of gold particles in the profile view in the DP urea sample reduced under H_2: (a) 200 °C; (b) at 500 °C (the view axis, perpendicular to the image, is in the [110] direction); and (c) average lattice distance in the successive layers starting from the interface.

particles were mainly found to be metallic Au on the surface after the calcination treatment.[29,32,42] The metallic Au particles or clusters could chemosorb CO molecules or activate oxygen molecules through gold–oxygen complexes formed by gold supported on a reducible oxide (TiO_2).[32] Park and co-workers recently reported that Pt-TiO_2 catalyst prepared by APD showed superior catalytic activity for CO oxidation compared with impregnated Pt nanoparticles due to the catalytically active nature of the mild surface oxidation and the active Pt metal, rather than inactive Pt bulk oxide prepared by impregnation.[36]

Recently the higher catalytic activity of tri-component Au-M (Ag, Fe, Ir, *etc.*)-TiO_2 catalysts compared with TiO_2-supported gold has been

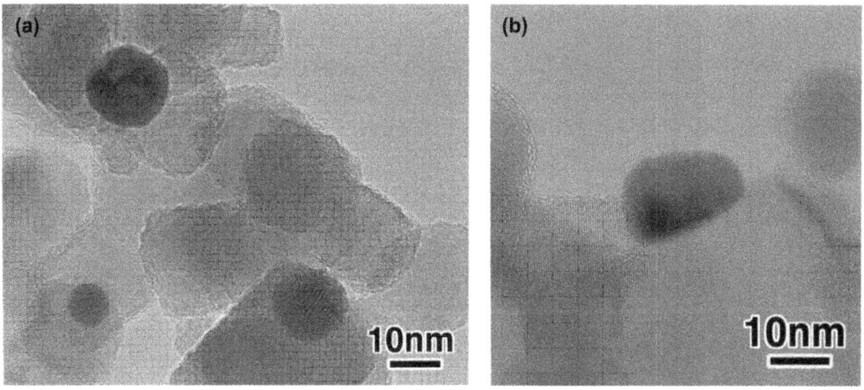

Figure 12.3 High-resolution TEM images of a mechanical mixture of Au colloids with TiO_2 powder calcined in air at different temperatures: (a) 200 °C; and (b) 600 °C. Au loading is 3.0 wt%.
Reproduced with permission from ref. 37. Copyright 1998 J.C. Baltzer AG, Science Publishers.

reported.[30,31,38] For example, iron doping enhanced the catalytic activity of the Au-TiO_2 catalyst as well as stabilizing the catalyst at high temperatures.[38] Iron doping helps in the creation of new adsorption sites at the Au-TiO_2 interface for the adsorption and activation of O_2 molecules. SMSI could be also responsible for these synergetic effects and presented higher catalytic activity compared with bi-component Au-TiO_2. With respect to Fe, Ir or Ag could form bimetallic particles with Au and exhibit a strong interaction between Au and the second metal. This would modify the adsorption properties of catalyst surface, resulting in enhanced catalytic performance of the catalysts.[30,31]

Goodman has presented a good review of the SMSI effect involving group VIII metals and catalytically active Au, and it is apparent that these two phenomena should be closely related.[43] In the late 1970s and 1980s considerable attention was given by Tauster and co-workers[44–46] to the unusual properties of Pt and Ir supported on TiO_2 after reduction at relatively high temperatures. Associated with such SMSI effects were the following: (1) CO and H_2 chemisorption suppression; (2) an increase in methanation activity; and (3) a decrease in hydrogenolysis activity. Similarly, the essential features of the interaction of Au with TiO_2 according to the recent theoretical and experimental studies that lead to enhanced catalytic activity are: (1) wetting of the support by the cluster; (2) strong bonding between the Au atoms at the interface with surface defects (reduced Ti sites); (3) electron-rich Au; (4) annealing at temperatures in excess of 477 °C; and (5) oxidation leads to deactivation *via* sintering of Au.

In any case, future work on catalytically active Au on TiO_2 and other supports known to activate Au should be considered to uncover many of the remaining mysteries associated with the SMSI effects.

12.3.2 Au-CeO_2 Interaction

Similar to TiO_2, CeO_2 has been commonly used as a reducible oxide support for gold nanoparticles to enhance catalytic activity in CO oxidation. Several preparation methods including deposition–precipitation,[47–53] coprecipitation,[47,48] wetness impregnation,[54,55] solvated metal atom dispersion,[47] and solution–combustion[55] have been successfully applied to synthesize active Au (Ag, Pd)-CeO_2 composite catalysts. Using density functional theory (DFT) studies, in most cases it has been proved that CO oxidation by Au-CeO_2 is similar to that by Au-TiO_2.[56,57] The following implications have been adduced:

(1) The Au dopant prefers to interact with Ce-defect sites on the surface rather than go into the bulk.
(2) The Au dopant weakens the bond between the surface oxygen atoms and the doped oxide, making it more reactive.
(3) These oxygen atoms react with gas-phase CO to form a CO_2 molecule that readily desorbs.
(4) The oxygen vacancy created by CO_2 removal adsorbs an oxygen molecule that reacts with CO to form a carbonate.
(5) The carbonate decomposes readily to produce gaseous CO_2 and heals the oxygen vacancy.
(6) Doping with Ag, Cu, Ni, Pd, or Pt weakens the oxygen bond to the surface more than doping with Au.

Recently, our group systematically investigated the relationship between the structure and CO oxidation activity of Au-CeO_2 catalysts.[47,53] The structural and surface analyses evidenced the presence of a modified CeO_2 phase in the case of the DP sample, and the presence of pure CeO_2 and metallic gold crystallites in the case of the SMAD and CP samples. By comparing the activities of the different catalysts, it was proposed that the small gold particle size was not the main requisite for the achievement of the highest CO conversion. The strong interaction between ionic gold and CeO_2, by enhancing the CeO_2 surface oxygen reducibility, may determine the particularly high activity. The presence of ionic gold in intimate contact with CeO_2, even forming a mixed oxide $Au_xCe_{1-x}O_{2-\delta}$, contributed to the weakening of the Ce–O bond, differently from the case of metallic gold dispersed on the support surface.[47]

More recently, our group has further studied the structure of the metal support interface and oxidation state of gold nanoparticles supported on CeO_2.[53] The result demonstrated the existence of a large interface between the gold particle and the support oxide, characterized by well-defined Au-O, and Au-Ce interactions extending up to $\sim$6.4 Å observed in the oxidized and CO-treated samples (Figure 12.4). From the X-ray absorption near-edge structure (XANES) analysis (data not shown), it was argued that a fraction of the gold atoms, most likely those at the interface with the support, are

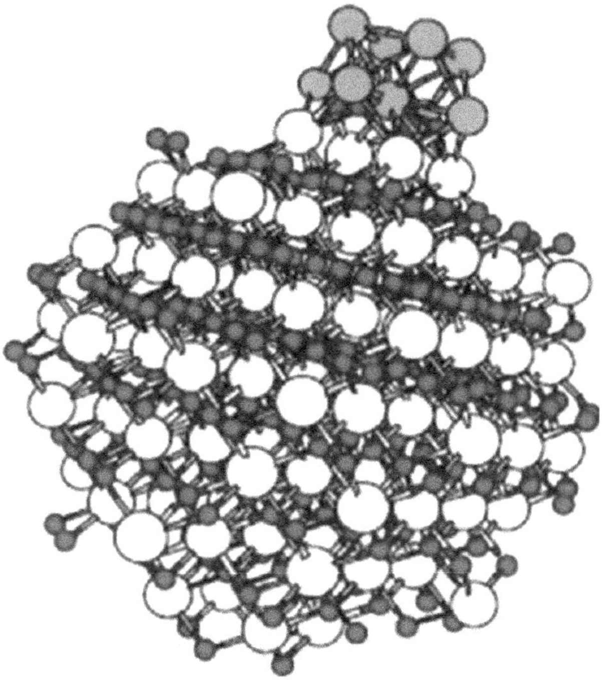

Figure 12.4 Pictorial view of a gold cluster supported on ceria. Gold atoms in yellow, oxygen atoms in red, and cerium atoms in white.
Reproduced with permission from ref. 53. Copyright 2012 American Chemical Society.

oxidized and in an oxidation state lower then Au^{3+}. It was proposed that the gold–ceria interface structure involved a contact oxygen layer between the gold and the support, giving rise to roughly linear Au-O-Ce bonds extending over a wide interface region, basically stable in oxidizing and reducing cycles. The gold–ceria strong interaction enhanced the CeO_2 surface oxygen mobility by weakening the Ce–O bond.[53]

Chen and co-workers have also studied the structure, morphology, and electronic properties of Au-CeO_2 catalysts prepared by DP.[49] The existence of Au^0 and Au^{3+} states were detected on the catalysts, whereas only metallic gold was detected on Au-MnO_2. The Au-CeO_2 demonstrated higher catalytic activity for CO oxidation than Au-MnO_2. The co-existence of metallic and oxidized Au species on Au-CeO_2 seemed to be the main reason for the higher activity with respect to Au-MnO_2. However, Flytzani-Stephanopoulos and co-workers have reported that well dispersed gold nanoparticles in reduced form provided the active sites for the CO oxidation reaction on supports.[50] They investigated various gold nanoparticles interacting with supports to different extents in the CO oxidation and water gas shift (WGS) reaction. They found that the reduced gold nanoparticles without strongly bound $Au^{\delta+}$ species on the surface of supports could promote CO oxidation activity but not WGS activity. This means that the weakly bounded metallic Au species on the surface of supports can only enhance the catalytic activity of CO oxidation rather than the WGS reaction. This result is absolutely in disagreement with the results reported by Chen and co-workers and by our group.

Recently, Yoshida and co-workers used a visualized method to understand gas molecules interacting with supported nanoparticulate catalysts at reaction conditions by means of aberration-corrected environmental TEM (ETEM).[51,52] They found that adsorbed CO molecules caused the {100} facets of gold nanoparticles to reconstruct during CO oxidation at room temperature (Figure 12.5). The CO molecules adsorbed at the 'on-top' sites of gold atoms in the reconstructed surface; the energetic favorability of this reconstructed structure was confirmed by *ab initio* calculations and image simulations.

Besides, Huang and co-workers have also investigated the shape-dependent interplay in Ag-CeO_2 catalysts prepared by IMP.[54] The silver–nanorod–CeO_2 interaction was stronger than the silver–nanocube–CeO_2

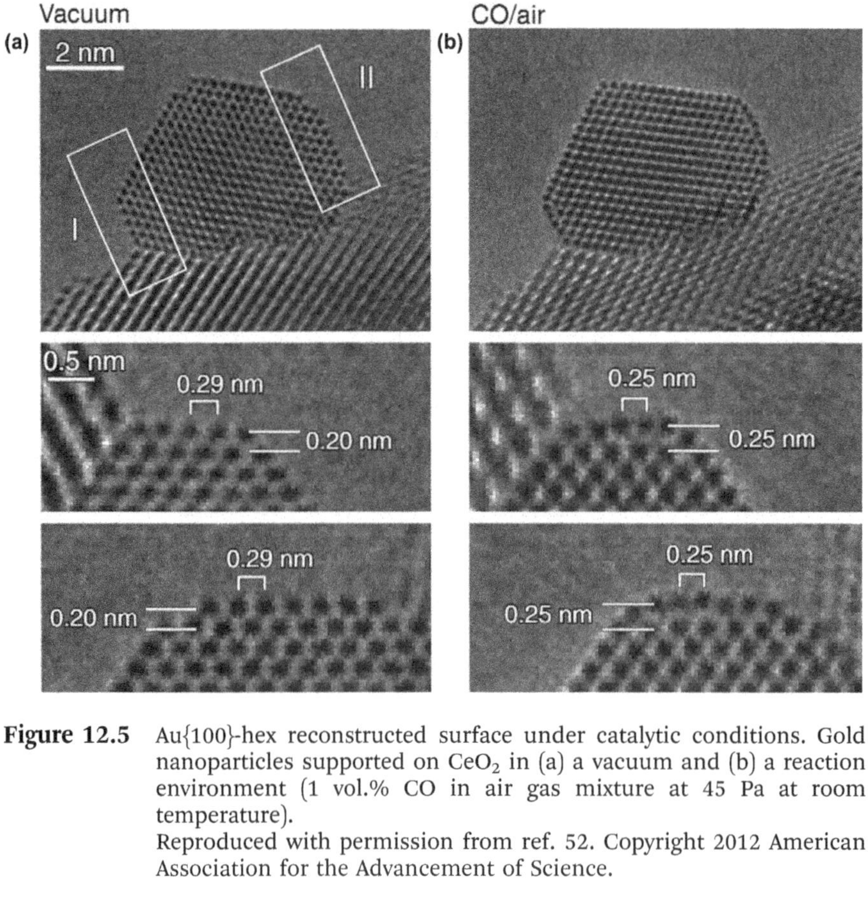

Figure 12.5 Au{100}-hex reconstructed surface under catalytic conditions. Gold nanoparticles supported on CeO_2 in (a) a vacuum and (b) a reaction environment (1 vol.% CO in air gas mixture at 45 Pa at room temperature).
Reproduced with permission from ref. 52. Copyright 2012 American Association for the Advancement of Science.

interaction. Positively Ag_n^+ clusters dominated in 1%-Ag-nanorod-CeO_2, but they were not capable of creating oxygen vacancies in CeO_2 compared with fine Ag nanoparticles. Ag nanoparticles greatly promoted the reduction and catalytic activity in CO oxidation over CeO_2 supports but positively charged Ag_n^+ clusters did not. These results demonstrated the concept that the interplay between oxygen vacancies and Ag-CeO_2 interaction controlled the structure of silver and CeO_2 in Ag-CeO_2 catalysts, and thus their surface reactivity and catalytic activity. The Ag clusters, CeO_2 supports, and their interplay between oxygen vacancies and Ag–CeO_2 interaction determined their catalytic activity in CO oxidation.

Moreover, Luo and co-workers have prepared, by a solution–combustion method, a $PdO/Ce_{1-x}Pd_xO_{2-\delta}$ catalyst containing free surface PdO species and PdO species in $Ce_{1-x}Pd_xO_{2-\delta}$ solid solution.[55] The contributions of the PdO species to catalytic CO oxidation were quantitatively evaluated. The free PdO species in the catalyst were shown to have the highest activity in CO oxidation, while the PdO species in the $Ce_{1-x}Pd_xO_{2-\delta}$ solid solution had the lowest activity. The presence of both species enhances the catalyst's activity due to synergetic effects, as the free surface PdO species provided CO chemisorption sites and the solid solution generated more oxygen vacancies for oxygen activation (Figure 12.6).

12.3.3 Au-Fe_xO_y Interaction

Considering the so-called strong metal-support interactions (SMSI) which have been the subject of a number of catalytic studies since their discovery in the late of 1970s, the main task of the support has to shift from ensuring sufficient particle dispersion to participating in the catalytic cycle in CO oxidation.[58] Many different transition metal oxides are known to show a SMSI effect which can strongly influence the electronic structure of noble metal catalysts and even lead to thin oxide layers covering the noble metal nanoparticles. Such synergism can result from the fact that the reducible oxide provides special adsorption sites at the perimeter of the particles, induces different particle morphology, or delivers active species. Indeed, this type of SMSI effect was recently demonstrated by a Pt-Fe_3O_4 catalyst.[58] The

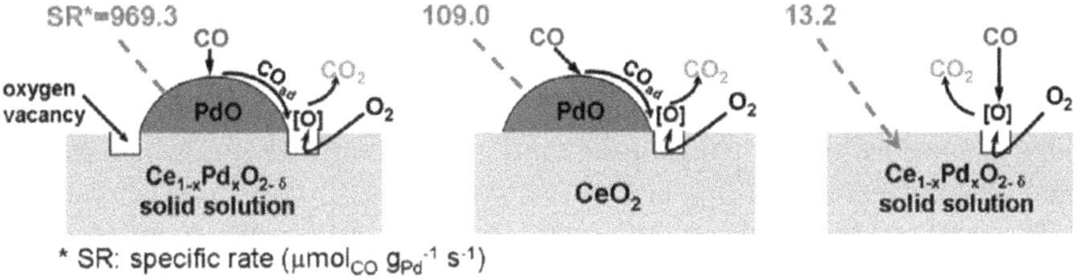

Figure 12.6 Reaction models of CO oxidation over the various catalysts. Reproduced with permission from ref. 55. Copyright 2011 American Chemical Society.

interaction of Pt with Fe_3O_4, which delivers oxygen at the border between the two constituents, was exploited in the development of a very efficient catalyst for PROX (preferential oxidation of CO in the presence of H_2). Moreover, the active site was proposed to be a highly reduced Fe-O layer on Pt, while H_2 is necessary to replenish these reduced sites. Bäumer and co-workers recently reported that they employed organic ligands to mediate the interaction between the Pt and the Fe_3O_4; the great advantage of using organic ligand-capped colloidal nanoparticles was the structure control achieved in the colloidal synthesis.[58]

As expected, the main preparation methods for gold-supported Fe_xO_y (Fe_2O_3 or Fe_3O_4) catalyst, including deposition–precipitation,[59] coprecipitation,[59–63] wetness impregnation,[60] colloidal deposition,[58] grafting,[64] double impregnation (DIM),[65] liquid phase reductive deposition (LPRD),[65] and ultrasonication,[65] are similar to those for Au-TiO_2 and Au-CeO_2 catalysts. Carabineiro *et al.* used three different methods for gold deposition and compared their catalytic activities in CO oxidation. The best catalytic result was obtained for the DIM samples, which showed the relatively large Au particle size compared with LPRD ones. This was most likely related to the oxidation state of gold (Au^+) found in the DIM catalysts, in contrast with LPRD and ultrasonication ones, where metallic gold (Au^0) was present (Figure 12.7).[65] Hoflund and co-workers and Hao and co-workers also reported that a small amount of $Au^{\delta+}$ was present on the surface of Au-Fe_2O_3 catalyst.[61,62] These catalysts had relatively higher low-temperature activities and the possible catalytic active center was the gold in the partially oxidized state ($Au^{\delta+}$).[61]

The results described above suggest that there is strong interaction between the Au nanoparticles and the Fe_xO_y, resulting in the strongly electronic structure change of Au nanoparticles in intimate contact with Fe_xO_y. Even the Fe-O oxide layer would cover the surface of Au due to the SMSI effect and leads to the new active sites favorable for possible reaction pathway.

12.3.4 Au-MnO_x Interaction

More recently, much attention has been paid to Au-MnO_x catalysts because of their exceptionally high activity and stability in low-temperature CO oxidation. Different aspects have been investigated in order to explain such extraordinary catalytic performances, including the nature of gold nanoparticles, the gold particle size and physical and chemical state, the influence of support, the preparation method, the preparation condition, and the catalyst pre-treatment.[66]

In a comparative study on CeO_2 and MnO_2 supported gold catalysts for CO oxidation at low temperature, Chen and co-workers reported that Au-CeO_2 showed higher catalytic activity for CO oxidation than Au-MnO_2.[49] The authors attributed this difference to the co-existence of metallic and oxidized Au species on the CeO_2-based catalyst, whereas only metallic Au species were

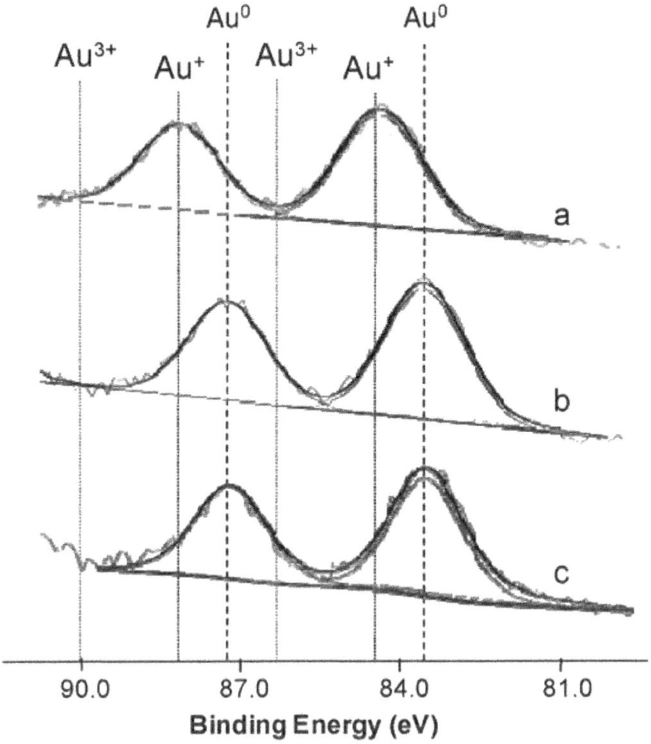

Figure 12.7 Au 4f XPS spectra with peak deconvolution of Au-Fe_2O_3 samples prepared by: (a) double impregnation; (b), liquid phase reductive deposition; and (c) ultrasonication.
Reproduced with permission from ref. 65. Copyright 2012 Royal Society of Chemistry.

recognized on the MnO_2 one.[49] Despite this, recently some authors have reported that Au-MnO_x (Mn_2O_3, MnO_2, or Mn_3O_4) catalysts for the oxidation of VOCs exhibited high catalytic performance in the oxidation of ethanol.[67] The metal–support interaction was claimed to enhance the concentration of oxygen vacancies, thereby promoting the adsorption/dissociation of the VOCs at the gold sites and improving the oxidation reaction. The increase in oxygen vacancies was attributed to a structural change of MnO_x towards a more reduced phase, evidenced by a strong contraction of the Mn–O first shell coordination number. The strong metal–support interaction found as a function of the structural and chemical state changes of the gold and the support at the Au-MnO_x interface was confirmed.

Similarly, the Au-MnO_x catalysts were usually prepared by deposition precipitation,[49,66–68] coprecipitation,[69] and direct thermal decomposition (DTD),[70,71] *etc.* for CO oxidation at low temperatures. Our group recently reported that gold promoted the reduction of MnO_2 and that this effect was more pronounced in the gold (10 wt%) catalyst, which exhibited an enhanced CO oxidation activity with respect to the gold (1wt%) catalyst.[66] The higher activity of Au-MnO_2 catalyst was attributed to the strong metal–support interaction between gold nanoparticles and support, as demonstrated by the presence of a component Au-O at 2.1 Å in the Fourier transform of the extended X-ray absorption for fine structure (EXAFS) signal of both the fresh and the CO-treated samples (proved by XANES, see Figure 12.8). This means that the interaction between Au clusters and the MnO_2, similarly to the CeO_2 supported catalysts, occurs *via* oxygen vacancy formation, in agreement with previous reports. However, no evidence of metal–support interaction was found for gold over inert silica.

Lee *et al.* reported an Au-MnO_x catalyst prepared by CP and found to contain only metallic Au.[69] Subsequent heat treatment at 400 °C enhanced the activity. This enhancement was associated with the shift of Au 4f binding energy and the presence of metallic Mn caused by strong interaction between metallic Au and the support, the transformation of $MnCO_3$ to MnO_x, and the high surface Na concentration.

Cao and co-workers systematically investigated the α-Mn_2O_3 materials as supports for anchoring gold nanoparticles which exhibited high activity for low-temperature CO oxidation in the absence or presence of excess H_2.[70,71] They found that comparing the specific rates for CO oxidation with the

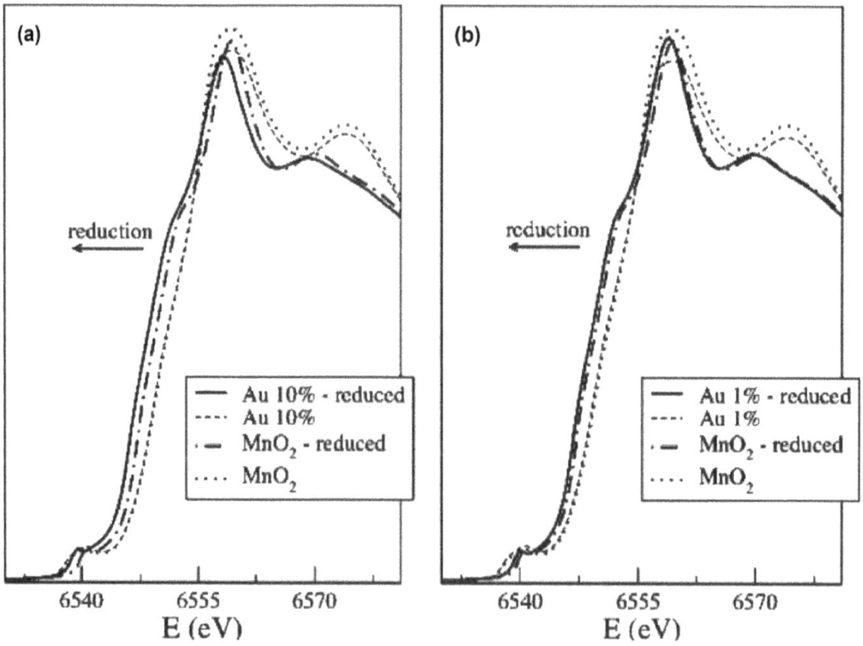

Figure 12.8 XANES spectra for as-prepared and reduced manganese oxide supports and gold catalysts: (a) Au 10 wt%; and (b) Au 1 wt%. Reproduced with permission from ref. 66. Copyright 2010 American Chemical Society.

highly active Au catalysts in the current literature confirmed the exceptionally high activity of these new materials. XPS combined with *in situ* CO adsorption studies revealed the presence of both metallic and positively charged Au species in the catalyst under reaction conditions. The superior activities of Au-α-Mn_2O_3 catalysts were attributed to their unique redox properties and the facile formation of activated oxygen species on the surface.[70] Furthermore, Cao and co-workers reported the effect of the pre-treatment atmosphere on the structure and catalytic properties of the Au-α-Mn_2O_3 catalyst in the low-temperature CO oxidation reaction.[71] XPS revealed a higher fraction of positively charged Au species in the catalyst pre-treated with O_2. Compared with He and H_2 pre-treated samples, the superior stability of O_2-Au-α-Mn_2O_3 catalyst was attributed to its enhanced metal–support synergy rather than suppressed accumulation of carbonate-like surface species as revealed by *in situ* diffuse reflectance infrared Fourier transform spectroscopy (DRIFTS) studies. The enhanced metal–support synergy would lead to the creation of a specific oxygen-enriched interface between the gold and support, and the abundant oxygen vacancies in turn influence the metal support interaction.[70,71] Finally, Cao and co-workers compared the catalytic activities of a series of gold catalysts deposited on various single phase manganese oxides (MnO_x, such as MnO_2, Mn_2O_3 and Mn_3O_4) and found their activities to be in the order: Au-Mn_2O_3 > Au-MnO_2 > Au-Mn_3O_4. XPS suggested both metallic and cationic Au species existed in the Au-Mn_2O_3, while only metallic gold species were present in the other catalysts (Figure 12.9). The superior performance of the Au-Mn_2O_3 catalyst could be attributed to the presence of highly dispersed gold species and the unique redox properties originating from the intimate metal-support interaction.[67]

More interestingly, Dai and co-workers recently found that a the Au-β-MnO_2 catalysts were active for the complete oxidation of CO, benzene, and toluene.[68] The loading of gold could greatly modify the reducibility of Au-β-MnO_2 catalysts *via* a strong metal–support interaction. It was concluded that higher Au^{3+}, oxygen adspecies concentrations, and low-temperature reducibility due to the SMSI effect might be responsible for the excellent catalytic performance of Au-β-MnO_2 catalysts.

12.3.5 Au-M(OH)$_x$ (M = Fe^{3+}, Mn^{2+}, Ti^{4+}, Ce^{4+}, *etc.*) Interaction

Molecular metal complexes (Au-phosphine complexes) and clusters supported on hydroxides have also been extensively studied as precursors for active metal particle catalysts.[72,73] Both selections of appropriate gold complexes as metal precursors and application of as-precipitated wet metal hydroxides as support precursors allowed us to obtain highly active Au catalysts for low-temperature CO oxidation.[72,73] It was suggested that effective interaction of the $Au(PPh_3)(NO_3)$ precursor with OH groups and defects of the support [*e.g.* amorphous $Fe(OH)_3$] surface was responsible for the

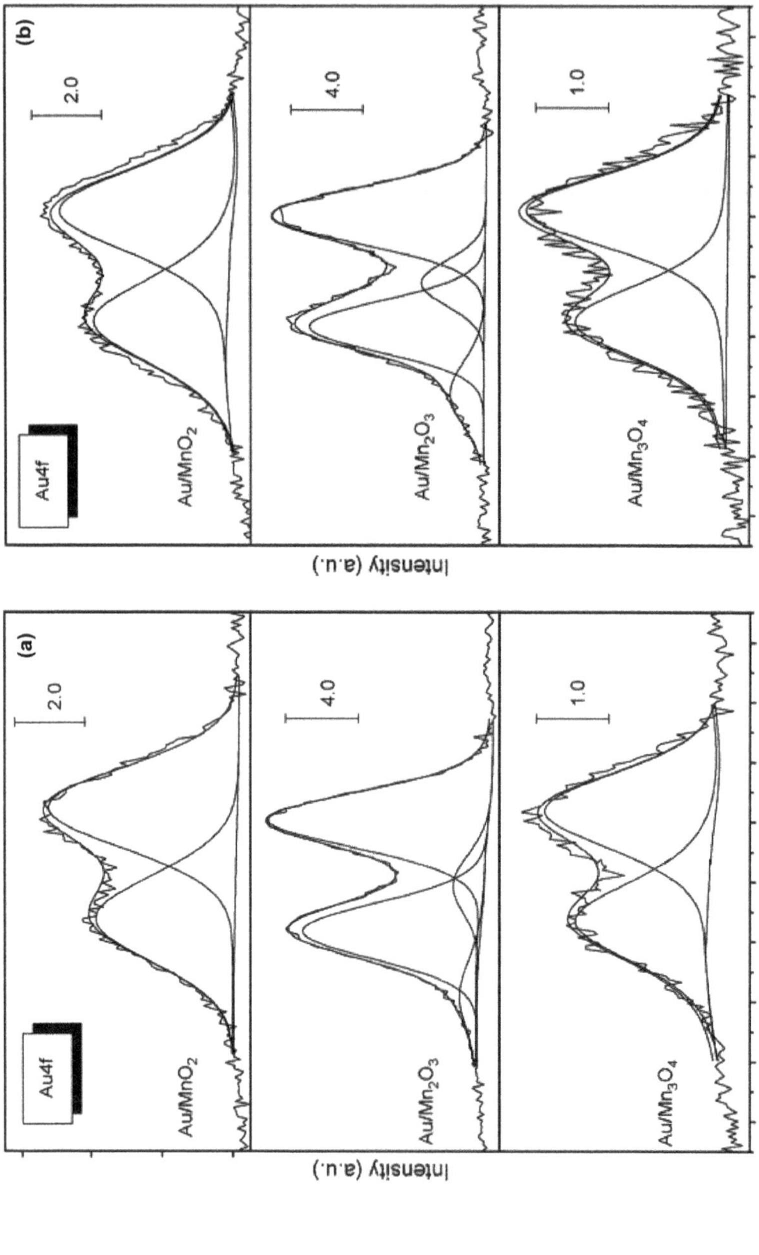

Figure 12.9 Au 4f XPS spectra of various MnO_x supported gold catalysts before and after reaction: (a) fresh catalysts; and (b) used catalysts. Reproduced with permission from ref. 67. Copyright 2009 Elsevier BV.

formation of the high active catalyst. The supported Au catalyst derived from $Au(PPh_3)(NO_3)$ and $Fe(OH)_3$ showed high CO conversion even at −70 °C.[72,73]

On the other hand, the support effect and particle size effect on the gold catalysis have been substantiated with a number of gold catalysts. Gold catalysts showed an overall trend of CO oxidation activity increasing with decreasing size of the Au nanoparticles. Ultrafine Au nanoparticles may be a prerequisite for observation of a gold–support interaction. On this basis, the activity of supported Au catalysts can be tailored by a preparation method which controls Au particle size and gold–support interaction. The choice of suitable gold complexes such as $[Au(PPh_3)(NO_3)]$ or $[Au_9(PPh_3)_8](NO_3)_3]$, and the use of as-precipitated wet metal hydroxides, such as $Ti(OH)_4$ or $Fe(OH)_3$, as precursors for the oxide support, is a promising way to prepare tailored gold catalysts with efficient metal–support interaction as reported by Kozlova *et al.*[72] and Yuan *et al.*[73] They found that catalysts obtained by attaching the Au complexes onto commercial available metal hydroxides showed negligible catalytic activity, while gold complexes interacting with as-prepared metal hydroxides produced highly active catalysts for CO oxidation at low-temperature.[72,73]

The obtained Au catalysts showed high catalytic activities in low-temperature CO oxidation.[73] Among the obtained Au catalysts, $Mn(OH)_2$ and $Co(OH)_2$ were the most highly active even at −70 °C. $Fe(OH)_3$ and $Ti(OH)_4$ also catalyzed CO oxidation at low temperatures from −70 to 0 °C, whereas Fe_2O_3 and TiO_2 prepared by supporting $Au(PPh_3)(NO_3)$ on conventional Fe_2O_3 and TiO_2 showed negligible activity under similar reaction conditions. It was estimated by TEM and X-ray diffraction (XRD) that the mean diameter of Au particles in $Fe(OH)_3$ was about 2.9 nm, which was 10 times smaller than that for Fe_2O_3. EXAFS for $Ti(OH)_4$ revealed that the coordination number of the Au–Au bond was 8–10, while that for TiO_2 was 11.0, which also indicates that Au particle size for $Ti(OH)_4$ is smaller than that for TiO_2.

12.3.6 Au-Other Oxide Interaction

With respect to the commonly used reducible oxides such as TiO_2, CeO_2, Fe_xO_y, MnO_x, $M(OH)_x$, *etc.*, as supports for gold catalysts in CO oxidation at low temperatures, other oxides like ZnO, Co_3O_4, CuO, SnO_2, $FePO_4$, spinel $MgFe_2O_4$, *etc.*, have also been found to enhance the catalytic activity of gold catalysts in the complete oxidation of CO or VOCs due to the SMSI effect.[74–79] We recently compared the Au catalysts over the different supports (*e.g.* ZnO, Co_3O_4, Al_2O_3 and MgO) prepared by the colloidal deposition (CD) method according to the literature.[80] Using this synthesis method, we can easily control the particle size of gold colloids as the same size by using a protecting agent, *e.g.* polyvinylalcohol (PVA), and thus isolate the effect of the support on catalytic activity since the deposition process does not change the particle sizes of the gold particles. Therefore, only the influence of the support could be studied in our work.[74,75] The possible mechanism for the formation of Au-ZnO catalyst is shown in Figure 12.10.

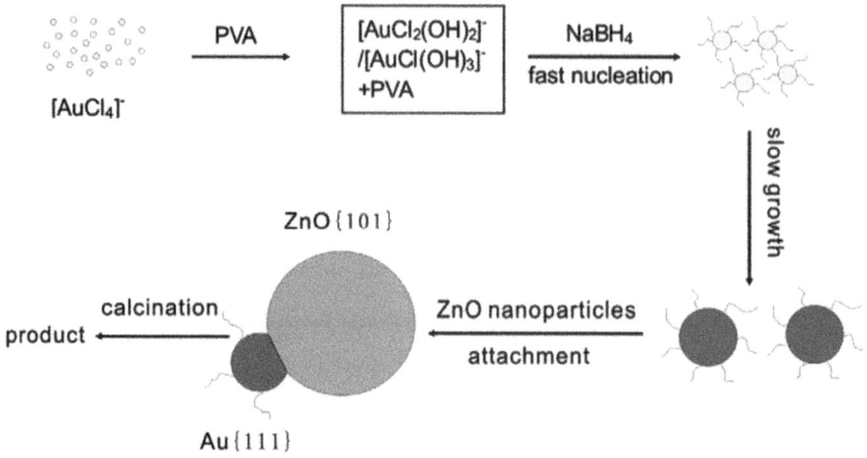

Figure 12.10 A suggested mechanism for the formation of Au-ZnO catalyst. Reproduced with permission from ref. 74. Copyright 2011 Elsevier BV.

In our paper, we suggested that the higher catalytic activity of the Au-ZnO catalyst might be attributed to the effect of SMSI effect which possibly originates from the small lattice parameter different between Au{111} and ZnO{101} lattice planes (Figure 12.11).[74] Schüth and co-workers reported that the activity for CO oxidation is strongly dependent on the gold particle size.[80] The most important finding of their study was the fact that almost identical gold particle size distribution on different supports resulted in different activities for CO oxidation. This clearly demonstrated that the metal–support interaction strongly influenced the catalytic properties of the final materials. However, this influence did not follow the reducibility and oxygen activation ability of the supports. Although the gold particle size was not changed by deposition, the support influenced the shape of the deposited particles, leading to faceting and possibly creating defect sites. The exact mechanism of how the different supports influence the activity is still unclear, but different faceting may be related to the catalytic activity. Their results are similar to our recent reports. However, the facets of gold particles in our work are different from those reported in the literature, which means that a different synthetic mechanism may exist in our study.

Overbury and co-workers have reported that Au supported on $FePO_4$ could be stable and active for CO oxidation and that oxygen from the $FePO_4$ could participate in the CO oxidation.[78] They found that cationic Au was present after oxidative treatment, and that metallic Au dominated after reductive treatment. In addition, Au was accompanied by anionic Au in the case of Au-$LaPO_4$, not seen on Au-$FePO_4$, which accumulated during CO exposure, even after an oxidative pre-treatment. The activities of the Au catalysts on the two supports were roughly comparable when the Au was in metallic form. However, the activity of the Au-$FePO_4$ was distinctly lower after oxidative

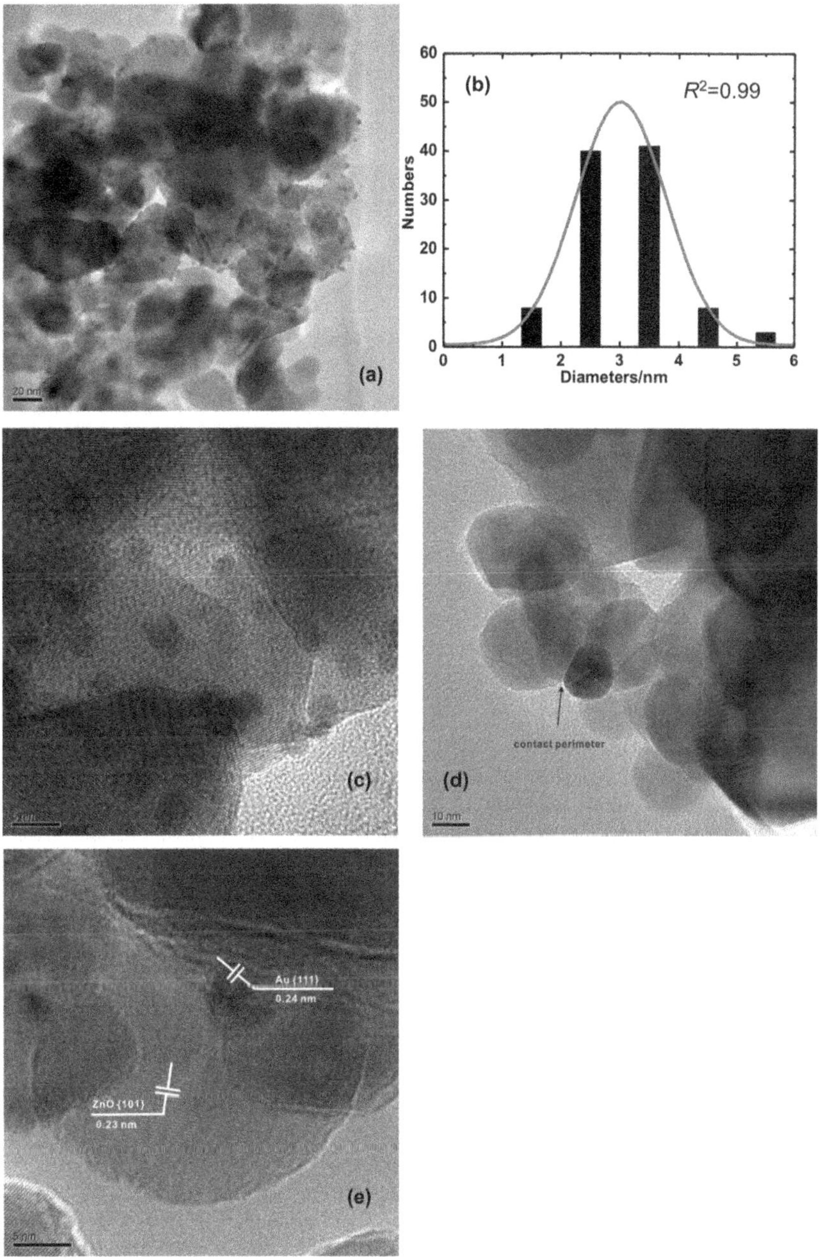

Figure 12.11 TEM images of: (a) Au-ZnO; (b) particle size distribution for gold nanoparticles; and (c–e) high-resolution images of Au-ZnO. Reproduced with permission from ref. 74. Copyright 2011 Elsevier BV.

treatment, due to less active cationic Au.[78] In spite of these support differences and different pathways identified to be contributing, in the end, the most important reaction pathway appeared to be direct reaction pathway catalyzed on reduced Au. Similarly, oxidative Au-$MgFe_2O_4$ catalyst exhibited superior activity towards CO oxidation, indicating that catalysts have high capabilities to activate oxygen molecules and supply active oxygen species for CO oxidation.[79] Based on the results above, we conclude that the capability to supply active oxygen and the chemical state of Au species are very important for these catalysts in CO oxidation at low temperatures.

12.4 Summary and Perspectives

In this chapter, we first introduced the syntheses of heterogeneous gold supported reducible oxide catalysts (Au-MO_x) prepared by chemical routes. Subsequently the emphasis has been on the catalytic performance of various heterogeneous composite catalysts enhanced by the interaction effects between the gold nanoparticles and the host supports, depending on the varied interfacial surfaces among them. Several kinds of possible synergetic/cooperation effects or interaction effects and the corresponding structural properties based on heterogeneous composite catalysts have been proposed. To understand the interaction effects more clearly, catalytic reactions catalyzed by composite catalysts both reported in the literature and the subject of our own investigations have been presented as typical examples and discussed as to their specific behaviors and properties in terms of the interaction effects between the noble metal(s) and support(s). CO oxidation was used as a sensitive probe reaction in all the cases because it is easy to carry out.

It is generally agreed that the catalytic activity of gold catalysts depends on the size of the gold particles, but the nature of the supports, the preparation methods, and the activation procedures have also been suggested to play a key role.[81] In this chapter, exceptionally high activities for oxidation of CO have been reported for highly dispersed Au catalysts on reducible oxides. It has been proposed that the reducible oxides supply oxygen to form oxidic gold sites exceptionally active in CO oxidation. In addition, positively charged gold particles or negatively charged clusters of a few nanometres have been shown to be catalytically active. Other explanations focus on the charge transfer between the support, particularly negatively charged defects (F centers), and the Au particles. Strain in the Au particle due to the mismatch of the lattices at the interface with the support has also been suggested. Besides, the presence of a metal to non-metal transition in very small, two-dimensional Au particles may exist in the gold catalysts. More importantly, several groups of researchers have pointed to the effect of low-coordinated sites and the roughness of Au clusters on the reducible oxides.

Effects related to the interaction with the support may contribute to a considerably smaller extent compared with the effect of a low-coordination number or gold particle size as reported by Nørskov and co-workers.[81] However, the effect of the interaction with the reducible supports is generally much stronger than that with the irreducible supports. Therefore, the latter are generally less reactive than the former, even by a factor of 2–4. This means that effects related to exchange of oxygen between the gold and the support, and effects due to charge transfer from oxygen vacancies to the gold particles can be the main contributor at a given same particle size. Furthermore, support-specific interactions between adsorbates or reaction intermediates on gold close to the perimeter of the support are apparently the main contributors to the activity in gold catalysts at the same level of gold particle size.

This chapter presents only an introductive study of the interaction effects possible in composite catalysts, and the corresponding behaviors and properties related to each type of heterogeneous composite catalysts are also explored very preliminarily based on limited reaction examples and literature reports. More work is needed in future investigations to obtain a clearer understanding of the interaction effects in the composite catalysts for the invention of better composite catalysts with enhanced performance.

We feel that aspects deserving further attention in the near future are the following:

(1) Despite the reported interface control *via* building up and tuning the metal–metal and metal–oxide interactions, metal–organic interactions of Au-oxide composite catalysts could optimize the catalytic performance of this kind of catalysts.
(2) Since the interactions between Au particles and reducible oxide supports usually arise from steric or electronic effects, characterization techniques such as *in situ* ETEM or IR-CO probe should be applied to accurately recognize the interfacial interactions.
(3) Studies on other interfaces, such as metal–hydroxide, bi-metal–oxide, and metal–bi-oxide should be envisaged.
(4) A combination of DFT calculations and structural/surface characterization techniques could be used to obtain more insight into the metal–support interaction effects in catalysis.

Acknowledgements

The authors acknowledge support by COST Action CM1104. The China Scholarship Council is sincerely acknowledged for supporting H.J.W.'s scholarship. H.J.W. thanks the Excellent Doctorate Foundation, the Doctorate Foundation of the Northwestern Polytechnical University, and the Scholarship Award for Excellent Doctoral Student granted by Ministry of Education, PR China.

References

1. J. L. Shi, *Chem. Rev.*, 2013, **113**, 2139.
2. C. Boissiere, D. Grosso, A. Chaumonnot, L. Nicole and C. Sanchez, *Adv. Mater.*, 2011, **23**, 599.

3. D. P. Debecker and P. H. Mutin, *Chem. Soc. Rev.*, 2012, **41**, 3624.
4. J. E. Mondloch, E. Bayram and R. G. Finke, *J. Mol. Catal. A*, 2012, **355**, 1.
5. G. A. Somorjai and J. Y. Park, *Angew. Chem., Int. Ed.*, 2008, **47**, 9212.
6. A. Corma and H. Garcia, *Chem. Soc. Rev.*, 2008, **37**, 2096.
7. J. B. Hansen and P. E. Nielsen, *Handbook of Heterogeneous Catalysis*, ed. G. Ertl, H. Knozinger, F. Schüth and J. Weirkamp, Wiley-VCH, Weiheim, 2008, p 2920.
8. C. Hess and R. Schlogl, in *Nanostructured Catalysts: Selective Oxidation*; ed. C. Hess and R. Schlogl, RSC Publishing, Cambridge, UK, 2011, vol. 19, p. 355–397.
9. H. L. Jiang and Q. Xu, *J. Mater. Chem.*, 2011, **21**, 13705.
10. B. H. Wu and N. F. Zheng, *Nano Today*, 2013, **8**, 168.
11. G. Ertl, *Angew. Chem., Int. Ed.*, 2008, **47**, 3524.
12. G. Ertl and H. J. Freund, *Phys. Today*, 1999, **52**, 32.
13. G. Ertl, *Chem. Rec.*, 2001, **1**, 33.
14. D. W. Goodman, *Surf. Sci.*, 1994, **299**, 837.
15. G. A. Somorjai, A. M. Contreras, M. Montano and R. M. Rioux, *Proc. Natl. Acad. Sci. U. S. A.*, 2006, **103**, 10577.
16. Q. Fu, W. X. Li, Y. X. Yao, H. Y. Liu, H. Y. Su, D. Ma, X. K. Gu, L. M. Chen, Z. Wang, H. Zhang, B. Wang and X. H. Bao, *Science*, 2010, **328**, 1141.
17. J. Zhang, H. Z. Yang, J. Y. Fang and S. Z. Zou, *Nano Lett.*, 2010, **10**, 638.
18. Y. Li, Q. Liu and W. Shen, *Dalton Trans.*, 2011, **40**, 5811.
19. M. Haruta, *CATTECH*, 2002, **6**, 102.
20. I. X. Green, W. J. Tang, M. Neurock and J. T. Yates, *Science*, 2011, **333**, 736.
21. X. Liu, M. H. Liu, Y. C. Luo, C. Y. Mou, S. D. Lin, H. K. Cheng, J. M. Chen, J. F. Lee and T. S. Lin, *J. Am. Chem. Soc.*, 2012, **134**, 10251.
22. S. H. Zhou, K. McIlwrath, G. Jackson and B. Eichhorn, *J. Am. Chem. Soc.*, 2006, **128**, 1780.
23. Z. Ma and S. Dai, *Nano Res.*, 2012, **4**, 3.
24. L. W. Zhang, H. Y. Cheng, R. L. Zong and Y. F. Zhu, *J. Phys. Chem. C*, 2009, **113**, 2368.
25. J. T. Tian, L. J. Chen, Y. S. Yin, X. Wang, J. H. Dai, Z. B. Zhu, X. Y. Liu and P. W. Wu, *Surf. Coat. Technol.*, 2009, **204**, 205.
26. H. B. Fu, T. G. Xu, S. B. Zhu and Y. F. Zhu, *Environ. Sci. Technol.*, 2008, **42**, 8064.
27. S. Bonanni, K. Aït-Mansour, H. Brune and W. Harbich, *ACS Catal.*, 2011, **1**, 385.
28. J. M. C. Soares, M. Hall, M. Cristofoini and M. Bowker, *Catal. Lett.*, 2006, **1**, 103.
29. R. Zanella, S. Giorgio, C. H. Shin, C. R. Henry and C. Louis, *J. Catal.*, 2004, **222**, 357.
30. A. Sandoval, A. Aguilar, C. Louis, A. Traverse and R. Zanella, *J. Catal.*, 2011, **281**, 40.
31. A. Gómez-Cortés, G. Díaz, R. Zanella, H. Ramírez, P. Santiago and J. M. Saniger, *J. Phys. Chem. C*, 2009, **113**, 9710.

32. N. Weiher, A. M. Beesley, N. Tsapatsaris, L. Delannoy, C. Louis, J. A. van Bokhoven and S. L. M. Schroeder, *J. Am. Chem. Soc.*, 2007, **129**, 2240.
33. M. Okumura, J. M. Coronado, J. Soria, M. Haruta and J. C. Conesa, *J. Catal.*, 2001, **203**, 168.
34. B. K. Chang, B. W. Jang, S. Dai and S. H. Overbury, *J. Catal.*, 2005, **236**, 392.
35. N. J. Ossipoff and N. W. Cant, *Top. Catal.*, 1999, **8**, 161.
36. K. Qadir, S. H. Kim, S. M. Kim, H. Ha and J. Y. Park, *J. Phys. Chem. C*, 2012, **116**, 24054.
37. S. Tsubota, T. Nakamura, K. Tanaka and M. Haruta, *Catal. Lett.*, 1998, **56**, 131.
38. K. M. Parida, N. Sahu, P. Mogapatra and M. S. Scurrell, *J. Mol. Catal. A*, 2010, **319**, 92.
39. S. H. Kim, C. H. Jung, N. Sahu, D. Park, J. Y. Yun, H. Ha and J. Y. Park, *Appl. Catal. A*, 2013, **454**, 53.
40. S. H. Wu, X. C. Zheng, S. R. Wang, D. Z. Han, W. P. Huang and S. M. Zhang, *Catal. Lett.*, 2004, **97**, 17.
41. Z. Zhao, T. Diemant, D. Rosenthal, K. Christmann, J. Bansmann, H. Rauscher and R. J. Behm, *Surf. Sci.*, 2006, **600**, 4992.
42. T. V. Choudhary, C. Sivadinarayana, C. C. Chusuei, A. K. Datye, J. P. Fackler and D. W. Goodman, *J. Catal.*, 2002, **207**, 247.
43. D. W. Goodman, *Catal. Lett.*, 2005, **99**, 1.
44. S. J. Tauster, S. C. Fung and R. L. Garten, *J. Am. Chem. Soc.*, 1978, **100**, 170.
45. S. J. Tauster and S. C. Fung, *J. Catal.*, 1978, **55**, 29.
46. S. J. Tauster, *Acc. Chem. Res.*, 1987, **20**, 389.
47. A. M. Venezia, G. Pantaleo, A. Longo, G. D. Carlo, M. P. Casaletto, L. F. Liotta and G. Deganello, *J. Phys. Chem. B*, 2005, **109**, 2821.
48. X. Y. Wang, S. P. Wang, S. R. Wang, Y. Q. Zhao, J. Huang, S. M. Zhang, W. P. Huang and S. H. Wu, *Catal. Lett.*, 2006, **112**, 115.
49. L. H. Chang, N. Sasirekha, B. Rajesh and Y. W. Chen, *Sep. Purf. Technol.*, 2007, **58**, 211.
50. W. L. Deng, C. Carpenter, N. Yi and M. Flytzani-Stephanopoulos, *Top. Catal.*, 2007, **44**, 199.
51. T. Uchiyama, H. Yoshida, Y. Kuwauchi, S. Ichikawa, S. Shimada, M. Haruta and S. Takeda, *Angew. Chem., Int. Ed.*, 2011, **50**, 10157.
52. H. Yoshida, Y. Kuwauchi, J. R. Jinschek, K. Sun, S. Tanaka, M. Kohyama, S. Shimada, M. Haruta and S. Takeda, *Science*, 2012, **335**, 317.
53. A. Longo, L. F. Liotta, G. Pantaleo, F. Giannici, A. M. Venezia and A. Martorana, *J. Phys. Chem. C*, 2012, **116**, 2960.
54. S. Chang, M. Li, Q. Hua, L. Zhang, Y. Ma, B. Ye and W. Huang, *J. Catal.*, 2012, **293**, 195.
55. L. Meng, A. P. Jia, J. Q. Lu, L. F. Luo, W. X. Huang and M. F. Luo, *J. Phys. Chem. C*, 2011, **115**, 19789.
56. V. Shapovalov and H. Metiu, *J. Catal.*, 2007, **245**, 205.
57. H. Y. Kim, H. M. Lee and G. Henkelman, *J. Am. Chem. Soc.*, 2012, **134**, 1560.

58. P. Sonström, D. Arndt, X. D. Wang, V. Zielasek and M. Bäumer, *Angew. Chem., Int. Ed.*, 2011, **50**, 3888.
59. M. Khoudiakov, M. C. Gupta and S. Deevi, *Nanotechnology*, 2004, **15**, 987.
60. A. M. Visco, A. Donato, C. Milone and S. Galvagno, *React. Kinet. Catal. Lett.*, 1997, **61**, 219.
61. Z. Hao, L. An and H. Wang, *Sci. Chin.*, 2001, **44**, 596.
62. W. S. Epling, G. B. Hoflund, J. F. Weaver, S. Tsubota and M. Haruta, *J. Phys. Chem.*, 1996, **100**, 9929.
63. T. T. M. Nguyet, N. C. Trang, N. Q. Huan and N. Xuan, *J. Korean Phys. Soc.*, 2008, **52**, 1345.
64. A. P. Kozlova, S. Sugiyama, A. I. Kozlov, K. Asakura and Y. Iwasawa, *J. Catal.*, 1998, **176**, 426.
65. S. A. C. Carabineiro, N. Bogdanchikova, P. B. Tavares and J. L. Figueiredo, *RSC Adv.*, 2012, **2**, 2965.
66. A. Longo, L. F. Liotta, G. D. Carlo, F. Giannici, A. M. Venezia and A. Martorana, *Chem. Mater.*, 2010, **22**, 3952.
67. A. K. Sinha, K. Suzuki, M. Takanara, H. Azuma, T. Nonaka and K. Fukumoto, *Angew. Chem., Int. Ed.*, 2007, **46**, 2891.
68. Q. Ye, J. S. Zhao, F. Huo, D. Wang, S. Y. Cheng, T. F. Kang and H. X. Dai, *Micropor. Mesopor. Mater.*, 2013, **172**, 20.
69. S. Lee, A. Gavriilidis, Q. A. Pankhurst, A. Kyek, F. E. Wagner, P. C. L. Wong and K. L. Yeung, *J. Catal.*, 2001, **200**, 298.
70. L. C. Wang, X. S. Huang, Q. Liu, Y. M. Liu, Y. Cao, H. Y. He, K. N. Fan and J. H. Zhuang, *J. Catal.*, 2008, **259**, 66.
71. L. C. Wang, L. He, Y. M. Liu, Y. Cao, H. Y. He, K. N. Fan and J. H. Zhuang, *J. Catal.*, 2009, **264**, 145.
72. A. P. Kozlova, A. I. Kozlov, S. Sugiyama, Y. Matsui, K. Asakura and Y. Iwasawa, *J. Catal.*, 1999, **181**, 37.
73. Y. Z. Yuan, A. P. Kozlova, K. Asakura, H. L. Wan, K. Tsai and Y. Iwasawa, *J. Catal.*, 1997, **170**, 191.
74. H. J. Wu, L. D. Wang, J. Q. Zhang, Z. Y. Shen and J. H. Zhao, *Catal. Commun.*, 2011, **12**, 859.
75. H. J. Wu, L. D. Wang, Z. Y. Shen and J. H. Zhao, *J. Mol. Catal. A*, 2011, **351**, 188.
76. G. J. Hutchings, M. R. H. Siddiqui, A. Burrows, C. J. Kiely and R. Whyman, *J. Chem. Soc. Faraday Trans.*, 1997, **93**, 187.
77. Q. Ye, J. Wang, J. S. Zhao, L. N. Yan, S. Y. Cheng, T. F. Kang and H. X. Dai, *Catal. Lett.*, 2010, **138**, 56.
78. M. J. Li, Z. L. Wu and S. H. Overbury, *J. Catal.*, 2011, **278**, 133.
79. C. J. Jia, Y. Liu, M. Schwickardi, C. Weidenthaler, B. Spliethoff, W. Schmidt and F. Schüth, *Appl. Catal. A*, 2010, **386**, 94.
80. M. Comotti, W. C. Li, B. Spliethoff and F. Schüth, *J. Am. Chem. Soc.*, 2006, **128**, 917.
81. N. Lopez, T. V. W. Janssens, B. S. Clausen, Y. Xu, M. Mavrikakis, T. Bligaard and J. K. Nørskov, *J. Catal.*, 2004, **223**, 232.

CHAPTER 13

Role of Oxygen Vacancies in Gold Oxidation Catalysis

O. H. LAGUNA, M. I. DOMÍNGUEZ, F. ROMERO-SARRIA, J. A. ODRIOZOLA AND M. A. CENTENO*

Instituto de Ciencia de Materiales de Sevilla, Centro mixto CSIC-Universidad de Sevilla, Avda Americo Vespucio 49, 41092, Sevilla, Spain
*Email: centeno@icmse.csic.es

13.1 Overview

The importance and uses of supported gold nanoparticles in heterogeneous oxidation catalysis is continuously growing from the astonishing discovery of Haruta and co-workers of their exceptional activity in the oxidation of carbon monoxide at very low temperatures.[1,2] Since then, gold catalysts have demonstrated to be very active in a large number of oxidation reactions with academic and industrial concerns, such as total volatile organic compounds oxidation (VOCs), preferential CO oxidation in the presence of hydrogen (PROX), the water gas shift (WGS) reaction, alkenes epoxidation, and the selective oxidation of sugars and hydrocarbons. The prominence of gold catalysts in these applications can be foreseen from the increasing number of patents and research papers as well as recent reviews and books[3–15] focused on such topics.

Although the reaction mechanisms are still under debate, there is a good consensus concerning the determinant role not only of the physical and chemical properties of the gold nanoparticles (size, shape, coordination, oxidation state) but also of the nature of the support (specific surface area, presence of defects, reducibility, *etc.*). Specifically, the gold–support

RSC Catalysis Series No. 18
Heterogeneous Gold Catalysts and Catalysis
Edited by Zhen Ma and Sheng Dai

Published by the Royal Society of Chemistry, www.rsc.org

interaction in terms of charge transfer and stabilization of gold species is considered to be a key factor in the activity, and from here, several mechanisms give a special relevance to the perimeter of contact between gold nanoparticles and the support surface.

Although several supports without oxygen in their structure, such as carbons[16–18] and nitrides,[19] have been studied and shown to produce active oxidation gold catalysts, the most successful and used supports are those containing oxygen, typically metal oxides.

The surface chemistry and the reactivity of the metal oxides are directly or indirectly connected to the presence of defects, with the oxygen vacancies being the most important.[20] This type of defect controls the optical, electronic, and transport properties of the material. Oxygen vacancies are naturally present in every oxide in the form of Schottky or Frenkel defects, and their concentration can be increased or reduced in several ways. They can be situated at the surface or in the bulk of the oxide but, in any case, their presence modifies the geometric and electronic structure as well as the chemical properties of the whole system.

The creation of an oxygen vacancy in an oxide generates an excess of negative charge[21] because a pair of electrons remains in the vacancy. These high electronic density centers may act as nucleation sites for the gold, binding it strongly and hindering agglomeration.[22–24] Hence, metallic oxides containing oxygen vacancies result in catalysts with a very high gold dispersion and small particle sizes.[24,25]

However, the redistribution of this excess of electronic density depends strongly on the electronic structure of the metal oxide, changing dramatically from one metal oxide to another,[26] and controlling the extent of the electronic transfer from the support to the metallic particle (metal–support interaction strength).

The nature of the metallic oxide and the easiness of creating oxygen vacancies strongly influence the properties of the support and the prepared gold catalyst. In such sense, it has been described that doping transition metal and rare earth oxides with other metals weakens the metal–oxygen bonds around the dopant, thus reducing the energy of formation of oxygen vacancies and favoring their creation.[27,28] This strategy has been successfully used to produce more active gold catalysts by doping the support (for instance, ceria or titania) with transition metals (Zr, Fe, Zn, *etc.*) in order to increase the number of oxygen vacancies in the material.[29–31]

Thus, the type of oxidic support has been described as a determinant in the catalytic performances of the supported gold catalyst. In particular, the reducibility properties of the support are considered to be a key factor. The reducibility of some of these oxides is associated with the possibility of having metal cations in more than one oxidation state. The presence of a mixture of oxidation states induces the generation of oxygen vacancies in the oxide lattice to maintain the electroneutrality of the structure.[32]

Since the surface of the oxide marks the cutoff of its crystalline structure, the density of oxygen vacancies is higher here. Diminution of the support particle size to the nanometre range increases the surface-to-volume ratio, increasing in this way the proportion of surface defects, mainly oxygen vacancies, in the material.

In general, the reducible metal oxides defined by Schubert *et al.*[33] as 'active' supports produce more active gold catalysts. In these oxides, not only the linkage of small clusters or ionic gold species to their surface is ensured, but also the active participation of the support in the adsorption–desorption processes of reactants and products, or even direct involvement of the oxygen species of the oxide lattice in the oxidation mechanism, is proposed.

In particular, the activation of the oxygen species by these reducible supports seems to be one of the most relevant contributions to the mechanism. Since oxygen activation is considered to be the rate-determining step[34] of the CO oxidation reaction, the new oxygen activation routes opened by the support nature result in the observed enhanced catalytic activity.

These reducible supports are able to create oxygen vacancies, mainly in the perimeter of contact between the gold particle and the oxide support as a consequence of the Schottky function,[35] where oxygen activation proceeds generating very active species in the CO oxidation reaction—typically peroxo and superoxo species.[36,37]

The oxygen vacancies can also promote oxygen migration or be directly involved in the CO oxidation processes.[38,39] In fact, a direct correlation between the relative concentration of oxygen vacancies of ceria-doped solid solutions, with and without gold, and the specific reaction rates for the CO oxidation is commonly found.[25,40,41] This mechanism has been called the 'metal/oxide boundary mechanism'.

The utilization of nanocrystalline supports increment the number of oxygen vacancies, resulting in much more active gold catalysts.[42] The stabilization of cationic gold (active species according to some authors) and the reactive intermediate oxygen species at these oxygen vacancies generated at the gold–support interface has been claimed to be responsible for such an enhancement in activity.

On the other hand, the reducibility of these oxides provides an interesting property from a catalytic point of view, *i.e.* active surface dynamics. The reduction of these solids leads to the loss of oxygen atoms from the structure, forming oxygen vacancies and allowing in that way the stabilization of smaller particles. Since the external conditions (flow composition, temperature, *etc.*) may modify the number of oxygen vacancies on the surface, dynamic modification of the gold particle size may be induced under reactive conditions.[25,43]

On the contrary, non-reducible metal oxides or 'inert' supports, generally lead to catalysts with a weak gold–support interaction producing low dispersion and susceptibility to sintering, although as said before, the interaction depends strongly on the nature and properties of the material. In this case, the reaction proceeds only by an Eley–Rideal mechanism known as the 'gold-only mechanism',[5,34] whereby both CO adsorption and oxygen activation proceeds over the gold particle without support participation.

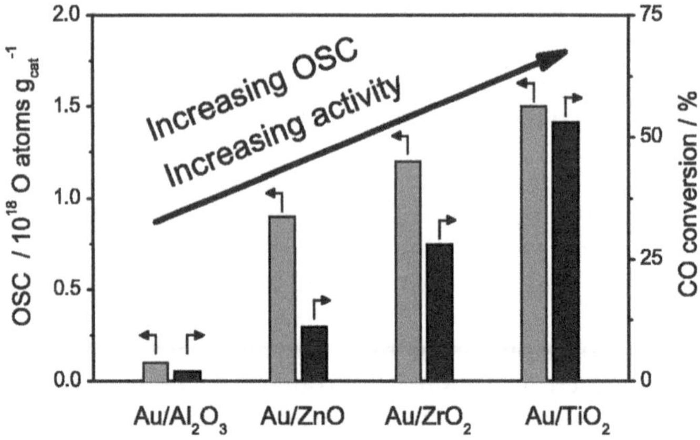

Figure 13.1 Influence of support reducibility on catalytic activity of gold-based catalysts
(Reproduced with permission of Elsevier from ref. 46.)

This mechanism can also work simultaneously with the 'metal/oxide boundary' one in reducible solids.[44,45] As a consequence, the factors affecting the oxidation activity are different for each type of support. As an example, Widmann *et al.*[46] found the following order for catalytic activity, Au/TiO_2 > Au/ZrO_2 > Au/ZnO_2 > Au/Al_2O_3, which coincides with the support reducibility characterized by oxygen storage capacity (OSC) (Figure 13.1).

In this chapter, we discuss the role of the oxygen vacancies present on supported gold catalysts in oxidation reactions, as a function of the reducible or irreducible nature of the support.

13.2 Reducible Supports

Many reducible metal oxides have been studied as supports for the preparation of supported gold catalysts: cobalt,[47,48] nickel,[49] manganese,[50,51] and probably the most studied, titanium,[24,52] iron,[53–55] and cerium oxides.[25,56]

The interest in oxygen vacancies on these oxides from experimental and theoretical points of view has motivated many researchers. For instance, Nolan *et al.*[30] analyzed the vacancy formation on gold-doped ceria surfaces through density functional theory (DFT) demonstrating that, in the case of CeO_2, there is a small oxygen vacancy formation energy at the surface because of the redox behavior of Ce^{4+}/Ce^{3+} (Figure 13.2).

In addition, these authors established that the energy for a vacancy formation in a specific zone of the CeO_2 framework is lower than that required for producing a second one near the first one. In this sense, the population of such punctual defects would depend on the redox potential of the cations in the oxide, and on the tolerance of the framework to stabilize the structural distortions produced by the modification of the symmetry, bond distances, and the electronic structure around the vacancies. This is closely related to

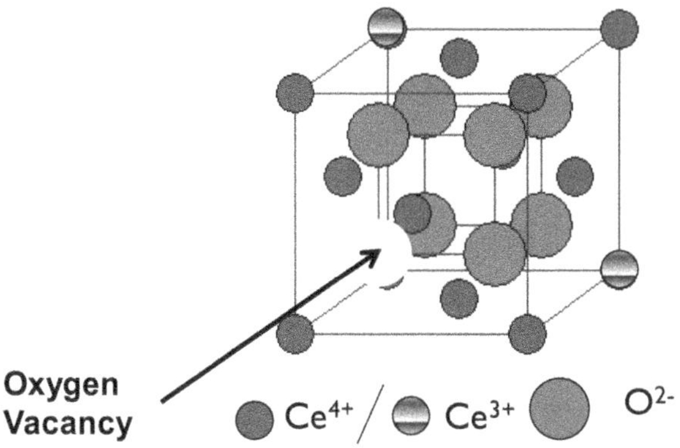

Figure 13.2 Oxygen vacancy formation in CeO_2 with appearance of two Ce^{3+}.

the ability of the oxide to stabilize structural defects in preferential crystallographic directions. In the study by Nolan *et al.*,[30] the presence of gold in the surface of CeO_2 promoted additional oxygen vacancies that may result in favoring a high oxidizing potential—very useful in oxidation reactions. This ability depends on the exposed crystallographic plane following the order: (111) < (110) < (100).

Therefore, the design of the synthesis method becomes relevant for producing supports with a specific morphology that exhibits the desired crystallographic planes in order to stabilize the oxygen population in the support.

In a similar work, Shojaee *et al.*[57] studied the oxygen stability and vacancy formation on Co_3O_4 model slabs with (100) and (110) terminations using spin-polarized periodic density functional simulations. These authors demonstrated that the removal of surface lattice oxygen by an on-surface oxygen producing molecular O_2 and leaving a slab with an oxygen vacancy site is, at least, 65 kJ mol^{-1} more activated in the (110) surface than in the (100) surface. In addition, they concluded that a surface oxygen atom in the (100)-B termination sitting above a cobalt atom located in the third layer is the most favorable site for the formation of an oxygen vacancy from a thermodynamic point of view. The formation of oxygen vacancies in the (100) and (110) terminations are less favorable, energetically speaking. The differences in binding and oxygen recombination energies and in oxygen vacancy formation suggest that (100) and (110) surfaces behave markedly differently in oxidative reactions over Co_3O_4. Therefore this work is in good agreement with the fact that the distortions produced by the presence of oxygen vacancies in a specific oxide are preferably stabilized in some crystallographic planes, making them potentially more active in oxidation reactions.

Within the discussion about how the vacancies in reducible supports improve the activation of oxygen for oxidation reactions, several reports demonstrate the formation of the peroxides and/or superoxides (very active species) by the reaction between oxygen defect sites of the support and O_2. For instance Shojaee *et al.*[57] demonstrated the strong interaction between on-surface and lattice oxygen, giving rise to an oxygen adsorbate with superoxide character in Co_3O_4. This interaction is closely related to the vacancy formation produced by the recombination of on-surface and lattice oxygen.

Carrettin *et al.*,[58] studying the CO oxidation reaction over supported gold catalysts employing a Fe-modified TiO_2, also observed the close relationship between the existence of oxygen vacancies on the support and the formation of peroxides and superoxides species by interaction with O_2. The Raman spectra of TiO_2 P25 showed a band at 1123 cm^{-1}, commonly assigned to superoxide species, with an intensity proportional to the amount of vacancies in the support. The interaction of gold with such peroxides and superoxides species at the support surface may improve the adsorption of the CO molecule, and then the reaction with structural and activated oxygen species. In this case, the superior activity of the gold catalyst is due to oxygen species such as η^1 superoxide and peroxide adspecies at one-electron defect sites.[58] In this study, it was also observed that the number of surface peroxide species increased when the gold loading was increased, resulting in high CO oxidation rates. The strong interaction between gold and the support was also observed through extended X-ray absorption fine structure (EXAFS) and X-ray absorption near-edge structure (XANES) spectroscopy experiments, demonstrating that the cationic gold species observed in the as-prepared catalysts were rapidly reduced to zero-valent gold under catalytic reaction conditions, resulting in very small clusters. This fact reflects the surface dynamic of this type of solids and the strong influence of the external conditions on the surface state.

Guzman *et al.*[59] used Raman spectroscopy to study gold supported on nanocrystalline CeO_2. They detected a band at 1571 cm^{-1} assigned to O_2 adsorption and other ones above 1300 cm^{-1} corresponding to $O_2^{\delta-}$ $(0<\delta<1)$ adspecies. Moreover, a band at 1123 cm^{-1} assigned to η^1 superoxide species was also observed. In the peroxide region, bands at 864, 871, and 831 cm^{-1} were reported. These features were attributed to peroxide adspecies at the one-electron defect site, to nonplanar bridging peroxide species, and to the η^2 peroxide species, respectively. Based on these results the authors suggested that nanocrystalline CeO_2 stabilizes O_2 as superoxide and peroxide species, thus resulting in the reactivity enhancement for the gold-supported catalyst (Figure 13.3).

Above, we have considered the ability of the reducible supports to produce intrinsic vacancies. However, it has been widely recognized that the promotion of oxygen vacancies in reducible materials can be enhanced by the modification of the structure through the inclusion of doping agents, even for oxides with low redox properties. In this sense, many researchers have studied doped systems demonstrating how different mixed oxides presented an improvement in the population of oxygen vacancies.[24,30,60–72] The interaction between the reducible oxide and the doping agent may form a solid

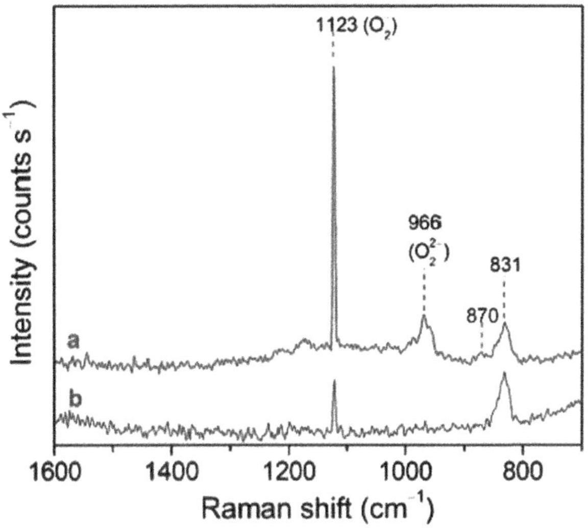

Figure 13.3 Raman bands characterizing peroxide and superoxide species. Reprinted with permission from ref. 59. Copyright 2005 American Chemical Society.

solution or generate segregated phases as a function of the properties of the doping element (solubility). The formation of a perfect solid solution could induce the formation of oxygen vacancies in the whole material, but even without solid solution formation, the promotion of oxygen vacancies may be achieved just on the surface by interaction of the segregated phases.[73,74]

Another interesting aspect to be considered from a catalytic point of view is the relationship between oxygen vacancies and catalytic properties towards oxidation reactions. It has been reported that oxygen vacancies may act as preferential nucleation sites for the formation of gold clusters,[73–76] in such a way that a high number of oxygen vacancies favors the obtaining of highly-dispersed gold catalysts and therefore high activity in oxidation reactions. Although some authors have proposed that cationic vacancies could be anchoring sites for gold nanoparticles (for instance, Ce vacancies[77]), it seems that, at least for ceria oxides, oxygen vacancies are much easier to form and much more stable, and that the gold addition does not alter the relative stabilities of these two different vacancies. Thus, it must be concluded that gold atoms strongly prefer the oxygen vacancy sites to be anchored.[23]

Considering that reducible supports are able to harbor higher populations of oxygen vacancies, these materials would potentially be better promoters of highly dispersed deposits of gold than inactive supports. The applications of this fact can be adapted to many technological fields in catalysis by gold. In a recent work, Pan and Xu[78] proposed a novel strategy for enhancing the number of oxygen vacancies in TiO_2 and then to reduce the size of the gold nanoparticles on the surface to improve the electronic transfer from the support to the gold cluster. This resulted in the increment of the photocatalytic activity of the obtained composite for the degradation of rhodamine B under visible light.[78] In this study the bands' structure model for TiO_2 with oxygen vacancies (TiO_2-OV) was constructed to explain the gold–support electronic transfer. The reduction potential of oxygen vacancy states is about 0.46 V *vs.* normal hydrogen electrode (NHE), which is much lower than that of $HAuCl_4$ (1.002 V *vs.* NHE). Therefore, the relative potential levels rationalize the spontaneous electron transfer from the oxygen vacancy states (oxidation) to the gold ions, leading to the formation of metallic gold nanoparticles. Also, Hernandez and co-workers[25,41] reported that the formation of a Ce-Eu solid solution induces the generation of oxygen vacancies able to anchor gold nanoparticles, as evidenced by the disappearance of the characteristic Raman band of oxygen vacancies in ceria lattice at 532 cm^{-1} (Figure 13.4).

From here, it is clear that a higher dispersion of gold on the surface of a support can be achieved by promoting the formation of oxygen vacancies. Thus, the existence of such point defects influences positively at least one of the key factors that determine the oxidation catalysis by gold, *i.e.* the gold particle size. Since the pioneering works of Haruta and co-workers,[1,2,79–81] it is widely accepted that gold must be deposited as ultrafine particles with a particle size below 5 nm to be highly active.[3,82] As an example, Santos *et al.*[83] reported that the doping of cryptomelane with Ce promoted the population of oxygen vacancies of the material resulting in a higher dispersion of gold

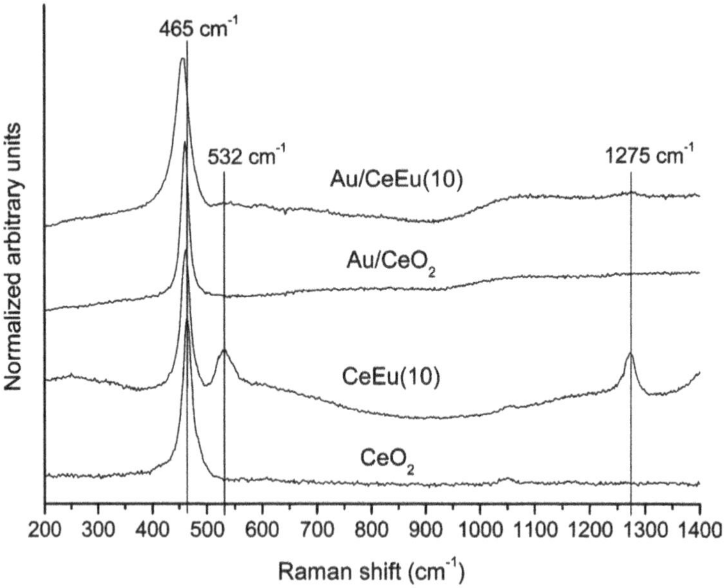

Figure 13.4 Raman Spectra of CeO_2 and CeO_2-Eu solid solution support and supported gold catalysts.
Reprinted with permission from ref. 25. Copyright 2010 American Chemical Society.

nanoparticles and, consequently, the gold catalysts prepared on cerium-modified cryptomelane (Au/Ce-K-OMS-2) exhibited a higher catalytic activity in CO oxidation compared with that of the other undoped systems.

The interaction between gold and a support containing oxygen vacancies may also improve the reducibility. For instance in the case of gold nanoparticles supported on β-MnO_2 nanorods used as catalysts for the selective oxidation of alcohols by Wang *et al.*,[84] the temperature-programmed reaction (TPR) studies clearly demonstrated that the presence of gold nanoparticles strongly promoted the reducibility of the MnO_2, and on increasing the gold loading, a progressive shift of the reduction peak to lower temperatures was observed without a significant enhancement of the percentage of reducibility (Figure 13.5). This was ascribed to the increment in the size of the gold nanoparticles at high loadings, which results in the diminishing of the gold–support interactions.

Obviously, the synthesis method of the gold-supported catalysts is a major determinant in the final size of the gold particles, and hence, of catalytic performance. Although there are a large variety of synthetic routes, according to Bond and Thompson,[3] they could be classified in two main categories: (1) those where the support and the gold species are produced at the same time, as coprecipitation or the sol-gel method; and (2) those where

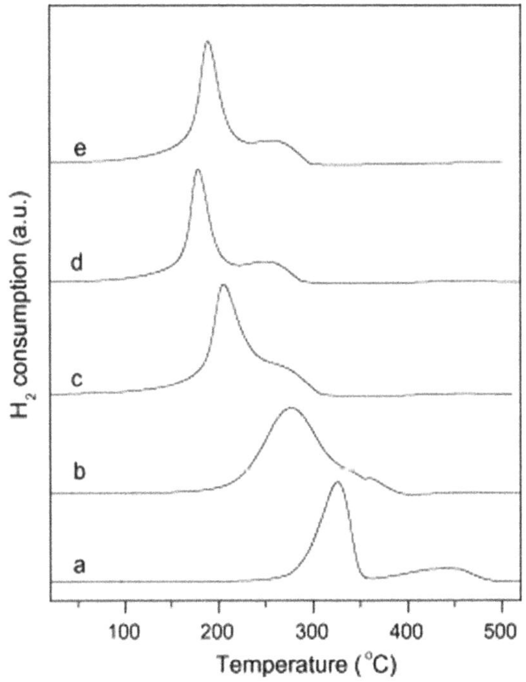

Figure 13.5 Reducibility properties of MnO_2 as a function of the gold content: (a) without gold; (b) 1%wt Au; (c) 3% Au; (d) 5%wt Au; and (e) 7.9%wt Au. Reproduced with permission of Elsevier from ref. 84.

gold is deposited onto the previously prepared support. The latter category includes: impregnation and the alternative, impregnation to incipient wetness, that generally do not achieve a high dispersion, and where some steps of the procedure such as the drying are determinant in the final gold particle size; and ion adsorption where cationic gold species are adsorbed on O^- sites on the surface of the support under pH values above the point zero charge. Also anionic gold precursors may be adsorbed under pH below point zero charge. Another possibility is the deposition–precipitation method that implies that the gold hydroxide or a hydrated oxide is deposited on the surface of the support through a gradual modification of the pH of a solution where the support is dispersed.[3]

In any case, there is a general agreement that no matter which support is selected or which synthesis procedure is used to obtain gold catalysts, the interaction between gold and oxygen vacancies is positive for promoting the catalytic activity in oxidation reactions. Nevertheless, some fundamental aspects of such interaction remain under discussion. Although some authors have described the improvement of the oxidizing ability in gold catalysts as an enhancement of the electronic transfer between the gold cluster and the surface of the support,[85] and theoretical studies have extensively examined the thermodynamic aspects of the behavior of the oxygen vacancies and gold,[28,30] there is no general agreement on the oxidation state of the really active species of such a noble metal. This discussion extends to the oxidation reaction mechanisms by considering the role of the support and the chemical nature of the gold species.

For instance, Chen and Goodman[86] created well-ordered gold monolayers and bilayers that completely covered titania; they demonstrated through high-resolution electron energy loss spectroscopy, CO adsorption measurements, and kinetic studies of the CO oxidation reaction that the gold bilayer structure is significantly more active than the monolayer. The major catalytic activity of Au^0 particles has also been suggested by Hickey *et al.*[87] They proposed that the CO oxidation in Au/CeO_2 catalysts depends strongly on the CO adsorption, which takes place over metal gold nanoparticles and not over ionic Au species. In a similar way, Sudarsanam *et al.*[88] revealed by X-ray photoelectron spectroscopy (XPS) studies the presence of only Au^0 species in Au/CeO_2 catalysts doped with Fe^{3+}, La^{3+}, and Zr^{4+}, concluding that such species are responsible for the observed CO oxidation activity.

The role of cationic gold species in oxidation reactions has been also discussed by many authors. For instance, Corma and co-workers[89] proposed the positive role of stabilized cationic gold species on Au/CeO_2 catalysts for CO oxidation, based on the detection of higher activities when the Au^{3+}/Au^0 ratio was increased. The influence of the electronic structure of $Au/CeO_2/Al_2O_3$ catalysts in the CO oxidation properties has been also evidenced, in such a way that the overlapping of the gold bands with the conduction band of the support results in the lowering of the bandgap and in the enhancement of the catalytic activities.[82] On the other hand, Fu and co-workers[32,90] studied the WGS reaction on Au-CeO_2 catalysts and found that cationic gold species strongly bounded to ceria, Au_n-O-Ce, are the catalytically relevant species, and that the coexisting metallic gold nanoparticles do not participate in the reaction, as demonstrated by the similar activity measured for cyanide-leached samples devoid of gold nanoparticles. Reducibility and XPS studies demonstrated the isolated presence of Au^+ or Au^{3+} cations, or the coexistence of both species, in different gold catalysts.[91–94]

In another work presented by Ousmane *et al.*,[95] the catalytic activity was evaluated in the total oxidation of propene and toluene, chosen as model VOC molecules. The activity trend was as follows: $Au/CeO_2 > Au/7.5Ce/Al_2O_3 > Au/TiO_2 > Au/Al_2O_3$. Au/CeO_2, containing highly dispersed gold species in intimate contact with ceria, was the best performing, achieving full propene and toluene conversions at 230 °C and 293 °C, respectively. Through XPS analysis, only Au^0 species were detected in the Au/Al_2O_3 solid, but in the case of the other systems, the presence of Au^+ and/or Au^{3+} was observed. Compared with other gold-supported catalysts, Au/CeO_2 is better performing in VOC oxidation at low temperature. The occurrence of a strong metal–support interaction between highly dispersed gold nanoparticles and ceria seems to be the main requisite for the achievement of high oxidation activity. This is due to the presence of oxygen vacancies which also contributes to enhancing the oxygen mobility and the metal–support interactions.

A very interesting contribution to the discussion about the really active gold species in oxidation reactions has been made by several authors who presented results about a more complex and dynamic behavior of the catalyst's surface, depending on the oxidant conditions of the atmosphere and the location of the gold species in the catalyst structure. Manzoli *et al.*[96] studied the spectroscopic features and the reactivity of CO adsorbed on different Au/CeO_2 catalysts with variable gold contents (0.6 and 0.7 at.%). They found an infrared adsorption (IR) band at 2130–2140 cm^{-1} due to CO adsorbed on cationic gold clusters, irreversible on outgassing and more resistant to oxidation than the usual carbonyl band on Au^0 sites. In addition, a broad absorption band in the 2000–2100 cm^{-1} range was observed after reduction in hydrogen, due to structural and electronic changes of gold. Moreover, the reduced gold species in ceria can be re-oxidized under mild conditions, and TPR and XPS analyses of the fresh and used catalysts corroborated the stability of ionic gold in ceria up to 120 °C in the reaction gas mixture.

In good agreement with the active role of reducible supports in gold catalysts for oxidation reactions, Liu and Rodriguez[97] recently reported that gold-supported systems based on defective structured materials can be considered as bifunctional catalysts in the WGS reaction; the dissociative adsorption of water takes place on the oxide, while CO adsorbs on the gold nanoparticles and all subsequent reaction steps occur at the metal–oxide interface. Moreover, these authors also demonstrated that the rate-limiting step of the WGS reaction is the water dissociation, explaining the higher activity of Au-CeO_2, Au-MoO_2, Au-TiO_2 and Au-CeO_x-TiO_2 compared with an Au-ZnO system by the direct participation from ceria, molybdena, titania, and mixed ceria–titania supports in the water dissociation. Taking into account the fact that the rate-limiting step of the WGS reaction is the dissociation of the water molecules and that this is favored by the presence of oxygen vacancies, the highest catalytic activity of Au/CeO_2 can be related to the higher population of oxygen vacancies in the ceria support.[98] In this sense, initially, the oxygen vacancies of the support allow a strong metal–support interaction, thus resulting in very small gold nanoparticles. Then, the gold species with ionic character in close contact with the support reduce the energy required for creating additional vacancies, which are crucial for the water dissociation. This agrees with the theoretical studies about the evolution of oxygen vacancies in reducible materials presented by Nolan *et al.*[30]

The easiness of formation of additional oxygen vacancies in gold catalysts induced by the presence of gold may even result in the redispersion of such gold species over the catalyst surfaces under reductive atmosphere. In this sense, Hernández *et al.*[25] observed that, in such reductive conditions, the population of oxygen vacancies in Au/CeO_2-Eu increases, thus creating new nucleation sites for gold nanoparticles. These authors analyzed by Raman spectroscopy this type of solid after treatment in different atmospheres. The evolution of the area ratio of the bands at 532 and 465 cm^{-1}, representative of the number of vacancies, showed the presence of a high number of oxygen vacancies on the catalysts if treated in a reductive atmosphere (Figure 13.6). Consequently the noble metal deposits became smaller by redispersion. This redispersion represents a significant improvement in the catalytic activity in the CO oxidation reaction.

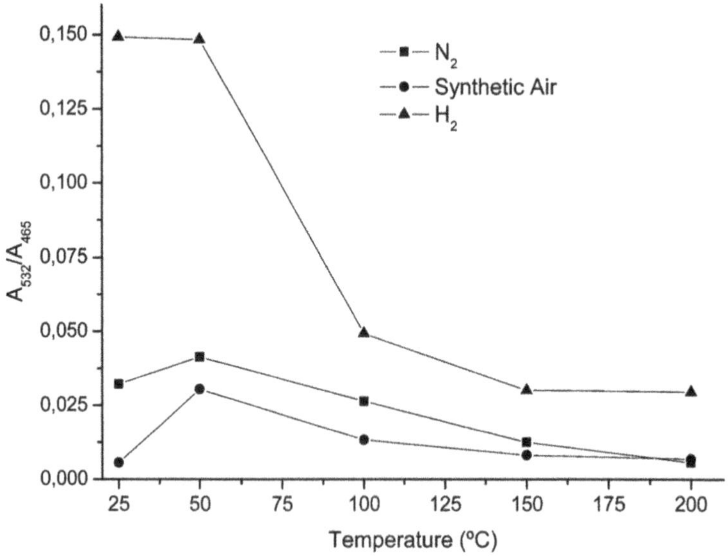

Figure 13.6 Relationship between A_{532}/A_{465} and temperature under different atmospheres for an Au/CeO_2-Eu catalyst.
Reprinted with permission from ref. 25. Copyright 2010 American Chemical Society.

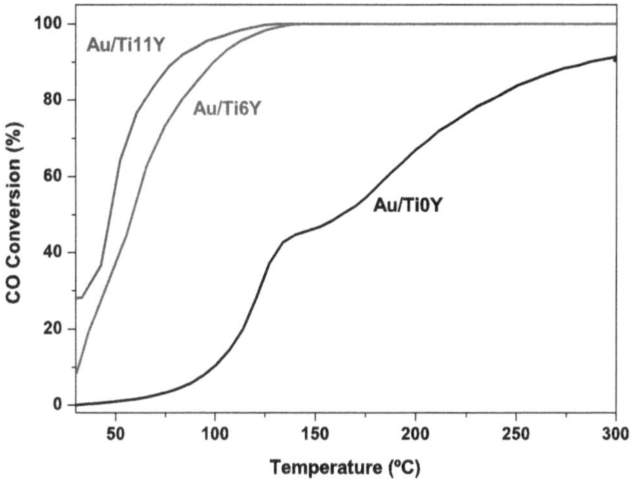

Figure 13.7 Light-off curves for Au/TiO_2-Y catalysts. The number indicates the percentage in weight of yttrium.
Adapted with permission of Elsevier from ref. 24.

A similar behavior has been observed in catalytic supports based on TiO_2 doped with different quantities of yttrium. In this case, the low solubility of the yttrium atoms in the oxide impedes the formation of a complete solid solution. According to XRD analysis and theoretical calculations,[24] yttrium atoms and oxygen vacancies are located on the surface, as expected by the properties of the doping atom and the tolerance limit of the framework. The addition of different quantities of yttrium (below the solubility limit) to the TiO_2 controls the number of created oxygen vacancies. The gold incorporation to the so-prepared supports evidenced the role of the oxygen vacancies as gold nucleation sites since, for similar gold contents, when yttrium content was increased, the average gold particle size was reduced and the CO oxidation activity was increased, as observed in the CO light-off curves (Figure 13.7). An in-depth study of these catalytic systems combining experimental and theoretical calculations demonstrated that the oxygen activation proceeds on the metal/oxide interface with a strong electronic transfer from gold to the oxygen (activation), while CO is adsorbed on the gold particle with a similar extent of the σ-donation and π-retro-donation. The high electronic transfer to the O_2 molecule permitted the proposal of a peroxo-type reaction intermediate.

13.3 Non-reducible Supports

Not many works devoted to study the role of oxygen vacancies in irreducible supports are available, the most studied being MgO, SiO_2, Al_2O_3, and phosphates. All studies agree in the necessity to have a very low gold particle size (<3.5 nm) to obtain good CO conversions because oxygen activation proceeds only on the metallic particle, without participation of the support. However, every support has a different influence on the morphology of the gold particles, the metal–support interaction, and the stabilization of ionic gold species, resulting in large differences in activity. Some of the most interesting aspects related to the use of irreducible supports in gold oxidation catalysis are summarized below.

13.3.1 Phosphates

In spite of the well-established 'gold-only mechanism', the generation of active species on the support has been detected when phosphate compounds (which are irreducible supports) are used for supporting gold. Domínguez *et al.*[99] studied the CO oxidation reaction on gold-supported calcium hydroxyapatite and observed an initial very high activity (~100%) at room temperature during a very short time. An *in operando* diffuse reflectance infrared Fourier transform spectroscopy–mass spectrometry (DRIFTS-MS) study of this catalytic system (3.4% CO, 21% O_2 balanced by N_2) evidenced a decrease in the amount of hydroxyl groups of the hydroxyapatite support (responsible for the IR band at 3572 cm^{-1}), probably due to reaction with the CO contained in the reactive flow. According to the stoichiometry of the hydroxyapatite, OH removal from the structure provokes the appearance of structural oxygen vacancies according to the following equation:[100]

$$Ca_{10}(PO_4)_6(OH)_2 \rightarrow Ca_{10}(PO_4)_6(OH)_{2-2x}O_x\square_x + xH_2O \qquad (13.1)$$

A detailed analysis of the 900–800 cm^{-1} spectral region showed the appearance of peroxide species (840 cm^{-1}) under these conditions. When an oxygen vacancy is created on an oxide, two electrons remain in this vacancy, and its interaction with gaseous O_2 generates species O_2^- or O_2^{2-}.[21,99] These species, strongly oxidizing and reactive, are responsible for the 100% CO conversion observed at room temperature for both the support and the gold catalyst. The formation of active species by interaction with the oxygen vacancies has also been detected by Matsumura and co-workers[100,101] for hydroxyapatite-type supports and by other authors[102] for reducible oxides. This fact suggests that one of the roles of the oxygen vacancies in oxidation reactions is similar for reducible and irreducible solids, *i.e.* the creation of species able to participate actively in the oxidation mechanism.

Irreducible and reducible phosphates were used as supports for gold and tested in the CO oxidation reaction.[103–105] In the case of the reducible phosphate, Liu and co-workers evidenced that the CO oxidation reaction proceeds by a direct pathway with the participation of oxygen atoms from the support.[46] The nature of gold species was explored by CO adsorption and only cationic gold species, which are reduced to metallic gold during the CO adsorption, were detected, in agreement with some other data reported for this type of supports.[91,106,107]

For the irreducible phosphate studied ($LaPO_4$), in addition to the direct pathway (with 'O' from the support), the participation of hydroxyl groups from the surface to form a carboxyl intermediate was also observed.[103] In this case, the CO adsorption allowed to determine the presence of metallic and anionic gold species, as well as formation of CO_2 and carbonates. Despite the detection of the products of the CO oxidation (CO_2 and carbonates), the loss of structural oxygen was not observed by Raman analysis. CO-TPR and Raman results permitted the authors to conclude that hydroxyl groups are the oxygen source for the oxidation reaction, according to the following equation:

$$CO + OH^- \rightarrow CO_2 + \frac{1}{2}H_2\square^- \quad (13.2)$$

An oxygen vacancy is created in the oxidation process. Given that oxygen vacancies are sites with high electronic density, their interaction with gold species results in an electronic transfer from these sites to the gold, explaining the observation of anionic species after CO adsorption. This is in good agreement with previous results evidencing that oxygen vacancies act as nucleation sites for the gold, similar to that observed with reducible supports.[24,43] From a mechanistic point of view, it is worth remembering that electronic transfer from the surface sites to the O_2 molecule is crucial for the gas activation and, therefore, the gold particles sited on oxygen vacancies may constitute centers for oxygen activation on irreducible supports.

13.3.2 MgO

Another irreducible support that has been widely treated in the literature is MgO. In this case, the active gold species in the CO oxidation reaction is under discussion because the charge transfer between the gold and the support is more noticeable in this oxide compared with other irreducible ones.[108] Some authors[22,109] detected negatively charged gold in the Au-MgO interface in studies of CO adsorption as the probe molecule. The formation of these anionic gold species was attributed to the placement of gold particles in the oxygen vacancies or F-sites of the support, in good agreement with the observations of Li *et al.*[103] for another irreducible support. The interaction of gold with the F-sites of MgO has been also studied by Sterrer *et al.*[110] In their work, MgO(001) films were studied by CO adsorption after different pre-treatments by electron bombardment and gold deposition. The authors concluded that gold nucleates in both terraces and F-sites. In the first case, the gold particle is neutral (IR band at 2120 cm^{-1}), while in the second one, negatively charged gold is detected (band at 2070 cm^{-1}). However, the presence of $Au^{\delta+}$ has also been observed in Au/MgO solids.[111,112] The work by Li *et al.*[113] pointed out the importance of the MgO exposed faces for the gold particles. The authors observed that gold strongly bonds to the (111) surfaces but not to (100) and (110) ones; this means that the charge transfer is more important in the first situation, leading to solids with a higher catalytic activity.

The participation of hydroxyl groups from the surface in the CO oxidation reaction is discussed above for the irreducible phosphates. In the case of MgO, its influence is still under discussion. On one hand, an improvement in the catalytic activity for hydroxylated surfaces of On the one hand Au/MgO and an increased thermal stability of the gold particles after adsorption of small quantities of water have been detected.[114] This stability is attributed to the formation of strong Au–O interactions by participation of type (II) OH groups. These interactions result in the inhibition of gold sintering and in higher thermal stability. On the other hand, it has been proposed that hydroxyl groups may interact with carbonate species *via* hydrogen bonds, blocking the active sites for CO oxidation and leading to catalyst deactivation.

However, Au/MgO catalysts show a high activity in the CO oxidation reaction and it is believed that both the participation of the hydroxyl groups and the presence of F-sites are the key factors that explain this behavior.

13.3.3 SiO_2

Another typical irreducible support is SiO_2, in which the presence of oxygen vacancies is not usual. The main problem with Au/SiO_2 catalysts is the difficulty in obtaining well-dispersed gold particles on the surface. The large negative charge on the silica surface (an isoelectric point of approximately 2) motivates a weak interaction between the support and the gold precursor (usually chloroauric acid) when conventional methods are used to prepare Au/SiO_2 catalysts.[115] Due to this, the catalysts obtained show a large gold particle size and very low activity in the CO oxidation reaction. To prepare solids with stronger gold–support interactions and smaller gold particles, more complex methods, such as chemical vapor deposition (CVD)[116] and grafting,[117] of support modification by active oxides[118] are being employed.

In the case of the solids with a low catalytic activity (prepared by conventional method), the active gold species have been investigated by *in situ* X-ray absorption spectroscopy (XANES); it was demonstrated that metallic gold is the active site in CO oxidation at room temperature. More active Au/SiO_2 catalysts, obtained by deposition–precipitation using $Au(en)_2Cl_3$ (en = ethylenediamine) as gold precursor, were studied by DRIFTS-MS to determine the active gold species.[119,120] The authors studied the catalytic activity after different pre-treatments (O_2, H_2, CO) that produced a similar number of gold sites on the surface but with different oxidation states. They proposed the metallic gold species as the active species in the CO oxidation reaction (at room temperature), and found a correlation between the reduction degree and the catalytic activity. Moreover, the positive effect of water, as previously observed in other irreducible supports, was evidenced.

13.3.4 Al_2O_3

In the case of Au/γ-Al_2O_3 catalyst, the active species have been investigated by DRIFTS.[121] In this study, the presence of Au^0 and Au^+ was detected and the authors proposed a reaction mechanism involving the OH groups from the surface. The authors suggested that the presence of neutral gold is needed, but the key role in the reaction is played by the cationic species. The mechanism of CO oxidation on Au/Al_2O_3 has also been investigated by Gavril *et al.*[122] The authors observed that CO adsorption in the absence of oxygen resulted in CO_2 formation, and concluded that the active sites must be composed of metallic gold atoms and cationic sites with an OH group. The positive effect of water on the catalytic activity in this solid agrees with the participation of hydroxyl groups in the mechanism of reaction. The decisive role of the hydroxyl groups in the mechanism of CO oxidation on Au/Al_2O_3 catalysts and the requirement of an Au^0 and Au^1-OH ensemble was first evidenced by Kung *et al.*[123]

Besides this, other authors proposed, after XANES and EXAFS studies, that small gold metallic particles can activate the molecular oxygen directly, being oxidized to Au_xO_y, which is further reduced to Au^0 by reaction with CO, resulting in CO_2 and closing the catalytic cycle.[124]

13.4 Comparison

The work reviewed above points to a similar role of oxygen vacancies in reducible and irreducible supports, *i.e.* generating of active species in the CO oxidation reaction and acting as sites for gold nucleation. However, the catalytic behavior of reducible and irreducible supports is markedly different in most cases. Some authors[46] attribute the differences to the minor contribution of oxygen vacancies to the formation of active species in the case of irreducible supports compared with the extent of the metal–support interface in the reducible ones.

In this way, Laursen and Linic,[21] using DFT and *ab initio* thermodynamic calculations, investigated the most probable oxidation state of gold and the role of the support on gold catalysts supported on one reducible (TiO_2) and one irreducible (SiO_2) oxide. From their calculations of local density of states the authors concluded that:

(1) In the case of TiO_2, characterized by a low bandgap, the electronic density freed by the formation of oxygen vacancies occupies the 3d bands of the Ti atoms, unoccupied in a solid free of defects. These levels are in the Fermi level, and the electronic density may be really transferred to the gold deposed on a reducible support containing a certain density of oxygen vacancies and explaining the detection of negatively charged gold species on TiO_2 supports.[125] The strong electronic transfer from the support to the gold means that oxygen adsorbed on these sites may be easily activated (sites with high electronic density).
(2) For an irreducible support like SiO_2, in which the bandgap is larger, the electronic density generated by the formation of oxygen vacancies is used to form Si–Si bonds (the bond distance is decreased compared

with the form free of vacancies), but not transferred to the gold deposited on the silica. This means that, in this case, $Au^{\delta-}$ species are not formed. This is due to the stabilization of the Si–Si bond below the Fermi level, and implies that the oxygen activation is more difficult on these sites.

13.5 Concluding Remarks

It is clear that the role of the oxygen vacancies in catalysis by gold seems to be similar in reducible and irreducible supports, *i.e.* they are able to generate active species for the oxidation reactions and act as sites for gold nucleation. However, the ease of creating oxygen vacancies and their increased mobility in reducible supports make the number and the reactivity of these sites higher in these solids than in the case of irreducible supports.

Electronic transfer from the support to the metallic particle occurs in both cases, but the extent of this transfer is determined by the properties of the oxide and dictates the strength of the gold–support interaction. Moreover, the good redox properties of the reducible ones facilitate the mobility of charged species, and metallic and oxidized gold species have been detected. In the case of the irreducible supports, the detected gold species are mainly negatively charged due to the limited redox properties. In this case, the participation of hydroxyl groups in the oxidation process seems to be proved by several authors, being the catalytic ensemble formed by Au^0 and Au^1-OH sites

Finally, it must be remarked that the atmosphere of reaction or pretreatment of the sample strongly influences the gold oxidation state. Therefore, studies by *in operando* techniques are fundamental tools to ensure the correct determination of the mechanism of the reaction and the behavior of the oxygen vacancies in each support.

Acknowledgements

Financial support for this work was obtained from the Spanish Ministerio de Economía y Competitividad (ENE2012-374301-C03-01) co-financed by FEDER funds from the European Union and from Junta de Andalucía (TEP-8196). All the authors acknowledge the Junta de Andalucía for the financing of the 'TEP 106 group'.

References

1. M. Haruta, T. Kobayashi, H. Sano and N. Yamada, *Chem. Lett.*, 1987, **16**, 405.
2. M. Haruta, N. Yamada, T. Kobayashi and S. Iijima, *J. Catal.*, 1989, **115**, 301.
3. G. C. Bond, C. Louis and D. T. Thompson, *Catalysis by Gold*, ed. G. J. Hutchings, Imperial College Press, London, 2006.

4. D. T. Thompson, *Top. Catal.*, 2006, **38**, 231.
5. G. C. Bond and D. T. Thompson, *Catal. Rev.: Sci. Eng.*, 1999, **41**, 319.
6. A. Corma and H. García, *Chem. Soc. Rev.*, 2008, **37**, 2096.
7. A. Corma and H. García, *Nanoparticles and Catalysis*, ed. D. Astruc, Wiley-VCH, Weinheim, 2008, ch. 12, pp. 389–426.
8. C. Louis, *Nanoparticles and Catalysis*, ed. D. Astruc, Wiley-VCH, Weinheim, 2008, ch. 15, pp. 475–503.
9. S. Scirè and L. F. Liotta, *Appl. Catal. B*, 2012, **125**, 222.
10. S. A. C. Carabineiro and D. T. Thompson, *Nanocatalysis*, ed. U. Heiz and U. Landman, Springer, Berlin, 2007, pp. 377–489.
11. A. S. K. Hashmi, *Chem. Rev.*, 2007, **107**, 3180.
12. R. Burch, *Phys. Chem. Chem. Phys.*, 2006, **8**, 5483.
13. S. A. C. Carabineiro and D. T. Thompson, *Gold Science and Applications*, ed. C. Corti and R. Holliday, CRC Press, Boca Raton, FL, 2010, ch. 6, pp. 89–122.
14. F. Tao and Z. Ma, *Phys. Chem. Chem. Phys.*, 2013, **15**, 15260.
15. G. Avgouropoulos and T. Tabakova, *Environmental Catalysis over Gold-Based Materials*, RSC Publishing, Cambridge, 2013.
16. S. Gil, M. Marchena, C. M. Fernández, L. Sánchez-Silva, A. Romero and J. L. Valverde, *Appl. Catal. A*, 2013, **450**, 189.
17. E. G. Rodrigues, M. F. R. Pereira, X. Chen, J. J. Delgado and J. J. M. Órfão, *J. Catal.*, 2011, **281**, 119.
18. X. Yu, Y. Huo, J. Yang, S. Chang, Y. Ma and W. Huang, *Appl. Surf. Sci.*, 2013, **280**, 450.
19. M. A. Centeno, M. Paulis, M. Montes and J. A. Odriozola, *Appl. Catal. B*, 2005, **61**, 177.
20. G. Pacchioni, *Solid State Sci.*, 2000, **2**, 161.
21. S. Laursen and S. Linic, *Phys. Rev. Lett.*, 2006, **97**, 026101–1.
22. B. Yoon, H. Häkkinen, U. Landman, A. S. Wörz, J. M. Antonietti, S. Abbet, K. Judai and U. Heiz, *Science*, 2005, **307**, 403.
23. C. Zhang, A. Michaelides, D. A. King and S. J. Jenkins, *J. Phys. Chem. C*, 2009, **113**, 6411.
24. J. J. Plata, A. M. Márquez, J. Fdez. Sanz, R. S. Avellaneda, F. Romero-Sarria, M. I. Domínguez, M. A. Centeno and J. A. Odriozola, *Top. Catal.*, 2011, **54**, 219.
25. W. Y. Hernández, F. Romero-Sarria, M. A. Centeno and J. A. Odriozola, *J. Phys. Chem. C*, 2010, **114**, 10857.
26. M. V. Ganduglia-Pirovano, A. Hofmann and J. Sauer, *Surf. Sci. Rep.*, 2007, **62**, 219.
27. V. Shapovalov and H. Metiu, *J. Catal.*, 2007, **245**, 205.
28. M. Nolan, *J. Phys. Chem. C*, 2009, **113**, 2425.
29. O. H. Laguna, F. Romero Sarria, M. A. Centeno and J. A. Odriozola, *J. Catal.*, 2010, **276**, 360.
30. M. Nolan, V. Soto and H. Metiu, *Surf. Sci.*, 2008, **602**, 2734.
31. M. Boronat and A. Corma, *Catal. Today*, 2011, **169**, 52.

32. Q. Fu, H. Saltsburg and M. Flytzani-Stephanopoulos, *Science*, 2003, **301**, 935.
33. M. M. Schubert, S. Hackenberg, A. C. van Veen, M. Muhler, V. Plzak and R. J. Behm, *J. Catal.*, 2001, **197**, 113.
34. G. C. Bond and D. T. Thompson, *Gold Bull.*, 2000, **33**, 41.
35. J. C. Frost, *Nature*, 1988, **334**, 577.
36. H. Liu, A. I. Kozlov, A. P. Kozlova, T. Shido, A. Asakura and Y. Iwasawa, *J. Catal.*, 1999, **185**, 252.
37. J. Guzman, S. Carrettin, J. C. Fierro-González, Y. Hao, B. C. Gates and A. Corma, *Angew. Chem., Int. Ed.*, 2005, **44**, 4778.
38. C. Li, K. Domen, K.-I. Maruya and T. Onishi, *J. Catal.*, 1990, **123**, 436.
39. G. Sedmak, S. Hõcevar and J. Levec, *J. Catal.*, 2004, **222**, 87.
40. S.-Y. Wang, N. Li, S.-F. Luo, W.-X. Huang, Z.-Y Pu, Y.-J. Wang, G.-S. Hu, M.-F. Luo and J.-Q. Lu, *Appl. Catal. B*, 2014, **144**, 325.
41. W. Y. Hernández, F. Romero-Sarria, M. A. Centeno and J. A. Odriozola, *J. Phys. Chem. C*, 2009, **113**, 5629.
42. S. Carrettin, P. Concepción, A. Corma, J. M. López-Nieto and V. F. Puntes, *Angew. Chem., Int. Ed.*, 2004, **43**, 2538.
43. F. Romero-Sarria, L. M. Martínez, T, M. A. Centeno and J. A. Odriozola, *J. Phys. Chem. C*, 2007, **111**, 14469.
44. I. N. Remediakis, N. López and J. K. Norskov, *Angw. Chem., Int. Ed.*, 2005, **44**, 1824.
45. I. N. Remediakis, N. López and J. K. Norskov, *Appl. Catal. A*, 2005, **291**, 13.
46. D. Widmann, Y. Liu, F. Schüth and R. J. Behm, *J. Catal.*, 2010, **276**, 292.
47. B. Solsona, E. Aylón, R. Murillo, A. M. Mastral, A. Monzonís, S. Agouram, T. E. Davies, S. H. Taylor and T. Garcia, *J. Hazard. Mater.*, 2011, **187**, 544.
48. B. Solsona, T. García, G. J. Hutchings, S. H. Taylor and M. Makkee, *Appl. Catal. A*, 2009, **365**, 222.
49. B. Solsona, T. Garcia, E. Aylón, A. M. Dejoz, I. Vázquez, S. Agouram, T. E. Davies and S. H. Taylor, *Chem. Eng. J.*, 2011, **175**, 271.
50. L.-C. Wang, X.-S. Huang, Q. Liu, Y.-M. Liu, Y. Cao, H.-Y. He, K.-N. Fan and J.-H. Zhuang, *J. Catal.*, 2008, **259**, 66.
51. A. Zhao, W. Ying, H. Zhang, M. Hongfang and D. Fang, *J. Nat. Gas Chem.*, 2012, **21**, 170.
52. M. A. Centeno, M. C. Hidalgo, M. I. Domínguez, J. A. Navío and J. A. Odriozola, *Catal. Lett.*, 2008, **123**, 198.
53. G. Avgouropoulos, T. Ioannides, C. Papadopoulou, J. Batista, S. Hõcevar and H. K. Matralis, *Catal. Today*, 2002, **75**, 157.
54. W. Deng, C. Carpenter, N. Yi and M. Flytzani-Stephanopoulos, *Top. Catal.*, 2007, **44**, 199.
55. A. K. Kandalam, B. Chatterjee, S. N. Khanna, B. K. Rao, P. Jena and B. V. Reddy, *Surf. Sci.*, 2007, **601**, 4873.
56. N. Bion, F. Epron, M. Moreno, F. Mariño and D. Duprez, *Top. Catal.*, 2008, **51**, 76.

57. K. Shojaee, A. Montoya and B. S. Haynes, *Comput. Mater. Sci.*, 2013, **72**, 15.
58. S. Carrettin, Y. Hao, V. Aguilar-Guerrero, B. C. Gates, S. Trasobares, J. J. Calvino and A. Corma, *Chem. – Eur. J.*, 2007, **13**, 7771.
59. J. Guzman, S. Carretín and A. Corma, *J. Am. Chem. Soc.*, 2005, **127**, 3286.
60. H. Bao, X. Chen, J. Fang, Z. Jiang and W. Huang, *Catal. Lett.*, 2008, **125**, 160.
61. D. Djurovic, M. Zinkevich and F. Aldinger, *Solid State Ionics*, 2008, **179**, 1902.
62. V. Idakiev, T. Tabakova, K. Tenchev, Z.-Y. Yuan, T.-Z. Ren and B.-L. Su, *Catal. Today*, 2007, **128**, 223.
63. V. Idakiev, T. Tabakova, Z.-Y. Yuan, T.-Z. Ren, X.-D. Zou and B.-L. Su, *Stud. Surf. Sci. Catal.*, 2006, **162**, 1017.
64. O. H. Laguna, M. A. Centeno, G. Arzamendi, L. M. Gandía, F. Romero-Sarria and J. A. Odriozola, *Catal. Today*, 2010, **157**, 155.
65. F. J. Pérez-Alonso, T. Herranz, S. Rojas, M. Ojeda, M. López Granados, P. Terreros, J. L. G. Fierro, M. Gracia and J. R. Gancedo, *Green Chem.*, 2007, **9**, 663.
66. F. J. Pérez-Alonso, M. López Granados, M. Ojeda, P. Terreros, S. Rojas, T. Herranz, J. L. G. Fierro, M. Gracia and J. R. Gancedo, *Chem. Mater.*, 2005, **17**, 2329.
67. F. J. Pérez-Alonso, I. Melián-Cabrera, M. López Granados, F. Kapteijn and J. L. G. Fierro, *J. Catal.*, 2006, **239**, 340.
68. F. J. Pérez-Alonso, M. Ojeda, T. Herranz, S. Rojas, J. M. González-Carballo, P. Terreros and J. L. G. Fierro, *Catal. Comm.*, 2008, **9**, 1945.
69. A. Razeghi, A. Khodadadi, H. Ziaei-Azad and Y. Mortazavi, *Chem. Eng. J.*, 2010, **164**, 214.
70. D. Terribile, A. Trovarelli, C. de Leitenburg, A. Primavera and G. Dolcetti, *Catal. Today*, 1999, **47**, 133.
71. G. Xiao, S. Li, H. Li and L. Chen, *Microporous Mesoporous Mater.*, 2009, **120**, 426.
72. S. Zhao and R. J. Gorte, *Appl. Catal. A*, 2003, **248**, 9.
73. O. H. Laguna, M. A. Centeno, F. Romero-Sarria and J. A. Odriozola, *Catal. Today*, 2011, **172**, 118.
74. A. Penkova, O. H. Laguna, M. A. Centeno and J. A. Odriozola, *J. Phys. Chem. C*, 2012, **116**, 5747.
75. O. H. Laguna, W. Y. Hernández, G. Arzamendi, L. M. Gandía, M. A. Centeno and J. A. Odriozola, *Fuel*, 2014, **118**, 176.
76. O. H. Laguna, F. Romero Sarria, M. A. Centeno and J. A. Odriozola, *J. Catal.*, 2010, **276**, 360.
77. D. Tibiletti, A. Amieiro-Fonseca, R. Burch, Y. Chen, J. M. Fisher, A. Goguet, C. Hardcare, P. Hu and D. Thompsett, *J. Phys. Chem. B*, 2005, **109**, 22553.
78. X. Pan and Y.-J. Xu, *Appl. Catal. A*, 2013, **459**, 34.

79. M. Haruta, *Catal. Today*, 1997, **36**, 153.
80. M. Haruta, *Gold Bull.*, 2004, **37**, 27.
81. H. Sakurai, A. Ueda, T. Kobayashi and M. Haruta, *Chem. Comm.*, 1997, 271.
82. M. A. Centeno, C. Portales, I. Carrizosa and J. A. Odriozola, *Catal. Lett.*, 2005, **102**, 289.
83. V. P Santos, S. A. C. Carabineiro, J. J. W. Bakker, O. S. G. P. Soares, X. Chen, M. F. R. Pereira, J. J. M. Órfão, J. L. Figueiredo, J. Gascon and F. Kapteijn, *J. Catal.*, 2014, **309**, 58.
84. L.-C. Wang, L. He, Q. Liu, Y.-M. Liu, M. Chen, Y. Cao, H.-Y. He and K.-N. Fan, *Appl. Catal. A*, 2008, **344**, 150.
85. D. Andreeva, P. Petrova, L. Ilieva, J. W. Sobczak and M. V. Abrashev, *Appl. Catal. B*, 2008, **77**, 364.
86. M. S. Chen and D. W. Goodman, *Science*, 2004, **306**, 252.
87. N. Hickey, P. Arneodo Larochette, C. Gentilini, L. Sordelli, L. Olivi, S. Polizzi, T. Montini, P. Fornasiero, L. Pasquato and M. Graziani, *Chem. Mater.*, 2007, **19**, 650.
88. P. Sudarsanam, B. Mallesham, P. S. Reddy, D. Großmann, W. Grünert and B. M. Reddy, *Appl. Catal. B*, 2014, **144**, 900.
89. P. Concepción, S. Carrettin and A. Corma, *Appl. Catal. A*, 2006, **307**, 42.
90. Q. Fu, W. Deng, H. Saltsburg and M. Flytzani-Stephanopoulos, *Appl. Catal. B*, 2005, **56**, 57.
91. A. Amieiro Fonseca, J. Fisher, D. Ozkaya, M. Shannon and D. Thompsett, *Top. Catal.*, 2007, **44**, 223.
92. G. Jacobs, S. Ricote, P. M. Patterson, U. M. Graham, A. Dozier, S. Khalid, E. Rhodus and B. H. Davis, *Appl. Catal. A*, 2005, **292**, 229.
93. A. Karpenko, R. Leppelt, V. Plzak and R. J. Behm, *J. Catal.*, 2007, **252**, 231.
94. C. H. Kim and L. T. Thompson, *J. Catal.*, 2005, **230**, 66.
95. M. Ousmane, L. F. Liotta, G .D. Carlo, G. Pantaleo, A. M. Venezia, G. Deganello, L. Retailleau, A. Boreave and A. Giroir-Fendler, *Appl. Catal. B*, 2011, **101**, 629.
96. M. Manzoli, F. Boccuzzi, A. Chiorino, F. Vindigni, W. Deng and M. Flytzani-Stephanopoulos, *J. Catal.*, 2007, **245**, 308.
97. P. Liu and J. A. Rodriguez, *J. Chem. Phys.*, 2007, **126**, 164705.
98. F. Vindigni, M. Manzoli, T. Tabakova, V. Idakiev, F. Boccuzzi and A. Chiorino, *Phys. Chem. Chem. Phys.*, 2013, **15**, 13400.
99. M. I. Domínguez, F. Romero-Sarria, M. A. Centeno and J. A. Odriozola, *Appl. Catal., B*, 2009, **87**, 245.
100. Y. Matsumura, H. Kanai and J. B. Moffat, *J. Mol. Catal. A*, 1997, **115**, L229.
101. Y. Matsumura, H. Kanai and J. B. Moffat, *J. Mol. Catal. A*, 1998, **135**, 317.
102. S. Rossignol, F. Gerard and D. Duprez, *J. Mater. Chem.*, 1999, **9**, 1615.
103. M. Li, Z. Wu and S. H. Overbury, *J. Catal.*, 2011, **278**, 133.

104. M. Li, Z. Wu, Z. Ma, V. Schwartz, D. Mullins, S. Dai and S. H. Overbury, *J. Catal.*, 2009, **266**, 98.
105. Z. Ma, H. Yin, S. H. Overbury and S. Dai, *Catal. Lett.*, 2008, **126**, 20.
106. S. Minico, S. Scire, C. Crisafulli, A. M. Visco and S. Galvagno, *Catal. Lett.*, 1997, **47**, 273.
107. F. Bocuzzi, A. Choirino and M. Manzoli, *Surf. Sci.*, 2002, **502–503**, 513.
108. X. Y. Liu, A. Wang, T. Zhang and C. Mou, *Nanotoday*, 2013, **8**, 403.
109. X. Lin, N. Nilius, M. Sterrer, P. Koskinen, H. Hakkinen and H. J. Freund, *Phys. Rev. B*, 2010, **81**, 153406.
110. M. Sterrer, M. Yulikov, E. Fischbach, M. Heyde, H. P. Rust, G. Pacchioni, T. Risse and H. J. Freund, *Angew. Chem., Int. Ed.*, 2006, **45**, 2630.
111. P. Frondelius, H. Hakkinen and K. Honkala, *Angew. Chem., Int. Ed.*, 2010, **49**, 7913.
112. Y. Hao, M. Mihaylov, E. Ivanova, K. Hadjiivanov, H. Knoezinger and B. C. Gates, *J. Catal.*, 2009, **261**, 137.
113. Z. Li, C. V. Ciobanu, J. Hu, J. P. Palomares-Baéz, J. L. Rodríguez-López and R. Richards, *Phys. Chem. Chem. Phys.*, 2011, **13**, 2582.
114. M. A. Brown, Y. Fujimori, F. Ringleb, X. Shao, F. Stavale and N. Nilius, *J. Am. Chem. Soc.*, 2011, **133**, 10668.
115. Y. Zhang, B. Zhaorigetu, M. Jia, C. Chen and J. Zhao, *Catal. Comm.*, 2013, **35**, 72.
116. M. A. P. Dekkers, M. J. Lippits and B. E. Nieuwenhuys, *Catal. Today*, 1999, **54**, 381.
117. Y. S. Chi, H. P. Lin and C. Y. Mou, *Appl. Catal. A*, 2005, **284**, 199.
118. A. Penkova, J. M. Martínez Blanes, S. A. Cruz, M. A. Centeno, K. Hadjiivanov and J. A. Odriozola, *Microporous Mesoporous Mater.*, 2009, **117**, 530.
119. Z. Wu, S. Zhou, H. Zhu, S. Dai and S. Overbury, *J. Phys. Chem. C*, 2009, **113**, 3726.
120. Z. Wu, S. Zhou, H. Zhu, S. Dai and S. Overbury, *Chem. Commun.*, 2008, 3308.
121. A. Leba, T. Davran-Candan, Z. I. Önsan and R. Yildirim, *Catal. Comm.*, 2012, **29**, 6.
122. D. Gavril, A. Georgaka, V. Loukopoulus, G. Karaiskakis and B. E. Nieuwenhuys, *Gold Bull.*, 2006, **39**, 192.
123. C. K. Costello, J. H. Yang, H. Y. Law, Y. Wang, J.-N. Lin, L. D. Marks, M. C. Kung and H. H. Kung, *Appl. Catal. A*, 2003, **243**, 15.
124. J. A. van Bokhoven, C. Louis, J. T. Miller, M. Tromp, O. V. Safonova and P. Glatzel, *Angew. Chem., Int. Ed.*, 2006, **45**, 4651.
125. T. Minato, T. Susaki, S. Shiraki, H. S. Kato, M. Kawai and K. Aika, *Surf. Sci.*, 2004, **566–568**, 1012.

CHAPTER 14

Infrared Spectroscopic Insights into the Role of the Support in Heterogeneous Gold Catalysis

ZILI WU* AND STEVEN H. OVERBURY*

Chemical Science Division and Center for Nanophase Materials Sciences, Oak Ridge National Laboratory, Oak Ridge, TN 37831, USA,
*Email: wuz1@ornl.gov; overburysh@ornl.gov

14.1 Introduction

The purpose of using a support in heterogeneous catalyst was originally to make efficient use of the active components, usually metals, by dispersing them as small (1–1000 nm) particles on a stable platform. Yet it turned out that the support materials are not just acting as a spectator, but often working together with the metals for the catalysis taking place upon them. It has now been recognized that the heterogeneous catalyst support is more like a ligand to the active components than an inert support, resembling the organic ligands coordinated to the active centers in homogeneous counterparts. It is thus the case that the support material may exhibit a profound effect on the overall catalytic performance.[1,2] No better example can be found for such a support effect than the intensively investigated heterogeneous catalysis on supported gold nanoparticles. Since the discovery of the exceptional low-temperature CO oxidation reactivity of gold nanoparticles by Haruta and co-workers,[3] oxidation catalysis by

RSC Catalysis Series No. 18
Heterogeneous Gold Catalysts and Catalysis
Edited by Zhen Ma and Sheng Dai

Published by the Royal Society of Chemistry, www.rsc.org

oxide-supported gold nanoparticles was found to be extremely sensitive to the nature of the support.[4,5] Although the size[6] and shape[7] of the support can impact the reactivity of gold nanoparticles, the composition of the support has been shown to be the dominant factor affecting gold nanoparticle catalysis.[4,7]

Many high surface area oxides have been tried as supports for gold. Metal oxides used as gold supports may be classified as inert or active according to their reducibility.[7,8] Thus, Al_2O_3, SiO_2, and MgO are included within the inert supports, while reducible transition metal oxides such as Fe_2O_3, MnO_2, TiO_2, and CeO_2 may be considered as active supports. The catalytic activity of the latter group may be hypothesized to depend on the ability of the support to provide reactive oxygen to the catalytic system. However, the relationship between activity and oxide reducibility is not straightforward. Support reducibility does not guarantee high activity, and if properly prepared and pre-treated, gold can be very active on non-reducible supports such as SiO_2[9,10] and $LaPO_4$.[11–13] Several mechanisms have been proposed to explain the enhanced activity at the metal–oxide interface including the presence of highly reactive surface gold atoms at growth steps, the ease of charge transfer at the interface, and in the case of reducible supports, the ability to provide activated oxygen that can be trapped at the reaction sites. Whatever the mechanism, it is evident that the interaction between the support and gold must be tailored for performance enhancement.

Due to the fact that gold nanoparticles catalysis can be affected by a variety of factors[4,14–17] such as the size of the gold nanoparticles, the oxidation state of Au sites, synthesis methods, pre-treatment conditions, dependence upon moisture, and the type of support, the origin of the interesting catalytic behavior of gold nanoparticles remains elusive. It is useful to look back upon this research field to summarize what has been achieved so that a comprehensive understanding of gold catalysis can be made possible. A number of reviews[4,14,15,18–20] have shed light on the catalysis by supported gold nanoparticles with an emphasis on the support effect.

In this chapter, we focus on non-reducible oxide and non-oxide supports which have been rarely discussed in the literature. Despite the dominant use of oxide supports, recent studies show that non-oxide supports such as phosphates can also be promising materials for supporting gold nanoparticles in various reactions. We summarize primarily our recent work[12,21–26] on SiO_2 and phosphate supported gold catalysts for CO oxidation reaction and demonstrate how one can make use of *in situ* infrared (IR) spectroscopy to gain insights into the intriguing support effect in gold nanoparticle catalysis. Together with the work reviewed in the literature on reducible oxide supported gold catalysis, we hope to provide a complementary picture on how the support materials impact gold catalysis by influencing the nature of Au active sites and the dynamics of Au sites during activation and reaction, and controlling the reaction pathways and mechanism.

14.2 SiO_2 Supported Au Catalysts

SiO_2 has long been employed as a support for a variety of metal catalysts owing to its high surface area, thermal stability, mechanical strength, and non-reducibility. Considering its inertness (non-reducibility), one may be motivated to utilize SiO_2 as a support for gold nanoparticles in order to separate the complex support effect from other factors affecting gold catalysis.

Initial trials[27,28] using the conventional deposition–precipitation method were not quite successful since this approach leads to inactive catalysts with large gold particles. This is because the low isoelectric point (IEP) of SiO_2 (IEP $\sim$ 2) does not allow a strong interaction between the support and the commonly used auric acid precursor during the preparation process.[29] More active Au/SiO_2 catalysts with small gold nanoparticles were successfully prepared *via* chemical vapor deposition (CVD),[30] grafting,[31] or by employing a different gold precursor for the DP process,[9] *e.g.* $Au(en)_2Cl_3$ (en = ethylenediamine), leading to a strong interaction between the deposited gold precursor and the SiO_2 support. It was shown that gold nanoparticles deposited on silica by the latter method leads to activity comparable with Au on TiO_2 for low temperature CO oxidation.[9] We were intrigued by this type of active Au/SiO_2 catalyst and motivated to reveal the nature of the Au sites and how they work for the low temperature CO oxidation on the inert SiO_2 surface.[21,23]

14.2.1 Oxygen-assisted Reduction of Au Species on SiO_2

Due to the strong interaction between gold nanoparticles and the ethylenediamine ligands in the $Au(en)_2Cl_3$ precursor, it was found that the Au/SiO_2 has to be calcined at temperature of 400 °C and above to remove the ligands so that gold nanoparticles could show CO adsorption from the *in situ* IR study.[21] The small size ($\sim$5 nm) of gold nanoparticles on SiO_2 after 500 °C calcination is rather unusual considering that gold nanoparticles supported on TiO_2 would increase in size to over 10 nm after similar thermal treatment.

This unusual high thermal stability of gold nanoparticles on SiO_2 from this preparation method is not yet quite understood. The stability is possibly due to partial capsulation of gold nanoparticles by amorphous silica.[9] The Au sites for CO adsorption are cationic (oxidation state between 0 and +1) after calcination of a 4 wt% Au/SiO_2 sample at 500 °C, as characterized by an IR band at 2141 cm^{-1} from initial CO adsorption at room temperature. At prolonged CO flow at room temperature for 20 h, the IR band of CO–Au red shifts slowly from 2141 to 2124 cm^{-1}. The CO band position is plotted in Figure 14.1 as a function of time. This red shift is due to the partial reduction of the cationic Au sites. But the Au sites are not completely reduced to the metallic state as the IR band at 2124 cm^{-1} is still attributed to CO adsorbed on positively charged Au sites. When this experiment was repeated in the presence of O_2, *i.e.* co-flow of CO–O_2 (1 : 4) at room temperature, the red shift of the CO–Au band was also observed (Figure 14.1) but at a much faster pace. The shift from 2141 to 2124 cm^{-1} takes 6 h and the band further shifts to

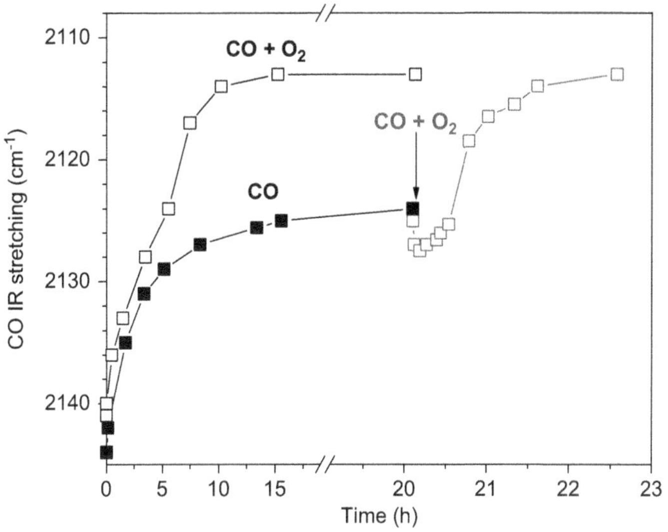

Figure 14.1 IR frequency of adsorbed CO on Au/SiO_2 pre-calcined at 500 °C as a function of time on stream at room temperature. Solid square: CO flow; open square: $CO + O_2$ flow. Curve and open squares above 20 h, indicate introduction of O_2 to CO after 20 h.
Adapted from ref. 23.

2113 cm^{-1} after 15 h time on stream, an indication of the complete reduction of cationic Au to metallic Au by $CO–O_2$ flow at room temperature. There is a clear contrast in the reduction degree of Au sites between CO flow only and $CO–O_2$ co-flow. Since it is normally assumed that the reduction of Au is done by just CO,[32,33] this unexpected observation suggests that CO alone is not able to completely reduce cationic Au to the metallic state on Au/SiO_2 at room temperature. Instead, the reduction of cationic Au to metallic Au during CO oxidation on Au/SiO_2 is assisted by the presence of oxygen.

This observation is further confirmed by another experiment. After CO flow for 20 h on calcined Au/SiO_2, O_2 was added into the stream and the reactant mixture was adjusted to the same conditions as in CO oxidation. The introduction of oxygen into the CO stream induces a shift in the CO band from 2124 cm^{-1} down to 2112 cm^{-1} within 2 h. The small blue shift of CO band from 2125 to 2128 cm^{-1} upon intial O_2 introduction is likely due to the co-adsorption of oxygen with CO on Au sites because adsorbed oxygen abstracts electrons from Au sites, making them more cationic.[34] Co-feeding $^{18}O_2$ with CO can also shift the CO band down to 2112 cm^{-1} but at a slower rate than using $^{16}O_2$, indicating that the presence of oxygen is directly related to this shift. Other experiments showed that an excess oxygen to CO ratio is not needed to induce the assisted reduction of Au, while co-existence of CO and O_2 is necessary. Co-feeding CO_2 during CO reduction does not change the CO band position at 2124 cm^{-1}, an indication that CO_2 formed during reduction (or reaction) is not relevant to the shift.

Based on the results above, we confirmed that the unusual CO band shift upon oxygen introduction is due to oxygen-assisted reduction of Au species by CO. This seems contradictory to the intuitive expectation that molecular oxygen should be excluded from the reaction media when reduction is to be performed. There exist only very few examples in which oxygen induces or accelerates reduction in certain organic reactions.[35] In our case of Au/SiO_2,[23] we propose that the reduction of cationic Au by CO involves hydroxyl groups on cationic Au sites and proceeds *via* a carboxyhydroxyl (COOH) intermediate as shown in Figure 14.2. The carboxyhydroxyl intermediate has been proposed in several other studies[36–38] for either CO oxidation or water gas shift reactions on Au and Cu catalysts. The presence of surface OH groups associated with Au is critical for the formation of COOH and they probably originate from water accumulated on the surface during CO flow (trace water, *ca.* 1 part per million, contained in the gases) as evident from the increase in intensity of OH stretching (broad feature from 2700 to 3700 cm^{-1} centered at *ca.* 3380 cm^{-1}) *vs.* time on stream. The promotional effect of water is further confirmed by the fact that the reduction of cationic Au is greatly enhanced (the shift from 2142 to 2112 cm^{-1} takes less than 2 h) as water, *ca.* 500 parts per million (ppm) was co-fed with CO/O_2. The broad OH stretching band may also contain OH stretching from COOH, which is calculated to be at 3530 cm^{-1} on Au(110).[39] The COOH species is somehow stable on Au/SiO_2 and thus prevents reduction of cationic Au by CO. However, when O_2 is present, the removal of H atoms is accelerated by forming water/hydroxyl, thus facilitating the decomposition of carboxyhydroxyl and the eventual reduction of cationic Au. Further theoretical investigation is

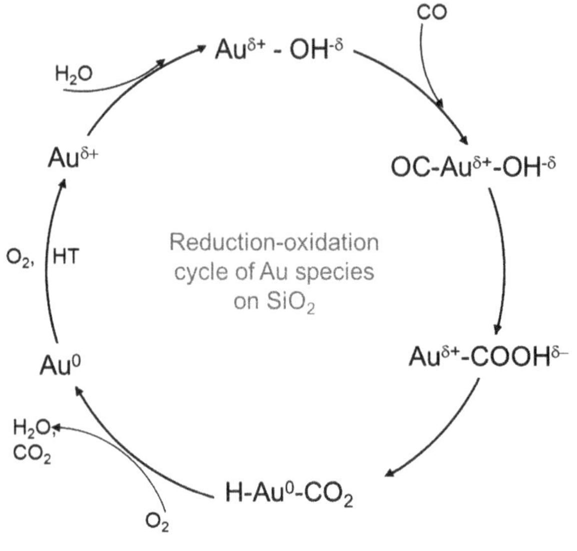

Figure 14.2 Mechanistic scheme showing the O_2 assisted reduction of cationic Au on SiO_2.

warranted to gain insight into this oxygen-assisted reduction of Au sites on SiO_2. We believe that this unusual finding is of fundamental importance in understanding the channels leading to the reduction of metal oxides by CO.

The presence of the carboxyhydroxyl species on Au/SiO_2 surface was also found during CO adsorption on H_2 pre-reduced Au/SiO_2. Shown in Figure 14.3(a) are the IR spectra collected during a CO pulse at room temperature on H_2-treated Au/SiO_2. Three bands at *ca.* 3400 cm^{-1} (OH stretching), 2350 cm^{-1} (CO_2 asymmetric stretching), and 2119 cm^{-1} (CO stretching) appear and increase in intensity simultaneously upon the increase in CO amount. During the later portion of the pulse (desorption process), the three bands decrease and disappear concurrently.

The broad OH band at around 3400 cm^{-1} is not observed when CO is pulsed over O_2 or CO-treated Au/SiO_2 sample. Therefore, it seems that the CO_2 production from a CO pulse is related to a surface species whose OH stretching is at *ca.* 3400 cm^{-1}. Considering the possible chemistry, we

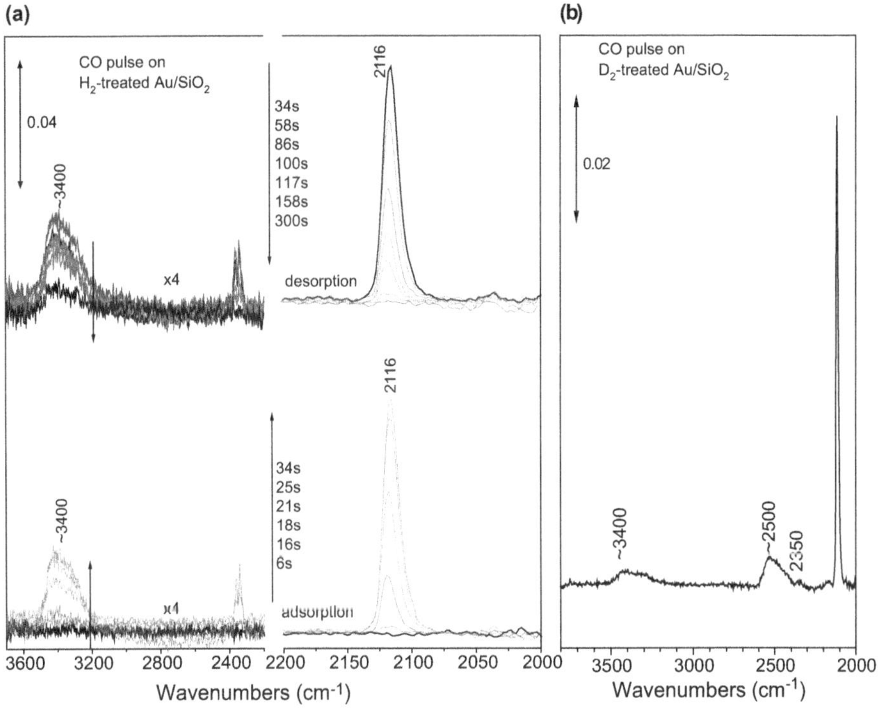

Figure 14.3 (a) IR spectra of CO adsorbed on 500 °C H_2-treated Au/SiO_2 during one CO pulse at room temperature during the onset adsorption of the pulse (bottom) and the desorption phase (top). (b) Representative IR spectrum of CO adsorbed on 500 °C D_2-treated Au/SiO_2 at room temperature.
Reprinted with permission from ref. 21. Copyright 2009 American Chemical Society.

propose that the band at *ca.* 3400 cm^{-1} is due to the OH stretching in adsorbed carboxyhydroxyl ($COOH_a$) an intermediate to the production of CO_2:

$$CO_a + OH_a \rightarrow COOH_a \rightarrow CO_2 + H_a \quad (14.1)$$

where (a) refers to an adsorbed species. This hypothesis is further confirmed by using D_2 as the reductant instead of H_2 during 500 °C treatment. A CO pulse on the D_2-treated (4% D_2/Ar) Au/SiO_2 shown in Figure 14.3(b) (one representative spectrum) yields a new broad IR band at around 2500 cm^{-1} in addition to the CO, CO_2, and OH bands in the H_2-treated case. A red shift of about 900 cm^{-1} is expected when the H atom in COOH is isotopically exchanged with a D atom. Thus it is very likely that COOH/COOD is the source of reactive O for CO_2 production observed after H_2/D_2 treatment. This observation substantiates our proposal of carboxyhydroxyl species as an intermediate during the oxygen-assisted reduction of cationic Au by CO on the calcined Au/SiO_2 catalyst.

It is notable that the nature of the support for gold catalysts seems to affect the reduction behavior of cationic Au by CO in addition to its known effect on CO oxidation behavior. The reduction is facile for gold supported on reducible oxides such as TiO_2,[33] but difficult on non-reducible oxides such as Al_2O_3[40] and SiO_2.[23] The reduction of cationic Au by CO was also found to be promoted by water on Au/Al_2O_3.[40] It is possible that a COOH intermediate could be at work in the reduction of cationic Au on both Au/Al_2O_3 and Au/SiO_2. Therefore, the finding of the oxygen-assisted reduction of cationic Au by CO may have general implications for gold supported on non-reducible metal oxide supports.

14.2.2 Role of Different Au Sites on SiO_2 for CO Oxidation

To determine the role of different Au species, cationic and metallic ones, in CO oxidation on Au/SiO_2, we used *in situ* IR spectroscopy coupled with online mass spectrometry (QMS) to follow both surface CO–Au species and the gas phase product. Shown in Figure 14.4(a) are the IR spectra collected during room temperature CO oxidation ($CO:O_2 = 1:4$) on a 500 °C O_2-treated Au/SiO_2 sample. As described in the previous section, the CO band shifts gradually downwards from 2142 to 2113 cm^{-1} at prolonged reaction time. This CO band shift *vs.* time is plotted in Figure 14.3(b) and is an indication of the reduction of cationic Au to metallic Au. Also plotted in Figure 14.4(b) are the integrated areas of the CO_2 (at 2342 cm^{-1}) and OH (2800–3700 cm^{-1}) IR signals, and the CO conversion as measured by QMS as a function of time on stream. It is interesting to note that, after the first 3 h, the activity of the catalyst correlates well with the red shift of the CO band. As the CO band moves to lower frequency, the CO oxidation activity and CO_2 IR area simultaneously increase and then stabilize as the CO band levels off at around 2113 cm^{-1}.

The excellent correlation among the three plots clearly indicates that metallic Au species play a major role in CO oxidation on Au/SiO_2 at room temperature, while cationic Au species alone are weakly or not active.

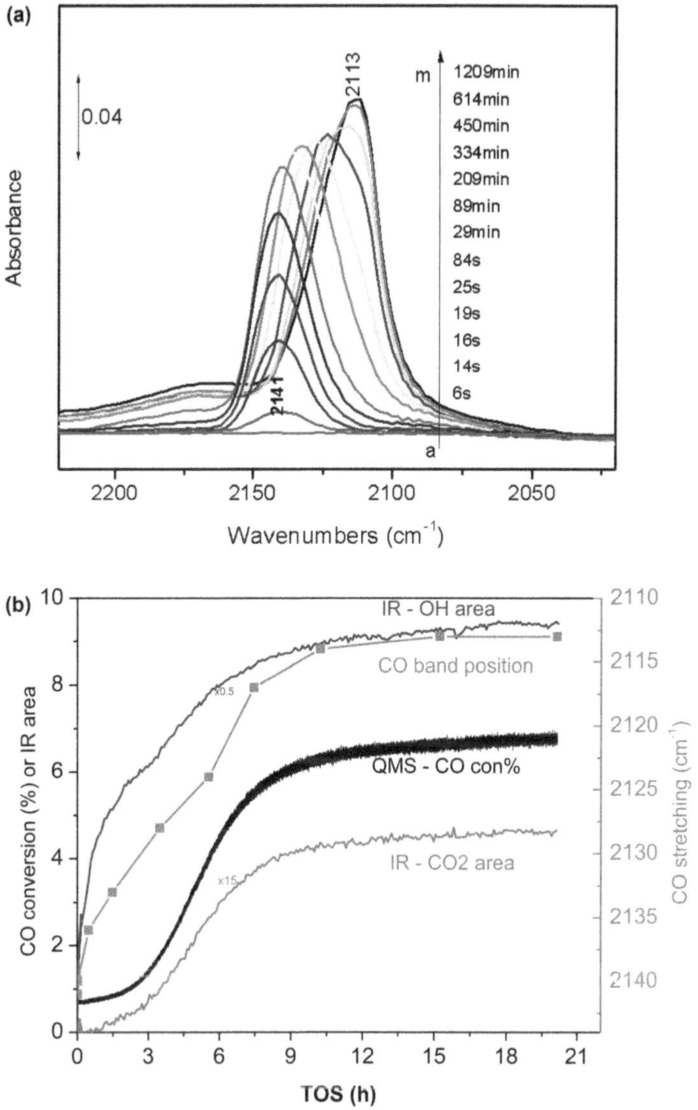

Figure 14.4 (a) IR spectra during room temperature CO oxidation on 500 °C O_2-treated Au/SiO_2. (b) Plots of IR CO_2 area, QMS CO conversion, IR CO band position, and IR OH band area as a function of reaction time during room temperature CO oxidation on 500 °C O_2-treated Au/SiO_2. Reprinted with permission from ref. 21. Copyright 2009 American Chemical Society.

This was the first report about the role of different Au species in low-temperature CO oxidation on a functioning Au/SiO_2 catalyst. During the first 3 h the CO conversion does not increase because the majority of the Au species is still cationic, but after 3 h reduced Au appears.

To determine if there is a general trend in CO oxidation activity *vs.* Au oxidation state, differently treated Au/SiO_2 samples were tested for room temperature CO oxidation. These different treatments include O_2, H_2, and CO treatment, and re-oxidation of the H_2-treated Au/SiO_2 at different temperatures in order to alter the oxidation state of Au species. The oxidation state of Au species is represented by the initial CO adsorption band position. When the CO conversion was plotted against the CO IR band frequency,[21] it showed clearly that the CO oxidation activity increased consistently as the CO band shifted to lower wavenumbers, implying that the activity correlates with the reduction degree of surface Au species. A similar correlation was observed for CO oxidation on Au/TiO_2,[41] but not on Au/Al_2O_3.[40] In the latter case, metallic Au is necessary but cationic Au was suggested to also play a role in CO oxidation. On an Au/SiO_2 catalyst, there is no indication of a positive role of cationic Au species in CO oxidation.

The dynamic nature of the Au sites on SiO_2 is not only reflected in the *in situ* reduction of cationic Au to metallic Au during CO oxidation, but is also found during steady state CO oxidation at room temperature. The Au sites are metallic in nature when the Au/SiO_2 catalyst was pre-reduced by either H_2 or CO.[21] On the reductively pre-treated Au/SiO_2, the CO oxidation activity at room temperature declines slightly as a function of time; meanwhile the IR spectra from adsorbed CO indicate that part of the Au species is oxidized. Three possible explanations for the slight deactivation might be: (1) aggregation of the gold particles; (2) accumulation of carbonate species; and (3) oxidation of Au species. The transmission electron microscopy (TEM) results show no sign of particle size change after CO oxidation on the differently pre-treated Au/SiO_2 samples. In all IR studies, no bands ascribed to carbonate species were observed. This is, on the one hand, due to the strong absorption of SiO_2 in the carbonate spectral region, and on the other hand, mostly due to the inert property of the SiO_2 surface. Temperature programmed oxidation of the catalyst after CO oxidation did not show appreciable production of CO_x (both CO and CO_2) from carbonate species decomposition. Thus it seems that the oxidation of Au species during CO oxidation is responsible for the slight deactivation, pointing to a detrimental effect of cationic Au species on CO oxidation. The dynamic change of the oxidation state of the Au sites on Au/SiO_2 underscores the importance of *operando* study of a working catalyst.

14.2.3 SiO_2 Supported Au-based Bimetallic Catalysts

Although supported gold nanoparticles have proven to be extremely active catalysts, bimetallic catalysts may have a remarkable ability to enhance the reactivity, selectivity, and especially the stability achieved by controlling their structure and composition. Bimetallic nanoparticle catalysts of AuM (M = Ag, Pt, Pd, In, Ni, Cu, *etc.*)[24–26,42,43] with heteroaggregate, core–shell, or alloy structures have shown an improvement toward catalytic activity and selectivity for a number of chemical reactions (including CO oxidation, alcohol oxidation, and hydrogenation). Our work[24–26,44] on NiAu and CuAu alloys supported on SiO_2 showed that the alloyed metals can become precursors to metal oxide components upon high-temperature oxidative treatment. The *in situ* generated oxide components have an intimate interaction with gold nanoparticles on SiO_2, and thus greatly improve the catalytic performance of Au nanoparticles in redox reactions.

The first example is a NiAu alloy supported on SiO_2 for CO oxidation.[25,26] The solution phase synthesized NiAu alloy and gold nanoparticles were loaded onto a SiO_2 support *via* a colloidal deposition method and tested in CO oxidation. The catalyst pre-treated in H_2 at 720 °C and then in O_2 at 300 °C showed the highest activity among differently pre-treated catalysts. For comparison, solely pre-treating the NiAu/SiO_2 sample in H_2 does not lead to a reasonably active catalyst. Under similar treatment conditions, the Au/SiO_2 sample made by the same synthesis method gave much lower activity for CO oxidation due to the large size of the gold nanoparticles. The actual working structure of the most active NiAu/SiO_2 catalyst under reaction conditions turned out to be Au-NiO/SiO_2 instead of alloy NiAu/SiO_2, a conclusion supported by various characterization data including high-resolution TEM, X-ray absorption spectroscopy and IR spectroscopy of CO probing. It was proposed that the nickel in the NiAu alloy nanoparticles diffused toward the surface during H_2 treatment and was then oxidized by O_2 treatment to NiO that is situated around the gold nanoparticles, as illustrated in Figure 14.5. The very small, amorphous NiO structures interacted with the Au nanoparticles in a non-uniform fashion, but clearly served an important role in preventing Au sintering and enhancing the catalytic activity for CO oxidation by forming a stable interface between Au and NiO. The influence of the Ni to Au ratio in the pre-treated catalyst was studied and the activity was highest when the Ni to Au ratio was 1. A lower activity of Ni_3Au/SiO_2-derived catalyst compared with NiAu/SiO_2-derived catalyst was explained as being caused by the gold nanoparticles being covered by too much NiO.

The oxidative transformation of supported alloy catalysts into oxide-decorated Au catalysts was later found to also occur for a CuAu/SiO_2 catalyst.[24,44] Similarly, the CuAu alloy supported on SiO_2 that undergoes an oxidative de-alloying process has increased activity for low-temperature CO oxidation and

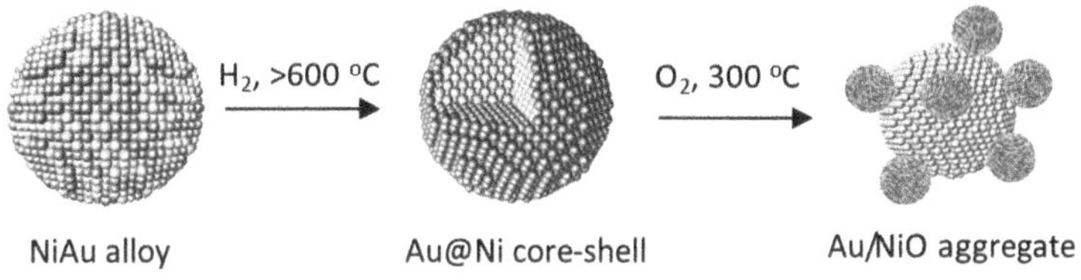

Figure 14.5 Schematic representation of the phase transformation from NiAu alloy to Au/NiO hetero-aggregate.
Reprinted with permission from ref. 26. Copyright 2008, John Wiley & Sons.

selective oxidation of ethanol to acetaldehyde. High-resolution TEM and *in situ* IR spectroscopy confirmed that the Cu component is oxidized and that the Au surface sites are exposed and decorated by the CuO_x after the oxidative treatment. When this de-alloyed catalyst was reduced by H_2 at 300 °C, the catalyst lost its room temperature CO oxidation activity due to re-alloying of Au and Cu, which prevents the adsorption of CO on Au sites as probed by IR spectroscopy. The alloy and de-alloy process is reversible and thus the catalytic performance of $CuAu/SiO_2$ catalyst can be turned on or off.

These two examples of Au alloy catalysts show that the presence of an oxide or another component near gold nanoparticles can work together with the traditional support to form the interface structure so that the catalytic property of Au nanoparticles can be preferably tuned. The examples also manifest the necessity of *operando* study of catalyst structure, as sometimes the presumed 'alloy catalysts' are not real alloy catalysts after oxidative pre-treatments or under reaction conditions.

14.3 Phosphate Supported Au Catalysts

Although there are numerous reports on oxide-supported Au catalysts, there are much fewer investigations of Au supported on salts. Salts used as supports for gold nanoparticles include metal carbonates,[45] hydroxyapatite,[46,47] orthovanadates,[48] and phosphates.[11,13] However, our interest was largely on phosphates since a variety of phosphates are available and some of them are reducible ($FePO_4$) and some are not ($LaPO_4$). It was shown that gold nanoparticles supported on nanosized $LaPO_4$ (6–8 nm) are active for CO oxidation below room temperature.[13] It has also been demonstrated that a number of other metal phosphates can be good supports for gold nanoparticles for low-temperature CO oxidation.[11]

The application of the phosphate-based supports opens up a new avenue in the search for highly active, stable, and selective gold catalysts for a number of catalytic reactions. Following these initial reports of a wide range of phosphate supported gold catalysts, we employed *in situ* IR spectroscopy to focus on the reaction mechanism of CO oxidation by gold nanoparticles supported on reducible $FePO_4$ *vs.* non-reducible $LaPO_4$.[12,22]

14.3.1 Participation of Phosphate Support in CO Oxidation

When the 200 °C O_2-treated $Au/FePO_4$ and $Au/LaPO_4$ samples were exposed to CO at room temperature, production of CO_2 was observed for both samples as evident from both IR spectroscopy and QMS. After careful characterization, we showed that the origin of the CO_2 production, *i.e.* the source of oxygen, was different on the two phosphate-supported gold catalysts.

14.3.1.1 Au/FePO$_4$

The O_2-treated sample leads to oxidized Au species as indicated by *in situ* X-ray absorption near edge spectroscopy (XANES). Room temperature exposure to CO results in reduction of oxidized Au to metallic Au as shown by both XANES and *in situ* IR spectroscopy, contributing partially to the CO_2 production. For example, in the *in situ* IR spectra[22] during CO flow over the O_2-treated sample, a band at 2169 cm^{-1} due to CO adsorbed on cationic Au sites was observed upon initial CO adsorption. Two other bands at 2129 cm^{-1} (CO on cationic Au) and 2114 cm^{-1} (CO on metallic Au) developed and dominated the spectra as more CO adsorbed on the surface and induced the reduction of the oxidized Au. Meanwhile, the $FePO_4$ support was also found to be reduced by CO at room temperature flow; this was confirmed by the *in situ* Raman spectra collected during CO flow. The Raman band at 1611 cm^{-1} caused by CO reduction at room temperature was similar to that caused by high-temperature H_2 reduction of the catalyst. Also, this band was found to disappear when a flow of O_2 is introduced over the sample.

Therefore CO_2 production from CO flow originates from the reduction of both cationic Au and the $FePO_4$ support. The reduced $FePO_4$ can be re-oxidized at room temperature by exposure to O_2, so the structural oxygen of $FePO_4$ can be catalytically involved in low-temperature CO oxidation. This is demonstrated in Figure 14.6(a) where IR spectra and QMS profiles are shown during repeated CO and O_2 exposure (sequential flow) at room temperature on O_2-treated Au/$FePO_4$. The IR spectra shown are for CO adsorption after first CO–O_2 exposure (*i.e.* during the second CO exposure). The IR spectra are dominated by 2114 cm^{-1} (CO on metallic Au) and 2129 cm^{-1} (CO on cationic Au) bands from the beginning of the adsorption, suggesting that the reduction of oxidized Au species is completed in the first CO flow cycle.

However, CO_2 production is still observed [inset, Figure 14.6(a)], indicating that there are active oxygen species on the surface of Au/$FePO_4$ that are not due to oxygen associated with cationic Au. Repeating the sequence of CO adsorption followed by room temperature O_2 exposure multiple times on the same sample gave almost identical growth of CO bands (position and relative intensity), and CO_2 was evolved during each CO adsorption cycle as shown in the inset of Figure 14.6(a).

Therefore, it appears that the active oxygen species that is consumed to evolve CO_2 can be replenished by gas-phase O_2 at room temperature without variation in the relative amounts of Au species. Evidently the surface of gold nanoparticles helps to activate both CO and O_2 so that the $FePO_4$ support can undergo cycles of reduction (by CO) and re-oxidation (by O_2) at room temperature. In the absence of gold nanoparticles, $FePO_4$ shows redox behavior only at temperatures above 250 °C in various oxidation and oxidative dehydrogenation reactions.[49]

14.3.1.2 Au/$LaPO_4$

Most surface Au sites on the O_2-treated Au/$LaPO_4$ sample are metallic as indicated by measurement of CO IR peak positions (CO–Au band at 2102 cm^{-1}), with a minority of sites being negatively charged (CO–Au bands at 2075 and 2050 cm^{-1}). The observed CO_2 production during room

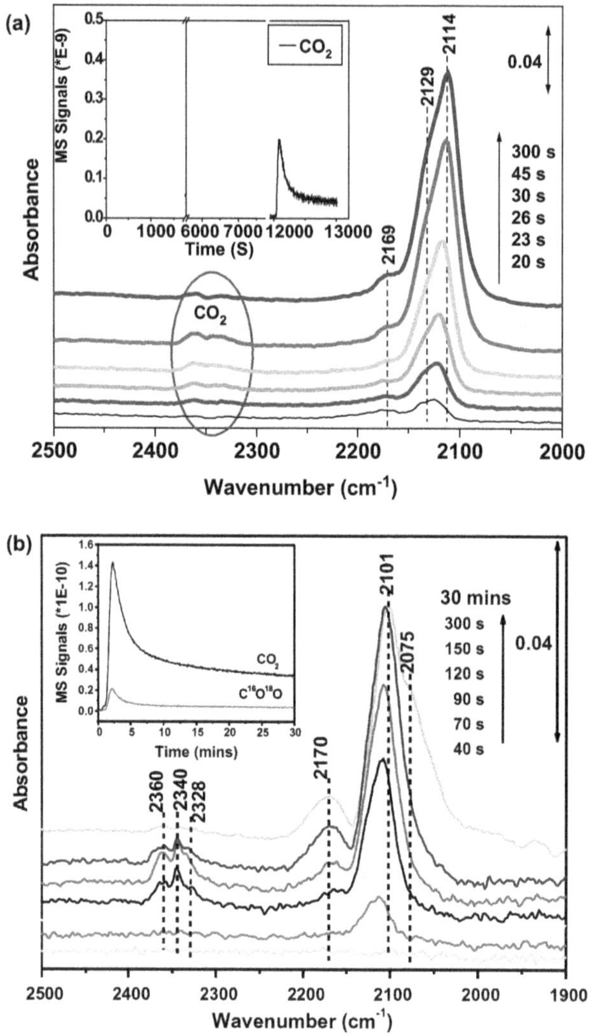

Figure 14.6 (a) IR spectra obtained during the second CO adsorption switch at room temperature on an O_2-pretreated Au/$FePO_4$ after an initial CO adsorption/desorption cycle. The inset shows the QMS (CO_2) response during this second adsorption and two subsequent adsorptions, each separated by an intermediate interval of room temperature O_2 exposure. Reprinted from ref. 22. Copyright 2009, with permission from Elsevier. (b) IR spectra and QMS (inset) results for CO adsorption at room temperature on $^{18}O_2$-pretreated Au/$LaPO_4$ as a function of adsorption time after switching from He to CO.
Reprinted from ref. 12. Copyright 2011, with permission from Elsevier.

temperature CO adsorption indicates that there is active oxygen available on the catalyst surface after O_2 pre-treatment. Since the surface Au sites are metallic or negatively charged, reduction of Au oxide cannot be responsible for the CO_2 formation. So, the active oxygen species on Au/$LaPO_4$ that is consumed to evolve CO_2 must be from either chemisorbed oxygen or structural oxygen from $LaPO_4$. However, it is unexpected that room temperature CO can remove structural oxygen from $LaPO_4$ because $LaPO_4$ is non-reducible. This was confirmed by H_2 temperature-programmed reduction (TPR) where no hydrogen consumption was observed when H_2 flowed across either $LaPO_4$ or Au/$LaPO_4$ during temperature ramps up to 500 °C.

To clarify the source of active oxygen in Au/$LaPO_4$, isotopic IR experiments were carried out as shown in Figure 14.6(a). When Au/$LaPO_4$ was pre-treated to 200 °C in $^{18}O_2$ and then exposed to $C^{16}O$ at room temperature, the majority of evolved CO_2 is unlabeled $C^{16}O^{16}O$ along with a small amount of $C^{16}O^{18}O$ as shown by both IR and QMS. This suggests the existence of active oxygen inherent to $LaPO_4$ in addition to either chemisorbed $^{18}O_2$, formation of peroxides, or incorporation of ^{18}O into the Au/$LaPO_4$ during the pre-treatment in $^{18}O_2$. Raman experiments were also conducted to clarify the source of oxygen as was performed in the case of Au/$FePO_4$. Room temperature CO exposure onto O_2 pre-treated Au/$LaPO_4$ showed no changes in the Raman spectrum, indicating no structural change of the Au/$LaPO_4$ sample, *i.e.* no active lattice oxygen. Therefore, we speculate that the active oxygen species responsible for the evolution of CO_2 must originate from hydroxyl groups in addition to a trace of chemisorbed oxygen on the Au/$LaPO_4$.

The reactivity of surface hydroxyl groups on Au/$LaPO_4$ was further elaborated by the comparison of CO-TPR for both the Fe and La phosphate supported Au catalysts as presented in Figure 14.7. The $FePO_4$ with 2.5 wt% Au loading showed an intense CO_2 formation peak around 230 °C and less

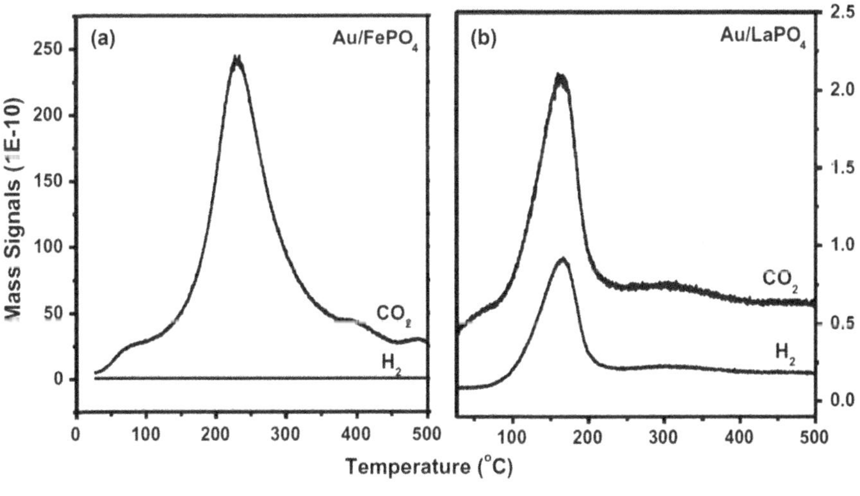

Figure 14.7 CO temperature-programmed reaction is compared for: (a) Au/$FePO_4$; and (b) Au/$LaPO_4$. The H_2 signal is scaled according to the CO_2 to H_2 QMS sensitivity ratio.
Reprinted from ref. 12. Copyright 2011, with permission from Elsevier.

pronounced reduction features peaking near 90, 400, and 480 °C. The CO_2 is produced from the reduction of cationic Au as well as $FePO_4$. Quantification of the TPR profile shows that the CO_2 formation is sufficient to imply complete reduction of not only the Au but also all of the $FePO_4$. It is evident that the presence of gold enhances the reduction of Fe^{III} by CO. The absence of H_2 evolved during CO-TPR suggests that the reduction of (surface) hydroxyl groups is negligible for Au/$FePO_4$. The CO-TPR profile of Au/$LaPO_4$ is shown for comparison in Figure 14.7(b). CO_2 formation peaked at 168 °C with 0.10 mol CO_2 mole^{-1} La, *ca.* 6 times less than that formed on Au/$FePO_4$. This ratio suggests that the consumption of O by CO is confined to a limited region, probably the surface region of the $LaPO_4$. Furthermore, the observation of an H_2 formation peak at the same temperature as the CO_2 formation suggests that the reaction involves attack of hydroxyl groups by CO according to the reaction:

$$CO + OH^- \rightarrow CO_2 + \frac{1}{2}H_2 + \square^{1-} \tag{14.2}$$

This reaction is confirmed by the fact that the amount of H_2 produced is very close to half the amount of the CO_2 produced. The result indicates that surface hydroxyl on $LaPO_4$ is reactive toward CO adsorbed on gold nanoparticles, yielding CO_2 and H_2, and leaving a hydroxyl vacancy site, $\square^{1-}$, on $LaPO_4$. This loss of oxygen (hydroxyl) from the phosphate anion may be related to the formation of negatively charged Au species. Because the evolution of H_2 was not observed over $LaPO_4$ alone during CO-TPR, the presence of Au species must enhance the removal of the hydroxyl group from Au/$LaPO_4$ sample. It is reasonable that only hydroxyl groups adjacent to Au species could be attacked by adsorbed CO on Au sites. It was also interesting to find that the active hydroxyl groups on Au/$LaPO_4$ can be readily replenished by impurity H_2O present in the gas streams, whereas the hydroxyl groups on Au/SiO_2 cannot be replenished once they reacted with CO.[21]

14.3.2 CO Oxidation Mechanism on Au/$FePO_4$ and Au/$LaPO_4$

Since both $FePO_4$ (lattice oxygen) and $LaPO_4$ (surface hydroxyls) can supply active oxygen for CO oxidation on the supported Au catalysts, we further investigated if there is direct participation of adsorbed oxygen species for CO oxidation on gold nanoparticles dispersed on phosphates.[12,22] To determine this, $^{16}O_2$-treated Au/$FePO_4$ was used to catalyze CO oxidation by labeled $^{18}O_2$ at room temperature. As shown in Figure 14.8, $C^{16}O_2$ appears immediately after the introduction of CO and $^{18}O_2$. This unlabeled CO_2 may derive from both the reduction of cationic Au by CO, and the reaction between CO and structural oxygen of $FePO_4$. As the reaction proceeds, IR bands at 2340 and 2325 cm^{-1}, due to $C^{16}O^{18}O$, grow in parallel with the band at 2114 cm^{-1} (CO–Au^0). The $C^{16}O^{18}O$ is produced from the gold-catalyzed reaction between CO and incoming $^{18}O_2$. The intensity of the gas phase band at

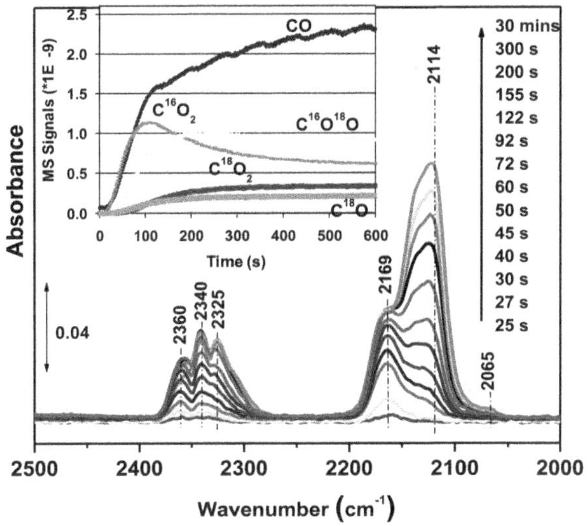

Figure 14.8 IR spectra and QMS (inset) results following switch from He to $CO + {}^{18}O_2$ at room temperature as a function of time after the switch. The Au/$FePO_4$ catalyst was pretreated in ${}^{16}O_2$ at 200 °C.
Reprinted from ref. 22. Copyright 2009, with permission from Elsevier.

2360 cm^{-1} ($C^{16}O_2$) maximizes at about 90 s and then decreases with time on-stream, due to the limited ${}^{16}O$ source on the surface. Correspondingly, the QMS signal (inset in Figure 14.8) of $C^{16}O_2$ increases simultaneously with the m/e = 28 signal (CO) and decreases after about 90 s. Appearance of $C^{16}O^{18}O$ (at m/e = 46) is delayed relative to $C^{16}O_2$, but dominates CO_2 production at prolonged reaction time. The parallel growth of the IR bands due to $C^{16}O^{18}O$ (2325 cm^{-1}) and CO–Au^0 (2114 cm^{-1}) suggests that the production of $C^{16}O^{18}O$ takes place only after metallic Au is formed from the reduction of cationic Au by CO. This points to the positive catalytic role of metallic Au in room temperature CO oxidation on Au/$FePO_4$, and is confirmed by a control experiment done over a 200 °C H_2 pre-treated sample whose surface is dominated by metallic Au.

The participation of adsorbed oxygen species in CO oxidation is better evidenced in a study carried out on a H_2 pre-treated Au/$FePO_4$. In this case, IR bands and QMS signals due to $C^{16}O_2$, $C^{16}O^{18}O$, and even $C^{18}O^{18}O$ were observed simultaneously from the very beginning of the reaction. This implies that CO reacts with structural oxygen of $FePO_4$ and activated ${}^{18}O_2$ from the onset when metallic Au exists on the surface. The simultaneous appearance of $C^{16}O_2$ and $C^{16}O^{18}O$ suggests that two reaction pathways exist: (1) a redox pathway in which $FePO_4$ supplies active O; and (2) a direct pathway in which gas phase O_2 provides the active O occurring on a metallic Au surface *via* a Langmuir–Hinshelwood mechanism.

CO oxidation with isotopically labeled ${}^{18}O_2$ was also conducted on an Au/$LaPO_4$ catalyst. Gaseous $C^{16}O_2$ and $C^{16}O^{18}O$ appeared simultaneously at room temperature after introducing CO and $^{18}O_2$ to an $^{16}O_2$-treated Au/$LaPO_4$ sample. The unlabeled CO_2 derives from the reaction between CO and chemisorbed oxygen and an active hydroxyl group on the Au/$LaPO_4$ catalyst. The $C^{16}O^{18}O$ is produced from the gold-catalyzed reaction between CO and the incoming $^{18}O_2$. As the reaction continues, the intensity of the gas phase $C^{16}O_2$ band at 2360 cm^{-1} decreases with time on stream due to the limited ^{16}O source on the surface. Simultaneous appearances of $C^{16}O^{18}O$ and $C^{16}O_2$ indicate that CO reaction with gas phase O_2 and hydroxyl groups initiate at the same time upon introduction of CO oxidation conditions. The immediate appearance of $C^{16}O^{18}O$ confirms the positive role of the metallic Au in catalyzing CO oxidation at room temperature on $LaPO_4$ supported Au catalysts. Therefore, there also exist two reaction pathways for CO oxidation on Au/$LaPO_4$, *i.e.* CO reaction with replenishable hydroxyl groups on $LaPO_4$ and the Langmuir–Hinshelwood channel. Figure 14.9 depicts the CO oxidation pathways on the Au nanoparticles supported on the two phosphates.

Although our results suggest two different CO oxidation pathways in which the phosphate support contributes to CO oxidation, quantification of the exact contribution of each is not straightforward. However, our reactivity measurements[12] suggest that the reaction pathway involving the phosphate support is not the primary pathway, as long as small particle, reduced Au is present. The catalytic performance results show that the H_2 pre-treated Au/$FePO_4$ catalyst exhibits higher rates of CO_2 formation than the O_2 pre-treated Au/$FePO_4$ at the temperatures investigated. After reductive treatment, there is less available structural oxygen on the $FePO_4$ but there are more available reduced Au sites. We conclude that it is the reduced Au, and not the availability of a reducible support, that is critical. Similarly, in spite of not having structural oxygen available, the Au/$LaPO_4$ catalyst gave higher CO conversion and turnover frequency (TOF) than the Au/$FePO_4$ catalyst, for both oxidative and reductive treatments. This information indicates that metallic Au is the active site for CO oxidation over phosphate-supported Au catalysts.

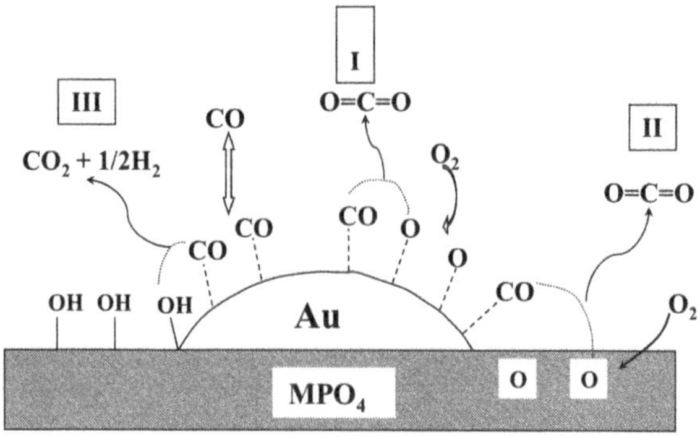

Figure 14.9 Schematic representation of three reaction pathways for low-temperature CO oxidation on gold nanoparticles supported on two different phosphates, $FePO_4$ (I + II) and $LaPO_4$ (I + III).

Based on these results, it is obvious that support reducibility does not guarantee high activity. The results for both Au/$LaPO_4$ and Au/$FePO_4$ suggest that neutral Au is more active than cationic Au, although it is not yet possible to compare the activity of the negatively charged Au apparent on the $LaPO_4$ surface.

To put the mechanism of CO oxidation on gold supported on phosphates into perspective, the participation of lattice oxygen and surface hydroxyls in CO oxidation has also been observed on oxide-supported Au catalysts. Although the differences in activities for reducible supports have been noted and their role in activating oxygen has been suggested,[27,50] there is still controversy as to how the support supplies active oxygen species in low-temperature CO oxidation catalyzed by gold.

Support-induced activation of O_2 has been reported to occur for Au on Fe-doped TiO_2[51] and nanocrystalline CeO_2.[6] In these cases, the presence of Au enhances the activation of oxygen through the formation of peroxide and superoxides at the metal–oxide interface. It is possible that the O_2 is bridge-bonded between Au and the reducible support cations, and such a configuration was proposed previously for Au or Pt supported on TiO_2[52–54] and CeO_2.[55] The Au-assisted Mars van Krevelen mechanism, operating in the case of CO oxidation on Au/$FePO_4$, was proposed for some reducible oxide-supported Au nanoparticles such as Au/FeO_x,[56] Au/TiO_2,[57] and Au/$CuMnO_x$[58] where the support lattice oxygen first reacts directly with CO and then gas phase O_2 heals the created oxygen vacancies. The role of surface hydroxyls in CO oxidation, as in the case of Au/$LaPO_4$, has also been previously suggested for other non-reducible oxide-supported metal catalysts. One case is demonstrated for Au/SiO_2[21,23] as described in the previous section. Another example is Pt supported on mesoporous SiO_2[59] where CO oxidation was assumed to be promoted by the attack of OH groups at the internal surface of mesoporous silica toward CO on Pt. Costello and co-workers[60,61] have shown that the presence of hydroxyls, as introduced by exposure to H_2 or H_2O, has a large positive effect on the activity of Au/Al_2O_3. They propose that the reaction between CO and OH to form a hydroxycarbonyl on the Au clusters is the primary reaction pathway for CO oxidation.

14.4 Conclusions

We have employed *in situ* spectroscopy to study CO oxidation on non-reducible oxides and non-oxide supported Au catalysts to understand the intriguing support effect in gold nanoparticle catalysis. Although a Langmuir–Hinshelwood mechanism may occur on the reduced gold nanoparticles, it was shown that non-reducible SiO_2 and $LaPO_4$ or reducible non-oxide $FePO_4$ supports can also participate in the CO oxidation reaction. Namely, surface hydroxyls on non-reducible supports and surface lattice oxygen on the reducible phosphate salt react with adsorbed CO on Au sites at the gold–support interface to form CO_2, pointing to a key role of the interface in CO oxidation on Au nanoparticles. It appears that the Langmuir–Hinshelwood mechanism plays a larger role in CO oxidation for gold nanoparticles supported on non-reducible materials than those on reducible ones. In all cases, the *in situ* spectroscopy results clearly imply that metallic Au is the active center for CO oxidation at low temperature, while the cationic Au plays a negligible role. The distribution of the Au sites appears to be dependent on the nature of the support, and the charge state of the Au sites can be dynamic under reaction conditions.

The results presented here highlight the role of non-traditional supports in gold nanoparticle catalysis and demonstrate the necessity of carrying out *in situ/operando* spectroscopy to identify the electronic state of the gold catalysts as the catalyst structure can be dynamic under reaction conditions.

Acknowledgements

This research is sponsored by the Division of Chemical Sciences, Geosciences, and Biosciences, Office of Basic Energy Sciences, US Department of Energy. We are grateful for contributions from co-workers including Dr Meijun Li, Dr Shenghu Zhou, Dr Haoguo Zhu, Dr Zhen Ma, Dr J. Chris Bauer, Dr David R. Mullins, and Dr Sheng Dai. Part of the work, including the infrared study, was conducted at the Center for Nanophase Materials Sciences, which is sponsored at Oak Ridge National Laboratory by the Scientific User Facility Division, Office of Basic Energy Science, US Department of Energy.

References

1. M. Cargnello, V. V. T. Doan-Nguyen, T. R. Gordon, R. E. Diaz, E. A. Stach, R. J. Gorte, P. Fornasiero and C. B. Murray, *Science*, 2013, **341**, 771.
2. J. Y. Liu, *ChemCatChem*, 2011, **3**, 934.
3. M. Haruta, T. Kobayashi, H. Sano and N. Yamada, *Chem. Lett.*, 1987, 405.
4. T. Takei, N. Iguchi and M. Haruta, *New J. Chem.*, 2011, **35**, 2227.
5. M. Haruta, *Faraday Discuss.*, 2011, **152**, 11.
6. J. Guzman, S. Carrettin, J. C. Fierro-Gonzalez, Y. L. Hao, B. C. Gates and A. Corma, *Angew. Chem., Int. Ed.*, 2005, **44**, 4778.
7. D. Widmann, Y. Liu, F. Schüth and R. J. Behm, *J. Catal.*, 2010, **276**, 292.
8. X. Y. Liu, A. Q. Wang, T. Zhang and C. Y. Mou, *Nano Today*, 2013, **8**, 403.
9. H. G. Zhu, Z. Ma, J. C. Clark, Z. W. Pan, S. H. Overbury and S. Dai, *Appl. Catal. A-Gen.*, 2007, **326**, 89.
10. M. Okumura, S. Nakamura, S. Tsubota, T. Nakamura, M. Azuma and M. Haruta, *Catal. Lett.*, 1998, **51**, 53.
11. Z. Ma, H. F. Yin, S. H. Overbury and S. Dai, *Catal. Lett.*, 2008, **126**, 20.
12. M. J. Li, Z. L. Wu and S. H. Overbury, *J. Catal.*, 2011, **278**, 133.
13. W. F. Yan, S. Brown, Z. W. Pan, S. M. Mahurin, S. H. Overbury and S. Dai, *Angew. Chem., Int. Ed.*, 2006, **45**, 3614.
14. A. Chiorino, M. Manzoli, F. Menegazzo, M. Signoretto, F. Vindigni, F. Pinna and F. Boccuzzi, *J. Catal.*, 2009, **262**, 169.

15. J. C. Fierro-Gonzalez, J. Guzman and B. C. Gates, *Top. Catal.*, 2007, **44**, 103.
16. M. Date, M. Okumura, S. Tsubota and M. Haruta, *Angew. Chem., Int. Ed.*, 2004, **43**, 2129.
17. M. C. Kung, R. J. Davis and H. H. Kung, *J. Phy. Chem. C*, 2007, **111**, 11767.
18. M. Haruta, *CATTECH*, 2002, **6**, 102.
19. F. Vindigni, M. Manzoli, A. Chiorino and F. Boccuzzi, *Gold Bull.*, 2009, **42**, 106.
20. Z. Ma and S. Dai, *Nano Res.*, 2011, **4**, 3.
21. Z. L. Wu, S. H. Zhou, H. G. Zhu, S. Dai and S. H. Overbury, *J. Phy. Chem. C*, 2009, **113**, 3726.
22. M. J. Li, Z. L. Wu, Z. Ma, V. Schwartz, D. R. Mullins, S. Dai and S. H. Overbury, *J. Catal.*, 2009, **266**, 98.
23. Z. L. Wu, S. H. Zhou, H. G. Zhu, S. Dai and S. H. Overbury, *Chem. Comm.*, 2008, 3308.
24. J. C. Bauer, D. Mullins, M. J. Li, Z. L. Wu, E. A. Payzant, S. H. Overbury and S. Dai, *Phys. Chem. Chem. Phys.*, 2011, **13**, 2571.
25. S. H. Zhou, Z. Ma, H. F. Yin, Z. L. Wu, B. Eichhorn, S. H. Overbury and S. Dai, *J. Phy. Chem. C*, 2009, **113**, 5758.
26. S. H. Zhou, H. F. Yin, V. Schwartz, Z. L. Wu, D. Mullins, B. Eichhorn, S. H. Overbury and S. Dai, *ChemPhysChem*, 2008, **9**, 2475.
27. S. H. Overbury, L. Ortiz-Soto, H. G. Zhu, B. Lee, M. D. Amiridis and S. Dai, *Catal. Lett.*, 2004, **95**, 99.
28. M. T. Bore, H. N. Pham, E. E. Switzer, T. L. Ward, A. Fukuoka and A. K. Datye, *J. Phy. Chem. B*, 2005, **109**, 2873.
29. R. Zanella, A. Sandoval, P. Santiago, V. A. Basiuk and J. M. Saniger, *J. Phy. Chem. B*, 2006, **110**, 8559.
30. M. Okumura, S. Tsubota and M. Haruta, *J. Mole. Catal. A Chem.*, 2003, **199**, 73.
31. Y. S. Chi, H. P. Lin and C. Y. Mou, *Appl. Catal. A-Gen.*, 2005, **284**, 199.
32. K. Hadjiivanov and G. Vayssilov, in *Advances in Catalysis*, ed. B. C. Gates and H. Knozinger, Academic Press, New York and London, 2002, vol. 47, pp. 307–511.
33. M. A. P. Dekkers, M. J. Lippits and B. E. Nieuwenhuys, *Catal. Lett.*, 1998, **56**, 195.
34. F. Boccuzzi, A. Chiorino, M. Manzoli, P. Lu, T. Akita, S. Ichikawa and M. Haruta, *J. Catal.*, 2001, **202**, 256.
35. T. Masawaki, Y. Uchida, A. Ogawa, N. Kambe, N. Miyoshi and N. Sonoda, *J.Phys. Org. Chem.*, 1988, **1**, 115.
36. C. K. Costello, M. C. Kung, H. S. Oh, Y. Wang and H. H. Kung, *Appl. Catal. A-Gen.*, 2002, **232**, 159.
37. G. C. Bond and D. T. Thompson, *Gold Bull.*, 2000, **33**, 41.
38. J. A. Rodriguez, S. Ma, P. Liu, J. Hrbek, J. Evans and M. Perez, *Science*, 2007, **318**, 1757.
39. T. E. Shubina, C. Hartnig and M. T. M. Koper, *Phys. Chem. Chem. Phys.*, 2004, **6**, 4215.

40. C. K. Costello, J. Guzman, J. H. Yang, Y. M. Wang, M. C. Kung, B. C. Gates and H. H. Kung, *J. Phys. Chem. B*, 2004, **108**, 12529.
41. J. H. Yang, J. D. Henao, M. C. Raphulu, Y. M. Wang, T. Caputo, A. J. Groszek, M. C. Kung, M. S. Scurrell, J. T. Miller and H. H. Kung, *J. Phys. Chem. B*, 2005, **109**, 10319.
42. Z. Ma and S. Dai, *ACS Catal.*, 2011, **1**, 805.
43. C. L. Bracey, P. R. Ellis and G. J. Hutchings, *Chem. Soc. Rev.*, 2009, **38**, 2231.
44. J. C. Bauer, G. M. Veith, L. F. Allard, Y. Oyola, S. H. Overbury and S. Dai, *ACS Catal.*, 2012, **2**, 2537.
45. H. L. Lian, M. J. Jia, W. C. Pan, Y. Li, W. X. Zhang and D. Z. Jiang, *Catal. Commun.*, 2005, **6**, 47.
46. N. Phonthammachai, Z. Ziyi, G. Jun, H. Y. Fan and T. J. White, *Gold Bull.*, 2008, **41**, 42.
47. A. Venugopal and M. S. Scurrell, *Appl. Catal. A-Gen.*, 2003, **245**, 137.
48. J. F. Liu, W. Chen, X. W. Liu, K. B. Zhou and Y. D. Li, *Nano Res.*, 2008, **1**, 46.
49. P. Nagaraju, N. Lingaiah, A. Balaraju and P. S. S. Prasad, *Appl. Catal. A-Gen.*, 2008, **339**, 99.
50. M. M. Schubert, S. Hackenberg, A. C. van Veen, M. Muhler, V. Plzak and R. J. Behm, *J. Catal.*, 2001, **197**, 113.
51. S. Carrettin, Y. Hao, V. Aguilar-Guerrero, B. C. Gates, S. Trasobares, J. J. Calvino and A. Corma, *Chem.- Eur. J.*, 2007, **13**, 7771.
52. S. N. Rashkeev, A. R. Lupini, S. H. Overbury, S. J. Pennycook and S. T. Pantelides, *Phys. Rev. B*, 2007, **76**, 035438.
53. I. X. Green, W. J. Tang, M. Neurock and J. T. Yates, *Science*, 2011, **333**, 736.
54. L. M. Molina, M. D. Rasmussen and B. Hammer, *J. Chem. Phys.*, 2004, **120**, 7673.
55. H. Y. Kim, H. M. Lee and G. Henkelman, *J. Am. Chem. Soc.*, 2012, **134**, 1560.
56. L. Li, A. Q. Wang, B. T. Qiao, J. Lin, Y. Q. Huang, X. D. Wang and T. Zhang, *J. Catal.*, 2013, **299**, 90.
57. D. Widmann and R. J. Behm, *Angew. Chem., Int. Ed.*, 2011, **50**, 10241.
58. K. Morgan, K. J. Cole, A. Goguet, C. Hardacre, G. J. Hutchings, N. Maguire, S. O. Shekhtman and S. H. Taylor, *J. Catal.*, 2010, **276**, 38.
59. A. Fukuoka, J. Kimura, T. Oshio, Y. Sakamoto and M. Ichikawa, *J. Am. Chem. Soc.*, 2007, **129**, 10120.
60. C. K. Costello, M. C. Kung, H.-S. Oh, Y. Wang and H. H. Kung, *Appl. Catal. A-Gen.*, 2002, **232**, 159.
61. C. K. Costello, J. H. Yang, H. Y. Law, Y. Wang, J. N. Lin, L. D. Marks, M. C. Kung and H. H. Kung, *Appl. Catal. A-Gen.*, 2003, **243**, 15.

CHAPTER 15

Model Catalysts for Au Catalysis: From Single Crystals to Supported Nanoparticles

WEIXIN HUANG

Hefei National Laboratory for Physical Sciences at the Microscale, Collaborative Innovation Center of Suzhou Nano Science and Technology, CAS Key Laboratory of Materials for Energy Conversion and Department of Chemical Physics, University of Science and Technology of China, Hefei, 230026, PR China
Email: huangwx@ustc.edu.cn

15.1 Introduction

Gold (Au) lies in the sixth period and the IB group of Periodic Table of the Elements and is of a $[Xe]4f^{14}5d^{10}6s^1$ electronic configuration. Due to the large size of its nucleus, gold chemistry is strongly affected by relativistic effects.[1–3] The first study of Au as a catalyst dates back to 1926 in which it was reported to catalyze the $H_2 + O_2$ reaction, but at a much lower rate than Ag.[4] Subsequently Au catalysts, either supported or unsupported, were also reported to exhibit low activity for other catalytic reactions.[5–10] The low catalytic activity of Au seemed reasonable because of its chemical inertness and consequent inability to chemisorb and activate molecules. Thus in classical catalysis, Au was considered to be catalytically inactive. However, the situation changed dramatically in the 1980s. In 1987 Haruta and co-workers[11,12] first reported that ultrafine Au particles of ~5 nm in size supported on α-Fe_2O_3, Co_3O_4, and NiO prepared by coprecipitation from an

RSC Catalysis Series No. 18
Heterogeneous Gold Catalysts and Catalysis
Edited by Zhen Ma and Sheng Dai

Published by the Royal Society of Chemistry, www.rsc.org

aqueous solution of chloroauric acid and the nitrate of transition metals were highly active in catalyzing CO and H_2 oxidation. In 1985 Hutchings[13] reported that a supported gold chloride catalyst could theoretically be the most active supported metal chloride catalyst for the hydrochlorination of ethylene, and in 1988 his prediction was confirmed experimentally.[14]

The pioneering works of Haruta and co-workers[11,12] and Hutchings and co-workers[14] have motivated many researchers to re-explore catalysis by Au. Their great successes eventually developed into a gold rush in heterogeneous catalysis.[15–24] Supported Au nanoparticles with the size below 10 nm have been demonstrated to actively catalyze a wide array of reactions including reactions involving CO, hydrogenation reactions, and selective oxidation reactions.

The catalyst preparation method strongly affects the size of supported Au nanoparticles.[24,25] Because chlorine contamination results in the rapid growth of Au particles, impregnation, the most traditional method for preparing supported metal catalysts, is not suitable to prepare active supported Au catalysts when $HAuCl_4$, the most popular source of Au, is used. Unfortunately, supported Au catalysts were all prepared by impregnation using $HAuCl_4$ as the Au source in the early studies,[4–10] and as a result, the acquired supported Au catalysts exhibited very poor catalytic activity. Coprecipitation and deposition–precipitation (DP) methods, however, are effective for the preparation of highly active supported Au catalysts.

Au catalysis exhibits several unique characteristics:

(1) Supported Au catalysts are highly active at mild reaction temperatures. Au nanoparticles of $\sim$5 nm supported on α-Fe_2O_3, Co_3O_4, and NiO can catalyze CO oxidation at temperatures as low as $-70\,^\circ$C.[11,12] Supported Au catalysts are also very active in catalyzing the low-temperature water gas shift (WGS) reaction.[26–30]
(2) Supported Au catalysts are very selective. Supported Au nanoparticles and clusters can catalyze gas-phase propylene epoxidation with molecular O_2 to produce propylene oxide with selectivities higher than 90%.[31–33]
(3) The catalytic performance of supported Au catalysts is sensitively dependent on the size of the supported Au nanoparticles. Haruta *et al.*[34] observed that the turnover frequency (TOF) of Au/TiO_2, Au/α-Fe_2O_3, and Au/Co_3O_4 based on Au atoms exposed on Au nanoparticles increased sharply with the decrease of the Au nanoparticle size to within 5 nm. Qi *et al.*[35] observed a clear Au nanoparticle size effect in the reaction of propylene with O_2 and H_2 on Au/Ti-based oxides. It is noteworthy that the calculation and comparison of TOF among a series of supported Au catalysts with different Au nanoparticle sizes are meaningful only when the active site of the supported Au nanoparticles can be determined and does not vary with their size. Fujitani and Nakamura[36] found that the TOF of Au/TiO_2(110) model catalysts calculated based on Au atoms exposed on Au nanoparticles and based on Au atoms at the Au-TiO_2 perimeter interfaces varied in different trends as a function of the mean size of the Au nanoparticles.

(4) The catalytic performance of supported Au catalysts depends on the support. The support not only can affect the Au-support interaction, and subsequently the structure and size of the supported Au nanoparticles, but also can directly participate in catalytic reactions. Schubert *et al.*[37] grouped oxide supports into 'active' supports (TiO_2, α-Fe_2O_3, CoO_x, and NiO_x) and 'inert' supports (SiO_2, Al_2O_3, and MgO) with respect to CO oxidation. Reactive oxygen species could facilely form on the 'active' supports and supply the catalytic CO oxidation but not on the 'inert' supports; thus Au nanoparticles supported on the 'active' supports exhibited a significantly enhanced activity for CO oxidation than those supported on the 'inert' supports.

(5) The catalytic performance of supported Au catalysts in reactions involving molecular O_2 is affected by moisture (gas phase reactions) and basicity (liquid phase reactions). Daté *et al.*[38] reported that the moisture enhanced the catalytic activity of Au/TiO_2, Au/Al_2O_3, and Au/SiO_2 in CO oxidation at 273 K for no less than two orders of magnitude and that the effect of moisture depended on the type of metal oxide support. Kung and co-workers[39–41] observed the promotional effect of H_2O or H_2 in the reaction mixture on Au/Al_2O_3 for low-temperature CO oxidation. Sanchez-Castillo *et al.*[42] reported that support-free Au nanotubes in polycarbonate membranes exhibited catalytic activity of CO oxidation at room temperature and that the activity was enhanced by liquid H_2O and further promoted by increasing the pH of the solution. They also observed the catalytic activity in CO oxidation at room temperature in aqueous solutions of polyoxometalates over carbon-supported Au nanoparticles that are catalytically inactive in gas-phase CO oxidation at room temperature.[43]

The gold rush in heterogeneous catalysis and the unique catalytic behaviors of supported Au catalysts have greatly inspired relevant fundamental investigations Most efforts have been devoted to low-temperature CO oxidation catalyzed by supported Au catalysts. This reaction seems simple, but its complexity has turned out to be far beyond expectation. Arguments still exist about the active site and reaction mechanism. The perimeter mechanism is popular for Au nanoparticles supported on 'active' oxide supports, in which CO is considered to adsorb on the Au surface (most likely at edges and corners) and the oxygen molecule is activated at the Au–metal oxide perimeter interfaces.[44,45] Another popular reaction mechanism involves both metallic and cationic Au, in which CO chemisorbed on metallic Au surface is proposed to insert into the cationic Au–OH bond to form a hydroxycarbonyl intermediate, the hydroxycarbonyl intermediate is oxidized to bicarbonate, and then bicarbonate decomposes into Au–OH and CO_2.[46,47] Low-coordinated metallic Au atoms were also proposed to be able to catalyze low-temperature CO oxidation themselves, and the support acted as a structural and electronic promoter.[48,49]

The size-dependent catalytic activity of supported Au nanoparticles in low-temperature CO oxidation is also a core issue in fundamental investigations, but the structure of supported Au nanoparticles with the optimal size and the interpretation of size-dependent catalytic activity of supported Au catalysts remain controversial. The decrease in the size of supported Au nanoparticles increases the Au-metal oxide perimeter interface length and thus enhances the catalytic activity of supported Au catalysts following the perimeter mechanism. The size of supported Au nanoparticles also significantly affects their structure and morphology.

Tai *et al.*[50] revealed that lattice contraction and structural changes of Au nanoparticles supported on TiO_2-coated silica aerogel became prominent at Au diameters less than 4–5 nm, consistent with the enhanced catalytic activity. Fine Au nanoparticles generally expose a high density of low-coordinated Au atoms at their corners, steps, and edges. Zanella *et al.*[51] attributed the decrease in catalytic activity of Au nanoparticles supported on TiO_2 with increasing calcination temperature to the change in particle shape from a shape containing a large proportion of low-coordinated sites to truncated octahedra with smooth facets.

The reduction in the size of Au nanoparticles could also change their electronic structure, in which the quantum size effect, *i.e.* the metal-to-nonmetal transition, is of particular interest. The critical thickness of supported Au nanoparticles for the metal-to-nonmetal transition was reported to be two atom layer thickness.[52,53] Goodman's group correlated the size-dependent catalytic activity of supported Au nanoparticles on $TiO_2(110)$ to the quantum size effect and proposed that supported Au nanoparticles with two layers of Au were most catalytically active.[52] Later they also demonstrated that a two-layered Au film vacuum-deposited on a thin layer of TiO_2 on Mo(112) was the most active structure for low-temperature CO oxidation.[54]

Au clusters with a bilayer structure was also proposed to be responsible for the high catalytic activity of coprecipitated Au/FeOOH dried at 393 K in low-temperature CO oxidation.[55] However, in a highly active Au/Fe_2O_3 catalyst prepared by the colloidal deposition method, Liu *et al.*[56] reported that supported Au nanoparticles had diameters larger than 1 nm and that bilayer structures and/or diameters of about 0.5 nm were not mandatory to achieve the high activity. Simultaneous reactivity measurements and grazing incidence small-angle X-ray scattering characterizations of an Au/$TiO_2(110)$ model catalyst revealed that supported Au nanoparticles with the optimal size (2.1 ± 0.3 nm) had a height of about six atomic layers.[57]

The complexity of catalysis by supported Au catalysts exemplifies the complexity of heterogeneous catalysis. As illustrated in Figure 15.1, the catalytic performance of a single-component solid catalyst particle microscopically is affected by its surface composition, surface electronic structure, and surface geometric structure because heterogeneous catalysis always occurs on the solid catalyst surface.[58] The surface composition of a solid particle is determined by the composition and crystal planes exposed on the

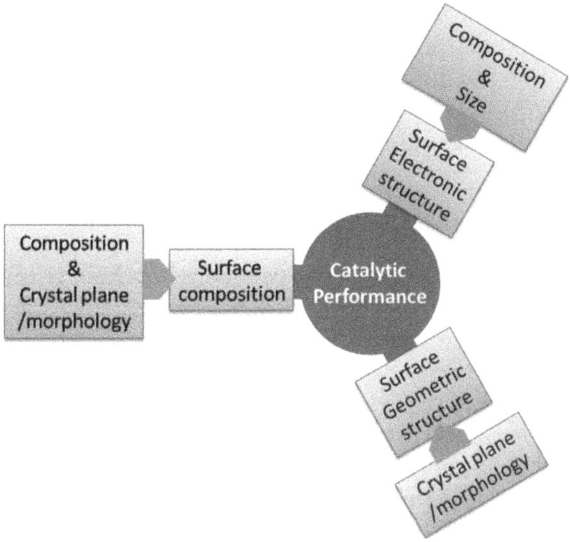

Figure 15.1 Schematic illustration of microscopic and macroscopic structural parameters that affect the catalytic performance of a solid particle.

surface (for oxides), the surface electronic structure is determined by the composition and size, and the surface geometric structure is determined by the crystal planes exposed on the surface.

According to Wulff's rule,[59] the crystal planes exposed on the surface of a crystal are determined by its morphology. Therefore, macroscopically the catalytic performance of a solid catalyst particle varies with its composition, size, and morphology. For multicomponent solid catalysts, the interaction and interfacial structure between different components are also vital for catalytic performance. During a heterogeneous catalytic reaction, the catalyst surface provides various types of active sites with different structures to chemisorb and activate reactants, and subsequently catalyze their surface reactions following different pathways. For each type of active site and the corresponding surface reaction pathway, the reaction rate depends on the chemisorption energy of involved surface adsorbates and the activation energy of involved elementary surface reactions.

The measured activity of the catalyst is the sum of the rates of surface reactions catalyzed by various types of active sites, while the measured selectivity of the catalyst is due to the different products produced by the surface reactions catalyzed over various types of active sites. The variation in the catalyst structure changes the types and concentrations of various active sites and thus the catalytic performance, while the variation in the reaction temperature changes the rate of each surface reaction and thus the catalytic performance.

The unique behaviors of supported Au catalysts such as size-, support-, and moisture-dependent catalytic performance and high activity at mild reaction temperatures offer an appropriate platform to comprehensively address fundamental issues in heterogeneous catalysis. Therefore, supported Au catalysts not only are promising novel catalysts with potential applications for industrial catalytic processes, but also constitute a novel system for the fundamental understanding of heterogeneous catalysis.

15.2 Model Catalyst Approach

Fundamental understanding of heterogeneous catalysis is indispensible to realize the ultimate goal of catalyst design, but remains a great challenge. Although we are able to comprehensively characterize the catalyst surface structure, surface adsorbate structure and catalytic performance, the inhomogeneity of catalyst particles in their sizes and morphologies makes the unambiguous correlation between the catalyst surface structure and the catalytic performance very difficult. With this respect, a successful approach has been developed to employ the uniform and well-defined surface as the model surface to establish the structure–catalytic property relation.[60]

Traditional model catalysts are mainly based on single crystals including single crystal thin films and nanoparticles supported on single crystals/single crystal thin films.[61–66] As illustrated in Figure 15.2, single crystals-based model catalysts can address the size effect, morphology effect and interfacial structure effect in the structure–catalytic performance relation of catalysts. However, there exists a so-called 'materials gap' and 'pressure gap' between single crystals-based model catalysts and corresponding powder catalysts. Powder catalysts consist of supported three-dimensional nanoparticles exhibiting a large specific surface area and the catalytic reactions are carried out at atmospheric or higher pressure,

Structure complexity of model catalysts (↓)

		Morphology effect	Size effect	Interfacial structure effect	Reaction pressure
Single crystals-based model catalysts	Metal/oxide single crystals	☑	☒	☒	UHV ~ mbar
	Oxide single crystal thin films	☑	☒	☒	UHV ~ mbar
	Nanoparticles on single crystals/single crystal thin films	☑	☑	☑	UHV ~ mbar
Nanocrystals-based model catalysts	Uniform nanocrystals	☑	☑	☒	≥ 1atm
	Uniform nanocomposites	☑	☑	☑	≥ 1atm
Powder catalysts		☑	☑	☑	≥ 1atm

Figure 15.2 Schematic illustration of correlations among single crystals-based model catalysts, nanocrystals-based model catalysts, and powder catalysts.

whereas single crystals-based model catalysts are extended two-dimensional surfaces exhibiting a low specific surface area and the reaction pressure usually does not exceed several millibar (*i.e.* several hundred Pa). Thus the structure–catalytic property relation derived from single crystals-based model catalysts sometimes cannot be simply extended to practical heterogeneous catalytic reactions, and several issues in practical heterogeneous catalytic reactions cannot be adequately approached using single crystals-based model catalysts.[67–71]

Benefiting from the recent significant progress of nanotechnology, nanocrystals and even powder catalysts with a uniform composition and structure (size and morphology) can be successfully prepared,[72–75] and constitute a novel type of model catalysts to explore the structure–catalytic performance relation.[76–82] The study of structure–catalytic performance relation employing nanocrystals/powder catalysts-based model catalysts can be carried out under the same experimental condition as practical heterogeneous catalytic reactions (Figure 15.2) and thus can bridge the 'materials gap' and 'pressure gap' between single crystals-based model catalysts and corresponding powder catalysts. Therefore, comprehensive investigations of single crystals-based and nanocrystals/powder catalysts-based model catalysts are able to turn fundamental understanding of heterogeneous catalysis at the molecular level into a reality.

As a novel and interesting system in heterogeneous catalysis, Au catalysis has been extensively studied employing the single crystals-based model catalyst approach.[83–91] Supported Au catalysts are exceptionally active in catalyzing oxidation reactions at mild conditions, and thus the activation of O_2 and the reactivity of chemisorbed oxygen species on Au surfaces have been a central issue of fundamental investigations. However, due to the chemical inertness of bulk gold, O_2 does not molecularly or dissociatively chemisorb on Au single crystal surfaces no matter how large the O_2 exposure is.[92]

Several methods have been developed to prepare oxygen adatom and surface oxide, including use of an atomic oxygen source[93,94] or ozone,[95,96] electron or photon irradiation of Au surfaces with physisorbed O_2 at 28 K,[97,98] and electron bombardment of chemisorbed NO_2.[99] Surface oxide was observed to be less thermally stable than oxygen adatoms.[97,98] CO does not chemisorb on closely-packed Au(111) at temperatures as low as 78 K under ultra high vacuum (UHV), but chemisorbs on more open Au single crystal surfaces such as Au(110) and Au(100).[94,100–102] This can be attributed to the presence of low-coordinated Au atoms on Au single crystals with open surfaces. At elevated pressures, CO can chemisorb on Au(111) even at 300 K.[103] Once formed on Au single crystal surfaces, oxygen adatoms react facilely with CO to produce CO_2 at temperatures far below room temperature,[85,88,94,104–107] but CO_2 production was observed to depend on the nature of oxygen adatoms following the order of chemisorbed oxygen > oxygen in a surface oxide > oxygen in a bulk gold oxide.[85,88,106,107] The reactivity of oxygen adatoms on Au single crystal surfaces toward other reactants including H_2O,[105,107,108–111] CO_2,[94,105,112] NO_X,[113–115] NH_3,[116,117] propylene,[118] formic acid[119] and alcohols[105,108,120–127] has also been studied. The studies of Au single crystal surfaces have greatly deepened fundamental understanding of the reactivity of oxygen adatoms on the Au surface, but these studies have failed to address the key step in the oxidation reactions catalyzed by supported Au catalysts, namely, the chemisorption and activation of molecular O_2.

Since supported Au catalysts exhibit a very strong size-dependent catalytic performance, Au deposited on oxide single crystals or single crystal thin films has been extensively fabricated by physical vapor deposition (PVD). Among them Au deposited on rutile $TiO_2(110)$ remains the most widely studied system.[83,84,85,128] Au preferentially nucleates on bridging oxygen vacancies, and experiences a two-dimensional (2D) growth mode or quasi-2D growth for very low coverage and then subsequently changes to a three-dimensional (3D) growth mode on rutile $TiO_2(110)$.[129–134]

Matthey *et al.*[135] studied the growth and stability of Au on three kinds of rutile $TiO_2(110)$ surfaces: reduced $TiO_2(110)$ with bridging oxygen vacancies; hydrated $TiO_2(110)$ with bridging hydroxyl groups; and oxidized $TiO_2(110)$ with oxygen adatoms. They found that the stability of Au clusters follows the order: oxidized $TiO_2(110)$ > reduced $TiO_2(110)$ > hydrated $TiO_2(110)$. These results have useful implications for supported Au catalysts that are usually prepared by wet chemistry methods with hydrated oxide supports and often suffer from the sintering of supported Au nanoparticles. Thus the stability of supported Au catalysts is likely to be improved by the use of oxidized or reduced oxide supports described in ref. 135, but novel preparation method similar to the PVD deposition procedure of Au on planar oxide supports under UHV condition needs to be developed for Au on high specific surface area powder oxide supports. At elevated temperatures, small Au nanoparticles on rutile $TiO_2(110)$ were observed to undergo Ostwald ripening in which large Au nanoparticles formed at the expense of small ones,[136] and the Ostwald ripening process becomes accelerated under reactive environments.[137–140]

The size of Au nanoparticles on oxide single crystals or single crystal thin films influences their structure. With the decrease in the size of supported Au nanoparticles, the quantum size effect,[141] *i.e.* the metal to non-metal transition, is expected and thus the exact point at which this transition occurs is the subject of much interest. Scanning tunneling microscopy (STM) combined with scanning tunneling spectroscopy (STS) that can simultaneously probe the size/morphology and bandgap of supported Au nanoparticles is the most powerful technique. A bandgap of 0.2–0.6 eV was measured for Au nanoparticles on rutile $TiO_2(110)$ at a diameter of 3.2 nm and height of 1.0 nm ($\sim$300 atoms), and increased up to 1.4 eV for the smallest particles (diameter $<$ 2.0 nm). For Au nanoparticles on highly ordered pyrolytic graphite (HOPG), the metal-non-metal transition was measured to take place at $\sim$1 nm.[142]

X-ray photoelectron spectroscopy (XPS) is a routine technique with which to probe the electronic structure of supported Au nanoparticles. The Au 4f core level binding energy of supported Au nanoparticles was generally observed to shift positively as their size decreased.[143–149] This might be taken as an indication for the occurrence of metal-non-metal transition, but the contribution of final state effects to the core level binding energy shift needs to be considered.[83] The size of supported Au nanoparticles not only influences their electronic structure, but also their morphology and geometric structure. In their study of an Au particle grown on a MgO(100) step, Kizuka and Tanaka[150] observed that the particle could rearrange continuously to maintain the lowest energy structure when holding the sample substrate at room temperature. Costanzo *et al.*[151] made a comparative study of Au nanoparticles with approximately the same total number of atoms but different morphologies, and observed that two-dimensional Au nanoparticles on HOPG experienced smaller binding energy shifts compared with more spherical nanoparticles deposited on amorphous graphite. The differences were interpreted as initial state effects arising from the higher percentage of surface atoms in the two-dimensional Au nanoparticles on HOPG.

It is noteworthy that the observed size-dependent structure of supported Au nanoparticles could originate either from the size-dependent structure of Au nanoparticles or from the size-dependent Au-support interaction. With this respect unsupported size-selective Au clusters have been much examined to look at the intrinsic properties of small Au particles.[86,152] Another effective approach is to comparatively investigate Au on various oxide supports. Chusuei *et al.*[153] measured Au 4f core level binding energy shifts of Au nanoparticles on SiO_2 and TiO_2, and observed rather large differences between the two substrates (0.8 eV for Au/TiO_2, 1.6 eV for Au/SiO_2). These results strongly indicate that interactions of the support with Au nanoparticles must affect the electronic structure of the Au nanoparticles.

As expected, supported Au nanoparticles on oxide single crystals or single crystal thin films exhibit the size-dependent chemisorption and catalytic activities that originate from the size-dependent geometric and electronic structures. Shaikhutdinov *et al.*[154] comparatively investigated the adsorption of CO on Au nanoparticles with various sizes supported on model surfaces of FeO, Fe_3O_4, Fe_2O_3 and Al_2O_3, and proposed the size effect for CO adsorption as such that small particles contain a higher percentage of highly uncoordinated Au atoms which favor CO adsorption. They also observed that, regardless of substrate, Au nanoparticles of a threshold diameter (3.0 nm) desorbed CO at the same maximum temperature.

Meier and Goodman employed infrared reflection absorption spectroscopy to study CO adsorption on Au clusters ranging in size from 1.8 to 3.1 nm supported on rutile TiO_2(110).[155] The vibrational frequency of adsorbed CO blue-shifted slightly compared with that adsorbed on bulk Au, whereas the heats of adsorption increased sharply with decreasing cluster size. The Au cluster size corresponding to the largest heat of adsorption for CO was very close to that exhibiting the maximum catalytic activity for CO oxidation. However, no direct evidence was reported for the adsorption of molecular O_2 on supported Au nanoparticles.

Mullins' group reported the evidence for molecularly chemisorbed oxygen on TiO_2 supported Au nanoparticles exposed to an oxygen plasma jet molecular beam, and further observed the reaction between molecularly chemisorbed oxygen and CO to produce CO_2 at 77 K.[156,157] But the mechanisms for formation of molecularly chemisorbed oxygen were not understood and the size-dependence of the reaction between CO and molecularly chemisorbed oxygen was not explored. In order to observe the catalytic activity of supported Au model catalysts in CO oxidation, a high-pressure reaction cell with the pressure of the reactants up to Torr was used.

Goodman's group observed a strong particle size effect with Au nanoparticles supported on rutile $TiO_2(110)$ in the range of 2.5–3.5 nm showing a maximum in activity, and related this maximum in activity to the transition between metallic and non-metallic particles.[52] However, simultaneous reactivity measurements and grazing incidence small-angle X-ray scattering characterizations of an Au/$TiO_2(110)$ model catalyst revealed that supported Au nanoparticles with the optimal size (2.1 ± 0.3 nm) in catalyzing CO oxidation were with a height of about six atomic layers.[57] It was argued that the structure of Au nanoparticles supported on rutile $TiO_2(110)$ in the work by Goodman's group[52] was not determined under reactive conditions. Goodman's group also demonstrated that a two-layered Au film vacuum-deposited on a thin layer of TiO_2 on Mo(112) was the most active structure for low-temperature CO oxidation.[54] This indicates the high intrinsic catalytic activity of Au in CO oxidation.

During the exploration of size-dependent property of Au nanoparticles, the idea of 'magic number' Au clusters that exhibit extraordinary activity was proposed.[158] One of the first examples of a magic cluster is Au_{55}, which has been extensively studied.[159–163] Boyen *et al.*[163] confirmed that Au_{55} clusters are oxidation-resistant. They attributed this inertness to the closed-shell structure of such magic clusters rather than the size-induced metal-to-nonmetal transition, but unfortunately did not measure the catalytic activity of Au_{55} clusters.

To summarize, although the results from model catalyst study employing Au single crystals and Au nanoparticles supported on oxide single crystals or single crystal thin films have greatly advanced the fundamental understanding of Au catalysis, the conclusions are somewhat diverse and even contradictory. Moreover, several key issues in Au catalysis, such as activation of molecular O_2 and reaction mechanisms, are not understood. The following reasons are put forward for this.

(1) Au catalysis is complicated and ultra structure-sensitive. Thus slightly different structures of as-synthesized supported Au nanoparticles for the same system will result in quite different chemisorption behaviors and catalytic activities.
(2) The chemisorption of small molecules such as O_2, H_2, and CO is weak on Au surfaces so that the 'pressure gap' and 'materials gap' are more evident on Au surfaces than on other metal surfaces.

(3) The extensively investigated size effect in Au catalysis could actually arise from size-dependent geometric structure (type and density of uncoordinated Au atoms), size-dependent electronic structure (the metal-to-nonmetal transition), or size-dependent gold–oxide interaction and interfacial structure.
(4) The contribution of 'active' supports, *e.g.* the frequently-employed rutile $TiO_2(110)$ surface, to supported Au model catalysts whose extent remains uncertain adds additional complexity to fundamental understanding.
(5) A comprehensive study of the size–structure–chemisorption–catalytic activity relation for any given system needs to be conducted for unambiguous conclusions on Au catalysis, which is generally lacking in previous studies.

On the basis of above considerations, we propose to perform a step-by-step and comprehensive study of Au catalysis employing single crystals-based model catalysts, nanocrystals-based model catalysts, and powder catalysts with well-defined structures. Single crystals-based model catalysts are used to address basic concepts in Au catalysis, whereas nanocrystals-based model catalysts and powder catalysts with well-defined structures are used to bridge the 'materials gap' and 'pressure gap' between single crystals-based model catalysts and powder catalysts. Such a study is challenging and time-consuming, and in the following section, the progress we have achieved so far is summarized.

15.3 Model Catalysts from Single Crystals to Well-defined Powder Catalysts

Au catalysis can be divided into two parts: one is intrinsic Au catalysis; the other is support-involved Au catalysis. Accordingly our model catalysts are classified into two groups. One group is for the study of the intrinsic Au catalysis, including Au single crystals, Au nanocrystals, and Au/SiO_2 catalyst. SiO_2 is an 'inert' support and the observed catalytic activity of Au/SiO_2 catalysts can be associated with the structure of supported Au nanoparticles. The other group is for the study of support-involved Au catalysis, including Au nanoparticles supported on oxide single crystals or single crystal thin films, oxide nanoparticles supported on Au single crystals, Au nanoparticles supported on oxide nanocrystals, oxide nanoparticles supported on Au nanocrystals, and Au/MO_x catalysts.

The (111) single crystal surface is the most common and stable surface for metals with *fcc* structure, but an Au(111) surface exposing 9-coordinated Au atoms has been demonstrated to be very inert and not able to chemisorb small molecules such as CO and NO even at 80 K. Therefore, Au(111) is not a suitable model surface for the study of Au catalysis. Low-coordinated Au atoms exhibit enhanced reactivity and catalytic activity, and can be modeled with stepped single crystal surfaces.

In our study, we choose stepped Au(997) and Au(110)-(1×2) single crystals as model surfaces. As schematically illustrated in Figure 15.3(a), the (997) vicinal surface is created by cutting an Au(111) crystal at an angle of approximately 7° offset to obtain atomic height steps along the dense [110] separated by (111) terraces, and is composed of close-packed (111) terraces and a mono-atomic step with (111) microfacet. The coordination numbers of atoms on the step and terrace of Au(997) surface are 7 and 9, respectively.

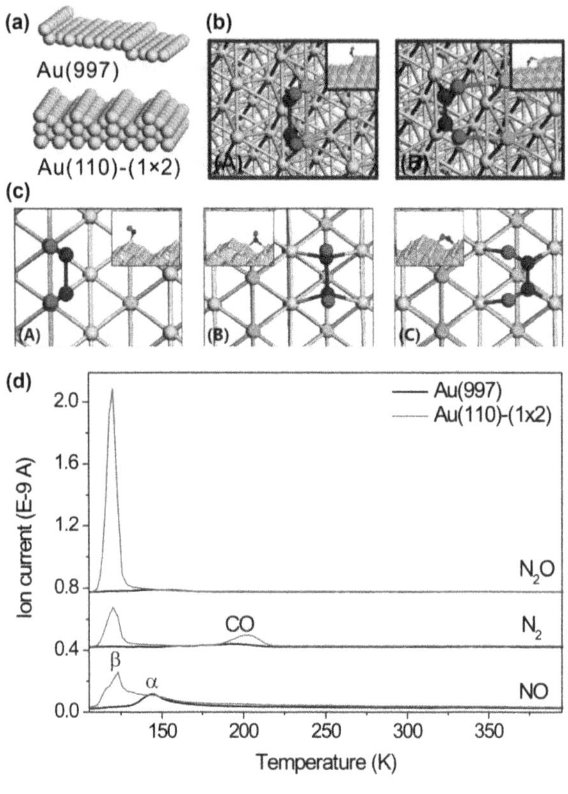

Figure 15.3 (a) Schematic illustration of Au(997) and Au(110)-(1×2) single crystal surfaces. (b) Top-view of optimized structures of $(NO)_2$ dimer on Au(997) bonded to 7-coordinated Au atoms *via* N atom (A) and bonded to both step and terrace Au atoms *via* O and N atoms (B). The inset shows the side view of optimized structures. Green, blue, and red balls represent 7-coordinated Au atom, N atom, and O atom, respectively. (c) Top-view of optimized structures of $(NO)_2$ dimer on Au(110)-(1×2) bonded to 7-coordinated Au atoms *via* N atom (A), bonded to the trench Au atoms *via* N atoms (B) and bonded to the trench Au atoms *via* N and O atoms (C). The inset shows the side view of optimized structures. Green, blue, and red balls represent 7-coordinated Au atom, N atom, and O atom, respectively. (d) Thermal desorption spectroscopy (TDS) spectra of NO, N_2O, and N_2 after saturating exposure of NO on Au(997) and Au(110)-(1×2) single crystal surfaces at 105 K.
Reproduced/adapted from ref. 165.

The Au(110)-(1×2) surface can be described by a missing-row model with paired rows in the second layer and buckled rows in the third layer, and gives rise to on-top-of-row atoms, side-of-row atoms, and trench atoms whose co-ordination numbers are 7, 9, and 11, respectively. Therefore, the Au(997) surface offers a model surface to compare the reactivity and catalytic activity of Au atoms with the same coordination environment but different coordination numbers, while the Au(997) and Au(110)-(1×2) surfaces offer model surfaces to compare the reactivity and catalytic activity of Au atoms with the same coordination number but different coordination environments.

The adsorption and reaction of CO, NO, and NO_2 on Au(997) and Au(110)-(1×2) surfaces have been studied;[164–166] 7-coordinated Au atoms on the step of Au(997) exhibit a stronger chemisorption ability than 9-coordinated Au atoms on the terrace of Au(997), clearly demonstrating the importance of low-coordinated Au atoms in Au catalysis. Interestingly, a comparative study of NO adsorption and reaction on Au(997) and Au(110)-(1×2) surfaces reveals the origin of excellent Au catalysis at low temperatures, and points out the importance of coordination environments of Au atoms in Au catalysis.[165]

As shown in Figure 15.3(d), following saturating NO exposure at 105 K, a similar NO desorption peak at 144 K appears for both surfaces and can be attributed to the desorption of species chemisorbed on 7-coordinated Au atoms; however, for Au(110)-(1×2) surface, stronger desorption peaks at 123 K with a shoulder at 116 K also appear. More strikingly, the desorption peaks of N_2O and N_2, the products of NO decomposition, are very obvious for the Au(110)-(1×2) surface but negligible for the Au(997) surface, and the N_2O and N_2 desorption peaks are clearly associated with the NO desorption peaks at lower temperatures.

Thus, although 7-coordinated Au atoms bind NO most strongly, surface sites other than 7-coordinated Au atoms on the Au(110)-(1×2) surface are most catalytically active in decomposing NO. Density functional theory (DFT) calculation results suggest that the chemisorbed $(NO)_2$ dimer and its decomposition dominate the surface chemistry of NO on the Au(997) and Au(110)-(1×2) surfaces. The NO desorption peak at 144 K arises from the most stable $(\underline{N}O)_2$ dimer bonded to 7-coordinated Au atoms *via* the N atom [Figure 15.3(b)(A) and (c)(A)], and the NO desorption peaks at lower temperatures arise from less stable $(\underline{N}O)_2$-II and $(\underline{NO})_2$ dimer, respectively, bonded to the trench Au atoms of Au(110)-(1×2) surface *via* an N atom, and N and O atoms [Figure 15.3(c)(B) and (c)(C)]. On the Au(997) surface, the less stable $(\underline{NO})_2$ dimer bonded to both step and terrace Au atoms *via* O and N atoms [Figure 15.3(b)(B)] needs to compete with more stable $(\underline{N}O)_2$ dimer for the step Au atoms, and thus its formation is suppressed.

DFT calculation results found that the decomposition activity of various $(NO)_2$ dimers on the Au(997) and Au(110)-(1×2) surfaces into N_2O varied in a trend contrasting with their adsorption energy. The most stable $(\underline{N}O)_2$ dimer bonded to 7-coordinated Au atoms exhibited a high activation barrier ($\sim$1.0 eV), whereas the less stable $(\underline{N}O)_2$-II and $(\underline{NO})_2$ dimers bonded to the trench Au atoms of the Au(110)-(1×2) surface exhibited a low activation barrier (~0.5 eV). Therefore, DFT calculation results agree well with experimental results.

The observation that the decomposition activity of various $(NO)_2$ dimers on Au surfaces into N_2O varied in a trend contrasting with their adsorption energy sheds light on the nature of the exceptional high activity of supported Au nanoparticles at low reaction temperatures. During the catalytic reaction, the coverage of a surface species on the catalyst surface is determined by the gas-phase pressure (liquid-phase concentration), the reaction temperature, and its adsorption energy on the catalyst surface. The larger the adsorption energy of a surface species is, the higher its equilibrium coverage. The reactivity of every type of surface species present on the catalyst surface will contribute to the observed catalytic activity. For a catalyst surface on which surface species with small adsorption energies exhibit high or better reactivity than the surface species with the largest adsorption energy, the catalyst will exhibit a high low-temperature catalytic activity, or a better catalytic activity at low temperatures than at high temperatures. NO decomposition on Au surfaces is a nice example and we believe that such a concept is also applicable for other catalytic reactions in which Au surface exhibits exceptionally high activity at low reaction temperatures.

To explore the size-dependent electronic structure of Au nanoparticles, Au colloids with different sizes were prepared by the reduction of aqueous $HAuCl_4$ solution at different pH values by ascorbic acid (vitamin C) in the presence of sodium benzenesulfonate (SDBS).[167] Aqueous $HAuCl_4$ solution consists of $[AuCl_x(OH)_{4-x}]^-$ ($x \geq 2$) at low pH but $[AuCl_x(OH)_{4-x}]^-$ ($x < 2$) at high pH. Reduction of aqueous $HAuCl_4$ solution at low pH leads to the synthesis of well-dispersed uniform fine Au colloids, whereas at high pH, large Au colloids and ensembles of fine Au colloids are formed [Figure 15.4(a)].

The S 2p XPS peaks of Au colloids synthesized at different pH values [Figure 15.4(b)] are similar and can be fitted by two components with the S $2p_{3/2}$ binding energy at 169.5 and ~168.0 eV. The former could be assigned to S in free SDBS molecules[168] and the latter to S in SDBS molecules interacting with Au colloids. Thus the SDBS molecules bind to the surface of Au colloids through the S atom, and charge transfer from Au atoms to S atoms occurs. However, as shown in Figure 15.4(c), the Au 4f XPS features of Au colloids synthesized at different pH values are size-dependent. Large Au colloids and Au ensembles of 3–4 nm Au colloids synthesized at pH = 10.35 exhibited a single feature with the Au $4f_{7/2}$ binding energy at 85.7 eV. The much higher Au $4f_{7/2}$ binding energy than that of bulk Au (84.0 eV) indicates that these Au ensembles and large Au colloids are positively charged, consistent with the occurrence of charge transfer from Au to S atom of SDBS. The Au 4f XPS peak of well-dispersed 3–4 nm Au colloids synthesized at pH = 2.91 consists of two components with the Au $4f_{7/2}$ binding energy at 85.7 and 86.1 eV. The component with the Au $4f_{7/2}$ binding energy at 86.1 eV arises from Au atoms on fine Au colloids directly interacting with S of SDBS and the other component with the Au $4f_{7/2}$ binding energy at 85.7 eV from Au atoms on fine Au colloids not interacting with the S atom. Fine Au

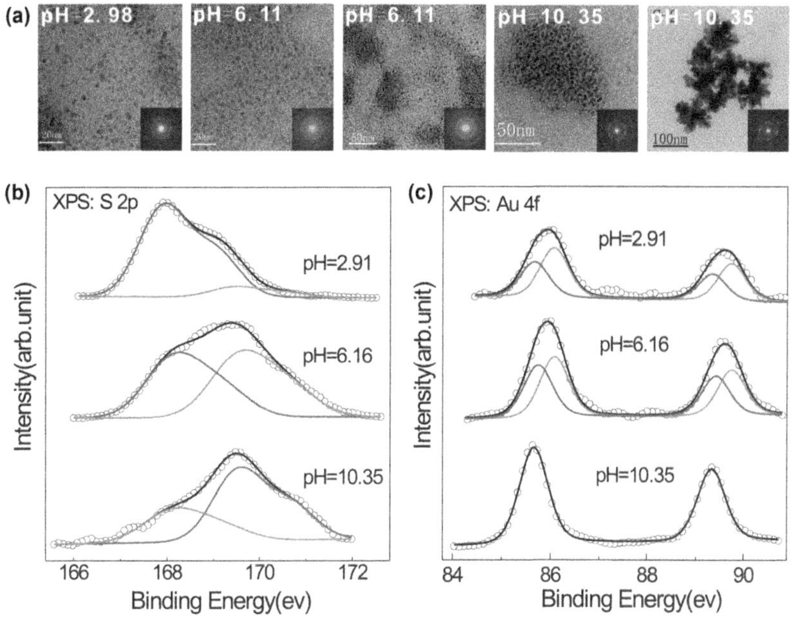

Figure 15.4 (a) Representative TEM images, (b) S 2p and (c) Au 4f XPS spectra of Au colloids synthesized by the reduction of aqueous $HAuCl_4$ solution at different pH values by ascorbic acid in the presence of sodium benzenesulfonate. The scatter data and the solid lines represent the original and fitted spectra, respectively. The grey lines represent the spectra arising from different components, and the black lines represent the sum spectra.
Reproduced/adapted from ref. 167.

nanoparticles exhibit a much higher Au 4f binding energy than bulk Au metals, whose value depends on the size of Au nanoparticles.[83] Due to the charge transfer from Au to S atom of SDBS, Au atoms on fine Au colloids directly interacting with S are positively charged and thus exhibit a higher Au 4f binding energy than those not interacting with S atom.

The Au 4f XPS peak of well-dispersed 3–4 nm Au colloids and their ensembles synthesized at pH = 6.16 also shows two components, with the Au $4f_{7/2}$ binding energy at 85.7 and 86.1 eV. Since both Au atoms on well-dispersed 3–4 nm Au colloids are not interacting with the S of SDBS and Au ensembles contribute to the Au 4f XPS component at 86.1 eV, the peak area (86.1 eV) to peak area (85.7 eV) ratio of well-dispersed 3–4 nm Au colloids and their ensembles synthesized at pH = 6.16 is smaller than that of well-dispersed 3–4 nm Au colloids synthesized at pH = 2.91.

On the basis of these XPS results, the charges are localized around the Au atoms directly participating in the charge transfer process when charge transfer occurs between well-dispersed 3–4 nm Au colloids and SDBS. This indicates that 3–4 nm Au colloids exhibit more or less characteristic of semiconductors. However, the transferred charges are delocalized to the entire Au colloids for those with a large diameter (*i.e.* characteristic of metals). This is the case at pH = 10.35, when Au ensembles and large Au colloids exhibit only a single XPS peak even though charge transfer also occurs. These results provide solid experimental results for the quantum size effect of Au nanoparticles.

We employed Au/SiO_2 catalysts to study the structure-intrinsic catalytic activity relation of Au nanoparticles in CO oxidation.[169–172] Au/SiO_2 catalysts were prepared by the traditional DP method employing aqueous $HAuCl_4$ solution as the Au precursor at pH 9–10, followed by calcination in air or reduction in H_2. As shown in Figure 15.5, for Au/SiO_2 catalysts with the same Au loading (1.8% weight ratio), when calcined in air at 300 °C (denoted as Au/SiO_2-Air-300), the acquired supported Au nanoparticles were mostly larger than 4.5 nm (~95%).[169–172] But when reduced in H_2, the acquired supported Au nanoparticles were smaller and exhibited reduction temperature-dependent size distribution.[172] In the catalyst reduced in H_2 at 300 °C (denoted as Au/SiO_2-H_2-300), ~23% of the Au nanoparticles were larger than 4.5 nm, ~70% were between 3 and 4.5 nm, and ~7% were between 2 and 3 nm. In Au/SiO_2-H_2-200, ~24% of the Au nanoparticles were larger than 4.5 nm, ~54% were between 3 and 4.5 nm, and ~22% were between 2 and 3 nm. In Au/SiO_2-H_2-120, ~35.6% of the Au nanoparticles were larger than 4.5 nm, ~39% were between 3 and 4.5 nm, ~23% were between 2 and 3 nm, and ~2.4% were smaller than 2 nm. Good resolution of the Au nanoparticles smaller than 2 nm was not possible using the TEM facility employed in our study and, therefore, it is likely that there were more Au nanoparticles smaller than 2 nm in Au/SiO_2-H_2-120 than we counted.

In the Au 4f XPS spectra [Figure 15.6(a)], both Au/SiO_2-Air-300 and Au/SiO_2-H_2-300 exhibit a single Au 4f component at 84.0 eV. Au/SiO_2-H_2-200 exhibits two Au 4f components at 84.0 eV (fraction: 60%) and 84.3 eV (fraction: 40%). Au/SiO_2-H_2-120 exhibits three Au 4f components at 84.0 eV (fraction: 55.56%), 84.3 eV (fraction: 33.33%), and 85.3 eV (fraction: 11.11%). Comparing the Au 4f binding energy and size distribution of supported Au nanoparticles in various Au/SiO_2 catalysts, it can be concluded that supported Au nanoparticles larger than 3 nm exhibit the Au 4f binding energy at 84.0 eV, while those between 2 and 3 nm and those smaller than 2 nm exhibit the Au 4f binding energy at 84.3 and 85.3 eV, respectively. Bulk Au exhibits its Au $4f_{7/2}$ binding energy at 84.0 eV.[168] These observations demonstrate the size-dependent electronic structure and the quantum size effect of Au nanoparticles supported on SiO_2. Au nanoparticles supported on SiO_2 larger than 3 nm exhibit the bulk Au-like electronic structure whereas those smaller than 3 nm exhibit electronic structures deviating, more or less, from that of bulk Au.

The chemisorption ability of Au nanoparticles in various Au/SiO_2 catalysts toward CO was studied by *operando* diffuse reflectance Fourier transform infrared spectroscopy (DRIFTS) spectra of saturated pulse chemisorption of CO at room temperature [Figure 15.6(b)]. Au/SiO_2-Air-300 does not chemisorb CO, while Au/SiO_2-H_2 chemisorbs CO reversibly. The observed

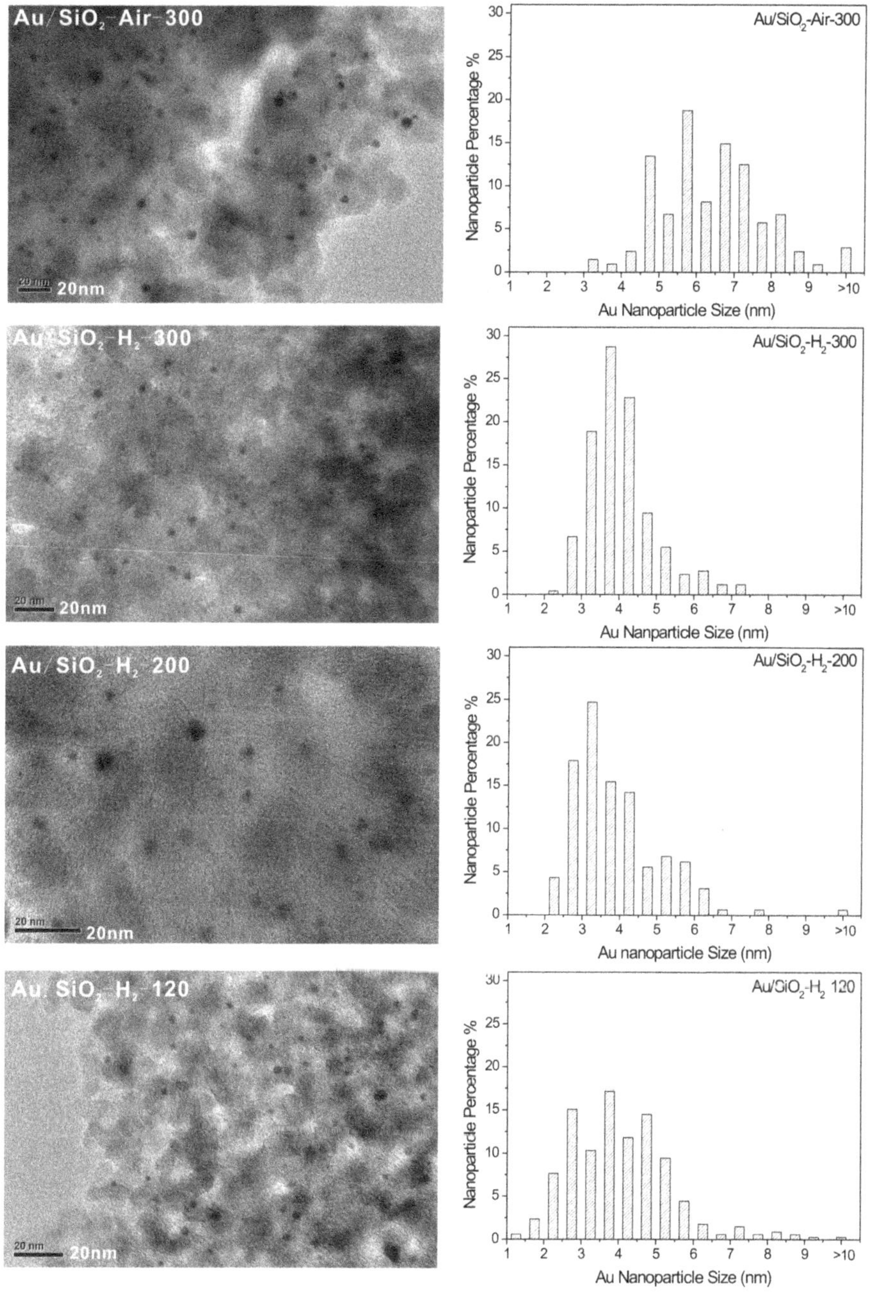

Figure 15.5 Representative TEM images and corresponding size distribution of Au/SiO_2-Air-300, Au/SiO_2-H_2-120, Au/SiO_2-H_2-200, and Au/SiO_2-H_2-300 catalysts.
Reproduced/adapted from ref. 172.

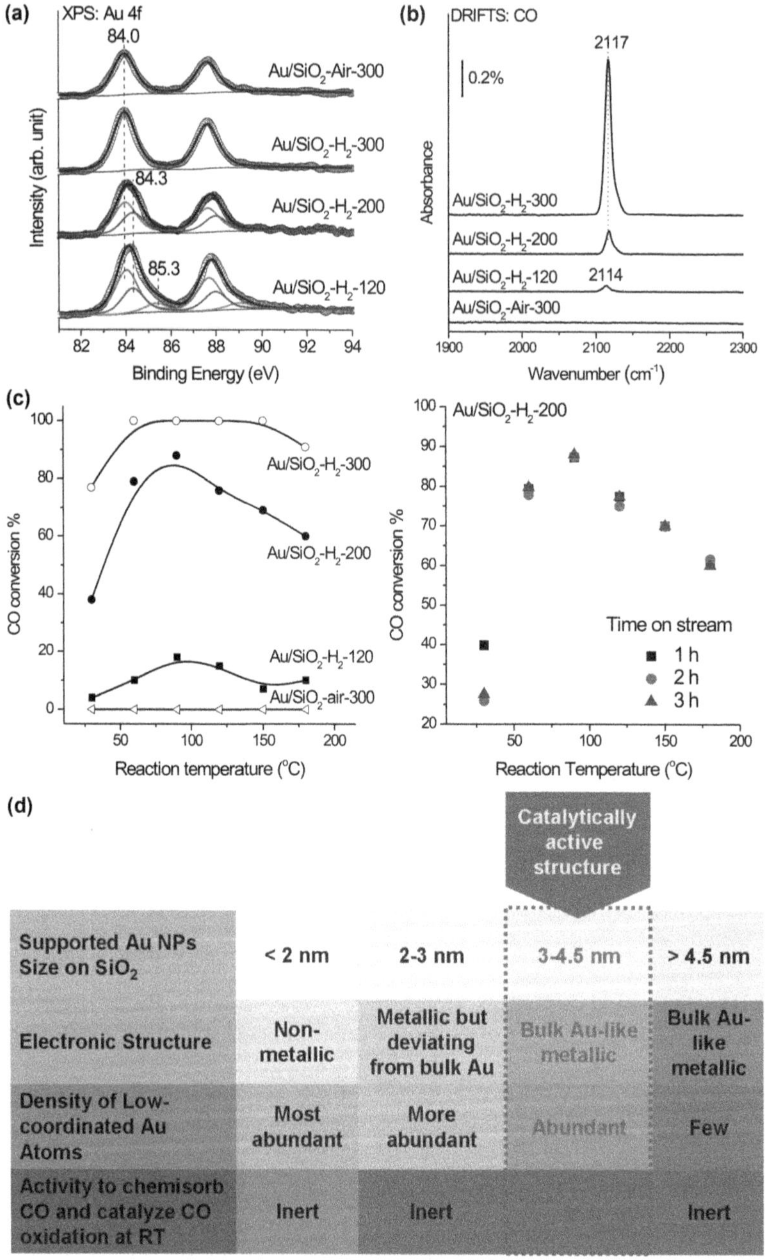

Figure 15.6 (a) Au 4f XPS spectra of Au/SiO_2-Air-300, Au/SiO_2-H_2-120, Au/SiO_2-H_2-200, and Au/SiO_2-H_2-300 catalysts. (b) DRIFTS spectra of CO chemisorption on Au/SiO_2-Air-300, Au/SiO_2-H_2-120, Au/SiO_2-H_2-200, and Au/SiO_2-H_2-300 catalysts at room temperature. (c) Catalytic activity and stability of Au/SiO_2-Air-300, Au/SiO_2-H_2-120, Au/SiO_2-H_2-200, and Au/SiO_2-H_2-300 catalysts in CO oxidation. (d) Schematic illustration of the size–structure–chemisorption–catalytic; activity relation and the catalytically active structure of Au nanoparticles supported on SiO_2 in low-temperature CO oxidation.
Reproduced/adapted from ref. 172.

vibrational frequencies (2114 and 2117 cm^{-1}) of chemisorbed CO agree well with those reported for CO chemisorbed on the top of low-coordinated Au atoms.[155,173,174] The amount of CO chemisorbed on various Au/SiO_2-H_2 catalysts increases with increasing reduction temperature. Thus, Au nanoparticles in Au/SiO_2-Air-300 expose few low-coordinated Au atoms, an observation which could reasonably be associated with their large size and thus likely smooth surfaces.

The experimental observation that the number of chemisorption sites for CO in Au/SiO_2-H_2 increases with the reduction temperature is consistent with their fractions of supported Au nanoparticles between 3 and 4.5 nm, but contrasts with the general assumption that smaller Au nanoparticles expose more low-coordinated Au atoms because the population of small Au nanoparticles in various Au/SiO_2-H_2 increases with decreasing reduction temperature. This infers that Au nanoparticles smaller than 3 nm in Au/SiO_2-H_2 should not actively chemisorb CO even though they expose a high density of low-coordinated Au atoms, which could reasonably be associated with their electronic structure which deviates from that of bulk Au. Since CO chemisorption on Au nanoparticles is required for catalytic CO oxidation, the observed catalytic activity of Au/SiO_2 catalysts at room temperature [Figure 15.6(c)] follows the same order as their ability to chemisorb CO: Au/SiO_2-H_2-300 > Au/SiO_2-H_2-200 > Au/SiO_2-H_2-120 ≫ Au/SiO_2-Air-300.

These results provide unprecedented solid and comprehensive experimental evidence for the size–structure–chemisorption–intrinsic catalytic activity relation of supported Au nanoparticles in low-temperature CO oxidation. As schematically illustrated in Figure 15.6(d), both the ability to chemisorb CO and the activity in catalyzing CO oxidation at room temperature of supported Au nanoparticles clearly exhibit a volcano-shaped dependence on their size. Supported Au nanoparticles between 3 and 4.5 nm with the bulk Au-like electronic structure effectively chemisorb CO and catalyze CO oxidation at room temperature; this could be attributed to the existence of abundant low-coordinated Au atoms. Larger supported Au nanoparticles with few low-coordinated Au atoms neither chemisorb CO nor catalyze CO oxidation at room temperature; however, smaller supported Au nanoparticles with more abundant low-coordinated Au atoms whose electronic structure deviates from that of bulk Au also neither chemisorb CO nor catalyze CO oxidation at room temperature. Therefore, without the involvement of support, the catalytically active structure of supported Au nanoparticles in low-temperature CO oxidation is low-coordinated Au atoms on Au nanoparticles with a bulk Au-like electronic structure. With the decrease in the size of Au nanoparticles, the density of low-coordinated Au atoms generally increases while the electronic structure deviates from bulk Au-like electronic structure. Therefore, the volcano-shaped dependence of intrinsic catalytic activity of Au nanoparticles on their size is an inevitable outcome of the fact that their size-dependence of geometric structure effect is opposite to their size dependence of electronic structure effect.

It is of interest that Au/SiO_2 catalysts active in low-temperature CO oxidation can be simply prepared by the conventional DP method employing $HAuCl_4$ as the Au precursor followed by H_2 reduction. SiO_2 is an inert support and it is quite difficult to prepare Au/SiO_2 catalysts active in low-temperature CO oxidation. Previous successful examples involve complicated methods or unusual Au precursors, including Au/SiO_2 catalysts prepared by chemical vapor deposition,[175] by using gold–inorganic–organic high surface area materials as precursors,[176] by the deposition of fine Au sols followed by oxidation and reduction treatments,[177] by the *in situ* formation of gold amine cation complexes,[178] and by using cationic Au gold complex cation, $[Au(en)_2]^{2+}$ (en = ethylenediamine) as Au precursor.[179]

We used *in situ* X-ray absorption spectroscopy (XAS) to understand the compositional and structural change of the catalyst precursor during calcination in air and reduction in H_2 (Figure 15.7).[172] The white line peak of Au L_{III}-edge X-ray absorption near-edge structure (XANES) spectrum at ~11 923 eV is associated with the $2p_{3/2} \rightarrow 5d_{5/2,3/2}$ dipole transition probing the unoccupied densities of d states of Au species. $HAuCl_4$ exhibits a strong white line peak, while Au foil does not show an obvious white line peak but exhibits a characteristic absorption peak at 11 948 eV. The Au L_{III}-edge XANES spectra during calcination in air vary very differently from those during reduction in H_2. The white line peak height (h_1) and the peak height at 11 948 eV (h_2) were plotted as a function of treatment temperature during different treatment processes. The catalyst precursor prepared by drying at 60 °C for 12 h mainly consists of SiO_2 with deposited Au(III) hydroxides[167,180,181] and exhibits a quite strong white line peak whose peak height is lower than that of $HAuCl_4$, but no absorption peak at 11948 eV. Additional calcination of the catalyst precursor in air at 60 °C for 30 min only results in a further decrease in the white line peak but also without the appearance of the peak at 11 948 eV. This observation infers that calcination of the catalyst precursor in air at 60 °C might reduce Au(III) to Au(I), but not to Au(0). Increasing the calcination temperature to 90 °C leads to the simultaneous quick decrease of h_1 and increase of h_2, demonstrating the simultaneous reduction of Au cations and formation of Au(0). The peak at 11 948 eV further grows at the expense of the white line peak with the increase of the calcination temperature. After the catalyst precursor was reduced in H_2 at 60 °C, the white line peak attenuates rapidly; meanwhile, the peak at 11 948 eV obviously appears, indicating the reduction of Au(III) cations to Au(0). The peak at 11 948 eV further grows at the expense of the white line peak when reduced at 90 °C. When the reduction was increased to 120 °C, the peak at 11 948 eV still grows, but interestingly, the white line peak height also increases. The fraction of Au cations surely decreases with the increase in reduction temperature, and thus the weakening of the white line peak should be expected.

Zhang and Sham[182] reported that, resulting from the interplay of quantum size and surface effect, the lattice contracts and the *d* charge at the Au atom site depletes relative to bulk Au as the size of the Au nanoparticle decreases,

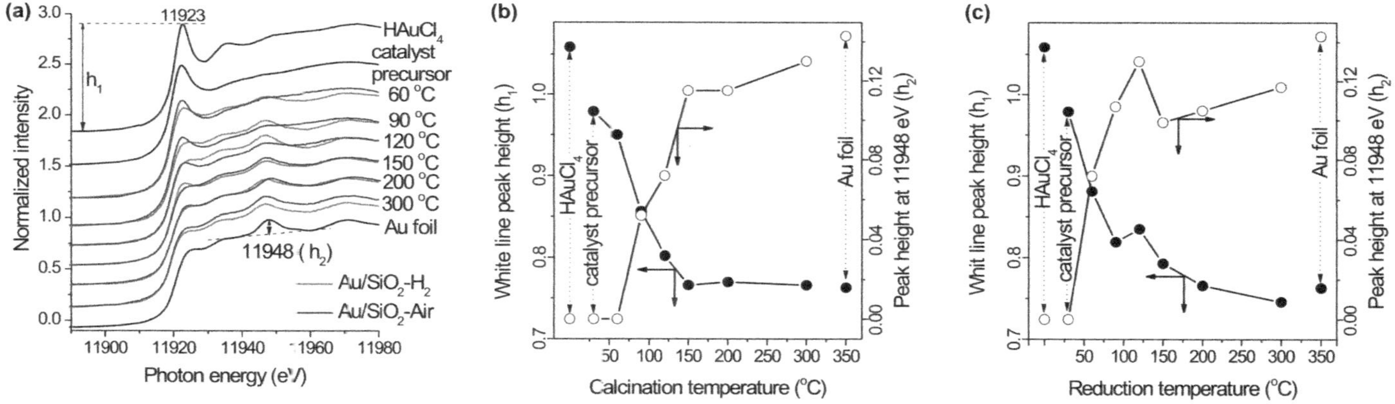

Figure 15.7 (a) *In situ* Au L_{III}-edge XANES spectra during the calcination of catalyst precursor in air and during the reduction of catalyst precursor in H_2. (b) Variations in the white line peak height (h_1) and the peak height at 11 948 eV (h_2) in the Au L_{III}-edge XANES spectra during the calcination of catalyst precursor in air. (c) Variations in the white line peak height (h_1) and the peak height at 11 948 eV (h_2) in the Au L_{III}-edge XANES spectra during the reduction of catalyst precursor in H_2. Reproduced/adapted from ref. 172.

leading to enhanced white line peak intensity. Therefore, the observation of simultaneous intensity growth of the white line peak and the peak at 11948 eV indicates the formation of dominance Au clusters with depleted *d* charge ($Au_n^{\delta+}$) when the catalyst precursor was reduced in H_2 at 120 °C for 30 min. With the further increase in the reduction temperature, the white line peak intensity keeps decreasing; however, the peak height at 11948 eV drops considerably when the reduction temperature increased from 120 to 150 °C, and then increases slowly.

The large differences among these Au L_{III}-edge XANES spectra indicate that the electronic structure of dominant Au clusters formed after reduction at 120 °C for 30 min is not in line with those formed after reduction at higher temperatures. In other words, the dominant Au clusters formed by reduction at 120 °C for 30 min are likely to have the non-metallic electronic structure resulting from the quantum size effect. These Au clusters with the non-metallic electronic structure are likely to be Au nanoparticles smaller than 2 nm with the Au 4f binding energy at 85.3 eV in $Au/SiO_2/H_2$-120. The change in the Au L_{III}-edge XANES spectrum when the reduction temperature was increased from 120 to 150 °C likely corresponds to the transition from non-metallic electronic structure to metallic electronic structure resulted from the growth in size and thickness. Therefore, when calcined in air, Au(III) hydroxides in the catalyst precursor seem to only decompose into Au(I) hydroxides or oxides at low temperatures; then Au(I) hydroxides or oxides decompose to Au(0) and nucleate at elevated temperatures, and the Au nuclei grow to form large Au nanoparticles. When reduced in H_2, Au(III) hydroxides in the catalyst precursor seem to be facilely reduced to Au(0) at low temperatures, forming Au nuclei likely with depleted *d* charge. Then these Au nuclei aggregate at elevated temperatures, forming Au nanoparticles with H_2 reduction temperature-dependent size distributions. These XAS results explain well the different size distributions and electronic structures of supported Au nanoparticles in various Au/SiO_2 catalysts.

The catalytic activity of low-coordinated Au atoms on large Au nanoparticles with bulk Au-like electronic structure in low-temperature CO oxidation was also reported in $NaNO_3$-modified Au/SiO_2 catalysts calcined in air at 200 °C.[170] Large Au nanoparticles supported on SiO_2 are poor in catalyzing CO oxidation. With the addition of $NaNO_3$ to the catalyst precursor before calcination, the acquired Au nanoparticles supported on SiO_2 become active in catalyzing low-temperature CO oxidation. Adding the $NaNO_3$ additive before calcination does not lead to the formation of detectable small Au nanoparticles, nor does it alter the electronic structure of the Au nanoparticles. The Au 4f peak intensity of Au/SiO_2 with $NaNO_3$ additive is much larger than that of Au/SiO_2. At a fixed total volume of particles, the XPS peak intensity depends on the surface-to-bulk atomic ratio of the particles; the larger the surface-to-bulk atomic ratio, the stronger the XPS peak intensity. It was proposed that, during catalyst preparation, $NaNO_3$ could interact with the precursor of large Au nanoparticles, inducing surface roughening and restructuring of acquired large Au nanoparticles. Therefore, Au nanoparticles in Au/SiO_2 with $NaNO_3$ additive exhibit a larger surface-to-bulk atomic ratio than those in Au/SiO_2; meanwhile low-coordinated Au atoms are created on the surface of large Au nanoparticles and catalyze low-temperature CO oxidation.

Au/SiO_2 catalysts with poor catalytic activity in CO oxidation are a good candidate to study the effect of promoters on catalytic activity. The catalytic performance of supported Au catalysts in reactions involving molecular O_2 is affected by moisture (gas phase reactions) and basicity (liquid phase reactions).[38–43] Employing an isotope labeling technique, Zope *et al.*[183] demonstrated that, in selective oxidation of alcohols in aqueous phase over supported Au catalysts at high pH conditions, oxygen atoms originating from hydroxide ions instead of molecular oxygen are incorporated into the alcohol, while molecular oxygen is proposed to participate in the catalytic cycle by regenerating hydroxide ions formed *via* the catalytic decomposition of a peroxide intermediate. Qian *et al.*[184] reported that NaOH additive substantially enhances the catalytic activity of Au/SiO_2 catalyst inert in catalyzing CO oxidation at temperatures below 150 °C and that $Au/NaOH/SiO_2$ catalyst with a NaOH to Au atomic ratio of 6 is active at room temperature [Figure 15.8(a) and (b)]. Both the particle size distribution and the electronic structure of Au nanoparticles were found to be similar in the Au/SiO_2 and $Au/NaOH/SiO_2$ catalysts, unambiguously proving that hydroxyls on 'inert' Au nanoparticles can induce the activation of O_2 for CO oxidation at room temperature.

Au(111) surface was used to model the surface of large Au nanoparticles. DFT calculation results reveal two likely reaction mechanisms for CO oxidation on Au(111) pre-covered with adsorbed OH(OH(a)) [Figure 15.9(a)]. The key elementary step is the reaction of CO(a) with OH(a) to form COOH(a) on Au(111) that proceeds with an activation energy of 0.44 eV and a reaction enthalpy of −1.15 eV.

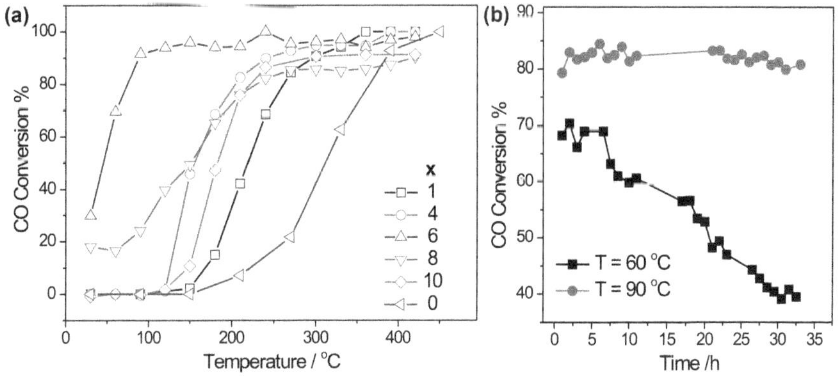

Figure 15.8 (a) Catalytic activity of $Au/NaOH/SiO_2$ catalysts in CO oxidation. X denotes the NaOH to Au molar ratio. (b) Catalytic stability of $Au/NaOH/SiO_2$ in CO oxidation at 60 and 90 °C. Reproduced/adapted from ref. 184.

In mechanism I in Figure 15.9(a), COOH(a) first decomposes to CO_2 and H(a) on Au(111) with an activation energy of 0.98 eV and a reaction enthalpy of −0.54 eV. The formed H(a) on Au(111) can readily react with O_2 to form HO_2(a) with an activation energy of 0.34 eV and a reaction enthalpy of −0.42 eV. HO_2(a) dissociates to OH(a) and O(a) on Au(111) with an activation energy of 0.57 eV and a reaction enthalpy of −0.26 eV, and O(a) then readily reacts with CO(a) to form CO_2. In this reaction mechanism, all the elementary surface reactions are exothermic, and the rate-determining step is the decomposition of COOH(a), and the activation of O_2 is accomplished by its reaction with H(a) on Au(111). This reaction mechanism is similar to the water gas shift reaction mechanism catalyzed by TiO_{2-x}/Au(111) and CeO_{2-x}/Au(111) catalysts proposed by Rodriguez *et al.*[185]

In mechanism II in Figure 15.9(a), COOH(a) reacts directly with O_2 to form OOCOOH(a) with a hydrogen-bonded five-membered ring structure on Au(111) with an activation energy of 0.94 eV and a reaction enthalpy of −0.81 eV. OOCOOH(a) can undergo a decomposition reaction to form O(a) and *di*-CO_3H(a) with an activation energy of 0.78 eV and a reaction enthalpy of −0.87 eV. The formed O(a) reacts with CO(a) to form CO_2. *di*-CO_3H(a) follows the reaction pathway schematically illustrated in Figure 15.9(b): *di*-CO_3H(a) first transforms into *mono*-CO_3H(a) with an activation energy of 0.60 eV and a reaction enthalpy of 0.41 eV, and then *mono*-CO_3H(a) decomposes into CO_2 and OH(a) with an activation energy of 0.39 eV and a reaction enthalpy of 0.04 eV. Mechanism II is a novel mechanism, and the rate-determining step is the reaction of COOH(a) with O_2 to form OOCOOH(a) on Au(111), which is also the activation step of O_2. In this mechanism, the transformation of *di*-CO_3H(a) into *mono*-CO_3H(a) and the decomposition of *mono*-CO_3H(a) into CO_2 and OH(a) are endothermic; other elementary surface reactions are exothermic. This indicates that *di*-CO_3H(a) is thermodynamically favored and will accumulate on the surface at low reaction temperatures, which might eventually block this mechanism. Furthermore, it was found that the reaction product CO_2 can facilely react with hydroxyls to form *di*-CO_3H(a) on the Au(111) surface with an activation energy of 0.35 eV and a reaction enthalpy of −0.45 eV.

These results suggest that *di*-CO_3H(a) on the Au surface could be the poison that leads to the deactivation of Au catalysts in CO oxidation at low temperatures. These DFT calculation results well explain the experimental observation of the deactivation of Au/NaOH/SiO_2-6 at 60 °C but not at 90 °C, which could be attributed to the formation and accumulation of *di*-CO_3H(a) on the Au surface that is favored thermodynamically at low reaction temperatures [Figure 15.9(b)]. Such a poisoning route for CO oxidation on OH(a)-covered Au(111) is thermodynamically controlled instead of kinetically controlled. In both mechanisms, the overall surface reaction is $2\ CO(a) + O_2(g) + OH(a) \rightarrow 2CO_2(g) + OH(a)$, with OH(a) acting as a co-catalyst. These results successfully demonstrate the influence of hydroxyls on the intrinsic activity of Au nanoparticles in CO oxidation and elucidate the underlying mechanism.

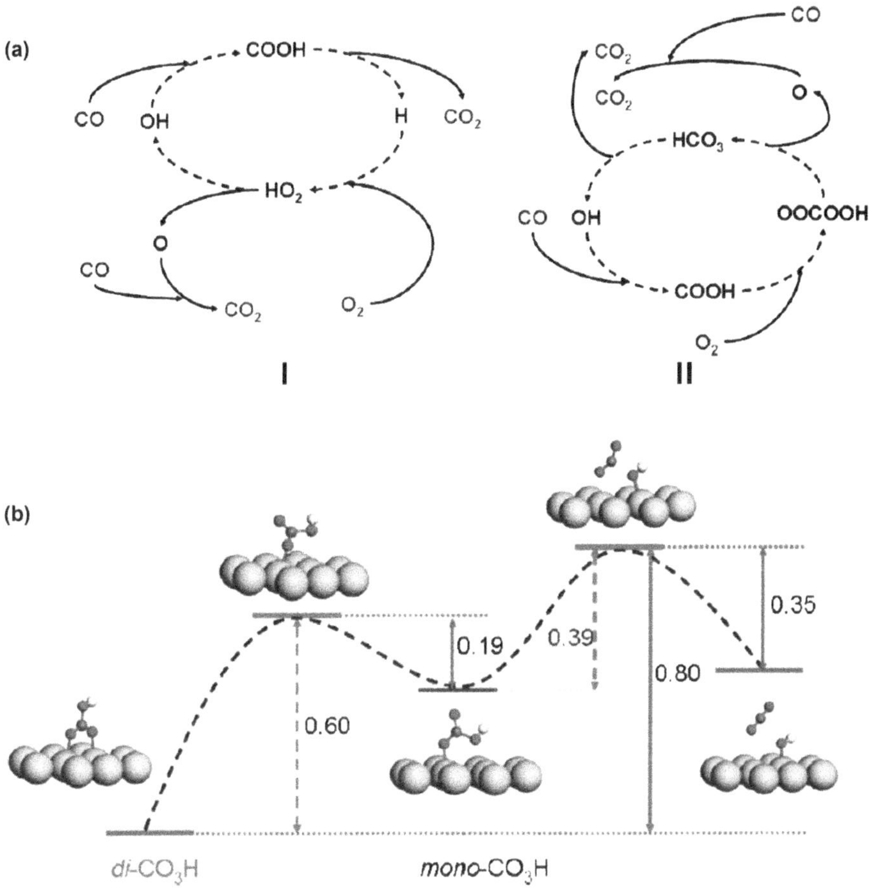

Figure 15.9 (a) Schematic illustration of two different surface reaction cycles of hydroxyl-co-catalyzed low-temperature CO oxidation on Au(111): (I) dissociation of COOH(a) and (II) reaction of COOH(a) with O_2. (b) Schematic illustration of the dissociation process of *di*-CO_3H(a) on Au(111) (from left to right) and the deactivation process of Au(111) (from right to left). The activation energy of each elementary step is also indicated.
Reproduced/adapted from ref. 184.

The structure and catalytic performance of supported Au nanoparticles sensitively depend on the employed oxide support. The gold–'active' oxide support interaction is a core issue. Charge transfer between 'active' oxide support and supported Au nanoparticles is also considered an important factor. Employing Au/rutile-TiO_2(110) as model catalysts, we have successfully provided solid photoelectron spectroscopic evidence for the charge transfer process (Figure 15.10).[186,187] Stoichiometric and reduced rutile-TiO_2(110) surfaces were prepared, and the growth and electronic structure of Au clusters on both surfaces were comparatively studied by XPS.[186] As shown in Figure 15.10(a), on a stoichiometric rutile-TiO_2(110) surface, the $4f_{7/2}$ XPS peak of supported Au clusters at the lowest coverage locates at 84.66 eV; it shifts monotonically toward lower binding energy with increasing Au coverage and reaches the binding energy value of the bulk Au at a deposited thickness of 10.0 Å. On a reduced rutile-TiO_2(110) surface, the $4f_{7/2}$ XPS peak of supported Au clusters at the lowest coverage locates at 84.32 eV; it shifts toward lower binding energy with increasing Au coverage and reaches 84.06 eV at a deposited thickness of 1.5 Å; however, the Au $4f_{7/2}$ binding energy shifts unexpectedly from 84.06 eV to 84.13 eV when the deposited Au thickness increases from 1.5 Å to 2.0 Å, and then does not change much with further increase of deposited Au thickness up to 10.0 Å.

The variation in Au 4f binding energy of Au clusters on stoichiometric rutile-TiO_2(110) surface agrees well with the size effect that the Au core level binding energy shifts to a higher binding energy with decreasing size of Au clusters on the substrates,[83] but that on a reduced rutile-TiO_2(110) surface can only result from both a size effect and charge transfer effect. Au clusters preferentially nucleate on the oxygen vacancy sites of the reduced rutile-TiO_2(110) surface and charge transfer occurs from the oxygen vacancy sites to Au clusters. As schematically illustrated in Figure 15.10(b), the charge transfer from the oxygen vacancy sites to Au clusters decreases the Au 4f binding energy, and thus the charge transfer effect compensates the size effect. Therefore, at the lowest thickness, Au clusters on a reduced rutile-TiO_2(110) surface exhibit a lower Au 4f binding energy than those on stoichiometric surface.

The charge transfer effect is also size-dependent and there is a critical size. Before reaching the critical size, the charge transfer effect can effectively compensate for the size effect, but after reaching the critical size, it cannot. When the Au particle size is increased over the critical size, the Au 4f binding energy will increase rather than decrease. This is exactly what we observed in the case of the growth of Au clusters on a reduced rutile-TiO_2(110) surface. The observation of the upward binding energy shift of Au clusters on a reduced rutile-TiO_2(110) surface with an increase in their size provides solid proof for the charge transfer from the oxygen vacancy sites of a reduced rutile-TiO_2(110) surface to Au clusters.

The charge transfer from the oxygen vacancy sites of a reduced rutile-TiO_2(110) surface to Au clusters was also evidenced by ultraviolet photoemission spectroscopy (UPS) results.[187] The Ti^{3+} feature of the reduced rutile-TiO_2(110) surface, ~0.9 eV below the Fermi level, attenuates considerably upon the deposition of a 0.1 Å-thick Au film [Figure 15.10(c)]. Thus, Au clusters preferentially nucleate on the oxygen vacancy sites of reduced rutile-TiO_2(110) surface and the charge transfer occurs from the oxygen vacancy sites to Au clusters, resulting in the oxidation of Ti^{3+} to Ti^{4+}. The charge transfer from the oxygen vacancy sites on reduced rutile-TiO_2(110) surface to Au clusters not only modifies the electronic structure of supported Au clusters, but also greatly enhances the thermal stability of Au clusters. As shown in Figure 15.10(d), Au clusters at a deposited thickness of

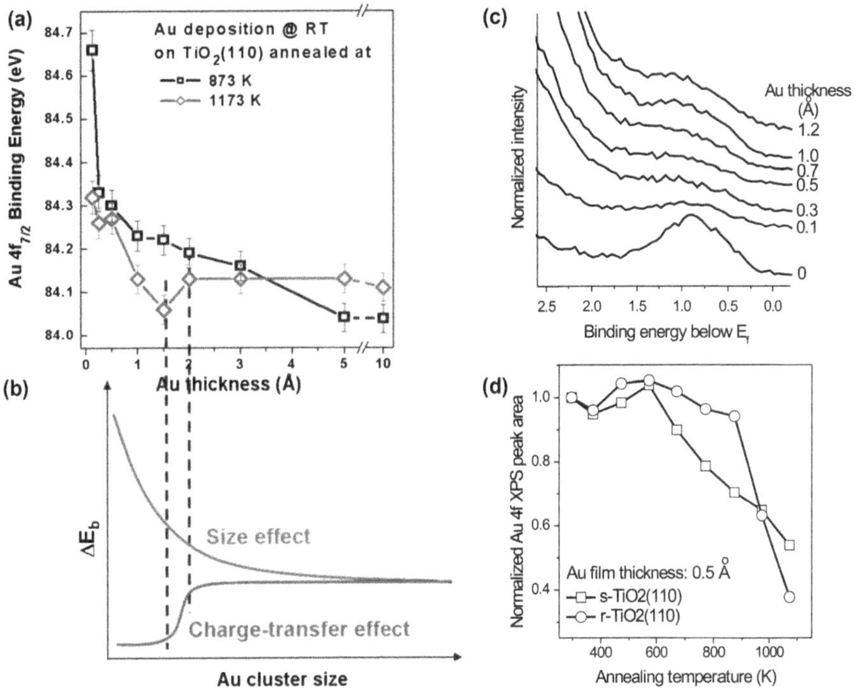

Figure 15.10 (a) Au $4f_{7/2}$ binding energy of Au clusters deposited on rutile Tio2(110) surface annealed at 873 k (stoichiometric surface) and 1173 K (reduced surface) as a function of Au thickness. (b) Schematic illustration of the size effect and size-dependent charge transfer effect on the Au $4f_{7/2}$ binding energy of Au clusters. (c) Valence-band photoemission spectra of Au clusters deposited on reduced rutile $TiO_2(110)$ surface with indicated thicknesses. hν = 28 eV. (d) Normalized Au 4f SR-PES peak intensity of Au clusters with a thickness of 0.5 Å on stoichiometric and reduced rutile $TiO_2(110)$ surfaces. hν = 150 eV. Reproduced/adapted from ref. 186 and ref. 187.

0.5 Å are thermally stable up to 673 K on the stoichiometric rutile-$TiO_2(110)$ surface but up to 873 K on the reduced rutile-$TiO_2(110)$ surface.

To understand the gold–'active' oxide support interaction in powder catalysts, we employed $Au/MO_x/SiO_2$ model catalysts in which both Au and 'active' oxide are supported on 'inert' SiO_2 so that their loadings can be varied to reflect the gold–'active' oxide support interaction.[188–191] CoO_x/SiO_2 was prepared by the incipient wetness impregnation method followed by drying at 60 °C, and then used directly to prepare $Au/CoO_x/SiO_2$ catalysts by the DP method using $HAuCl_4 \cdot 4H_2O$ as the precursor followed by calcination in air at 200°C for 4 h.[188,189] Only metallic Au exists in the catalysts. As shown in Figure 15.11(a), 2%-Au/SiO_2 catalyst mainly contains Au nanoparticles between 7 and 15 nm, and 2%-Au/6%-CoO_x/SiO_2 catalyst contains Au nanoparticles with a bimodal size distribution: small Au nanoparticles between 5 and 7 nm; and large Au aggregates.

These observations directly prove that Au-CoO_x interaction is stronger than Au-SiO_2 interaction so that small Au nanoparticles can be stabilized on CoO_x but not on SiO_2.[188] By varying the Au to CoO_x ratio in Au/CoO_x/SiO_2 catalysts, we found that the size distribution of supported Au nanoparticles in Au/CoO_x/SiO_2 catalysts and their catalytic performance depend on the Au to CoO_x ratio.[189] Large Au aggregates preferentially form in Au/CoO_x/SiO_2 catalysts with low Au to CoO_x ratios; with an increase in the Au CoO_x ratio, small Au nanoparticles appear and eventually dominate in the catalysts, though large Au aggregates are always present. As shown in Figure 15.11(b), Au/CoO_x/SiO_2 catalysts containing only large Au aggregates exhibit better catalytic activity than Au/SiO_2 in CO oxidation but are not active at room temperature, and Au/CoO_x/SiO_2 catalysts containing small Au nanoparticles are active in catalyzing CO oxidation at room temperature.

These results clearly demonstrate the size effect of supported Au nanoparticles on catalytic activity. During our preparation procedure, the Co precursor supported on SiO_2 is $Co(OH)_2$, and hydrogen-bonded hydroxyls in $Co(OH)_2$ can bind the Au precursor more strongly than isolated hydroxyls. Thus the Au precursor preferentially deposits on the support surface by interacting with hydrogen-bonded hydroxyls in $Co(OH)_2$ on SiO_2, resulting in the formation of clusters of Au species on the support surface and eventually large Au aggregates in the catalyst. Only after the occupation of hydrogen-bonded hydroxyls in $Co(OH)_2$ on SiO_2 can the Au precursor deposit on the support surface by interacting with isolated hydroxyls in $Co(OH)_2$, resulting in the formation of isolated Au species on the support surface and eventually small Au nanoparticles in the catalyst [Figure 15.11(c)]. These results reveal that the distribution of surface hydroxyls on oxide supports could be an important factor to affect the size and catalytic activity of supported Au nanoparticles prepared by the DP method. The distribution of surface hydroxyls on oxides is difficult to control, and thus it might be an important reason why the reproducible preparation of highly active Au/MO_x catalysts was sometimes reported by different research groups to be difficult, even though the employed materials and preparation methods were similar.[47]

We also studied the effect of CeO_2 microstructure on the structure and catalytic activity of supported Au nanoparticles in CO oxidation employing Au/CeO_2/SiO_2 model catalysts.[191] CeO_2/SiO_2 supports with different CeO_2 microstructures were prepared by the incipient wetness impregnation and DP methods, and used to prepare Au/CeO_x/SiO_2 catalysts by the DP method. Independent of the microstructure of CeO_2, the Ce(III)/Ce(IV) atomic ratio in CeO_2 increases greatly with Au loading. This demonstrates that the loading of Au on CeO_2/SiO_2 facilitates the formation and stabilization of surface oxygen vacancies in CeO_2. Both Au(I) and metallic Au exist in the catalysts. The formation of Au(I) species is closely associated with Ce(III) sites, *i.e.* surface oxygen vacancy sites of CeO_2. The Au(I) to Au atomic ratio varies proportionally with the Ce(III) to Ce(IV) atomic ratio in Au/CeO_2/SiO_2 catalysts. Au nanoparticles also nucleate on the surface oxygen vacancy sites of CeO_2. Therefore, Au precursors on CeO_2 surface can form either Au(I) species

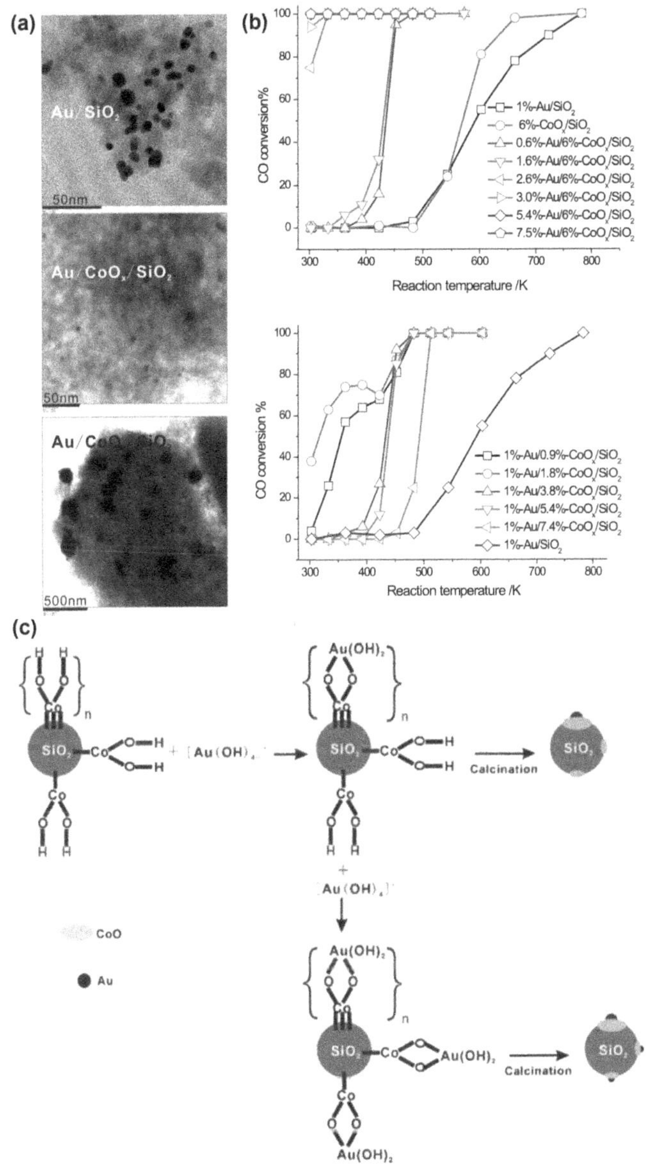

Figure 15.11 (a) Representative TEM images of 2%-Au/SiO_2 and 2%-Au/6%-CoO_x/SiO_2 catalysts. (b) Catalytic activity of Au/CoO_x/SiO_2 catalysts with various Au and CoO_x loadings in CO oxidation. (c) Schematic illustration of the deposition of Au precursor on $Co(OH)_2$/SiO_2 and its influence on the structure of supported Au nanoparticles in Au/CoO_x/SiO_2 catalysts. $Co(OH)_2$ and $\{Co(OH)_2\}_n$ represent isolated $Co(OH)_2$ and $Co(OH)_2$ clusters with hydrogen-bonded hydroxyls on SiO_2, respectively.
Reproduced/adapted from ref. 188 and ref. 189.

or Au, and the available surface oxygen vacancy sites and their distribution on CeO_2, the surface oxygen vacancy-Au(I) interaction, and the surface oxygen vacancy–gold cluster interaction cooperatively determine the resulting Au species in the catalysts. Highly dispersed CeO_2 in CeO_2/SiO_2 supports facilitates the formation and stabilization of Au(I) species. $Au/CeO_2/SiO_2$ catalysts containing both Au(I) species and Au nanoparticles are much more active in CO oxidation than those containing only Au(I) species. This indicates that Au(I) species on CeO_2 alone is not active in CO oxidation and that the presence of Au nanoparticles on CeO_2 is important. Meanwhile, the activity of $Au/CeO_2/SiO_2$ catalysts in CO oxidation could correlate well with the reducibility of CeO_2 and the activity of the corresponding CeO_2/SiO_2 catalyst in CO oxidation, indicating the participation of CeO_2 in the catalytic reaction.

ZnO does not belong to reducible oxides such as TiO_2, CeO_2, and CoO_x, but Au/ZnO catalysts were also reported to be active in low-temperature CO oxidation.[192,193] We thus studied the Au-ZnO interaction employing $Au/ZnO/SiO_2$ model catalysts.[190] 6%-ZnO/SiO_2 was prepared by the incipient wetness impregnation method followed by calcination at 200 °C, and then directly used to prepare 2%-Au/6%-ZnO/SiO_2 catalysts by the DP method using $HAuCl_4 \cdot 4H_2O$ as the precursor followed by calcination in air at 200°C for 4 h. Ammonium hydroxide and an aqueous solution of Na_2CO_3 were used as the DP agent, and the resulting catalysts were designated $Au/ZnO/SiO_2$-NH_3 and $Au/ZnO/SiO_2$-Na_2CO_3.

Both Au(I) and metallic Au exist in the catalysts. Au nanoparticles in $Au/ZnO/SiO_2$-Na_2CO_3 are smaller than those in $Au/ZnO/SiO_2$-NH_3 (Figure 15.12),

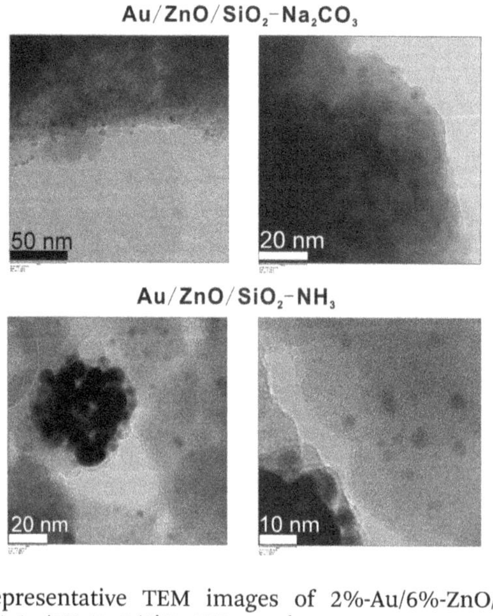

Figure 15.12 Representative TEM images of 2%-Au/6%-ZnO/SiO_2-Na_2CO_3 and 2%-Au/6%-ZnO/SiO_2-NH_3 catalysts.
Reproduced/adapted from ref. 190.

and the Au(I) to Au(0) atomic ratio in Au/ZnO/SiO_2-Na_2CO_3 is larger than that in Au/ZnO/SiO_2-NH_3 [Figure 15.13(a)]. Interestingly, we observed strong photoluminescence of ZnO in ZnO/SiO_2 arising from oxygen vacancies and grain boundary defects becoming significantly quenched after the loading of Au [Figure 15.13(b)]. This demonstrates that Au interacts with oxygen vacancies and grain boundary defects of ZnO. The luminescence intensity of ZnO in Au/ZnO/SiO_2-Na_2CO_3 is weaker than that ZnO in Au/ZnO/SiO_2-NH_3. These results indicate that the Au–ZnO interaction in Au/ZnO/SiO_2-Na_2CO_3 is stronger than that in Au/ZnO/SiO_2-NH_3. As shown in Figure 15.13(c), Au/ZnO/SiO_2-Na_2CO_3 is more active in catalyzing CO oxidation than Au/ZnO/SiO_2-NH_3. ZnO does not participate in catalytic CO oxidation and the observed catalytic activity can be only related to supported Au nanoparticles. Thus, these results demonstrate that small Au nanoparticles with low-coordinated Au atoms and bulk Au-like electronic structure are able to catalyze low-temperature CO oxidation without the involvement of oxide supports, agreeing with the results of Au/$NaNO_3$/SiO_2[170] and Au/SiO_2-H_2 catalysts.[172]

As shown in Figure 15.6(c), Figure 15.11(b) and Figure 15.13(c), several catalysts were observed to exhibit an unusual dependence of catalytic activity on the reaction temperature. As the reaction temperature increases, CO conversion initially increases, then decreases, and eventually increases again to reach the 100% conversion. As representatively shown in Figure 15.13(c), such a behavior is reproducible when the used catalyst was tested once again; thus it is not due to the structural change of supported Au nanoparticles during the catalytic activity evaluation.

We have proposed that this unusual activity–reaction temperature dependence indicates different mechanisms of CO oxidation catalyzed by supported Au nanoparticles at low and high reaction temperatures.[190] At high reaction temperatures, O_2 can directly dissociate on the Au surface to form oxygen adatoms that react with CO to form CO_2. In this high-temperature reaction mechanism, oxygen dissociation is the rate-limiting step and its barrier reduces as the Au particle size decreases and the Au atoms become low-coordinated.[194] The critical temperature at which the high-temperature reaction pathway does not contribute much to the catalytic activity depends mainly on the structure and size of the Au nanoparticles. At low reaction temperatures ($\sim$300 to 450 K), CO oxidation catalyzed by Au nanoparticles proceeds *via* a different mechanism. We proposed that a characteristic of low-temperature reaction mechanism is that O_2 does not need to undergo direct dissociation on the Au surface.

The observed 'volcano' shape of the activity–reaction temperature dependence strongly suggests that a weakly chemisorbed species is the key active species in the low-temperature reaction mechanism. With an increase in reaction temperature, the coverage of the weakly chemisorbed species initially does not change much and thus the reaction rate increases. When the reaction temperature continues to increase, however, this results in the desorption of the weakly chemisorbed species, its coverage on Au surface decreases rapidly and eventually the weakly chemisorbed species cannot

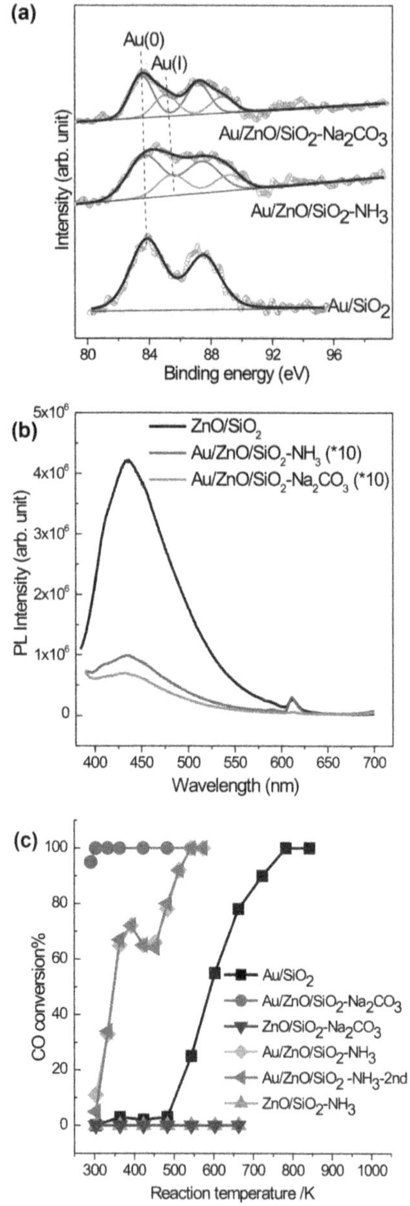

Figure 15.13 (a) Au 4f XPS spectra of Au/SiO_2, 2%-Au/6%-ZnO/SiO_2-Na_2CO_3, and 2%-Au/6%-ZnO/SiO_2-NH_3 catalysts. (b) Photoluminescence (PL) spectra of Au/SiO_2, 2%-Au/6%-ZnO/SiO_2-Na_2CO_3, and 2%-Au/6%-ZnO/SiO_2-NH_3 catalysts excited by the laser at 366 nm. (c) Catalytic activity of Au/SiO_2, ZnO/SiO_2, and $Au/ZnO/SiO_2$ catalysts in CO oxidation. Reproduced/adapted from ref. 190.

form on the surface and, consequently, the reaction rate decreases. Therefore, it is the competition between desorption and surface reaction of the weakly chemisorbed species on the Au surface that leads to the observation of a local maximum reaction rate with an increase in reaction temperature at low temperatures. The low-temperature reaction mechanism is facilitated over small Au nanoparticles because small Au nanoparticles are active in catalyzing low-temperature CO oxidation. The nature of the weakly chemisorbed species requires further investigation. Likely candidates include molecularly chemisorbed O_2(a), chemisorbed water/hydroxyls, and chemisorbed COOH(a) species.[83,184] Similar experimental results were reported in Au-catalyzed low-temperature CO oxidation by other groups.[38,195,196] Daté *et al.*[38] attributed the activity decrease to desorption of moisture from the catalyst surface which plays an important role in the catalytic activity of Au/SiO_2 catalyst in low-temperature CO oxidation.

The key role of weakly chemisorbed species on Au surface in low-temperature CO oxidation also supports our conclusion drawn from NO decomposition on Au single crystal surfaces that the excellent catalytic activity of supported Au nanoparticles is due to the presence of weakly chemisorbed and highly reactive species.[165] Therefore, the apparent catalytic activity of supported Au nanoparticles in CO oxidation consists of contributions from CO oxidation following both the low-temperature and high-temperature reaction mechanisms. Following either reaction mechanism, the catalytic activity of supported Au nanoparticles increases with a decrease in the Au particle size within reasonable ranges. With the increase in the reaction temperature, the activity of supported Au nanoparticles following the low temperature reaction mechanism will intrinsically exhibit a 'volcano' dependence on the reaction temperature, and the activity of Au nanoparticles following the high-temperature reaction mechanism the activity will rise and dominate. Thus there should be a temperature window in which the reaction mechanism switches from a dominant low-temperature one to a dominant high-temperature one. However, whether or not this temperature window can be observed experimentally depends on the catalyst. For highly active Au catalysts such as Au/ZnO/SiO_2-Na_2CO_3, the activity loss with the increase in reaction temperature resulting from the low-temperature reaction mechanism can be effectively compensated for by the activity gain contributed by the high-temperature reaction mechanism, and no loss of apparent activity can be observed. But for Au/ZnO/SiO_2-NH_3, such a temperature window was observed, providing experimental evidence for the existence of reaction temperature-dependent mechanisms of CO oxidation catalyzed by supported Au nanoparticles.

15.4 Summary and Outlook

Gold catalysis is very complex and structure-sensitive, which adds great difficulty in fundamental studies. The model catalyst approach is effective, but has to be comprehensive. Our strategy of model catalysts from single crystals to well-defined powder catalysts can achieve this goal, and so far we have provided solid experimental evidence for several important issues in Au catalysis:

(1) The high catalytic activity of the Au surface at low temperatures results from the formation of weakly chemisorbed but highly reactive surface species on the Au surface. Such weakly chemisorbed species can effectively accumulate on the Au surface only at low temperatures and contribute to the catalytic activity. Resulting from the reaction temperature-dependent competition between desorption and the surface reaction of weakly chemisorbed species, the activity of the Au surface contributed by weakly chemisorbed species will intrinsically exhibit a 'volcano' dependence on the reaction temperature.
(2) The active Au sites in low-temperature CO oxidation without the involvement of a support are low-coordinated Au atoms of Au nanoparticles with a bulk Au-like electronic structure. With a decrease in the size of Au nanoparticles, the density of low-coordinated Au atoms generally increases while the electronic structure deviates from bulk Au-like electronic structure. Therefore, the volcano-shaped dependence of intrinsic catalytic activity of Au nanoparticles on their size is an inevitable outcome resulting from the fact that their size-dependence on the geometric structure effect is opposite to their size-dependence on the electronic structure effect.
(3) Hydroxyls on the inert Au surface can act as co-catalyst to catalyze low-temperature CO oxidation in which chemisorbed COOH(a) species are responsible for the activation of molecular O_2. *di*-CO_3H(a) species on the Au surface whose formation is favored thermodynamically at low reaction temperatures are the poison for low-temperature CO oxidation catalyzed by the Au surface with hydroxyls.
(4) The gold–oxide interaction sensitively depends on the microstructure of the oxide. Oxygen vacancies and isolated surface hydroxyls on the oxide facilitate the formation and stabilization of Au(I) and small Au nanoparticles. H_2 reduction is a useful method to prepare small Au nanoparticles supported on SiO_2.

Future efforts will be devoted to: (1) fabrication of model catalysts for supported Au catalysts with precise structural control of both Au nanoparticles and support over a wide range. and (2) *in situ* spectroscopic and microscopic studies of Au catalysis under working condition to identify the active structure and the active surface species.

Recent advances in the fabrication of model supported Au catalysts include the deposition of well-defined metal colloids on the support[197] and the use of oxide nanocrystals exposing well-defined crystal planes.[198,199] Experimental evidence for active oxygen species on supported Au catalysts for oxidation reactions is still rare, although supported Au catalysts are versatile in catalyzing various oxidation reactions with molecular O_2 at mild conditions.

Meanwhile, supported Au nanoparticles were reported to restructure under working conditions.[200,201] Thus the ultimate goal of a thorough fundamental understanding of Au catalysis is still far off, but with the model catalyst strategy from single crystals to well-defined powder catalysts, hope is there.

Acknowledgements

Financial support from the National Natural Science Foundation of China, Ministry of Science and Technology of China, Chinese Academy of Sciences, and the Max Planck Society *via* the MPG-CAS partner group program are gratefully acknowledged. I am indebted to my students and co-workers (Kun Qian, Zhiquan Jiang, Xin Yu, Yulin Zhang, Zongfang Wu, Liangfeng Luo, Shilong Chen, and Yunsheng Ma) for their contributions to the Au catalysis project in my research group.

References

1. K. Pitzer, *Acc. Chem. Res.*, 1979, **12**, 271.
2. P. Pyykkö and J. P. Desclaux, *Acc. Chem. Res.*, 1979, **12**, 276.
3. P. Pyykkö, *Angew. Chem., Int. Ed.*, 2002, **41**, 3573.
4. A. F. Benton and J. C. Elgin, *J. Am. Chem. Soc.*, 1926, **48**, 3027.
5. J. G. Firth, *Trans. Faraday Soc.*, 1966, **62**, 2566.
6. H. R. Gerberich, N. W. Cant and W. K. Hall, *J. Catal.*, 1970, **16**, 204.
7. D. Y. Cha and G. Parravano, *J. Catal.*, 1970, **18**, 200.
8. M. Boudart and L. D. Ptak, *J. Catal.*, 1970, **16**, 90.
9. W. R. Patterson and C. Kemball, *J. Catal.*, 1963, **2**, 465.
10. N. W. Cant and W. K. Hall, *J. Phys. Chem.*, 1971, **75**, 2914.
11. M. Haruta, T. Kobayashi, H. Sano and N. Yamada, *Chem. Lett.*, 1987, **16**, 405.
12. M. Haruta, N. Yamada, T. Kobayashi and S. Iijima, *J. Catal.*, 1989, **115**, 301.
13. G. J. Hutchings, *J. Catal.*, 1985, **96**, 292.
14. B. Nkosi, J. Coville and G. J. Hutchings, *Appl. Catal.*, 1988, **43**, 33.
15. G. C. Bond, C. Louis and D. Thompson, *Catalysis by Gold*, Imperial College Press, London, 2006.
16. C. Corti and R. Holliday, *Gold Science and Applications*, CRC Press, Boca Raton, FL, 2010.
17. *Faraday Discussions*, Royal Society of Chemistry, Cambridge, 2011, vol. 152.
18. C. Louis and O. Pluchery, *Gold Nanoparticles for Physics, Chemistry and Biology*, World Scientific, Singapore, 2012.
19. G. C. Bond and D. Thompson, *Cat. Rev. Sci. Eng.*, 1999, **41**, 319.
20. M. Haruta and M. Daté, *Appl. Catal. A*, 2001, **222**, 427.
21. G. J. Hutchings, *Catal. Today*, 2005, **100**, 55.
22. A. S. K. Hashmi and G. J. Hutchings, *Angew. Chem., Int. Ed.*, 2006, **45**, 7896.

23. Z. Ma and S. Dai, *Nano Res.*, 2011, **4**, 3.
24. T. Takei, T. Akita, I. Nakamura, T. Fujitani, M. Okumura, K. Okazaki, J. Huang, T. Ishida and M. Haruta, *Adv. Catal.*, 2012, **55**, 1.
25. M. Haruta, *Catal. Today*, 1997, **36**, 153.
26. D. Andreeva, V. Idakiev, T. Tabakova and A. Andreev, *J. Catal.*, 1996, **158**, 354.
27. H. Sakurai, A. Ueda, T. Kobayashi and M. Haruta, *Chem. Commun.*, 1997, 271.
28. Q. Fu, A. Weber and M. Flytzani-Stephanopoulos, *Catal. Lett.*, 2001, **77**, 87.
29. Q. Fu, H. Saltsburg and M. Flytzani-Stephanopoulos, *Science*, 2003, **301**, 935.
30. T. Hayashi, K. Tanaka and M. Haruta, *J. Catal.*, 1998, **178**, 566.
31. M. Ojeda and E. Iglesia, *Chem. Commun.*, 2009, 352.
32. S. Lee, L. M. Molina, M. López, J. A. Alonso, B. Hammer, B. Lee, S. Seifert, R. E. Winans, J. W. Elam, M. J. Pellin and S. Vajda, *Angew. Chem., Int. Ed.*, 2009, **48**, 1467.
33. J. Huang, T. Akita, J. Faye, T. Fujitani, T. Takei and M. Haruta, *Angew. Chem., Int. Ed.*, 2009, **48**, 7862.
34. M. Haruta, S. Tsubota, T. Kobayashi, H. Kageyama, M. J. Genet and B. Delmon, *J. Catal.*, 1993, **144**, 175.
35. C. Qi, J. Huang, S. Bao, H. Su, T. Akita and M. Haruta, *J. Catal.*, 2011, **281**, 12.
36. T. Fujitani and I. Nakamura, *Angew. Chem., Int. Ed.*, 2011, **50**, 10144.
37. M. M. Schubert, S. Hackenberg, A. C. van Veen, M. Muhler, V. Plzak and R. J. Behm, *J. Catal.*, 2001, **197**, 113.
38. M. Daté, M. Okumura, S. Tsubota and M. Haruta, *Angew. Chem., Int. Ed.*, 2004, **43**, 2129.
39. C. K. Costello, M. C. Kung, H.-S. Oh, Y. Wang and H. H. Kung, *Appl. Catal. A*, 2002, **232**, 159.
40. C. K. Costello, J. H. Yang, H. Y. Law, Y. Wang, J.-N. Lin, L. D. Marks, M. C. Kung and H. H. Kung, *Appl. Catal. A*, 2003, **243**, 15.
41. H. H. Kung, M. C. Kung and C. K. Costello, *J. Catal.*, 2003, **216**, 425.
42. M. A. Sanchez-Castillo, C. Couto, W. B. Kim and J. A. Dumesic, *Angew. Chem., Int. Ed.*, 2004, **43**, 1140.
43. W. B. Kim, G. J. Rodriguez-Rivera, S. T. Evans, T. Voitl, J. J. Einspahr, P. M. Voyles and J. A. Dumesic, *J. Catal.*, 2005, **235**, 327.
44. M. Haruta, *Chem. Rec.*, 2003, **3**, 75.
45. I. X. Green, W. Tang, M. Neurock and J. T. Jr Yates, *Science*, 2011, **333**, 736.
46. G. C. Bond and D. T. Thompson, *Gold Bull.*, 2000, **33**, 41.
47. M. C. Kung, R. J. Davis and H. H. Kung, *J. Phys. Chem. C*, 2007, **111**, 11767.
48. N. Lopez, T. V. W. Janssens, B. S. Clausen, Y. Xu, M. Mavrikakis, T. Bligaard and J. K. Nørskov, *J. Catal.*, 2004, **223**, 232.
49. H. Falsig, B. Hovalbæk, I. S. Kristensen, T. Jiang, T. Bligaard, C. H. Christensen and J. K. Nørskov, *Angew. Chem., Int. Ed.*, 2008, **47**, 4836.

50. Y. Tai, W. Yamaguchi, K. Tajiri and H. Kageyama, *Appl. Catal. A*, 2009, **364**, 143.
51. R. Zanella, S. Giorgio, C.-H. Shin, C. R. Henry and C. Louis, *J. Catal.*, 2004, **222**, 357.
52. M. Valden, X. Lai and D. W. Goodman, *Science*, 1998, **281**, 1647.
53. Y. Maeda, M. Okumura, S. Tsubota, M. Kohyama and M. Haruta, *Appl. Surf. Sci.*, 2004, **222**, 409.
54. M. S. Chen and D. W. Goodman, *Science*, 2004, **306**, 252.
55. A. A. Herzing, C. J. Kiely, A. F. Carley, P. Landon and G. J. Hutchings, *Science*, 2008, **321**, 1331.
56. Y. Liu, C.-J. Jia, J. Yamasaki, O. Terasaki and F. Schüth, *Angew. Chem., Int. Ed.*, 2010, **49**, 5771.
57. I. Laoufi, M.-C. Saint-Lager, R. Lazzari, J. Jupille, O. Robach, S. Garaudée, G. Cabailh, P. Dolle, H. Cruguel and A. Bailly, *J. Phys. Chem. C*, 2011, **115**, 4673.
58. G. A. Somorjai, *Introduction to Surface Chemistry and Catalysis*, John Wiley & Sons, New York, 1994.
59. G. Wulff, *Z. Kristallogr.*, 1901, **34**, 449.
60. G. Ertl, Nobel Lecture. Reactions at Surfaces: From Atoms to Complexity [online]. http://nobelprize.org/nobel_prizes/chemistry/laureates/2007/ertl-lecture.html.
61. H. J. Freund, *Chem. Eur. J.*, 2010, **16**, 9384.
62. J. H. Sinfelt, *Surf. Sci.*, 2004, **500**, 923.
63. G. Ertl, *Angew. Chem., Int. Ed.*, 2008, **19**, 3524.
64. F. H. Ribeiro, G. A. Somorjai, G. Wedler, B. C. Gates, A. Baiker, C. T. Campbell, C. Xu, D. W. Goodman, A. Zecchina, D. Scarano, S. Bordiga, A. Cimino, F. S. Stone, Y. Iwasawa, M. A. Barteau, J. M. Vohs and J. P. Vigneron, *Handbook of Heterogeneous Catalysis*, ed. G. Ertl, H. Knözinger, F. Schüth and J. Wietkamp, Wiley-VCH, Weinheim, 2nd edn, 2008, vol. 2, ch. 4, pp. 1259–1398.
65. C. T. Campbell, *Surf. Sci. Rep.*, 1997, **227**, 1.
66. C. R. Henry, *Surf. Sci. Rep.*, 1998, **31**, 231.
67. V. Gorodetskii, J. Lauterbach, H. H. Rotermund, J. H. Block and G. Ertl, *Nature*, 1994, **370**, 276.
68. W. X. Huang, R. S. Zhai and X. H. Bao, *Langmuir*, 2001, **17**, 3629.
69. W. X. Huang, X. H. Bao, H. H. Rotermund and G. Ertl, *J. Phys. Chem. B*, 2002, **106**, 5645.
70. G. A. Somorjai and J. Y. Park, *Chem. Soc. Rev.*, 2008, **37**, 2155.
71. F. Gao and D. W. Goodman, *Annu. Rev. Phys. Chem.*, 2012, **63**, 265.
72. O. Masala and R. Seshadri, *Annu. Rev. Mater. Res.*, 2004, **34**, 41.
73. C. Burda, X. Chen, R. Narayanan and M. A. El-Sayed, *Chem. Rev.*, 2005, **105**, 1025.
74. Y. N. Xia, Y. J. Xiong, B. Lim and S. E. Skrabalak, *Angew. Chem., Int. Ed.*, 2009, **48**, 60.
75. G. R. Patzke, Y. Zhou, R. Kontic and F. Conrad, *Angew. Chem., Int. Ed.*, 2011, **50**, 826.

76. J. A. van Bokhoven, *ChemCatChem*, 2009, **1**, 363.
77. G. A. Somorjai, H. Frei and J. Y. Park, *J. Am. Chem. Soc.*, 2009, **131**, 16589.
78. K. B. Zhou and Y. D. Li, *Angew. Chem., Int. Ed.*, 2012, **51**, 602.
79. N. Musselwhite and G. A. Somorjai, *Top. Catal.*, 2013, **56**, 1277.
80. W. X. Huang, *Top. Catal.*, 2013, **56**, 1363.
81. Z.-A. Qiao, Z. Wu and S. Dai, *ChemSusChem*, 2013, **6**, 1797.
82. F. Zaera, *ChemSusChem*, 2013, **6**, 1821.
83. R. Meyer, C. Lemire, Sh. K. Shaikhutdinov and H.-J. Freund, *Gold Bull.*, 2004, **37**, 72.
84. M. S. Chen and D. W. Goodman, *Acc. Chem. Res.*, 2006, **39**, 739.
85. B. K. Min and C. M. Friend, *Chem. Rev.*, 2007, **107**, 2709.
86. T. Risse, Sh. Shaikhutdinov, N. Nilius, M. Sterrer and H.-J. Freund, *Acc. Chem. Res.*, 2008, **41**, 949.
87. M. S. Chen and D. W. Goodman, *Chem. Soc. Rev.*, 2008, **37**, 1860.
88. J. L. Gong and C. B. Mullins, *Acc. Chem. Res.*, 2009, **42**, 1063.
89. S. A. C. Carabineiro and B. E. Nieuwenhuys, *Gold Bull.*, 2010, **43**, 252.
90. J. L. Gong, *Chem. Rev.*, 2012, **112**, 2987.
91. K. J. Stowers, R. J. Madix and C. M. Friend, *J. Catal.*, 2013, **308**, 131.
92. C. W. Olsen and R. I. Masel, *Surf. Sci.*, 1988, **201**, 444.
93. N. D. S. Canning, D. Outka and R. J. Madix, *Surf. Sci.*, 1984, **141**, 240.
94. D. A. Outka and R. J. Madix, *Surf. Sci.*, 1987, **179**, 351.
95. N. Saliba, D. H. Parker and B. E. Koel, *Surf. Sci.*, 1998, **410**, 270.
96. D. H. Parker and B. E. Koel, *J. Vac. Sci. Tech. A*, 1990, **8**, 2585.
97. J. M. Gottfried, K. J. Schmidt, S. L. M. Schroeder and K. Christmann, *Surf. Sci.*, 2003, **525**, 184.
98. J. M. Gottfried, K. J. Schmidt, S. L. M. Schroeder and K. Christmann, *Surf. Sci.*, 2002, **511**, 65.
99. X. Deng, B. K. Min, A. Guloy and C. M. Friend, *J. Am. Chem. Soc.*, 2005, **127**, 9267.
100. J. M. Gottfried, K. J. Schmidt, S. L. M. Schroeder and K. Christmann, *Surf. Sci.*, 2003, **536**, 206.
101. D. C. Meier, V. Bukhtiyarov and D. W. Goodman, *J. Phys. Chem. B*, 2003, **107**, 12668.
102. G. McElhiney and J. Pritchard, *Surf. Sci.*, 1976, **60**, 397.
103. K. F. Peters, P. Steadman, H. Isern, J. Alvarez and S. Ferrer, *Surf. Sci.*, 2000, **467**, 1007.
104. J. M. Gottfried and K. Christmann, *Surf. Sci.*, 2004, **566**, 1112.
105. M. A. Lazaga, D. T. Wickham, D. H. Parker, G. N. Kastanas and B. E. Koel, *ACS Symp. Ser.*, 1993, **523**, 90.
106. B. K. Min, A. R. Alemozafar, D. Pinnaduwage, X. Deng and C. M. Friend, *J. Phys. Chem. B*, 2006, **110**, 19833.
107. J. L. Gong, R. A. Ojifinni, T. S. Kim, J. D. Stiehl, S. M. McClure, J. M. White and C. B. Mullins, *Top. Catal.*, 2007, **44**, 57.
108. D. A. Outka and R. J. Madix, *J. Am. Chem. Soc.*, 1987, **109**, 1708.

109. T. S. Kim, J. Gong, R. A. Ojifinni, J. M. White and C. B. Mullins, *J. Am. Chem. Soc.*, 2006, **128**, 6282.
110. R. A. Ojifinni, N. S. Froemming, J. Gong, M. Pan, T. S. Kim, J. M. White, G. Henkelman and C. B. Mullins, *J. Am. Chem. Soc.*, 2008, **130**, 6801.
111. R. G. Quiller, T. A. Baker, X. Deng, M. E. Colling, B. K. Min and C. M. Friend, *J. Chem. Phys.*, 2008, **129**, 064702.
112. R. A. Ojifinni, J. L. Gong, N. S. Froemming, D. W. Flaherty, M. Pan, G. Henkelman and C. B. Mullins, *J. Am. Chem. Soc.*, 2008, **130**, 11250.
113. M. E. Bartram and B. E. Koel, *Surf. Sci.*, 1989, **213**, 137.
114. J. Wang and B. E. Koel, *Surf. Sci.*, 1999, **436**, 15.
115. S. M. McClure, T. S. Kim, J. D. Stiehl, P. L. Tanaka and C. B. Mullins, *J. Phys. Chem. B*, 2004, **108**, 17952.
116. J. L. Gong, R. A. Ojifinni, T. S. Kim, J. M. White and C. B. Mullins, *J. Am. Chem. Soc.*, 2006, **128**, 9012.
117. N. Lopez, M. Garcia-Mota and J. Gomez-Diaz, *J. Phys. Chem. C*, 2008, **112**, 247.
118. K. A. Davis and D. W. Goodman, *J. Phys. Chem. B*, 2000, **104**, 8557.
119. D. A. Outka and R. J. Madix, *Surf. Sci.*, 1987, **179**, 361.
120. J. L. Gong, D. W. Flaherty, R. A. Ojifinni, J. M. White and C. B. Mullins, *J. Phys. Chem. C*, 2008, **112**, 5501.
121. J. L. Gong and C. B. Mullins, *J. Am. Chem. Soc.*, 2008, **130**, 16458.
122. J. L. Gong, D. W. Flaherty, T. Yan and C. B. Mullins, *ChemPhysChem*, 2008, **9**, 2461.
123. X. Liu, B. Xu, J. Haubrich, R. J. Madix and C. M. Friend, *J. Am. Chem. Soc.*, 2009, **131**, 5757.
124. B. Xu, X. Liu, J. Haubrich, R. J. Madix and C. M. Friend, *Angew. Chem., Int. Ed.*, 2009, **48**, 4206.
125. B. Xu, X. Liu, J. Haubrich and C. M. Friend, *Nat. Chem.*, 2009, **2**, 61.
126. B. Xu, R. J. Madix and C. M. Friend, *J. Am. Chem. Soc.*, 2010, **132**, 16571.
127. B. Xu, J. Haubrich, T. A. Baker, E. Kaxiras and C. M. Friend, *J. Phys. Chem. C*, 2011, **115**, 3703.
128. F. Cosandey and T. Madey, *Surf. Rev. Lett.*, 2001, **8**, 73.
129. L. Zhang, R. Persaud and T. E. Madey, *Phys. Rev. B*, 1997, **56**, 10549.
130. X. Lai, T. P. St. Clair, M. Valden and D. W. Goodman, *Progr. Surf. Sci.*, 1998, **59**, 25.
131. Q. Guo, K. Luo, K. A. Davis and D. W. Goodman, *Surf. Interface Anal.*, 2001, **32**, 161.
132. N. Spiridis, J. Haber and J. Korecki, *Vacuum*, 2001, **63**, 99.
133. S. C. Parker, A. W. Grant, V. A. Bondzie and C. T. Campbell, *Surf. Sci.*, 1999, **441**, 10.
134. E. Wahlström, N. Lopez, R. Schaub, P. Thostrup, A. Rønnau, C. Africh, E. Lægsgaard, J. K. Nørskov and F. Besenbacher, *Phys. Rev. Lett.*, 2003, **90**, 026101.
135. D. Matthey, J. G. Wang, S. Wendt, J. Matthiesen, R. Schaub, E. Lægsgaard, B. Hammer and F. Besenbacher, *Science*, 2007, **315**, 1692.

136. C. E. J. Mitchell, A. Howard, M. Carney and R. G. Egdell, *Surf. Sci.*, 2001, **490**, 196.
137. X. Lai and D. W. Goodman, *J. Mol. Catal. A: Chem.*, 2000, **162**, 33.
138. A. Kolmakov and D. W. Goodman, *Catal. Lett.*, 2000, **70**, 93.
139. A. Kolmakov and D. W. Goodman, *Surf. Sci.*, 2001, **490**, L597.
140. K. Fukui, S. Sugiyama and Y. Iwasawa, *Phys. Chem. Chem. Phys.*, 2001, **3**, 3871.
141. W. P. Halperin, *Rev. Mod. Phys.*, 1986, **58**, 533.
142. C. P. Vinod, G. U. Kulkarni and C. N. R. Rao, *Chem. Phys. Lett.*, 1998, **289**, 329.
143. S. B. DiCenzo, S. D. Berry and E. H. Hartford, *Phys. Rev. B*, 1988, **38**, 8465.
144. H. G. Boyen, T. Herzog, G. Kastle, F. Weigl, P. Ziemann, J. P. Spatz, M. Möller, R. Wahrenberg, M. G. Garnier and P. Oelhafen, *Phys. Rev. B*, 2002, **65**, 075412.
145. M. G. Mason, *Phys. Rev. B*, 1983, **27**, 748.
146. K. S. Liang, W. R. Salaneck and I. A. Aksay, *Solid State Commun.*, 1976, **19**, 329.
147. O. Cordes and M. Harsdorff, *Appl. Surf. Sci.*, 1988, **33**, 152.
148. S. Zafeiratos and S. Kennou, *Surf. Sci.*, 1999, **443**, 238.
149. D. C. Lim, I. Lopez-Salido, R. Dietsche, M. Bubek and Y. D. Kim, *Angew. Chem., Int. Ed.*, 2006, **45**, 2413.
150. T. Kizuka and N. Tanaka, *Phys. Rev. B*, 1997, **56**, 10079.
151. E. Costanzo, G. Faraci, A. R. Pennisi, S. Ravesi, A. Terrasi and G. Margaritondo, *Solid State Commun.*, 1992, **81**, 155.
152. H. Häkkinen, *Chem. Soc. Rev.*, 2008, **37**, 1847.
153. C. Chusuei, X. Lai, K. Luo and D. W. Goodman, *Top. Catal.*, 2001, **14**, 71.
154. S. Shaikhutdinov, R. Meyer, M. Naschitzki, M. Bäumer and H. J. Freund, *Catal. Lett.*, 2003, **86**, 211.
155. D. C. Meier and D. W. Goodman, *J. Am. Chem. Soc.*, 2004, **126**, 1892.
156. J. D. Stiehl, T. S. Kim, S. M. McClure and C. B. Mullins, *J. Am. Chem. Soc.*, 2004, **126**, 1606.
157. J. D. Stiehl, T. S. Kim, S. M. McClure and C. B. Mullins, *J. Am. Chem. Soc.*, 2004, **126**, 1606.
158. W. D. Knight, K. Clemenger, W. A. de Heer, W. A. Saunders, M. Y. Chou and M. L. Cohen, *Phys. Rev. Lett.*, 1984, **52**, 2141.
159. G. Schmid, R. Pfeil, R. Boese, F. Bandermann, S. Meyer, G. H. M. Calis and J. W. A. van der Velden, *Chem. Ber.*, 1981, **114**, 3634.
160. M. A. Marcus, M. P. Andrews, J. Zegenhagen, A. S. Bommannavar and P. Montano, *Phys. Rev. B*, 1990, **42**, 3312.
161. D. van der Putten and R. Zanoni, *Phys. Lett. A*, 1995, **208**, 345.
162. H. Zhang, G. Schmid and U. Hartmann, *Nano Lett.*, 2003, **3**, 305.
163. H. G. Boyen, G. Kästle, F. Wiegl, B. Koslowski, C. Dietrich, P. Ziemann, J. P. Spatz, S. Riethmüller, C. Hartmann, M. Moller, G. Schmid, M. G. Garnier and P. Oelhafen, *Science*, 2002, **297**, 1533.

164. Z. F. Wu and W. X. Huang, unpublished results.
165. Z. F. Wu, L. S. Xu, W. H. Zhang, Y. S. Ma, Q. Yuan, Y. K. Jin, J. L. Yang and W. X. Huang, *J. Catal.*, 2013, **304**, 112.
166. Z. F. Wu, Y. S. Ma, Y. L. Zhang, L. S. Xu, B. H. Chen, Q. Yuan and W. X. Huang, *J. Phys. Chem. C*, 2012, **116**, 3608.
167. S. Wang, K. Qian, X. Z. Bi and W. X. Huang, *J. Phys. Chem. C*, 2009, **113**, 6505.
168. J. F. Moulder, W. F. Stickle, P. E. Sobol and K. D. Bomben, *Handbook of X-ray Photoelectron Spectroscopy*, Perkin-Elmer Corporation, Minnesota, MN, 1992.
169. K. Qian, Z. Q. Jiang and W. X. Huang, *J. Mol. Catal. A: Chem.*, 2007, **264**, 26.
170. K. Qian, H. X. Sun, W. X. Huang, J. Fang, S. S. Lü, B. He, Z. Q. Jiang and S. Q. Wei, *Chem. Eur. J.*, 2008, **14**, 10595.
171. K. Qian, L. F. Luo, C. B. Chen, S. F. Yang and W. X. Huang, *ChemCatChem*, 2011, **3**, 161.
172. K. Qian, L. F. Luo, H. Z. Bao, Q. Hua, Z. Q. Jiang and W. X. Huang, *Catal. Sci. Technol.*, 2013, **3**, 679.
173. M. L. Kottke, R. G. Greenler and H. G. Tompkins, *Surf. Sci.*, 1972, **32**, 231.
174. C. Ruggiero and P. Hollins, *J. Chem. Soc. Faraday Trans.*, 1996, **92**, 4829.
175. M. Okumura, S. Tsubota and M. Haruta, *J. Mol. Catal. A: Chem.*, 2003, **199**, 73.
176. G. Budroni and A. Corma, *Angew. Chem., Int. Ed.*, 2006, **45**, 3328.
177. G. Martra, L. Prati, C. Manfredotti, S. Biella, M. Rossi and S. Coluccia, *J. Phys. Chem. B*, 2003, **107**, 5453.
178. F. Somodi, I. Borbáth, M. Hegedűs, A. Tompos, I. E. Sajó, Á. Szegedi, S. Rojas, J. L. G. Fierro and J. L. Margitfalvi, *Appl. Catal. A*, 2008, **347**, 216.
179. H. G. Zhu, C. D. Liang, W. F. Yan, S. H. Overbury and S. Dai, *J. Phys. Chem. B*, 2006, **110**, 10842.
180. J. H. Yang, J. D. Henao, C. Costello, M. C. Kung, H. H. Kung, J. T. Miller, A. J. Kropf, J. G. Kim, J. R. Regalbuto, M. T. Bore, H. N. Pham, A. K. Datye, J. D. Laeger and K. Kharas, *Appl. Catal. A*, 2005, **291**, 73.
181. F. Moreau, G. C. Bond and A. O. Taylor, *J. Catal.*, 2005, **231**, 105.
182. P. Zhang and T. K. Sham, *Phys. Rev. Lett.*, 2003, **90**, 245502.
183. B. N. Zope, D. D. Hibbitts, M. Neurock and R. J. Davis, *Science*, 2010, **330**, 74.
184. K. Qian, W. H. Zhang, H. X. Sun, J. Fang, B. He, Y. S. Ma, Z. Q. Jiang, S. Q. Wei, J. L. Yang and W. X. Huang, *J. Catal.*, 2011, **277**, 95.
185. J. A. Rodriguez, S. Ma, P. Liu, J. Hrbek, J. Evans and M. Pérez, *Science*, 2007, **318**, 1757.
186. Z. Q. Jiang, W. H. Zhang, L. Jin, X. Yang, F. Q. Xu, J. F. Zhu and W. X. Huang, *J. Phys. Chem. C*, 2007, **111**, 12434.
187. Y. Yu, L. S. Xu, W. H. Zhang, Z. Q. Jiang, J. F. Zhu and W. X. Huang, *Chin. J. Chem. Phys.*, 2009, **22**, 339.
188. K. Qian, W. X. Huang, Z. Q. Jiang and H. X. Sun, *J. Catal.*, 2007, **248**, 137.

189. K. Qian, J. Fang, W. X. Huang, B. He, Z. Q. Jiang, Y. S. Ma and S. Q. Wei, *J. Mol. Catal. A: Chem.*, 2010, **320**, 97.
190. K. Qian, W. X. Huang, J. Fang, S. S. Lv, B. He, Z. Q. Jiang and S. Q. Wei, *J. Catal.*, 2008, **255**, 269.
191. K. Qian, S. S. Lv, X. Y. Xiao, H. X. Sun, J. Q. Lu, M. F. Luo and W. X. Huang, *J. Mol. Catal. A: Chem.*, 2009, **306**, 40.
192. F. Boccuzzi, A. Chiorino, S. Tsubota and M. Haruta, *Catal. Lett.*, 1994, **29**, 225.
193. G. Y. Wang, W. X. Zhang, H. L. Lian, D. Z. Jiang and T. H. Wu, *Appl. Catal. A*, 2003, **239**, 1.
194. N. Lopez and J. K. Norskov, *J. Am. Chem. Soc.*, 2002, **124**, 11262.
195. H. G. Zhu, Z. Ma, J. C. Clark, Z. W. Pan, S. H. Overbury and S. Dai, *Appl. Catal. A*, 2007, **326**, 89.
196. A. Q. Wang, Y. P. Hsieh, Y. F. Chen and C. Y. Mou, *J. Catal.*, 2006, **237**, 197.
197. M. Cargnello, V. V. T. Doan-Nguyen, T. R. Gordon, R. E. Diaz, E. A. Stach, R. J. Gorte, P. Fornasiero and C. B. Murray, *Science*, 2013, **341**, 771.
198. R. Si and M. Flytzani-Stephanopoulos, *Angew. Chem., Int. Ed.*, 2008, **47**, 2884.
199. S. J. Chang, M. Li, Q. Hua, L. J. Zhang, Y. S. Ma, B. J. Ye and W. X. Huang, *J. Catal.*, 2012, **293**, 195.
200. Y. Kuwauchi, H. Yoshida, T. Akita, M. Haruta and S. Takeda, *Angew. Chem., Int. Ed.*, 2012, **51**, 7729.
201. N. Ta, J. Y. Liu, S. Chenna, P. A. Crozier, Y. Li, A. L. Chen and W. J. Shen, *J. Am. Chem. Soc.*, 2012, **134**, 20585.

Subject Index